THE THERMAL-HYDRAULICS OF A BOILING WATER NUCLEAR REACTOR

THE THERMAL-HYDRAULICS OF A BOILING WATER NUCLEAR REACTOR

R. T. Lahey, Jr.
Rensselaer Polytechnic Institute, Troy, New York

and

F. J. Moody
General Electric Company, San Jose, California

Second Edition

American Nuclear Society
La Grange Park, Illinois USA

Library of Congress Cataloging-in-Publication Data

Lahey, Richard T.
The thermal-hydraulics of a boiling water nuclear reactor / R. T. Lahey, Jr. and F. J. Moody. — 2nd ed.
p. cm.
Includes bibliographical references and index.
ISBN 0-89448-037-5
1. Boiling water reactors. 2. Heat—Transmission. 3. Nuclear reactors—Fluid dynamics. I. Moody, F. J. (Frederick J.) II. Title.
TK9203.B6L33 1993
621.48'34—dc20 93-12785
CIP

ISBN: 0-89448-037-5
Library of Congress Catalog Card Number: 93-12785
ANS Order Number: 300026

555 N. Kensington Avenue
La Grange Park, Illinois 60525 USA

Second Edition, 1993
Second Printing, 1979
First Printing, 1977
Printed in the United States of America

CONTENTS

PREFACE

The purpose of the second edition of this monograph is to present an up-to-date overview of the thermal-hydraulic technology that underlies the design, operation, and safety assessment of boiling water nuclear reactors (BWRs). Significantly, BWRs represent a large fraction of the world's installed nuclear power capacity. Some important new material on pressure suppression containment technology has been added in order to provide the reader with a more comprehensive understanding of BWR power plants.

This monograph is not intended to be a reactor design manual. Rather, emphasis is placed on the physical understanding of thermal-hydraulic phenomena, as opposed to tabulating correlations and design methods. Our strong conviction is that once the basic physical principles involved are understood, the synthesis of design techniques is rather straightforward.

This monograph has been divided into four main parts:

I. *Description of BWR Systems*
   - The evolution of BWR technology
   - A description of BWR systems and hardware

II. *Basic Thermal-Hydraulic Analyses*
   - Applied thermodynamics
   - Boiling heat transfer and two-phase flow technology

III. *Performance of the Nuclear Steam Supply System*
   - Thermal-hydraulic performance of reactor components and systems
   - Stability analysis

IV. *Performance of BWR Safety Systems*
   - Accident analysis
   - Critical flow
   - Piping loads
   - Pressure suppression system performance

We sincerely hope that the information contained herein will be of use to our colleagues and to students of nuclear engineering. Indeed, it is intended to be both a reference book and a text.

The revision of this monograph was a major undertaking that required a great amount of time and effort. We appreciate the help and advice we were given by our colleagues and students and gratefully acknowledge the excellent secretarial work done by Ms. Elaine Verrastro and the continued patience and understanding of our wives, Ellee and Phyllis.

Dr. R. T. Lahey, Jr.
Rensselaer Polytechnic Institute
Troy, New York

Dr. F. J. Moody
General Electric Company
San Jose, California

February 1993

NOTATION

a_v = vapor absorption coefficient
a_l = liquid absorption coefficient
A_{HT} = heat transfer area
A_{x-s} = cross-sectional flow area
A_i = cross-sectional flow area occupied by phase, i
B = Biot number
B_i = radiosity of surface, i
c_p = specific heat at constant pressure
c_f = Fanning fraction factor
C_0 = void concentration parameter
C_α = $\partial k/\partial\langle\alpha\rangle_{\mathrm{ave}}$ = core-average void reactivity coefficient
c = sonic speed
D_{H} = $4A_{x-s}/P_f$ = hydraulic diameter
D_b = bubble diameter
E = total stored energy of a system
e = total specific convected energy of a system
f = Darcy-Weisbach friction factor
$\langle f\rangle$ = cross-section average of a function, f
$\langle f\rangle_i$ = the average of a function, f, across the flow area of phase i
F = force, or Tong F factor
F_L = leakage flow fraction
F_{QL} = leakage power fraction
F_{i-j} = radiation view factor
G = mass flux
$g_c = 32.17\,\dfrac{\mathrm{lb}_m\text{-ft}}{\mathrm{lb}_f\text{-sec}^2} = 1.0\,\dfrac{\mathrm{kg-m}}{\mathrm{N-sec}^2}$
g = gravity, or h-Ts = Gibb's free energy function
Gr = Grashoff number
$G(S)$ = forward loop transfer function
h = $\mu + pv$ = specific enthalpy
$h_0 = e - \dfrac{gZ}{g_cJ}$ = specific stagnation enthalpy
$\bar{h}$ = density weighted enthalpy
H = convective heat transfer coefficient
H_0 = condensation parameter
Δh_{sub} = $h_f - h_i$ = inlet subcooling
H_g = gap conductance
$H(S)$ = feedback loop transfer function
H_i = radiant energy flux incident on surface, i
J = 778 ft-lb$_f$/Btu = 1.0 mn/J
j_g = volumetric flux of vapor phase
j_f = volumetric flux of liquid phase
j = $j_f + j_g$ = volumetric flux (Homogenous velocity) of two-phase mixture
K = local loss coefficient, or c_p/c_v = specific heat ratio

k = reactivity
K_i = parameters used in transient analysis
L_B = boiling length
L_H = heated length
L_i = flow length of segment, i
L_m = mean beam length
M = mass
m_e'' = liquid entrainment mass flux
m_{co}'' = liquid carryover mass flux
m_d'' = liquid deposition mass flux
Nu = Nusselt number
p = static pressure
p_0 = stagnation pressure
P_H = heated perimeter
Pr = Prandtl number
P_f = friction perimeter
P_i = interfacial perimeter
p_I = impingement pressure
P = Peclet number
q = heat rate
q' = linear heat generation rate
q'' = heat flux
$\bar{q}''$ = axial-average heat flux
q''' = volumetric heat generation rate
Q_i = volumetric flow rate of phase, i
Q_{active} = active core power
R_v = thermodynamic gas constant
r = radial distance
R_j = jet reaction force
S = Ms = entropy, or Slip ratio, or Laplace transform variable
s = specific entropy
t = time
t_0 = time for a particle to cross the boiling boundary
T = temperature
ΔT = temperature difference
T_0 = rewetting temperature
$T_i(S)$ = transfer functions for core-average void fraction perturbations
T_{sat} = saturation temperature
u_i = velocity of phase, i, or location, i
U = velocity of a falling film, or μM = total internal energy
u_r = relative velocity
U_r = one-dimensional-averaged relative velocity
U_m = one-dimensional-averaged velocity of the center-of-mass
U_p = one-dimensional-averaged velocity of the center-of-momentum
U_e = one-dimensional-averaged velocity of the center-of-energy
U_t = terminal rise velocity
V = volume
v = V/M = specific volume
V_{gj} = drift velocity
$\dot{V}_B$ = volumetric rate of vapor formation
$\hat{v}$ = Laplace transform of the variable, $v(t)$
$\delta v(t)$ = perturbation of variable, $v(t)$

$\delta \hat{v}$ = Laplace transformation of the perturbation of variable, $v(t)$
w = flow rate
$\dot{W}_k$ = work rate (mechanical power)
$\delta w'$ = liquid phase evaporation rate per unit axial length
x_S = static quality
x_e = thermodynamic equilibrium quality
x = flow quality
X_{tt} = Martinelli parameter
α = void fraction
α_T = coefficient of thermal expansion
α_k = absorptivity of body, k
β = volumetric flow fraction
$\beta(S)$ = a transfer function for boiling boundary dynamics
β_c = contact angle
β_v = volumetric expansivity
γ_i = angle, i
ρ = density
$\bar{\rho}$ = two-phase density
ρ_H = homogeneous (no-slip) density
ρ' = momentum density
ρ'' = temporal energy density
ρ''' = spatial energy density
ρ_k = reflectivity of body, k
η = efficiency
η_a = actual efficiency
η_e = mechanical efficiency
η_m = phase-to-phase momentum transfer parameter
μ = U/M = specific internal energy
M_i = viscosity of phase i
$\Gamma_i(S)$ = transfer functions for single-phase hydrodynamics
$\pi_i(S)$ = transfer functions for two-phase hydrodynamics
$\Lambda_i(S)$ = transfer functions for boiling boundary dynamics
τ_i = interfacial shear stress
τ_w = wall shear stress
τ_k = transmissivity of body, k
ϵ = $e - pv$ = total specific stored energy
ϵ = ratio of pumping to evaporation heat flux
ϵ_k = emissivity of body, k
σ = surface tension, or Stefan-Boltzmann constant
δ_m = thickness of metal plate
Ω_T = Tong's memory parameter
Ω = characteristic frequency of phase change
θ = $(T - T_{sat})/(T_0 - T_{sat})$, or angle of inclination
$\theta(S)$ = a transfer function for boiling boundary dynamics
ϕ_{l0}^2 = two-phase friction loss multiplier
Φ = two-phase local loss multiplier
κ = thermal conductivity, or Armand's void parameter
ν = time for a particle to lose its subcooling, or kinematic viscosity
λ = nonboiling length (boiling boundary)
ψ = two-phase dissipation function
ζ = slope of the pump head-flow curve
τ = $t - t_0$ = boiling time.

SUBSCRIPTS

av = average
ave = axial average
b = boiling
B = blowdown
c = critical
cond = condensation
cont = contraction
crit = thermodynamic critical point
℄ = center line
cr = crud layer
D = discharge
d = departure point
eq = equilibrium point
eff = effective
evap = evaporative
exp = expansion
ext = external
f = saturated liquid
fg = the difference between saturated vapor and liquid properties
F = feedback
g = saturated vapor
gap = pellet-clad gap
h = heater
HT = heat transfer
i = inlet, or interfacial, or ideal
in = inflow
l = liquid phase
le = entrained liquid
m = metal
n = nucleation point
o = initial conditions
ox = oxide layer
out = outflow
pump = liquid agitation (pumping)
QL = power loss due to leakage
R = radiation, or riser, or recover
Rw = rewet
sat = saturation
S = shear
T = total
t = throat
v = vapor phase
w = wall, or wave
1 = jet pump drive flow
2 = jet pump suction flow
I = Section I of duct
II = Section II of duct
1ϕ = single-phase
2ϕ = two-phase

PART ONE

Description of BWR Systems

CHAPTER ONE

Reactor and Containment Configurations

In the late 1950s, two distinct light-water-cooled nuclear reactor steam supply systems became commercially available: the pressurized water reactor (PWR) system and the boiling water reactor (BWR) system. The commercial PWR systems largely grew out of the technology developed for naval nuclear submarine reactors, while early BWR technology was developed mainly at Argonne National Laboratory (ANL) and the Nuclear Energy Division (NED) of General Electric Company (GE). The BWR steam supply system is attractive due to its basic simplicity and potential for greater thermal efficiency, better reliability, and lower capital cost than other competing light water reactor (LWR) systems.

Two reactor containment concepts were also developed during the 1950s. The first and earliest type was *dry containment,* which surrounded the reactor with a large, passive, leakproof shell, capable of withstanding the pressures caused from a pipe break and complete discharge of the reactor coolant. The *pressure suppression containment* concept soon followed, in which the reactor coolant energy, released from a postulated pipe rupture, is contained by a surrounding drywell shell that discharges through vents to a large water pool. The pressure suppression containment concept is the one that has been adopted for modern BWRs.

Various steps in the pressure suppression containment design evolution, while incorporating significant economic and functional improvements, also introduced hydrodynamic phenomena associated with submerged air discharge and steam condensation. Extensive analytical and experimental studies were performed to help understand the unsteady phenomena and to verify the prediction of the associated forces. Recent efforts have focused on refining predictive models for the analysis of postulated severe accidents involving core melting and fission product retention in the containment.

The purpose of this chapter is to give a concise overview of the evolution of BWRs and containment from the early experimental stage to current modern BWR power plants. It is recognized that many organizations have made substantial contributions to early BWR technology. However, since all domestic and most foreign BWRs are currently manufactured by either General Electric (GE) or are manufactured under a GE license agreement, this monograph is primarily concerned with BWRs of the GE design.

Kramer (1958) previously documented the pioneering work done at ANL, including the BORAX experiments, the Experimental Boiling Water Reactor (EBWR), and the Argonne Low-Power Reactor (ALPR) features and operating experiences. In addition, Kramer has presented an excellent treatment of the SPERT experiments, GE's Vallecitos Boiling Water Reactor (VBWR), and the Dresden Nuclear Power Station (D-1). Although some of the salient features of VBWR and D-1 are discussed in subsequent sections of this chapter, the reader interested in the details of these plants and the earlier work done in the development of BWR technology is referred to Kramer's excellent book (Kramer, 1958).

The initial strategic plan followed by GE in the development of BWR technology was known as Operation Sunrise (Cohen and Zebroski, 1959). This plan involved the parallel development of natural and forced circulation plant concepts and nuclear superheat technology. In accordance with the original action plan, natural and forced circulation BWRs have been designed and are currently in operation, whereas the nuclear superheat concept is not currently being pursued. The increased efficiency and reduced plant size theoretically possible with superheat steam were evaluated through several major developmental programs. Extensive fuel development work was done in VBWR and the Esada Vallecitos Experimental Superheat Reactor (EVESR). In addition, many out-of-core heat transfer experiments were performed to quantify the thermal-hydraulic performance of typical superheat fuel bundles. Both separate and integral superheating concepts were evaluated during the early 1960s. The program was finally discontinued due to problems with fuel integrity, the relatively low power density achievable, and the marginal economic advantages to be gained through nuclear superheat.

Pioneering work in pressure suppression containment technology moved forward through a cooperative effort by GE and Pacific Gas and Electric in the late 1950s. Condensing tests with vertical vents in water pools showed that condensation was effective for a broad range of vent sizes, submergence, and steam discharge rates. Scaled-down transient flow tests, consisting of a pressure vessel with steam discharged into a simulated drywell and vented to a water pool, confirmed the effectiveness of pressure suppression.

Additional foreign and domestic tests were performed to further understand phenomena associated with pressure suppression, including broken pipe discharge of steam/water mixtures, system pressurization, and condensation in water pools.

Joint design efforts between GE and the Japanese have furthered the technology for an advanced BWR (ABWR) and a simplified BWR (SBWR), based on state-of-the-art components and concepts. These concepts for the ABWR include:

- internal recirculation pumps
- fine motion control rod drives
- digital control and instrumentation
- multiplexed fiber optic cabling networks
- cylindrically reinforced pressure suppression containments
- horizontal vents
- structural integration of the containment/reactor building
- severe accident mitigation capability
- state-of-the-art fuel
- advanced turbine/generators (e.g., 52-in. last-stage buckets)
- advanced radwaste technology.

1.1 Boiling Water Reactor Cycles

The BWR is characterized by the fact that bulk boiling takes place in the core. However, BWRs can be further characterized by whether the steam generated in the core passes directly to the turbine or is used to make steam for the turbine in a primary/secondary heat exchanger; i.e., a steam generator. When the steam generated in the core passes directly to the turbine, the plant is classified as a direct-cycle BWR. When the steam/water mixture generated in the core passes through a steam generator, such that the steam going to the turbine is produced on the secondary side of the steam generator, the plant is classified as an indirect-cycle BWR. Thus, the indirect-cycle BWR is quite similar to a PWR; the basic difference is that bulk boiling is allowed in the core of a BWR. The final cycle of interest is the so-called "dual-cycle BWR," in which part of the steam going through the turbine is produced directly in the core and part is produced in a steam generator. In addition to cycle classification, BWRs are normally classified as either forced circulation or natural circulation plants, depending on whether the coolant is pumped through the core (forced circulation) or whether it flows through the core due to density differences between the fluid in the downcomer and the core region (natural circulation). Figure 1-1 is a simplified schematic of several possible BWR cycles.

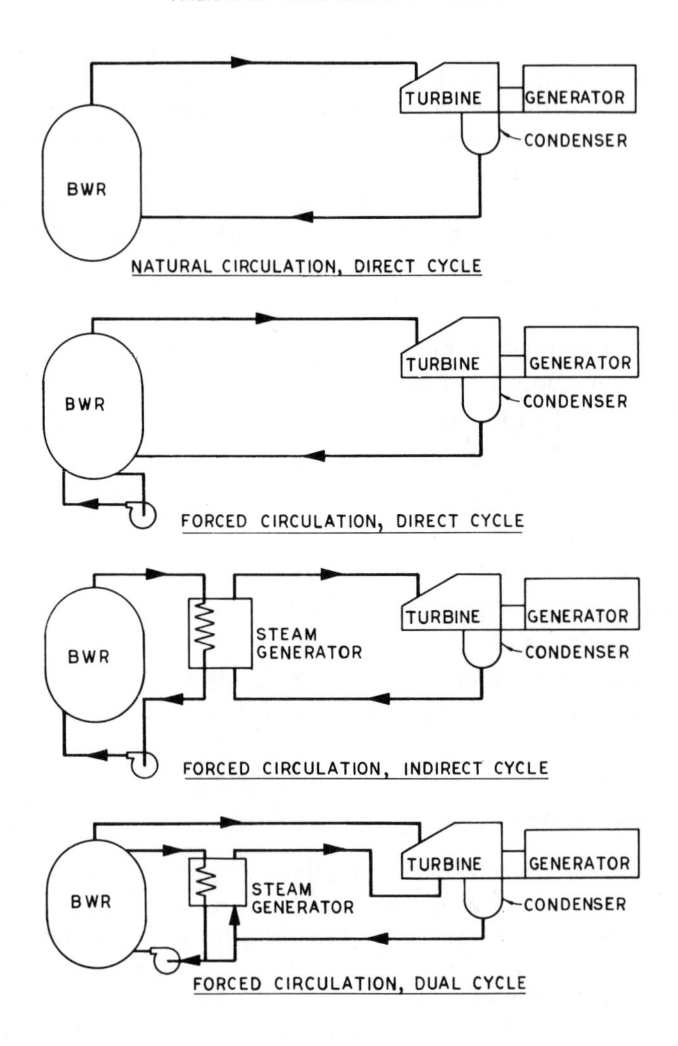

Fig. 1-1 Various boiling water reactor cycles.

The indirect-cycle BWR is not a very attractive concept, since it involves the increased cost and maintenance associated with large steam generators and requires relatively high operating pressures on the primary side. Only one plant of this type, the KAHL Reactor in West Germany, has ever been built by GE. In contrast, there are numerous operating plants of the dual- and direct-cycle types that have been designed and built by GE.

The first commercial nuclear power plant to be licensed by the U.S. Atomic Energy Commission (USAEC) was the VBWR. It was basically an experimental reactor used to develop BWR nuclear fuel technology and to demonstrate various BWR cycle concepts, although a small amount of power (~5 MW) was supplied to the Pacific Gas and Electric Company grid. The VBWR was constructed so that it could be operated in the following modes:

- natural circulation, direct cycle
- forced circulation, direct cycle
- natural circulation (through the core), dual cycle
- forced circulation, dual cycle.

The operating experience gained with the VBWR showed that BWRs have low-turbine radiation levels and are stable in the various operating modes. This was an important step in the evolution of modern BWRs since earlier low-pressure experiments conducted by ANL had indicated that nuclear-coupled instability and radioactive contamination of the turbine were potential problems. Although VBWR demonstrates stable operation in all modes, it was observed that the natural circulation mode of operation was less stable than forced circulation. This was primarily due to the fact that the natural circulation driving head was relatively small in the VBWR, thus only limited inlet orificing of the core could be accomplished. In addition, it was found that the method of steam separation in the VBWR was not very efficient since the steam carry-under criterion limited the downcomer velocity during the various modes of operation. These and other experimental observations aided in improving the BWR designs of all subsequent plants, such that current BWRs are quite reliable, stable, and efficient.

The first BWR specifically constructed as a commercial reactor was D-1. As shown in Fig. 1-2, D-1 was a dual-cycle plant and was the prototype for several other dual-cycle plants including Tarapur, Garigliano (Senn), and Gundremmingen (KRB). These dual-cycle plants have been designated as the BWR/1-type design, indicating that they were in the first "product line" marketed by NED. Basically, BWR/1-type plants were a series of prototypes of both dual and early direct-cycle (e.g., Big Rock Point, Humboldt Bay) plants, demonstrating key improvement features.

The dual-cycle design was chosen as the basis for the first commercial BWR since it combined reliability, stability, and fairly high power density with a well-controlled reactor response to load changes. The unique features of a dual-cycle BWR can best be described with reference to some operational transients. The control concept is basically one of the turbine being slaved to the reactor; i.e., the turbine responds to reactor conditions. Referring to Fig. 1-2, if the load on the turbine increases, it tends to slow down; however, in a dual-cycle plant, a signal is sent to the secondary

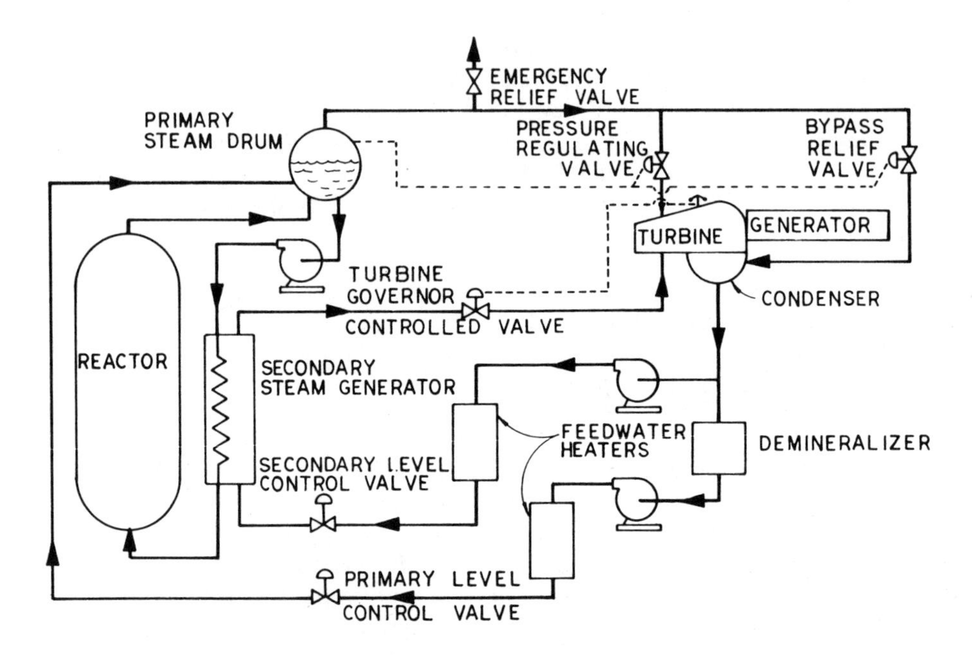

Fig. 1-2 Dresden-1 circulation system.

steam generator's control valve, which causes sufficient steam to be admitted to a lower stage of the turbine to carry the increased load. Thus, in most cases the load change is taken up by the secondary steam generator such that the primary pressure and steaming rate from the reactor are essentially unchanged. As the steaming rate in the steam generator increases, it causes the reactor inlet coolant to become more subcooled, which in turn causes some of the in-core voids to collapse, thereby increasing the reactor power level to meet the new load requirements. The opposite occurs for turbine load rejections, with the secondary steam generator again handling the load change. In this manner, the dual-cycle plant could handle many operational transients with minimum perturbation to the primary system and a minimum of control rod action. It did, in fact, provide a very convenient means for an essentially base-loaded plant to "load follow" using the principle of subcooling control.

Dual-cycle power plants proved that the basic BWR concept was sound. However, these plants required the additional capital cost and maintenance expense associated with large steam generators and large containment buildings. Significantly, the inherent simplicity and thermodynamic potential of the BWR concept can be realized fully only in a direct-cycle plant. Thus, starting in 1963, with GE's BWR/2 product line, all subsequent plant offerings were exclusively of the direct-cycle type.

TABLE 1-1
GE/BWR Product Lines

Product Line Number	Year of Introduction	Characteristic Plants and Salient Features
BWR/1	1955	D-1, Big Rock Point, Humboldt Bay, Tarapur, Senn, KRB, LaCrosse Initial commercial BWRs Dual cycle Direct-cycle prototype Natural circulation (Humboldt Bay) Pressure suppression containment First internal steam separation (KRB)
BWR/2	1963	Oyster Creek, Nine Mile Point First domestic turnkey plant Direct cycle M/G flow control 7×7 fuel bundle
BWR/3	1965	D-2, Millstone, Monticello, D-3, Quad Cities 1, 2, Pilgrim Turnkey plants Internal jet pumps Improved emergency core cooling system (ECCS)
BWR/4	1966	Browns Ferry (TVA) 1, 2, 3, Vermont Yankee, Duane Arnold, Cooper, Peach Bottom 2, 3, Fitzpatrick 1, Brunswick 1, 2, Hatch 1, 2, Fermi 2, Hope Creek 1, Susquehanna, Limerick 1, 2 Increased power density
BWR/5	1969	Zimmer, LaSalle 1, 2, WNP 2, Nine Mile Point 2 Improved safeguards Valve flow control
BWR/6	1971	Grand Gulf 1, Perry 1, 2, Clinton 1, River Bend 1, Skagit 1 8×8 fuel bundle Added fuel bundles; increased power output Improved ECCS performance Reduced fuel duty (lower kW/ft)

1.2 The Evolution of BWR Nuclear Steam Supply Systems (NSSS)

Table 1-1 summarizes the various GE/BWR product line plants from BWR/1 through the current BWR/6 plants. Some of the site names commonly associated with a given product line and some of the salient plant characteristics are also tabulated for convenient reference. Basically, the BWR/1 vintage plants were demonstration plants and were custom-made to the individual utility's specifications. Oyster Creek was the first real attempt to establish a standard BWR product line. These BWR/2 product-line plants featured the internal[a] steam separation system proven previously on KRB.

[a]Steam separators are located entirely inside the pressure vessel rather than in high-rise steam drums as on earlier BWR/1 plants.

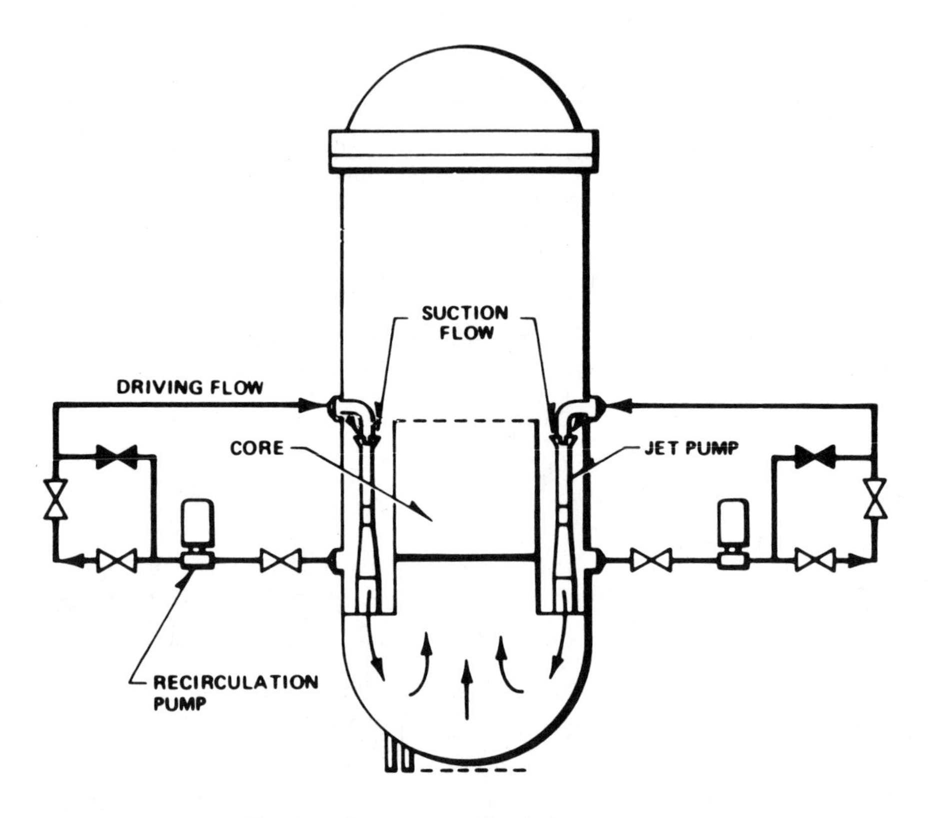

Fig. 1-3 Jet pump recirculation system.

They had square-pitch 7×7 fuel rod bundles and were of a forced circulation type design in which all the core flow was pumped through five external recirculation loops. The recirculation pumps had variable speed capability through motor-generator (M/G) sets to accomplish load following through flow control. This concept of flow control is described more fully later, since it is common to all direct-cycle BWR plants.

In 1965, GE introduced the BWR/3 product line. Dresden-2 (D-2) typifies this design class, which was similar to the BWR/2 design in many respects, but featured internal jet pumps in place of the external recirculation pumps used in BWR/2-type plants. The use of high-efficiency jet pumps reduced the size and number of external recirculation loops, since only about one-third of the core flow was needed as the driving flow; the rest was entrained as suction flow. Figure 1-3 shows this type of recirculation system. The principles of jet pump operation and more detail on the actual system are given in Chapter 2.

In 1966, the Brown's Ferry design was introduced by GE as the BWR/4 product line. These plants were quite similar to BWR/3 plants except the power density of the core was increased by 20%, to 51 kW/liter. In the standard 7×7 fuel bundle, this rating gave a peak fuel duty of 18.5 kW/ft. This increase in power density enabled a given size reactor to produce more usable energy, thus improving the overall economic evaluation.

The last year GE marketed the so-called "turnkey" plants was 1966, i.e., the last year that GE was the prime contractor and thus was responsible for the construction of site buildings, obtaining USAEC licensing, and furnishing the nuclear steam supply system and the balance of plant equipment. Since 1966, NED has marketed only the nuclear steam supply system. The balance of plant and site construction is now being done by various architect/engineering (A/E) firms or by the utilities making the purchase. Table 1-2 tabulates some of the salient information on the twelve turnkey plants that were sold during the period from 1962 through 1966. During the era of turnkey projects, with their fixed price but ever-changing construction and licensing requirements, nuclear energy was not a profit-making venture. Nevertheless, these plants demonstrated successful performance and essentially "turned the key" toward putting the BWR on a solid commercial basis.

In 1969, the Zimmer class of plant was marketed by GE as BWR/5. Although these plants were quite similar to BWR/4 plants, they have several unique features. The ECCS and recirculation system were redesigned to improve reactor safeguards to any hypothetical accident conditions, and flow control was accomplished by valve control rather than controlling the speed of the recirculation pumps. This latter innovation allows BWR/5

TABLE 1-2
Turnkey Plants

Station	Size [MW(*e*)]	Type	Order Date	Year of Commercial Operation
Domestic Plants:				
Oyster Creek	650	BWR/2	1963	1969
D-2	809	BWR/3	1965	1970
D-3	809	BWR/3	1966	1971
Millstone-1	652	BWR/3	1965	1971
Monticello	545	BWR/3	1966	1971
Quad Cities-1	809	BWR/3	1966	1972
Quad Cities-2	809	BWR/3	1966	1972
Foreign Plants:				
Tarapur-1	210	BWR/1	1962	1969
Tarapur-2	210	BWR/1	1962	1969
Tsuruga	342	BWR/2	1965	1970
Fukushima-1	440	BWR/3	1966	1971
Nuclenor	440	BWR/3	1965	1971

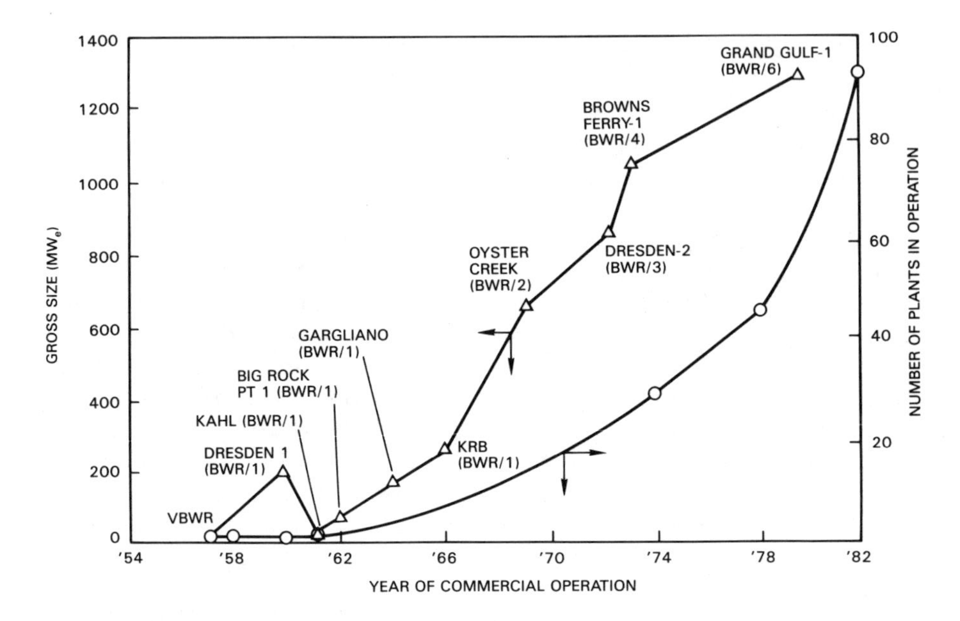

Fig. 1-4 Year of commercial operation versus number of plants in operation and gross size in megawatts.

plants to follow more rapid load variations and reduced the capital cost of the overall control system.

The current product line marketed by GE is BWR/6. As in previous design changes, BWR/6 is similar in many respects to the previous product line (BWR/5); however, it employs improved steam separators, higher efficiency multihole jet pumps, and enhanced power flattening through better coolant distribution and zone loading of burnable gadolinia poison. The most significant change from BWR/5 is that a square-lattice 8×8 fuel rod bundle is used in BWR/6 rather than the previous standard 7×7 lattice. This lattice change allowed the power density to be increased to 56 kW/liter while, due to the increased heat transfer surface area, the peak fuel duty was reduced to 13.4 kW/ft. This increase in power density coupled with loading more subassemblies into a given size pressure vessel enables BWR/6 to deliver 20% more power than BWR/5 for a given pressure vessel size. Figure 1-4 shows how plant ratings and numbers evolved with time.

This discussion of the design evolution of BWRs from VBWR through BWR/6 clearly indicates that the overall design philosophy has been a fairly conservative approach of building on proven technology. Any major innovations are invariably given thorough developmental and proof testing before they are put into commercial power plants. In this way, significant

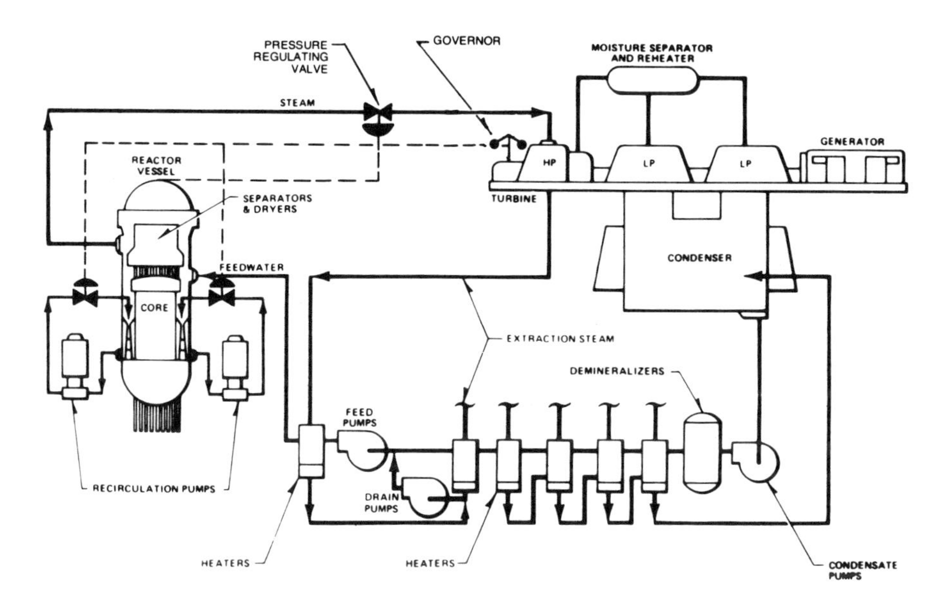

Fig. 1-5 Direct-cycle reactor system.

technological improvements can be made in each new product line with a high degree of certainty for successful operation.

The generic classification of BWRs into product-line types serves to focus attention on the concept of design standardization used by GE in the BWR/2 through BWR/6 power plants. In particular, these product lines all have been of the direct-cycle type as opposed to earlier BWR/1 dual-cycle plants. As in the description of dual-cycle plants, the unique operational features of direct-cycle plants can best be described in terms of operational transients. Figure 1-5 shows a typical BWR/6 power plant.[b]

All large nuclear power plants are designed for base-loaded service. Hence, the overall control concept of a BWR is that the reactor pressure is maintained approximately constant by slaving the turbine to the reactor. To follow load changes, a direct-cycle BWR uses the inherent feature of varying the reactor power level by reactor coolant flow control.

For instance, for an increase in generator load, more reactor power is required. To obtain an increase in reactor power level, the core flow is increased causing the core-average void fraction to decrease, which in turn causes the neutron moderation, and thus core power, to increase. An equilibrium condition is achieved at a higher power and flow rate in which

---

[b]Earlier plants are quite similar but may use pump speed control rather than valve flow control.

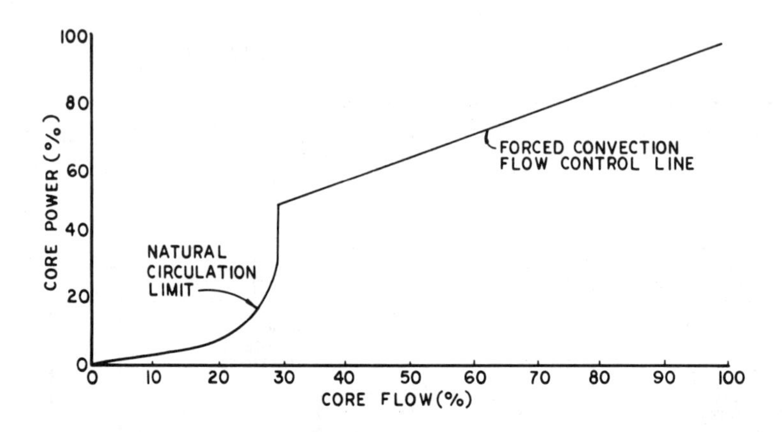

Fig. 1-6 Typical BWR flow control line.

the core-average void fraction is approximately the same as for the previous condition.

In contrast, for a partial load rejection, the turbine tends to speed up sending a signal to the flow control valves to close to a new position. The increased throttling of the flow control valves lowers the recirculation flow through the reactor core, thus tending to increase the core-average void fraction. This increase in steam content reduces the moderating ability of the coolant and, thus, the reactor power level is lowered to a new operating condition consistent with the reduced load. Therefore, for most operational transients, the principle of flow control is used to follow load changes. This method of load following assures that the primary system pressure is essentially constant and minimizes the need for control rod action. A typical BWR flow control line is shown in Fig. 1-6.

This classification and short historical review of the development of GE BWRs was meant to give the reader an overview of the evolution of domestic BWR technology. Subsequent chapters discuss the details of a particular product-line reactor so that more in-depth understanding can be achieved. The bulk of these discussions is concerned with features of modern BWRs, such as those contained in BWR/4-, BWR/5-, and BWR/6-type plants.

1.3 The Evolution of BWR Containment Systems

The major function of a containment system is to protect the public from injury resulting from a design-basis loss-of-coolant accident (LOCA). The basic containment design provides a barrier that prevents fission product release to the atmosphere.

The containment designs of the early 1950s featured a large dry shell structure, surrounding the primary system, which could withstand the pressure increase resulting if all reactor fluid mass and energy were released through a rupture of the largest pipe. This design was followed in the mid-1950s by pressure suppression containment system development, in which the reactor was surrounded by a large vessel that was vented to a water pool capable of condensing all steam released during a LOCA.

1.3.1 Dry Containment

The dry containment design employed in the Dresden I, Consumers Power, and LaCrosse reactors is illustrated in Fig. 1-7. Also shown is a typical containment pressure-time curve resulting from a large pipe break accident and complete loss of reactor coolant. The pressure rises to a maximum

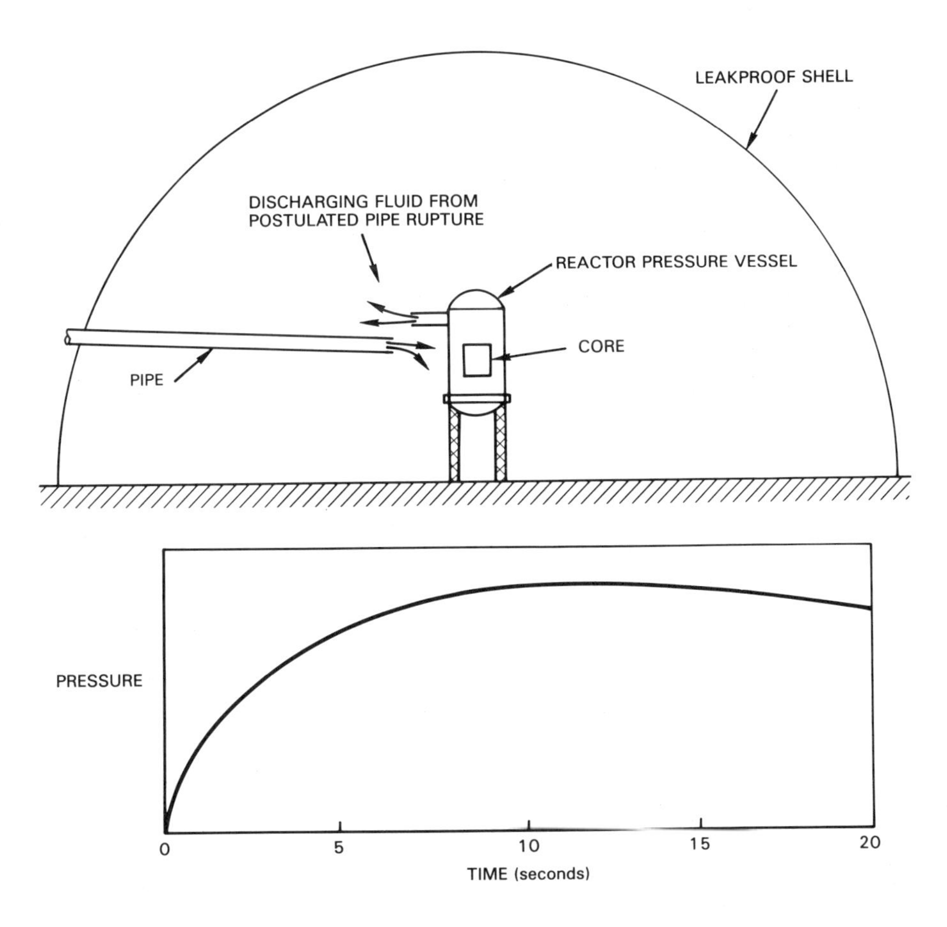

Fig. 1-7 Dry containment.

value, from which it decreases slowly as thermal energy is removed by heat transfer to the shell, concrete, and other equipment surfaces.

Further pressure reduction is obtained by containment sprays that condense steam and cool surfaces for continued heat transfer. Moreover, core cooling by in-vessel sprays transports additional decay heat energy to the containment, which must be accommodated. It is desirable to maintain the lowest possible containment pressure after a LOCA to minimize leakage of radioactive contaminated products to the environment.

1.3.2 Pressure Suppression Containment Systems

Pressure suppression containment designs have substantially smaller volumes than dry containments because of the high energy absorption capacity of the water pool. The basic design components of a pressure suppression containment are shown in Fig. 1-8.

The drywell surrounds the reactor and provides a primary barrier to steam/water release during a LOCA and to subsequent fission product releases, which can occur from the reactor under postulated inadequate cooling conditions.

Discharge of drywell air and LOCA steam/water components occurs through a vent system, which is sized to limit the maximum drywell pressure. Figure 1-8 also shows typical containment pressure transients for the drywell and wetwell resulting from a large break LOCA and complete discharge of the reactor coolant. Comparison with Fig. 1-7 shows that the drywell pressure quickly reaches a maximum value, which decreases rapidly after the vent water is expelled and steam/air discharge to the pool begins. The drywell region of the containment system must allow adequate room for the steam and water released from the reactor during a LOCA. Pressure suppression in the wetwell region then takes the steam and rapidly condenses it in the water pool.

The suppression pool is a large, passive heat sink, which can absorb all the steam released during a LOCA plus an additional amount that is generated by decay heat in the core and is transferred to the containment with the emergency core coolant. The pool also absorbs steam releases from the safety/relief valves (SRVs) during operation of the automatic depressurization system (ADS), and other occasional steam releases to control reactor pressure. Another passive feature of the water pool is its absorption and retention capability for filtering radioactive material released during a severe core accident when insufficient core cooling water is available. The pressure suppression pool also provides a large source of primary coolant makeup or emergency core coolant (ECC). Since the vents are submerged in the pool, the air purge from the drywell during a LOCA causes the pool surface to move upward until the air reaches the rising surface and breakthrough occurs to the wetwell air space. Subsequent steam discharge condenses and is retained by the pool.

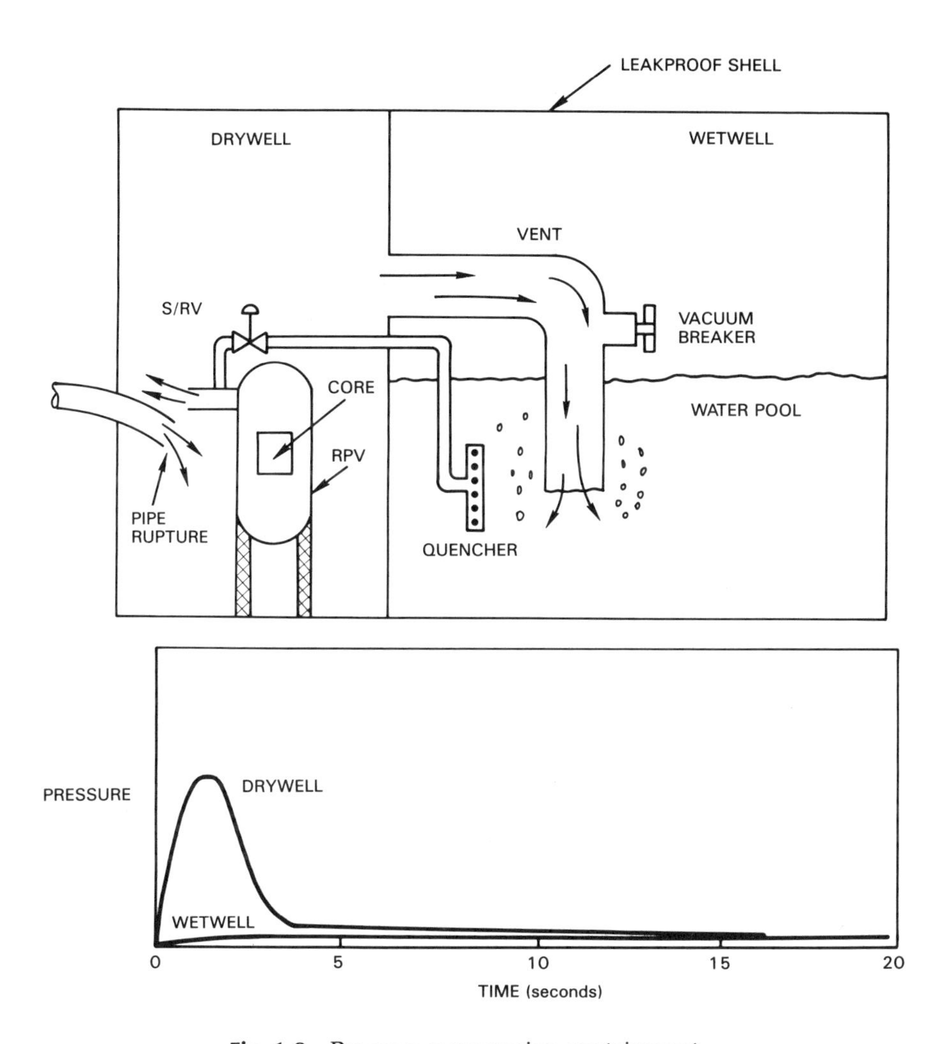

Fig. 1-8 Pressure suppression containment.

Safety/relief valves are discharged into the pool when it is necessary to reduce pressure in the reactor during accident and nonaccident conditions.

The wetwell is a large volume that contains the pool and an air space large enough to receive all drywell air plus additional heating and evaporative mass transfer from the pool without exceeding a specific maximum pressure.

The vacuum breakers permit air flow back from the wetwell to the drywell if the drywell pressure is reduced by spray cooling and steam condensation.

Steam quenchers are pipes with many small holes, which are attached to the submerged ends of SRV discharge lines for the purpose of improving overall steam condensation, pool mixing, and reducing condensation forces exerted on pool structures.

The residual heat removal (RHR) system provides cooling of the pool when its temperature has been increased by a LOCA or a long period of SRV discharge.

The fission product removal capability of a pressure suppression system is enhanced by the water pool, which is an effective filter of radioactive aerosols released from a postulated severe accident. A large fraction of airborne radioactive aerosols, which are discharged through the vents or relief valve system, are scrubbed as the bubbles rise through the water pool. Also, these aerosol particles may be deposited on containment and structural surfaces.

Early testing of the pressure suppression concept confirmed the capability both for absorbing the energy of the steam released and for quickly lowering the containment pressure resulting from a LOCA. Other tests also exhibited short-term vibrations and impulsive pressures, which were attributed to submerged air discharge and expansion and to unsteady steam condensation. Although these short-term phenomena were not predicted by analytical models in use at the time, they resulted in pressure loads that were bounded and were lower than the dominant pressure transient load. It was judged that the experiments were sufficiently broad to embrace all pertinent loads and could be used in the specification of the pressure suppression design requirements.

Three commercial pressure suppression designs were developed by GE and are shown in Fig. 1-9. The first is called the *MARK-I design,* which consists of an inverted light-bulb-shaped steel drywell surrounding the reactor and vented to a horizontal torroidal vessel encircling the drywell and containing a water pool. The second is called the *MARK-II design* or over-under containment system. The third containment is called the *MARK-III design,* which has a much larger drywell, connected by a weir and a horizontal submerged vent arrangement to an annular pool of water. Table 1-3 gives a list of GE BWR plants and their respective containment types.

Typical Japanese containment designs, shown in Fig. 1-10, incorporate modifications to the MARK-I and -II systems to give more working space in the drywell and wetwell. The larger volumes also reduce the magnitude of LOCA pressure loads.

The German design shown in Fig. 1-11 is similar to a MARK-II system with the vents and pool surrounding the reactor, largely to simplify construction and reduce costs.

The ASEA-ATOM pressure suppression system in Fig. 1-12 also incorporates a modified MARK-II design, with accommodation for external and internal recirculation pumps, respectively.

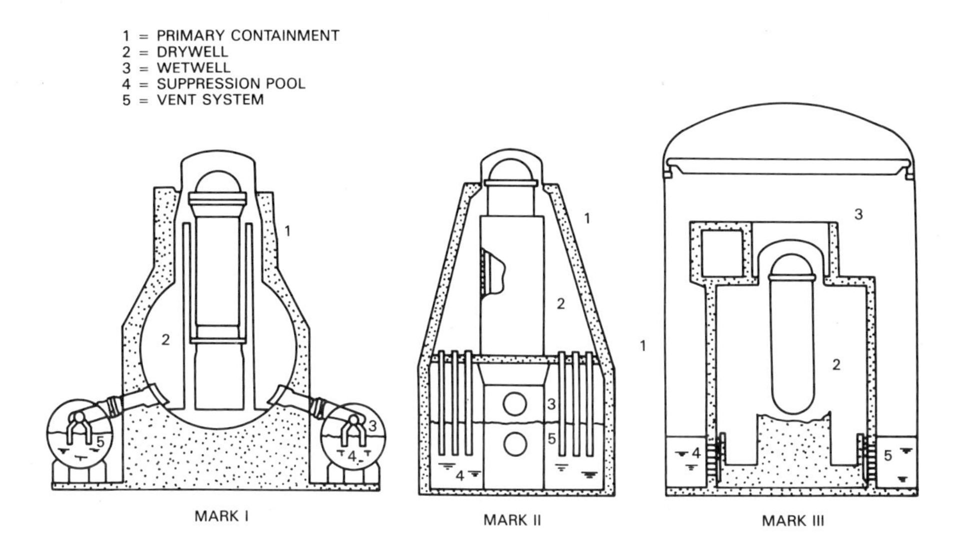

Fig. 1-9 GE pressure suppression system designs.

1.3.3 Pressure Suppression Tests

Early testing of the pressure suppression containment concept was performed for the Humboldt Bay Power Plant in a cooperative effort by GE and Pacific Gas and Electric in the late 1950s. One test involved the full-scale condensation of steam, discharged through vents into rectangular water pools in the facility shown in Fig. 1-13.

The condensing tests were performed by discharging 100-psi saturated steam at flow rates up to 100,000 lbm/h through pipes ranging from 4 to 14 in. in diameter, and having submergences from 1.0 in. to 6 ft, into water at temperatures of 50 to 150°F. Condensation in all these tests was rapid and complete.

Another test at approximately the same time was constructed to simulate transient operation of a complete pressure suppression system in a small-scale facility, shown in Fig. 1-14.

A LOCA was simulated by a rupture disk, which allowed a discharge of saturated steam and water into the simulated drywell for venting into the pool. The simulated reactor pressure vessel contained saturated water up to 1000 psig that was discharged through orifices to simulate various break sizes. Figure 1-15 gives an example of typical pressures recorded during one of these tests. The simulated reactor pressure is seen to decay slowly, relative to the time of pressure rise in the drywell, which decays rapidly when the vent water is fully expelled, after which the simulated wetwell pressure rises. This behavior confirmed the pressure suppression

TABLE 1-3
GE BWR Plants and Containment Types

	Plant Identity	Design		Plant Identity	Design
1	Dresden 1 (INOP)	Dry	46	Bailly[a]	MARK-II-C
2	Humboldt Bay 3 (INOP)	Wet	47	La Salle 1	MARK-II-C
3	Garigliano (INOP)	Dry	48	La Salle 2	MARK-II-C
4	Big Rock Point	Dry	49	Susquehanna 1	MARK-II-C
5	KRB-A (INOP)	Dry	50	Susquehanna 2	MARK-II-C
6	Tarapur 1	Wet	51	Fukushima 6	MARK-II
7	Tarapur 2	Wet	52	Tokai 2	MARK-II
8	Dodewaarde	Wet	53	Hanford 2	MARK-II
9	Nine Mile Point 1	MARK-I	54	Nine Mile Point 2	MARK-II-C
10	Oyster Creek 1	MARK-I	55	Grand Gulf 1	MARK-III-C
11	Dresden 2	MARK-I	56	Grand Gulf 2	MARK-III-C
12	Dresden 3	MARK-I	57	Perry 1	MARK-III
13	Millstone 1	MARK-I	58	Perry 2	MARK-III
14	Tsuruga	MARK-I	59	River Bend 1	MARK-III
15	Nuclenor	MARK-I	60	River Bend 2[a]	MARK-III
16	Monticello	MARK-I	61	Laguna Verde 1	MARK-II
17	Quad Cities 1	MARK-I	62	Laguna Verde 2	MARK-II
18	Quad Cities 2	MARK-I	63	Clinton 1	MARK-III-C
19	Fukushima 1	MARK-I	64	Clinton 2	MARK-III-C
20	Brown's Ferry 1	MARK-I	65	Hartsville A1[a]	MARK-III
21	Brown's Ferry 2	MARK-I	66	Hartsville A2[a]	MARK-III
22	Brown's Ferry 3	MARK-I	67	Hartsville B1[a]	MARK-III
23	Vermont Yankee	MARK-I	68	Hartsville B2[a]	MARK-III
24	Peach Bottom 2	MARK-I	69	Phipps Bend 1[a]	MARK-III
25	Peach Bottom 3	MARK-I	70	Phipps Bend 2[a]	MARK-III
26	KKM	MARK-I	71	Kuo Sheng 1	MARK-III-C
27	Fitzpatrick 1	MARK-I	72	Kuo Sheng 2	MARK-III-C
28	Shoreham	MARK-II-C	73	Cofrentes	MARK-III
29	Cooper	MARK-I	74	Leibstadt	MARK-III
30	Pilgrim	MARK-I	75	Skagit/Hanford[a]	MARK-III
31	Fukushima 2	MARK-I	76	Skagit 2[a]	MARK-III
32	Hatch 1	MARK-I	77	Black Fox 1 (Can)	MARK-III
33	Hatch 2	MARK-I	78	Black Fox 2 (Can)	MARK-III
34	Brunswick 2	MARK-I-C	79	Alto Lazio 1	MARK-III
35	Brunswick 1	MARK-I-C	80	Alto Lazio 2	MARK-III
36	Duane Arnold	MARK-I	81	CNV 1	MARK-III
37	Enrico Fermi 2	MARK-I	82	CNV 2	MARK-III
38	Limerick 1	MARK-II-C	83	Graben[b]	MARK-III
39	Limerick 2	MARK-II-C	84	Kaiseraugst[b]	MARK-II-C
40	Hope Creek 1	MARK-I	85	Allens Creek 1[a]	MARK-III
41	Hope Creek 2 (Can)	MARK-I	86	Santillan[b]	MARK-III
42	Wm. H. Zimmer 1	MARK-II-C			
43	Chinshan 1	MARK-I			
44	Chinshan 2	MARK-I			
45	Caorso	MARK-II-C			

[a]Cancelled

[b]Suspended

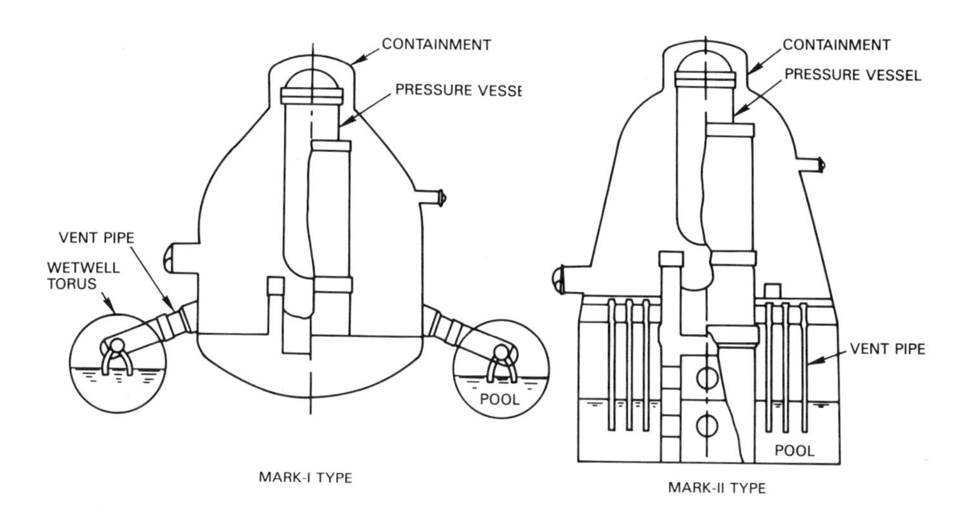

Fig. 1-10 Japanese modified MARK-I and -II pressure suppression system designs.

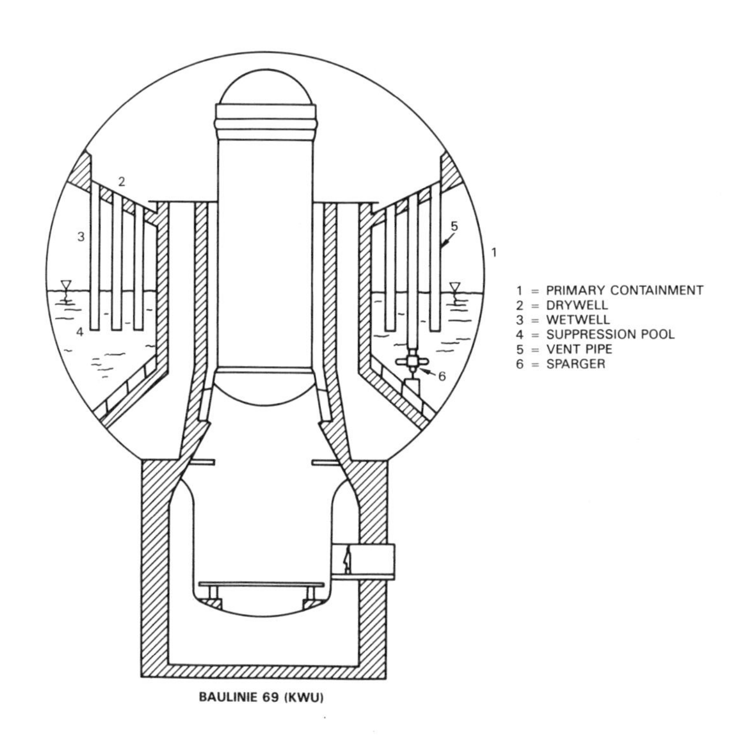

Fig. 1-11 German pressure suppression system designs.

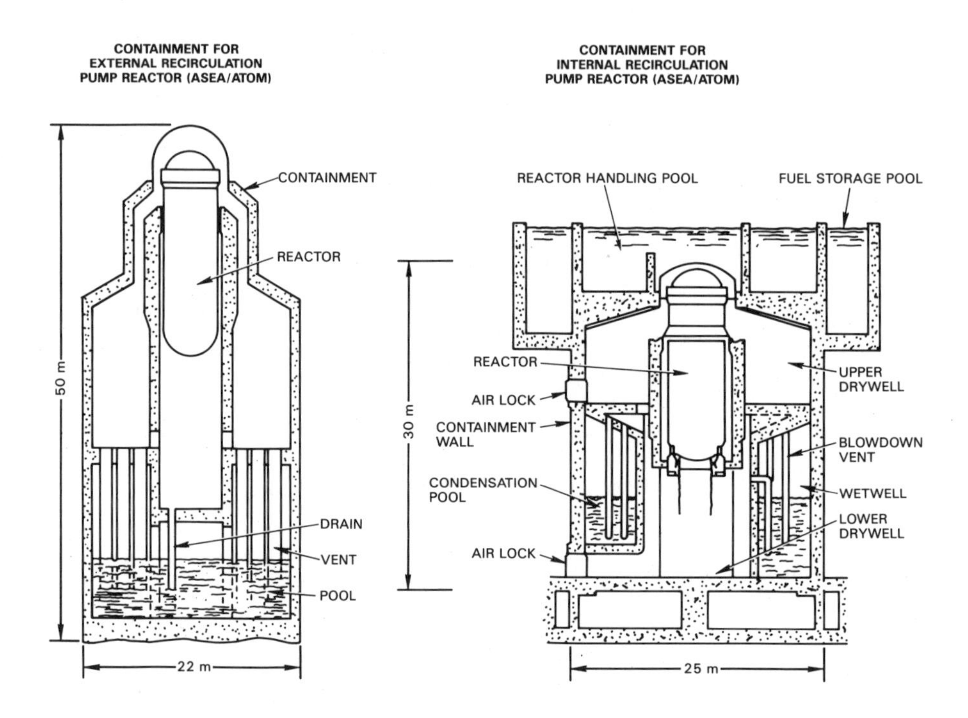

Fig. 1-12 ASEA-ATOM pressure suppression system designs.

design concept for limiting maximum containment pressure, while absorbing large amounts of energy from steam condensation.

Additional confirmatory pressure suppression tests were performed in a full-scale 1/48 segment of the Humboldt Bay and a 1/112 segment of the Bodega Bay containment designs. Both tests included simulated reactor vessels with pressure up to 1250 psig, a double rupture-disk/orifice assembly to simulate a pipe rupture and discharge to the drywell, with venting to a water pool. Variations in pool water level, temperature, reactor water subcooling, and orifice and vent sizes were found to have only minor effects on the pressure suppression behavior. The ranges of parameters tested provided data for the verification of analytical models.

Continued study and testing of pressure suppression designs made it possible to identify and quantify short-term containment loads, which may occur during a LOCA or after vessel depressurization is almost complete, and during SRV discharge. These short-term loads are associated with several dynamic phenomena, including the following:

1. *Pool swell:* When the drywell air is discharged through the vents, a pool swell occurs, which may create impact loads on various wetwell structures.

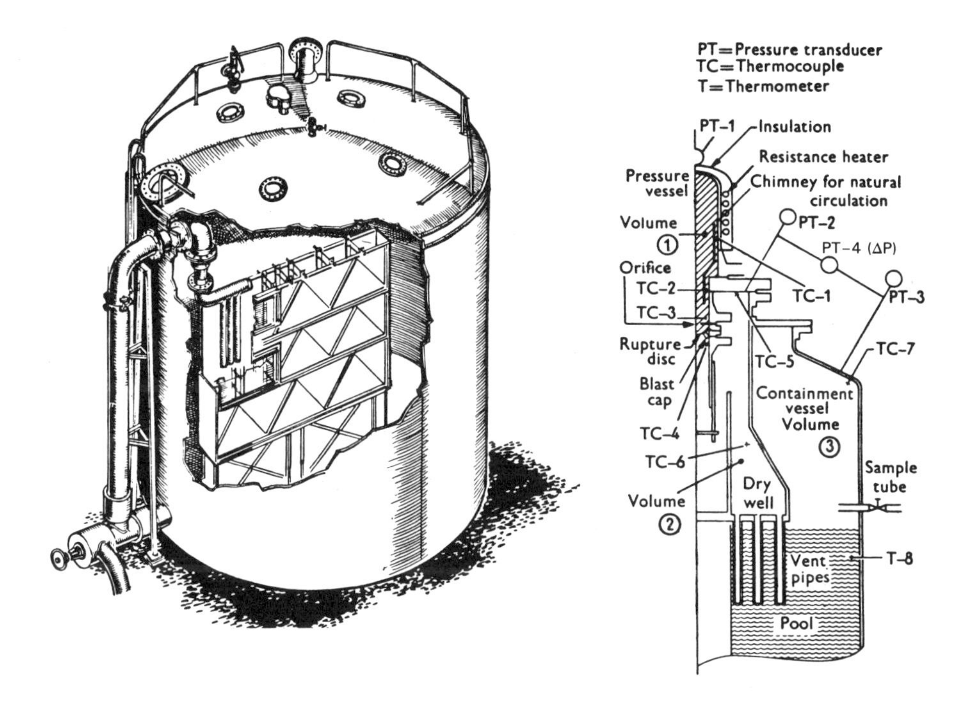

Fig. 1-13 Condensing test facility showing test compartment with vents.

Fig. 1-14 Arrangement of transient test facility.

2. *Condensation oscillations:* This phenomenon occurs during early steam discharge from the vents, and resembles a flickering flame of steam as the condensation boundary oscillates and creates acceleration of the surrounding water, which imposes unsteady loads on submerged structures.
3. *Chugging:* This occurs at low steam flow, and results in the cyclic rise and expulsion of pool water in the vents like a large percolator, resulting in dynamic loads on submerged structures.
4. *SRV air clearing:* Steam discharges into the submerged relief pipes and compresses the initial air prior to water seal expulsion, causing the introduction of a compressed air bubble into the pool. This bubble undergoes large amplitude oscillations and creates dynamic loads on wetwell structures.
5. *High-temperature condensation instability:* This phenomenon can occur if an SRV steam discharge warms local regions of the pool water, delaying steam condensation until bubbles move into regions of cold water, where they collapse and produce dynamic loads on wetwell structures.

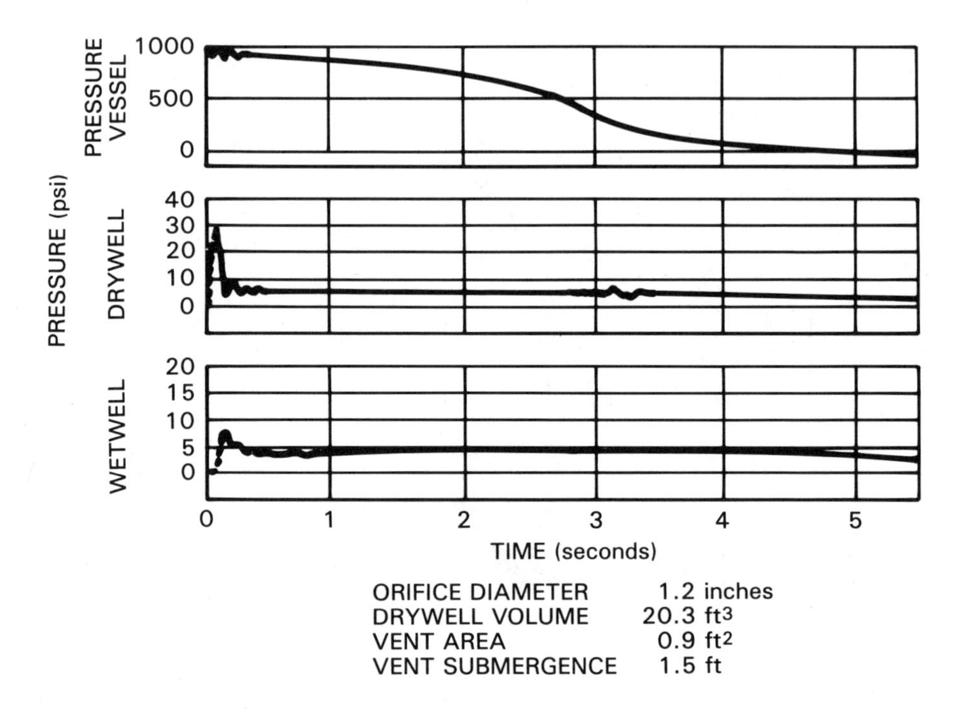

Fig. 1-15 Pressures recorded during a transient test.

Research programs were conducted for MARK-II-type pressure suppression designs in Europe at the Marviken Laboratory in Sweden. The test facility had four large vents for the study of both dynamic response and overall containment behavior. An eight vent test for the MARK-I containment was also performed at GE's Full-Scale Test Facility to obtain a better understanding of multivent condensation phenomena. Multivent steam blowdown tests were also conducted at the GKSS and KWU plants in Germany to study condensation and vent pipe loads. Also, four to seven vent pipe tests of a 20-deg sector of a full-size plant were conducted at the JAERI laboratory in Japan. A full-scale, 1/1000 volume segment of the MARK-III containment was studied by GE to provide a comprehensive data base for vent clearing, condensation, pool swell, pool impact with structures, and dynamic loads on the pool and vent boundaries. Other single vent tests were focused on a study of condensation oscillation and chugging phenomena.

The multitude of tests performed have confirmed the pressure suppression containment concept, defined the various loads, and provided a substantial data base for containment design and the verification of analytical models. Theoretical models have not yet been refined to a degree that describes all aspects of unsteady steam condensation. While state-of-the-art modeling of unsteady condensation continues to improve the overall understanding of this unsteady phenomena, pressure suppression designs are currently based on experimental results with well-accepted design factors.

References

Ashworth, C. P., D. B. Barton, E. Janssen, and C. H. Robbins, "Predicting Maximum Pressure in Pressure Suppression Containment," ASME Paper 61-WA-222 (1961).

Ashworth, C. P., D. B. Barton, and C. H. Robbins, "Pressure Suppression," *Nucl. Eng.* (Aug. 1962).

Cohen, K., and E. Zebroski, "Operation Sunrise," *Nucleonics*, **17** (Mar. 1959).

"Final Hazards Summary Report: Humboldt Bay Power Plant Unit 3," Pacific Gas and Electric (Sep. 1, 1961).

Kramer, A. W., *Boiling Water Reactors*, Addison-Wesley Publishing Company, Reading, Massachusetts (1958).

"Marviken Full Scale Containment Experiment," Summary Reports MxA-301, MxA-402, MxB-302, MxB-402, Studsvik Energiteknik AB, Sweden (Dec. 1974; Mar. 1977).

McIntyre, T., et al., "The MARK III Confirmatory Test Program," *Proc. Int. Specialty Mtg. BWR Pressure Suppression Containment Technology*, GKSS 81/E/27 (1981).

Miller, D. R., "Pressure Suppression Containment Design—Current State of the Art," *J. Eng. Power*, **91**(A:1) (Jan. 1969).

Namatame, K., et al., "Full-Scale MARK II Containment Response Test Program: Test Facility Description," JAERI-M 8780, Japan Atomic Energy Research Institute (1980).

"Pressure Suppression Test Program, Appendix I, Preliminary Hazards Summary Report: Bodega Bay Atomic Park Unit 1," General Electric Company and Pacific Gas and Electric Company (Dec. 1962).

Voight, O., et al., "Consequences Drawn from a Stuck Open Relief Valve Incident at the Wurgassen Power Plant," Bereicht Kernreactoren, KWU-Frankfurt, Germany.

Welchel, C. C., "Pressure Suppression Approved for Humboldt Bay," *Electrical World* (Nov. 21, 1960).

Welchel, C. C., and C. H. Robbins, "Pressure Suppression Containment for Nuclear Power Plants," ASME Paper 59-A-215 (1959).

CHAPTER TWO

The Nuclear Boiler Assembly

The nuclear boiler assembly consists of the equipment and instrumentation necessary to produce, contain, and control the steam flow required by the turbine-generator. The principal components of a BWR/6 nuclear boiler are (BWR/6, 1975):

- *Reactor vessel and internals:* reactor pressure vessel, jet pumps for reactor water recirculation, steam separators, steam dryers, and core support structure
- *Reactor water recirculation system:* pumps, valves, and piping used in providing and controlling core flow
- *Main steam lines:* main steam line valves, piping, and pipe supports from reactor pressure vessel up to and including the isolation valves outside the primary containment barrier
- *Control rod drive (CRD) system:* control rods, CRD mechanisms, and hydraulic system for insertion and withdrawal of the control rods
- *Nuclear fuel and instrumentation:* fuel rods, channels, and in-core neutron flux monitors.

2.1 Reactor Assembly

The reactor assembly (Fig. 2-1) consists of the reactor vessel, internal components of the core, shroud, top guide assembly, core plate assembly, shroud head, steam separator, dryer assemblies, and jet pumps. Also included in the reactor assembly are the in-core neutron flux monitors, control rods, CRD housings, and the CRDs. However, due to the importance of these components, they are considered separately in later sections.

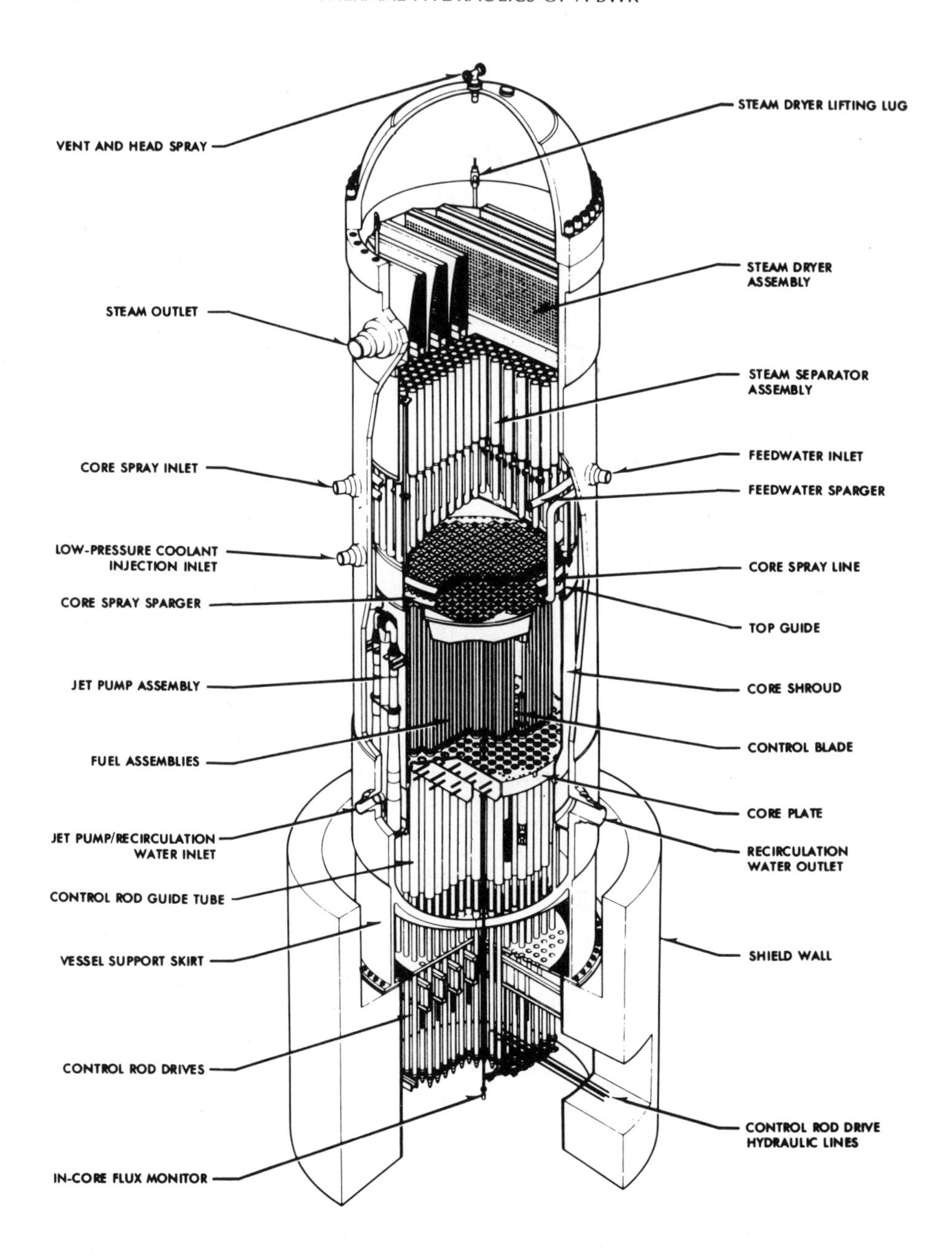

Fig. 2-1 BWR/6 reactor assembly.

Most of the fuel assemblies that make up the core rest on orificed fuel supports mounted on top of the control rod guide tubes. Each guide tube, with its fuel support piece, bears the weight of four assemblies and is supported by a CRD penetration nozzle in the bottom head of the reactor vessel. The core plate provides lateral guidance at the top of each control rod guide tube. The remaining fuel assemblies, located at the periphery of the core, are supported by orificed fuel supports that are welded to the core plate to provide both vertical and lateral support at the lower end of the fuel. The top guide provides lateral support for the top of each fuel assembly.

Control rods occupy alternate spaces between fuel assemblies and can be withdrawn into the guide tubes below the core during plant operation. The rods are coupled to CRDs mounted within housings that are welded to the bottom head of the reactor vessel.

Figures 2-1 and 2-2 show that there are many penetrations in the lower head to accommodate the control rod drive housings and the in-core instrument housings. In particular, there are instrument tubes, control rod drive housings, and a drain plug (in the center of the lower head). Details of the in-core neutron detectors are shown in Fig. 2-3. All of these penetrations involve members with a tube wall thickness that is significantly less than that of the lower head. As a consequence, these members may preferentially fail during severe accidents, such as those discussed in Sec. 8.5.

2.1.1 Reactor Vessel

The reactor vessel is a pressure vessel with a single full-diameter removable head. The base material of the vessel is low-alloy steel that is clad on the interior, except for the top head and nozzles, with stainless-steel weld overlay to provide corrosion resistance.

The vessel head closure seal consists of two concentric metal O-rings. To monitor seal integrity, a leak detection system is provided.

Vessel supports, internal supports, their attachments, and adjacent shell sections are designed to take maximum combined loads, including CRD reactions, earthquake loads, and pipe break reaction thrusts. The vessel is mounted on a supporting skirt that is bolted to a concrete and steel cylindrical vessel pedestal, which is integral with the reactor building foundation.

2.1.2 Core Shroud

The core shroud is a cylindrical, stainless-steel structure that surrounds the core and provides a barrier to separate the upward flow through the core from the downward flow in the annulus. A flange at the top of the shroud mates with a flange on the top guide which, in turn, mates with a flange on the shroud head and steam separator assembly to form the core outlet plenum. The jet pump diffusers penetrate the peripheral shelf

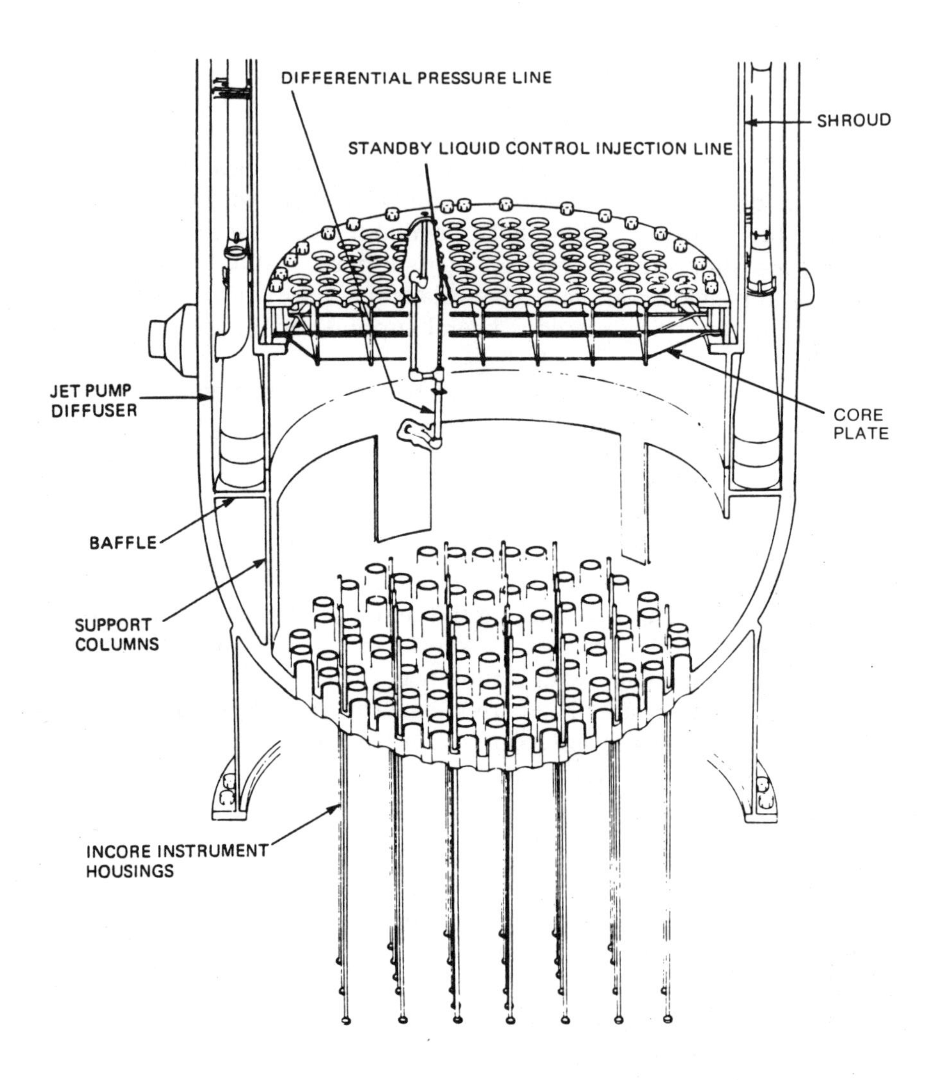

Fig. 2-2 Reactor pressure vessel lower plenum.

of the shroud support below the core elevation to introduce the coolant into the inlet plenum. The peripheral shelf of the shroud support is welded to the vessel wall to prevent the jet pump outlet flow from bypassing the core and to form a chamber around the core that can be reflooded in the event of a loss-of-coolant accident (LOCA). The shroud support is designed to carry the weight of the shroud, the steam separators, the jet pump system, and the seismic and pressure loads both in normal and abnormal conditions of operation.

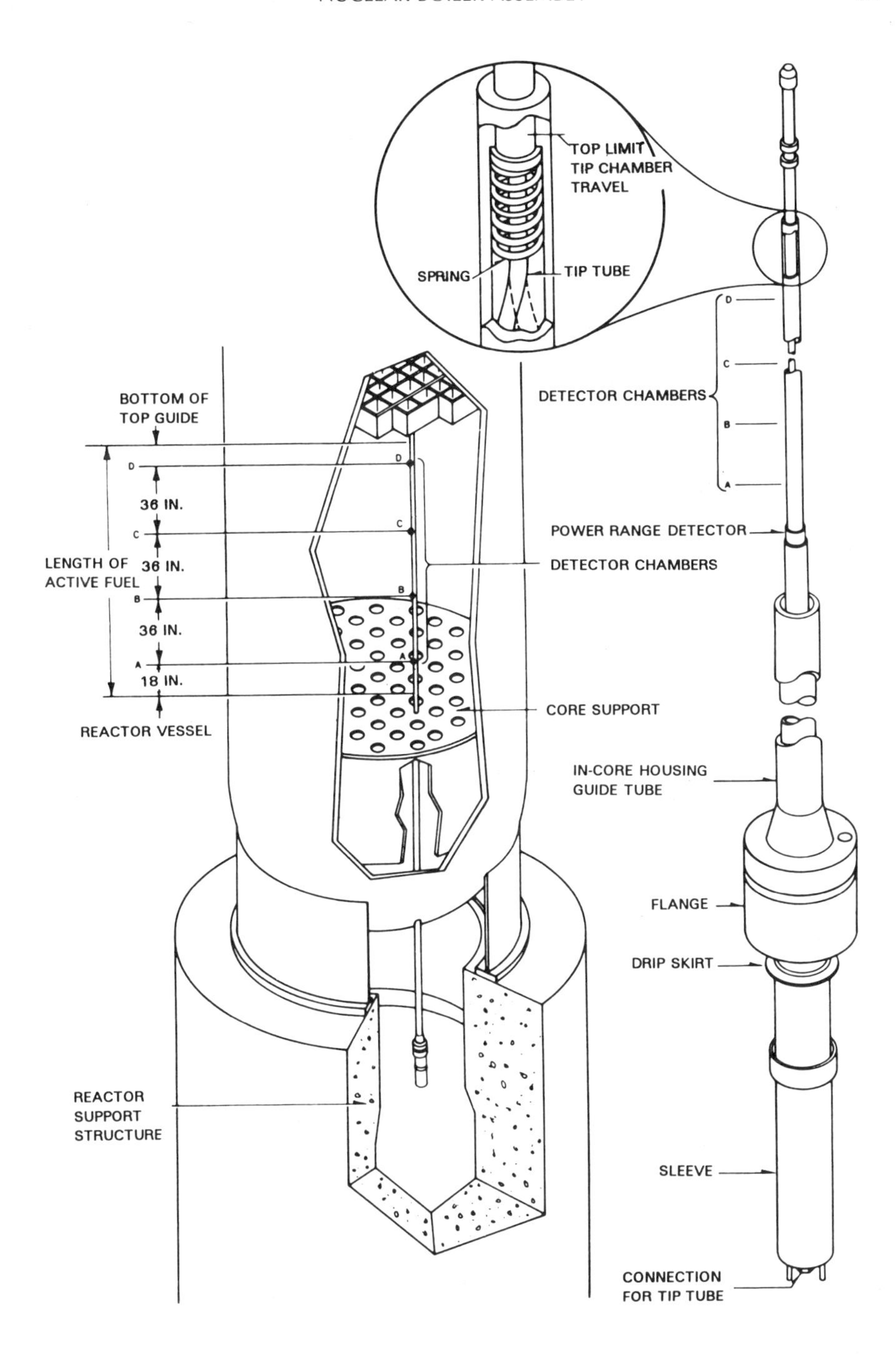

Fig. 2-3(a) In-core neutron detector assembly.

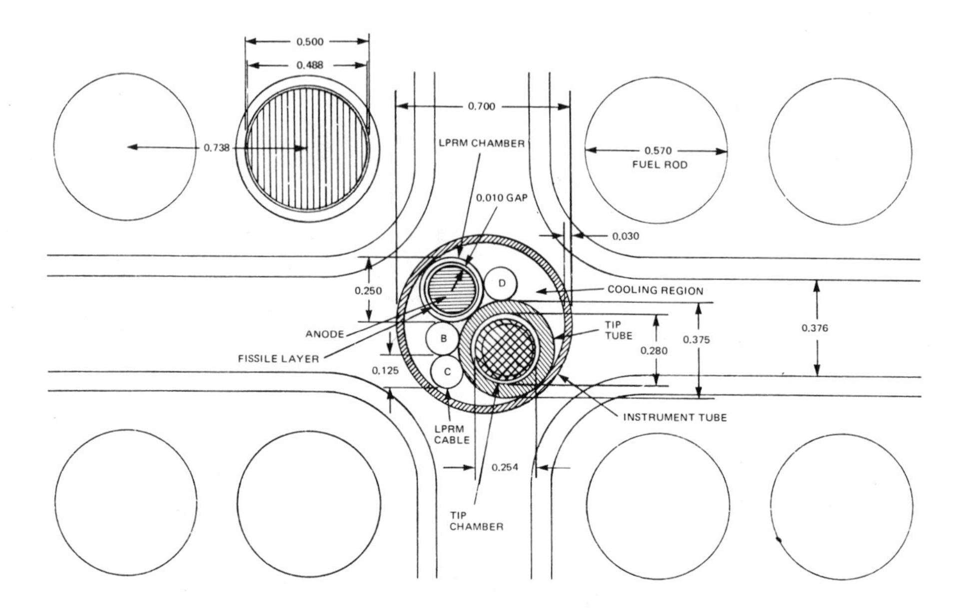

Fig. 2-3(b) Cross section of a typical in-core neutron detector.

Two ring spargers, one for low-pressure core spray and the other for high-pressure core spray, are mounted inside the core shroud in the space between the top of the core and the steam separator base. The core spray ring spargers are provided with spray nozzles for the injection of cooling water. The core spray spargers and nozzles have been designed not to interfere with the installation or removal of fuel from the core.

2.1.3 Shroud Head and Steam Separator Assembly

As can be seen in Fig. 2-4, the shroud head and steam separator assembly consists of a domed base, on top of which is welded an array of standpipes with a three-stage steam separator located at the top of each standpipe. The shroud head and steam separator assembly rests on the top flange of the top guide grid and forms the cover of the core outlet plenum region. The fixed axial-flow-type steam separators have no moving parts and are made of stainless steel.

In each separator, the steam-water mixture rising through the standpipe impinges on vanes that give the mixture a spin to establish a vortex such that the centrifugal forces separate the water from the steam in each of three stages. Steam leaves the separator at the top and passes into the wet steam plenum below the dryer. The separated water exits from the lower end of each stage of the separator and enters the pool that surrounds the

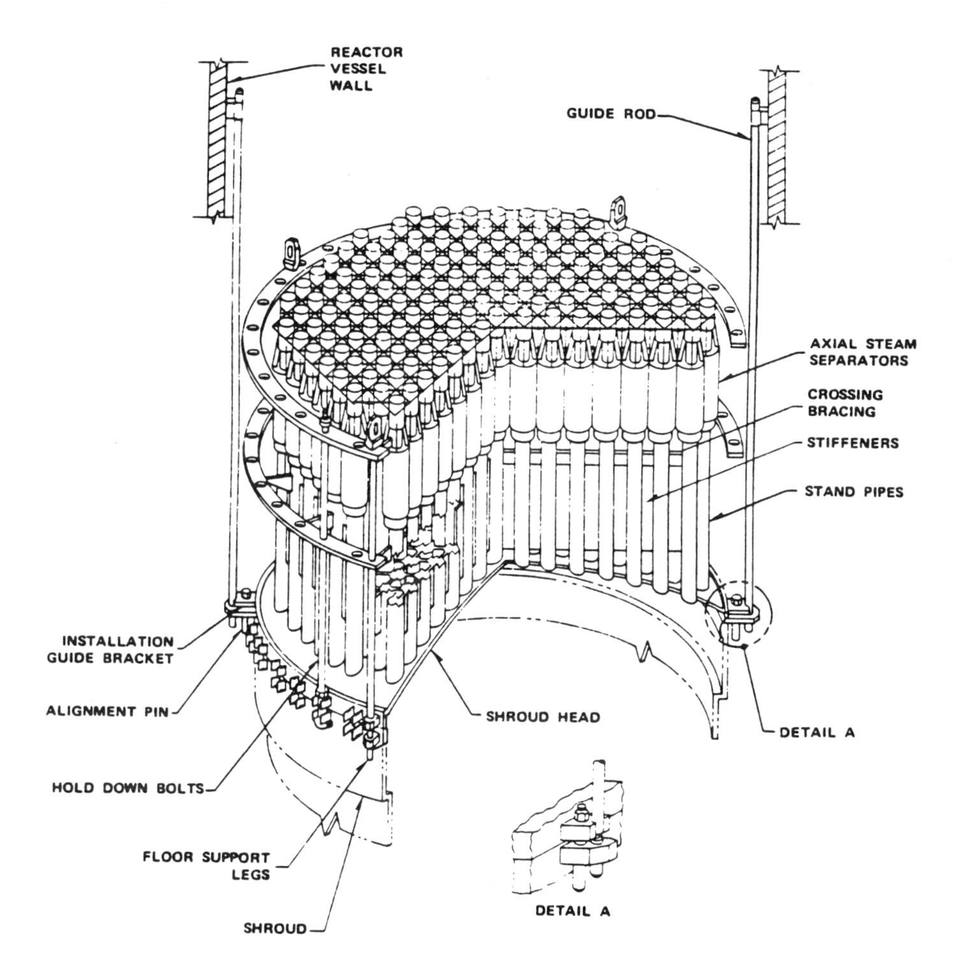

Fig. 2-4 Shroud head and steam separator assembly.

standpipes to join the downcomer annulus flow. An internal steam separator is shown schematically in Fig. 2-5.

2.1.4 Steam Dryer Assembly

The steam dryer assembly is mounted in the reactor vessel above the shroud head and steam separator assembly and forms the top and sides of the wet steam plenum. Vertical guides on the inside of the vessel provide alignment for the dryer assembly during installation. The dryer assembly is supported by brackets extending inward from the vessel wall. There are brackets attached to the vessel head, which would limit upward motion of the dryer assembly during seismic or pipe break conditions. Steam from

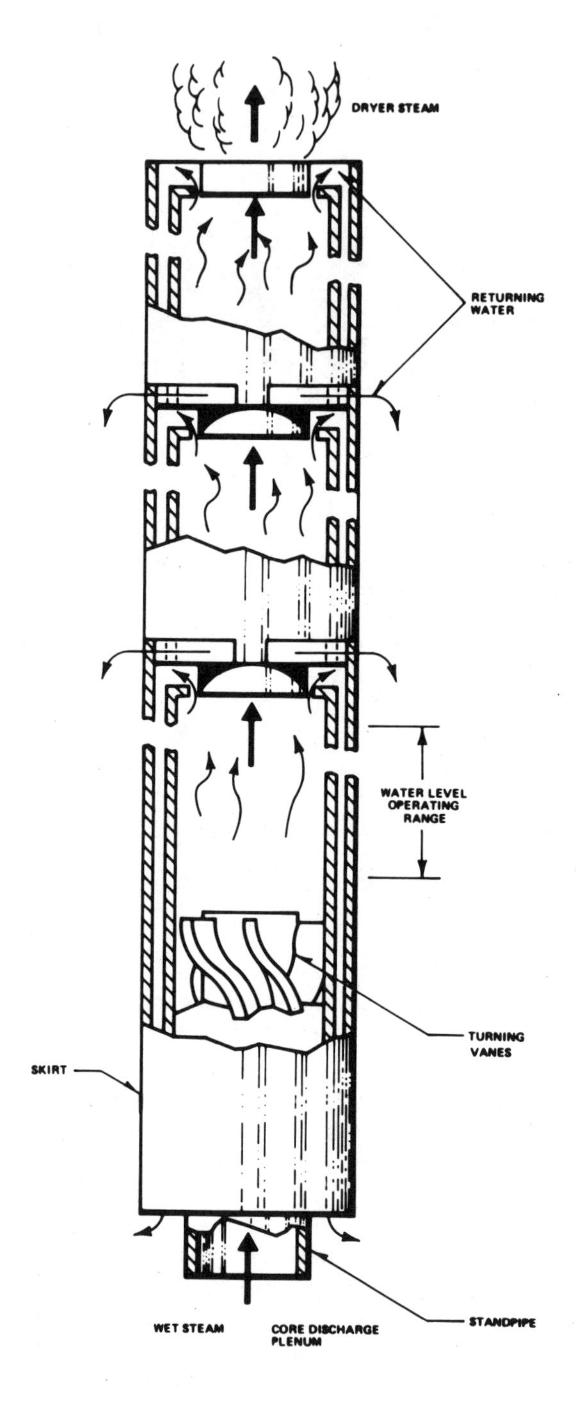

Fig. 2-5 Internal steam separator.

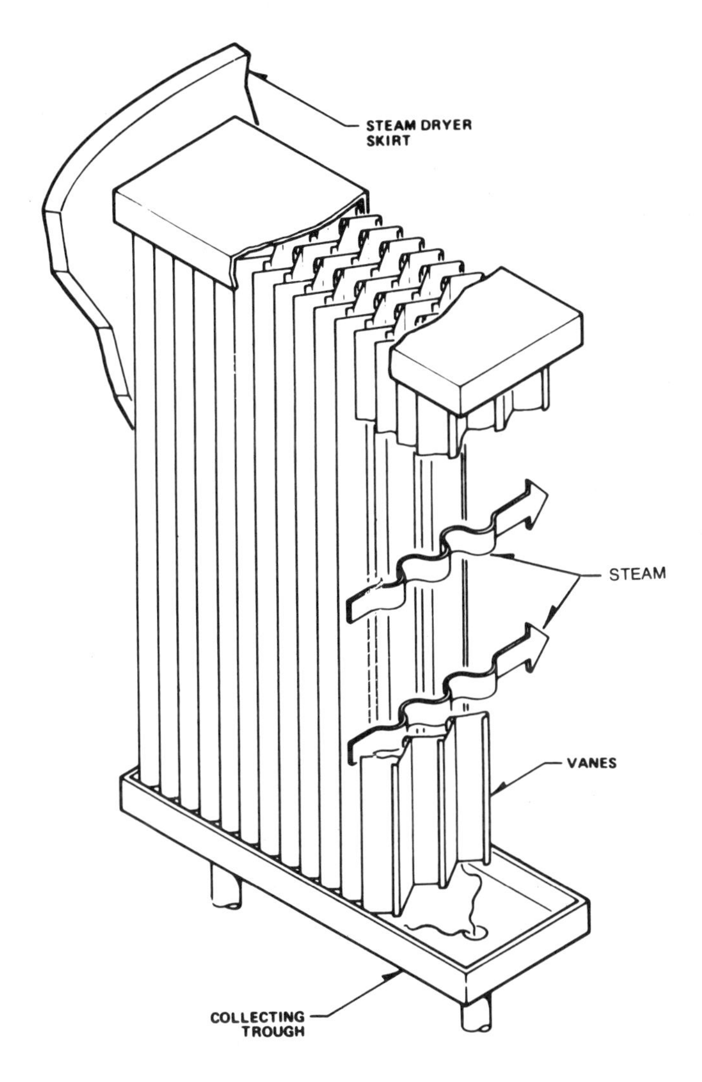

Fig. 2-6 Steam dryer.

the separators flows upward and outward through the chevron-type drying vanes. These vanes are attached to a top and bottom supporting member forming a rigid, integral unit. Moisture is removed and carried by a system of troughs and drains to the pool surrounding the separators and then into the recirculation downcomer annulus between the core shroud and reactor vessel wall. Figure 2-6 shows a typical steam dryer panel.

2.2 Reactor Water Recirculation System

The function of the reactor water recirculation system is to circulate the required coolant through the reactor core. The system consists of two loops external to the reactor vessel, each containing a pump with a directly coupled water-cooled motor, a flow control valve, two shutoff valves, and a bypass valve.

High-performance jet pumps located within the reactor vessel are used in the BWR/6 recirculation system. The jet pumps, which have no moving parts, provide a continuous internal circulation path for a major portion of the core coolant flow.

2.2.1 Jet Pump Recirculation System

The jet pump recirculation system provides forced circulation flow through BWR cores. As shown in Fig. 2-7, the recirculation pumps take suction from the downward flow in the annulus between the core shroud and the vessel wall. Approximately one-third of the core flow is taken from the vessel through two recirculation suction nozzles. There, it is pumped to a higher pressure, distributed through a manifold to which a number of riser pipes are connected, and returned to the vessel inlet nozzles. This flow is discharged from jet pump nozzles into the initial stage of the jet pump throats where, due to momentum exchange, it induces the surrounding water in the downcomer region to be drawn into the jet pump throats where these two flows mix and flow through the diffuser, into the lower plenum of the reactor pressure vessel.

The jet pump diffusers are welded into openings in the core shroud support shelf, which forms a barrier between the lower plenum and the annular downcomer region where the jet pumps are located. The flow of water from the jet pumps enters the lower plenum, flows between the CRD guide tubes, and enters into the fuel support where the flow is individually directed to each fuel bundle through the nose piece. Orifices in each fuel support piece provide the desired flow distribution among the fuel assemblies. The coolant water passes along the individual fuel rods inside the fuel channel where it boils and becomes a two-phase steam/water mixture. The steam/water mixture enters a plenum located directly above the core and bounded by the shroud head, which opens to the separator array of fixed steam separators. As discussed previously, the steam is separated from the water and passes through a dryer where any remaining water is removed. The saturated steam leaves the vessel through steam line nozzles located near the top of the vessel body and is piped to the turbine. Water collected in the support tray of the dryer is routed through drain lines, joins the water leaving the separators, and flows downward in the annulus between the core shroud and the vessel wall. Feedwater is added to the system through spargers located above the annulus

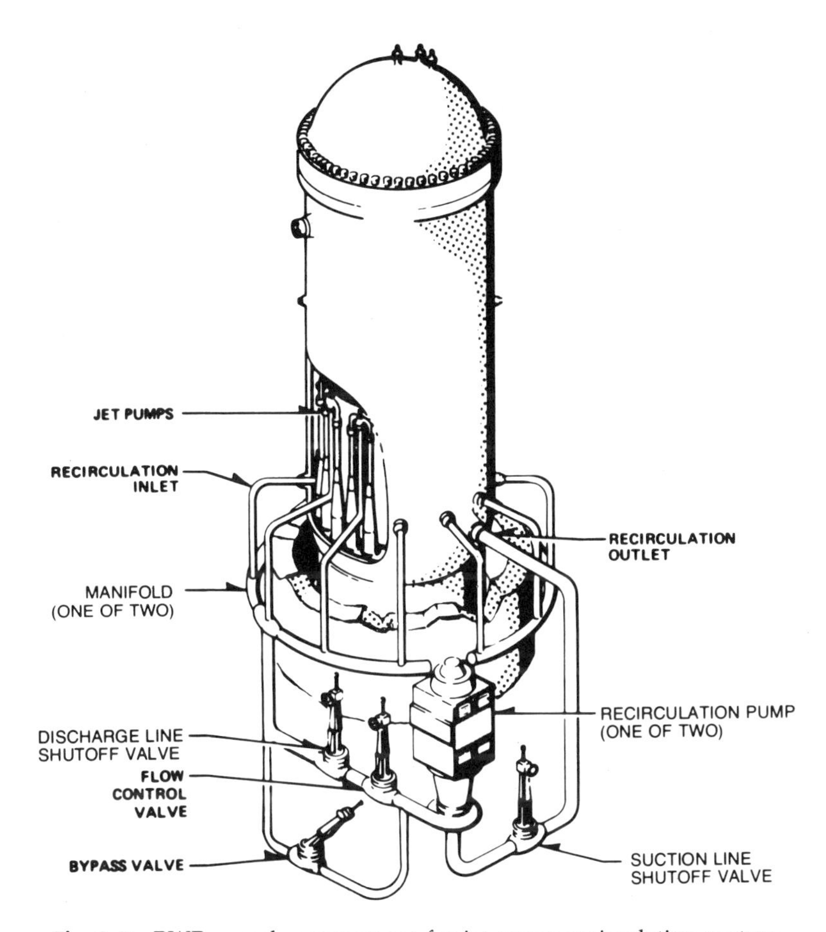

Fig. 2-7 BWR vessel arrangement for jet pump recirculation system.

and joins the downward flow of water. A portion (about two-thirds) of this downward flow enters the jet pumps and the remainder exits from the vessel as recirculation flow.

Figure 2-8 shows that each jet pump assembly is composed of two jet pumps and contains no moving parts. Each BWR/6 jet pump consists of an inlet mixer, a nozzle assembly with five discharge nozzles, and a diffuser.

The inlet mixer assembly is a constant-diameter section of pipe with a diffuser entrance section at the lower end and the drive nozzle at the upper end.

The jet pump diffuser is a gradual conical section terminating in a straight cylindrical section at the lower end, which is welded into the shroud support.

The overall length of the jet pumps is about 19 ft. Each pair of jet pumps is supplied driving flow from a single riser pipe. These risers have indi-

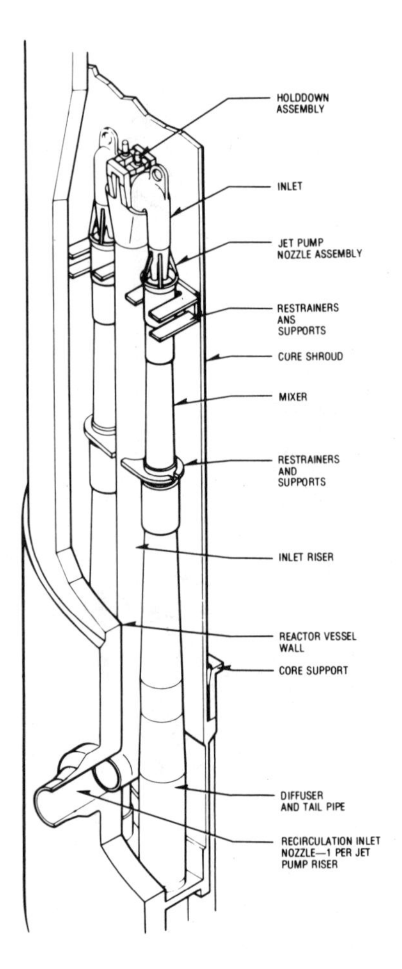

Fig. 2-8 Jet pump assembly.

vidual vessel penetrations and receive flow from one of two external manifolds. Driving flow to each distribution manifold is furnished by its associated centrifugal pump. The recirculation system includes 20 to 24 jet pumps, depending on the size of the nuclear boiler system.

Instrumentation monitors jet pump flow to determine individual and collective flow rates under varying operating conditions.

2.2.2 Jet Pump Operating Principles and Features

The jet pump drive flow enters the nozzle section at a high pressure and is accelerated to a high velocity because of the constriction at the nozzle

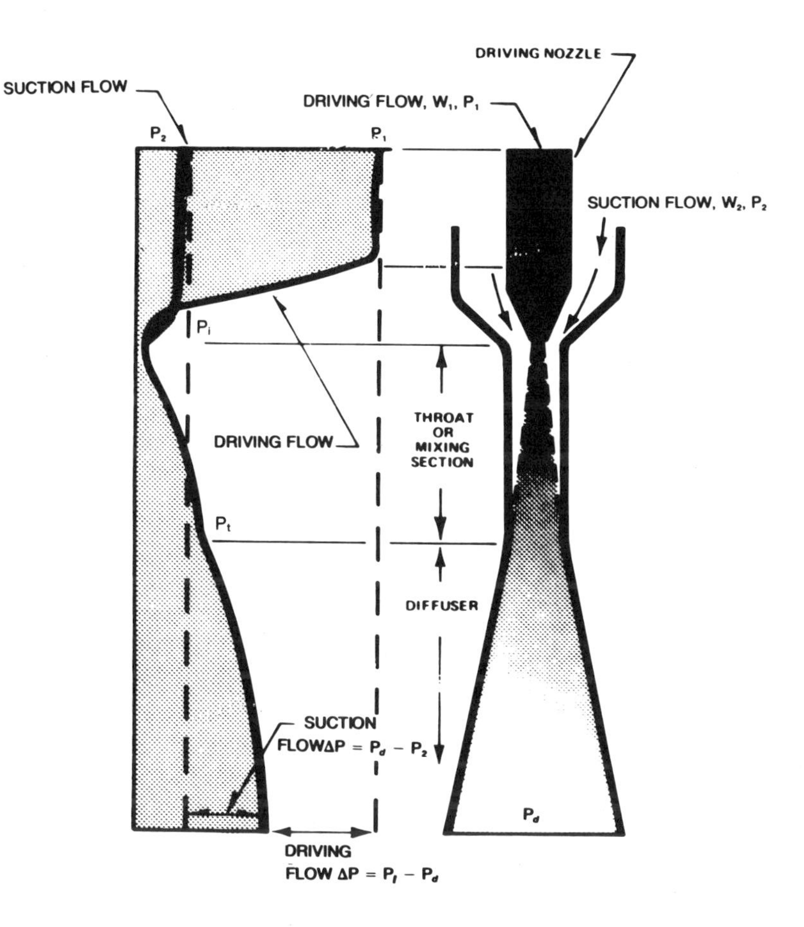

Fig. 2-9 Jet pump principle.

outlet. The suction flow enters at a low pressure, which is reduced further as the flow is accelerated through the converging suction inlet nozzle. These two streams merge in the mixing section where a pressure rise occurs because of velocity profile rearrangement and the momentum transfer caused by mixing. The rate of pressure rise decreases near the end of the mixing section because mixing essentially is completed. A diffuser is located downstream from the mixing section to slow the relatively high-velocity mixed streams. This diffuser converts the dynamic head into static head. These processes are illustrated in Fig. 2-9.

It is instructive to perform a simple analysis of a jet pump. The steady one-dimensional mass conservation equation shows that the velocity at the exit of the throat section, v_t, is:

$$v_t = (v_1 A_1 + v_2 A_2)/A_t \ . \tag{2.1}$$

Similarly, neglecting all irreversible hydraulic losses, the steady one-dimensional momentum conservation equation shows that the difference in the static pressures between the inlet, p_i, and the exit, p_t, of the throat section is given by:

$$p_i(A_1 + A_2) - p_t A_t = \frac{\rho_f}{g_c}(A_t v_t^2 - A_1 v_1^2 - A_2 v_2^2) \ , \tag{2.2}$$

where, as can be noted in Fig. 2-6, we have assumed that at the discharge of the drive nozzle both the suction and driving flow fluid pressures are equal to p_i. Thus, Eq. (2.2) yields,

$$p_t = p_i(A_1 + A_2)/A_t - \frac{\rho_f}{A_t g_c}[A_t v_t^2 - A_1 v_1^2 - A_2 v_2^2] \ . \tag{2.3}$$

To obtain the static pressure at the exit of the diffuser, p_d, we may use the ideal Bernoulli equation,

$$\frac{p_t g_c}{\rho_f g} + \frac{v_t^2}{2g} = \frac{p_d g_c}{\rho_f g} + \frac{v_d^2}{2g} \ , \tag{2.4}$$

where,

$$v_d = \frac{\rho_f A_t v_t}{\rho_f A_d} \ . \tag{2.5}$$

Using Eqs. (2.3) and (2.4) we can calculate the pressure rise in the suction flow due to the jet pump, $p_d - p_2$, and the pressure loss in the drive flow, $p_1 - p_d$.

The operating efficiency, η, of a jet pump is normally defined as the ratio of the energy increase of the suction flow to the energy decrease of the driving flow,

$$\eta \triangleq MN \cdot 100 \ , \tag{2.6}$$

where the M ratio is defined as the ratio of the suction flow divided by the driving flow,

$$M \triangleq \frac{w_2}{w_1} \ , \tag{2.7}$$

and the N ratio is defined as the specific energy increase of the suction flow divided by the specific energy decrease of the driving flow,

$$N \triangleq \frac{\left(\frac{p_d}{J\rho_l} + \frac{u_d^2}{2g_c J}\right) - \left(\frac{p_2}{J\rho_l} + \frac{u_2^2}{2g_c J}\right)}{\left(\frac{p_1}{J\rho_l} + \frac{u_1^2}{2g_c J}\right) - \left(\frac{p_d}{J\rho_l} + \frac{u_d^2}{2g_c J}\right)} \ . \tag{2.8}$$

In actual reactor applications, the N ratio is normally modified to account for nozzle (inlet) and diffuser (exit) losses and, thus, a net efficiency for reactor application, η_a, can be defined (Kudirka and Gluntz, 1974). This efficiency is related to the classical mechanical efficiency, η_e, in the following way:

$$\eta_e = \eta_a \frac{(M+1)}{(M+\eta_a/100)}\ , \tag{2.9}$$

although Eq. (2.6) is frequently used in jet pump technology. Note that modern BWR jet pumps are quite efficient, with values of $\eta = 45\%$ ($\eta_a = 41.5\%$) not uncommon.

An important safety feature inherent in a jet pump system is that the sizes of the recirculation lines do not need to be large enough to pass the total core flow; i.e., only about one-third of the flow is used as drive flow. Thus, the maximum pipe break area is reduced. In addition, as shown in Fig. 2-10, the long-term core flooding capability of a jet pump design is

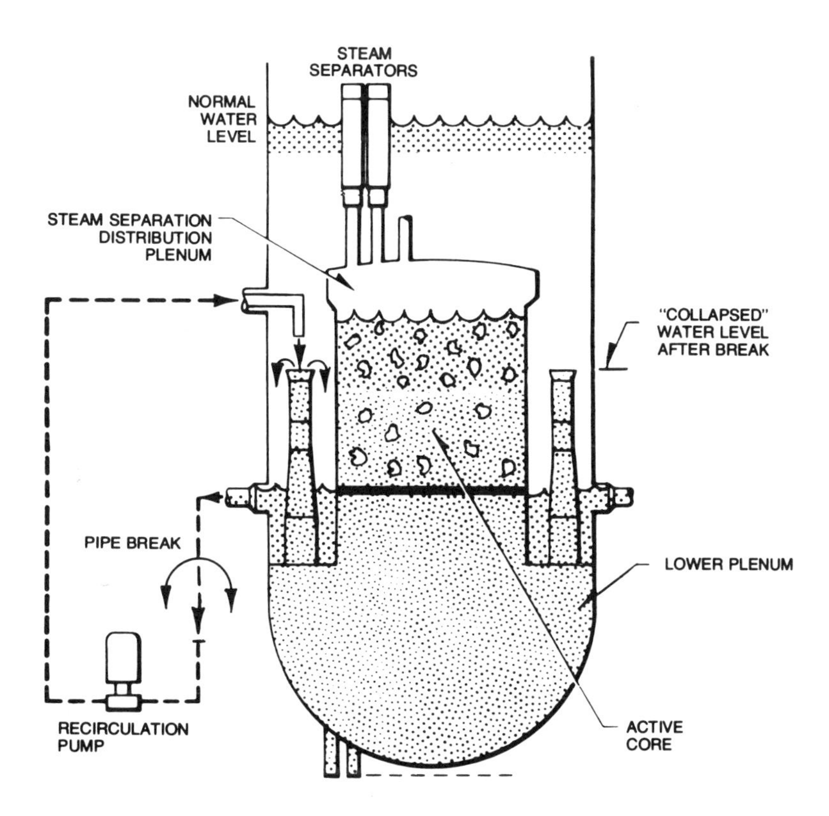

Fig. 2-10 Core submersion capability of jet pump system.

quite good. This design allows a "collapsed" water level to be maintained at two-thirds of the core height, which swells in the core due to boiling, such that the core can be kept covered to accomplish long-term cooling (Kamath and Lahey, 1980; Fakory and Lahey, 1985).

2.3 Main Steam Lines and Valves

Steam exits from the vessel several feet below the reactor vessel flange through four nozzles. Carbon-steel steam lines are welded to the vessel nozzles and run parallel to the vertical axis of the vessel, downward to the elevation, where they emerge from the containment. Two air-operated isolation valves are installed on each steam line, one inboard and one outboard of the primary containment penetration. The combination safety/relief valves are flange connected to the main steam line for ease of removal for test and maintenance.

A flow restricting nozzle is included in each steam line as an additional engineered safeguard to protect against rapid uncovering of the core in case of a main steam line break.

2.3.1 Safety/Relief Valves

The safety/relief valves are dual-function valves discharging directly to the pressure suppression pool. The safety function includes protection against overpressure of the reactor primary system.

The relief function provides power-actuated valve opening to relieve steam during transients resulting in high system pressure or during postulated accident conditions to depressurize the reactor primary system. The valves are sized to accommodate the most severe pressurization transients:

1. turbine trip from turbine design power, failure of direct scram on turbine stop-valve closure, failure of the steam bypass system, and reactor scrams from an indirect scram
2. closure of all main steam line isolation valves, failure of direct scram based on valve position switches, and reactor scrams from an indirect scram.

For the safety function, the valves open at spring-set point pressure and close when inlet pressure falls to 96% of spring-set point pressure.

For the pressure relief function, the valves are power actuated manually from the control room or power actuated automatically on high pressure (70 to 90 psi above the rated operating pressure). Each valve is supplied by separate power circuits. Valves that are power actuated automatically on high pressure are closed when pressure falls to a preset pressure of 35 to 55 psi below the pressure switch setpoint. Selected valves are associated with the automatic depressurization system (ADS) and, in addition to the capabilities discussed above, are power actuated automatically on coincident LOCA signals of high drywell pressure and low reactor water level. These selected valves also can be manually power actuated to open at any pressure. The signal for manual power actuation is from redundant control room switches from different power sources. The signal for automatic depressurization is from redundant channels, powered from different power sources.

2.3.2 Isolation Valves

The isolation valves for the main steam line are spring-loaded pneumatic piston-operated globe valves designed to fail closed should loss of pneumatic pressure or loss of power to the pilot valves occur. Each valve has an air accumulator to assist in the closure of the valve on loss of the air supply, electrical power to the pilot valves, and failure of the loaded spring. Each valve has an independent position switch initiating a signal into the reactor protection system scram trip circuit when the valve closes. Lights in the control room also indicate the position of the valves.

The signal for closure comes from two independent channels; each channel has two independent tripping sensors for each measured variable. Once isolation is initiated, the valves continue to close and cannot be opened except by manual means.

2.4 Control Rod Drive System

Positive core reactivity control is maintained by using movable control rods interspersed throughout the core. These control rods control the overall reactor power level and provide the principal means of quickly and safely shutting down the reactor. The rods are vertically moved by hydraulically actuated locking-piston-type drive mechanisms. The drive mechanisms perform both a positioning and latching function and a scram function, which overrides any other signal. The drive mechanisms are bottom entry, upward scramming drives that are mounted on a flanged housing on the reactor vessel bottom head. Here they cause no interference during refueling and are readily accessible for inspection and servicing. Moreover, control rod entry from below provides the best power shaping and, thus, fuel economy for a BWR.

The CRD system, shown in Fig. 2-11, consists of a number of locking piston CRD mechanisms, a hydraulic control unit for each drive mechanism, a hydraulic power supply for the entire system, and instrumentation and controls with necessary interconnections. The locking-piston-type CRD mechanism is a double-acting hydraulic piston that uses condensate water as the operating fluid. In addition, this water, which comes from the condensate storage tank, acts to cool the CRD mechanisms. (Typical flow rates are given in Fig. 3-3.)

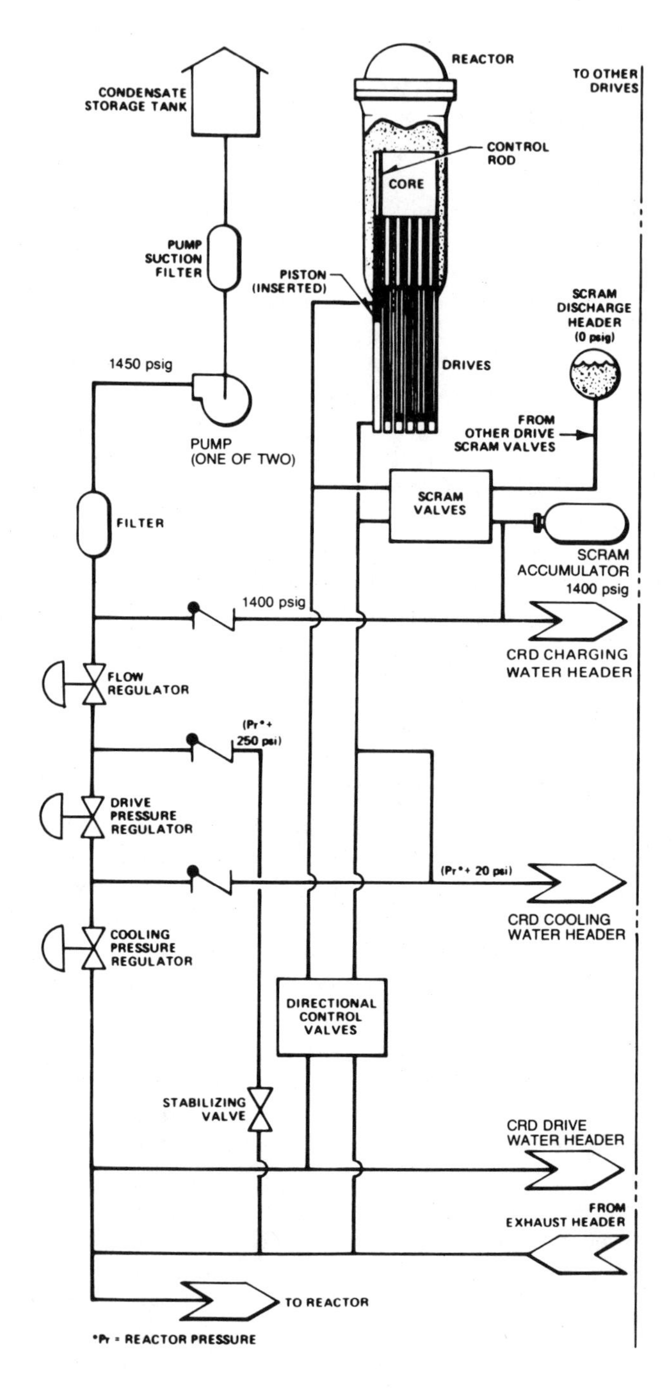

Fig. 2-11 Basic CRD system.

An index tube and piston, coupled to the control rod, are locked at fixed increments by a collet mechanism. The collet fingers engage notches in the index tube to prevent unintentional withdrawal of the control rod, but without restricting insertion. The drive mechanism can position the rods at intermediate increments over the entire core length. The CRDs can be uncoupled and removed from below the vessel without removing the reactor vessel head. Also, the control rods can be uncoupled and removed, with the vessel head removed for refueling, without removing the drive mechanism.

The number of drives supplied with a particular reactor is selected to optimize the power distribution in the core and to give the operator the maximum degree of control flexibility during startup, maneuvering, and flux shaping.

2.4.1 Control Rod Drive Hydraulic Supply System

The CRD hydraulic system pump and filters, shown in Fig. 2-11, are located outside the primary containment. The remaining major elements of the CRD system are located within the primary containment. This system supplies pressurized demineralized water, on demand, for operation of the CRDs, and cooling water for the drive mechanisms. The system is made up of high-head low-flow pumps and the necessary piping, filters, control valves, and instrumentation. Two pumps are provided, one of which is used as a standby spare. The pumps take suction from the condensate storage tank and discharge excess water (not used for CRD operation or cooling) directly into the reactor vessel.

Water from the condensate storage tank is pumped to a nominal pressure of 1450 psig by a multistage centrifugal pump. After filtering at ~1400 psig, the water charges accumulators that store the high-pressure water for the scram function. The flow regulator automatically maintains a constant flow to the system. During normal operation periods, when rod drive movement is not required, adequate flow is provided for cooling each of the drives by way of the cooling water header, with the remainder of the flow going to the reactor vessel through the cooling water pressure regulator valve. The cooling water pressure regulator valves are manually adjusted to give the desired pressure at the cooling water header to maintain proper temperatures within the drive mechanisms.

The manually adjustable drive pressure regulator valve maintains the correct pressure on the drive water header for control rod positioning. During periods when a drive mechanism is in motion, stabilizing valves are automatically closed to reduce the flow bypassing the drive water pressure regulator valve by the amount required to move a drive, thus maintaining the pressure balance of the system. During multiple drive operation (ganged rods) an appropriate additional number of stabilizing valves are closed to accommodate the increased flow requirements.

The basic components of the hydraulic system for controlling the drive mechanism during rod positioning and scram operation are shown in Fig. 2-12. All components shown are typical for each rod drive.

The main movable element of the system consists of the main drive piston, the index tube, and the control rod coupled to the index tube. The movable element is held in any chosen position by a collet that engages one of several notches in the index tube. Gravity holds the tube notch against the latch since the entire mechanism is essentially at reactor pressure. The control rod is moved by applying a pressure greater than reactor pressure to either the top or bottom of the main drive piston. When the reactor protection system calls for a reactor scram, all control rods are driven into the core at the maximum rate of speed.

When a scram signal is initiated, control air is vented from the scram valves allowing them to open by spring action. Opening of the exhaust scram valve vents the pressure above the drive piston to the scram discharge volume, which is maintained at atmospheric pressure prior to scram, and opening of the inlet scram valve applies the accumulator pressure to the bottom of the piston. Since the notches in the index tube are tapered on the lower edge, the latch is forced open by cam action, allowing the

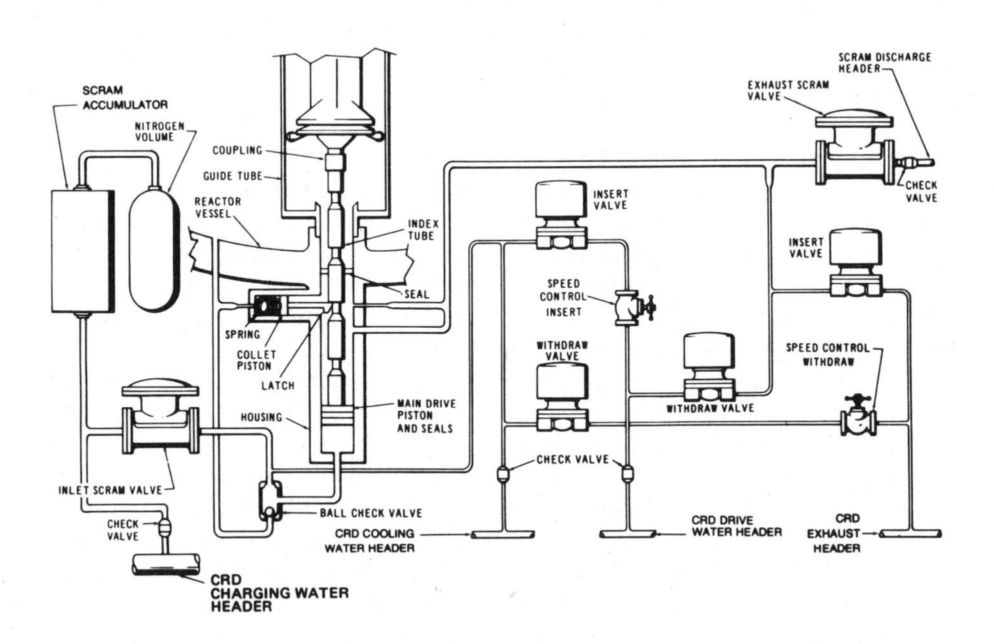

Fig. 2-12 Locking piston drive system.

index tube to move upward without restriction under the influence of the high-pressure differential across the drive piston. As the drive moves upward and the accumulator pressure reduces to the reactor pressure, the ball check valve changes position, permitting reactor water to complete the scram action. If reactor pressure is low, such as during startup, the accumulator fully inserts the rod in the required time without assistance from reactor pressure.

The actual mechanical arrangement of the drive mechanism is illustrated in Fig. 2-13, which shows the important elements in the drive unit. In comparing this with Fig. 2-12, note the following:

1. The area above the main piston is inside the index tube. The moving piston is hollow and moves in an annulus between the stationary cylinder and the stationary piston tube.
2. The inside of the stationary piston tube is connected to the scram discharge volume when the exhaust scram valve opens. It also is connected to the area above the drive piston by orifices in the piston tube. These orifices are cut off progressively by the main piston seals at the end of the scram stroke to decelerate the control rod.
3. The latch mechanism is made of six collet fingers attached to an annular collet piston operating in an annular cylinder surrounding the index tube. The spring action of the fingers holds them against the index tube. To unlatch the rod, pressure must be applied to the collet piston, which drives the fingers up against the guide cap and cams them outward, free of the index tube.
4. The annular passage between the drive and housing is the hydraulic passage that connects reactor pressure to the bottom of the ball check valve and, hence, to the bottom of the main piston. This construction ensures that the reactor pressure is always available to the lower end of the piston for a scram and that the drive can be actuated in the withdraw direction only by a pressure higher than reactor pressure.

2.4.2 Control Rod Positioning

For normal manual positioning of the control rods, the operator selects one rod to be positioned. Then, by means of a control switch, appropriate relays and valves are activated to "jog" (i.e., move) the rod in the selected direction. Actuating the control for the insert direction, for example, actuates the two insert valves. This connects the drive water header to the chamber below the piston and the exhaust header to the chamber above the piston. As described under scram operation, the latch is cammed out of the way as the index tube moves upward. To move one notch only, the index tube moves slightly above the new latching position, then is allowed to settle downward until the latch engages the notch in the index tube. A timing

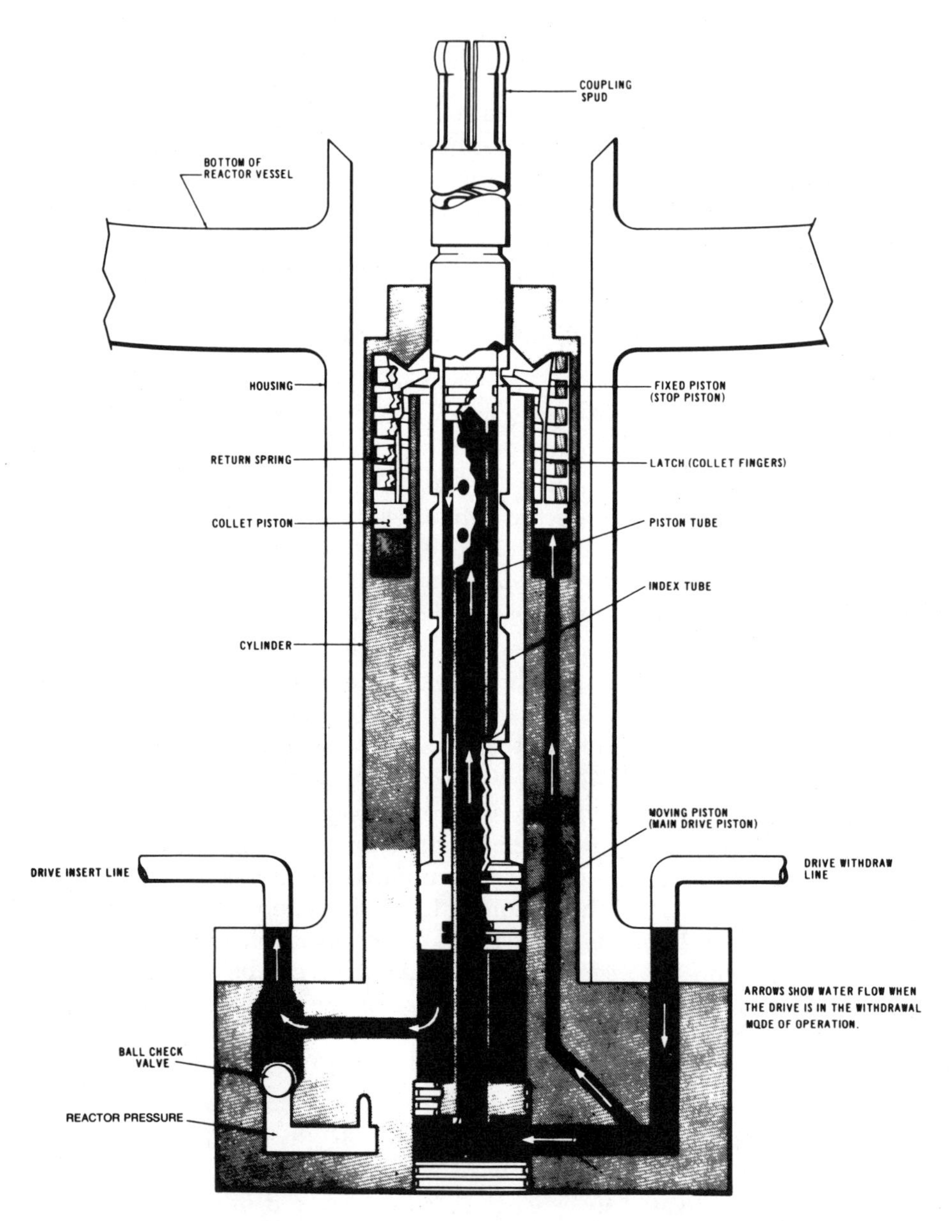

Fig. 2-13 Locking piston hydraulic drive.

relay automatically interrupts the "insert" signal and actuates a "settle" signal. The settle signal opens a "withdraw" valve, venting the chamber below the piston to the exhaust header, allowing the piston and rod to lower to the latch position without delay.

The rod withdrawal sequence occurs automatically through use of timing relays after a rod is selected and the control switch is turned to the withdraw position. The two insert valves open, connecting the drive water header to the area below the piston and the exhaust water header to the area above the piston. The piston and tube move upward far enough to free the latch. Next, the insert valves are closed and the withdraw valves are opened. This connects the drive water header to the area above the piston and the exhaust header to the area below the piston. At the same time, drive water pressure is also applied to the collet piston, which holds the latch in the retracted position. The piston and rod move downward. When sufficient time has elapsed for the piston to move past the previous latch point, the withdraw valve connected to the top of the drive piston is automatically closed. This also removes the pressure from the collet piston and the drive settles to the next notch where latch engagement occurs.

To withdraw a rod by several notches continuously, the notch override switch, which maintains both withdraw valves open, is operated, thus holding the latch in the retracted position. When the override switch is released, the index tube latches at the next notch.

Control rods cannot be withdrawn when any input from redundant sources indicates a block condition. Control circuitry prevents the simultaneous withdrawal of multiple control rods unless the ganged rod mode has been selected; interlocks prevent more than one control rod from being withdrawn from its fully inserted position at any time during refueling.

Rod position is sensed by a series of sealed glass-reed-type switches. They are spaced every 3 in. and are contained within a tube inside the drive mechanism. The switches are actuated by a magnet located in the drive piston. The entire switch assembly can be removed from the drive mechanisms without removing the drive from the reactor. The switches located between latching positions energize an alarm in the event that the drive moves from any latched position when not selected for operation.

The status of all scram valves, accumulators, and CRD "full in" and "full out" limit positions is indicated by panel lights in the control room. The position of all control rods can be displayed on a digital readout indicator (one for each control rod) located on the main operating console in the control room. The position indicator system indicates the numerical position of the drive at each notch position. Midway positions between notches are indicated nonnumerically. The positions of all drives not selected for movement are monitored continuously for motion. A change in drive position (i.e., drive drifting) initiates an audible alarm and an indication of which drive mechanism is drifting.

2.4.3 Description of Control Rods

The cruciform control rods contain stainless-steel tubes filled with boron carbide (B_4C) powder compacted to ~70% of theoretical density. The tubes are seal welded with end plugs on either end. Stainless-steel balls are used to separate the tubes into individual longitudinal compartments. The stainless-steel balls are held in position by a slight double crimp above and below each ball in the tube. The individual tubes act as pressure vessels to contain the helium gas released by the boron-neutron capture reaction.

The tubes are held in cruciform array by a stainless-steel sheath extending the full length of the tubes. A top casting and handle, shown in Fig. 2-14, aligns the tubes, provides structural rigidity, and contains positioning rollers and a parachute-shaped velocity limiter to mitigate rod-drop or ejection-type accidents. The castings are welded into a single structure by a small cruciform post located in the center of the control rod. The control rods can be positioned at 6-in. steps and have a nominal withdrawal and insertion speed of 3 in./sec.

2.5 Nuclear Fuel and Instrumentation

The reactor core of a BWR is arranged as an upright cylinder containing a large number of fuel assemblies, and is located within the reactor vessel. The coolant flows upward through the core. A typical core arrangement of a large BWR and the lattice configuration are shown in Figs. 2-15 and 2-16. It is seen that the BWR core is comprised essentially of only three components: fuel assemblies, control rods, and in-core neutron flux monitors.

2.5.1 Nuclear Fuel

A fuel rod consists of UO_2 pellets in a Zircaloy-2 cladding tube. The UO_2 pellets are manufactured by compacting and sintering UO_2 powder into cylindrical pellets and grinding to size. The nominal density of the pellets is ~95% of the theoretical UO_2 density.

A fuel rod is made by stacking pellets into a Zircaloy-2 cladding tube, which is evacuated, back-filled with helium to atmosphere pressure, and sealed by welding Zircaloy end plugs in each end of the tube. A fission gas plenum spring, shown in Fig. 2-17, is provided in the plenum space to exert a downward force on the pellets; this plenum spring keeps the pellets in place during the preirradiation handling of the fuel bundle.

The BWR fuel rod is designed as a pressure vessel. The rod is designed to withstand the applied loads, both externally and internally. The fuel pellet is sized to provide sufficient volume within the fuel tube to accommodate differential expansion between fuel and cladding.

Typical BWR/6 fuel bundles contain 62 active fuel rods and two hollow "water" rods, which are spaced and supported in a square (8×8) array by

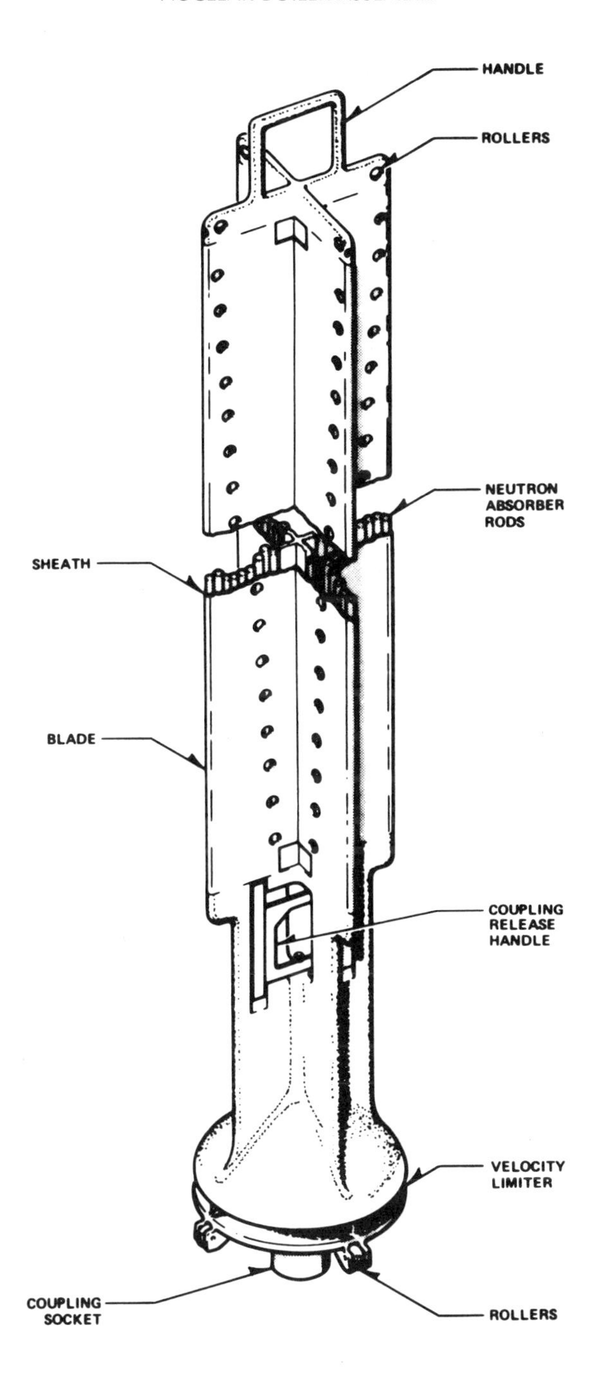

Fig. 2-14 Control rod.

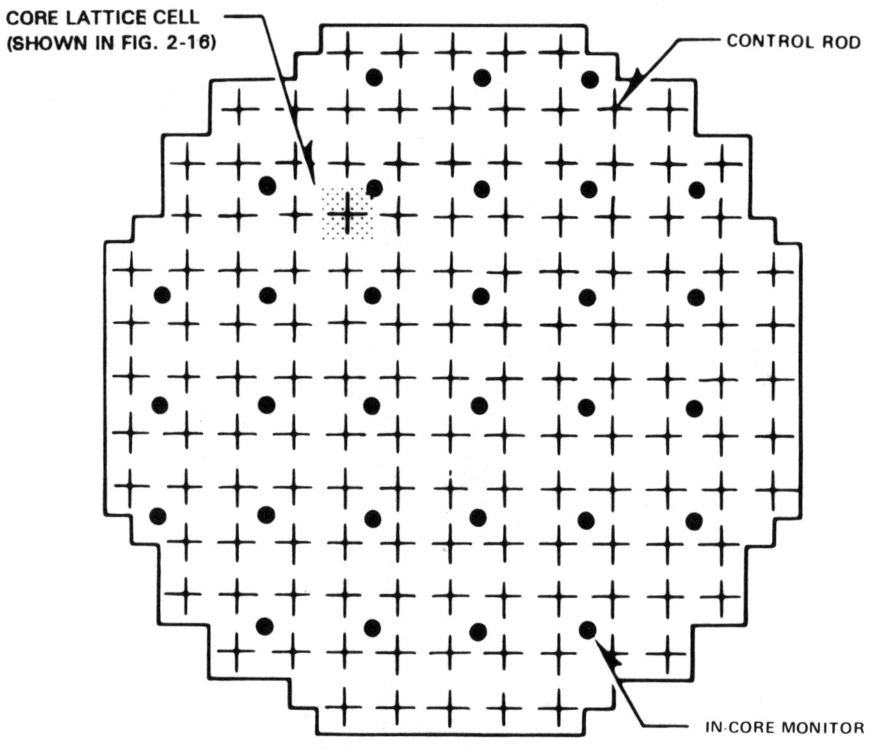

Fig. 2-15 Typical core arrangement.

a lower and upper tie plate. The lower tie plate has a nosepiece that fits into the fuel support piece and distributes coolant flow to the fuel rods. The upper tie plate has a handle for transferring the fuel bundle.

Three types of rods are used in a fuel bundle: tie rods, water rods, and standard fuel rods. The third and sixth fuel rods along each outer edge of a bundle are tie rods. The eight tie rods in each bundle have threaded end plugs that screw into the lower tie plate casting and extend through the upper tie plate casting. A stainless-steel hexagonal nut and locking tab are installed on the upper end plug to hold the assembly together. These tie rods support the weight of the assembly only during fuel handling operations when the assembly hangs by the handle; during operation, the fuel rods are supported by the lower tie plate. Two rods in the interior of typical bundles are water rods; i.e., hollow Zircaloy-2 tubes. Small holes are provided at both the lower and upper ends allowing water to flow through these rods, thus introducing moderating material within the bundle interior. The water rods also serve as spacer-capture rods, mechanically locked to each of the seven grid spacers, thereby fixing the axial position of each

(1) TOP FUEL GUIDE
(2) CHANNEL FASTENER
(3) UPPER TIE PLATE
(4) EXPANSION SPRING
(5) LOCKING TAB
(6) CHANNEL
(7) CONTROL ROD
(8) FUEL ROD
(9) SPACER
(10) CORE PLATE ASSEMBLY
(11) LOWER TIE PLATE
(12) FUEL SUPPORT PIECE
(13) FUEL PELLETS
(14) END PLUG
(15) CHANNEL SPACER
(16) PLENUM SPRING

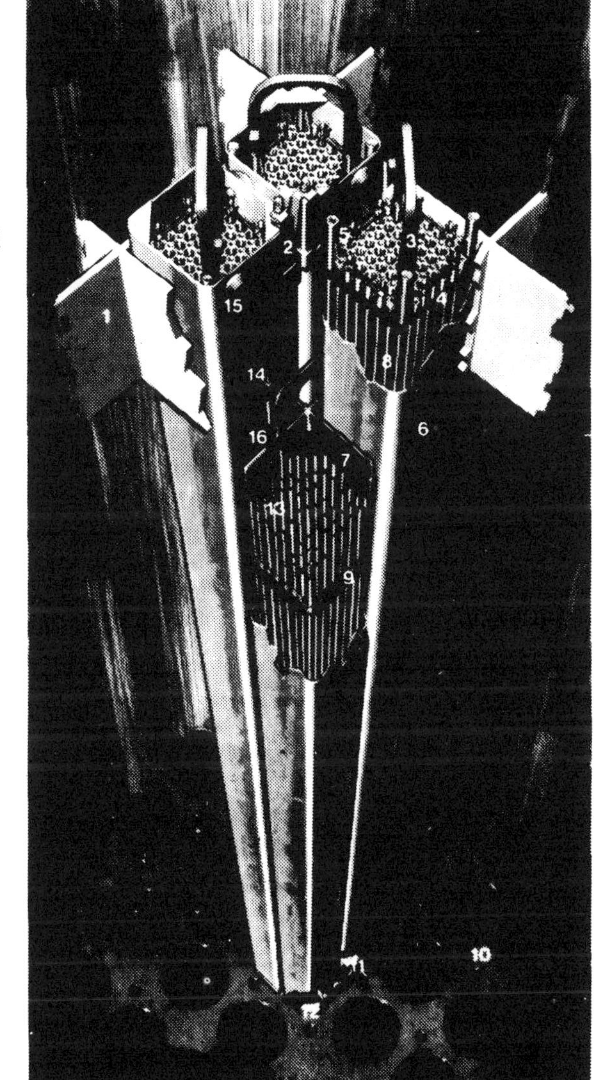

Fig. 2-16 Typical BWR/6 core lattice cell.

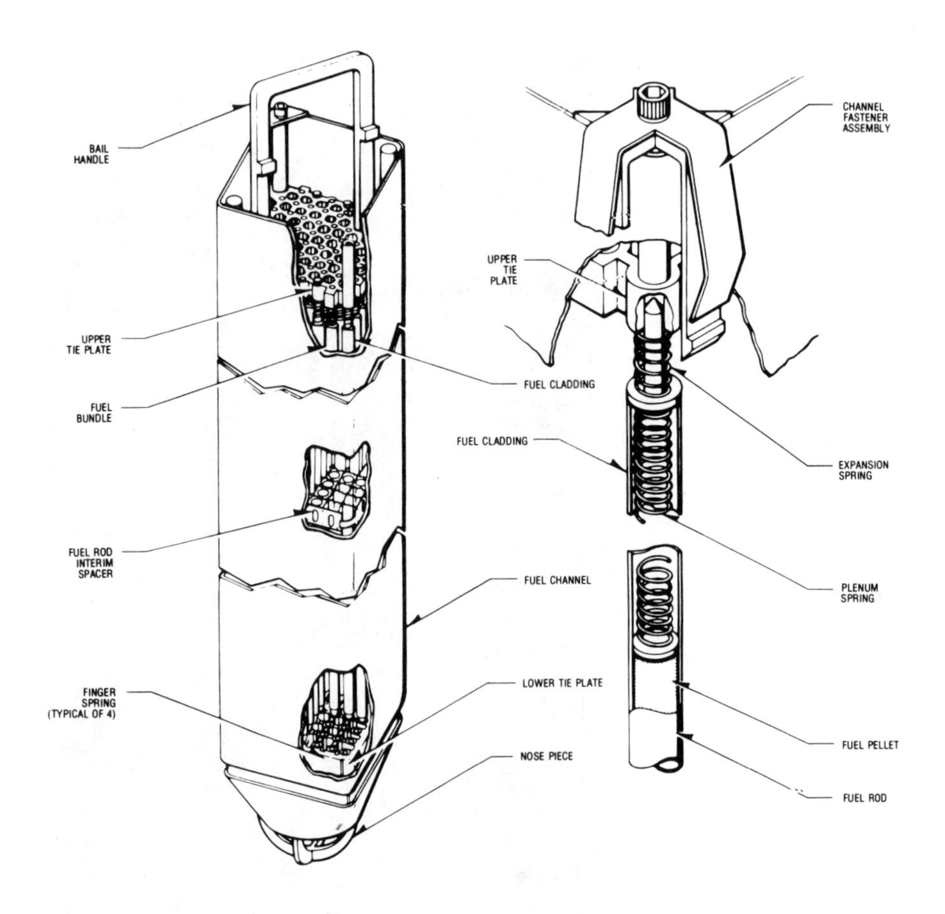

Fig. 2-17 Fuel assembly.

spacer. The fuel rod spacers are equipped with Inconel-X springs to maintain rod-to-rod spacing. The remaining 54 fuel rods in a bundle are standard rods and thus contain many fuel pellets. The end plugs of both the spacer-capture rods and the standard rods have pins that fit into anchor holes in the tie plates. An Inconel-X expansion spring located over the top end plug pin of each fuel rod keeps the fuel rods seated in the lower tie plate, while allowing them to expand axially by sliding within the holes in the upper tie plate to accommodate differential axial thermal expansion.

Different ^{235}U enrichments are used in fuel assemblies to reduce the local power peaking. In addition, selected rods in each assembly are blended with gadolinium, a burnable poison, to achieve optimum power flattening. Low-enrichment uranium rods are used in the corner rods and in the rods nearer the water gaps, while rods with higher enrichment are used in the central part of the fuel bundle. The fuel rods are designed with characteristic mechanical end fittings, one for each enrichment. End fittings are designed so that it is not mechanically possible to complete assembly of a fuel bundle with any high-enrichment rods in a position specified to receive a lower enrichment.

The structural design of the BWR fuel bundle permits the removal and replacement, if required, of individual fuel rods, and ensures that each fuel rod is free to expand in the axial direction, thus minimizing external forces on the rods.

A square Zircaloy-4 fuel channel encloses the fuel bundle. The combination of a fuel bundle and a fuel channel is called a fuel assembly. A typical BWR/6 fuel assembly is shown in Fig. 2-17. These reusable channels make a sliding seal fit on the lower tie plate surface. They are attached to the upper tie plate by the channel fastener assembly, consisting of a spring, guard, and capscrew secured by a lock washer. Spacer buttons are located on two sides of the channel to space properly the four assemblies within a core cell. The fuel channels direct the core coolant flow through each fuel bundle and also serve to guide the control rods.

The use of individual fuel channels greatly increases operating flexibility because the fuel bundles can be orificed separately and, thus, the reload fuel design can be changed to meet the newest requirements and technology. The channels also permit fast in-core sampling of the bundles to locate possible leaking fuel rods.

2.5.2 Core Power Distribution

The design power distribution is divided for convenience into several components: the relative assembly power, the local peaking factor, and the axial peaking factor. The relative assembly power (i.e., radial peaking factor, F_R) is the maximum fuel assembly average power divided by the reactor core-average assembly power. It is a direct measure of the gross radial peaking. The local power peaking factor, F_L, is the maximum fuel rod average heat flux in an assembly at a particular axial position divided by the assembly average fuel rod heat flux at the same axial position. The axial power peaking factor, F_A, is the maximum heat flux of a fuel rod divided by the axial average heat flux of that rod. Peaking factors vary throughout an operating cycle, even at steady-state full-power operation, since they are affected by withdrawal of control rods to compensate for fuel burnup. The design peaking factors represent the values of the most limiting power distribution that exists in the core throughout its life such that the peak linear power generation of 13.4 kW/ft is not exceeded.

The design peaking factors of a BWR/6 are approximately:

1. relative assembly (i.e., radial) peaking, $F_R = 1.40$
2. axial peaking, $F_A = 1.40$

3. local peaking, $F_L = 1.13$
4. total maximum-to-average peaking, $F_T = F_R F_A F_L = 2.22$.

These design peaking factors have been selected on the basis of analysis and performance data from large operating BWRs. Operation within these limits is assured by design and prescribed operating procedures.

Because of the presence of relatively high steam voids in the upper part of the core, there is a natural characteristic for a BWR to have the axial power peak in the lower part of the core. During the early part of an operating cycle, bottom entry control rods permit a partial reduction of this axial peaking by locating a larger fraction of the control rods in the lower part of the core. At the end of an operating cycle, the higher accumulated exposure and greater depletion of the fuel in the lower part of the core reduce the axial peaking. The basic operating procedure is to locate control rods so that the reactor operates with approximately the same axial power shape throughout an operating cycle (Haling strategy). In addition to control rods, the initial core of BWR/6 uses gadolinia in some fuel rods for temporary reactivity control. The magnitudes of gadolinia concentrations are adjusted according to the enrichment requirements of initial core fuel. Axial variation of the gadolinia is by zone, with the concentration held constant within each zone. Typically there are three zones for each gadolinium-bearing rod; the result is to lower the axial power peaking factor.

The relative assembly power distribution (i.e., radial distribution) is normally fairly flat in a BWR because of greater steam voids in the center bundles of the core. A control rod operating procedure is also used to maintain approximately the same radial power shape throughout an operating cycle.

2.5.3 In-Core Neutron Monitoring System

Reactor power is monitored from the source range up through the power operating range by suitable neutron monitoring channels. All detectors are located inside the reactor core. This location of detectors provides maximum sensitivity to control rod movement during the startup period and provides effective monitoring in the intermediate and power ranges. As shown in Fig. 2-18, three types of neutron monitoring are provided: source range (SRM), intermediate range (IRM), and local power range (LPRM). A traversing in-core probe system provides for periodic calibration of these neutron detectors.

In the SRM, the neutron flux is monitored by fission counters which, as shown in Fig. 2-19, are inserted to about the midplane of the core by the drive mechanisms that move each chamber into the core through inverted thimbles. A range from below the source level to 10^9 n/cm^2-sec is covered.

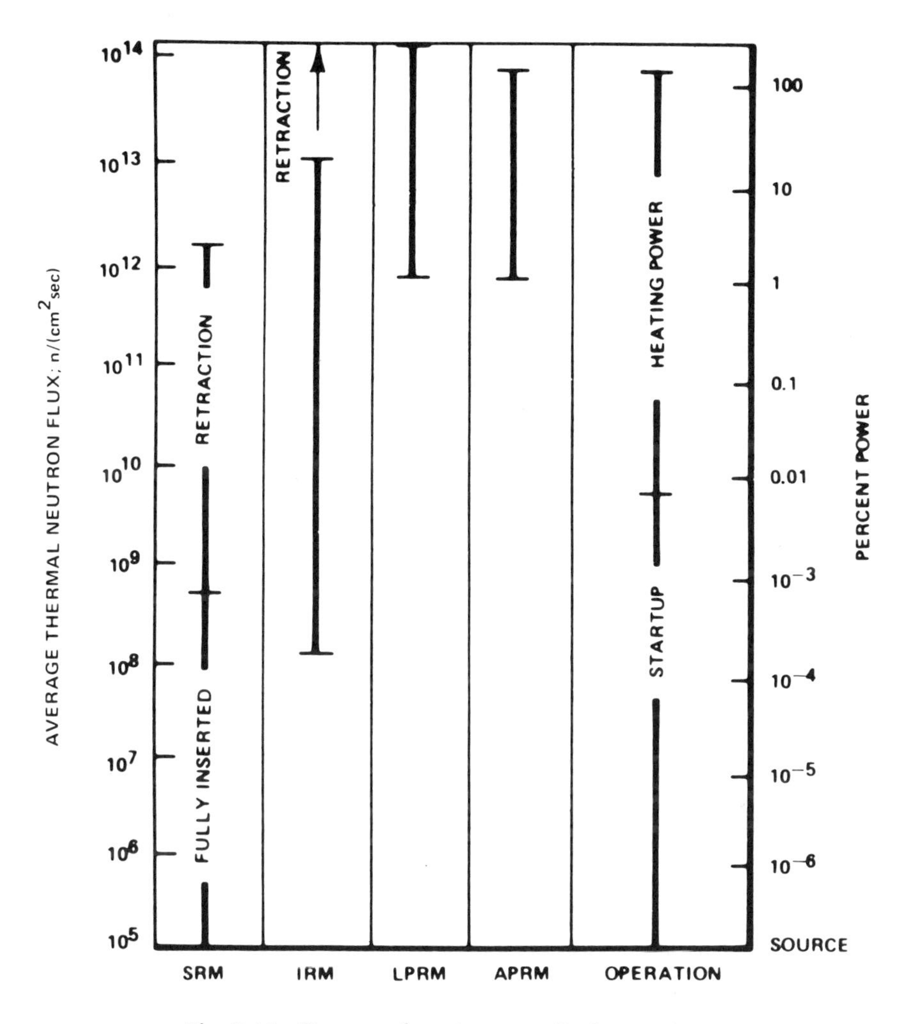

Fig. 2-18 Ranges of neutron monitoring system.

As startup progresses and the count rate approaches the top of the meter range ($\sim 10^6$ cps), the counters are withdrawn to give a drop in apparent count rate. Criticality normally occurs before movement of the counters is necessary. The counters can be motor driven to any position within their limits of travel; however, two or three selected positions provide the necessary range to achieve criticality and provide overlap with the intermediate range monitors.

When the reactor reaches the power range, the counters are moved to a position ~2 ft below the core. This places the counters in a low neutron

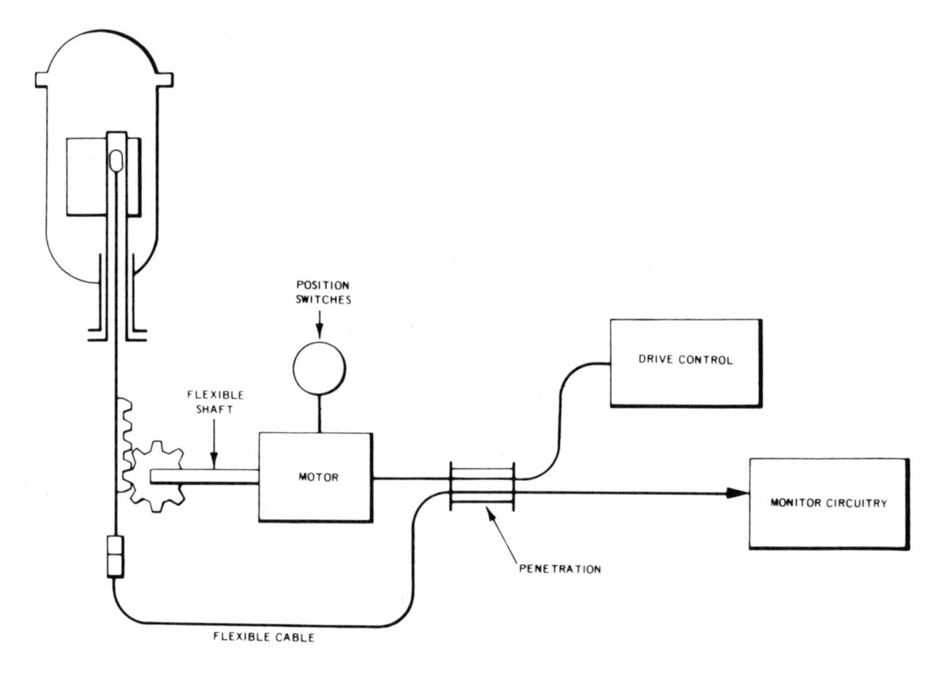

Fig. 2-19 Detector drive system.

flux region so that unnecessary burnup and activation of the counters are minimized.

The intermediate range is considered to be from $\sim 10^8$ to 1.5×10^{13} n/cm^2-sec. In this range, the neutron flux is monitored by a system using a voltage variance method (also known as MSV or the Campbell method). This method makes use of the ac component of voltage, which is due to the random nature of neutron pulses generated in a detection chamber. With small chambers located in the high-temperature environment of the reactor core, the ac component is used to measure neutron flux at lower power levels because cable leakage and gamma radiation have relatively little effect on the signal.

These fission chambers also are withdrawn during full-power operation to maintain their expected life and to reduce activation. They are positioned with drive mechanisms similar to those used for the source range fission counters.

In the power range, neutron flux is monitored by fixed in-core ion chambers that are arranged in a uniform pattern throughout the core. These chambers cover a range of ~ 1 to 125% of rated power on a linear scale.

Detector assemblies each contain four fission chambers and a calibration guide tube for a traversing ion chamber. The chambers are spaced uniformly in an axial direction and lie in four horizontal planes. Each ion chamber is connected to a dc amplifier with a linear output. Internal controls permit adjustment of the amplifier gain to compensate for the reduction of chamber sensitivity caused by burnup of its fissionable material.

These detectors are individually replaced as necessary through the bottom of the reactor vessel. The design lifetime of the ion chambers in the average core flux is 3 yr or a fluence of 1.1×10^{22} n/cm^2 before the neutron-to-gamma current ratio drops to 5:1.

The average power level is measured by four average power range monitors (APRMs). Each monitor measures bulk power in the core by averaging signals from as many as 24 LPRMs distributed throughout the core. The output signals from these monitors are displayed and also are used to operate trips in the reactor protection system.

The calibration guide tube included in each fixed in-core assembly permits the insertion of a traversing ion chamber to obtain axial flux profiles and to calibrate the ion chambers. Each calibration guide tube extends nearly to the top of the active portion of the core and is sealed at the upper end. The tubes pass through the nozzles and seals beneath the reactor vessel and connect to an indexing mechanism located inside the containment. The indexing mechanism permits the traversing ion chamber to be directed to many different detector assemblies.

Flux readings along the axial length of the core are obtained by fully inserting the traversing ion chamber into one of the calibration guide tubes, then taking data as the chamber is withdrawn. The data go directly to the reactor's process computer. One traversing chamber and its associated drive mechanism is provided for each group of seven to ten fixed in-core assemblies (depending on reactor size).

Now that the BWR nuclear boiler assembly has been described, we are ready to consider the thermal-hydraulic analyses of these systems.

References

"BWR/6—General Description of a Boiling Water Reactor," NED Document, General Electric Company (1975).

Fakory, M. R., and R. T. Lahey, Jr., "An Analytical Model for the Analysis of BWR/4 Long Term Cooling with Either Intact or Broken Jet Pump Seals," *J. Nucl. Eng. Des.*, **85** (1985).

Kamath, P. S., and R. T. Lahey, Jr., "The Analysis of Boiling Water Reactor Long-Term Cooling," *Nucl. Technol.*, **49** (1980).

Kudirka, A. A., and D. M. Gluntz, "Development of Jet Pumps for Boiling Water Reactor Recirculation Systems," *J. Eng. Power* (1974).

PART TWO

Basic Thermal-Hydraulic Analyses

CHAPTER THREE

Applied Thermodynamics

3.1 The First Law

The basic principle of energy conservation is called the First Law of Thermodynamics. The first law simply states that "energy is neither created nor destroyed." For thermal-hydraulic considerations, it is not necessary to include mass-energy equivalence as must be done for energy analyses of nuclear processes.

The first law can be expressed as follows with regard to an imaginary system contained by a well-defined control volume: "Total outflow rate of energy less total inflow rate of energy plus total storage rate of energy is zero."

Before this statement can be expressed in a convenient operational form, it is necessary to define some basic concepts:

1. *Energy* is the capacity of matter to do work or to supply heat in a given reference frame.
2. *Work* is energy in transit by means of a force acting through a distance.
3. *Heat* is energy in transit by means of a temperature difference.

Work and heat are not storable quantities, but their effect is to increase or decrease the total energy of a system. Several of the most important storable energy forms, which normally must be considered in thermal-hydraulic analyses, are:

1. kinetic energy (KE)
2. potential energy (PE)
3. internal thermal energy.

These storable forms of energy are additive and the summation is designated by,

$$E = \underset{\text{(KE)}}{E_{KE}} + \underset{\text{(PE)}}{E_{PE}} + \underset{\text{(internal)}}{U} = \varepsilon M \ , \tag{3.1}$$

where M is the system mass and ε is the total stored energy per unit mass,

$$\varepsilon \stackrel{\Delta}{=} \underbrace{\frac{u^2}{2Jg_c}}_{\text{(KE)}} + \underbrace{\frac{gZ}{g_cJ}}_{\text{(PE)}} + \underbrace{\mu}_{\text{(internal)}} \ . \tag{3.2}$$

Energy flow across a system boundary occurs by mass flow, work, and heat transfer. When matter crosses a system boundary, it not only carries its stored energy forms, but also provides energy transfer by flow work. That is, the transfer of a mass, m, also causes an associated energy transfer given by,

$$m(\varepsilon + pv) = m\left(\frac{u^2}{2Jg_c} + \frac{gZ}{g_cJ} + \mu + pv\right) \ , \tag{3.3}$$

or per unit mass, the total convected energy is,

$$h_o = h + \frac{u^2}{2Jg_c} + \frac{gZ}{g_cJ} \ , \tag{3.4}$$

where h_o is the so-called specific stagnation enthalpy, and the specific enthalpy (Btu/lb) has been defined as,

$$h \stackrel{\Delta}{=} \mu + pv \ . \tag{3.5}$$

The foregoing definitions lead to an operational formulation of the first law. Figure 3-1 shows a system contained by a control volume across which energy flows in the form of work, heat, or mass transfer.

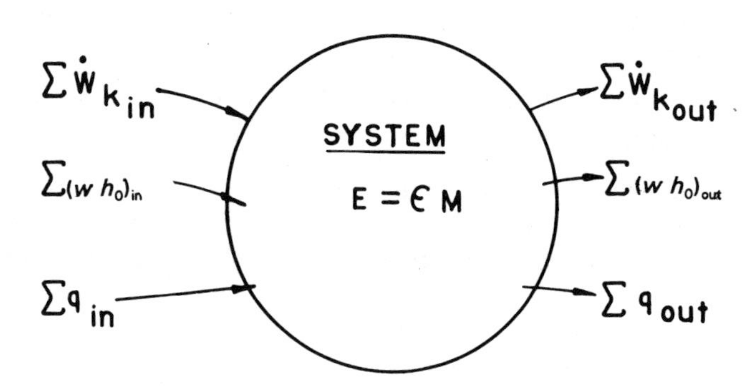

Fig. 3-1 The first law system.

The summations in Fig. 3-1 indicate numerous energy flows. Operationally, the first law can be written as,

$$\Sigma\dot{W}_{k_{out}}+\Sigma(wh_o)_{out}+\Sigma q_{out}-\Sigma\dot{W}_{k_{in}}-\Sigma(wh_o)_{in}-\Sigma q_{in}+\frac{dE}{dt}=0 \ . \quad (3.6)$$

Similarly, mass conservation is given by,

$$\Sigma w_{out}-\Sigma w_{in}+\frac{dM}{dt}=0 \ . \quad (3.7)$$

Equations (3.6) and (3.7) are used for the mass and energy "bookkeeping" required for system state and blowdown analyses.

3.1.1 System Blowdown Analysis

The term "blowdown" refers to a relatively rapid loss of coolant mass and energy from the reactor system. Reactor internal pressure and external containment pressure depend on blowdown mass and energy rates, which must be predicted to ensure an adequate safeguard design.

The overall primary system can be viewed as a collection of subsystems, interconnected by moving boundaries, mass flow paths, and heat flow boundaries. A reactor vessel is shown in Fig. 3-2 with different subsystems defined for studying various hypothetical accidents. Generally, the internal thermal energy dominates the total energy of a subsystem that is near saturation. It follows that kinetic and potential energy terms are normally negligible. A proper accounting of the mass and energy contained and the subsystem volume is adequate to determine its thermodynamic state, or specifically, its pressure and enthalpy.

Although a subsystem can contain liquid, vapor, or a two-phase equilibrium mixture, its static pressure can be expressed functionally as,

$$p=p(\mu,\ v)=p\left(\frac{U}{M},\ \frac{V}{M}\right) \ . \quad (3.8)$$

It is often desirable to determine the rate of pressure change of a subsystem. Since kinetic and potential energies in a subsystem are usually negligible, Eqs. (3.1) and (3.2) lead to the approximation,

$$E\cong U=\mu M \ . \quad (3.9)$$

It follows that Eq. (3.8) can be differentiated with respect to time and combined with Eqs. (3.6), (3.7), and (3.9) to give the subsystem pressure rate in the form,

$$\frac{dp}{dt}=\left(\frac{\partial p}{\partial \mu}\right)_v\frac{1}{M}\left[\Sigma w_{in}(h_{o,in}-\mu)-\Sigma w_{out}(h_{o,out}-\mu)\right.$$

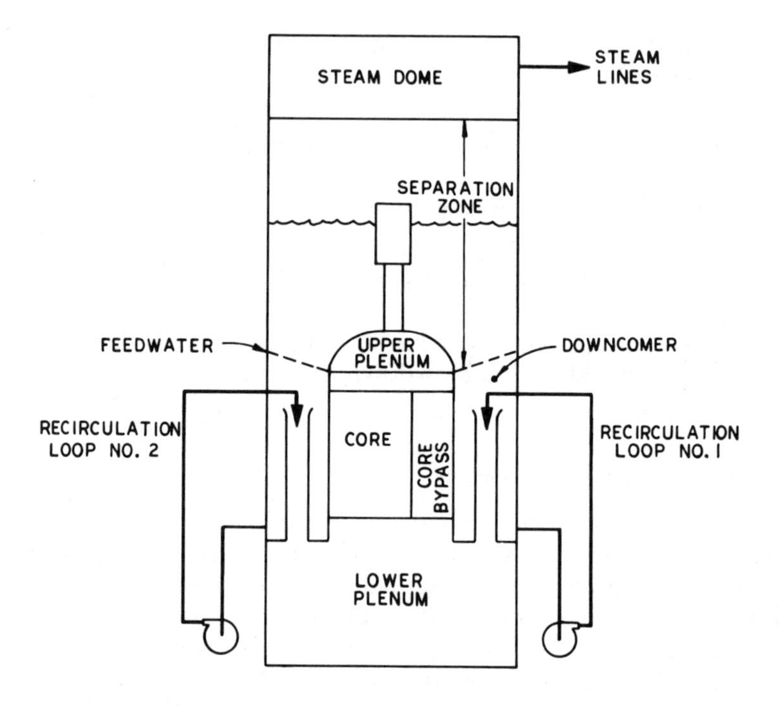

Fig. 3-2 Schematic diagram of typical BWR pressure vessel subsystems.

$$+\Sigma q_{\text{in}} - \Sigma q_{\text{out}} + \dot{W}_{k_{\text{in}}} - \dot{W}_{k_{\text{out}}}]$$

$$+\left(\frac{\partial p}{\partial v}\right)_{\mu} \frac{1}{M}\left[\frac{dV}{dt} - v(\Sigma w_{\text{in}} - \Sigma w_{\text{out}})\right] . \tag{3.10}$$

Equation (3.10) can be used for any simple compressible substance (SCS).

3.1.2 Useful Property Differentials

Enthalpy, h, rather than internal energy, μ, is readily available in tabulated form or in computational state property subroutines. Therefore, it is convenient to employ the state form, $p = p(\mu, v)$, whose differential at constant pressure can be combined with Eq. (3.5) to give

$$\left(\frac{\partial p}{\partial \mu}\right)_v = \frac{1}{\left[\left(\frac{\partial h}{\partial p}\right)_v - v\right]} \tag{3.11}$$

$$\left(\frac{\partial p}{\partial v}\right)_\mu = -\frac{\left[\left(\frac{\partial h}{\partial v}\right)_p - p\right]}{\left[\left(\frac{\partial h}{\partial p}\right)_v - v\right]} \quad . \tag{3.12}$$

Other thermodynamic relationships, useful in computation with Eqs. (3.11) and (3.12), are (Zemansky, 1955),

$$\begin{aligned}
\beta_v &\triangleq \frac{1}{v}\left(\frac{\partial v}{\partial T}\right)_p && \text{(volumetric expansivity)} \\
c_p &\triangleq T\left(\frac{\partial s}{\partial T}\right)_p && \text{(specific heat at constant pressure)} \\
\left(\frac{\partial v}{\partial s}\right)_p &\triangleq \frac{T\beta_v v}{c_p} \\
\left(\frac{\partial h}{\partial p}\right)_v &= v + \frac{g_c c_p v}{\beta_v c^2} \\
\left(\frac{\partial h}{\partial v}\right)_p &= \frac{c_p}{\beta_v v} \\
c^2 &= g_c\left(\frac{\partial p}{\partial \rho}\right)_s && \text{(sonic or acoustic speed)} \quad .
\end{aligned} \tag{3.13}$$

3.1.2.1 *Subcooled Liquid Subsystem*

For a subsystem containing subcooled liquid only, Eqs. (3.11) and (3.12) can be combined with Eq. (3.13) to yield,

$$\left(\frac{\partial p}{\partial \mu}\right)_v = \frac{\beta_v c^2}{g_c v c_p} \tag{3.14}$$

$$\left(\frac{\partial p}{\partial v}\right)_\mu = -\frac{\left(\frac{c_p}{v\beta_v} - p\right)}{v\left(\frac{g_c c_p}{\beta_v c^2} - 1\right)} = -\frac{c^2}{g_c v^2} + \frac{p\beta_v c^2}{g_c c_p v} \quad . \tag{3.15}$$

These two derivatives have large magnitudes and, thus, it follows from Eq. (3.10) that *dp/dt* is extremely sensitive to slight changes in flow rate, heat rate, power, or the rate of change of volume of a liquid subsystem. This feature can lead to computational instabilities during numerical integration of pressure, unless a rather tight error criterion is employed.

Instead, it may be preferable to use Eq. (3.10) for computing the unknown flow rate in a liquid subsystem with pressure rate considered an independent variable.

3.1.2.2 *Saturated Equilibrium Two-Phase Mixtures*

It is convenient to use the following state equation for a subsystem that contains an equilibrium mixture of saturated liquid and vapor:

$$h = h_f(p) + \frac{h_{fg}(p)}{v_{fg}(p)}[v - v_f(p)] \ . \tag{3.16}$$

Moreover, the derivatives $(\partial h/\partial v)_p$ and $(\partial h/\partial p)_v$ are readily found to be

$$\left(\frac{\partial h}{\partial v}\right)_p = \frac{h_{fg}(p)}{v_{fg}(p)} \tag{3.17}$$

$$\left(\frac{\partial h}{\partial p}\right)_v = \frac{dh_f}{dp} - \frac{d}{dp}\left(\frac{h_{fg}v_f}{v_{fg}}\right) + v\frac{d}{dp}\left(\frac{h_{fg}}{v_{fg}}\right) \ . \tag{3.18}$$

If the specific volume, v, is expressed in terms of the static quality, x_s, as,

$$v = v_f(p) + x_s v_{fg}(p) \ , \tag{3.19}$$

then Eqs. (3.17) and (3.18) can be used in Eqs. (3.11) and (3.12) to provide $(\partial p/\partial \mu)_v$ and $(\partial p/\partial v)_\mu$ for various pressures and static qualities.

An alternate form for dp/dt is given in Sec. 9.2.2 for saturated water systems.

3.1.2.3 *Superheated Vapor or Ideal Gas*

If a subsystem contains superheated vapor or gas, it is convenient to express its specific enthalpy by the ideal gas approximation,

$$h = \frac{k}{(k-1)}pv \ , \tag{3.20}$$

from which,

$$\left(\frac{\partial h}{\partial v}\right)_p = \frac{k}{(k-1)}p \ , \tag{3.21}$$

and,

$$\left(\frac{\partial h}{\partial p}\right)_v = \frac{k}{(k-1)}v \ . \tag{3.22}$$

The blowdown relationships presented here are general for subcooled liquid, saturated mixture, and superheated vapor subsystems. They can be combined with the critical flow rate expressions given in Chapter 9 to estimate time-dependent system pressures during liquid, vapor, and saturated mixture blowdowns. The steady form of Eq. (3.6) provides the basis for an overall system heat balance, which is considered next.

3.1.3 Boiling Water Reactor System Heat Balance

A heat balance is a systematic accounting of all the important energy flows that occur in a system or subsystem. A nuclear reactor core, for example, adds thermal energy to the working fluid at a rate of q_{core}. The turbine extracts energy for turning the generator armature at the rate of $\dot{W}_{k_{turb}}$. Thermal energy is extracted by the condenser at a rate of q_{cond}. There are numerous other energy transfers in a BWR nuclear steam supply system, which include pumping power and heat losses.

The steady-state form of Eq. (3.6) is used for tracking energy flows from one subsystem to another. An example of a simplified BWR system heat balance is shown in Fig. 3-3.

A steady-state heat balance is obtained by first choosing values for some of the system parameters and then solving for the others. Generally, that part of the system, including the turbine, condenser, and feedwater heaters, is described by a main feed enthalpy—steam flow characteristic curve similar to that shown in Fig. 3-4, which is based on a turbine "heat balance." The load demand at the turbine implies a given steam flow rate and corresponding enthalpy of water leaving the feedwater heaters. It follows from the steady form of Eq. (3.6) that since the kinetic and potential energy terms are relatively small, core heat can be estimated from the equation,

$$\begin{aligned} q_{core} = &(wh)_{main\ steam} + (wh)_{recirculation\ suction} \\ &- (wh)_{recirculation\ discharge} - (wh)_{main\ feed} \\ &- (wh)_{control\ rod\ drive} - q_{losses} \ . \end{aligned} \tag{3.23}$$

Energy rates associated with recirculation suction and discharge approximately cancel. Control rod drive flow and reactor heat loss energy rates are relatively small, usually $<1\%$ of the core thermal power. Therefore, a reasonably accurate q_{core} can be obtained in terms of turbine load demand.

Before a reactor heat balance is completed, the core thermal hydraulics must be considered. Briefly, the reactor core can be viewed as numerous heated parallel passages through which coolant flows. Boiling occurs at reactor pressure and provides the required turbine steam flow. However, core reactivity, pressure loss characteristics, the local void fraction, core and recirculation flow rates, and core temperature all are interrelated and must be considered together. Core and recirculation flow rates finally must be determined, for which thermal-hydraulic performance specifications are satisfied. A discussion of core thermal hydraulics is given in Chapter 6.

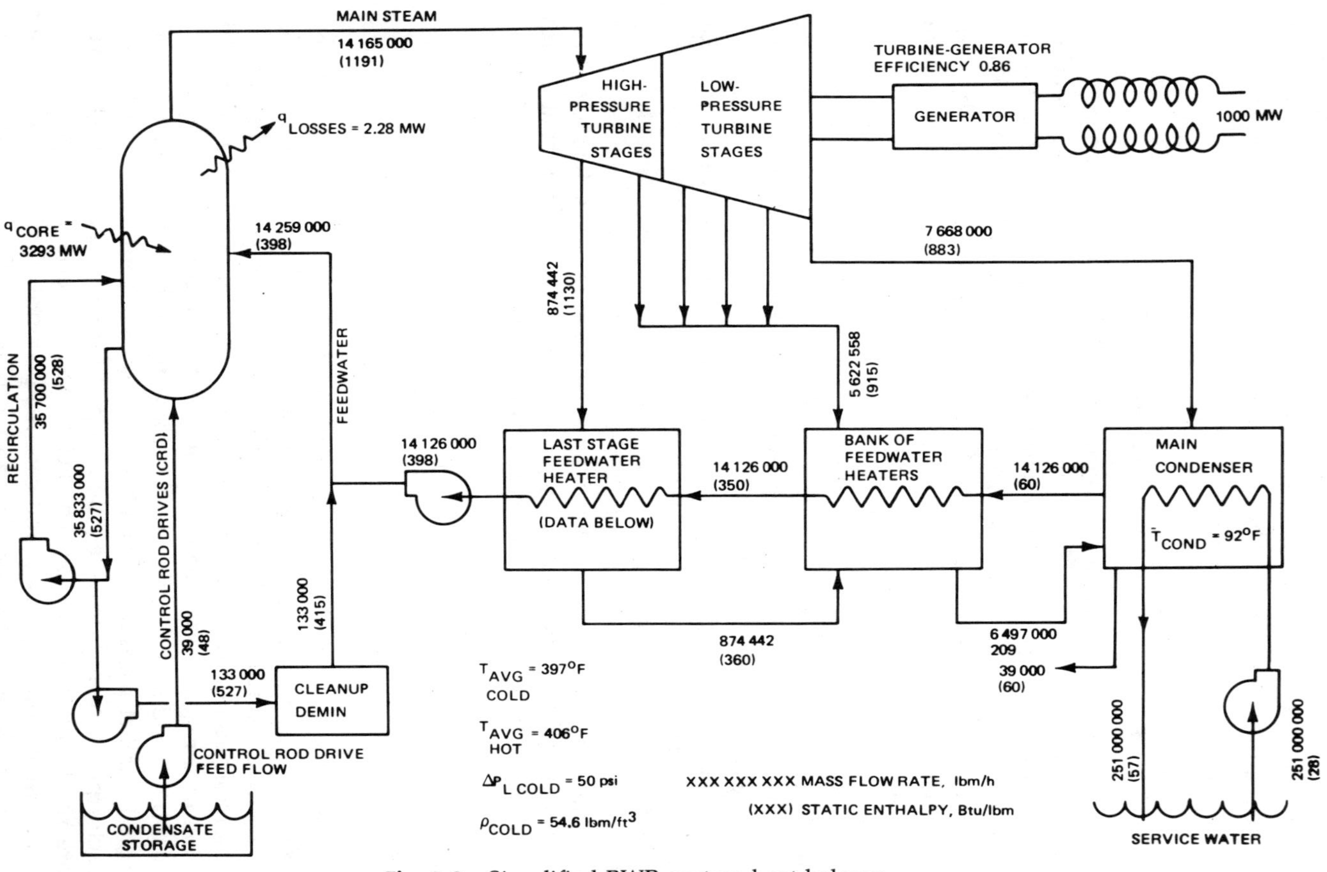

Fig. 3-3 Simplified BWR system heat balance.

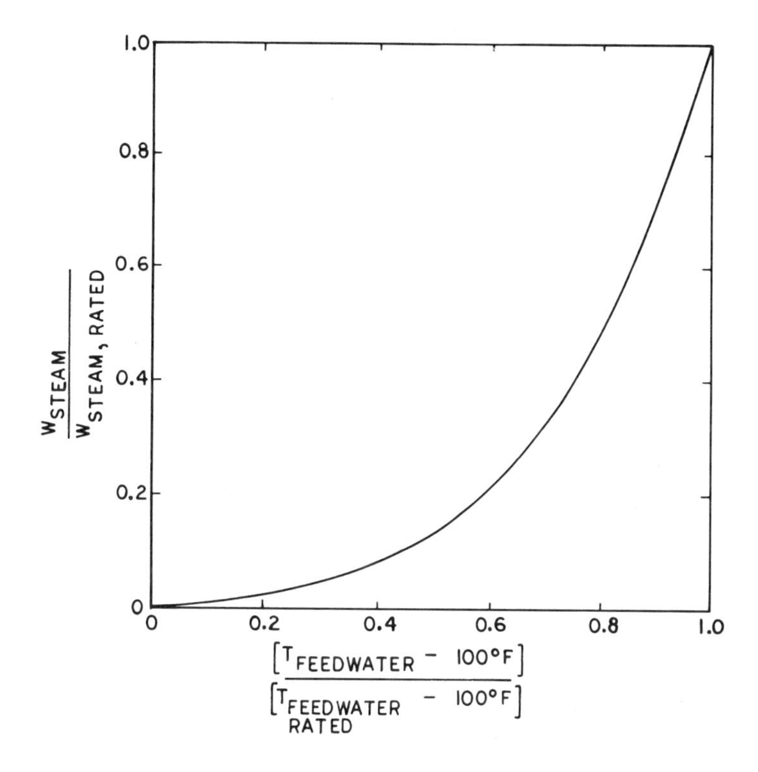

Fig. 3-4 Typical steam flow, main feed enthalpy operating curve.

Steady-state heat balances are satisfied for every interconnected subsystem in a BWR power plant, including the reactor, steam lines, turbine, condenser, feedwater lines, and recirculation loops. Although energy is conserved by each subsystem, various irreversible mechanisms exist, which transform part of the total energy to forms in which it becomes unavailable for power production. These irreversibilities are identified and quantified by a Second Law analysis.

3.2 The Second Law

A BWR/turbine-generator power system receives heat through the process of nuclear fission and converts it into electrical energy. The most ideal machine for energy conversion would provide a useful energy outflow exactly equal to the energy input. Unfortunately, no real processes exist where 100% energy conversion is realized. Bearing friction, turbine windage, heat, and pressure losses, and other irreversible processes associated with heat transfer, mixing, evaporation, and condensation, all act to reduce the useful power output. It is economically desirable to minimize these power losses in energy conversion. The Second Law of Thermodynamics can be used to identify power losses and to suggest ways to improve the overall thermal performance.

The efficiency, η, of an energy conversion process is defined as,

$$\eta \triangleq \frac{\text{useful power output}}{\text{power input}} \quad . \tag{3.24}$$

An overall power plant efficiency, expressed in terms of electrical power output and thermal power input, or equivalently, the reciprocal of the plant heat rate (Btu/kW-h), can be obtained from the product of efficiencies associated with the reactor system, turbine, and generator. Since this monograph pertains to BWR thermal-hydraulic design, only the reactor system efficiency is considered.

The concept of useful power output from the reactor system requires proper definition to help identify possible design improvements. If the useful power of the reactor system and total power input of the core are known, a reactor efficiency could be obtained from Eq. (3.24). To determine the power output available for use, the concept of an ideal machine is employed. This ideal machine makes use of all available power and rejects heat to a sink at some appropriate temperature. The concept of available power requires second law considerations, which are summarized next.

The Second Law of Thermodynamics is often expressed in terms of entropy, which is a property that cannot be physically sensed. Consequently, relatively few practicing engineers use second law ideas during the analysis of a process or the design of an apparatus. Since there is much emphasis today on conservation, efficient use of resources, and optimizing performance, this section is included to present the essential ideas required in performing a second law analysis of reactor systems.

Ideas involving entropy, reversible processes, and the Second Law of Thermodynamics provide a means of identifying, quantifying, and partially controlling losses of available energy in various processes.

The reason entropy is not more widely used is that it is an abstract property which is not measurable like force, length, velocity, temperature, and other common quantities that are easily perceived. Entropy has been presented in a classical scientific framework as a defined property of matter from both macroscopic and statistical viewpoints (Zemansky, 1955; Reynolds, 1965; Sonntag and VanWylen, 1965). It is then used in formulating the second law and shown to be related to available energy losses. Nevertheless, the macroscopic definition of entropy is still abstract. Some authors have tried to minimize its abstractness by reversing the classical presentation to show that losses of available energy can be quantified with something defined as entropy (Hatsopoulos and Keenan, 1965; Huang, 1976). Regardless of how entropy and related ideas are introduced, most thermo-science workers view entropy as something fictional, and this clouds their view of its usefulness.

Entropy is an abstract property much like the velocity potential ϕ used to predict velocity $v = -\nabla\phi$ in irrotational fluid flow. The velocity potential itself is an abstract mathematical quantity that exists only for ideal flow and has no meaningful physical interpretation. However, its gradient is equal to the velocity vector, which is readily understood. Therefore, workers in fluid mechanics are not bothered by an abstract velocity potential because they can use it in solving velocity fields. Similarly, entropy alone does not convey macroscopic meaning. But its change can be used to determine the loss of available power from a system, thus pointing the way for economic and design improvements.

It is assumed that readers already accept the definition of entropy, at least as an abstract property. A brief discussion follows, which shows how entropy is used to obtain an operational formulation of the second law. An application of the second law is included for determining the loss of available power, followed by analysis of several processes found in boiling water nuclear reactor systems.

3.2.1 Entropy and the Second Law

Figure 3-5 shows a fixed mass system, which contains an amount of entropy, S, with heat inflow and outflow at similarly identified temperatures. The differential change of entropy is given from its definition as,

$$dS = \frac{dQ_{R,\text{in}}}{T_{\text{in}}} - \frac{dQ_{R,\text{out}}}{T_{\text{out}}} , \tag{3.25}$$

where dQ_R designates idealized reversible heat transfer without temperature gradients. It should be noted that dQ_R is an inexact differential which depends on the state process path of the system during heat transfer.

The second law, or principle of entropy increase, can be demonstrated by a simple analysis of the isolated system shown in Fig. 3-6. Let us assume that one region of the system is at uniform temperature, T, and the other region is at $T(1-\varepsilon)$, where ε is a small quantity. This temperature difference causes a differential amount of heat transfer, dQ, to occur from the region at higher temperature to the lower one over an unspecified period of time. Since entropy is an extensive property, the total system entropy is,

$$S = S_1 + S_2 \ . \tag{3.26}$$

Its differential, dS, can be combined with Eq. (3.25) to give, for an isolated system,

$$dS = dS_1 + dS_2 = -\frac{dQ}{T} + \frac{dQ}{T(1-\varepsilon)} = \frac{dQ}{T}(\varepsilon + \varepsilon^2 \ldots) \geqslant 0 , \tag{3.27}$$

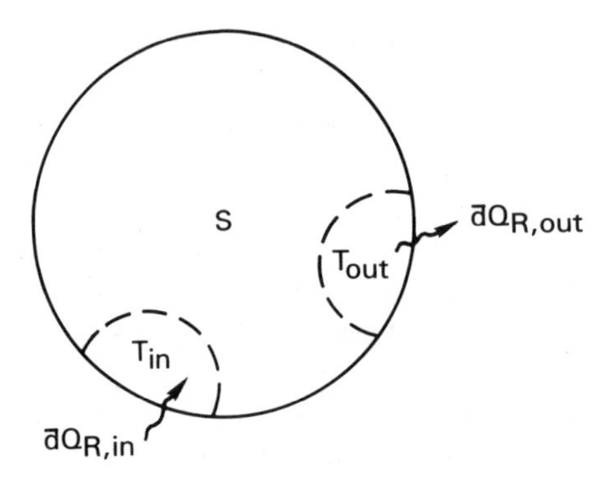

Fig. 3-5 Definition sketch for entropy.

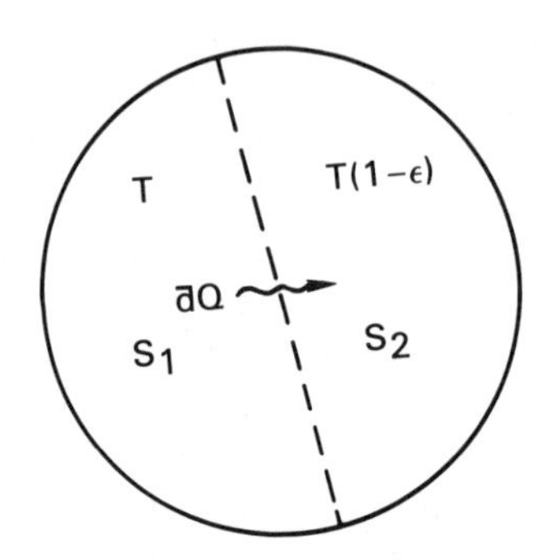

Fig. 3-6 Isolated systems and the second law.

which shows that the system's total entropy increases. This is one simple demonstration of the second law, which basically depends on the fact that heat transfer occurs from a region of higher temperature to a region of lower temperature. An isolated system with uniform temperature has no change of entropy. However, if any macroscopic disturbance occurred, such as the falling of an object from a shelf or the rattling of contained components, slight temperature changes would occur with associated internal heat transfer, causing the system's entropy to increase. When viewed from the microscopic standpoint, the system becomes more disorderly as entropy increases. It will be shown that entropy also provides a way to quantify the loss of available power in a system.

If a system of mass M has total entropy S, its specific entropy is $s = S/M$. Thus, when mass flows across a boundary at a rate w, the entropy flow rate is ws.

Consider the system enclosed by a dashed line in Fig. 3-7 with contained entropy S, which lies inside the solid boundary of an isolated system having entropy S_∞. Equation (3.27) can be employed to write the second law for the entire system of Fig. 3-7 in terms of the entropy rate as,

$$\frac{d(S+S_\infty)}{dt} \geqslant 0 \ . \tag{3.28}$$

Inflow and outflow mass rates are shown crossing the dashed boundary. Reversible heat transfer rates, q_{in} and q_{out}, are also shown between regions at temperatures T_{in}, T_{out}, and T_∞. Entropy storage rates dS/dt and dS_∞/dt are expressed in terms of entropy production rates associated with heat transfer and mass transfer, which gives for the system enclosed by the dashed lines,

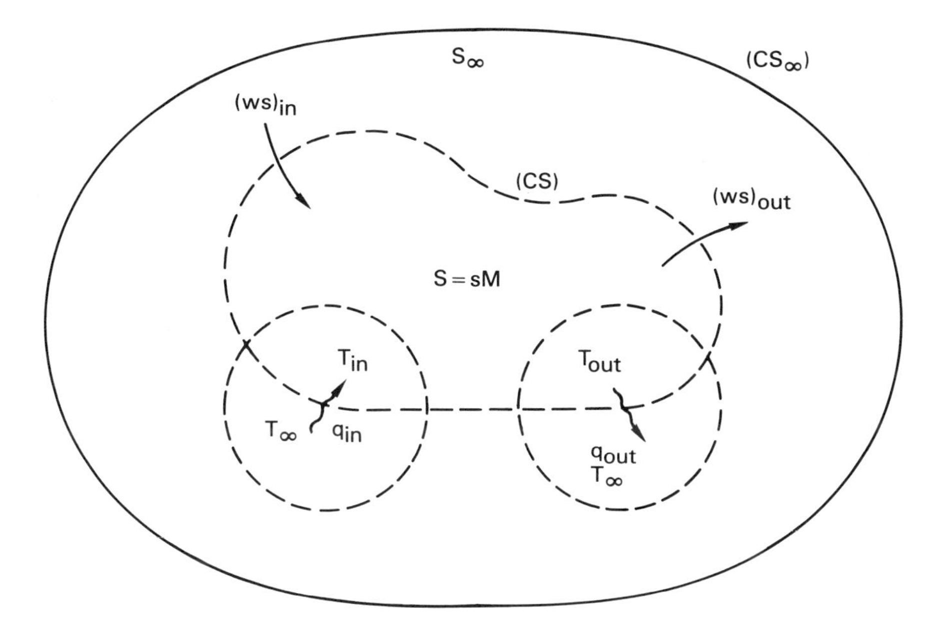

Fig. 3-7 Operation at formulation of the second law.

$$\frac{dS}{dt} = \frac{q_{\rm in}}{T_{\rm in}} - \frac{q_{\rm out}}{T_{\rm out}} + (sw)_{\rm in} - (sw)_{\rm out} \ . \tag{3.29}$$

Heat and entropy inflows and outflows across the dashed boundary have the opposite meaning when considering the outer system. That is,

$$\frac{dS_\infty}{dt} = \frac{q_{\rm out}}{T_\infty} - \frac{q_{\rm in}}{T_\infty} + (sw)_{\rm out} - (sw)_{\rm in} \ . \tag{3.30}$$

Equations (3.28), (3.29), and (3.30) yield:

$$\frac{q_{\rm out}}{T_\infty} - \frac{q_{\rm in}}{T_\infty} \geqslant \frac{q_{\rm out}}{T_{\rm out}} - \frac{q_{\rm in}}{T_{\rm in}} \tag{3.31}$$

which can be used to write Eq. (3.30) as:

$$\frac{dS_\infty}{dt} \geqslant (sw)_{\rm out} - (sw)_{\rm in} + \frac{q_{\rm out}}{T_{\rm out}} - \frac{q_{\rm in}}{T_{\rm in}} \ . \tag{3.32}$$

It follows from Eq. (3.28) that

$$\dot{I} = (sw)_{\rm out} - (sw)_{\rm in} + \frac{dS}{dt} - \left(\frac{q_{\rm in}}{T_{\rm in}} - \frac{q_{\rm out}}{T_{\rm out}}\right) \geqslant 0 \ , \tag{3.33}$$

where $\dot{I}$ is called the *irreversibility rate* (London, 1982). Equation (3.33) is an operational form of the second law which can be employed to assess the overall efficiency of a process. It will be shown next that the loss of available power is given by,

$$\mathcal{P}_{\text{loss}} \leqslant \dot{I}\overline{T} \ , \tag{3.34}$$

where $\overline{T}$ is an appropriate temperature.

3.2.2 Available Power Loss

If it were possible to provide an electronic computer that displayed a readout of the available power loss when it was connected to the inputs and outputs of a system, more engineers would probably be motivated to reduce irreversibilities. The idea of an available power loss computer is discussed next, and it is shown that its readout of power loss, $\mathcal{P}_{\text{loss}}$, is equivalent to the loss of available power expressed by Eq. (3.34).

Consider the general steady-state system of Fig. 3-8a. A mass flow rate, w, enters and leaves with the properties indicated. Mechanical energy and heat transfer rates also cross the system boundary. Note that this general system can be used to describe either a nonflow system by setting $w=0$, or other systems that may lack one or more of the mechanical energy and heat transfer rates shown by setting them equal to zero in Fig. 3-8a. The actual system is replaced by an ideal system in Fig. 3-8b, which performs all internal processes reversibly. Its mass inflow and outflow properties, as well as $\mathcal{P}_{\text{in}}$, $\mathcal{P}_{\text{out}}$, and q_{in}, are equal to those of the actual system. However, to perform all internal processes reversibly, various increments of reversible heating, cooling, compression, and expansion must be per-

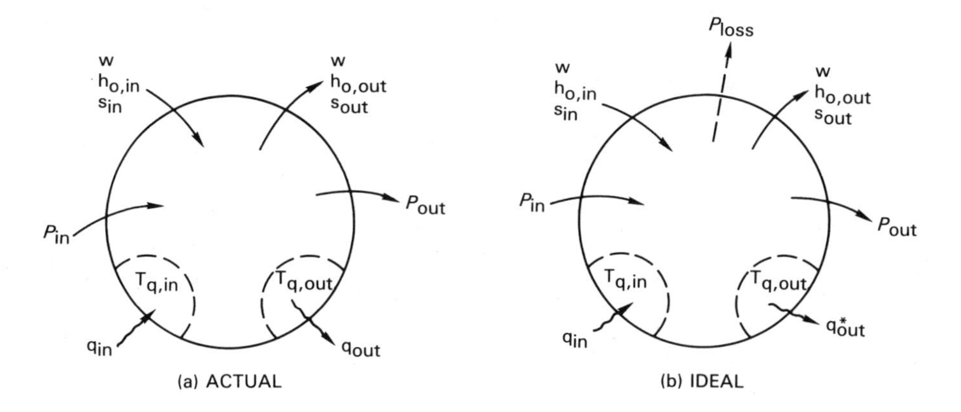

Fig. 3-8 Steady-state thermodynamic system.

formed on the fluid inside the system. The result is that heat rejection q^*_{out} from the ideal system would not necessarily be equal to q_{out} of the actual system. Furthermore, there may be an additional reversible power outflow from the ideal system, identified as $\mathcal{P}_{loss}$, which would be left over if the internal processes were performed reversibly. The extra power that could be ideally extracted, $\mathcal{P}_{loss}$, is imagined to be displayed as an electronic readout on the available power loss computer.

To quantify $\mathcal{P}_{loss}$, and recognizing that internal processes can never equal reversible performance, energy conservation and the second law are written for the ideal steady-state system of Fig. 3-8b as,

$$\mathcal{P}_{loss}+\mathcal{P}_{out}-\mathcal{P}_{in}+w(h_{o,out}-h_{o,in})+q^*_{out}-q_{in}=0 \tag{3.35}$$

and,

$$w(s_{out}-s_{in})\geqslant\frac{q_{in}}{T_{q,in}}-\frac{q^*_{out}}{T_{q,out}}\ . \tag{3.36}$$

Eliminating q^*_{out} yields,

$$\mathcal{P}_{loss}\leqslant\mathcal{P}_{in}\left(1-\frac{\mathcal{P}_{out}}{\mathcal{P}_{in}}\right)+q_{in}\left(1-\frac{T_{q,out}}{T_{q,in}}\right)+wh_{o,in}(a_{in}-a_{out}) \tag{3.37}$$

where $a_{in}-a_{out}$ is the change in a normalized steady flow availability function, defined by:

$$a_{in}-a_{out}\triangleq[(h_{o,in}-T_{q,out}s_{in})-(h_{o,out}-T_{q,out}s_{out})]/h_{o,in}\ . \tag{3.38}$$

The form of Eq. (3.37) applies to a system where both inflow and outflow rates of mechanical, thermal, and convected energy exist. If the system is a turbine for which $\mathcal{P}_{in}=0$, the first term on the right side of Eq. (3.37) becomes $-\mathcal{P}_{out}$. If the system has no thermal energy inflow rate, q_{in} is zero in Eq. (3.37). Furthermore, Eq. (3.37) characterizes a nonflow system when $w=0$.

When a system has both inflow and outflow of mechanical power, the term $\mathcal{P}_{out}/\mathcal{P}_{in}$ in Eq. (3.37) is a mechanical efficiency like that associated with the input and output drive shafts of a gear box. Therefore, $(1-\mathcal{P}_{out}/\mathcal{P}_{in})$ is the fraction of mechanical power that is dissipated. A reduction of bearing gear and fluid friction would make $\mathcal{P}_{out}$ approach $\mathcal{P}_{in}$, and the associated power loss would be reduced. The term $(1-T_{q,out}/T_{q,in})$ can be recognized as the Carnot efficiency (Reynolds, 1965) for thermal energy transfer. Internal thermal power losses would be diminished if the system temperature was more uniform, with $T_{q,out}$ approaching $T_{q,in}$. The thermal power loss is that which could be extracted by a Carnot engine running between $T_{q,out}$ and $T_{q,in}$. Finally, in the absence of thermal and mechanical power terms, the quantity, $(a_{in}-a_{out})$, is a fluid flow efficiency that quantifies the loss of available energy from a fluid stream by flow irreversibilities, which include frictional and geometrical pressure losses.

Generally when two or more energy transfer processes occur simultaneously in a system, their overall effects on property changes are interdependent. Nevertheless, the total available power loss is the sum indicated by Eq. (3.37).

Energy conservation for the actual system, shown in Fig. 3–8a, can be written as,

$$w(h_{o,\text{out}} - h_{o,\text{in}}) + \mathcal{P}_{\text{out}} - \mathcal{P}_{\text{in}} + q_{\text{out}} - q_{\text{in}} = 0 \ . \tag{3.39}$$

If $(\mathcal{P}_{\text{in}} - \mathcal{P}_{\text{out}})$ is eliminated between Eqs. (3.37) and (3.39), $\mathcal{P}_{\text{loss}}$ can be expressed as:

$$\mathcal{P}_{\text{loss}} \leq T_{q,\text{out}} \left[w(s_{\text{out}} - s_{\text{in}}) + \frac{q_{\text{out}}}{T_{q,\text{out}}} - \frac{q_{\text{in}}}{T_{q,\text{in}}} \right] = T_{q,\text{out}} \dot{I} \ . \tag{3.40}$$

A comparison shows that Eqs. (3.34) and (3.40) are the same if the temperature $\overline{T}$ is the system's heat rejection temperature, $T_{q,\text{out}}$. This analysis shows that the available power loss given by Eq. (3.34) is an expression of power losses associated with mechanical, thermal, and flow processes.

If $\mathcal{P}_{\text{loss}}$ was visibly displayed on an electronic computer connected to a given system, it probably would attract attention and issue a silent challenge to identify irreversibilities and reduce them by careful design. Although no such computer exists yet, the challenge still remains, which should encourage more second law thinking.

Several simple examples follow that give estimates of the available power loss from selected thermodynamic processes found in nuclear reactor and containment technology.

3.2.2.1 *Geometric or Frictional Pressure Loss*

Fluid flow in an adiabatic pipe may undergo a pressure loss from pipe geometry effects and friction. The available power loss is to be estimated for the orifice pressure loss of Fig. 3-9. Incompressible liquid flow is characterized by equal inflow and outflow values of h, h_o, μ, and v. Since the pipe is adiabatic, $q = 0$ and $T_{q,\text{out}} \approx T_1 \approx T_2$. The power loss of Eq. (3.40) is $T_1 w(s_2 - s_1)$. The entropy change is obtained from Gibbs equation,

$$T ds = dh - v dp = d\mu + p dv \ , \tag{3.41}$$

giving $T_1(s_2 - s_1) = v(p_1 - p_2)$ so that the available power loss is,

$$\mathcal{P}_{\text{loss}} \leq wv(p_1 - p_2) = Q \Delta p_{\text{loss}} \ . \tag{3.42}$$

A pressure loss of 100 kPa (14.5 psi) and a volumetric flow rate, Q, of 1000 liter/sec (35.3 ft^3/sec) give a power loss of as much as 100 kW. Since pressure loss is proportional to the square of fluid velocity, a larger pipe, less orificing or an additional pipe to carry a given total flow rate would reduce this power loss considerably.

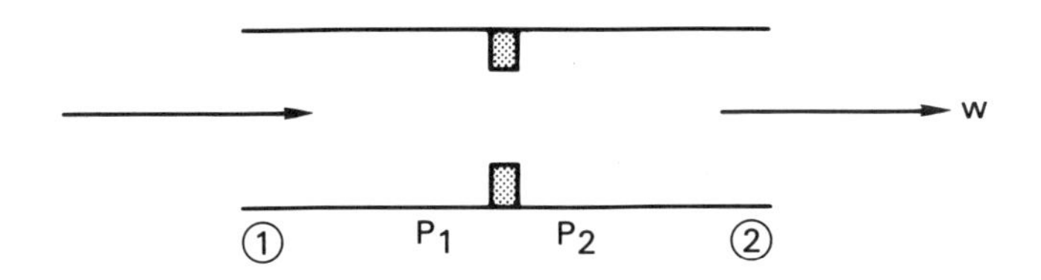

Fig. 3-9 Orifice pressure loss.

If perfect gas undergoes an adiabatic pressure loss $\Delta p_{\text{loss}} = p_1 - p_2$, the available power loss of Eq. (3.40) is again $T_{q,\text{out}} w(s_2 - s_1)$. The entropy increase is obtained by integrating Eq. (3.41), giving

$$s_2 - s_1 = c_p \ln\frac{T_2}{T_1} - R \ln\frac{p_2}{p_1} \ . \tag{3.43}$$

The ratio T_2/T_1 can be obtained for a known p_2/p_1 from the mass and energy conservation laws. The stagnation enthalpies first are expanded, with the help of Eq. (3.20), to give,

$$h_{o1} = \frac{k}{(k-1)} p_1 v_1 + \frac{u_1^2}{2g_c J} = h_{o2} = \frac{k}{(k-1)} p_2 v_2 + \frac{u_2^2}{2g_c J} \ .$$

Mass conservation yields,

$$\frac{u_1}{v_1} = \frac{u_2}{v_2} \ .$$

Now, $p_1 v_1 = RT_1$ and $p_2 v_2 = RT_2$ are employed next to obtain,

$$\frac{k}{(k-1)} p_1 v_1 \left(1 - \frac{T_2}{T_1}\right) = \left(\frac{v_2^2}{v_1^2} - 1\right) \frac{u_1^2}{2g_c J} = \left[\left(\frac{p_1}{p_2}\right)^2 \left(\frac{T_2}{T_1}\right)^2 - 1\right] \frac{u_1^2}{2g_c J} \ .$$

But since $kg_c p_1 v_1 = c_1^2$ and $u_1/c_1 = M_1$, the last equation can be put into the form,

$$\left(\frac{p_1}{p_2}\right)^2 \left(\frac{T_2}{T_1}\right)^2 + \frac{2}{(k-1)M_1^2}\left(\frac{T_2}{T_1}\right) - \left[\frac{2}{(k-1)M_1^2} + 1\right] = 0 \ .$$

This quadratic equation has a positive and a negative solution for T_2/T_1. Only the positive solution has physical meaning, and is written as:

$$\frac{T_2}{T_1} = \frac{\left\{\left[1 + 2(k-1)M_1^2 \left(\frac{p_1}{p_2}\right)^2 \left(1 + \frac{k-1}{2} M_1^2\right)\right]^{1/2} - 1\right\}}{(k-1)M_1^2 \left(\frac{p_1}{p_2}\right)^2} \ , \tag{3.44}$$

where M_1 is the entrance Mach number, $u_1/(kg_cRT_1)^{1/2}$.

Consider a stream flow with $k=1.4$ and $M_1=0.2$. If the upstream and downstream pressures across an orifice are $p_1=6.9$ MPa (1000 psia) and $p_2=6.55$ MPa (950 psia), the temperature ratio of Eq. (3.44) is $T_2/T_1=0.999$, and the normalized power loss from Eq. (3.42) is $\mathcal{P}_{loss}/wc_pT_1 \leqslant 0.014$. In other words, the orifice imposes an available power loss equal to 1.4% of the stream's energy flow rate.

3.2.2.2 *Heat Transmission from a Fin*

The available power loss is to be expressed for a fin like that shown in Fig. 3-10, which conducts heat from a boundary at temperature T_H. Convection to the surrounding fluid at temperature T_∞ is governed by the heat transfer coefficient, H. The differential control volume of Fig. 3-10 shows heat outflows $q+dq$ and dq_h at temperatures $T+dT$ and T, whereas heat inflow is q at temperature T. The film actually is in contact with the fin, but is shown removed for this analysis. Heat inflow to the film is dq_h at T, and heat outflow is dq_h at T_∞. Recognizing that $dq=-dq_h$, and setting $T_{q,\text{out}}=T_\infty$, Eq. (3.40) yields the following expression for differential power loss:

$$d\mathcal{P}_{loss} \leqslant -\left[\frac{T_\infty}{T}\frac{q}{T}\frac{dT}{dz}+\left(1-\frac{T_\infty}{T}\right)\frac{dq}{dz}\right] dz \ . \tag{3.45}$$

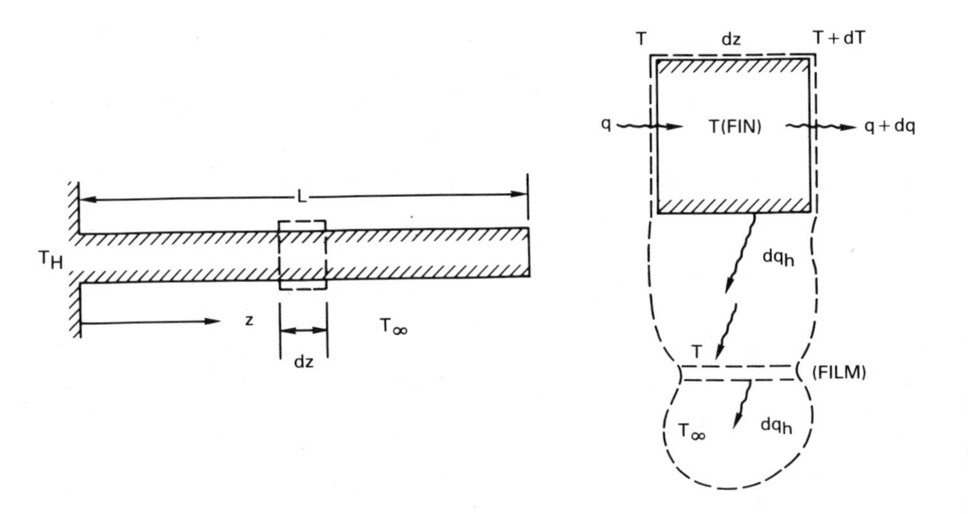

Fig. 3-10 Fin heat transfer.

This example is simplified by choosing a long fin ($L \to \infty$) for which the temperature distribution (Krieth, 1973) is,

$$T(z) - T_\infty = (T_H - T_\infty)e^{-mz} , \tag{3.46a}$$

where

$$m = \frac{HP_h}{KA} , \tag{3.46b}$$

and P_h is the heated perimeter.

Employing,

$$q(z) = -KA\frac{dT}{dz} = KAm(T_H - T_\infty)e^{-mz}$$

and,

$$\frac{dq}{dz} = -KA\frac{d^2T}{dz^2} = -KAm^2(T_H - T_\infty)e^{-mz} ,$$

the available power loss differential of Eq. (3.45) can be written as,

$$d\mathcal{P}_{\text{loss}} \leq KAm^2(T_H - T_\infty)^2 e^{-2mz} \frac{[2T_\infty + (T_H - T_\infty)e^{-mz}]}{[T_\infty + (T_H - T_\infty)e^{-mz}]^2} dz .$$

Integration of this expression along the fin from $z = 0$ to $z = \infty$, and introducing the heat transfer rate at the base of the fin,

$$q(0) = KAm(T_H - T_\infty)$$

yields,

$$\mathcal{P}_{\text{loss}} \leq \left(1 - \frac{T_\infty}{T_H}\right) q(0) . \tag{3.47}$$

Equation (3.47) shows that the available power loss from fins is the Carnot efficiency ($1 - T_\infty/T_H$) times the heat transfer rate from the hot reservoir at T_H. For example, if the temperature at the base of the fin, T_H, is 810 K (1000°F) and the surrounding fluid temperature is 558 K (545°F), the available power loss of $q(0)$ is 31%.

3.2.2.3 *Heat Transfer to a Flowing Liquid*

Liquid enters a nuclear fuel channel at temperature T_1 and pressure p_1, as shown in Fig. 3-11. It is heated by a known $q'(z)$ over length L to temperature T_2 where the pressure is p_2. The liquid temperature at z is $T(z)$, although its temperature at the heating surface is the slightly higher value $T_H = T + \Delta T$ where ΔT is a relatively constant temperature difference. Ne-

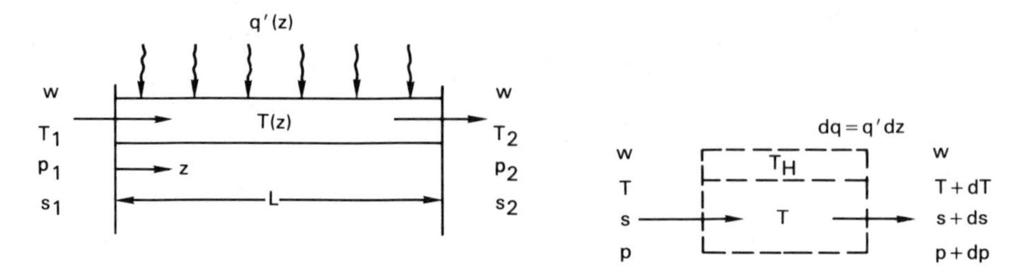

Fig. 3-11 Liquid heating.

glecting heat conduction in the flow direction, the available power loss of Eq. (3.40) can be written for the incremental control volume as,

$$d\mathcal{P}_{\text{loss}} \leqslant T\left(wds - \frac{dq}{T_H}\right) . \tag{3.48}$$

The entropy change is expressed from Gibbs equation,

$$Tds = dh - vdp$$

and the differential heat transfer is given by,

$$dq = wdh = wc_p dT .$$

Also, employing

$$T_H = T + \Delta T ,$$

Eq. (3.48) becomes

$$d\mathcal{P}_{\text{loss}} \leqslant w\left(\frac{\Delta T}{T+\Delta T} c_p dT - vdp\right) ,$$

which can be integrated over the channel length to give,

$$\mathcal{P}_{\text{loss}} \leqslant c_p w \Delta T \ln\left(\frac{T_2 + \Delta T}{T_1 + \Delta T}\right) + wv(p_1 - p_2) . \tag{3.49}$$

The two components of power loss in Eq. (3.49) are seen to be caused from an elevated temperature increment at the heating surface, and the pressure loss.

Suppose that water, $c_p = 4177$ J/kg-K (1.0 Btu/lbm-F), enters a channel at $T_1 = 477$ K (400°F) with a mass flow rate of 50 kg/sec (110 lbm/sec). It is heated with a temperature excess of $\Delta T = 10$ K (18°F) to $T_2 = 558$ K (545°F). If the pressure loss is negligible, the available power loss from Eq. (3.49) is 321 kW.

3.2.2.4 *The Boiling Process*

Bulk boiling occurs when saturated liquid flows through a heated fuel channel of a boiling water nuclear reactor. The fluid is superheated to temperature $T_s = T_f + \Delta T_s$ at the heating surface (McAdams, 1954). Although the boiling mixture will have vapor-liquid slip under most conditions, homogeneous flow is assumed for simplicity in this example. The control volume of Fig. 3-12 can be employed to write the differential power loss as,

$$d\mathcal{P}_{\text{loss}} \leq T_f \left(w ds - \frac{dq}{T_s} \right) . \tag{3.50}$$

Equation (3.34) is used with the saturated state equations

$$h = h_f + x h_{fg} \quad \text{and} \quad v = v_f + x v_{fg} \tag{3.51}$$

and Gibbs equation $T ds = dh - v dp$ to express $T_f w ds$. The mixture kinetic and potential energies are usually negligible, so that energy conservation yields $dq = q' dz = w h_{fg} dx$. The power loss of Eq. (3.50) becomes,

$$d\mathcal{P}_{\text{loss}} \leq (w dh - w v dp) - \frac{T_f}{T_s} w h_{fg} dx \leq \frac{\Delta T_s}{T_f + \Delta T_s} w h_{fg} dx - w v dp . \tag{3.52}$$

The pressure loss term can be treated separately. The remaining term is integrated with an average h_{fg} to give the power loss from boiling as,

$$\mathcal{P}_{\text{loss}} \leq w h_{fg} \left(\frac{\Delta T_s}{T_f + \Delta T_s} \right) x_2 . \tag{3.53}$$

If a 50 kg/sec (110 lbm/sec) mass flow rate of saturated water at 6.9 MPa (1000 psia) with $h_{fg} = (1.51) \times 10^6$ J/kg (650.4 Btu/lbm) and $T_f = 557$ K (544°F) is boiled with a wall superheat of $\Delta T_s = 10$ K (18°F) and leaves the channel at a quality of $x = 0.20$, the available power loss is 266 kW.

Heat transfer irreversibilities can be reduced by roughening the heating surface, although this procedure increases pressure loss irreversibilities.

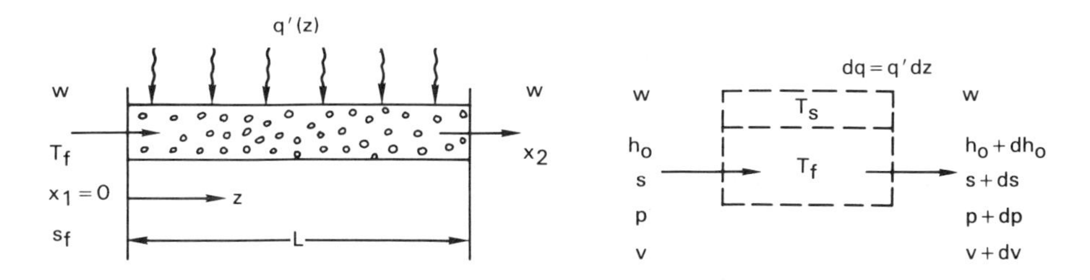

Fig. 3-12 Channel boiling.

Procedures have been formulated for evaluating the effectiveness of surface roughness and other design changes on the second law performance of heat transfer apparatus (Bejan, 1977, 1980).

3.2.2.5 *Available Power Loss from Heat Exchange*

Figure 3-3 indicates a number of feedwater heaters where steam, which is extracted from various turbine stages, is used to heat returning feedwater flow. This arrangement increases overall plant efficiency. The loss of available power association with heat exchange will now be considered to assist the BWR system designer.

Figure 3-13 shows the hot and cold sides of a heat exchanger. Inlet and outlet temperatures are indicated for each side, with hot and cold side average temperatures T_H and T_c. It is assumed that the external boundary of the heat exchanger is adiabatic and that pressure losses of both the hot and cold side flows are negligible. If Eq. (3.40) is applied to the hot and cold sides, and the available power losses of each side are added, the total loss is:

$$\mathcal{P}_{\text{loss}} \leqslant T_2\left[w_H(s_2-s_1)+\frac{q}{T_H}\right]+T_3\left[w_c(s_4-s_3)-\frac{q}{T_c}\right], \tag{3.54}$$

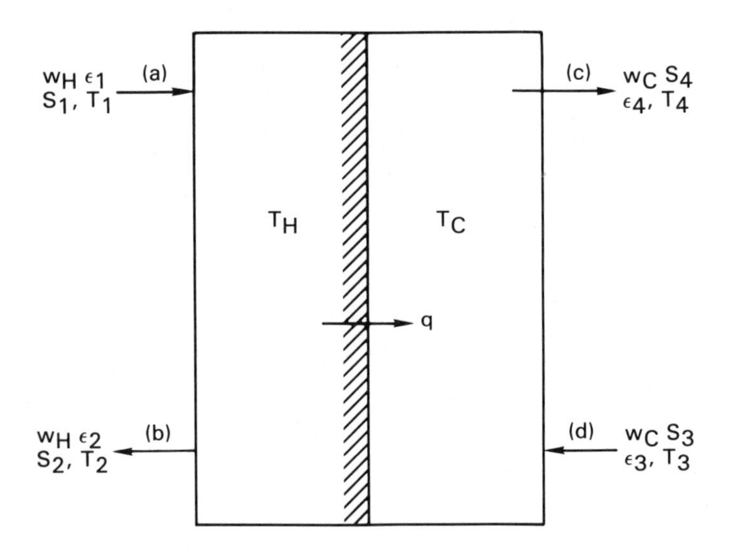

Fig. 3-13 Heat exchanger.

where T_2 and T_3 are the lowest temperatures on the hot and cold sides, respectively. A first law analysis on each side yields,

$$q = w_H c_p (T_1 - T_2) = w_c c_p (T_4 - T_3) \ . \tag{3.55}$$

The Gibbs equation for approximately constant pressure on each side can be integrated to yield:

$$s_2 - s_1 = c_p \ln\frac{T_2}{T_1} \ ; \quad s_4 - s_3 = c_p \ln\frac{T_4}{T_3} \ . \tag{3.56}$$

It follows that,

$$\frac{\mathcal{P}_{\text{loss}}}{q} \leqslant \frac{T_2}{T_1 - T_2} \ln\frac{T_2}{T_1} + \frac{T_3}{T_4 - T_3} \ln\frac{T_4}{T_3} + \frac{T_2}{T_H} - \frac{T_3}{T_c} \ . \tag{3.57}$$

An example case with the following parameters,

$T_1 = 887$ R	$T_H = 866$ R
$T_2 = 845$ R	$T_c = 857$ R
$T_3 = 837$ R	$q = 200$ MW
$T_4 = 877$ R	

has a maximum available power loss of 179 kW.

Available Power Loss from Steam Condensation. Steam condenser properties are indicated in Fig. 3-14, for which saturated steam enters and saturated water leaves the hot side, which is at uniform temperature T_{sat}. Cooling water enters at T_1 and leaves at T_2, and the average cold side temperature is $\overline{T}_c$. Application of Eq. (3.40) to both the hot and cold sides yields,

$$\mathcal{P}_{\text{loss}} \leqslant T_{\text{sat}} \left[w(s_f - s_g) + \frac{q}{T_{\text{sat}}} \right] + T_1 \left[w_c(s_2 - s_1) - \frac{q}{T_c} \right] \ . \tag{3.58}$$

Energy conservation for both sides gives,

$$q = w h_{fg} = w_c c_p (T_2 - T_1) \ . \tag{3.59}$$

Also, use of the Gibbs equation for the hot and cold sides results in,

$$T_{\text{sat}} s_{fg} = h_{fg} \ , \tag{3.60a}$$

$$s_2 - s_1 = c_p \ln\frac{T_2}{T_1} \ . \tag{3.60b}$$

We note from Eqs. (3.59) and (3.60a) that the first bracketed term on the right side of Eq. (3.58) is zero, which leads to the available power loss fraction,

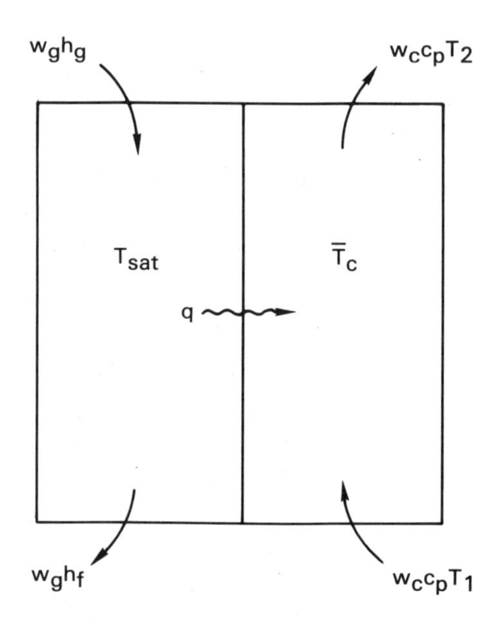

Fig. 3-14 Steam condensation.

$$\frac{\mathcal{P}_{\text{loss}}}{q} \leqslant \frac{T_1}{T_2 - T_1} \ln\frac{T_2}{T_1} - \frac{T_1}{\overline{T}_c} \ , \tag{3.61}$$

where $\overline{T}_c = (T_1 + T_2)/2$. A case with $T_1 = 500$ R and $T_2 = 520$ R gives a maximum available power loss fraction of,

$$\mathcal{P}_{\text{loss}}/q \leqslant 0.00013 \ .$$

3.2.3 Design to Minimize Available Power Loss

The procedure for achieving the minimum loss of available power in the reactor's thermal-hydraulic systems involves a careful examination of each component and process. If the associated available power losses are displayed in terms of system parameters and operating properties, design or operational changes often can be interpreted relative to their effect on system performance. Sometimes the available power loss reduction in a heat transfer process is counterbalanced by an increase from pressure drop. The problem often results in an optimization exercise. Some parameters are beyond design control, such as the axial power distribution in a fuel channel. Although there are notable differences in available power losses from cosine, uniform, and skewed distributions, the requirements of uniform burnup and optimum control rod patterns do not permit arbitrary shifting of power profiles.

Even fractions of a percent decrease of available power loss can result in huge energy savings over the life of a plant. These savings directly benefit the utility operating the plant. It follows that second law analyses can yield economic improvements.

References

Bejan, A., "The Concept of Irreversibility in Heat Exchanger Design: Counterflow Heat Exchangers for Gas-to-Gas Applications," *Trans. ASME,* **99**, 374 (1977).

Bejan, A., "Evaluation of Heat Transfer Augmentation Techniques Based on Their Impact on Entropy Generation," *Heat Mass Transfer,* **7**, 97–106 (1980).

Hatsopoulos, G. N., and J. H. Keenan, *Principles of General Thermodynamics,* John Wiley & Sons, New York (1965).

Huang, F. F., *Engineering Thermodynamics: Fundamentals and Applications,* Macmillan Publishing Company, New York (1976).

Huang, F. F., "Let Us De-mystify the Concept of Entropy," Paper 3257, ASEE 87th Annual Conference (1979).

Krieth, F., *Principles of Heat Transfer,* Harper & Row, New York (1973).

London, A. L., "Economics and the Second Law: An Engineering View and Methodology," *Int. J. Heat Mass Transfer,* **25**(6), 743–751 (1982).

McAdams, W., *Heat Transmission,* McGraw-Hill Book Company, New York (1954).

Reynolds, W. C., *Thermodynamics,* McGraw-Hill Book Company, New York (1965).

Sonntag and VanWylen, *Statistical Thermodynamics,* John Wiley & Sons, New York (1965).

Zemansky, M. W., *Heat and Thermodynamics,* McGraw-Hill Book Company, New York (1955).

CHAPTER FOUR

Boiling Heat Transfer

This chapter deals with the heat transfer mechanisms and processes that occur due to boiling. Boiling heat transfer has a rather vast literature, which includes such things as classical pool boiling, bubble dynamics, and forced convection boiling. Here, we try to restrict ourselves to those aspects of boiling heat transfer that are directly related to BWR technology. However, before discussing product-oriented applications, a brief introduction to some of the jargon, thermodynamics, and mechanistic considerations is given. Those readers interested in a more general treatment of the subject should consult other works, such as the reviews and/or references given in the *Handbook of Heat Transfer* (Rohsenow and Hartnett, 1973), *The Handbook of Multiphase Systems* (Hetsroni, 1982), and books by Hsu and Graham (1976), Van Stralen and Cole (1979), and Hahne and Grigull (1977).

4.1 Introduction to Boiling

This characteristic that distinguishes the boiling heat transfer process from other mechanisms is the change of phase that occurs in the coolant. This change in phase can occur within the fluid, due to a process known as homogeneous nucleation, or more commonly, it can occur at nucleation sites within the fluid (e.g., suspended foreign material in the fluid), or on heated surfaces.

The boiling process is a very efficient method for cooling heated surfaces since the enhanced agitation of the surrounding liquid, caused by boiling, apparently creates a "pumping" action leading to more effective convective cooling. This mechanism is known in the literature as microconvection, as opposed to the normal convective heat transfer mode due to turbulence. For boiling of low-pressure water at a low heat flux, only a few percent of the total energy transferred during boiling is due to latent heat flux; the remainder is due to microconvection and evaporation of the microlayer of liquid that is trapped beneath a growing bubble (Judd and Hwang, 1976).

Before going into the theories and mathematical modeling of boiling phenomena, let us review some of the thermodynamic and physical properties of water and the concepts of phase equilibrium that are necessary to describe the behavior of phase change.

An important property of the vapor-liquid interface is the surface tension, σ. This is the energy per unit area ($F \cdot L/L^2$) required to maintain the interface. Strictly speaking, surface tension is a property of the interface and is dependent on the nature of the gas as well as the liquid. However, for all practical purposes, it can be considered a property of the liquid phase. A plot of the surface tension of water, showing its dependence on temperature, is given in Fig. 4-1.

Another important property is the so-called contact angle. It is not a true property of the substance alone, but describes the wetting characteristics of a liquid in relation to a given solid surface and gas environment. The contact angle, β_c, is defined as the angle between the solid surface and the gas-liquid interface (see Fig. 4-2). By convention, it is always drawn through the liquid phase.

The next property of interest, the Gibbs free-energy function, or chemical potential, is a true thermodynamic property. It is defined as,

$$g \triangleq h - Ts \ . \tag{4.1}$$

If we differentiate Eq. (4.1) we obtain:

$$dg = dh - Tds - sdT \ . \tag{4.2a}$$

Thus, using the so-called Gibbs equation,

$$dg = \frac{1}{\rho} dp - sdT \ . \tag{4.2b}$$

For saturated equilibrium,

$$dp = 0 \text{ and } dT = 0 \ ;$$

thus, Eq. (4.2b) yields,

$$dg = 0 \ . \tag{4.2c}$$

Hence, we find that,

$$g_v = g_l \ . \tag{4.3}$$

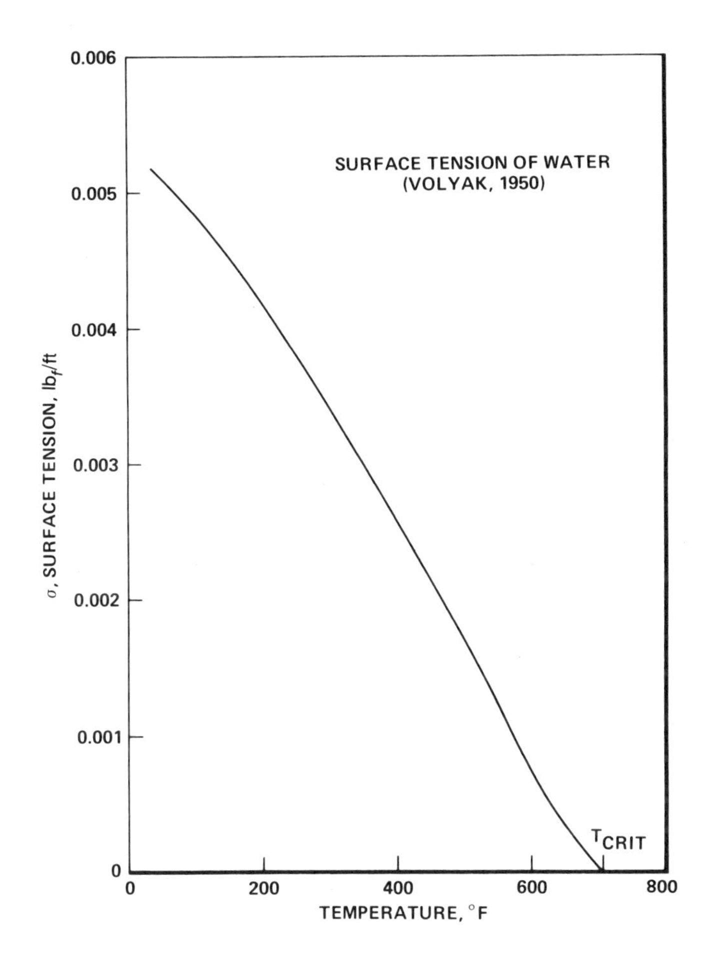

Fig. 4-1 The surface tension of water versus T_{sat}.

Fig. 4-2 Liquid-solid contact angles, β_c; left, a nonwetting solid and, right, a wetting solid.

That is, for equilibrium between two phases of the same substance, the Gibbs function must be the same for the liquid and vapor phase (Hatsopoulos and Keenan, 1965). A system is said to be in phase equilibrium if, when isolated from the surroundings, it does not go through any further change of state. In addition to phase equilibrium, we must also have mechanical and thermal equilibrium. That is, the net sum of the forces on the system must be zero and the temperature of each phase must be the same.

There are two cases of interest: equilibrium across a plane (flat) interface, and equilibrium across a curved interface. These two cases are shown in Fig. 4-3. For the first case (see Fig. 4-3a), the condition of mechanical equilibrium requires that,

$$p_v = p_l \ . \tag{4.4a}$$

For thermal equilibrium, the temperature in both phases must be equal.

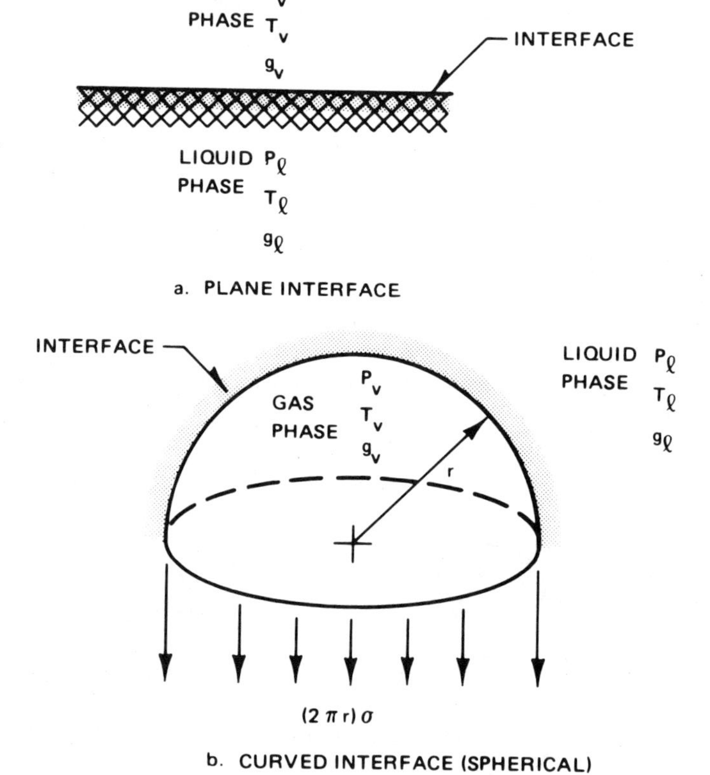

Fig. 4-3 Curved and plane interface configurations.

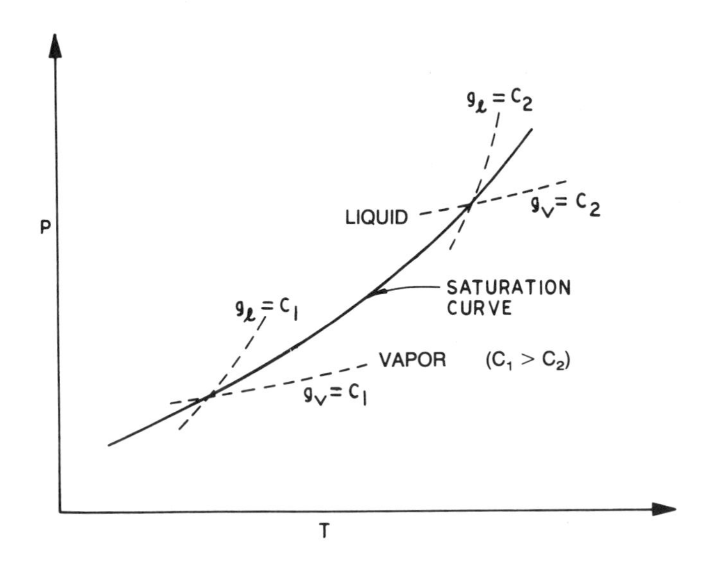

Fig. 4-4 Graphic interpretation of relationship between Gibbs function and saturation curve.

Thus,

$$T_v = T_l \ . \tag{4.4b}$$

Finally, for phase equilibrium, i.e., no condensation or evaporation,

$$g_v = g_l \ . \tag{4.4c}$$

The physical significance of Eq. (4.4c) is the well-known fact that for two phases of the same substance to be in equilibrium, the pressure and temperature are uniquely related. That is, besides the requirement that the temperature and pressure in the phases be equal, there is an additional requirement that they follow a saturation curve. This curve is defined graphically in Fig. 4-4, where it is seen to be the intersection of lines of constant g_l and g_v. In summary, for equilibrium of two phases separated by a plane interface, the appropriate thermodynamic and mechanical conditions require the conditions given in Eqs. (4.4). However, the condition given by Eq. (4.4c) frequently is replaced by the equivalent condition,

$$p = p_{\text{sat}}(T_{\text{sat}}) \ . \tag{4.5}$$

The conditions across a curved interface are somewhat more complex. For the case of a stationary bubble (Fig. 4-3b), the condition of mechanical equilibrium is obtained by a force balance,

$$p_v(\pi r^2) = p_l(\pi r^2) + (2\pi r)\sigma \ ,$$

which yields the well-known Laplace equation,

$$p_v - p_l = \frac{2\sigma}{r} \ . \tag{4.6}$$

As before, the necessary condition for thermal equilibrium is,

$$T_v = T_l \ . \tag{4.7}$$

Finally, the condition for phase equilibrium again implies,

$$g_v = g_l \ . \tag{4.8}$$

Here again, it is interesting to replace the condition for equilibrium given in Eq. (4.8) by an appropriate relation between the saturation pressures and temperatures. This can be done best in the p–g plane by drawing isotherms. An example of this procedure is shown in Fig. 4-5. In this figure, it is assumed that p_l is known; i.e., measured. If the interface separating the phases were flat, the three equilibrium conditions would be satisfied by state ⓓ, which falls on the normal saturation line at temperature, $T_1 = T_{sat}(p_l)$. For the curved interface under consideration, that is, an isolated bubble of radius, r, the equilibrium conditions are satisfied by states ⓐ and ⓑ.

It is evident that these equilibrium states are found at a temperature, T_3, which is higher than the saturation temperature, T_1, associated with the liquid pressure, p_l, and the saturation temperature, $T_2 = T_{sat}(p_v)$, associated with the gas pressure, p_v. In practice, since the slope of the isotherms in the liquid phase is much larger[a] than in the vapor phase, points ⓑ, ⓒ, and ⓒ′ are very close together. To a good first approximation, we may assume that the equilibrium temperature is the saturation temperature at the pressure of the gas phase. Thus, the conditions for equilibrium across a curved interface are given by Eqs. (4.6), (4.7), and (4.8). In accordance with Fig. 4-5, Eq. (4.7) can be written as,

$$T_v = T_l = T_3 \ .$$

---

[a]From the definition of the Gibbs function,

$$g \triangleq h - Ts = u + p/\rho - Ts$$

$$\therefore \ dg = du + dp/\rho - p\frac{d\rho}{\rho^2} - Tds - sdT \ ,$$

recalling the classic Gibbs equation,

$$Tds = du - p\frac{d\rho}{\rho^2} \ .$$

Combining these equations, $dg = dp/\rho - sdT$. Along an isotherm, $dT = 0$ and thus,

$$(\partial p/\partial g)_T = \rho \ .$$

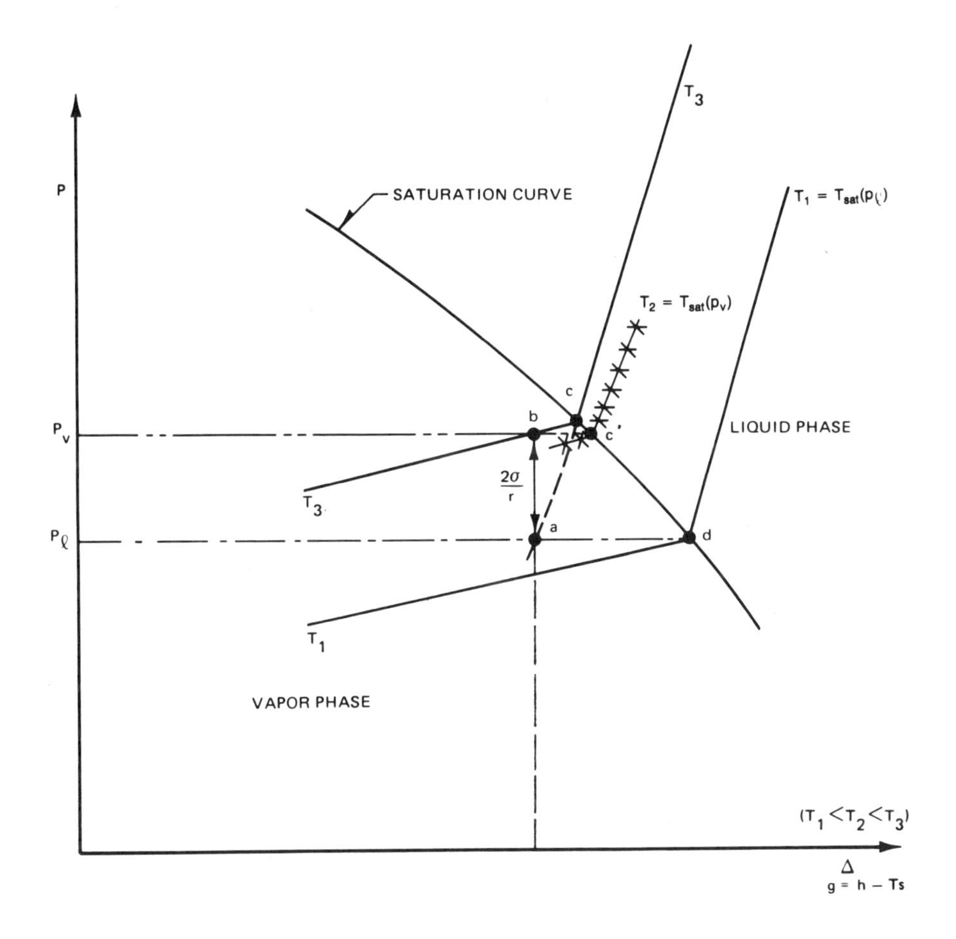

Fig. 4-5 Equilibrium diagram for an isolated bubble.

The equality expressed by Eq. (4.8) is approximately,

$$T_3 \simeq T_{sat}(p_v) = T_2 \quad . \tag{4.9}$$

It is emphasized that for equilibrium across a curved interface, the liquid phase is superheated and the amount of superheat increases as the radius of the bubble, r, decreases. The infinite radius case becomes the plane interface where no superheat is required for equilibrium.

Now we turn our attention to considerations of metastability in two-phase mixtures; that is, conditions that are stable to small disturbances, but unstable to large ones. This subject is best discussed in terms of the appropriate equation-of-state of the mixture. The equation-of-state is the relation that allows us to evaluate any property of a pure substance once two independent properties of that substance are given. For our purposes, we consider the van der Waals equation as the appropriate equation-of-state. This equation is given by,

$$p = \frac{RT}{[v - (v_{crit}/3)]} - \frac{3p_{crit}v_{crit}^2}{v^2} \ . \qquad (4.10)$$

It is well known (Hatsopoulos and Keenan, 1965) that this equation predicts the qualitative characteristics of water, even though it does not represent all observed trends with quantitative precision. Nevertheless, for our purposes it can be used to help us understand metastability effects in two-phase mixtures.

In the familiar p–v plane, the plot of an isotherm that satisfies the van der Waals equation is shown in Fig. 4-6. Branches a to b and e to f are the isotherms in the liquid and vapor regions, respectively. In the two-phase

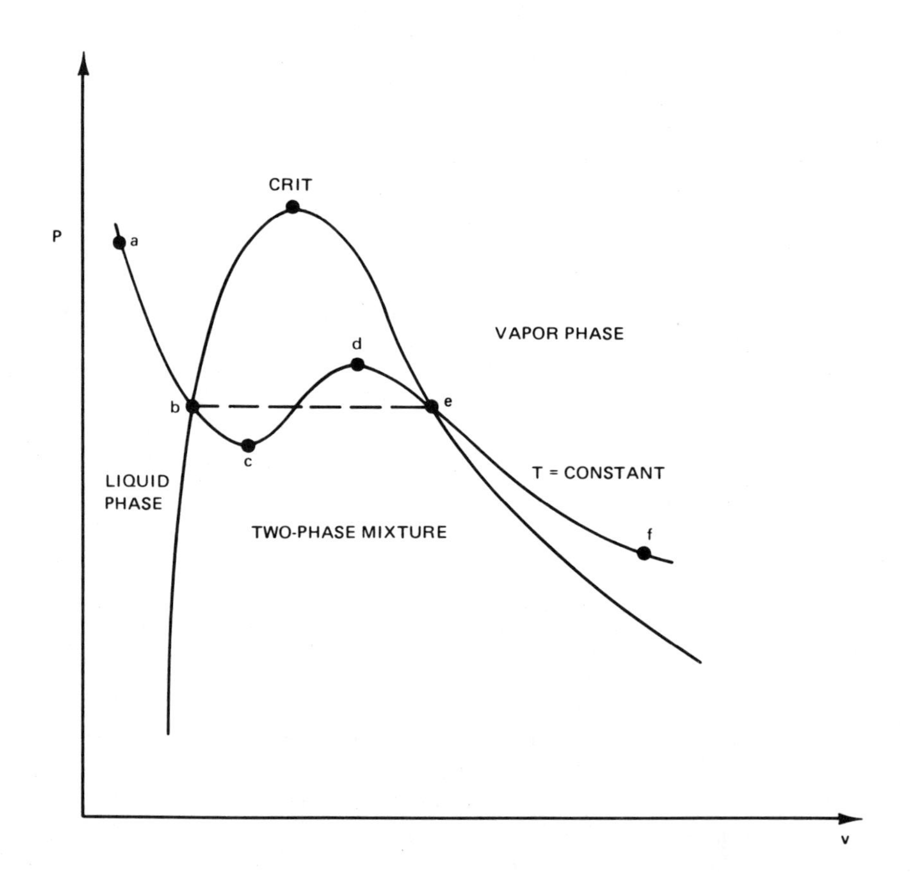

Fig. 4-6 The van der Waals equation.

region, liquid and vapor can coexist in equilibrium. For the case of a pure substance, which implies no surface tension and, thus, a plane interface, this condition of equilibrium must occur at the same pressure. Furthermore, since the two phases can be present in any proportion, a change of state can be achieved at constant pressure and temperature between state points *b* and *e*. However, in accordance with our assumed equation-of-state, it is also thermodynamically possible to follow a constant temperature path, *b–c–d–e*. The *b–c* branch represents a condition of metastability in which the liquid phase is superheated. This condition can be achieved experimentally if sufficient care is taken to avoid mucleation. Point *c* is called a spinoidal point and represents the point at which $(\partial p/\partial v)_T = 0$. The *d–e* branch represents another condition of metastability in which the vapor phase is supersaturated; i.e., at a temperature lower than the saturation temperature of the vapor. Again, this condition can be demonstrated experimentally if sufficient care is taken to avoid condensation. As before, point *d* is also a spinoidal point. The remaining branch, *c–d*, is clearly unstable since $(\partial p/\partial v)_T > 0$ in this region. Considerations of metastable two-phase flow conditions are of interest in BWR technology, particularly with respect to explaining bubble nucleation and transient accident phenomena such as pressure undershoot during system blowdown and the critical flow of flashing liquid.

No discussion of boiling is complete without some reference to the nucleation processes involved. A great deal of work has been done on this problem and we do not attempt to review it in detail. Nevertheless, it is worthwhile to discuss briefly the salient physical mechanisms involved in homogeneous and surface (cavity) nucleation.

First we discuss homogeneous nucleation. This type of bulk process rarely occurs in practice because surface nucleation occurs at much lower superheats. Homogeneous nucleation is an explosive type of nucleation process that results when the liquid phase becomes superheated before vapor formation. In this metastable condition, the liquid phase is subject to rapid phase change once the superheat has exceeded the rupture characteristics of the fluid. Figure 4-7 indicates the situation in terms of the saturation curve and the spinoidals of the van der Waals equation-of-state. Also shown are typical data trends observed by Wismer (Volmer, 1939), when a well-controlled experiment was performed with ether in a smooth glass test section. Although the trends are right, metastability is somewhat overpredicted by the spinoidals of the van der Waals equation-of-state.

The energy required to form a vapor bubble is composed of the surface energy, the latent energy of evaporation, and the thermodynamic *pdV* work term:

$$\frac{4\pi R^2}{J}\sigma + \frac{4}{3}\pi R^3\left[\rho_g h_{fg} - \frac{1}{J}(p_g - p_l)\right] . \tag{4.11}$$

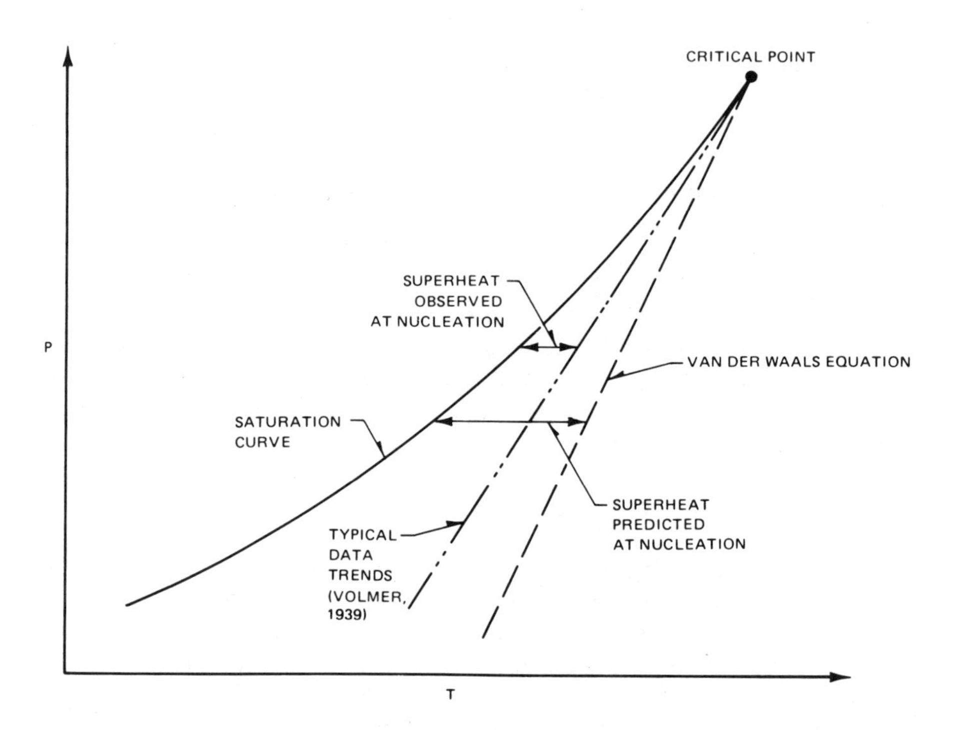

Fig. 4-7 Graphic representation of superheat required for homogeneous nucleation.

This energy can be obtained from the liquid superheat, as described above, or it can be imparted to the fluid through a process such as radiation attenuation or neutron thermalization. Note that if a surface is present spontaneous nucleation can occur on it at lower liquid superheats than for homogeneous nucleation (Van Stralen and Cole, 1979).

Although homogeneous and spontaneous nucleation is of some academic interest, it is of limited practical concern in BWR technology. In contrast, surface or cavity nucleation is a mechanism of considerable practical importance. It is, in fact, the process by which nucleate boiling occurs in fuel rod bundles. Briefly, the mechanism of surface nucleation can be considered a process of nucleation in the cavities that exist in all commercial surfaces. A key feature of this process is that these minute cavities entrap noncondensible gases, which in turn act as nucleation sites for the vapor bubbles. To understand the process involved, we must consider the relationship between equilibrium bubble size and liquid superheat.

In accordance with Fig. 4-3b, the condition for the mechanical equilibrium of a spherical bubble that contains a noncondensible gas of partial pressure p_G is,

$$p_v + p_G - p_l = \frac{2\sigma}{r} . \tag{4.12}$$

To relate the vapor-liquid pressure differential to liquid superheat, we must employ the Clausius-Clapeyron equation,

$$\frac{dT}{dp} = \frac{Tv_{fg}}{Jh_{fg}} . \tag{4.13}$$

This can be integrated along the saturation curve assuming that the quantity, $T_{sat}v_{fg}/h_{fg}$, is constant, yielding,

$$T_v - T_{sat} = (p_v - p_l)\frac{T_{sat}v_{fg}}{Jh_{fg}} . \tag{4.14}$$

By combining Eqs. (4.12), (4.14), and (4.7) we obtain,

$$T_l - T_{sat} = \left(\frac{2\sigma}{r} - p_G\right)\frac{T_{sat}v_{fg}}{Jh_{fg}} , \tag{4.15}$$

which is the liquid superheat required for the spherical vapor bubble of radius, r, to remain in equilibrium. That is, for a liquid superheat larger than that given by Eq. (4.15), the bubble will grow; for a smaller superheat, it will condense. Note that the required superheat varies as $1/r$ and is reduced by the partial pressure of an inert gas.

Equation (4.15) is frequently written as,

$$T_l - T_{sat} = \left(\frac{2\sigma}{r} - p_g\right)\frac{R_v T_{sat}^2}{Jp_v h_{fg}} , \tag{4.16}$$

in which the perfect gas law has been used to eliminate v_{fg}; i.e.,

$$v_{fg} \simeq v_g \simeq \frac{R_v T_{sat}}{p_v} . \tag{4.17}$$

Now we are in a position to discuss the process of cavity nucleation. In Fig. 4-8a, several parts of the nucleation and initial growth phase of a typical vapor bubble are shown. It is seen that the entrapped gas in the bottom of the cavity forms the nucleation site for the vapor bubble. Equation (4.15) yields the minimum superheat required for a bubble of radius, r_1. This requirement is shown in Fig. 4-8b. In the hypothetical case shown, it is assumed that the existing superheat exceeds the initial requirement and, thus, the bubble grows. This growth is rather fast since as the bubble grows, $1/r$ decreases and the excess superheat increases. At position 2, the bubble interface has reached the mouth of the cavity. From this point on, the

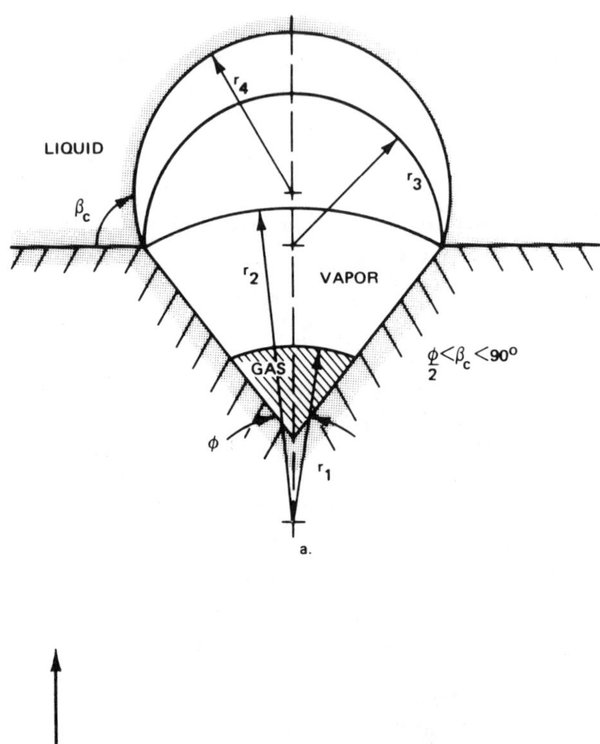

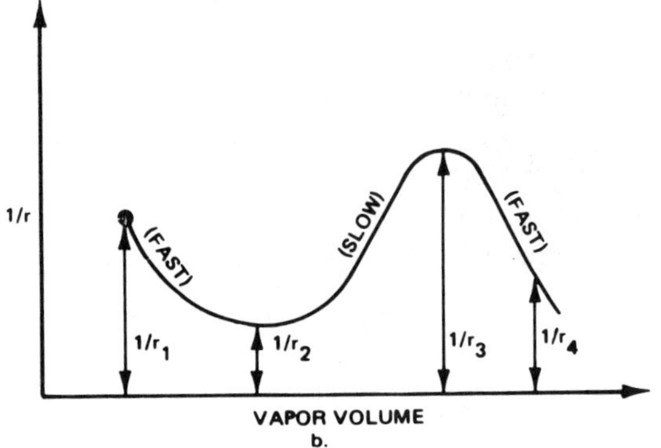

Fig. 4-8 Surface (cavity) nucleation considerations.

bubble growth is somewhat slower, since the center of curvature begins to shift upward. Thus, as shown in Fig. 4-8b, $1/r$ (and thus the required superheat) increases with bubble size. The critical point of bubble growth is position 3. As seen in Fig. 4-8, the center of curvature of the bubble has now shifted to the top of the cavity and $1/r$ (and thus the required superheat) is a maximum. If the local liquid superheat exceeds that required by Eq. (4.15) with $r = r_3$, then the bubble continues to grow rapidly (see Fig. 4-8b). Thus, we find that the superheat required for bubble growth is uniquely determined by r_3, which is the radius of the cavity.

Although the physical description given above has been highly simplified, it does capture the essence of the phenomena involved. Various studies have been performed to support the phenomenological explanation we have presented here. For instance, it has been shown experimentally that fluid-surface combinations, which produce large contact angles, β_c, such as that shown in Fig. 4-2a, tend to capture more inert gas in the surface cavities and, thus, require less superheat to nucleate. In addition, other experiments have been performed in which the entrapped inert gases have been removed. In these cases, nucleation was effectively suppressed. Finally, special surfaces have been prepared with a given cavity size to verify nucleation models.

In practice, there is a whole spectrum of cavity sizes present. Hence, depending on the local liquid superheat, some may nucleate while others remain dormant. Using this observation, we derive a nucleation criterion for heated commercial surfaces. Assuming liquid phase in the laminar sublayer, Fourier's law gives,

$$q'' = -\kappa_f \frac{\partial T}{\partial y} \; . \tag{4.18}$$

Now, in accordance with Fig. 4-8b, we assume that the condition for bubble growth is when the local liquid temperature is tangent to the required vapor temperature given by Eq. (4.15). By combining Eqs. (4.18) and (4.15), we obtain,

$$q'' = -\kappa_f \frac{\partial T}{\partial y}\bigg|_{y=r_*} = \frac{2\kappa_f \sigma}{r_*^2}\left(\frac{T_{\text{sat}} v_{fg}}{J h_{fg}}\right) \; . \tag{4.19}$$

Thus, the radius of the first cavity to nucleate is given by Eq. (4.19) as,

$$r_* = \left(\frac{2\sigma T_{\text{sat}} v_{fg} \kappa_f}{J h_{fg} q''}\right)^{1/2} \; . \tag{4.20}$$

If we assume a linear temperature gradient through the liquid in the laminar sublayer,

$$T_w - T_v = \frac{q'' r_*}{\kappa_f} \; , \tag{4.21}$$

then,

$$T_w - T_{\text{sat}} = T_v - T_{\text{sat}} + \frac{q'' r_*}{\kappa_f} \; . \tag{4.22}$$

By combining Eqs. (4.15) and (4.22), and by neglecting the partial pressure of any noncondensible gases,

$$T_w - T_{sat} = \frac{2\sigma}{r_*}\frac{T_{sat}v_{fg}}{Jh_{fg}} + \frac{q''r_*}{\kappa_f} \quad . \tag{4.23}$$

If only one cavity size is present (i.e., $r_* = r_o$) then Eq. (4.23) yields,

$$q'' = \kappa_f \frac{(T_w - T_{sat})}{r_o} - \frac{2\sigma\kappa_f T_{sat} v_{fg}}{J r_o^2 h_{fg}} \quad . \tag{4.24a}$$

In contrast, if a distribution of cavity sizes is present, Eqs. (4.20) and (4.23) yield the following nucleation criterion:

$$q'' = \frac{Jh_{fg}\kappa_f}{8\sigma T_{sat} v_{fg}} (T_w - T_{sat})^2 \quad . \tag{4.24b}$$

Equation (4.24b) is a simplified version of a more general result previously obtained (Davis and Anderson, 1966). This criterion assumes that cavities of all sizes are present. In practice, large cavities rarely occur and, thus, Eq. (4.24b) should be considered a lower bound for nucleation. That is, nucleation and bubble growth do not occur for conditions below the values given in Eq. (4.24b), and may not occur even for higher conditions (heat flux, wall temperature) if the necessary cavity sizes are not present.

The situation represented by the above analysis is shown in Fig. 4-9. In this figure, three cavity sizes and three temperature profiles are shown; Cases a, b, and c. In Case a, the liquid temperature is less than the required vapor phase temperature given by Eq. (4.15); thus, none of the bubbles grows. In Case b, the wall temperature has increased such that the liquid temperature is just tangent to the required vapor temperature. In this case, bubbles forming in cavities of radius, r_2, grow, while bubbles for smaller and larger cavities do not grow. In the final situation, Case c, all three cavity sizes are active.

Finally, let us consider an overview of the ebullition (i.e., surface boiling) process. After some dwell time, T_D, the liquid adjacent to a cavity on a heated surface reaches the conditions given by Eqs. (4.24). At this point nucleation and bubble growth may occur. During the bubble growth and departure period, T_G, energy will be extracted from the heated surface at a high rate due to microlayer evaporation. Subsequently, when the bubble departs from the surface, microconvection will occur where an equal volume of liquid will replace the departing bubble and begin to heat up. This starts the dwell period again. During this period the heat transfer from the heated surface to the liquid will fall off as the liquid superheats.

The ebullition period, T_B, is obviously given by,

$$T_B = T_D + T_G \quad .$$

Mechanistic models for boiling heat transfer (Hetsroni, 1982) attempt to quantify the wall heat transfer during the dwell period, T_D, and the bubble

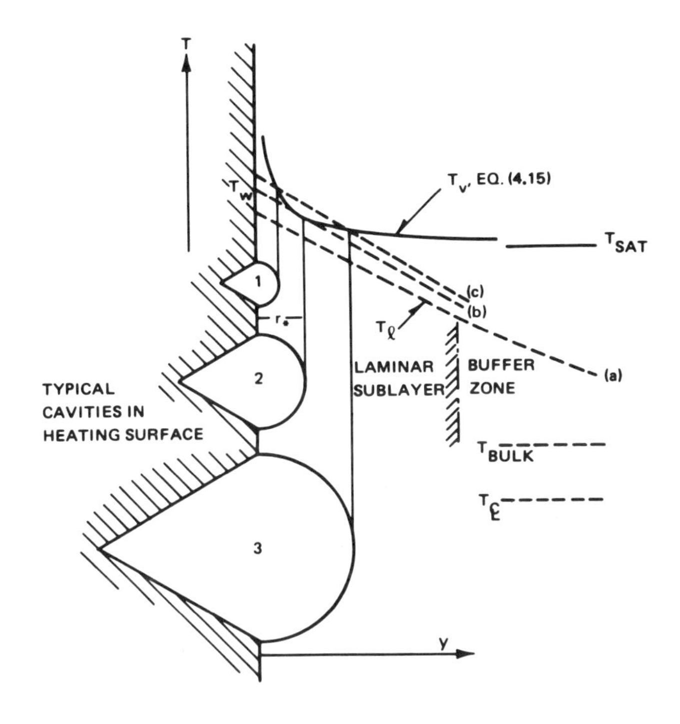

Fig. 4-9 Initiation of bubble growth in forced convection surface boiling.

growth period, T_G. Unfortunately, there are no mechanistic models that can be used to predict boiling heat transfer accurately. As a consequence, empirical correlations are widely used.

This brief introduction to boiling has prepared us for consideration of the boiling phenomena that occur in BWRs. In Sec. 4.2, we are concerned with pool-boiling mechanisms as they apply to nuclear fuel rod bundles.

4.2 Multirod Pool Boiling

Pool boiling is, as its name implies, a situation in which boiling takes place in a stagnant pool of liquid. Natural convection, microconvection (i.e., pumping), and microlayer evaporation are the important heat transfer mechanisms that occur when a heated surface experiences nucleate boiling. Classical pool boiling is characterized by the fact that the heated surface can freely communicate with the surrounding pool of liquid. This problem has been thoroughly investigated over the years and is rather well understood. The general shape of the so-called boiling curve for pool-boiling

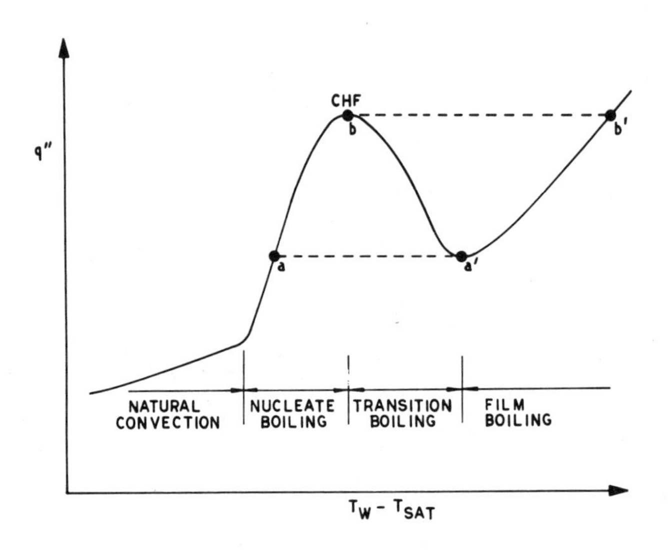

Fig. 4-10 Typical pool-boiling curve.

heat transfer is shown in Fig. 4-10. This curve is normally divided into four regions: natural convection, nucleate boiling, transition boiling, and film boiling. In the natural convection region, the heat flux is normally proportional[b] to $(T_w - T_{sat})^{5/4}$ and, thus, has a slope of 5/4 on a log-log plot. In contrast, in the nucleate boiling region, the slope ranges from 2 to 4 depending on the conditions of the heated surface.

A typical pool boiling heat transfer coefficient for water is given by (Rohsenow, 1952):

$$H_{2\phi} \triangleq \frac{q''}{(T_w - T_{sat})} = \frac{1}{C_{sf}^3} \left\{ \frac{(T_w - T_{sat})^2 \kappa_f^3}{h_{fg}^2 \mu_f^2} \left[\frac{g(\rho_f - \rho_g)}{g_c \sigma} \right]^{1/2} \right\}, \tag{4.25}$$

where representative values of the fluid/surface parameter are (Hetsroni, 1982):

C_{sf}	Fluids
0.014	water/stainless steel
0.013	water/copper
0.006	water/nickel

[b] $\mathrm{Nu} = 0.55(G_r P_r)^{1/4}$
$\therefore H_{1\phi} \propto (T_w - T_{sat})^{1/4}$
$\therefore q'' \propto (T_w - T_{sat})^{5/4}$.

As we continue to increase the power to the heated surface, a condition known as critical heat flux (CHF) is experienced. This phenomenon is basically hydrodynamic in nature and is frequently modeled as a localized flooding phenomenon. Basically, localized flooding occurs when the velocity of the vapor phase is sufficient to prevent an adequate amount of liquid from reaching the heated surface. Kutateladze (1951) found that the critical vapor volumetric flux was given by,

$$(j_g)_c = K\left[\frac{\sigma g g_c(\rho_f - \rho_g)}{\rho_g^2}\right]^{1/4}, \tag{4.26a}$$

where $K \cong 0.16$. In an independent experiment, Wallis (1969) found that the critical vapor volumetric flux for liquid droplet fluidization also was given by Eq. (4.26a) with $K \cong 0.2$. Finally, in an important analytical study on pool-boiling CHF, Zuber (1959) showed that the Taylor and Helmholtz hydrodynamic instability modes imply that for countercurrent vapor and liquid jets, the critical vapor volumetric flux is given by,

$$(j_g)_c = K\left[\frac{\sigma g g_c(\rho_f - \rho_g)}{\rho_g^2}\right]^{1/4}\left(\frac{\rho_f}{\rho_f + \rho_g}\right)^{1/2}. \tag{4.26b}$$

For many cases of interest, $\rho_f >> \rho_g$ and, thus, Eqs. (4.26a) and (4.26b) become essentially equivalent.

Once the critical vapor volumetric flux is known, the CHF can be readily determined. In a saturated pool-boiling situation, the vapor produced at the heated surface is given by,

$$j_g = \frac{q''}{\rho_g h_{fg}}. \tag{4.27}$$

Hence, at CHF, Eqs. (4.27) and (4.26) yield, respectively,

$$q''_c = K h_{fg} \rho_g^{1/2} [\sigma g g_c(\rho_f - \rho_g)]^{1/4} \tag{4.28}$$

$$q''_c = K h_{fg} \rho_g \left[\frac{\sigma g g_c(\rho_f - \rho_g)}{\rho_g^2}\right]^{1/4}\left(\frac{\rho_f}{\rho_f + \rho_g}\right)^{1/2}. \tag{4.29}$$

Equation (4.28), with $K = 0.16$, is the saturated pool-boiling CHF correlation recommended by Kutateladze (1951). Equation (4.29), with $K = \pi/24 \cong 0.13$, is the saturated pool-boiling CHF correlation recommended by Zuber (1959). More recently, Rohsenow (Rohsenow and Hartnett, 1973) has recommended a value of $K = 0.18$ for Eq. (4.29).

Not all successful pool-boiling CHF correlations are formulated in terms of phenomenological localized flooding mechanisms. A widely used correlation, synthesized from a completely different point of view, was formulated by Rohsenow and Griffith (1956) as,

$$q''_c = 143 h_{fg} \rho_g \left(\frac{\rho_f - \rho_g}{\rho_g}\right)^{0.6}\left(\frac{g}{g_c}\right)^{1/4}. \tag{4.30}$$

The correlations given by Eqs. (4.28), (4.29), and (4.30) are compared in Fig. 4-11. It is seen that the trends with system pressure are similar, indicating a peak in CHF for saturated pool boiling around 1000 psia.

Now we turn our attention to the regions of the boiling curve beyond the CHF point in Fig. 4-10. In a temperature-controlled experiment, we can trace out the transition boiling branch of the curve. This branch is characterized by unstable regions of both nucleate boiling and film boiling; the latter is a situation in which the heated surface is blanketed with vapor. For the interesting case of heat flux control,[c] this branch is unstable and, once CHF is experienced, the wall temperature increases rapidly from point b to b'. The film-boiling temperature at point b' frequently is high enough to cause melting of the heater material and, thus, classical pool-boiling CHF is frequently referred to as the burnout heat flux.

If the heater can withstand the elevated temperature at point b', a reduction in heater power can reduce the temperature along the film-boiling branch of the boiling curve to point a', the so-called rewetting temperature. At this point, the vapor film blanketing the heated surface begins to collapse and a rapid excursion of the transition and nucleate boiling branches, from point a' to point a, is experienced. Note the hysteresis that occurs when we move down the film-boiling branch of the curve toward the rewetting point, a'.

A typical film-boiling heat transfer coefficient is of the form:

$$H_{\text{film}} \triangleq \frac{q''}{(T_w - T_{\text{sat}})} = C_1 \left[\frac{\kappa_v^3 g \rho_v (\rho_f - \rho_v) h'_{fg}}{\mu_v L (T_w - T_{\text{sat}})} \right]^{1/4} , \tag{4.31}$$

where,

$$h'_{fg} = h_{fg}[1 + C_2 c_{p_g}(T_w - T_{\text{sat}})/h_{fg}] \ .$$

For a flat horizontal plate (Berenson, 1961):

$$C_1 = 0.425$$

$$C_2 = 0.5$$

$$L = \left[\frac{g_c \sigma}{g(\rho_f - \rho_v)} \right]^{1/2} \ .$$

For a vertical fuel rod of radius R (Bailey et al., 1973):

$$C_1 = 0.4$$

$$C_2 = 0.68$$

$$L = R \ .$$

---

[c]A nuclear reactor fuel element is heat flux controlled.

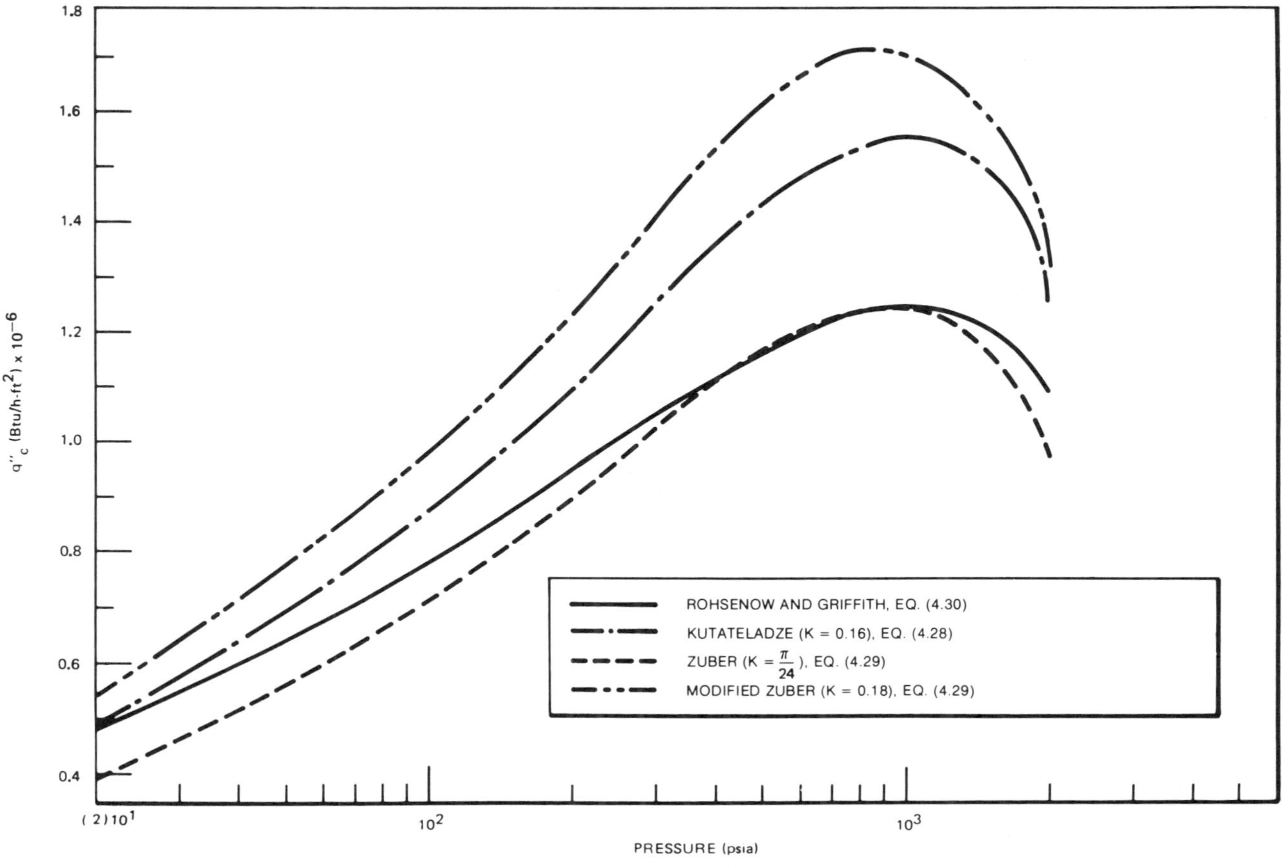

Fig. 4-11 Saturated water pool-boiling CHF correlations.

Although the heat flux at rewetting, $q''_{a'}$, is not well known, a typical expression for the Leidenfrost point is given by:

$$q''_{a'} = C_3 h_{fg} \rho_v \left[\frac{\sigma g(\rho_f - \rho_v)}{(\rho_f + \rho_v)^2} \right]^{1/4} , \tag{4.32}$$

where, for a horizontal flat plate, values of C_3 range from $\pi/24$ (Zuber, 1958) to 0.09 (Berenson, 1961). It should be obvious that the breakdown of film boiling is closely related to localized flooding.

Previously, we mentioned that pool boiling is an interesting and rather well understood mode of boiling heat transfer. Unfortunately, the conditions that exist during low- or no-flow situations in nuclear reactor fuel rod bundles are somewhat different from those that exist in an open pool. Specifically, the close spacing of the rods limits circulation of the coolant and, thus, could be expected to decrease the thermal performance; i.e., critical power. Moreover, it has been found that in BWR fuel rod bundles, which are enclosed by individual channels, a global countercurrent flow limitation (CCFL) normally determines the pool boiling critical power. In Sec. 4.2.1, this phenomenon is discussed.

4.2.1 Countercurrent Flow Limiting (Flooding) Considerations

The phenomenon known as flooding was discussed in Sec. 4.2 in connection with the hydrodynamic explanation of the pool-boiling CHF. In that discussion, the local or microscopic flooding phenomenon was of interest. In this section, we are interested in the same phenomenon, but on a more macroscopic scale. That is, we are concerned with the global CCFL (i.e., flooding phenomenon) in BWR fuel rod bundles.

The situation normally under consideration is shown in Fig. 4-12. For sufficiently low inlet mass flux, G_i, the water in the upper (outlet) plenum tries to drain down into the fuel rod bundle and replace the mass inventory lost due to evaporation. When the superficial velocity of the exiting vapor, $\langle j_g \rangle$, is sufficiently high, CCFL conditions occur at an upper area restriction, and the water in the outlet plenum is unable to drain back into the bundle fast enough to replenish the fluid lost. Flooding does not need to occur at the exit, rather, it may occur within the fuel bundle. That is, it is not unusual to find transients with co-current upward flow at the exit of the fuel bundle that have flooded countercurrent liquid-vapor flow in the lower part of the fuel bundle.

CCFL in BWR fuel rod bundles is a hydrodynamic phenomenon, which for conditions of practical concern, is adequately correlated by the following CCFL correlation (Wallis, 1969),

$$(j_g^*)^{1/2} + (|j_f^*|)^{1/2} = C , \tag{4.33a}$$

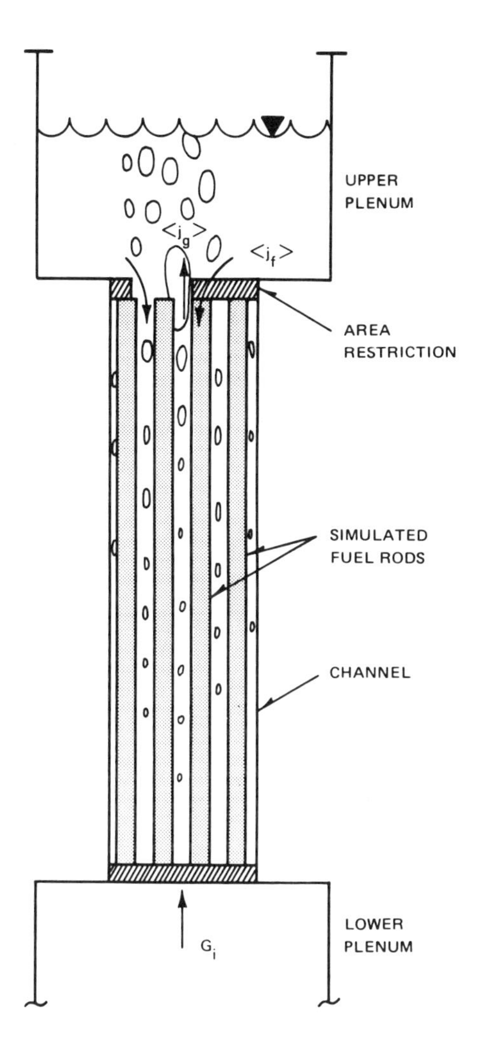

Fig. 4-12 Schematic of CCFL conditions in a BWR fuel rod bundle.

where,

$$j_g^* \triangleq \frac{G_g}{[gD_H\rho_g(\rho_f-\rho_g)]^{1/2}} \tag{4.33b}$$

$$j_f^* \triangleq \frac{G_f}{[gD_H\rho_f(\rho_f-\rho_g)]^{1/2}} \quad . \tag{4.33c}$$

Shiralkar et al. (1972) found that for geometries typical of BWR fuel rod bundles, the hydraulic diameter dependence is not important and, hence, Eq. (4.33a) is normally rewritten as,

$$(j_g')^{1/2} + (|j_f'|)^{1/2} = C' \ , \tag{4.34}$$

where,

$$j_g' \triangleq j_g^* D_H^{1/2}$$

$$j_f' \triangleq j_f^* D_H^{1/2}$$

$$C' \triangleq C D_H^{1/4} \ .$$

A typical value for the parameter, C', is $C' \cong 0.6$ (Shiralkar et al., 1972). Equation (4.34) is shown in Fig. 4-13. Typical loci of bundle exit conditions for the case of increasing power (i.e., evaporation rate) in a rod bundle that is assumed to have no liquid inflow at the bottom are shown in this figure. As bundle power is increased, the operating states at the exit of the bundle move along a locus from point ① to point ② as defined by the equation, $G_g = |G_f|$. This equation simply means that there is a balance between the vapor outflow and the liquid inflow. The CCFL, or flooding, line is intercepted at point ②. Any further increase in bundle power causes an increase in vapor outflow, but a decrease in liquid inflow due

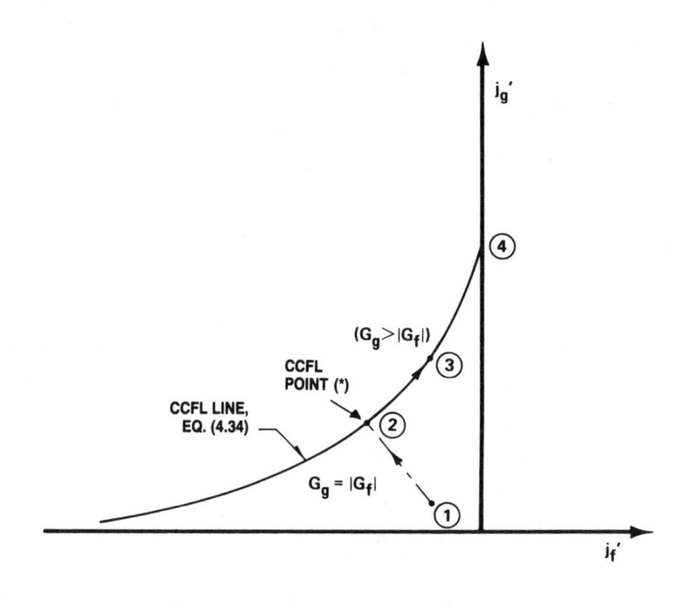

Fig. 4-13 Typical CCFL flooding locus.

to CCFL conditions. The locus of operating states from point ② to point ③ illustrates this trend. As power is further increased, point ④ can be reached where no liquid is able to enter the rod bundle. This operating state is frequently called the "hanging film" state. It should be obvious that continued operation at operating states along the CCFL line will result in overheating the fuel rod bundle, since the mass inventory in the bundle will be depleted with time.

For all practical purposes, the critical power, q_c, of a rod bundle under conditions of pool boiling can be obtained by determining the intersection of the mass balance with the CCFL line, that is, by determining point ② in Fig. 4-13. The most general analysis would include the effect of an inlet flow, G_i, in Fig. 4-12, which is low enough so that flooding is the controlling mechanism. Normally, this is satisfied for $G_i \leq 0.05 \times 10^6$ lbm/h-ft^2.

For steady-state operating conditions, a mass balance yields,

$$G_g A_0 = |G_f| A_0 + G_i A_{x-s} \ , \tag{4.35a}$$

where A_0 is the cross-sectional flow area of the most limiting restriction in the top of the heated bundle; e.g., a tie plate or a grid spacer. Assuming saturated fluid in the upper plenum, an energy balance yields,

$$G_i A_{x-s} h_i + |G_f| A_0 h_f + q = G_g A_0 h_g \ . \tag{4.35b}$$

Equations (4.35) can be combined to yield,

$$G_g = \frac{(q - G_i A_{x-s} \Delta h_{\text{sub}})}{A_0 h_{fg}} \ . \tag{4.36}$$

Thus, Eqs. (4.36) and (4.35a) yield,

$$|G_f| = \frac{(q - G_i A_{x-s} \Delta h_{\text{sub}})}{A_0 h_{fg}} - G_i \left(\frac{A_{x-s}}{A_0} \right) \ . \tag{4.37}$$

In Eqs. (4.36), (4.37), and (4.34), the empirical value for C', and the definitions of j_g', j_f' can now be combined to yield the critical bundle power, q_c, for the specified subcooling and imposed inlet flow. Note that for the case of no inlet flow, Eqs. (4.36) and (4.37) reduce to,

$$G_g = |G_f| = \frac{q}{A_0 h_{fg}} \ . \tag{4.38}$$

A CCFL condition ensures that overheating of the rod bundle will eventually occur. However, in the detailed analysis of hypothetical BWR accidents, which involve low- or no-flow situations, we also are interested in predicting when and where, in the bundle, overheating will occur. To do this, we must combine the applicable elements of the CCFL, multirod pool boiling, and the two-phase conservation equations (Chapter 5) into a transient prediction scheme.

For example, let us consider a specific case in which the inlet flow is suddenly reduced to zero due to loss of pumping capability or flow blockage. The resulting transient is seen in Fig. 4-14 in the $\langle j_f \rangle$–$\langle j_g \rangle$ plane. It is seen that as time, t, increases, the lower part of the bundle experiences countercurrent flow and flooding. Below the CCFL point, point *, the mass of liquid draining into the bundle just matches the vapor generated due to evaporation (i.e., $\rho_f |\langle j_f \rangle| = \rho_g \langle j_g \rangle$) and, thus, a quasi-steady-state situation is established. Above the flooding point, either the liquid downflow is insufficient to replace the rate of mass evaporated or, as seen in the upper region of the bundle, no liquid drains back at all.

Since the CCFL locus is the envelope of lines of constant void fraction, we can replot the information shown in Fig. 4-14 into the more familiar coordinates shown in Fig. 4-15. In this plane, it is seen that at time, t_6, CCFL is achieved at point *. From then on, the void fraction increases rapidly in the upper portion of the rod bundle, while the void fraction in the lower (unflooded) portion is time invariant. Naturally, as the liquid is depleted in the upper part of the rod bundle, the CCFL point, point *, moves lower into the bundle until the vapor updraft, i.e., $\langle j_g \rangle$, is reduced to the point where liquid can drain back in from the outlet plenum. The process then repeats itself in a somewhat cyclic manner provided the rods do not overheat excessively.

To determine the transient location of the dryout front during conditions of countercurrent flow, a method similar to the one developed by Griffith (Walkush and Griffith, 1975) can be employed. The essence of this method is seen in Fig. 4-16, which is a plot of the normalized CHF conditions in a rod bundle versus the local bundle-average void fraction. The normalizing factor on the ordinate, $q''_{c(z)}$, is the pool-boiling CHF correlation given by Zuber; i.e., Eq. (4.29) with $K = \pi/24$. Note that at very small void fractions, the ratio of the rod bundle CHF to that given by Zuber's flat plate pool-boiling correlation is ~0.9. This is in general agreement with Lienhard's results (Lienhard and Dhir, 1972), which indicated that the pool-boiling CHF of a vertical heated surface was ~85% of that of a horizontal plate, and that this ratio was insensitive to the geometry of the vertical surface. Note that in Fig. 4-16 the value for CHF falls off monotonically with increasing void fraction.

To use Fig. 4-16 in accident analysis, we must know $\langle \alpha(z, t) \rangle$. That is, we must have predictions of the transient void fraction as shown in Fig. 4-15. Once the transient void profile has been determined, Fig. 4-16 can be entered to determine $q''_c(z, t)$. Alternatively, we may use:

$$q''_c = q''_c \times \begin{cases} 0.9, & (\langle \alpha \rangle) \leq 0.2 \\ 0.9(1 - \langle \alpha \rangle), & \text{otherwise} \end{cases} . \tag{4.39}$$

If the local transient heat flux in the rod bundle exceeds this critical value, dryout and overheating would be predicted. While it is possible that dryout

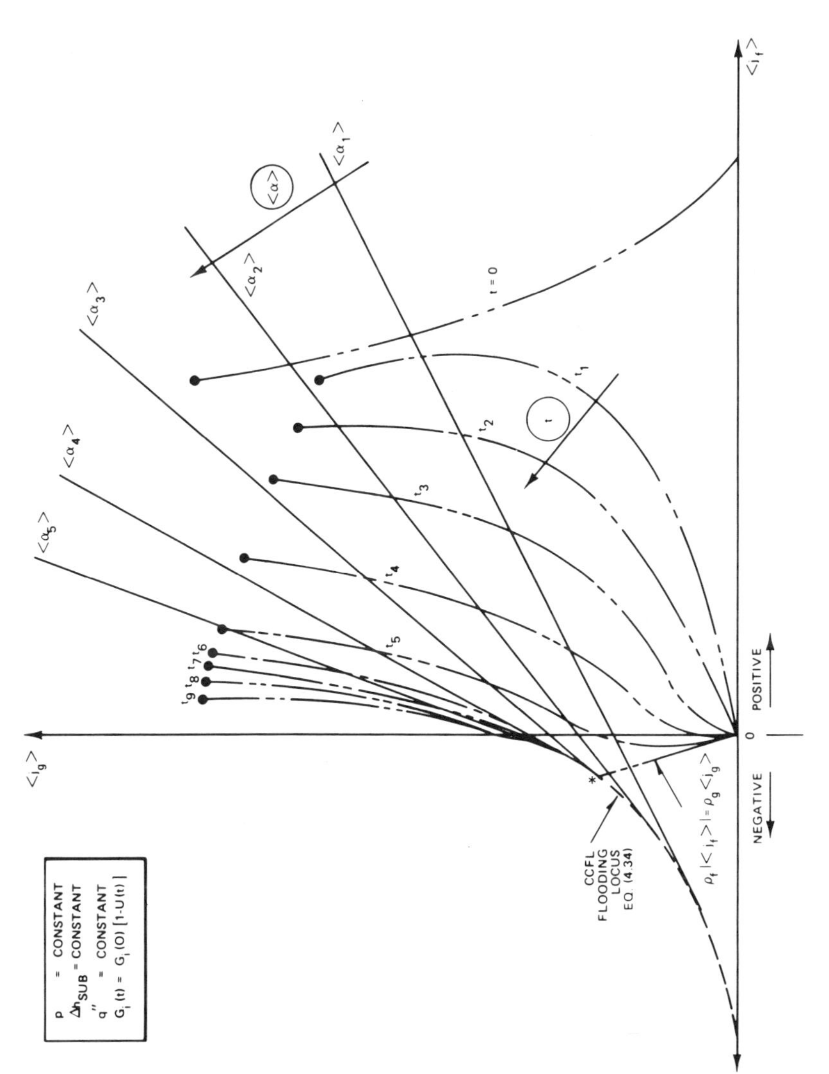

Fig. 4-14 Typical step change response of a BWR fuel rod bundle ($\langle j_g \rangle - \langle j_f \rangle$ plane).

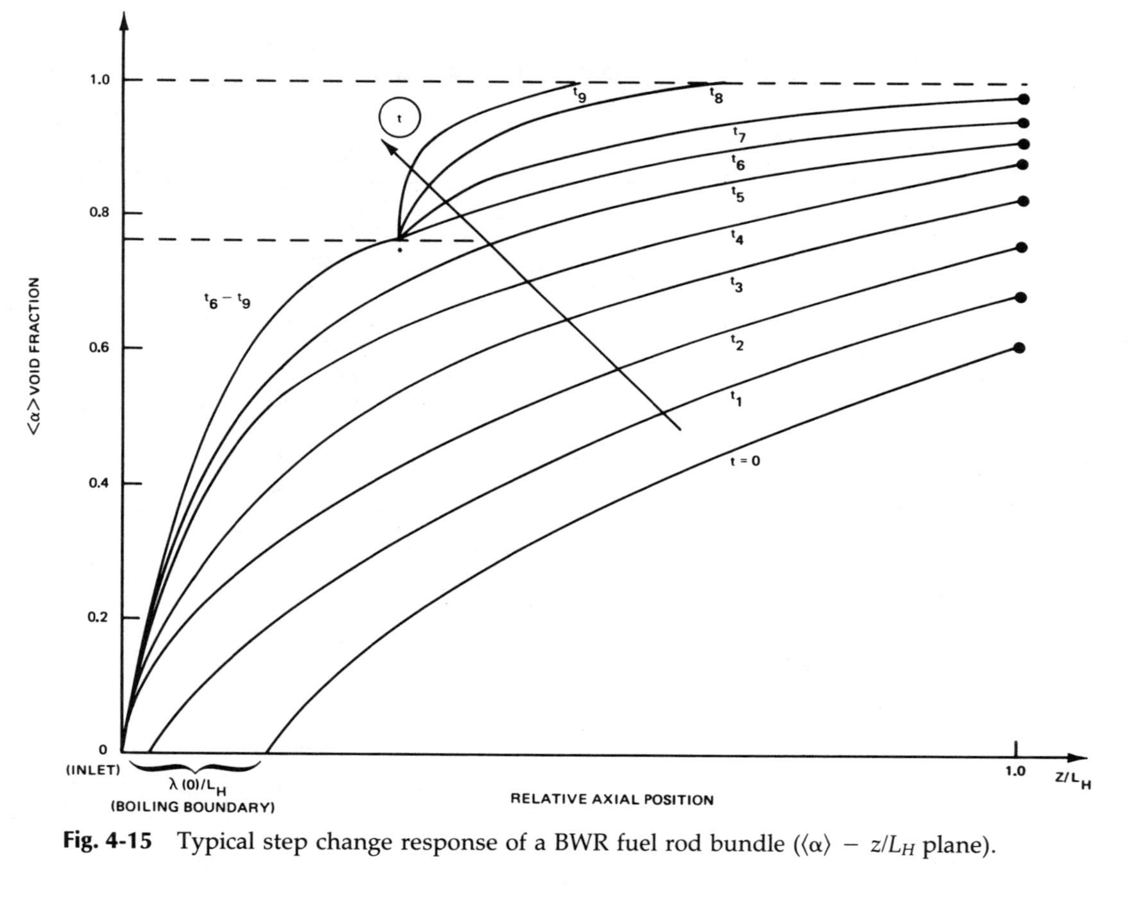

Fig. 4-15 Typical step change response of a BWR fuel rod bundle ($\langle\alpha\rangle$ – z/L_H plane).

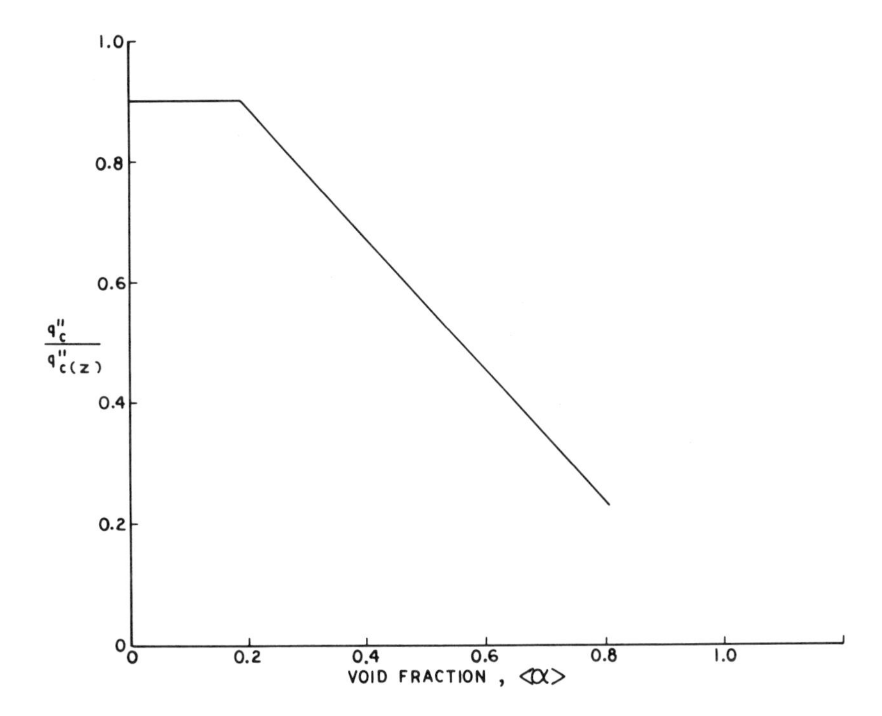

Fig. 4-16 The relation between counterflow BT, void fraction, and pool-boiling CHF for a vertical surface (Walkush and Griffith, 1975).

may be indicated in Fig. 4-16 before a CCFL is reached, it is unlikely, since Zuber's pool-boiling correlation indicates rather high values of allowable heat flux compared to typical BWR heat fluxes (see Fig. 4-11).

The calculational technique just described can be used to predict many items of interest including how the dryout front propagates into a rod bundle during transients. Section 4.4 will be concerned more specifically with the dryout phenomena (i.e., boiling transition) in BWR rod bundles.

4.3 Multirod Flow Boiling

The process of convective heat transfer is a very important mechanism for the transport of energy from nuclear reactor fuel rods. This section is concerned with representative single- and two-phase convective heat transfer modes in multirod bundles. Readers interested in a more general treatment of these topics are referred to other works more specifically concerned with convective boiling heat transfer; e.g., Collier (1972), Rohsenow and Hartnett (1973), and Hetsroni (1982).

To increase appreciation of the various modes of convective heat transfer that can exist in a nuclear fuel rod bundle, consider the two boiling curves shown in Fig. 4-17. The lower curve is a typical pool-boiling curve, previously discussed in Sec. 4.1. The upper curve is a typical forced convection boiling curve drawn for a heater with uniform axial heat flux and a given quality, flow rate, and pressure in the two-phase region of the curve. It is this upper curve that we discuss in this section, since it determines the convective heat transfer characteristics of a typical nuclear fuel rod bundle.

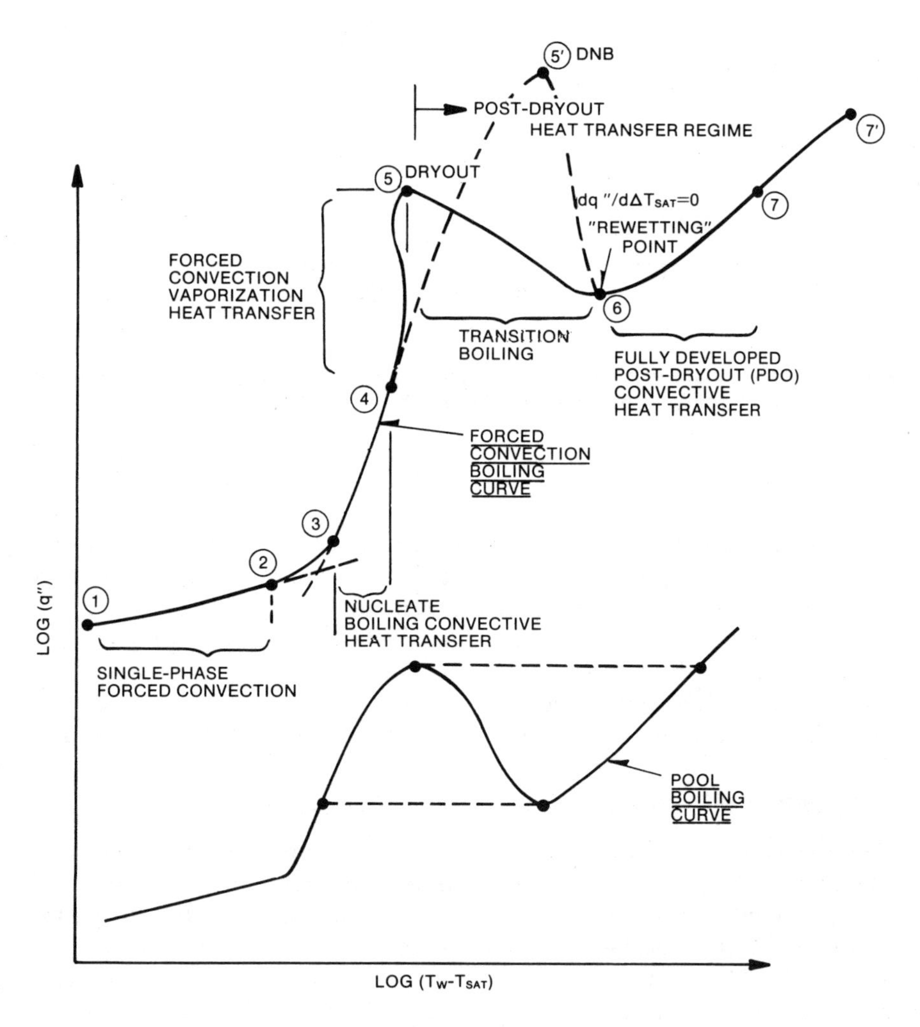

Fig. 4-17 Comparison of typical convective and pool-boiling curves.

Note that if subcooled boiling still occurs at the boiling transition (BT) point then a departure from nucleate boiling (DNB) will take place at a relatively high heat flux (point ⑤′). This is the situation of interest for a pressurized water nuclear reactor (PWR). In contrast, in a BWR, liquid film dryout will take place at the BT point (point ⑤). In this book we will concentrate on boiling heat transfer phenomena of interest in BWR technology.

4.3.1 Convective Heat Transfer in Rod Bundles

The first mode of convective heat transfer that we consider is single-phase convection, shown as branch ① to ② in Fig. 4-17. Single-phase convective heat transfer in multirod bundles can be correlated in the standard form (McAdams, 1954)

$$\mathrm{Nu} = C_1 \, \mathrm{Re}^{n_1} \, \mathrm{Pr}^{n_2} \; , \tag{4.40}$$

where the Nusselt number, Nu, Reynolds number, Re, and Prandtl number, Pr, are given by

$$\mathrm{Nu} \triangleq \frac{H_{1\phi} D_H}{\kappa} \tag{4.41a}$$

$$\mathrm{Re} \triangleq \frac{G D_H}{\mu} \tag{4.41b}$$

$$\mathrm{Pr} \triangleq \frac{c_p \mu}{\kappa} \; . \tag{4.41c}$$

For moderate temperature differences, the single-phase heat transfer coefficient, $H_{1\phi}$, for heating a liquid is often given by either the classical Dittus-Boelter equation,

$$\left(\frac{H_{1\phi} D_H}{\kappa_{\text{bulk}}}\right) = 0.023 \left(\frac{G D_H}{\mu_{\text{bulk}}}\right)^{0.8} \left(\frac{c_p \mu}{\kappa}\right)_{\text{bulk}}^{0.4} , \tag{4.42}$$

where the bulk temperature, $T_{\text{bulk}} = \langle T \rangle$, of the fluid is used to determine the thermodynamic properties, or the Colburn equation,

$$\left(\frac{H_{1\phi} D_H}{\kappa_{\text{film}}}\right) = 0.023 \left(\frac{G D_H}{\mu_{\text{film}}}\right)^{0.8} \left(\frac{c_p \mu}{\kappa}\right)_{\text{film}}^{1/3} \left(\frac{c_{p_{\text{bulk}}}}{c_{p_{\text{film}}}}\right) , \tag{4.43}$$

in which the mean film temperature, $(T_{\text{wall}} + T_{\text{bulk}})/2$, is used to determine the properties denoted by the subscript "film."

The Colburn equation frequently is rewritten in terms of the Stanton number, St, as,

$$\mathrm{St} = \frac{\mathrm{Nu}}{\mathrm{Re}\,\mathrm{Pr}} = \frac{H_{1\phi}}{G c_p} \; , \tag{4.44}$$

in the form,

$$j \triangleq \left(\frac{H_{1\phi}}{Gc_p}\right)_{\text{bulk}} \left(\frac{c_p \mu}{\kappa}\right)_{\text{film}}^{2/3} = 0.023 \left(\frac{GD_H}{\mu}\right)_{\text{film}}^{-0.2} . \tag{4.45}$$

Note that by using the Reynolds analogy, the so-called Colburn j-factor is uniquely related to the smooth tube (Darcy-Weisbach) friction factor by,

$$j = f/8 .$$

Either the Dittus-Boelter or the Colburn correlation can be used to evaluate the single-phase transfer coefficient, $H_{1\phi}$. Note that the Dittus-Boelter correlation requires only knowledge of the bulk temperature for property evaluations and, thus, is easier to use. However, when it is used, some authors (Kays, 1966) recommend a property correction of the form,

$$\frac{\text{Nu}}{\text{Nu}_{\text{bulk}}} = \begin{cases} \left(\dfrac{T_w}{\langle T \rangle}\right)^{-0.5} , & \text{for steam} \\ \left(\dfrac{\mu_w}{\langle \mu \rangle}\right)^{-0.2} , & \text{for water ,} \end{cases} \tag{4.46}$$

which also requires that we iterate on the wall temperature T_w, as with the Colburn correlation. Fortunately, for many practical cases of interest the Dittus-Boelter correlation, Eq. (4.42), can be used without the property correction given in Eq. (4.46), to yield a reasonably accurate estimation of the single-phase heat transfer coefficients in rod bundles.

The next mode of convective heat transfer considered is that associated with the nucleate boiling branch of Fig. 4-17. Section ② to ③ is associated with partial (subcooled) nucleate boiling. This mode of convective heat transfer is a complicated mixture of single-phase convection and nucleate boiling modes of heat transfer. In many cases of practical concern, it can be ignored. For those cases (e.g., transients) where it must be considered, an approximate treatment is possible (Lahey, 1974). Section ③ to ④ represents fully developed nucleate boiling and, for water, is normally correlated in the form,

$$q'' = \kappa (T_w - T_{\text{sat}})^m . \tag{4.47}$$

The Jens-Lottes correlation (Jens and Lottes, 1951) and the Thom correlation (Thom, 1966) are several frequently used correlations of this form for water. These correlations are given in the following table:

Correlation	Parameters [Eq. (4.47)]	Nucleate Boiling Heat Transfer Coefficient $H_{2\phi(NB)} \triangleq \frac{q''}{(T_w - T_{sat})}$
Jens-Lottes (1951)	$m=4$ $\kappa = \frac{\exp(4p/900)}{12.96}$	$\frac{\exp(4p/900)}{12.96}(T_w - T_{sat})^3$
Thom (1966)	$m=2$ $\kappa = \frac{\exp(2p/1260)}{0.005184}$	$\frac{\exp(2p/1260)(T_w - T_{sat})}{0.005184}$

where the heat flux, q'', is in Btu/h-ft^2, pressure, p, is in psia, and temperatures are in °F. Note that in the two correlations discussed above, fully developed nucleate boiling is insensitive to flow rate and flow quality, but is quite sensitive to system pressure.

For conditions of high-quality annular flow, such as may occur in BWR technology, the vapor velocity and interfacial turbulence level can become so high, and the liquid film can become so thin, that nucleation is suppressed and the heat is transferred by conduction through the thin liquid film on the heated surface. This convective heat transfer regime is known as forced convective vaporization and it is characterized by evaporation at the liquid-vapor interface and a flow effect similar to single-phase convective heat transfer. Note that in branch ④ to ⑤ in Fig. 4-17, this mode of heat transfer implies very high heat transfer coefficients resulting in decreasing wall temperature as the heat flux increases.

Forced convective vaporization heat transfer correlations generally are of the form,

$$\frac{H_{2\phi(FC)}}{H_{1\phi(lo)}} = A\left(\frac{1}{X_{tt}}\right)^n , \tag{4.48}$$

where the turbulent Martinelli parameter, X_{tt}, is normally given by,

$$\frac{1}{X_{tt}} = \left[\left(\frac{dp}{dz}\right)_g \bigg/ \left(\frac{dp}{dz}\right)_f\right]^{1/2} = \left(\frac{\langle x \rangle}{1-\langle x \rangle}\right)^{0.9} \left(\frac{\rho_f}{\rho_g}\right)^{0.5} \left(\frac{\mu_g}{\mu_f}\right)^{0.1} . \tag{4.49}$$

Equation (4.42) or (4.43) can be used to determine $H_{1\phi(lo)}$, where lo implies that the single-phase heat transfer correlation is evaluated assuming that the total mass flux of the system is saturated liquid. Several popular forced convective vaporization correlations are due to Dengler and Addoms (1956) and Bennett et al. (1961). These correlations are given as follows:

Correlation	Parameters [Eq. (4.48)]
Dengler and Addoms (1956)	$A=3.5$ $n=0.5$
Bennett et al. (1961)	$A=2.9$ $n=0.66$

Figure 4-18 shows a comparison of Bennett's correlation with some typical data. Note that in the right portion of the figure, for small values of X_{tt} (i.e., large values of flow quality) the forced convective vaporization heat transfer coefficient, $H_{2\phi(FC)}$, has the same mass flux dependence as $H_{1\phi(lo)}$. In contrast, for larger values of X_{tt} (i.e., smaller values of flow quality), the two-phase heat transfer coefficient is independent of mass flux and quality, which indicates a nucleate boiling mode of heat transfer.

One of the problems with the application of forced convective vaporization heat transfer correlations, such as that given in Eq. (4.48), is to know when to apply them. That is, we must have some criterion to determine when nucleation in the liquid film is suppressed. Although there are a number of approximate criteria available (Rohsenow and Hartnett, 1973),

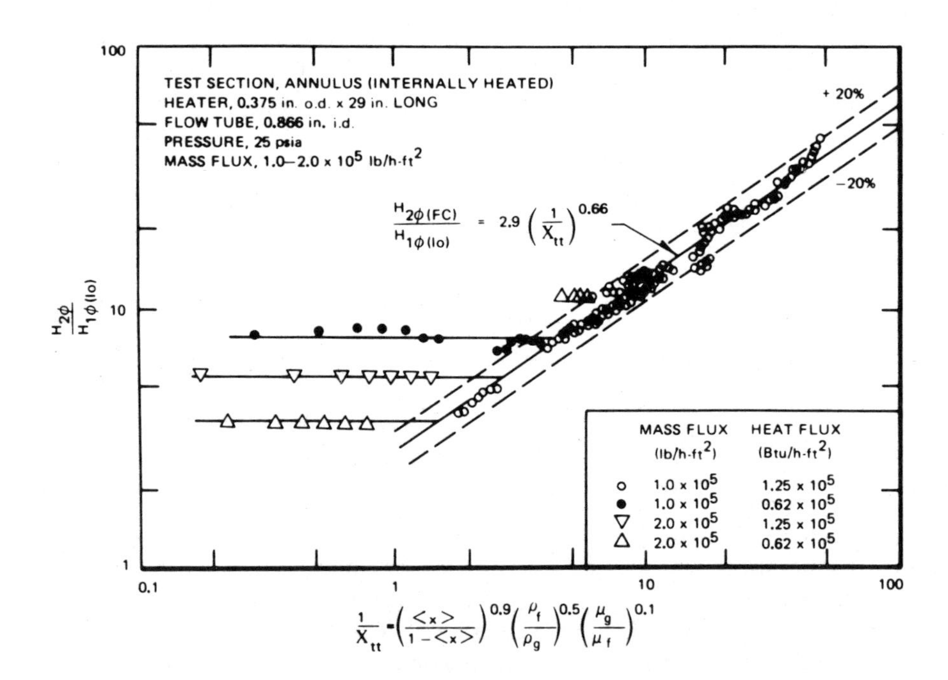

Fig. 4-18 Variation of heat transfer coefficient ratio with Martinelli parameter, X_{tt} (Bennett et al., 1961).

none has been proven accurate as yet for application to rod bundles. Fortunately, Chen (1963) has formulated a rather good correlation for both the fully developed nucleate boiling and forced convection vaporization regions; i.e., branch ③ to ⑤ of Fig. 4-17.

The Chen correlation is in the form of a superposition:

$$q''_{2\phi} = H_{1\phi(LP)}[(T_w - T_{sat}) + (T_{sat} - T_{bulk})] + H_{2\phi(NB)}(T_w - T_{sat}) \ , \quad (4.50)$$

where,

$$H_{1\phi(LP)} \triangleq 0.023\left[\frac{G(1-\langle x\rangle)D_H}{\mu_f}\right]^{0.8}\left(\frac{c_{pf}\mu_f}{\kappa_f}\right)^{0.4}\left(\frac{\kappa_f}{D_H}\right)F \quad (4.51)$$

$$H_{2\phi(NB)} \triangleq 0.00122\left[\frac{\kappa_f^{0.79} c_{p_f}^{0.45} \rho_f^{0.49} (g_c^*)^{0.25}}{\sigma^{0.5} \mu_f^{0.29} h_{fg}^{0.24} \rho_g^{0.24}}\right]\left(\frac{h_{fg}}{T_{sat} v_{fg}}\right)^{0.75} (T_w - T_{sat})^{0.99} S \quad (4.52)$$

$$g_c^* \triangleq 416.923 \times 10^6 \frac{\text{ft}}{h^2} \ .$$

It is often convenient to assume that the exponent 0.99 in Eq. (4.52) can be approximated by 1.0. In this case Eqs. (4.50), (4.51), and (4.52) yield:

$$q''_{2\phi} = H_{1\phi(LP)}[(T_w - T_{sat}) + \Delta T_{sub}] + \frac{H_{2\phi(NB)}}{[T_w - T_{sat}]}(T_w - T_{sat})^2 \ . \quad (4.53)$$

Equation (4.53) is a quadratic equation, which can be solved easily for the local wall superheat, $(T_w - T_{sat})$.

The single-phase heat transfer coefficient, $H_{1\phi(LP)}$, is the standard Dittus-Boelter correlation evaluated for saturated liquid conditions at the mass flux of the liquid phase, LP, and multiplied by a correction parameter, F, given by,

$$F \triangleq \left[\frac{Re_{2\phi}}{Re_{(LP)}}\right]^{0.8} \ , \quad (4.54)$$

where,

$$Re_{(LP)} \triangleq \frac{G(1-\langle x\rangle)D_H}{\mu_f} \ . \quad (4.55)$$

The parameter, F, can be recognized as the ratio of an effective two-phase Reynolds number to the Reynolds number used to evaluate $H_{1\phi(LP)}$. This parameter is plotted versus the Martinelli parameter, X_{tt}, in Fig. 4-19. A convenient fit for this parameter is given by,

$$F = 2.35\left[\frac{1}{X_{tt}} + 0.213\right]^{0.736} \ . \quad (4.56)$$

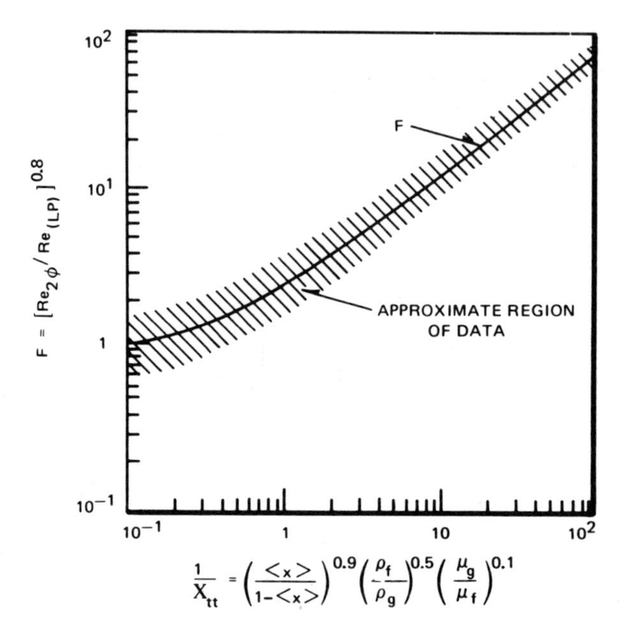

Fig. 4-19 Reynolds number factor, F (Chen, 1963).

The two-phase nucleate boiling heat transfer coefficient, $H_{2\phi(NB)}$, is a form of the Forster-Zuber correlation times a nucleation suppression factor, S, which is defined as,

$$S \triangleq \left[\frac{\Delta T_{ave}}{(T_w - T_{sat})}\right]^{0.99} , \tag{4.57}$$

where ΔT_{ave} is the effective radial average superheat in the liquid film. The suppression factor, S, is given in Fig. 4-20 as a function of the effective two-phase Reynolds number, $Re_{2\phi}$. A fit for this parameter is

$$S = 0.9622 - 0.5822\left[\tan^{-1}\left(\frac{Re_{2\phi}}{6.18 \times 10^4}\right)\right] . \tag{4.58}$$

Equations (4.50), (4.51), and (4.52) can be used in conjunction with Figs. 4-19 and 4-20 to evaluate the two-phase heat transfer; however, note that the correlation is implicit in $(T_w - T_{sat})$ and, thus, some iteration is required. The Chen correlation is widely used throughout the nuclear industry and is recommended for most convective boiling heat transfer evaluations.

To gain further appreciation of how several of the more popular boiling heat transfer correlations compare, the Chen, Jens-Lottes, and Dengler and Addoms correlations have been evaluated in Fig. 4-21 for a typical value of $\Delta T_{sat} = 10°F$. Note that the Chen correlation tends to merge with the Jens-Lottes correlation for nucleate boiling conditions and with the Dengler

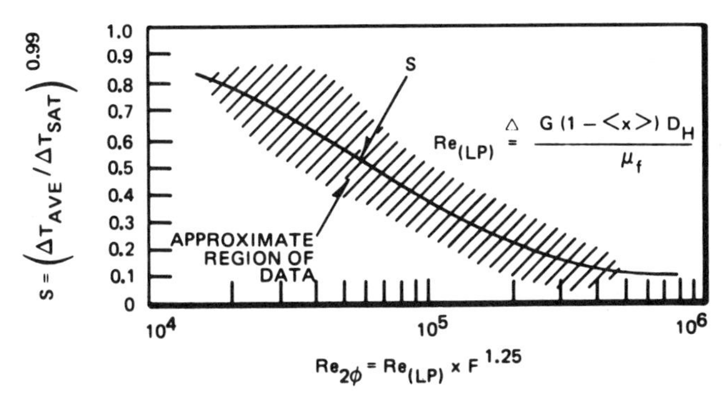

Fig. 4-20 Suppression factor, S (Chen, 1963).

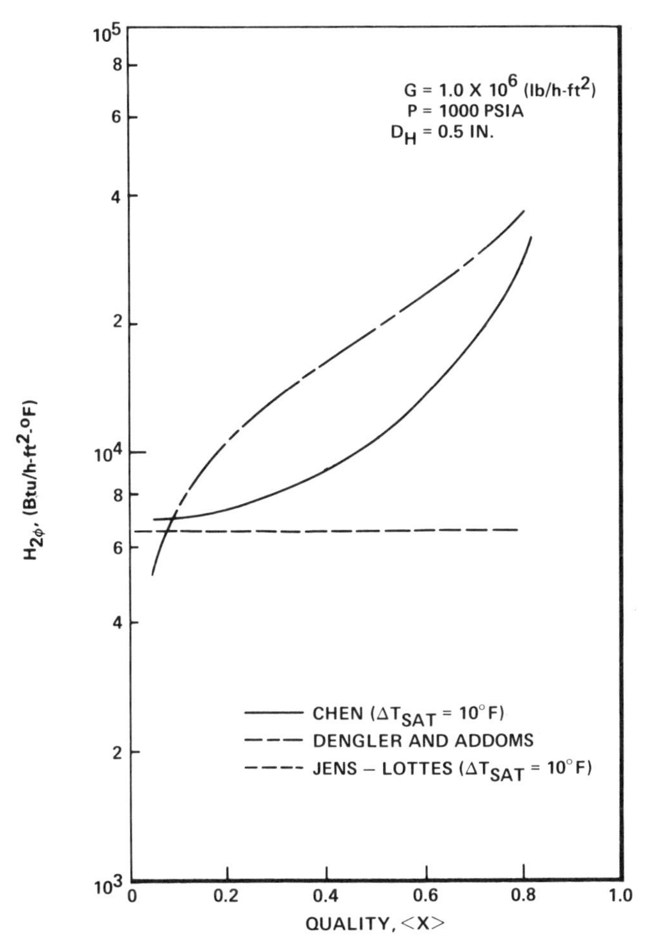

Fig. 4-21 A comparison of some convective boiling heat transfer correlations.

and Addoms correlation for forced convection vaporization conditions. Also note the rapid increase of the heat transfer coefficient with quality when nucleate boiling is suppressed, as discussed previously.

The next point of interest on the forced convection boiling curve, shown in Fig. 4-17, is the BT or dryout point, point ⑤. As discussed in Sec. 4.4, this point is uniquely defined in the q'' versus $(T_w–T_{sat})$ plane only for uniform axial heat flux. For nonuniform axial heat flux profiles, more typical of BWR fuel rods, there is a family of boiling curves which are a function of the axial heat flux profile and inlet subcooling; i.e., it is primarily the critical quality that determines dryout, not the local heat flux.

4.4 Boiling Transition

In Sec. 4.2, the phenomenon known as pool-boiling CHF was discussed. The CHF was characterized by a sudden deterioration in the boiling heat transfer mechanism resulting in a temperature excursion of the heated surface. A similar phenomenon can occur during flow-boiling situations of interest in BWR technology. The point at which the heat transfer coefficient deteriorates is known by various names, including

1. the boiling crisis
2. critical heat flux (CHF)
3. departure from nucleate boiling (DNB)
4. burnout (BO)
5. dryout.

In the American literature, CHF seems to be the preferred term, although DNB is frequently used to describe the high-pressure, high-flow phenomena that can occur in pressurized water reactor rod bundles. Unfortunately, all of these terms leave something to be desired. For instance, CHF has the connotation that it is the local heat flux that determines the onset of transition boiling. It will be shown subsequently that the so-called "local conditions hypothesis" is not generally valid for BWR conditions and, thus, unlike pool-boiling situations, the term CHF is not generally descriptive of the phenomena involved. Moreover, even in situations in which the local conditions hypothesis can be considered valid (i.e., test sections with uniform axial heat flux or very low qualities), it is well known that a temperature excursion can be achieved at constant power level by varying the system flow rate, inlet subcooling, or pressure and, thus, CHF is again misleading.

Of the terms tabulated above, dryout most accurately describes the physical mechanism that occurs at conditions of interest to BWR technology. However, as pointed out by Hewitt (1970), it also lacks generality. The term "boiling crisis" is frequently used in the European literature and, although it does accurately characterize the rapid temperature excursions typical of low-quality DNB, it is misleading for high-quality film dryout in which the temperature excursions are quite mild.

The thermal-hydraulic situation that we wish to describe is the onset of transition boiling. It is felt that the term boiling transition (BT) provides the correct connotations of the event and is generally applicable to the various phenomena and experimental techniques for achieving the event. In any case, this is the terminology that is used exclusively in BWR technology in the United States (GETAB, 1973).

Now that a terminology to describe the phenomenon of interest has been established, we are in a position to discuss the event more specifically. Note that there are numerous other works that classify and discuss the BT phenomena. Source material, for instance, includes Tong (1965), Hewitt and Hall-Taylor (1970), Collier (1972), Tong (1972), and Rohsenow and Hartnett (1973). These references discuss the analytical and experimental work that has been done in survey form. In most cases, these surveys provide material of generic interest rather than material for specific geometry and operating conditions. Here we restrict our attention to those geometries and operating conditions of interest in BWR technology.

4.4.1 Phenomenological Discussion

The BT, which occurs at qualities of interest in BWR technology, is primarily due to a dryout of the liquid film on the heated surface. This phenomenon is associated with two-phase annular flow conditions and has been studied rather extensively (Hewitt and Hall-Taylor, 1970). Once the film of liquid ceases to exist, the heated surface is cooled only by vapor and the impingement of liquid droplets. As subsequently discussed, this mode of heat transfer generally is rather good, particularly at high-quality conditions where vapor velocities can be quite high. Nevertheless, when the liquid film vanishes, a temperature excursion of the heated surface occurs. As shown in Fig. 4-22, the surface temperature normally oscillates due to oscillations of the dryout front. The magnitude of these temperature oscillations and the speed of the post-BT (PBT) temperature excursion depends on the operating conditions. However, for conditions of practical concern, the temperature excursion is normally rather slow and the oscillations are not excessive. A typical PBT temperature history is shown in Fig. 4-23.

4.4.1.1 *Film Dryout Modeling*

To gain some insight into the phenomena of film dryout, we formulate a simple mechanistic description of the process. Figure 4-24 is a representation of the incipient dryout of a liquid film. The liquid film is being depleted by evaporation, entrainment, i.e., shearing off liquid by the high-velocity vapor, causing wave action, etc., that results in an entrainment

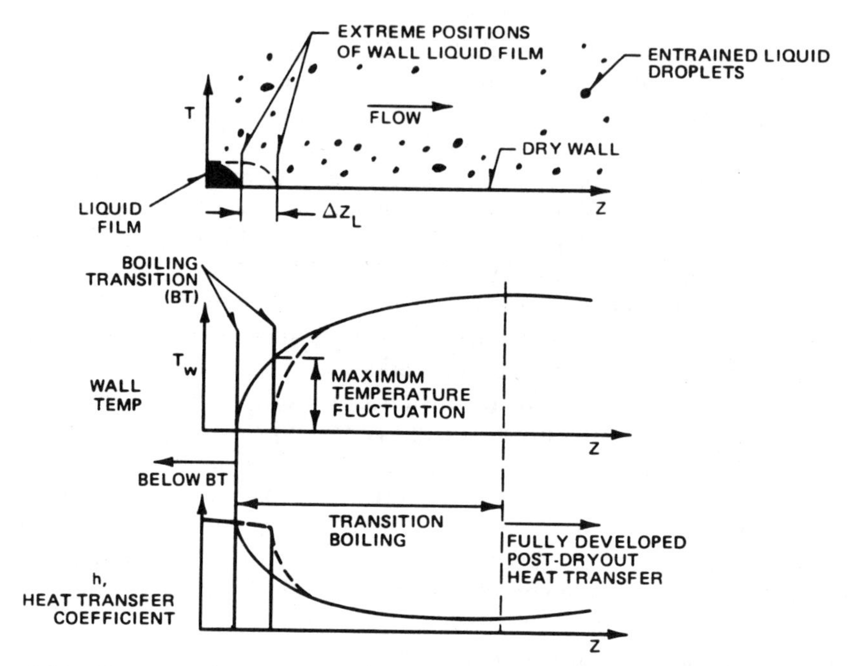

Fig. 4-22 Heat transfer behavior beyond BT for uniform axial heat flux (Quinn, 1966).

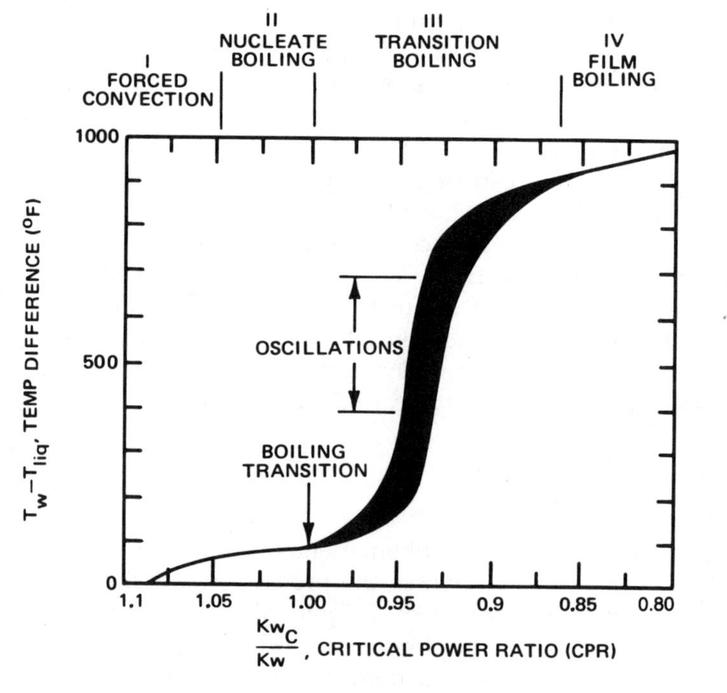

Fig. 4-23 Typical temperature history beyond BT (Levy and Bray, 1963).

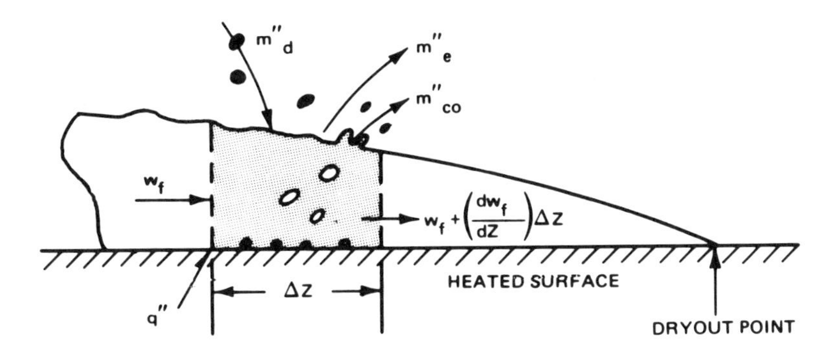

Fig. 4-24 A simple film dryout model.

mass flux, m_e'', and carryover; i.e., the liquid that is carried over by any vapor bubbles that burst through the liquid film and cause a carryover mass flux, m_{co}''. On the other hand, deposition, m_d'', of the liquid droplets tends to replenish the film.

The conservation of mass principle applied to a saturated liquid film implies that for steady state,

$$w_f + \left(\frac{dw_f}{dz}\right)\Delta z + \frac{P_H \Delta z q''}{h_{fg}} + m_e'' P_H \Delta z + m_{co}'' P_H \Delta z - w_f - m_d'' P_H \Delta z = 0 \ , \tag{4.59a}$$

which can be simplified to,

$$\frac{dw_f}{dz} = P_H\left(m_d'' - \frac{q''}{h_{fg}} - m_e'' - m_{co}''\right) \ . \tag{4.59b}$$

If we specify the functional form of m_d'', q'', m_e'', and m_{co}'' (i.e., how these parameters vary with z and w_f), then presumably we can integrate Eq. (4.59b) from the point where the annular flow regime begins to the point of dryout. Numerous investigators have attempted to specify these terms and integrate mechanistic film flow equations such as Eq. (4.59b). The works of Grace (1963), Tippets (1964), and Whalley et al. (1974) are typical of early efforts. A latter model, by Whalley (1987), is one of the most successful to date.

Some investigators (i.e., Hewitt and Hall-Taylor, 1970) have preferred to formulate Eq. (4.59b) in terms of the flow rate of the entrained liquid, w_{le}. The appropriate transformation is accomplished easily through use of the identity,

$$w_f = GA_{x-s}(1 - \langle x \rangle) - w_{le} \ . \tag{4.60}$$

The work done to date indicates that understanding the entrainment process is the key to understanding the dryout phenomena. Figure 4-25 shows that there is a significant difference between the developing diabatic entrainment curves and the equilibrium adiabatic entrainment curve. Moreover, there appears to be a tendency for the diabatic curves to approach the equilibrium situation. This tendency is seen more dramatically in Fig. 4-26, which indicates the strong tendency toward equilibrium when "cold patches" are installed in various axial positions. It is seen that the critical quality (i.e., the quality at the BT point) increases when the tendency is to decrease entrainment and decreases when the tendency is to increase it. Since equilibrium can never be achieved in a diabatic situation (i.e., the quality increases with length), the actual entrainment curve depends on many factors and is quite difficult to calculate accurately.

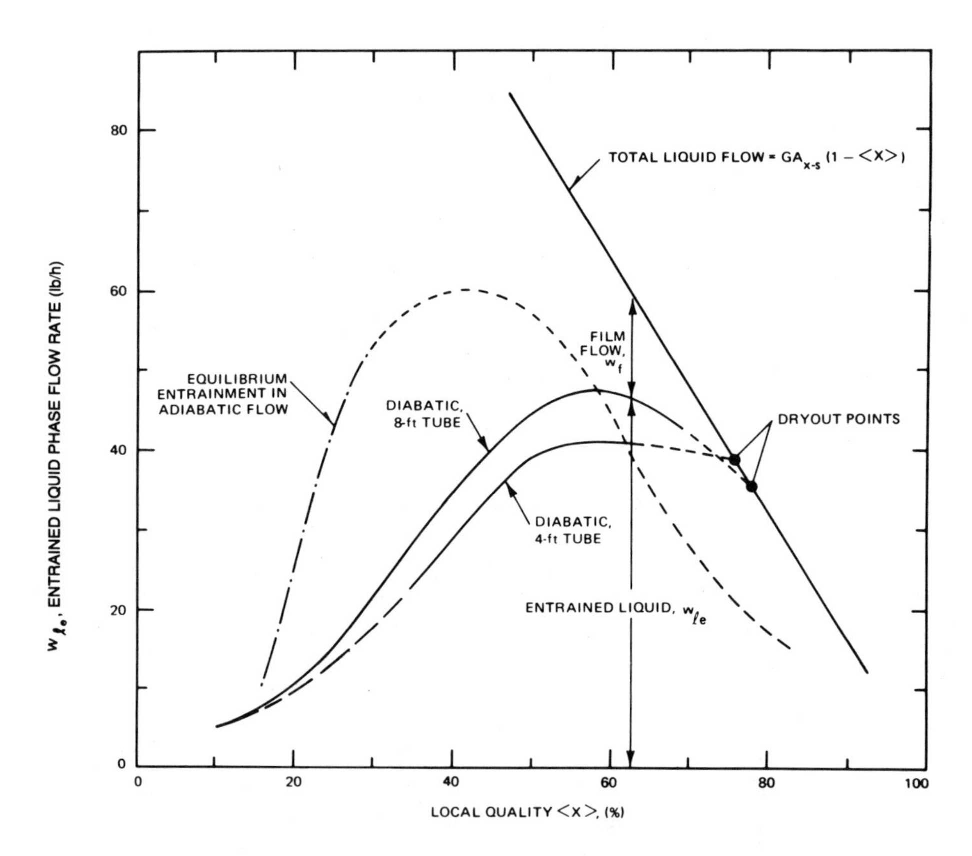

Fig. 4-25 Entrained liquid phase flow rate as a function of local quality (Hewitt, 1970).

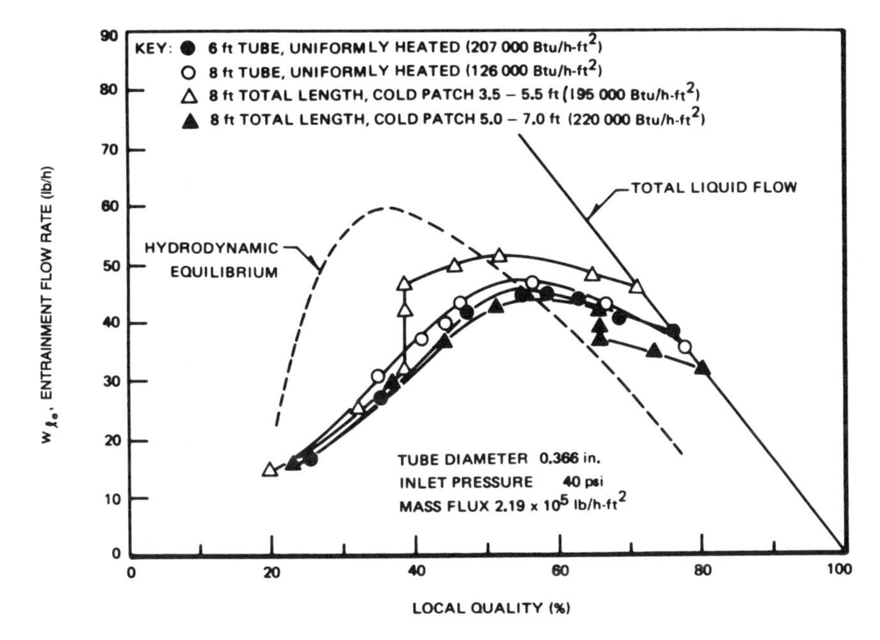

Fig. 4-26 Effect of unheated zones on the entrainment curve for the evaporation of low-pressure water in vertical tubes (Bennett et al., 1966).

At the present time, it is virtually impossible to perform an accurate film dryout calculation in complex geometries such as BWR fuel rod bundles. Nevertheless, valuable and interesting insight can be obtained from consideration of the mechanisms involved. As an example, Hewitt and Hall-Taylor (1970) consider Eq. (4.59b) at the dryout point. As shown in Fig. 4-27a, various possibilities occur. For the situation of uniform axial heat flux, in which dryout normally occurs at the end of the heated length (EHL), the film flow gradient can be either negative or zero. For the situation in which it is zero (i.e., $dw_f/dz=0$), Eq. (4.59b) yields,

$$q_c''=h_{fg}(m_d''-m_e''-m_{co}'') \quad . \tag{4.61}$$

For very thin liquid films, typical of those at incipient dryout, wave action and nucleation are normally suppressed. Thus, $m_e''=m_{co}''=0$. Equation (4.61) yields the CHF for so-called deposition controlled dryout,

$$q_c''=h_{fg}m_d'' \quad . \tag{4.62}$$

This situation is rarely achieved in practice, since normally rivulets are present, and $dw_f/dz\neq 0$.

For nonuniform axial heat flux situations, shown in Fig. 4-27b, in which upstream dryout may occur, the film flow gradient must be zero. Moreover, assuming at dryout $m_e''=m_{co}''=0$, the spatial derivative of Eq. (4.59b) yields,

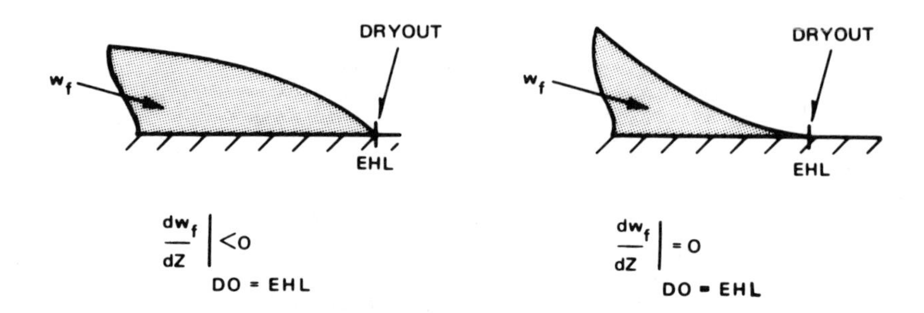

Fig. 4-27a Dryout situations with uniform axial heat flux.

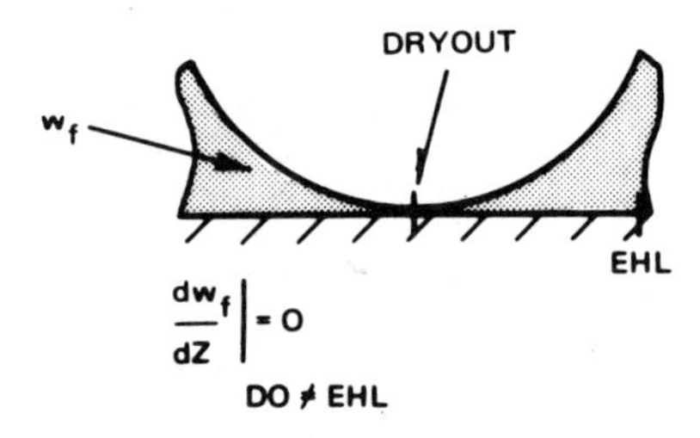

Fig. 4-27b Upstream dryout with nonuniform axial heat flux (Hewitt and Hall-Taylor, 1970).

$$\frac{d^2 w_f}{dz^2} = P_H \left(\frac{dm''_d}{dz} - \frac{1}{h_{fg}} \frac{dq''}{dz} \right) > 0 \ . \tag{4.63}$$

Normally, $dm''_d/dz<0$ as dryout is approached since the amount of entrained liquid is decreasing. Thus, $dq''/dz<0$ for upstream dryout to occur. This explains the experimentally observed fact that when upstream dryout occurs, it normally occurs in a region of decreasing heat flux.

Another interesting example of the insight that we obtain from consideration of film flow models is concerned with the effect of obstacles (Shiralkar and Lahey, 1973). It is well known that in BWR rod bundles the onset of BT always occurs initially upstream of grid spacers (Janssen, 1971). As shown in Fig. 4-28, this phenomenon is primarily due to impacting of the relatively high-velocity vapor stream on bluff spacer components, causing a horseshoe vortex that "scrubs" out the liquid film ahead of the obstruction, thus causing a premature dryout.

Obstacles of various shape and size were investigated in an adiabatic air/water experiment (Shiralkar and Lahey, 1973) to determine the critical liquid film flow rate (i.e., the rate below which upstream dryout occurs). It is seen in Fig. 4-29 that some shapes are much better than others. In

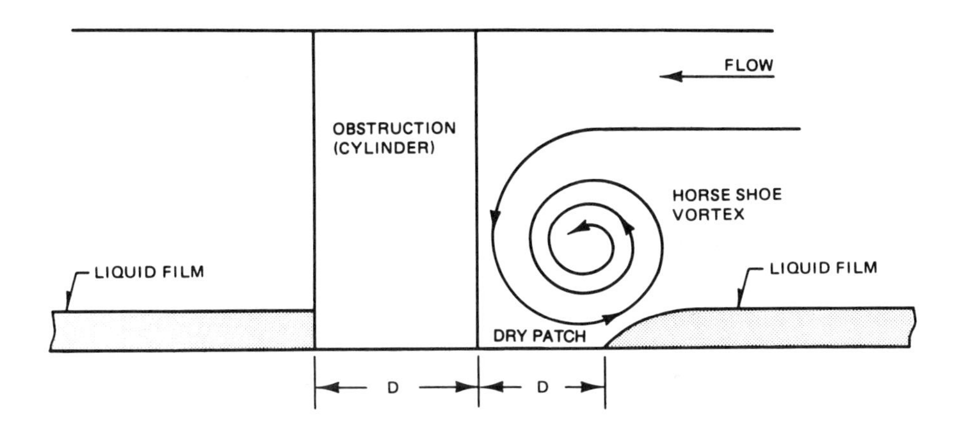

Fig. 4-28 Formation of an upstream dry patch (Shiralkar and Lahey, 1973).

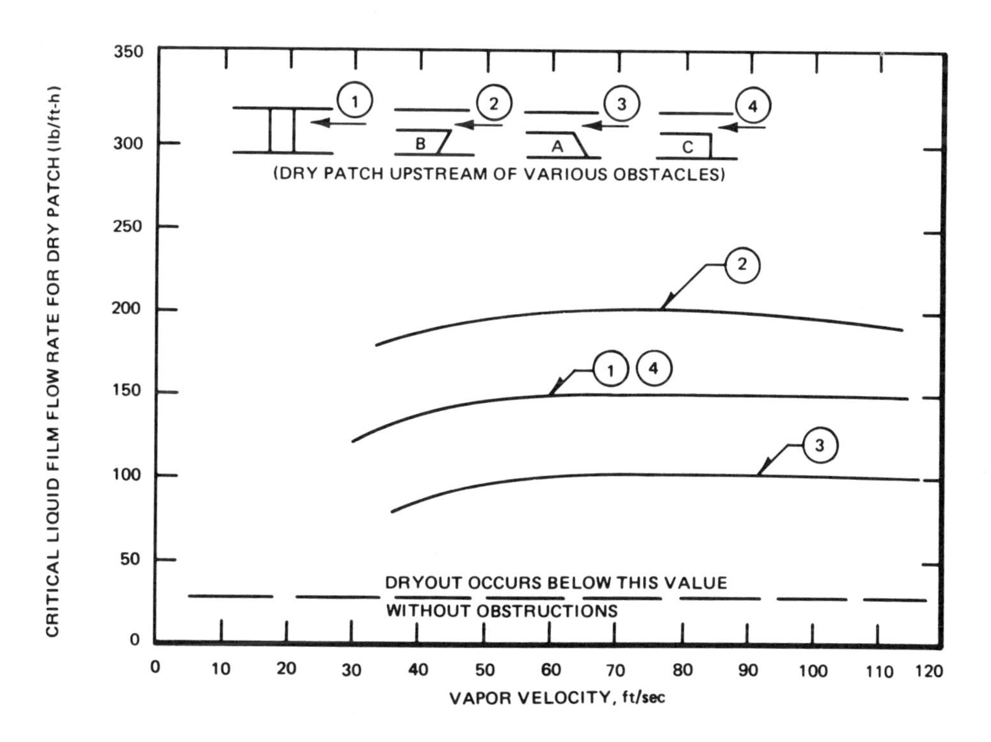

Fig. 4-29 Comparison of critical film flow rates for various obstacles.

general, the more streamlined the obstacle, e.g., shape ③, the lower the critical film flow rate and, thus, the more difficult it is to cause an upstream dryout.

The hydrodynamic situation that occurs in a BWR fuel rod bundle is completely analogous. Hence, the components contained in a grid-type spacer can strongly affect the thermal performance of a fuel rod bundle, and the spacer designer is well advised to make these components as streamlined as possible.

4.4.1.2 *Interpretation of Boiling Transition Data*

Boiling transition in BWR fuel rod bundles is rather complicated, not only due to the complex geometry involved, but also due to the nonuniform axial and local (i.e., rod-to-rod) heat flux profiles that exist. Rather than discuss the data trends observed in full-scale rod bundle experiments, it is instructive first to consider the effects separately. First we consider the effect of nonuniform axial heat flux on BT in simple geometry experiments, since it turns out that the effects seen in the more complex rod bundle experiments are exactly the same.

Collier (1972) has summarized the various techniques currently used in correlating nonuniform axial heat flux data. Essentially they are based either on the local conditions hypothesis or on an integral approach. The so-called local conditions hypothesis states in essence that it is only the local heat flux and local quality that control the BT; that is, the upstream history is not important. In contrast, the integral concept implies that the upstream history is normally quite important. The integral concept can take many analytic forms, all of which imply that it is not sufficient just to know the local thermal-hydraulic parameters, but we must also know how the local situation came about. For instance, the local flow quality, $\langle x(z) \rangle$, tells us the total liquid content but yields no information on how much of this liquid is actually on the heated surface.

In general, it has been found that the integral concept is essentially correct for heat flux profiles of practical concern. In particular, for operating conditions of interest to BWR technology, the local conditions hypothesis is not valid and some integral-type correlation of the data is required. This can be seen clearly in Fig. 4-30, which presents Freon-114 BT data taken in an internally heated annulus (Shiralkar, 1972). If only the local heat flux and quality were important, as implied by the local conditions hypothesis, then the data for the various axial heat flux profiles would coalesce. Obviously they do not. In fact, a strong dependence on axial heat flux profile is evident.

The effect of the axial heat flux profile, sometimes referred to as the "upstream memory effect," strongly depends on the flow regime and, thus, the quality. The data of De Bortoli et al. (1958), shown in Fig. 4-31, provide

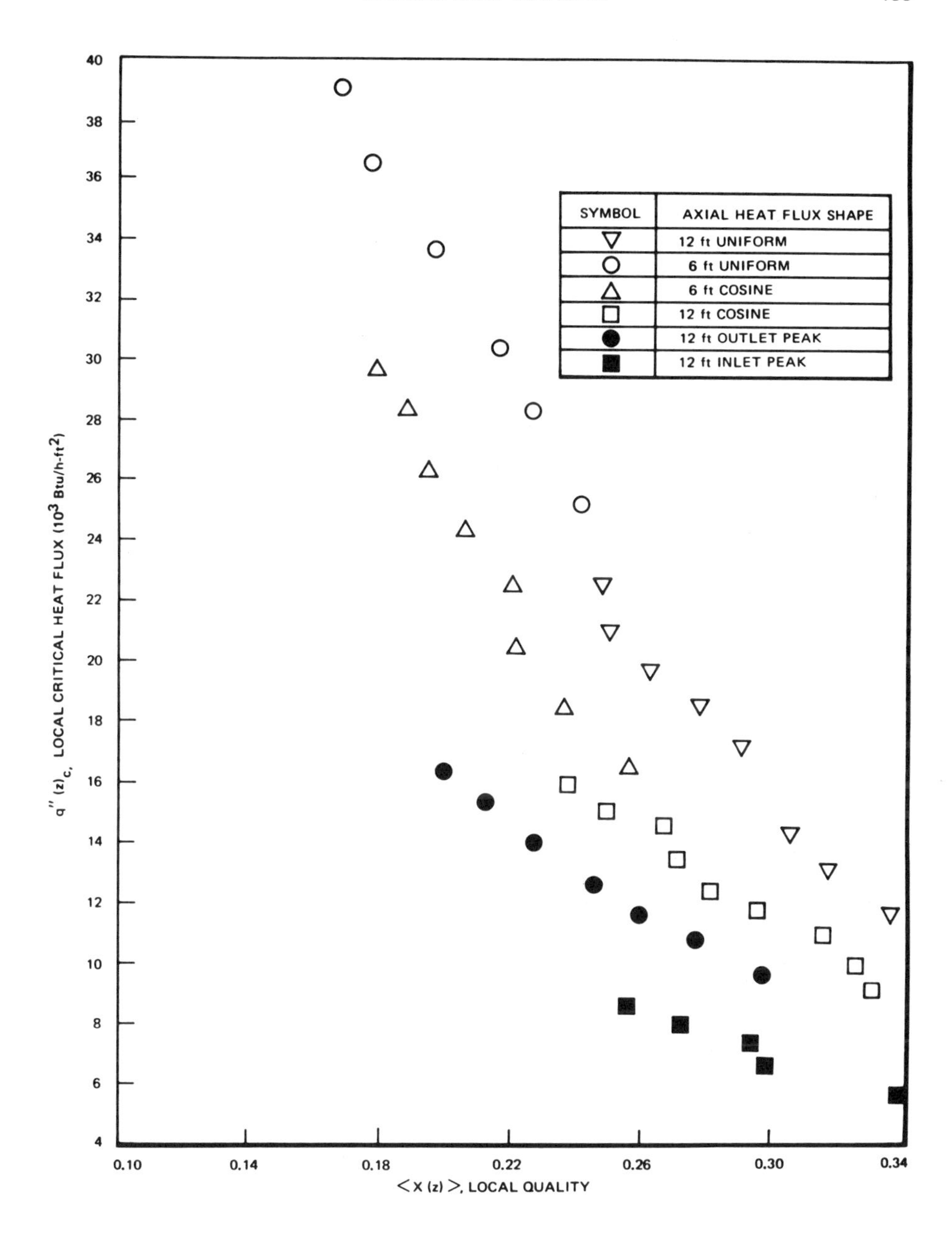

Fig. 4-30 Heat flux versus quality at location of BT, Freon-114 annulus data, $D_1 = 0.563$ in., $D_2 = 0.875$ in., 6- and 12-ft heated length, various axial profiles, $p = 123$ psia, $G = 0.54 \times 10^6$ lb/h-ft^2 (Shiralkar, 1972).

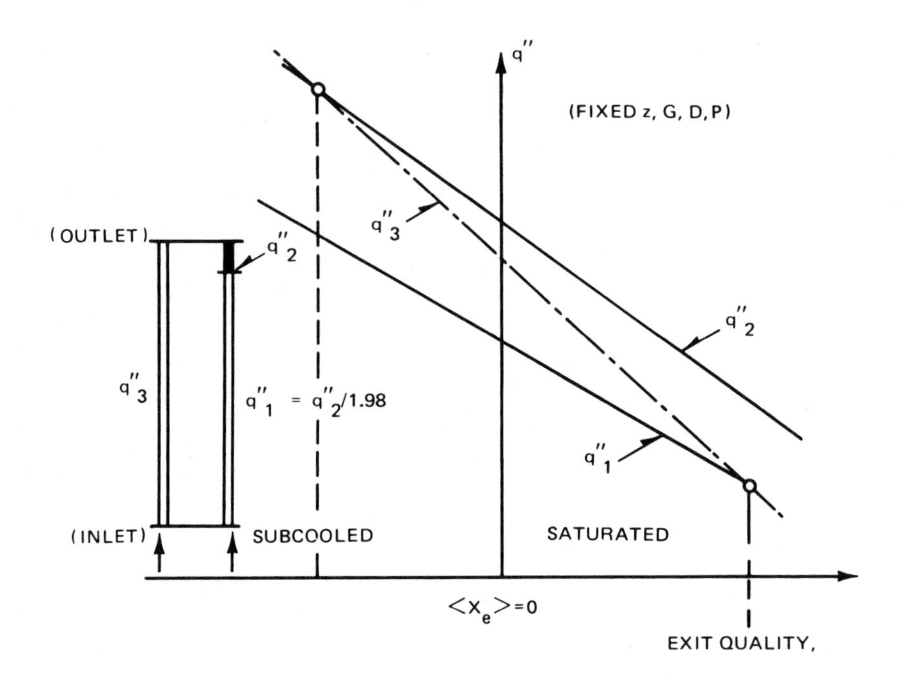

Fig. 4-31 Schematic display of hot patch data (De Bortoli et al., 1958).

interesting insight into the relationship between the quality and axial heat flux effects. This experiment was performed in high-pressure steam/water using two test sections. The first had a uniform axial heat flux profile, q_3''. The second had a uniform axial heat flux profile, q_1'', for ~99% of its length, with a small uniform axial hot patch at the end of magnitude q_2'', where $q_2''=1.98\ q_1''$. For a profile of this type, BT always occurs at the end of the heated length for flow rates of interest in BWR technology. Indeed, only for uniform axial profiles can we speak meaningfully in terms of a CHF.

Note that in Fig. 4-31 for highly subcooled conditions, the CHF data trends for both test sections become colinear. Hence, for subcooled DNB conditions, the local conditions hypothesis is approximately valid. In contrast, for the higher quality conditions more typical of BWR technology, the effect of the local hot patch washes out, which implies the validity of an integral concept.

There are various integral schemes for introducing the effect of upstream history. One of the most popular and widely used schemes is known as the Tong *F*-factor method (Tong, 1972). The Tong *F*-factor method is basically a flux-quality plane scheme in which uniform axial CHF test data are divided by the appropriate upstream weighting factor, *F*, to yield the appropriate thermal limit for the nonuniform axial heat flux profile under consideration. The functional form of the Tong F-factor is given by Tong (1972) and Smith et al. (1965) as,[d]

$$F \triangleq \frac{\overline{q''_{c(\text{EQ})}}}{q''_c(L_{B_c}+\lambda)} = \frac{\Omega_\text{T}}{q''_c(L_{B_c}+\lambda)[1-\exp(-\Omega_\text{T} L_{B_c})]} \times \int_\lambda^{L_{B_c}+\lambda} q''(z) \exp\{-\Omega_\text{T}[(L_{B_c}+\lambda)-z]\}\, dz \,, \qquad (4.64)$$

where

$\overline{q''_{c(\text{EQ})}} \triangleq$ equivalent uniform axial CHF

$q''_c(L_{B_c}+\lambda) \triangleq$ local nonuniform axial heat flux at the critical location

$L_{B_c} \triangleq$ critical boiling length (Fig. 4-32)

$\lambda \triangleq$ axial position of the boiling boundary; i.e., axial position where $\langle x_e \rangle = 0$.

It can be seen from Eq. (4.64) that the equivalent uniform axial heat flux is given by,

$$\overline{q''_{c(\text{EQ})}} = \frac{\Omega_\text{T}}{[1-\exp(-\Omega_\text{T} L_{B_c})]} \int_\lambda^{L_{B_c}+\lambda} q''(z) \exp\{-\Omega_\text{T}[(L_{B_c}+\lambda)-z]\}\, dz \,. \qquad (4.65)$$

For film dryout conditions, typical of BT in a BWR, the parameter, Ω_T, is very small, thus,

$$\Omega_\text{T} = C/D_H \cong 0.0 \,. \qquad (4.66)$$

Actually, the parameter, Ω_T, decreases as critical quality increases (Tong, 1972). To appreciate the implications of this variation of Ω_T with quality, let us consider the integrand of Eq. (4.64). For large values of the parameter Ω_T, the flux weighting term, $\exp\{-\Omega_\text{T}[(L_{B_c}+\lambda)-z]\}$, dies out strongly for axial positions less than the critical position; i.e., $z<(L_{B_c}+\lambda)$. Hence, for low-quality BT, there is little upstream memory effect and the local heat flux is very important. That is, the local conditions hypothesis is approximately valid. In contrast, for higher quality BT, typical of BWR technology, Ω_T is small and, thus, the exponential weighting term does not attenuate the effect of upstream heat flux as strongly. In the limit, as the parameter, Ω_T, goes to zero, there is equal weighting of the local and upstream heat fluxes, since, by using L'Hopital's rule, Eq. (4.64) becomes:

$$\lim_{\Omega_\text{T}\to 0} F \triangleq \frac{\overline{q''_{c(\text{EQ})}}}{q''_c(L_{B_c}+\lambda)} = \frac{1}{q''_c(L_{B_c}+\lambda) L_{B_c}} \int_\lambda^{L_{B_c}+\lambda} q''(z)\, dz \,, \qquad (4.67)$$

[d]The assumption is made in Eq. (4.64) that we can integrate from the boiling boundary, λ, rather than the axial position at which either subcooled boiling or annular flow begins.

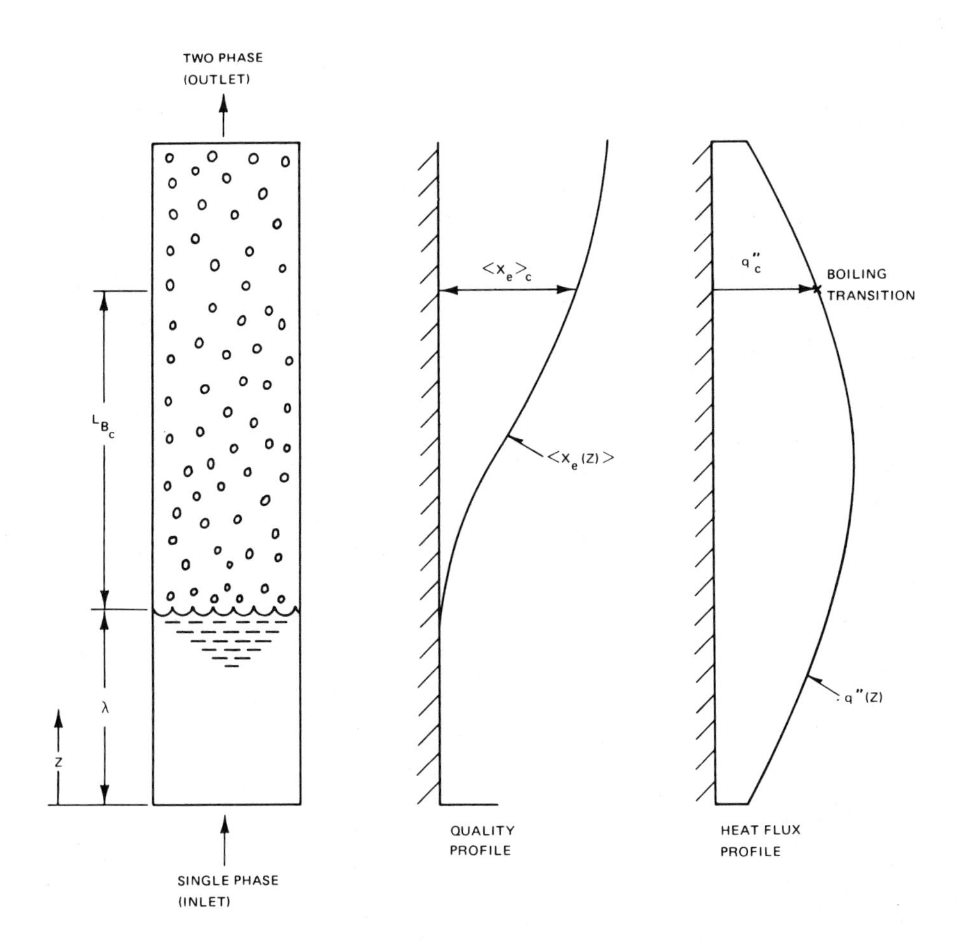

Fig. 4-32 Representation of boiling length, L_B.

which implies that the equivalent uniform axial CHF is given by the nonuniform axial heat flux profile averaged over the boiling length,

$$\overline{q''_{c(EQ)}} = \frac{1}{L_{B_c}} \int_{\lambda}^{L_{B_c}+\lambda} q''(z)\, dz \ . \tag{4.68}$$

This interesting observation will prove valuable shortly.

The Tong *F*-factor approach has been applied to data of interest in BWR technology and, subsequent to proper optimization of the parameter, Ω_T, has done a good job of correlating the nonuniform axial heat flux effect (Shiralkar,1972). There are, however, some fundamental drawbacks to the method. First, evaluation of Eq. (4.64) almost invariably requires numerical integration of an exponential integral. Thus, computer-based techniques are normally required. Second, as shown in Fig. 4-33, graphic interpretation of the technique in the flux-quality plane is very involved. This is because for a given flow rate, system pressure, and axial heat flux profile, the parameter, F, given by Eq. (4.64), and the energy balance curves, $q''(\langle x \rangle)$, are a function of inlet subcooling. Thus, rather than a single CHF line, there is a whole family of such lines depending on the inlet subcooling. Moreover, for a different axial heat flux profile, Eq. (4.64) implies that there is a new family of CHF lines for each inlet subcooling. Clearly a simple graphic interpretation is impossible. Moreover, as previously discussed, the exponential weighting of the upstream history, while important for low-quality conditions, is not nearly as important for BWR conditions of interest. Thus, logically we are led to explore alternate integral schemes that are more appropriate for BWR operating conditions.

There is a technique that has been developed specifically for BT in the higher quality annular flow regime. This technique originally was introduced into the literature by investigators (Bertoletti et al., 1965) at the Centre Informazioni Studi Esperienze (CISE) in Milan, Italy. The CISE-type correlation is basically of the critical quality-boiling length ($\langle x_e \rangle_c - L_{B_c}$) type, where the upstream history is implicit in the critical boiling length, L_{B_c}.

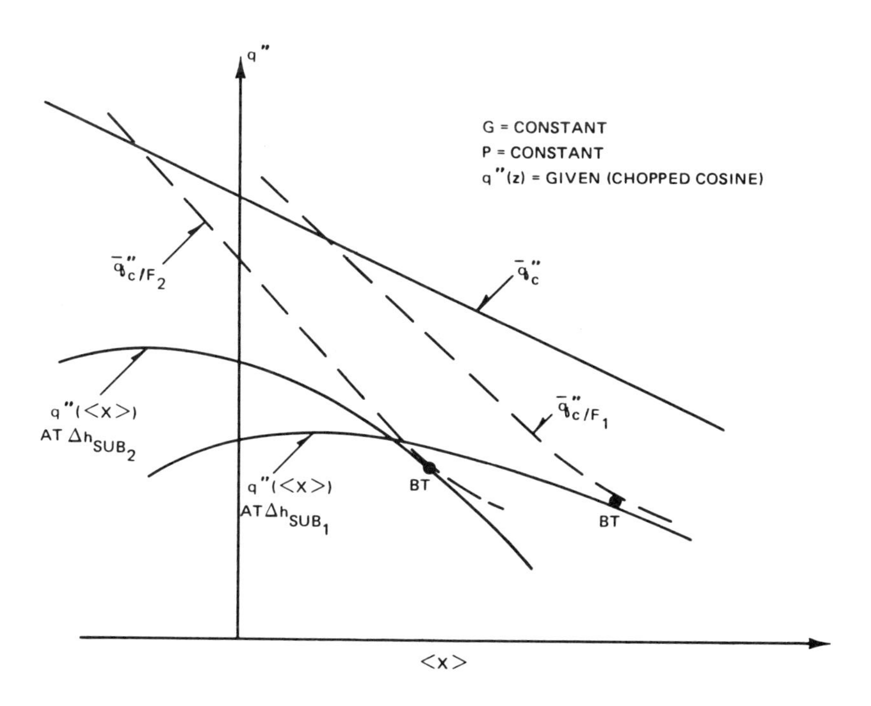

Fig. 4-33 Graphic interpretation of Tong F-factor technique.

To appreciate the $\langle x_e \rangle_c - L_{B_c}$ approach, consider the situation of a heated rod bundle with a chopped cosine axial heat flux profile operating with fixed inlet subcooling, flow rate, and system pressure. Moving from right to left in Fig. 4-32, the heat flux profile is shown with the typical location of a BT indicated; the axial quality distribution is shown with the critical quality, $\langle x_e \rangle_c$, indicated; and, on the left, the boiling boundary, λ, and critical boiling length, L_{B_c}, are indicated. It is seen that the boiling length is just the length over which bulk boiling occurs. It is, by convention, measured from the boiling boundary, λ. The critical boiling length is then just the length from the boiling boundary to the point at which a BT is observed.

Techniques that assume the validity of the local conditions hypothesis attempt to correlate in terms of local values of q_c'' and $\langle x_e \rangle_c$. Integral flux-quality techniques (i.e., the Tong F-factor approach) also attempt to correlate these local parameters through appropriate modification of uniform axial CHF correlations. As indicated in Fig. 4-32, rather than correlating q_c'' and $\langle x_e \rangle_c$, we could correlate $\langle x_e \rangle_c$ and L_{B_c}, where the boiling length is given implicitly by,

$$\int_{\lambda}^{L_B+\lambda} P_H q''(z) dz = G A_{x-s} h_{fg} \langle x_e \rangle \ , \tag{4.69}$$

and the position of the boiling boundary is given by,

$$\int_{0}^{\lambda} P_H q''(z)\ dz = G A_{x-s}(h_f - h_i) = G A_{x-s} \Delta h_{\text{sub}} \ . \tag{4.70}$$

The integral nature of the boiling length parameter is evident in Eqs. (4.69) and (4.70). To determine whether this integral weighting of the upstream heat flux history is appropriate, nonuniform axial heat flux data, including the flux-quality data given in Fig. 4-30, were replotted in Fig. 4-34 in the critical quality-boiling length plane. It is seen that this plane does an excellent job of collapsing the axial heat flux effect. This trend originally was noted by the CISE group, but has been observed by other investigators. Figure 4-35 is an example of British data (Keeys et al., 1971) plotted in the $\langle x_e \rangle_c$–L_{B_c} plane. Again, it is seen that the axial heat flux effect is well captured in this plane.

It should be evident that, in principle, we could develop a "best fit" line to the data and, thus, synthesize a critical quality-boiling length correlation for design purposes. The functional form of this correlation is quite arbitrary and depends completely on the ingenuity of the investigator. As an example, a well-known CISE correlation has a functional form given by,

$$\langle x_e \rangle_c = \frac{a(p,G) L_{B_c}}{[b(p,G,D_H) + L_{B_c}]} \ , \tag{4.71}$$

although more recent work at CISE (Gaspari et al., 1972) indicates that a quadratic form of the correlation may yield a better data fit. Nevertheless,

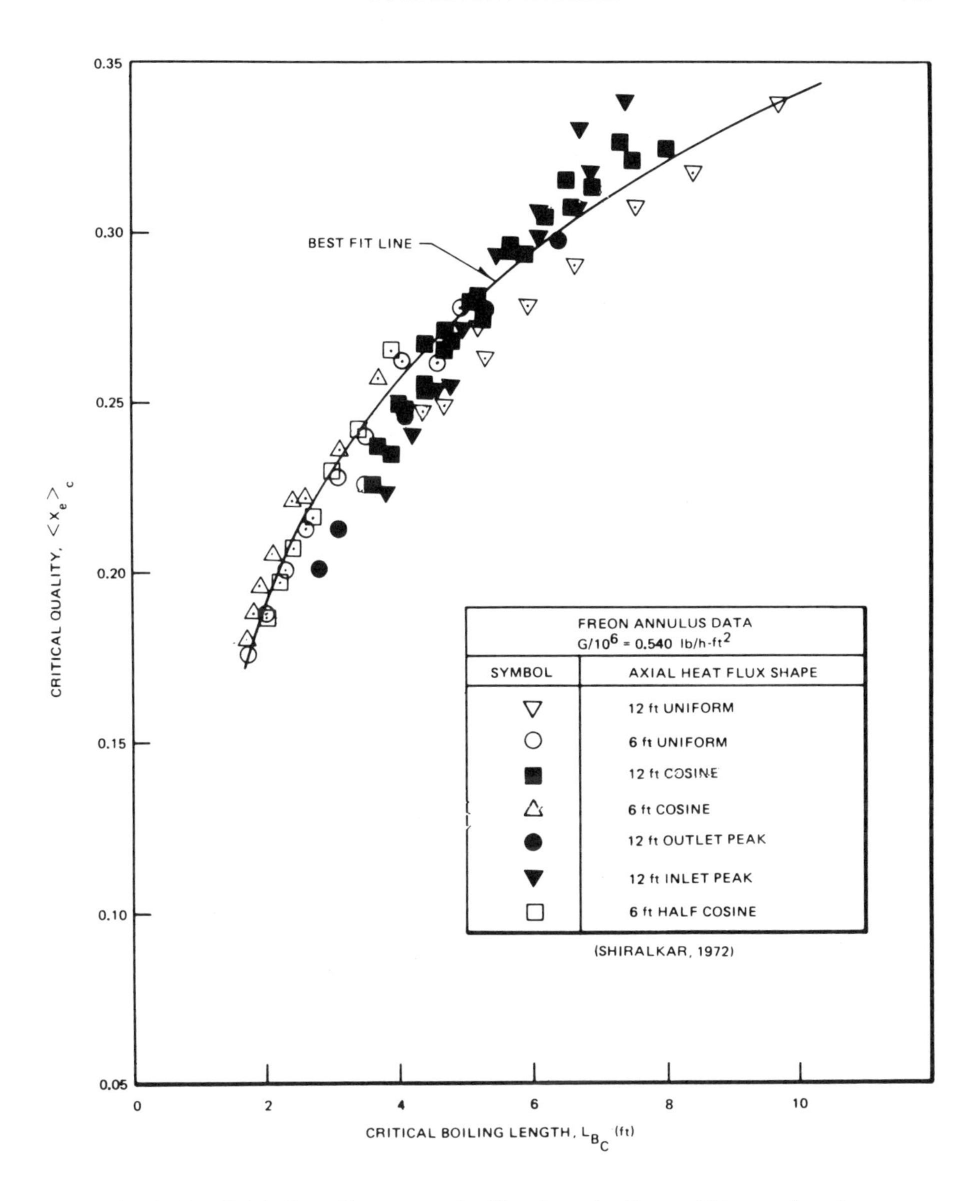

Fig. 4-34 Critical quality versus boiling length, Freon-114 annulus data.

as shown in Fig. 4-36, optimization of parameters a and b in Eq. (4.71) allows us to fit typical BT data.

We have discussed two techniques (i.e., the Tong F-factor and the critical quality-boiling length approach) for handling the effect of nonuniform axial heat flux. A systematic comparison of these techniques (Shiralkar, 1972) indicates that for conditions of interest in BWR technology, the accuracy

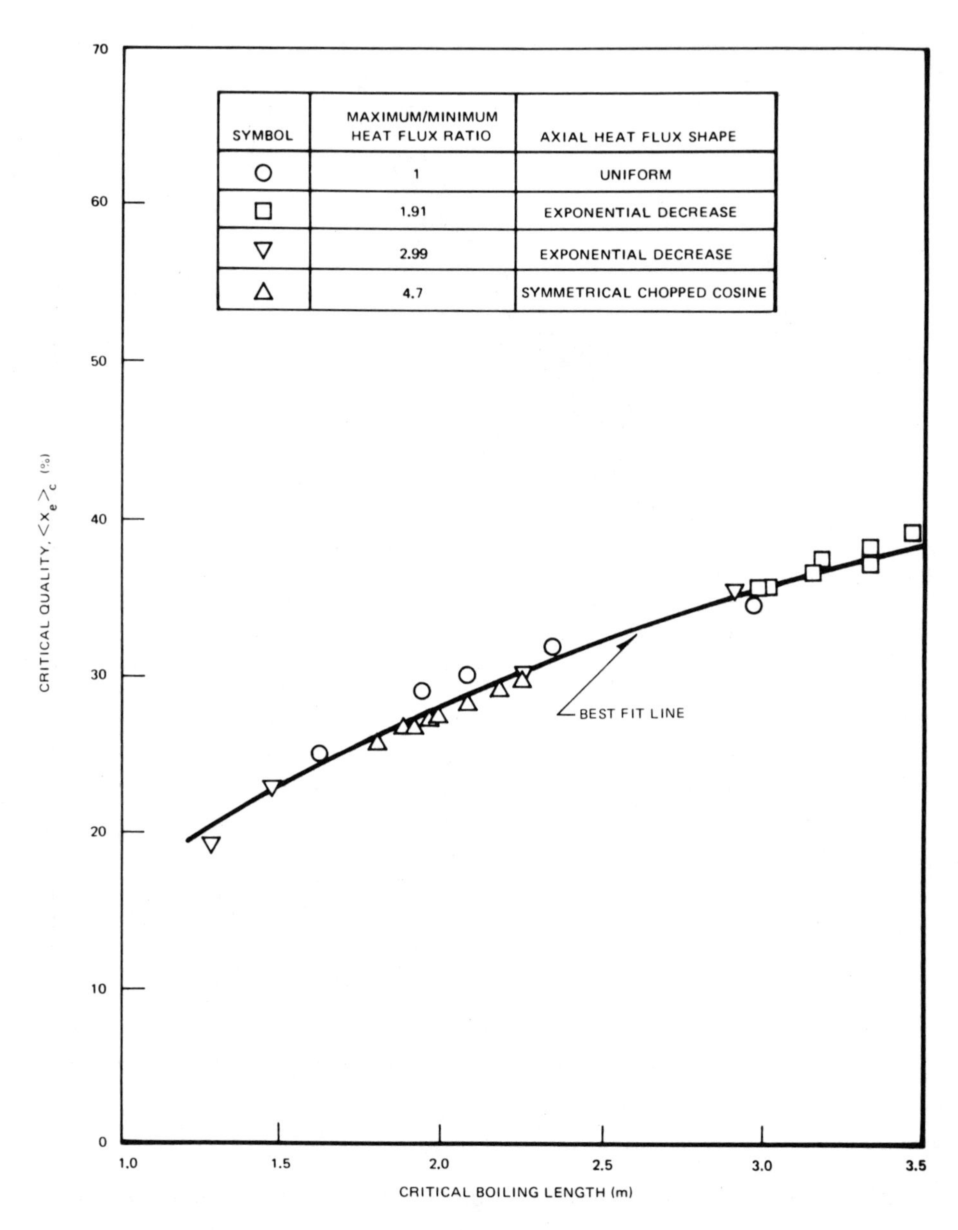

Fig. 4-35 Critical quality versus boiling length data, 12.6-mm-diam round tube, 3.66-m heated length, various axial power profiles, $G = 2.72 \times 10^3$ kg/sec-m^2 (Keeys et al., 1971).

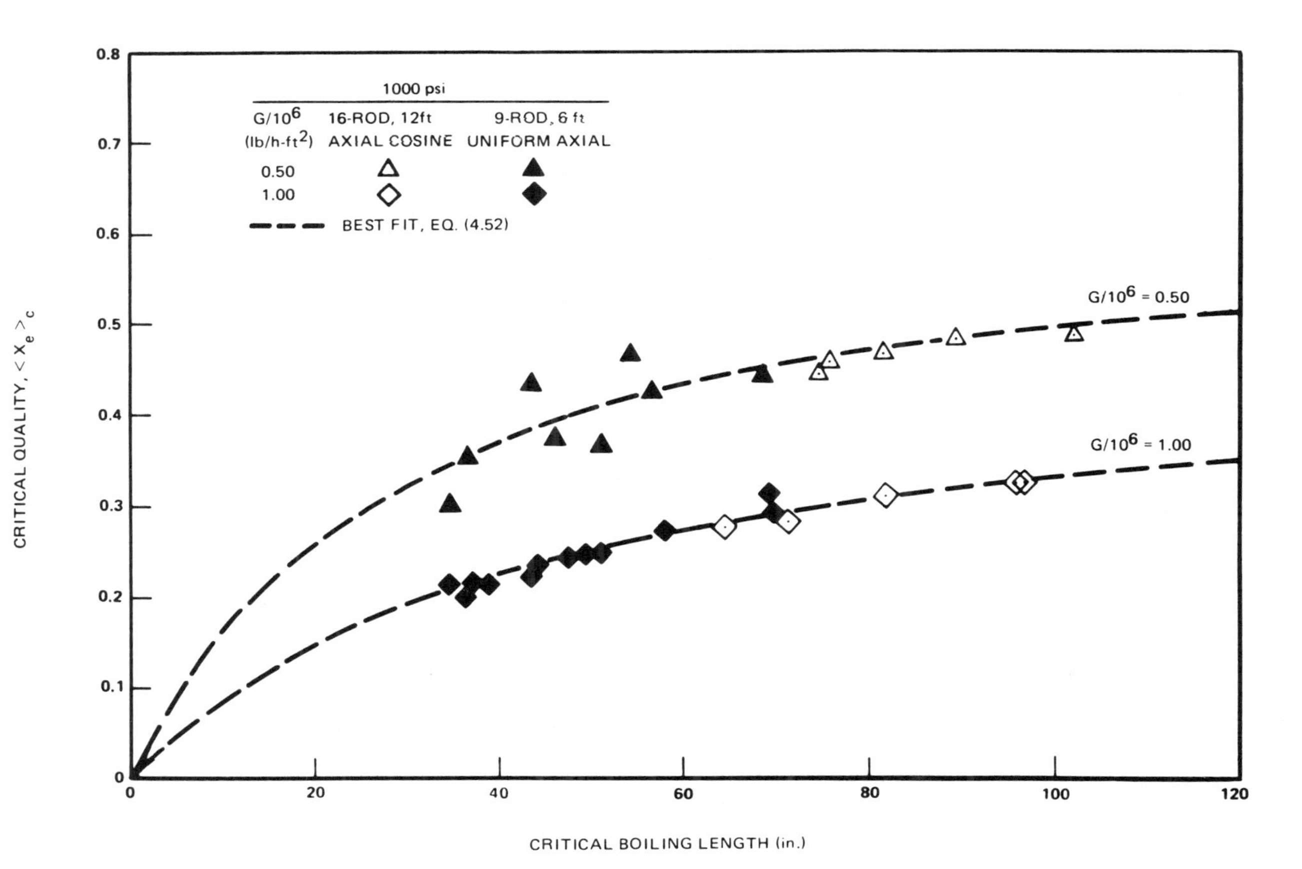

Fig. 4-36 Nine- and sixteen-rod critical quality versus boiling length, 1000 psi (Polomik et al., 1972).

with which the data trends are correlated is about the same. Thus, either technique can be used for design applications, although due to its relative simplicity, the critical quality-boiling length technique is preferred. However, before proceeding, it is informative to investigate further the relationship between flux-quality and the critical quality-boiling length approaches.

It is well known that, to a good first approximation, BT data for uniform axial heat flux profiles are linear in the flux-quality plane. This trend is seen in the data plotted in Fig. 4-30. Hence, the appropriate CHF correlation is of the form,

$$\overline{q''_c} = n(G, p) - m(G, p)\langle x_e \rangle_c \ . \tag{4.72}$$

To better understand the significance of the $\langle x_e \rangle_c - L_{B_c}$ correlation given in Eq. (4.71), we assume that, when appraising thermal limits, the appropriate heat flux to compare with $\overline{q''_c}$ is the heat flux averaged over the boiling length. That is, the equivalent uniform axial heat flux is, from Eq. (4.68),

$$\overline{q''_{c(\mathrm{EQ})}} \triangleq \frac{1}{L_{B_c}} \int_{\lambda}^{\lambda + L_{B_c}} q''(z)\, dz = \frac{1}{L_{B_c}} \int_{0}^{\langle x_e \rangle_c} q''(z) \frac{dz}{d\langle x_e \rangle}\, d\langle x_e \rangle \ . \tag{4.73}$$

By using Eq. (4.69) written in the form,

$$\langle x_e \rangle_c = \frac{\displaystyle\int_{\lambda}^{\lambda + L_{B_c}} P_H q''(z)\, dz}{G A_{x-s} h_{fg}} \ , \tag{4.74}$$

we obtain,

$$\langle x_e \rangle_c = \frac{(n/m) L_{B_c}}{[(G A_{x-s} h_{fg})/(m P_H) + L_{B_c}]} \ . \tag{4.75}$$

Comparison of Eqs. (4.71) and (4.75) indicates the unique relationship between the CISE correlation and the parameters, n (intercept) and m (slope), in the uniform axial CHF correlation. This is an interesting and important result since it implies that, at least in simple geometries, we only need to perform uniform axial heat flux experiments to obtain the slope and intercept. Then, by using Eq. (4.75), we can predict BT for conditions of nonuniform axial heat flux.

In more general situations in which the appropriate CHF correlation is no longer linear, as in Eq. (4.72), the same procedure can be followed to obtain the corresponding critical quality-boiling correlation. In this case, however, the correlation no longer has the same functional form as in Eq. (4.71).

It has been found (Bertoletti et al., 1965) that for cases of extreme axial heat flux profiles, the simple flux averaging given by Eq. (4.73) is inadequate. To account for this effect, Silvestri (1966) introduced an axial weighting factor, $\zeta(z)$, to obtain the expression for the equivalent uniform CHF,

$$\overline{q''_{c(\mathrm{EQ})}} = \frac{\int_{\lambda}^{\lambda + L_{B_c}} q''(z)\zeta(z)\, dz}{\int_{\lambda}^{\lambda + L_{B_c}} \zeta(z)\, dz} , \tag{4.76}$$

where,

$$\zeta(z) \triangleq [q''(z)]^{\eta} \exp\{-\Omega_{\mathrm{T}}[(L_{B_c} + \lambda) - z]\} , \tag{4.77}$$

and the equivalent critical boiling length is given by,

$$L_{B_c(\mathrm{EQ})} \triangleq \frac{1}{\overline{q''_{c(\mathrm{EQ})}}} \int_{\lambda}^{\lambda + L_{B_c}} q''(z)\, dz . \tag{4.78}$$

Note that for the case in which $\eta = 0$, Eqs. (4.76) and (4.77) yield,

$$\overline{q''_{c(\mathrm{EQ})}} = \frac{\Omega_{\mathrm{T}}}{[1 - \exp(-\Omega_{\mathrm{T}} L_{B_c})]} \int_{\lambda}^{\lambda + L_{B_c}} q''(z) \exp\{-\Omega_{\mathrm{T}}[L_{B_c} + \lambda) - z]\}\, dz . \tag{4.79}$$

By comparing Eqs. (4.65) and (4.79), we see that the Tong F-factor approach and the generalized critical quality-boiling length approach are completely equivalent.

For conditions of interest in BWR technology, the ratio of peak-to-average axial heat flux is normally less than 2 and, thus, to a good first approximation, standard $\langle x_e \rangle_c - L_{B_c}$ correlations are valid. This implies the validity of Eq. (4.73), which is entirely consistent with the Tong F-factor approach for very small Ω_{T}. The main difference between the two techniques has to do with how the upstream history is introduced. As shown in Fig. 4-37, the Tong F-factor approach modifies the uniform axial CHF correlation to yield the appropriate correlation for the existing nonuniform axial profile. In contrast, the critical quality-boiling approach[e] modifies the energy balance, $q''(\langle x_e \rangle)$, rather than the CHF correlation, $\overline{q''_c}$. Both techniques are equivalent and yield the same thermal margin.

The effect of nonuniform axial heat flux now has been considered and it has been shown that for BWR conditions there are at least two integral techniques that accurately correlate the observed data trends. Once a particular technique has been selected, we are faced with the problem of evaluating and interpreting the thermal margin. There are various figures of merit that can be used to quantify the thermal margin. Several popular figures of merit are indicated in the flux-quality plane in Fig. 4-38.

[e]Normally, the critical quality-boiling length correlation is not plotted in the flux-quality plane and is done here only for illustrative purposes.

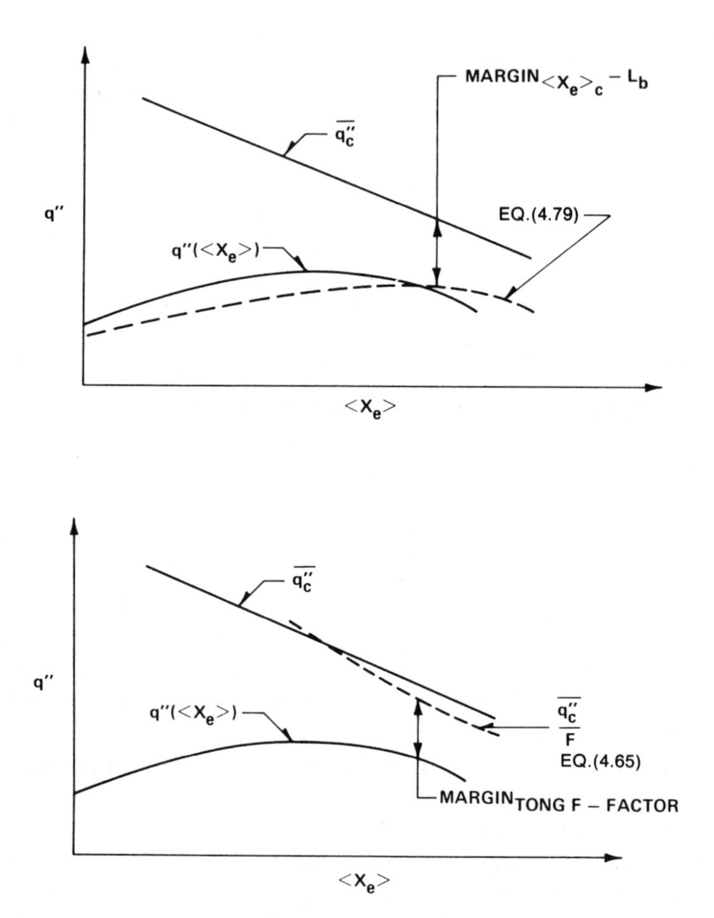

Fig. 4-37 Comparison of $\langle x_e \rangle_c - L_{B_c}$ and Tong F-factor methods in the flux-quality plane.

The most popular measure of the thermal margin has been the CHF ratio (CHFR). It is defined as the ratio, at a given quality, of CHF (given by the thermal limit, i.e., CHF correlation) divided by the local heat flux. This classical figure of merit does not give a true picture of the thermal margin. More specifically, if CHFR=1.5, this does not imply that the operating power can be increased 50%, nor does it imply that the local heat flux can be increased only 50%; i.e., it is the integral of the axial heat flux profile that is important rather than a local heat flux "spike."

The critical quality defect is a convenient figure of merit to use in transient analysis and is a natural measure of the thermal margin in the critical quality-boiling length plane. Unfortunately, from the operational point of view, this figure of merit is not very easy to interpret. In contrast, the

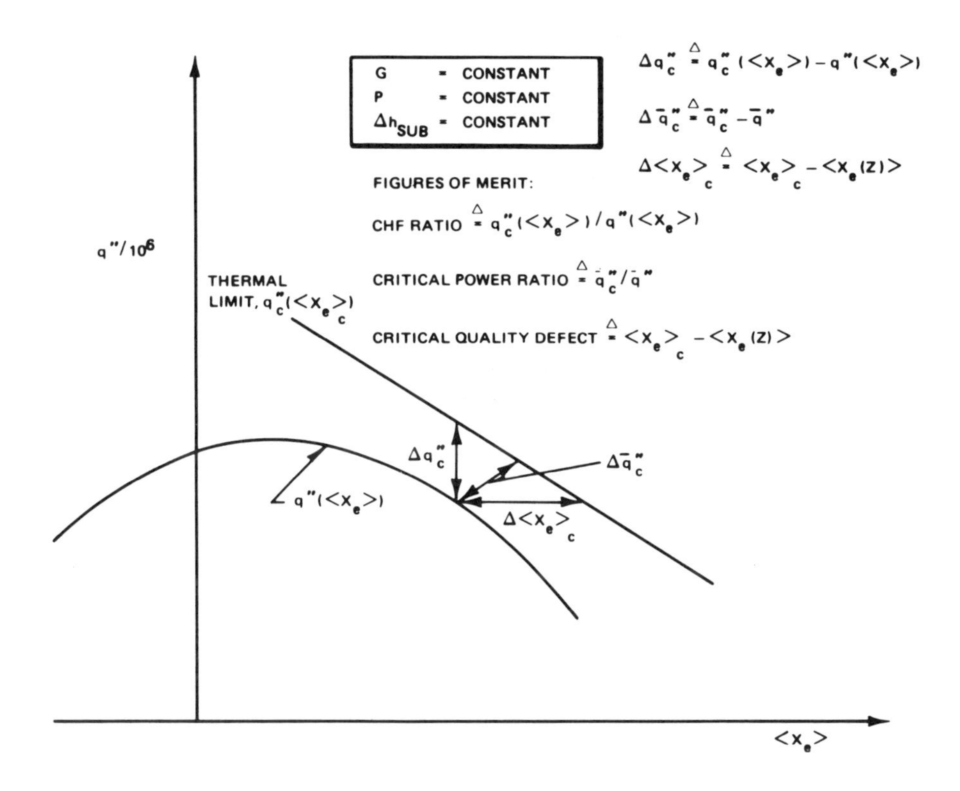

Fig. 4-38 Various figures of merit used to appraise thermal margins.

physical significance of the critical power ratio (CPR), defined as the critical power divided by the operating power, is very easy to appreciate. It is particularly useful to the experimentalist, since if CPR = 1.20, it implies that if system pressure, flow rate, and inlet subcooling are held constant, we should expect a BT when the test section power is increased 20%. An alternate interpretation of this figure of merit is in terms of a CHFR that is evaluated not at a constant quality, but (see Fig. 4-38) along the energy balance trajectory of each axial position on the heater.

Let us next consider the BWR design application of the techniques and concepts described in Sec. 4.4.1.1.

4.4.2 Boiling Water Reactor Design Application

The BWR fuel rod bundle has a rather complex geometry and power distribution. To determine its thermal limits, large-scale simulation experiments are required. Fortunately, because each rod bundle is separated from its neighbors by a channel box, we can use the "unit cell" concept and perform full-scale experiments using an electrically heated simulated fuel rod bundle. Figure 4-39 is typical of the results of an extensive experimental program (GETAB, 1973) performed in General Electric's (GE's) ATLAS heat transfer facility. Note that the critical power performance of the bundle increases approximately linearly with inlet subcooling and monotonically with mass flux.

Figure 4-40 indicates that the effect of increasing the local rod-to-rod peaking is generally to reduce the critical power of the rod bundle. This

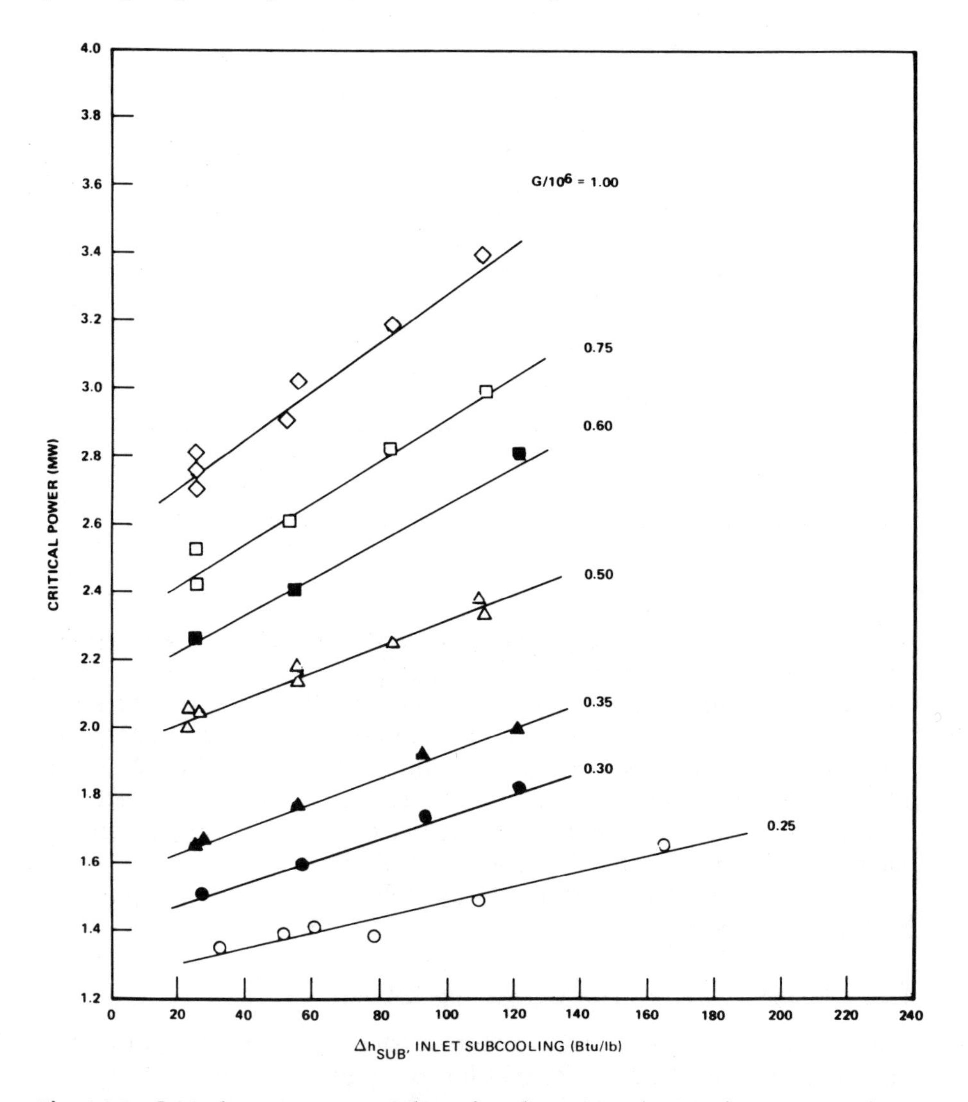

Fig. 4-39 Critical power versus inlet subcooling, 16-rod × 12-ft cosine, uniform local peaking, 1.38 *P*/*A* axial, 1000 psia, various flow rates (GETAB, 1973).

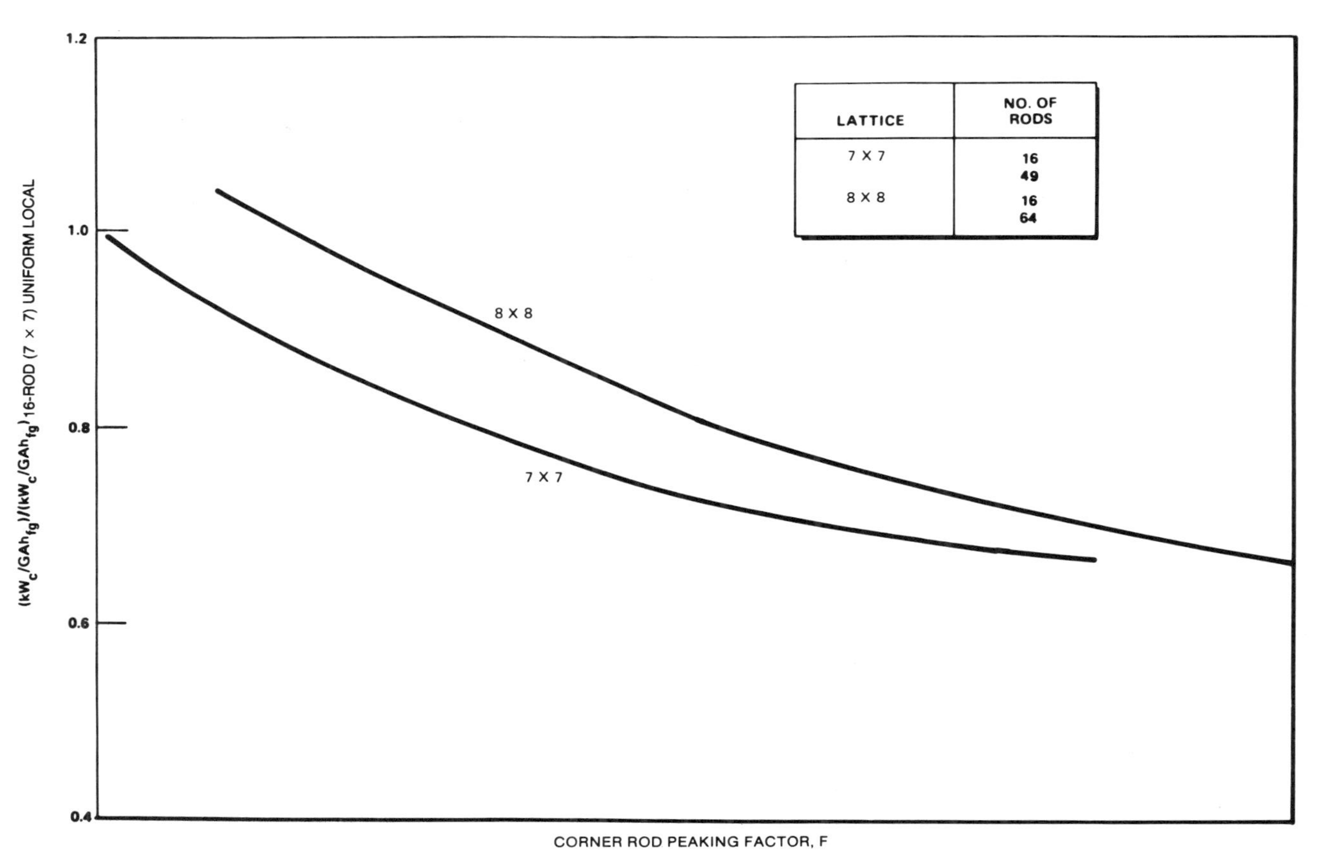

Fig. 4-40 Critical power per unit mass rate, normalized with respect to the value for uniform local peaking, versus corner rod peaking, $p = 1000$ psia, $\Delta h_{sub} = 20$ Btu/lb, $G = 0.5 \times 10^6$ and 1.0×10^6 lb/h-ft^2 (GETAB, 1973).

decrease is proportional to $1/\sqrt{F}$. For corner rod peaking, Fig. 4-40 indicates that 8×8 (64-rod) bundles are less sensitive to this effect than 7×7 (49-rod) bundles. Figure 4-41 indicates the effect of system pressure within the range $800 \leq p \leq 1400$ psia. Over this range, the critical power performance drops monotonically with increasing pressure. Based on other data (not shown), it appears that this effect turns around at lower system pressures, that is, $p<600$ psia.

Figure 4-42 indicates the effect of the axial heat flux profile on critical power. Note that the observed effect is generally within ±8% of the symmetric cosine data. As might be expected, the uniform axial and inlet peaked profiles have the highest thermal performance, while the outlet peaked and double humped (i.e., dromedary) profile have the lowest. To design and operate BWR fuel, we must have techniques (analytical models and/or correlations) with which to predict the critical power performance of the rod bundles. These techniques should be capable of following the experimentally observed data trends discussed above.

There are at least two approaches that can be taken to establish the required design techniques. The first approach that can be—and has been—used is to develop a limit line. That is, a conservative lower envelope to

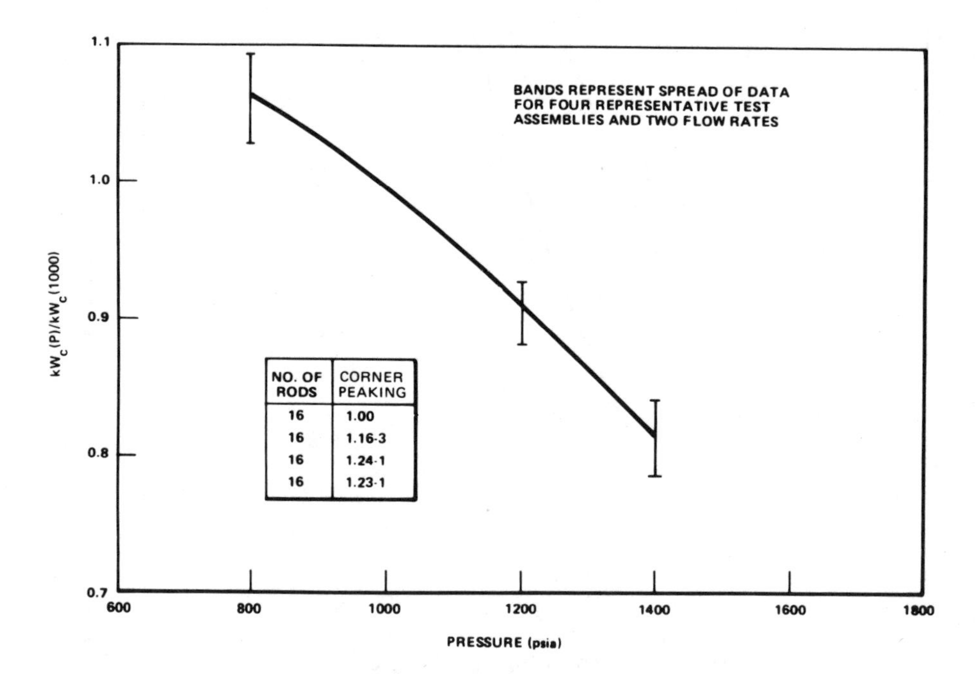

Fig. 4-41 Critical power, normalized with respect to value at 1000 psia, versus $\Delta h_{sub} = 20$ Btu/lb, $G = 0.5 \times 10^6$ and 1.0×10^6 lb/h-ft^2 (GETAB, 1973).

the appropriate BT data such that virtually no data points fall below this line. Historically, these limit lines have been constructed in the flux-quality plane. The first set of limit lines used by GE was based largely on single-rod annular BT data having uniform axial heat flux. These design lines were known as the Janssen-Levy limit lines (Janssen and Levy, 1962). They are considered valid for mass fluxes from 0.4×10^6 to 6.0×10^6 lb/h-ft^2, hydraulic diameters between 0.245 to 1.25 in., and system pressure from 600 to 1450 psia. For 1000 psia and a hydraulic diameter <0.60 in., these limit lines are given by,

$$(q_c''/10^6)=0.705+0.237(G/10^6),\ \frac{\text{Btu}}{\text{h-ft}^2} \tag{4.80}$$

for,

$$(\langle x_e\rangle_c)<0.197-0.108(G/10^6)$$

$$(q_c''/10^6)=1.634-0.270(G/10^6)-4.71\langle x_e\rangle_c,\ \frac{\text{Btu}}{\text{h-ft}^2} \tag{4.81}$$

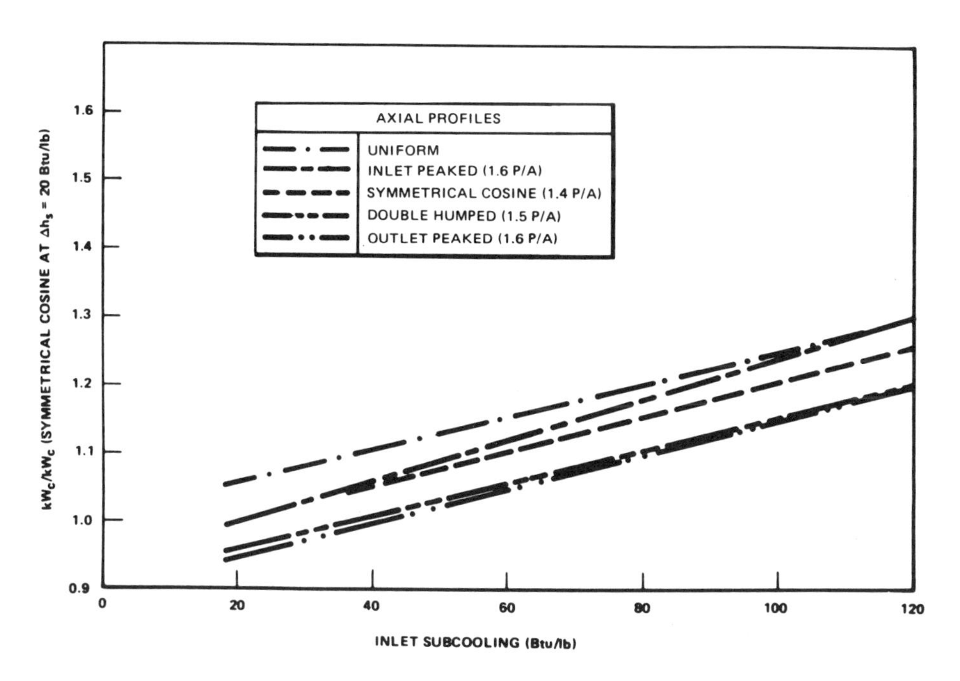

Fig. 4-42 Critical power, normalized with respect to value for symmetrical cosine axial profile at $\Delta h_{\text{sub}}=20$ Btu/lb, versus inlet subcooling, various axial power profiles, 1.23/1.26 three-rod corner peaking, 1000 psia, $G=1\times10^6$ lb/h-ft^2 (GETAB, 1973).

for,

$$0.197-0.108(G/10^6)<(\langle x_e\rangle_c)<0.254-0.026(G/10^6)$$

$$(q_c''/10^6)=0.605-0.164(G/10^6)-0.653\langle x_e\rangle_c,\ \frac{\text{Btu}}{\text{h-ft}^2} \tag{4.82}$$

for,

$$(\langle x_e\rangle_c)>0.254-0.026(G/10^6)\ .$$

For hydraulic diameters greater than 0.60 in., Eqs. (4.80), (4.81), and (4.82) should be modified by subtracting,

$$2.19\times10^6(D_H^2-0.36)\left[\langle x_e\rangle_c-0.0714\left(\frac{G}{10^6}\right)-0.22\right]\ . \tag{4.83}$$

At system pressures other than 1000 psia, the following pressure correction was recommended,

$$q_c''(p)=q_c''(1000)+400(1000-p)\ . \tag{4.84}$$

The validity of the Janssen-Levy limit lines was based on the hypothesis that since the corner rod in a multirod bundle is geometrically similar to an annular configuration, the thermal performance (i.e., BT) should be quite similar. In retrospect, this was a rather good assumption. Nevertheless, it was recognized at the time that more multirod BT data were needed and, thus, an experimental program using four- and nine-rod uniform axial heat flux bundles was conducted. These BT data indicated that some adjustment to the Janssen-Levy limit lines was in order, and, thus, the Hench-Levy limit lines were developed (Healzer et al., 1966).

The Hench-Levy lines are considered valid for mass fluxes from 0.2×10^6 to 1.6×10^6 lb/h-ft^2, hydraulic diameters from 0.324 to 0.485 in., system pressures from 600 to 1450 psia, and rod-to-rod and rod-to-wall spacings greater than 0.060 in. The mathematical expressions for this correlation at 1000 psia are given by,

$$(q_c''/10^6)=1.0,\ \frac{\text{Btu}}{\text{h-ft}^2} \tag{4.85}$$

for,

$$(\langle x_e\rangle_c)\leqslant0.273-0.212\ \tanh^2(3G/10^6)$$

$$(q_c''/10^6)=1.9-3.3\langle x_e\rangle_c-0.7\ \tanh^2(3G/10^6),\ \frac{\text{Btu}}{\text{h-ft}^2} \tag{4.86}$$

for,

$$0.273-0.212\ \tanh^2(3G/10^6)\leqslant(\langle x_e\rangle_c)\leqslant0.5-0.269\ \tanh^2(3G/10^6)$$
$$+0.0346\ \tanh^2\left(\frac{2G}{10^6}\right)$$

and,

$$(q_c''/10^6) = 0.6 - 0.7\langle x_e\rangle_c - 0.09 \tanh^2 (2G/10^6), \frac{\text{Btu}}{\text{h-ft}^2} \quad (4.87)$$

for,

$$(\langle x_e\rangle_c) \geqslant 0.5 - 0.269 \tanh^2 (3G/10^6) + 0.346 \tanh^2 \left(\frac{2G}{10^6}\right) .$$

At system pressures other than 1000 psia, the following pressure correction is recommended,

$$q_c''(p) = q_c''(1000) \left[1.1 - 0.1 \left(\frac{p - 600}{400}\right)^{1.25}\right] . \quad (4.88)$$

The 1000 psia limit lines are shown in Fig. 4-43. As with the Janssen-Levy limit lines, these design curves were constructed such that they fell below virtually all the data at each mass flux. Moreover, to allow for the fact that rod bundles of untested power distribution and configuration may yield BT data below these lines, they have always been applied in BWR design with a margin of safety. Historically, this margin of safety has been in terms of a minimum CHFR (MCHFR). For the Hench-Levy limit lines, the MCHFR was 1.9. That is, to allow for data uncertainty and anticipated transients, we must operate a BWR such that the MCHFR in any bundle is not less than 1.9.

The limit line approach allows for safe BWR operation, but leaves something to be desired in terms of following the data trends previously discussed. As an example, consider an experimentally observed situation shown in Fig. 4-44, in which a rod bundle with a cosine axial heat flux distribution is operated at three different power levels. At power level ①, no BT was observed experimentally and, since the heat balance line is everywhere below the limit line, none is predicted. The power was then increased to level ②, where the heat balance curve just touched the limit line. At this operating condition, a BT would be predicted, but none was observed. When the power level was increased further to level ③, a BT was measured although, in the flux-quality plane, the data point fell below the limit line. Thus, we have the paradox that at power level ③, BT is conservatively predicted along a whole axial region of the rod bundle, while further downstream, where BT was actually measured, the local conditions are below the limit line. Obviously, while the limit line concept can be used for design purposes, it does not capture the axial heat flux effect and, thus, the axial BT location normally is predicted incorrectly.

Figure 4-44 is another example of the fact that the local conditions hypothesis is not valid for BWR conditions of interest. In Sec. 4.4.1.2, it was shown that an appropriate integral technique was in terms of critical quality-boiling length. To eliminate the undesirable features inherent in the limit line approach, a new correlation known as the GE critical quality-boiling

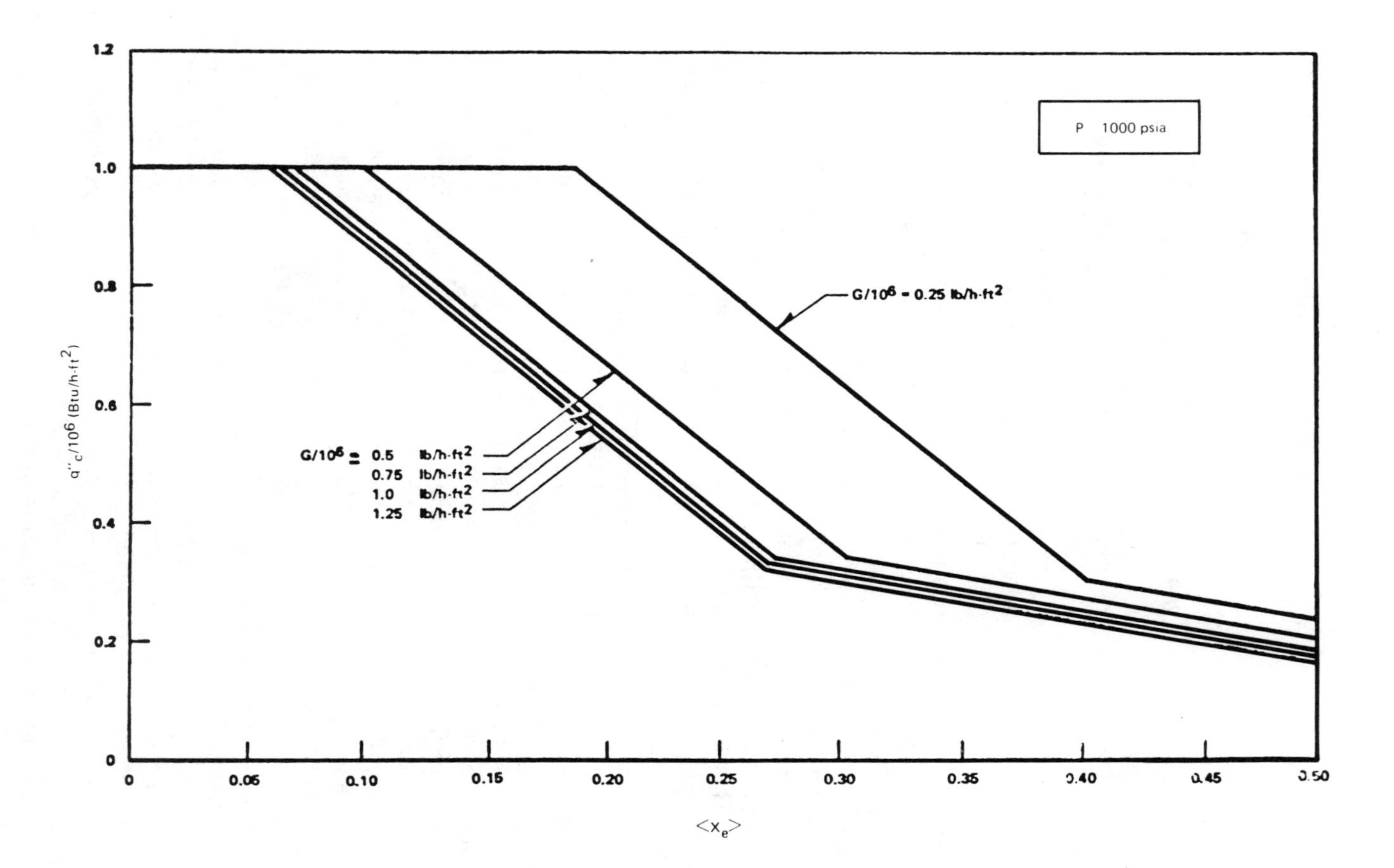

Fig. 4-43 Hench-Levy limit lines.

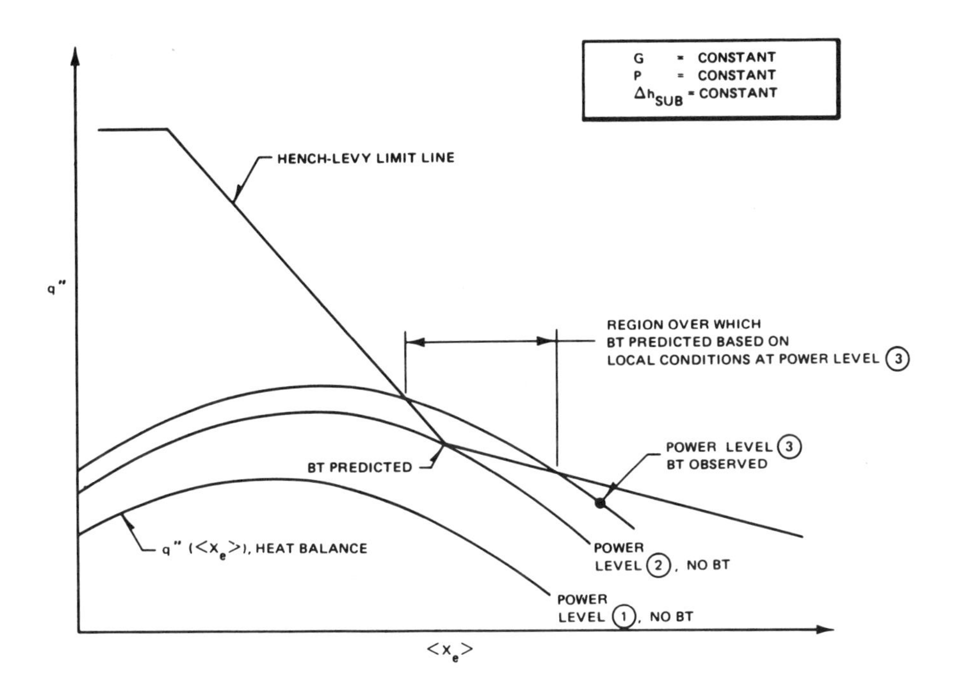

Fig. 4-44 Experimentally observed trend in BT data compared to Hench-Levy limit line.

length (GEXL) correlation was developed (GETAB, 1973). The GEXL correlation is based on a large amount of BT data taken in GE's ATLAS Heat Transfer Facility, which includes full-scale 49- and 64-rod data. The generic form of the GEXL correlation is

$$\langle x_e \rangle_c = f(L_{B_c}, G, p, L_H, D_q, R) \ , \tag{4.89}$$

where

D_q $\triangleq$ thermal diameter, $(4A_{x-s}/P_{\text{rods}})$
R $\triangleq$ generalized local peaking pattern factor
P_{rods} $\triangleq$ the perimeter of the active fuel rods.

The GEXL correlation, like the previous limit lines, uses cross-sectional bundle-average parameters rather than the local subchannel parameters discussed in Sec. 4.5. To predict accurately the local peaking effect, a generalized local peaking factor, R, was synthesized. The R parameter can be considered as the bundle-average analog of subchannel analysis, since it quantifies the effect of the power on the neighboring rods on the thermal performance of the critical rod.

Unlike the limit line approach, the GEXL correlation is a best fit to the experimental data and is able to predict a wide variety of data with a standard deviation of about ±3.5% (GETAB, 1973). Moreover, the GEXL correlation lends itself to a statistical treatment of the required thermal margin for safe plant operation. Figure 4-45 is a display of the GEXL correlation and several typical BWR (cosine axial heat flux) heat balance curves. The lower operating curve is typical of the operating conditions that would exist when the reactor is operated at a power level such that the plant-dependent statistically determined minimum CPR (MCPR) criterion is not exceeded. For instance, if the required CPR is 1.2, the plant must be operated such that the operating MCPR $\geqslant$ 1.2. The 20% margin includes uncertainties in the GEXL correlation (2σ), the reduction in CPR due to the most limiting anticipated transient (e.g., turbine trip without bypass), and some maneuvering room.

The energy balance curve in Fig. 4-45, which just touches the GEXL correlation, is associated with the critical power condition. For this operating condition, MCPR = 1.0, which implies that there is an ~50% probability that a BT would be experienced on some (~0.1%) of the fuel rods. The critical quality defect, $\Delta\langle x_e\rangle_c$, is also shown in Fig. 4-45. This parameter,

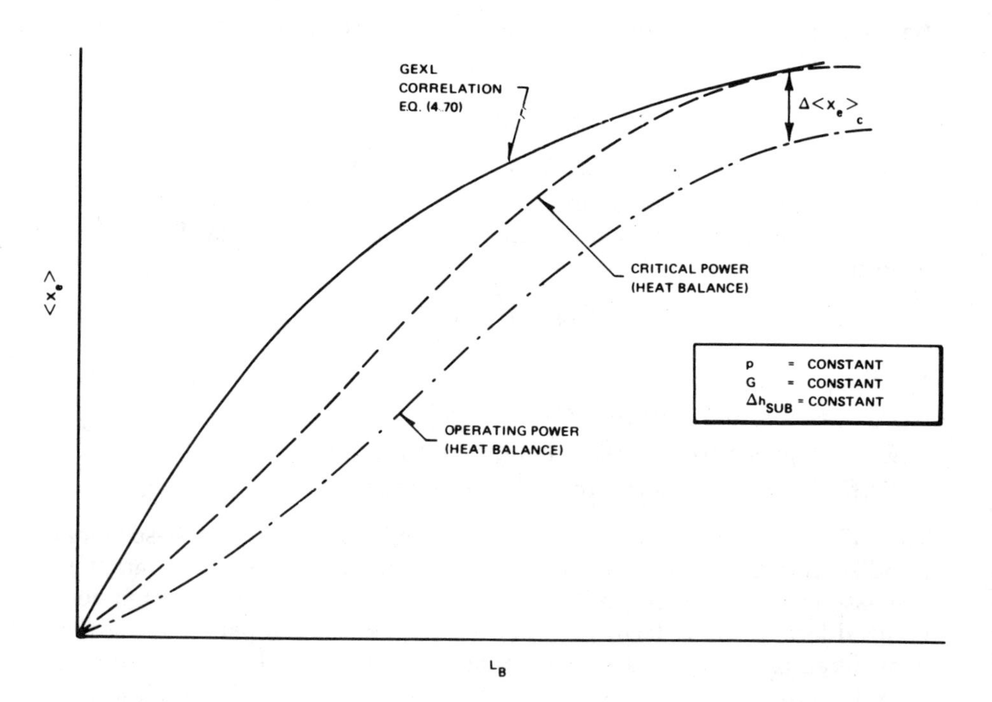

Fig. 4-45 Graphic display of GEXL correlation and BWR energy balance curves.

which is zero when MCPR = 1.0, can be used to approximate the operating CPR. That is, if we assume that the change in boiling length is small between the operating condition and the critical condition, then an energy balance, i.e., Eqs. (4.69) and (4.70), yields,

$$\mathrm{CPR} \triangleq \frac{KW_c}{KW} \cong 1 + \frac{\Delta\langle x_e(L_B)\rangle_c}{\langle x_e(L_B)\rangle + \Delta h_{\text{sub}}/h_{fg}} \ . \tag{4.90}$$

When $0.9 \leqslant \mathrm{CPR} \leqslant 1.1$, the approximation given in Eq. (4.90) has been found to be valid to within 1%. In any event, one can always iterate on the power level until the heat balance curve just touches the GEXL correlation to determine the critical power.

Although the bulk of the experimental data has been taken with simulated fuel rod bundles having nominal clearances, a detailed study was conducted (Lahey et al., 1973) in which it was demonstrated that pathological geometric conditions, such as reduced rod-to-wall clearance and rod bowing, have virtually no effect on the critical power performance of a BWR rod bundle. To extrapolate these results to design applications and to handle untested situations of concern, effective subchannel analyses are required. Such techniques will be considered in Sec. 4.6.

4.5 Post-Boiling-Transition Convective Heat Transfer

Branch ⑤ to ⑥ of the forced convection boiling curve shown in Fig. 4-17 is known as the transition boiling (TB) branch. This branch is highly unstable and is characterized by the intermittent physical rewetting of the heated surface by the liquid in the flow stream. In a temperature-controlled system, steady-state operation can be achieved on this branch of the curve, while for a heat flux controlled system, steady-state operation is possible only in some very special cases of little practical importance. Note that a nuclear fuel rod acts like a heat flux controlled system.

To better understand the basic heat transfer phenomena occurring during TB, it is convenient now to consider film boiling and post-dryout (PDO) heat transfer, since the basic heat transfer mechanisms are presumably the same as in the TB branch.

Branch ⑥ to ⑦ of the convective boiling curve is commonly known as the convective film boiling or fully developed PDO heat transfer branch, depending on the quality and flow rate at which BT occurs. As shown in Fig. 4-46, for low-quality/high-flow BT a rather severe temperature excursion is experienced at the departure from nucleate boiling (DNB) point. In addition, an inverse annular flow regime is established downstream of the DNB point such that an annular vapor film blankets the heated surface. Thus, for low-quality/high-flow conditions, classical film boiling exists on branch ⑥ to ⑦′. In contrast, for higher quality/low-flow conditions, the temperature excursion is milder and the heat transfer situation downstream

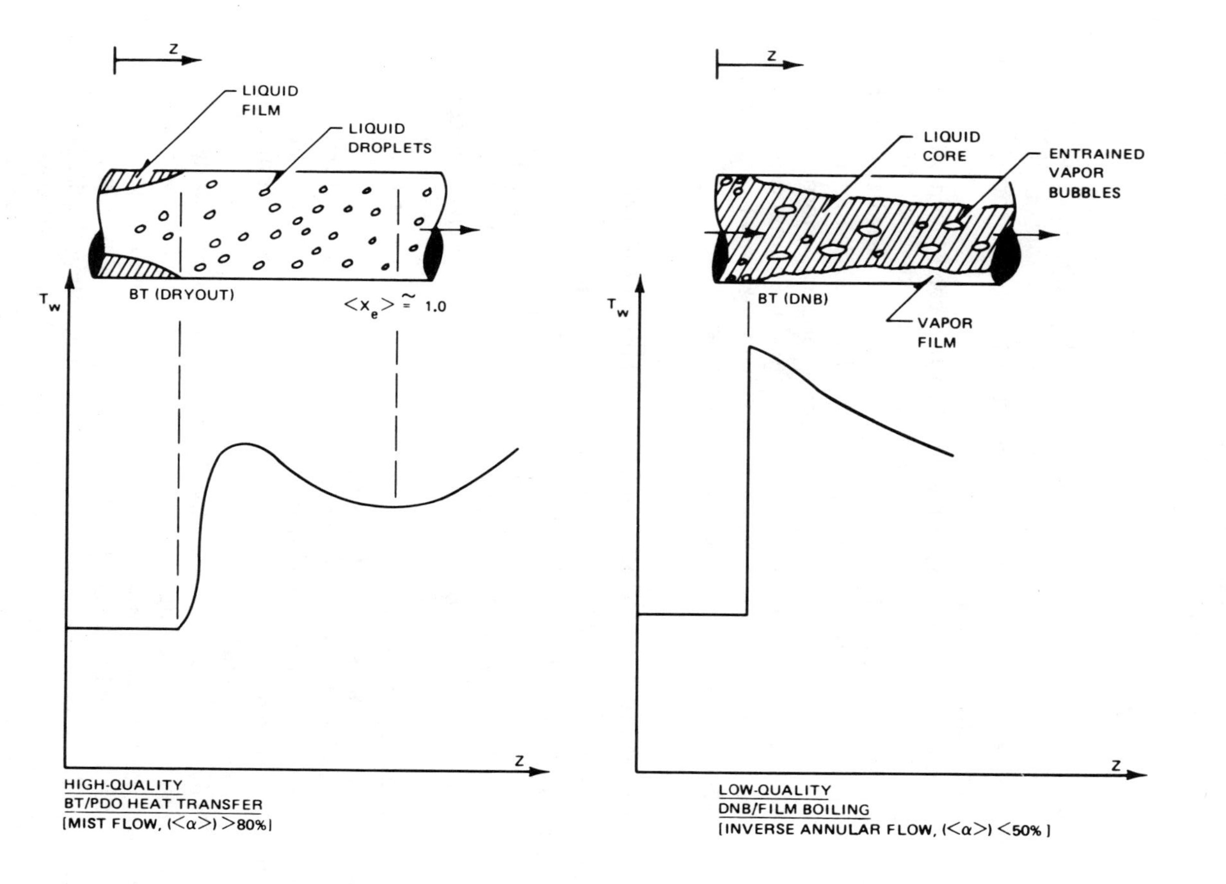

Fig. 4-46 Post-boiling transition heat transfer regimes.

of liquid film dryout is often referred to as a liquid deficient (mist flow) film boiling situation, although there is obviously no vapor film in the true sense of the word. To avoid confusion, we have adopted the terminology fully developed PDO heat transfer (Groeneveld, 1973) to describe the high-quality/low-flow heat transfer phenomena which occurs on branch (6) to (7). In BWR technology, we are concerned primarily with PDO heat transfer.

To gain a better mechanistic appreciation for PDO heat transfer, we consider the situation shown in Fig. 4-47. Note that subsequent to BT, the vapor temperature begins to rise as it becomes superheated. Calculation of the true vapor temperature is quite important since it is the key to the determination of accurate, fully developed PDO heat transfer coefficients. Before discussing various techniques for the determination of vapor superheat or, equivalently,[f] the actual flow quality, we consider some simple bounding cases that provide valuable and interesting insight into the observed data trends.

There are two bounding cases to be considered. First, we assume complete thermal equilibrium, which implies that the vapor temperature is always equal to the saturation temperature and, as discussed more thoroughly in Chapter 5, that the flow quality, $\langle x(z)\rangle$, is always equal to the thermodynamic equilibrium quality, $\langle x_e(z)\rangle$. This model implies no vapor superheat and assumes that all energy from the heated walls goes directly into evaporation of the liquid droplets. The other bounding case is for the assumption that, subsequent to BT, no further liquid droplet evaporation occurs. That is, the flow quality, $\langle x(z)\rangle$, is frozen at the critical quality, $\langle x_e(z)\rangle_c$. This case gives the most rapid heatup of the vapor phase that is possible since none of the energy from the heated walls goes into evaporation of the liquid droplets. This model is known as the frozen droplet or no-further-evaporation model.

The essence of the thermal equilibrium model is given by the steady-state energy balance,

$$\langle x_e(z)\rangle = \langle x_e\rangle_c + \frac{1}{GA_{x-s}h_{fg}}\int_{z_c}^{z} P_H q'' \, dz' \ , \tag{4.91}$$

and the assumptions,

[f]From a steady-state energy balance, neglecting kinetic and potential energy, and assuming a saturated liquid phase

$$\frac{d\langle T_v(z)\rangle}{dz} = \left[\frac{q''(z)P_H/A_{x-s} - G[\langle h_v(z)\rangle - h_f]\dfrac{d\langle x(z)\rangle}{dz}}{G\langle x(z)\rangle c_p}\right] .$$

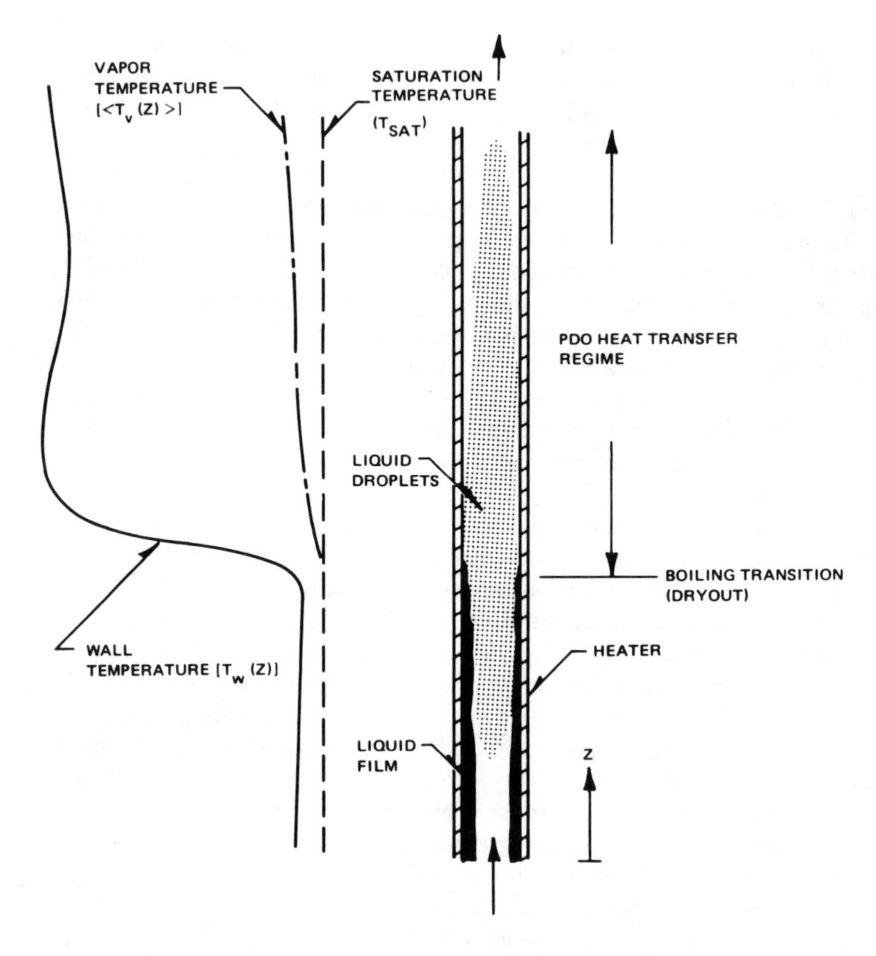

Fig. 4-47 Typical PDO heat transfer in BWR technology.

$$\langle x(z)\rangle = \begin{cases} \langle x_e(z)\rangle, & \text{for } [\langle x_c(z)\rangle] \leqslant 1.0 \\ 1.0, & \text{otherwise} \end{cases}$$

$$\langle T_v(z)\rangle = \begin{cases} T_{\text{sat}}, & \text{for } [\langle x_c(z)\rangle] \leqslant 1.0 \\ \text{superheated}, & \text{otherwise} . \end{cases} \tag{4.92}$$

The fully developed PDO heat transfer coefficient, H_{PDO}, is calculated from an appropriate single-phase vapor heat transfer correlation such as the Dittus-Boelter correlation, Eq. (4.42),

$$\left[\frac{H_{\text{PDO}}(z)D_H}{\kappa_g}\right] = 0.023\left[\frac{G\langle x(z)\rangle D_H}{\langle\alpha(z)\rangle\mu_g}\right]^{0.8}\left(\frac{c_{p_g}\mu_g}{\kappa_g}\right)^{0.4} , \tag{4.93}$$

and, thus, the fully developed PDO heat flux is given by,

$$q''_{\mathrm{PDO}}(z)=H_{\mathrm{PDO}}(z)[T_w(z)-\langle T_v(z)\rangle] \ , \tag{4.94}$$

where from Eq. (4.92) for,

$$(\langle x_e\rangle)\leq 1.0,\ \langle T_v(z)\rangle = T_{\mathrm{sat}} \ .$$

For $(\langle x_e\rangle)\geq 1.0$, we set $\langle x(z)\rangle$ and $\langle \alpha(z)\rangle$ in Eq. (4.93) equal to unity and evaluate the vapor temperature from the steady single-phase energy balance,

$$\langle T_v(z)\rangle = T_{\mathrm{sat}}+\frac{1}{GA_{x-s}c_{p_v}}\int_{z(\langle x_e\rangle=1.0)}^{z} P_H q'' dz' \ . \tag{4.95}$$

We now consider the other bounding case known as the frozen droplet model. This model is really just a single-phase heat transfer model in which the liquid phase is ignored; i.e., the liquid droplets are assumed not to participate in the heat transfer process. The vapor superheat is given by the steady-state energy balance,

$$\langle T_v(z)\rangle - T_{\mathrm{sat}}=\frac{1}{G\langle x_e\rangle_c A_{x-s}c_{p_v}}\int_{z_c}^{z} P_H q'' dz' \ , \tag{4.96}$$

where the assumption has been made that for $z\geq z_c$,

$$\langle x(z)\rangle = \langle x_e\rangle_c \ . \tag{4.97}$$

The fully developed PDO heat transfer coefficient again can be determined from an appropriate superheated vapor heat transfer correlation, such as that given by Heineman (1960) or through the use of the Colburn correlation, Eq. (4.43), written as,

$$\left(\frac{H_{\mathrm{PDO}}D_H}{\kappa_v}\right)_{\mathrm{film}}=0.023\left[\frac{G\langle x_e\rangle_c D_H}{\langle\alpha(z)\rangle\mu_v}\right]_{\mathrm{film}}^{0.8}\left(\frac{c_{p_v}\mu_v}{\kappa_v}\right)_{\mathrm{film}}^{1/3}\left(\frac{c_{p_{v\mathrm{bulk}}}}{c_{p_{v\mathrm{film}}}}\right) \ , \tag{4.98}$$

where the axial dependence of the void fraction is due to the effect of vapor superheat on ρ_v [see Eq. (5.25)]. Thus, the fully developed PDO heat flux is given by

$$q''_{\mathrm{PDO}}(z)=H_{\mathrm{PDO}}[T_w(z)-\langle T_v(z)\rangle] \ . \tag{4.99}$$

A comparison of the thermal equilibrium and the frozen droplet models with some typical data trends (Bailey et al., 1973) is shown in Fig. 4-48. Note that at lower mass flux, the data tend to approach the frozen droplet model, while at the higher mass fluxes, thermal equilibrium is approached. It is apparent that at high mass flux, the actual heat transfer process is somewhat better than predicted by the thermal equilibrium model, indicating that some nontrivial mechanism has been neglected in this model. It will be shown subsequently that this mechanism is the heat transfer due to wall-to-droplet interaction.

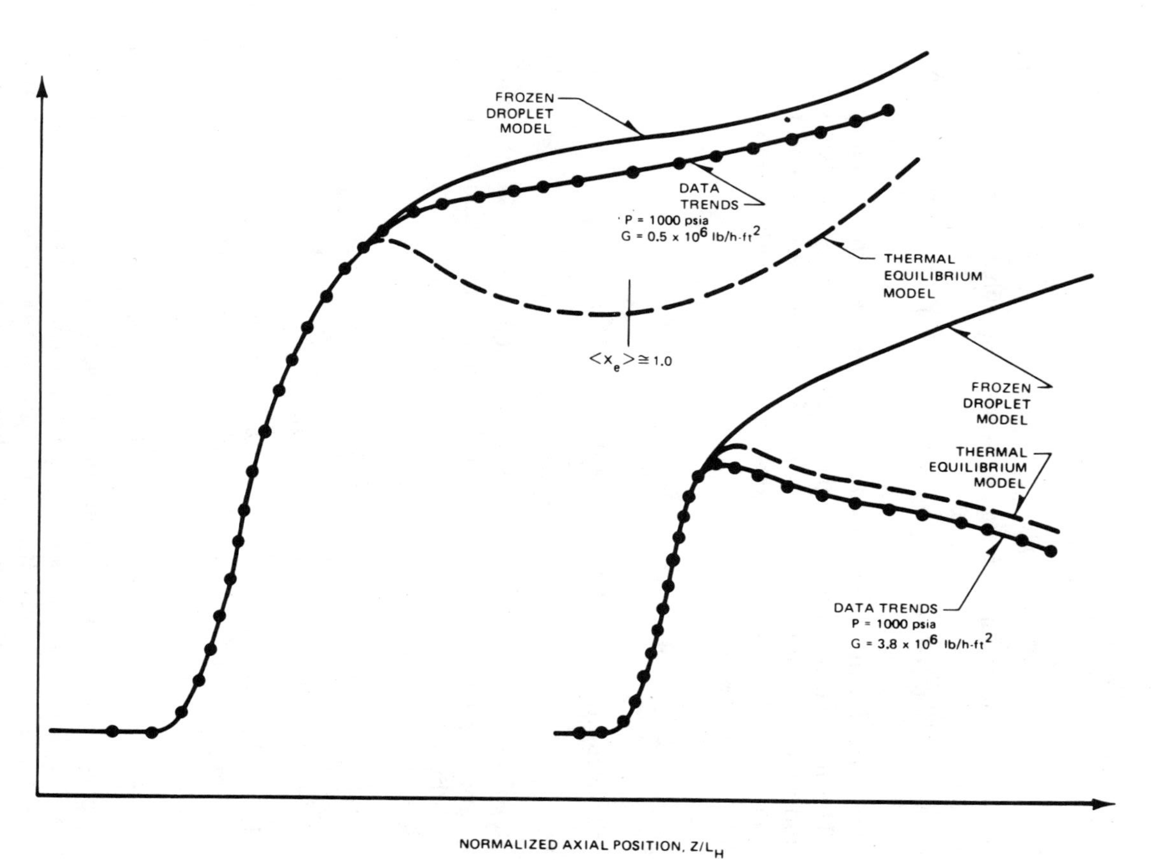

Fig. 4-48 Typical high-quality PDO heat transfer data trends (Bailey et al., 1973).

The observed trend with mass flux is primarily due to the fact that as the mass flux increases, the slip ratio increases. When the critical Weber number ($We_c \stackrel{\Delta}{=} \rho_v[\langle\mu_v\rangle_v - \langle\mu_l\rangle_l]^2 D_d/g_c\sigma = 7$ to 13) is exceeded, the liquid droplets are shattered into smaller droplets with increased interfacial area for heat transfer. In addition, the so-called ventilation factor (Bennett et al., 1967) increases and, thus, the vapor-to-droplet heat transfer coefficient improves as mass flux increases. These effects prevent the existence of large vapor superheats such as those observed at the lower mass fluxes. Naturally, any effective model for flow quality and vapor superheat must accurately account for the observed trend with mass flux.

There are two basic approaches that have been taken to synthesize techniques that can be used for prediction of PDO heat transfer. The empirical approach, that is, the use of global heat transfer correlations referenced to the saturation temperature, has been the most widely used to date. However, considerable progress recently has been made in the development of physically based models for PDO heat transfer. Typically, these models are concerned with the heat transfer mechanisms that occur during PDO conditions and the degree of superheat of the vapor phase.

A thorough review of the various global, fully developed PDO heat transfer correlation was previously given (Groeneveld, 1973) and is not repeated here. Nevertheless, since certain PDO correlations are currently important in BWR design and safety analysis, they are discussed briefly. The two most widely used, fully developed PDO heat transfer correlations accepted by the U.S. Nuclear Regulatory Commission for use in BWR design and analysis are the so-called Groeneveld correlation (Groeneveld, 1973) given by,

$$\frac{H_{PDO}D_H}{\kappa_g} = 0.052\left\{\left(\frac{GD_H}{\mu_g}\right)\left[\langle x_e\rangle + \frac{\rho_g}{\rho_f}(1-\langle x_e\rangle)\right]\right\}^{0.688} \times \left(\frac{c_{p_g}\mu_g}{\kappa_g}\right)_{\text{wall}}^{1.26}\left\{1.0 - 0.1\left[\left(\frac{\rho_f}{\rho_g}\right)-1\right]^{0.4}(1-\langle x_e\rangle)^{0.4}\right\}^{-1.06}, \quad (4.100)$$

and the Dougall-Rohsenow correlation (Dougall and Rohsenow, 1963) given by,

$$\frac{H_{PDO}D_H}{\kappa_g} = 0.023\left\{\frac{GD_H}{\mu_g}\left[\langle x_e\rangle + \frac{\rho_g}{\rho_f}(1-\langle x_e\rangle)\right]\right\}^{0.8}\left(\frac{c_p\mu}{\kappa}\right)_g^{0.4}. \quad (4.101)$$

The Dougall-Rohsenow correlation can be recognized as simple modification of the single-phase Dittus-Boelter equation given in Eq. (4.42). Both Eqs. (4.100) and (4.101) greatly underpredict fully developed PDO heat transfer for low qualities (<10%). In general, the Dougall-Rohsenow correlation yields fully developed PDO heat transfer coefficients that are somewhat higher than those given by the Groeneveld correlation, although both correlations tend to be conservative for BWR conditions.

For the higher system pressures, more typical PWR operation, Bishop (Bishop et al., 1965) has proposed the following fully developed PDO correlation:

$$\left(\frac{H_{\mathrm{PDO}} D_H}{\kappa_v}\right) =$$

$$0.0193 \left(\frac{G D_H}{\mu_v}\right)_{\mathrm{film}}^{0.8} \left(\frac{c_{p_v} \mu_v}{\kappa_v}\right)_{\mathrm{film}}^{1.23} \left(\frac{\rho_v}{\rho_f}\right)^{0.068} \left[\langle x_e \rangle + \frac{\rho_v}{\rho_f}(1 - \langle x_e \rangle)\right]^{0.68} . \quad (4.102)$$

Global fully developed PDO correlations, such as those just discussed, are frequently used to bound the temperature response of various hypothetical reactor accidents. Since they generally yield conservative fully developed PDO heat transfer coefficients, they are currently accepted by NRC for safety analysis. Unfortunately, while global correlations may be conservative, they do not accurately capture the data trends and they provide little insight into the heat transfer mechanisms involved. For instance, since fully developed PDO heat transfer coefficients are referenced arbitrarily to saturation temperature,

$$H_{\mathrm{PDO}} \triangleq \frac{q''}{(T_w - T_{\mathrm{sat}})} , \quad (4.103)$$

the effect of vapor superheat is not included. This explains why fully developed PDO heat transfer coefficients are frequently lower than corresponding single-phase vapor heat transfer coefficients.

To understand the observed data trends, we must consider the detailed PBT heat transfer mechanisms. Insight into these mechanisms has been provided by Iloeje et al. (1974). Iloeje partitioned the wall heat transfer into the heat transfer from:

1. the heated wall to the vapor, q''_{w-v}
2. the wall to the impinging liquid droplets, q''_{w-d}
3. the wall to the liquid droplets, which penetrate and agitate the thermal boundary layer but do not actually strike the wall, q''_{w-v-d}.

Inherent in this model is the assumption that the thermal energy, which is transferred to the vapor, tends to superheat the vapor phase, which in turn evaporates the liquid droplets in the flow stream.

Iloeje's model is basically an extension of the so-called "two-step" model (Bennett et al., 1967), in which wall-to-droplet interaction was neglected. As shown in Fig. 4-49, the PDO branch of a typical convective boiling curve can be considered composed of the superposition of various heat transfer mechanisms. The part due to single-phase vapor heat transfer increases

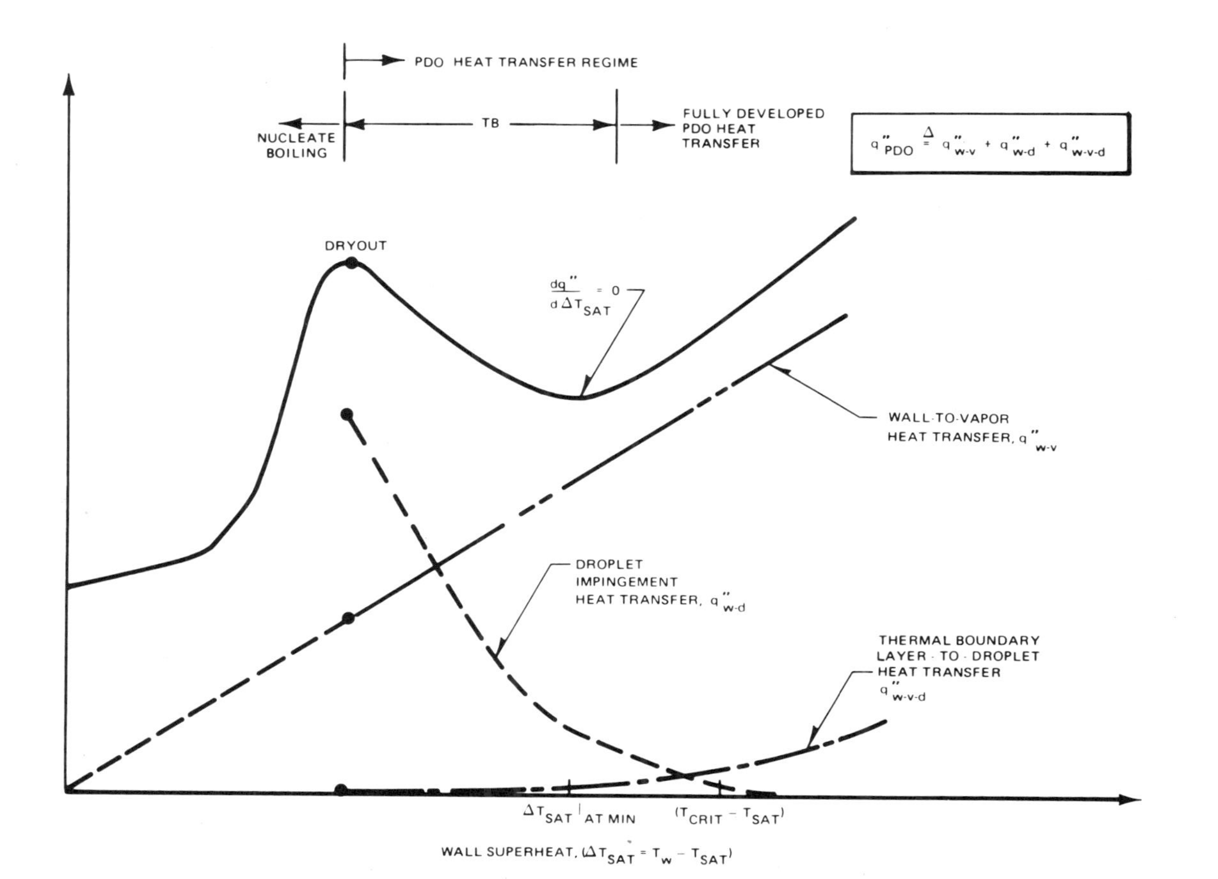

Fig. 4-49 A mechanistic model of PDO convective boiling curves.

monotonically with wall superheat, ΔT_{sat}. Similarly, the contribution due to droplet agitation of, and evaporation in, the thermal boundary layer increases with wall superheat and begins to be significant at the higher wall temperatures. The contribution due to droplet impingement on the heated wall is seen to decrease monotonically as the wall superheat increases. Note that the point at which $dq''/d\Delta T_{sat}$ is zero is determined by the relative value of the various heat transfer mechanisms and, thus, is largely thermal hydraulic, rather than thermodynamic, in nature. This minimum point is known frequently as the Leidenfrost or rewetting point; however, physical rewetting of the heated surface is not necessary for its existence.

At some local wall temperature, below the critical temperature of water ($T_{crit}=705.4°F$), no further direct wall-to-droplet heat transfer occurs since the liquid droplets can no longer wet the wall; i.e., surface tension goes to zero at the critical temperature and, thus, liquid droplets can no longer exist. Nevertheless, "droplet-like" heat transfer can occur at higher wall temperatures since, as the liquid droplets approach the heated surface, the pressure between the droplets and wall can build up sufficiently to form a very dense vapor adjacent to the wall. These dense vapor "droplets" can provide rather effective local heat transfer and, thus, affect the location and shape of the minimum (i.e., $dq''/d\Delta T_{sat}=0$) point.

Spiegler et al. (1963) have deduced from the spinoidals of the van der Waals equation-of-state the thermodynamic maximum temperature, T_{max}, at low pressure (i.e., the point at which homogeneous nucleation occurs). This result is given by Iloeje et al. (1974),

$$T_{max}(p)/T_{crit}=0.16\left(\frac{p}{p_{crit}}\right)+0.84 \ . \tag{4.104}$$

By surveying the available data, Iloeje has shown that, depending on the flow conditions, the ratio of the $\kappa\rho c_p$ product, the surface conditions, the contact angle, and the thermal diffusivity of the heated surface, localized physical wetting of the heated surface can occur even when the measured wall temperature is between $T_{max}(p)$ and T_{crit}. Moreover, the available data show that the minimum point, $dq''/d\Delta T_{sat}=0$, is sometimes above and sometimes below $T_{max}(p)$, confirming that the minimum is controlled by system thermal hydraulics rather than the thermodynamics of the fluid.

The importance of the $\kappa\rho c_p$ product can be best understood in terms of the contact temperature between the liquid and heated wall. It is a well-known fact in conduction heat transfer that the contact temperature is characterized by the parameter, $[(\kappa\rho c_p)_{wall}/(\kappa\rho c_p)_{liquid}]^{1/2}$. Thus, surfaces with a low $\kappa\rho c_p$ product can support local temperature depressions due to droplet impact and, thus, local rewetting can occur. While the $\kappa\rho c_p$ product is of some interest in PDO heat transfer in determining the effect of droplet heat transfer, it is of much more importance for inverse annular flow convective film-boiling conditions (Kalinin et al., 1969). For film-boiling conditions, it appears that repeated contact of the liquid interface with the heated wall eventually leads to the collapse of the vapor film and, thus, rewet of the heated surface.

Now that we have some phenomenological understanding of what are currently considered to be the important PDO heat transfer mechanisms, we shall consider the status of the various mechanistic models. The first and most widely known mechanistic model is the so-called two-step model (Bennett et al., 1967). The principal assumption behind models of this type is that the heat is transferred first to the vapor and then from the vapor to the liquid drops, in two steps. That is, direct wall-to-droplet heat transfer is ignored. Basically, these models involve the simultaneous numerical solution of four first-order ordinary differential equations for the characteristic liquid droplet diameter, D_d, and velocity, $\langle u_l \rangle_l$, the flow quality, $\langle x \rangle$, and the mean vapor temperature, $\langle T_v \rangle$.

In addition, the characteristic droplet diameter is constrained by a critical Weber number and the vapor velocity, $\langle u_v \rangle_v$, normally is obtained through a slip relationship. Once these parameters are known, fully developed PDO heat transfer is evaluated through the use of $q'' = H_{1\phi}(T_w - \langle T_v \rangle)$.

Two-step PDO heat transfer models take the form of digital computer codes and thus are rather awkward to evaluate, particularly when PDO heat transfer is only a small part of overall reactor safety analyses. A preferable model would be a correlation based on the mechanistic understanding of PDO heat transfer. Iloeje et al. (1974) have given the functional form of the various components of the PDO boiling curve as,

$$q''_{w-v} = H_{1\phi}\langle \alpha \rangle (T_w - \langle T_v \rangle) \tag{4.105}$$

$$q''_{w-d} = F_1(p, \langle x \rangle, G, \sigma, \text{surface roughness, fluid properties, } \kappa \rho c_p) \times \exp[-\kappa_1 (T_w - T_{sat}) \tag{4.106}$$

$$q''_{w-v-d} = (1 - \langle \alpha \rangle) F_2[p, \langle x \rangle, G, \kappa_1 (T_w - T_{sat}), \text{surface roughness, fluid properties}](T_w - T_{sat}) \;, \tag{4.107}$$

where,

$$\kappa_1 = \kappa_1(p, \langle x \rangle, G, \sigma, \text{surface roughness, fluid properties}) \;,$$

and, by superposition:

$$q''_{PDO} = q''_{w-v} + q''_{w-d} + q''_{w-v-d} \;. \tag{4.108}$$

Although quantitative models of this form have been derived (Iloeje et al., 1974), the precise definition of the functions, F_1, F_2, and κ_1, for steam/water is currently a state-of-the-art problem in PDO heat transfer. From Fig. 4-49 it should be obvious, however, that when these functions are well established, Eqs. (4.105) through (4.108) can be used to define the entire

PDO portion of the boiling curve. That is, the physics involved in the TB branch, the rewetting (i.e., $dq''/d\Delta T_{sat}=0$) point, and the fully developed PDO branch, are exactly the same.

Note that in Eqs. (4.105), (4.106), and (4.107), flow quality, $\langle x \rangle$, is required to evaluate accurately the PDO heat transfer terms. As discussed in more detail in Chapter 5, flow quality generally is not easily obtainable unless the system is in thermodynamic equilibrium. Unfortunately, PDO heat transfer is characterized by situations (e.g., superheated vapor) that normally are quite far from thermodynamic equilibrium.

For the representative case of high PDO wall temperatures, the effect of droplet heat transfer can be ignored (i.e., $q''_{w-d} \cong 0$) and, thus, all heat transfer to the entrained droplets is through the superheated vapor. In this case, the actual quality profile would look like the one shown in Fig. 4-50. This flow quality profile can be calculated, as in the two-step model, by solving the governing differential equations, or it can be estimated using a "profile fit" technique similar to that used in subcooled boiling models.

Plummer et al. (1974) assumed a linear relationship of the form,

$$\langle x \rangle = \langle x_e \rangle_c + K(\langle x_e \rangle - \langle x_e \rangle_c) \ . \tag{4.109}$$

While this fit agrees better with the data than the assumption of thermodynamic equilibrium, it does not satisfy all the required boundary conditions shown in Fig. 4-50,

$$\langle x \rangle \big|_{\langle x_e \rangle_c} = \langle x_e \rangle_c \tag{4.110}$$

$$\left. \frac{d\langle x \rangle}{d\langle x_e \rangle} \right|_{\langle x_e \rangle_c} = 0 \tag{4.111}$$

$$\left. \frac{d^2\langle x \rangle}{d\langle x_e \rangle^2} \right|_{\langle x_e \rangle_c} = F_3(G, p, \langle x_e \rangle_c, \text{ We, fluid and thermodynamic properties}) > 0 \tag{4.112}$$

$$\lim_{\langle x_e \rangle \to \infty} \langle x \rangle = 1.0 \tag{4.113}$$

$$\lim_{\langle x_e \rangle \to \infty} \left(\frac{d\langle x \rangle}{d\langle x_e \rangle} \right) = 0 \ . \tag{4.114}$$

The first two boundary conditions, at the location of the BT, imply that thermodynamic equilibrium initially exists; however, Eq. (4.112) implies that as the vapor becomes superheated, nonequilibrium conditions develop. The asymptotic conditions, Eqs. (4.113) and (4.114), imply that all liquid droplets eventually will be evaporated.

A profile fit that satisfies these boundary conditions (Saha, 1975) is given by,

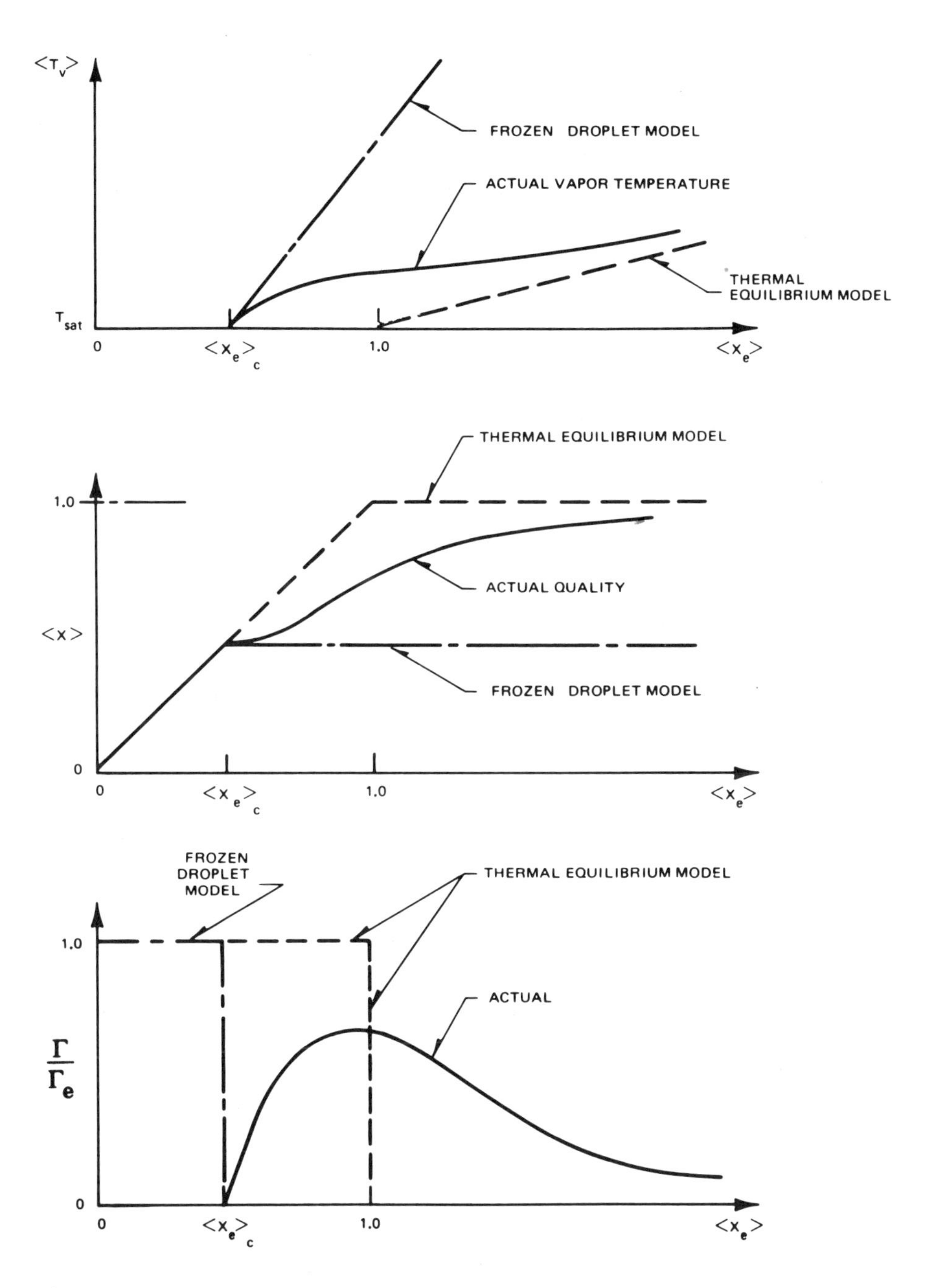

Fig. 4-50 Typical vapor temperature, flow quality, and nondimensional rate of vapor generation.

$$\langle x\rangle = 1.0 - a\exp\left[-\frac{1}{a}(\langle x_e\rangle - \langle x_e\rangle_c)\right] + b\exp\left[-\frac{1}{b}(\langle x_e\rangle - \langle x_e\rangle_c)\right] , \quad (4.115)$$

where the arbitrary parameters, a and b, can be evaluated from Eqs. (4.110) and (4.112),

$$a = \frac{1}{2}\{[(1-\langle x_e\rangle_c)^2 + 4(1-\langle x_e\rangle_c)/F_3]^{1/2} + (1-\langle x_e\rangle_c)\} \quad (4.116)$$

$$b = \frac{1}{2}\{[(1-\langle x_e\rangle_c)^2 + 4(1-\langle x_e\rangle_c)/F_3]^{1/2} - (1-\langle x_e\rangle_c)\} . \quad (4.117)$$

Figure 4-50 shows typical profiles for the two asymptotic cases previously discussed and the actual flow quality, $\langle x\rangle$, vapor temperature, $\langle T_v\rangle$, and nondimensional vapor generation rate, Γ/Γ_e. By using the relationships developed in Chapter 5, it can be shown that for steady-state conditions,

$$\frac{d\langle x\rangle}{d\langle x_e\rangle} = \frac{\Gamma}{\Gamma_e} , \quad (4.118)$$

and, thus, the flow quality profile is just the integral of the nondimensional vapor generation rate.

The whole thrust of the above discussion has been concerned with the proper definition of the forced convection boiling curves. Such curves are of prime importance in the analysis of hypothetical reactor accidents. Plummer et al. (1973) have proposed a tentative set of boiling curves for water. Several of these curves are shown in Fig. 4-51. Note that the flow effect is quite strong in the entire PBT region, while the quality effect is very pronounced only in the TB region. Note also that for high-quality conditions, there may be no point at which $dq''/d\Delta T_{sat}$ is equal to zero, which is in agreement with the mild PBT temperature excursions actually observed. Although research currently is quite active in this area of technology, no completely satisfactory set of boiling curves has been synthesized to date.

Now that the status of rod bundle heat transfer has been discussed, let's turn our attention to the use of subchannel techniques for the evaluation of the local thermal-hydraulic parameters in rod bundles.

4.6 Subchannel Analysis

For many decades, there has been considerable interest in the area of technology known as subchannel analysis. This work has been motivated largely by the desire of nuclear reactor designers to predict accurately BT and void distribution in fuel rod bundles. In Sec. 4.4.2, it was shown that a bundle-average-type correlation, such as GEXL, is able to predict accurately local BT, provided a generalized local peaking parameter, R, is properly formulated. Correlations of this type are empirical and, thus, generally

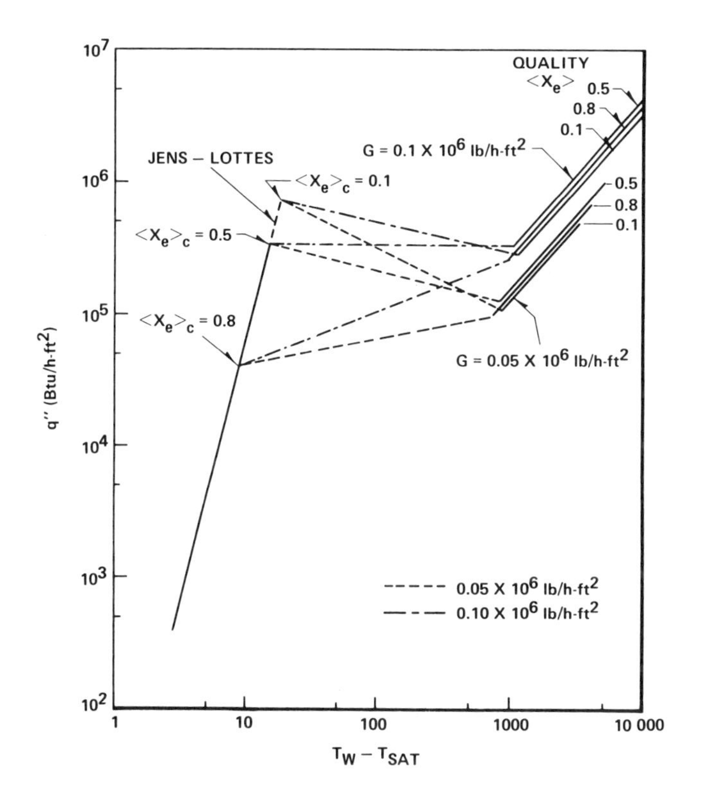

Fig. 4-51 Boiling curves for water (Plummer et al., 1973).

are valid only for conditions representative of the data on which they are based. If a designer is asked to investigate the thermal performance of a proposed new lattice type with different lattice dimensions and configuration, the empirical bundle-average correlation is normally of little help.

To optimize a given lattice, to investigate a proposed new fuel lattice, or to look at some abnormal lattice configurations, we normally must use subchannel techniques. The essence of the subchannel analysis approach is shown in Fig. 4-52. The rod bundle is divided into a number of ventilated flow tubes (subchannels). In the two cases shown in Fig. 4-52, it is assumed that the local peaking factor (i.e., power) on each rod is the same and, thus, only a small section of the bundle needs to be analyzed; the remainder is known by symmetry. By convention, subchannels 1, 2, and 3 in Fig. 4-52b are known as the corner, side, and center subchannel, respectively. In the nine-rod bundle shown, symmetry would yield eight corner subchannels, side, and center subchannels.

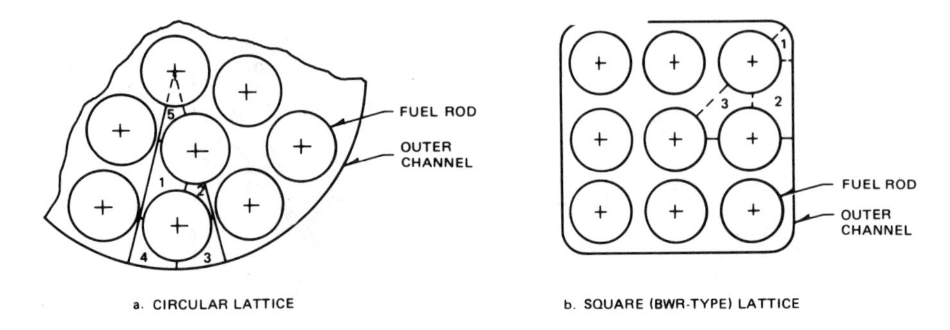

Fig. 4-52 Typical subchannel grids.

When choosing subchannels, it is important to select the right size. If we use subchannels that are too small (e.g., if subchannel 1 in Fig. 4-52b were subdivided into many smaller subchannels) then we must consider the details of the two-phase microstructure, a task that is beyond the current state-of-the-art. On the other hand, if the subchannels chosen are too large, then local conditions around the fuel rod of interest are not accurately predicted. The most popular subdivision is really a compromise and, as shown in Fig. 4-52, normally involves drawing the subchannel boundaries at the minimum rod-to-rod and rod-to-wall gaps. Once the subchannel grid has been established, the appropriate two-phase conservation equations can be integrated to yield the axial subchannel parameters of interest in the rod bundle.

Current generation subchannel techniques normally take the form of large digital computer codes, although rod-centered subchannel techniques are also available (Gaspari et al., 1974) with which first-order approximations to the local parameters can be obtained more easily. Here we discuss the classical approach, such as employed in the COBRA (Rowe, 1973) and HAMBO (Bowring, 1968) codes, although for BWR conditions, other formulations (Forti and Gonzalez-Santalo, 1973) may be preferable.

Before discussing the details of the various equations used in subchannel analysis, we will review the trends seen in the available data. Indeed, any effective subchannel technique must have analytical models capable of accurately predicting these data trends. A review of the available data (Lahey and Schraub, 1969) has indicated that there is an observed tendency for the vapor to seek the less obstructed, higher velocity regions of a BWR fuel rod bundle. This tendency can be seen in the adiabatic data shown in Fig. 4-53, where it is noted that the flow quality is much higher in the more open interior (center) subchannels than in the corner and side subchannels. This indicates the presence of a thick liquid film on the channel wall and an apparent affinity of the vapor for the more open subchannels. Diabatic

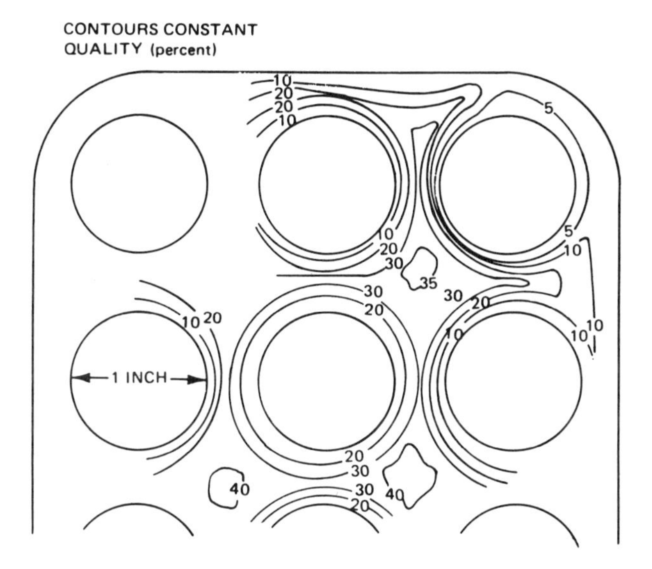

Fig. 4-53 Quality contours from isokinetic probe sampling of air-water flow in a nine-rod array (Schraub et al., 1969).

subchannel data (Lahey et al., 1971; Lahey et al., 1972) have confirmed this observation. The isokinetic steam/water data presented in Fig. 4-54 clearly show that even though the power-to-flow ratio of the corner subchannel is the highest of any subchannel in the bundle, the quality in the corner subchannel is the lowest, while that in the center subchannels is the highest. Moreover, the enhanced turbulent two-phase mixing that occurs near the slug-annular transition point (~10% quality at 1000 psia) is seen clearly in Figs. 4-54a and b. As shown in Fig. 4-55, the dependence of mixing on flow regime has also been observed by other investigators (Rowe and Angle, 1969).

It is interesting to speculate on the reasons for the observed data trends and, in particular, the observed distribution of subchannel void fraction. We can gain valuable insight by considering the data trends seen in experiments performed in much simpler geometries than rod bundles. As shown in Fig. 4-56, the data trends seen in adiabatic and diabatic experiments in eccentric annuli (Schraub et al., 1969; Shiralkar, 1970) show the same basic trend. That is, the vapor apparently has a strong affinity for the more open, higher velocity regions. It is important to note that this occurs for both developed adiabatic two-phase flows and developing diabatic two-phase flows. Hence, even in the developing situation, there is apparently a strong trend toward the equilibrium void distribution. This trend should not be too surprising since the same thing can be observed

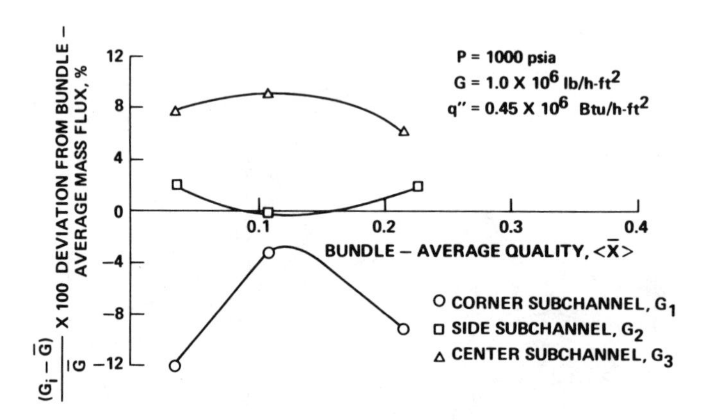

Fig. 4-54a Comparison of subchannel flows for the three subchannels (Lahey et al., 1971).

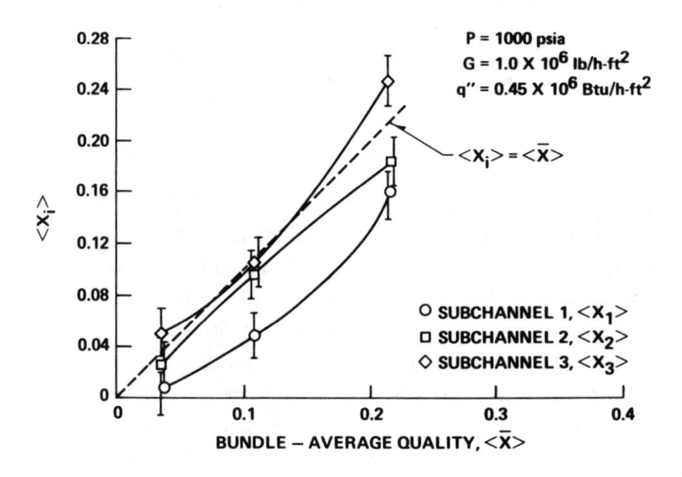

Fig. 4-54b Variation of subchannel qualities with average quality for three subchannels (Lahey et al., 1971).

in a tube during annular two-phase flow. That is, the vapor "drifts" to the higher velocity (center-line) region of the tube, while the liquid flows primarily as a film on the tube wall.

A complete understanding and quantification of this "void drift" phenomenon in two-phase flow remains one of the fundamental unsolved problems in two-phase flow today. Nevertheless, to develop accurate subchannel techniques, approximate void drift models must be synthesized.

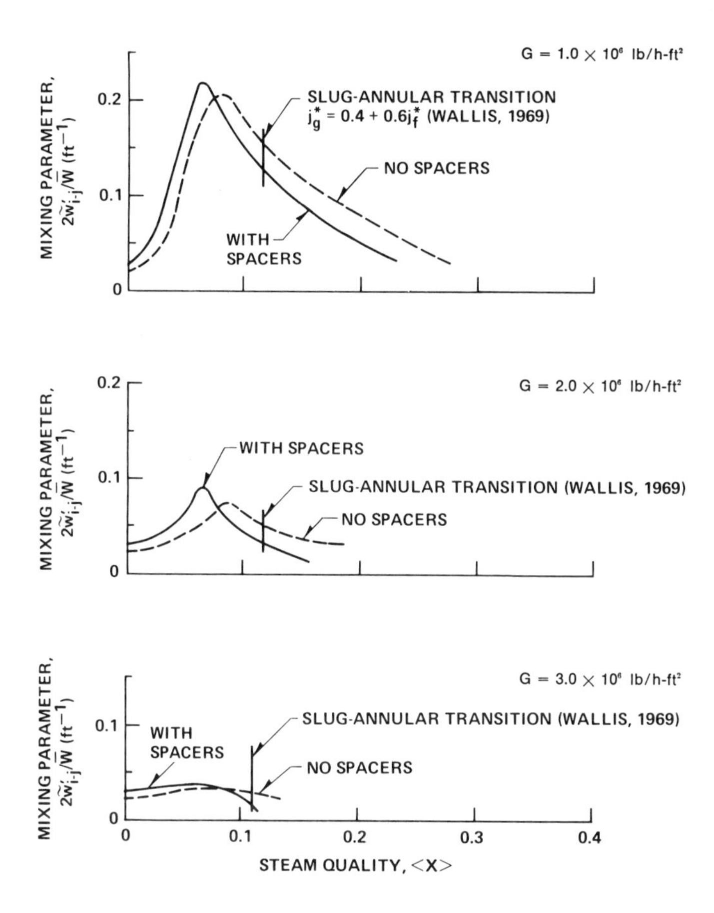

Fig. 4-55 The effect of flow regime on turbulent two-phase mixing (Rowe and Angle, 1969).

Now that the motivation for subchannel analysis has been established and the observed data trends discussed, we turn our attention to the development of a detailed analytical model. The specific model that is derived here is representative of current generation models that exist in the literature. To simplify this development and still retain the essence of the physics involved, a steady-state model is developed. The extension of this model to transient subchannel analysis is rather straightforward although, as might be expected, the numerics may become a problem.

In the derivation of our model, the basic approach is to consider the conservation of mass, momentum, and energy in each subchannel. Figure 4-52 shows that not only do axial effects need to be considered, but also the transverse (radial) interchange of mass, momentum, and energy across the imaginary interfaces dividing subchannels. These transverse inter-

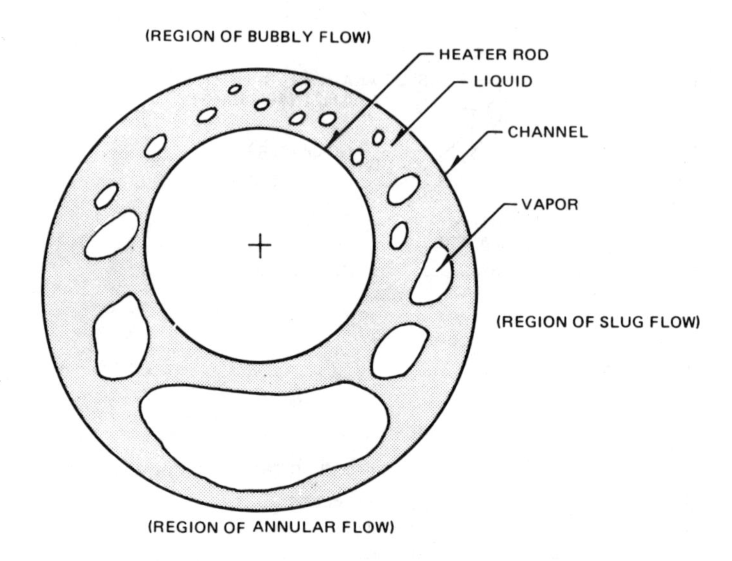

Fig. 4-56 Typical data trends seen in eccentric annuli.

changes are, in fact, the unique feature of subchannel analysis. The transverse exchange phenomena are quite complicated and difficult to decompose into more elementary interchange terms. Nevertheless, they normally are decomposed arbitrarily into three components:

1. flow diversions that occur due to imposed transverse pressure gradients
2. turbulent (eddy diffusivity) mixing that occurs due to stochastic pressure and flow fluctuations
3. lateral "void drift" that, as discussed previously, is apparently due to the strong tendency of the two-phase system to approach an equilibrium phase distribution.

Although this subdivision is arbitrary and not unique, it has proved useful in the development of subchannel models.

In the most general formulation, we must consider the conservation equations for mass, momentum, and energy in both the axial and transverse directions. Fortunately, in BWR-type fuel rod bundles, the rod-to-rod spacing is so large that only negligible transverse pressure gradients can exist (Lahey et al., 1970). Thus, to a good first approximation, we can set transverse pressure gradients equal to zero and solve the resulting set of sub-

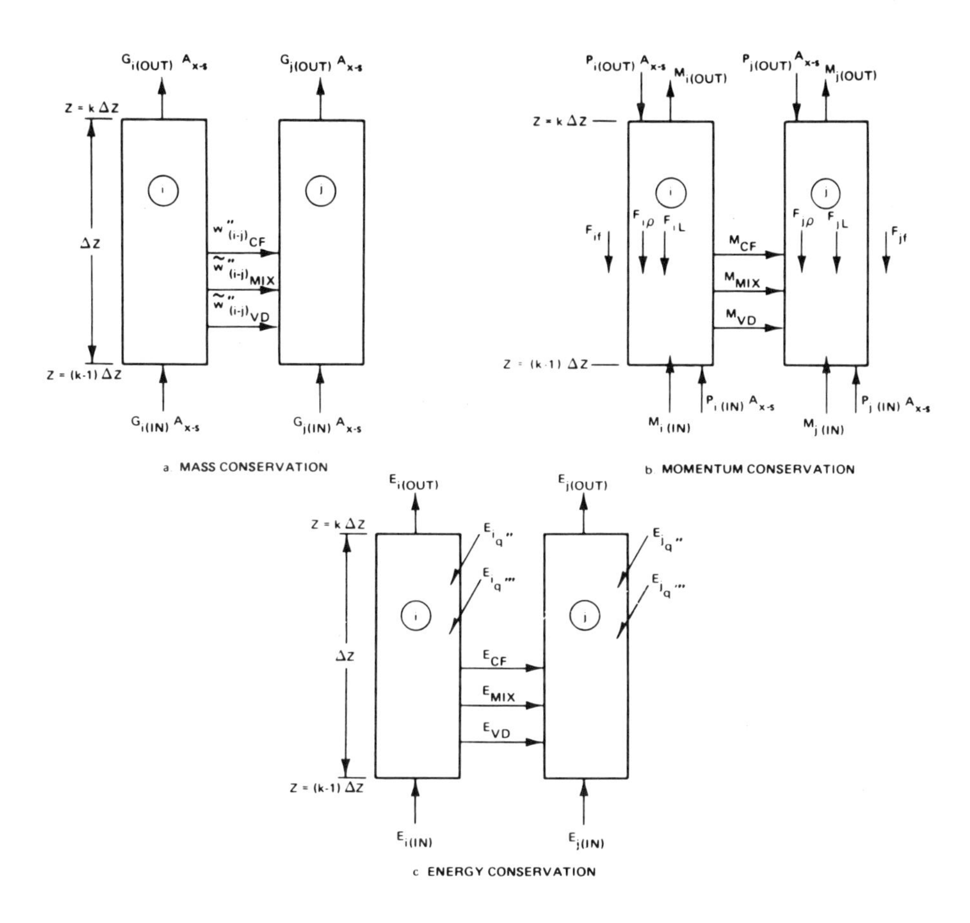

Fig. 4-57 Nodal conservation laws.

channel conservation equations with the physical constraint of no net circulation[g] around fuel rods.

The basic subchannel conservation equation can be formulated in various ways. Either we can write the appropriate conservation equations on a nodal basis, or we can finite difference the differential form of the appropriate conservation equations. For simplicity, the nodal derivation is presented here.

Consider axial node, k, of subchannel, i, shown in Fig. 4-57a. By applying the basic steady-state conservation principle—the inflows and outflows must be equal—for the nodal continuity equation, we obtain,

[g]Required to achieve unique solutions for the assumption of no transverse pressure gradients.

$$G_{i(\text{in})}A_{x-s} = G_{i(\text{out})}A_{x-s} + [(w''_{i-j})_{\text{CF}} + (\tilde{w}''_{i-j})_{\text{mix}} + (\tilde{w}''_{i-j})_{\text{VD}}]S_{i-j}\Delta z \ , \quad (4.119)$$

where the subscripts CF, mix, and VD denote diversion cross flow, mixing, and void drift, respectively. These three interaction terms are a unique feature of subchannel analysis. If flow tube, i, were not ventilated (i.e., if it does not communicate with flow tube, j), then Eq. (4.119) would yield the well-known result that the axial flow rate is constant for steady-state.

Next we consider momentum conservation in axial node, k, of subchannel, i, as shown in Fig. 4-57b. By applying Newton's Second Law, we obtain the nodal momentum conservation equation,

$$[p_{i(\text{in})} - p_{i(\text{out})}]A_{x-s} - F_{if} - F_{i\rho} - F_{iL} = M_{i(\text{out})} + M_{\text{CF}} + M_{\text{mix}} + M_{\text{VD}} - M_{i(\text{in})} \ , \quad (4.120)$$

where the two-phase friction pressure drop is given by,

$$F_{if}/A_{x-s} \triangleq \frac{f_i \Delta z}{D_{H_i}} \frac{G_i^2}{2g_c \rho_f} \phi_{lo}^2 \ , \quad (4.121)$$

the density head pressure drop by,

$$F_{i\rho}/A_{x-s} \triangleq \frac{g}{g_c} \langle \bar{\rho}_i \rangle \Delta z \ , \quad (4.122)$$

the local (e.g., grid spacer) pressure drop, if any, by,

$$F_{iL}/A_{x-s} \triangleq K_L \frac{G_i^2}{2g_c \rho_f} \left(1 + \frac{v_{fg}}{v_f} \langle x \rangle_i \right) \ , \quad (4.123)$$

the node exit and inlet momentum by,

$$M_{i(\text{out})} \triangleq \frac{G_{i(\text{out})}^2 A_{x-s}}{g_c \langle \rho'_{i(\text{out})} \rangle} \quad (4.124)$$

$$M_{i(\text{in})} \triangleq \frac{G_{i(\text{in})}^2 A_{x-s}}{g_c \langle \rho'_{i(\text{in})} \rangle} \ , \quad (4.125)$$

and the momentum transfer due to transverse flow diversion, mixing, and void drift by,

$$M_{\text{CF}} \triangleq \frac{(w''_{i-j})_{\text{CF}} S_{i-j} \Delta z}{2g_c} \left(\frac{G_i}{\langle \rho'_i \rangle} + \frac{G_j}{\langle \rho'_j \rangle} \right) \quad (4.126)$$

$$M_{\text{mix}} \triangleq \tau_{\text{Re}} S_{i-j} \Delta z \quad (4.127)$$

$$M_{\text{VD}} \triangleq \tau_{\text{VD}} S_{i-j} \Delta z \ . \quad (4.128)$$

The assumption of no transverse pressure gradient [i.e., $p_{i(\text{out})} = p_{j(\text{out})}$], the criterion of no net circulation, and the choice of an appropriate void-quality model (Chapter 5) yield a well-posed momentum equation.

The notation and models used in Eqs. (4.121) through (4.128) may be unfamiliar to some readers. These are discussed in Chapter 5, which can be referred to if clarification is required. The models for turbulent Reynolds stress, τ_{Re}, and the shear stress due to void drift, τ_{VD}, are discussed subsequently in this section.

The final nodal conservation principle that must be employed is the First Law of Thermodynamics; i.e., the conservation of energy. For axial node, k, and subchannel, i, Fig. 4-57c yields,

$$E_{i(\text{in})} + E_{i_{q''}} + E_{i_{q'''}} = E_{i(\text{out})} + E_{CF} + E_{\text{mix}} + E_{VD} \ , \tag{4.129}$$

where the convected energy flows are given by,

$$E_{i(\text{in})} \stackrel{\Delta}{=} G_{i(\text{in})} A_{x-s} h_{i(\text{in})} \tag{4.130}$$

$$E_{i(\text{out})} \stackrel{\Delta}{=} G_{i(\text{out})} A_{x-s} h_{i(\text{out})} \ . \tag{4.131}$$

The energy input to node k due to heat addition from the heated surfaces is given by,

$$E_{i_{q''}} \stackrel{\Delta}{=} q'' P_H \Delta z \ , \tag{4.132}$$

and by internal generation (i.e, gamma heating, etc.),

$$E_{i_{q'''}} \stackrel{\Delta}{=} q''' A_{x-s} \Delta z \ . \tag{4.133}$$

The energy transfers due to transverse flow diversion, mixing, and void drift are given by,

$$E_{CF} \stackrel{\Delta}{=} (w''_{i-j})_{CF} S_{i-j} \Delta z \ h_{CF} \tag{4.134}$$

$$E_{\text{mix}} \stackrel{\Delta}{=} \tilde{q}''_{\text{mix}} S_{i-j} \Delta z \tag{4.135}$$

$$E_{VD} \stackrel{\Delta}{=} \tilde{q}''_{VD} S_{i-j} \Delta z \ . \tag{4.136}$$

The normal choice for the cross-flow enthalpy, h_{CF}, is the enthalpy of the donor subchannel, that is, the average enthalpy of the subchannel from which the fluid is diverted. As shown in Fig. 4-58, the available data indicate that a more appropriate choice would be a flow regime dependent cross-flow enthalpy, which is in general larger than that of the donor subchannel.

The pseudo-heat-flux terms for energy exchanges due to turbulent mixing, $\tilde{q}''_{\text{mix}}$, and void drift, $\tilde{q}''_{VD}$, are very important terms. In fact, they represent the key to effective subchannel analyses. Hence, it is appropriate to consider models for the various mixing and void drift phenomena that have been postulated.

Two-phase turbulent mixing is frequently formulated as an extension of single-phase mixing. Hence, first we consider the nature of turbulent mixing in single-phase flows. The expression for the turbulent mass interchange divided by the mean subchannel flow rate, $\overline{W}$, can be expressed as,

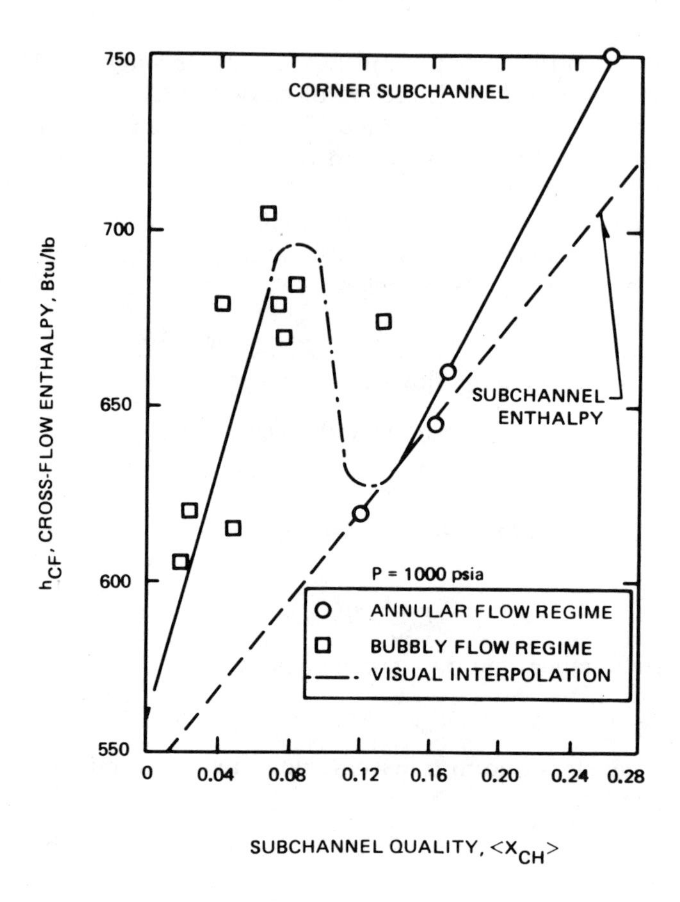

Fig. 4-58 Cross-flow enthalpy versus quality of donor subchannel (Lahey et al., 1971).

$$\frac{\tilde{w}'_{i-j}}{\overline{W}} = \frac{2\tilde{G}_{i-j}S_{i-j}}{(G_i + G_j)A_{x-s}} = \frac{\tilde{\beta}S_{i-j}}{A_{x-s}} , \tag{4.137}$$

where $\tilde{\beta}$ is recognized as the mixing Stanton number and S_{i-j} is the gap between subchannels i and j. The fluctuating transverse mass flux, $\tilde{G}_{i-j}$, is defined in terms of mixing length theory as,

$$\tilde{G}_{i-j} = \rho_l \tilde{u}_{i-j} = \rho_l l \frac{d\tilde{u}}{dy} = \rho_l \frac{\left(l^2 \dfrac{d\overline{u}}{dy}\right)}{l} = \frac{\rho_l \varepsilon}{l} . \tag{4.138}$$

The single-phase eddy diffusivity, ε, implicitly defined in Eq. (4.138) has been expressed by many independent investigators (Rogers and Todreas, 1968) as,

$$\varepsilon/\mu_l = K_1 \text{ Re } (f/2)^{1/2} = K_1 \left(\frac{GD_{H_{\text{ave}}}}{\rho_l \mu_l}\right) (f/2)^{1/2} \ . \tag{4.139}$$

By combining Eqs. (4.137), (4.138), and (4.139), we obtain,

$$\frac{\tilde{w}'_{i-j}}{\overline{W}} = K_1 \left(\frac{S_{i-j}}{l}\right) \frac{D_{H_{\text{ave}}}}{A_{x-s}} (f/2)^{1/2} \ . \tag{4.140}$$

The choice of what to use for the subchannel mixing length, l, is still debatable. By averaging Reichardt's data (Reichardt, 1951) over the cross-sectional flow area, we obtain,

$$l = K_2 D_{H_{\text{ave}}} = \frac{K_2}{2} (D_{H_i} + D_{H_j}) \ , \tag{4.141}$$

which, from Eq. (4.140), implies that the turbulent mixing, $\tilde{w}'_{i-j}/\overline{W}$, is directly proportional to the gap spacing, S_{i-j}. In contrast, other experimenters (Rowe and Angle, 1969) have suggested that the secondary flow patterns in the gap region are responsible for the scale of the turbulence and that the proper choice of mixing length is,

$$l = K_3 S_{i-j} \ , \tag{4.142}$$

which implies that the turbulent mixing, $\tilde{w}'_{i-j}/\overline{W}$, is completely independent of gap spacing but directly proportional to the mean hydraulic diameter. Finally, other studies (Rogers and Rosehart, 1972) have indicated that the mixing length should be a function of both gap spacing and hydraulic diameter. The true dependence of mixing length on geometric parameters is still to be established.

With such a divergence of opinion concerning the correct modeling of single-phase mixing in fuel rod bundles, it should not come as a great surprise that the more complicated case of two-phase mixing is not understood completely. Nevertheless, models based on realistic assumptions can be derived and employed in subchannel analysis.

First we consider turbulent mixing in two-phase systems. One way to conceptualize the process is, as shown in Fig. 4-59, to assume that all transfers are due to "globs" of fluid from one subchannel being exchanged with globs of fluid from the neighboring subchannels. In single-phase mixing, there is usually no net mass transfer due to this process since the density in neighboring subchannels is equal. In contrast, in two-phase systems, substantial mass transfer has been observed (Gonzalez-Santalo, 1971) and can be modeled as an exchange of globs of equal volume but of different density. This volume-for-volume exchange model implies that the

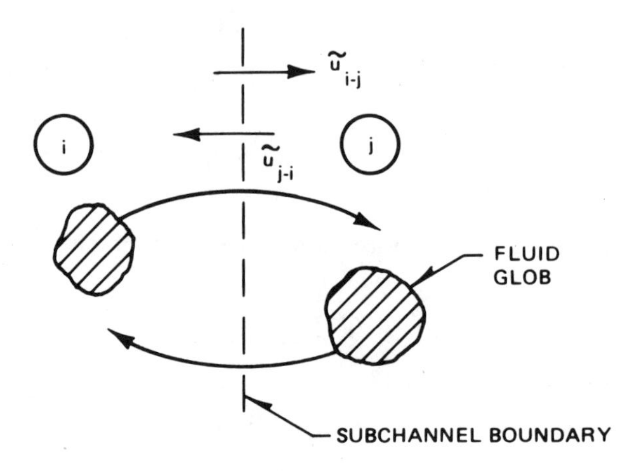

Fig. 4-59 A phenomenological model for two-phase mixing.

transverse fluctuating velocities for the vapor and liquid phase are equal and that,

$$\tilde{u}_{i-j}=\tilde{u}_{j-i}=\frac{\varepsilon}{l} \ . \tag{4.143}$$

The net fluctuating transverse mass flux is then expressed by,

$$\tilde{w}''_{i-j}=\frac{\varepsilon\rho_l}{l}[(1-\langle\alpha_i\rangle)-(1-\langle\alpha_j\rangle]+\frac{\varepsilon\rho_g}{l}(\langle\alpha_i\rangle-\langle\alpha_j\rangle) \ , \tag{4.144}$$

or,

$$\tilde{w}''_{i-j}=(\tilde{w}''_{i-j})_l+(\tilde{w}''_{i-j})_g \ , \tag{4.145}$$

where,

$$(\tilde{w}''_{i-j})_l \triangleq \frac{\varepsilon\rho_l}{l}(\langle\alpha_j\rangle-\langle\alpha_i\rangle) \tag{4.146}$$

$$(\tilde{w}''_{i-j})_g \triangleq \frac{\varepsilon\rho_g}{l}(\langle\alpha_i\rangle-\langle\alpha_j\rangle) \ . \tag{4.147}$$

Equation (4.144) also can be written in a more familiar form as,

$$\tilde{w}''_{i-j}=\frac{\varepsilon}{l}(\langle\overline{\rho}_i\rangle-\langle\overline{\rho}_j\rangle) \ , \tag{4.148}$$

which clearly implies that the mass transfer due to mixing is directly proportional to the transverse density gradient, $d\langle\overline{\rho}\rangle/dy$.

Phenomenologically, the energy transfer due to two-phase mixing, $\tilde{q}''_{mix}$, is due to the exchange of a volume of vapor in one subchannel for an equal volume of liquid in a neighboring subchannel. This process can be modeled easily using the results in Eq. (4.145). For instance, for conditions of bulk boiling,

$$\tilde{q}''_{mix} = (\tilde{w}''_{i-j})_f h_f + (\tilde{w}''_{i-j})_g h_g \ . \tag{4.149}$$

By using Eqs. (4.146) and (4.147), Eq. (4.149) can be expressed as,

$$\tilde{q}''_{mix} = \frac{\varepsilon}{l}(\rho_f h_f - \rho_g h_g)(\langle\alpha_j\rangle - \langle\alpha_i\rangle) \ . \tag{4.150}$$

Equation (4.150) can also be written as,

$$\tilde{q}''_{mix} = \frac{\varepsilon}{l}(\langle\overline{\rho}_i\rangle\langle\overline{h}_i\rangle - \langle\overline{\rho}_j\rangle\langle\overline{h}_j\rangle) \ , \tag{4.151}$$

where the center-of-mass enthalpy is given by,

$$\langle\overline{h}_i\rangle \triangleq \frac{1}{\langle\overline{\rho}_i\rangle}[\rho_f h_f(1 - \langle\alpha_i\rangle) + \rho_g h_g\langle\alpha_i\rangle] \ .$$

Equation (4.151) indicates that the energy transfer due to mixing is directly proportional to the mass-weighted transverse enthalpy gradient, $d(\langle\overline{\rho}\rangle\langle\overline{h}\rangle)/dy$.

Note that for the case in which $\langle\alpha_j\rangle$ is greater than $\langle\alpha_i\rangle$, there is a positive net mass transfer from subchannel i to subchannel j. That is, Eq. (4.144) implies,

$$\tilde{w}''_{i-j} = \frac{\varepsilon}{l}(\rho_f - \rho_g)(\langle\alpha_j\rangle - \langle\alpha_i\rangle) \ . \tag{4.152}$$

Equation (4.150) indicates a positive net energy exchange from subchannel i to subchannel j, even though subchannel j has a higher void fraction. This is because $\rho_f h_f > \rho_g h_g$ and, thus, the energy transported from subchannel i to j, in liquid form, more than offsets the energy transferred from j to i, in vapor form. This does not mean that mixing tends to increase further the void fraction in subchannel j, since the equivalent net mixing enthalpy, given from Eqs. (4.150) and (4.152) by,

$$\frac{\tilde{q}''_{mix}}{\tilde{w}''_{i-j}} = \frac{(\rho_f h_f - \rho_g h_g)}{(\rho_f - \rho_g)} \ , \tag{4.153}$$

is always less than the saturation enthalpy, h_f. Thus, as expected, mixing cools off the hot subchannel, j, since the mass exchange overshadows the energy exchange.

Now we determine the interfacial shear stress, i.e., the Reynolds stress, due to the mixing process. By using Eq. (4.143), the two-phase Reynolds stress can be written as,

$$\tau_{\mathrm{Re}} = -\frac{\varepsilon}{g_c l}\rho_g(\langle\alpha_i\rangle\langle u_g\rangle_{g_i} - \langle\alpha_j\rangle\langle u_g\rangle_{g_j}) - \frac{\varepsilon}{g_c l}\rho_l[(1-\langle\alpha_i\rangle)\langle u_l\rangle_{l_i} - (1-\langle\alpha_j\rangle)\langle u_l\rangle_{l_j}] \ . \tag{4.154}$$

Equation (4.154) can be regrouped and written more compactly as,

$$\tau_{\mathrm{Re}} = \frac{\varepsilon}{g_c l}(G_j - G_i) \ , \tag{4.155}$$

which is easily recognized as the classical Reynolds stress term.

All the necessary mixing terms have now been formulated. Unfortunately, these models alone will not allow us to predict the observed data trends; e.g., Figs. 4-54 and 4-56. That is, the magnitude of turbulent mixing can be varied from zero to as large as desired and the observed data trends will not be predicted. In fact, even infinite mixing fails to produce the required results since, as seen in Eq. (4.148), infinite mixing implies that subchannel void fractions must be the same for finite $\tilde{w}''_{i-j}$. Moreover, the same equation implies that for any finite fluctuating velocity, ε/l, the void distribution tends toward a uniform radial distribution rather than the observed nonuniform equilibrium distribution. Clearly, some important physical mechanism has been left out of our mixing model.

To develop the required phenomenological model, we first hypothesize that net two-phase turbulent mixing is proportional to the nonequilibrium void fraction gradient. This hypothesis implies that there is a strong trend toward an equilibrium distribution and that when this state has been achieved, the net exchange due to mixing ceases. It is convenient, but not necessary, to partition the net turbulent mass transfer into an eddy diffusivity mixing part and a void drift part, such that Eq. (4.148) can be written as,

$$\tilde{w}''_{i-j} = (\tilde{w}''_{i-j})_{\mathrm{mix}} + (\tilde{w}''_{i-j})_{\mathrm{VD}} = \frac{\varepsilon}{l}(\langle\bar{\rho}_i\rangle - \langle\bar{\rho}_j\rangle) - \frac{\varepsilon}{l}(\langle\bar{\rho}_i\rangle - \langle\bar{\rho}_j\rangle)_{\mathrm{EQ}} \ , \tag{4.156}$$

where the subscript, EQ, denotes the equilibrium lateral phase distribution. Note that net mass transfer due to mixing ceases when equilibrium conditions are achieved.

The net turbulent energy exchange can be written immediately in the form of Eq. (4.150) as,

$$\tilde{q}''_{\mathrm{net}} = \tilde{q}''_{\mathrm{mix}} + \tilde{q}''_{\mathrm{VD}} = \frac{\varepsilon}{l}(\rho_f h_f - \rho_g h_g)[(\langle\alpha_j\rangle - \langle\alpha_i\rangle) - (\langle\alpha_j\rangle - \langle\alpha_i\rangle)_{\mathrm{EQ}})] \ . \tag{4.157}$$

Note that when the equilibrium distribution is achieved, net turbulent energy transfer ceases.

Finally, the net interfacial shear stress due to turbulence can be written in the same form as Eq. (4.155),

$$\tau_{\mathrm{net}} = \tau_{\mathrm{Re}} + \tau_{\mathrm{VD}} = \frac{\varepsilon}{g_c l}[(G_j - G_i) - (G_j - G_i)_{\mathrm{EQ}}] \ . \tag{4.158}$$

As before, when equilibrium is achieved, the driving potential for momentum exchange vanishes.

The void drift models given implicitly in Eqs. (4.156), (4.157), and (4.158) represent the equilibrium turbulent mass, energy, and momentum transfer terms, respectively. The models are logical and simple to apply, provided we know the equilibrium void fraction and mass flux distribution. Unfortunately, correlations and/or models for predicting two-phase equilibrium distributions are not readily available.

A simple model (Levy, 1963) for the equilibrium void fraction distribution has been employed previously with some success (Lahey et al., 1972). This model implies that the equilibrium density distribution is related to the mass flux distribution by the expression,

$$\langle\bar{\rho}\rangle_{EQ} = aG_{EQ} + b \ , \tag{4.159}$$

where the constants, a and b, can be evaluated from the boundary conditions,

$$\left.\begin{aligned} \langle\bar{\rho}\rangle_{EQ} &= \rho_l, \text{ at } G_{EQ} = 0 \\ \langle\bar{\rho}\rangle_{EQ} &= \langle\bar{\rho}_{ave}\rangle_{EQ}, \text{ at } G = G_{ave} \end{aligned}\right\} , \tag{4.160}$$

where "no slip" at the wall implies $G_{EQ} = 0$ and ave denotes the average of the various communicating subchannels.

The equilibrium density difference between subchannels i and j is then given from Eqs. (4.159) and (4.160) as,

$$(\langle\bar{\rho}_i\rangle - \langle\bar{\rho}_j\rangle)_{EQ} = -(\rho_l - \rho_g)\frac{\langle\alpha_{ave}\rangle}{G_{ave}}(G_i - G_j)_{EQ} \ , \tag{4.161}$$

where the equilibrium mass flux distribution can be taken as the existing mass flux distribution, to a good first appropriation. Although this simple model leaves much to be desired, it does allow us to predict the observed data trends (Lahey et al., 1972). Moreover, it is supported by mechanistic models (Drew and Lahey, 1979) and by data for the fully developed void distribution in rod bundles (Lahey, 1986).

Some insight into why Eq. (4.161) allows us to predict the correct data trends is given by considering a typical case, such as the corner subchannel, 1, shown in Fig. 4-52b. The data shown in Fig. 4-54 indicate that the density is the highest (i.e., the quality is the lowest) and that the mass flux is the lowest in the corner subchannel. Equation (4.161) implies that since $(G_1 - G_2)$ is negative, $(\langle\bar{\rho}_1\rangle - \langle\bar{\rho}_2\rangle)_{EQ}$ is positive and, thus, $(\langle\alpha_2\rangle - \langle\alpha_1\rangle)_{EQ}$ is also positive. This agrees with the observed data trends. However, equilibrium conditions are never achieved in the diabatic case under consideration. Equations (4.146) and (4.147) indicate that while turbulent mixing causes a net flow of liquid from subchannel 1 to subchannel 2, and a net flow of vapor from subchannel 2 to subchannel 1, void drift works in the opposite direction. Thus, in this case, void drift and turbulent mixing oppose each other; one is trying to pump vapor out of subchannel 1, while the other is

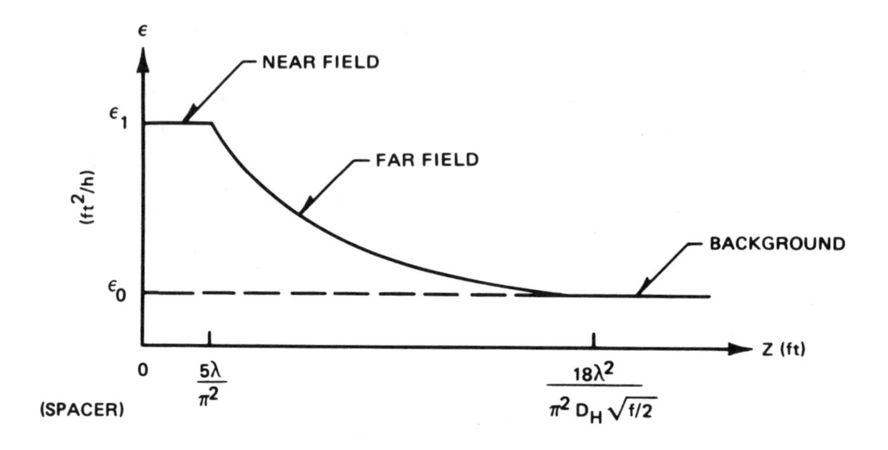

Fig. 4-60 Spacer downstream eddy enhancement model.

trying to pump it in. In the future, it is anticipated that a more physically based model for the equilibrium distribution will be developed. However, for the present, it should be clear that void drift, or something like it, is the key to accurate subchannel predictions.

The models discussed so far allow us to do a reasonably good job of predicting the observed flow and enthalpy distribution in a "clean" rod bundle (i.e., a rod bundle without grid-type spacers). However, in practice all BWR nuclear fuel rod bundles have grid spacers to maintain rod-to-rod clearance. To handle the effect of typical grid spacers, a model to quantify the eddy diffusivity enhancement observed downstream of these spacers must be employed. Such a model is shown in Fig. 4-60, in which λ is some characteristic dimension of the particular spacer under consideration. This simple model has been shown (Lahey et al., 1972) to do a good job of capturing the data trends seen in subchannel measurements taken at various distances downstream of typical grids.

An important potential application for subchannel analysis is in the area of fluid-to-fluid modeling of BT phenomena, which is considered in the next section.

4.6.1 Fluid-to-Fluid Modeling

To determine the thermal performance of nuclear reactor fuel rod bundles, large simulation experiments, such as those discussed in Sec. 4.4.2, are normally required. Since most modern nuclear reactors use water as the coolant and moderator, simulation experiments in large high-pressure steam-water test facilities are indicated. Unfortunately, these test facilities and experiments are normally quite expensive and difficult to perform. There are, in fact, only a few laboratories in the world that have the electrical power required for such experiments. Even when adequate facilities are available, it is much less expensive to conduct preliminary screening tests using other fluids. Because of the desire for less expensive experiments, there has been significant recent interest in the development of modeling relations to allow water conditions to be simulated by a fluid with a lower latent heat. Due to their low latent heat, safety, and well-established fluid properties, the fluids that typically have been chosen for this modeling are members of the fluorocarbon family (Freons).

The first attempts to model water BT phenomena with Freon were initiated in the United Kingdom (Barnett, 1963, 1964, 1965). These attempts were concerned with the identification of the appropriate dimensionless groups. Later, Stevens and Kirby (1964) developed a successful empirical correlation with which Freon-water modeling could be accomplished in simple geometries such as tubes and annuli. The current interest and confidence in Freon-water modeling of BT followed from this early work.

More recently, other investigators, Dix (1970), Bouré (1970), and Ahmad and Groeneveld (1972), have presented empirical modeling relations. In essence, all of these models are quite similar and involve scaling the mass flux, inlet subcooling, and pressure of the two fluids. A simple derivation that produces scaling laws typical of those currently used is presented to provide the reader with a more in-depth understanding of fluid-to-fluid modeling.

In Sec. 4.4.2, it was shown that for BWR technology, critical quality was of prime importance in determining BT. Consider a steady-state energy balance on a heated section having uniform axial heat flux. Equating the energy inflows to outflows,

$$GA_{x-s}h_{\text{in}} + \overline{q''}A_{\text{HT}} = GA_{x-s}(h_f + \langle x_e \rangle h_{fg}) \quad . \tag{4.162}$$

This equation can be rearranged to yield the following expression for quality,

$$\langle x_e \rangle = \left(\frac{\overline{q''}}{Gh_{fg}} \right) \frac{A_{\text{HT}}}{A_{x-s}} - \frac{\Delta h_{\text{sub}}}{h_{fg}} \quad , \tag{4.163}$$

where,

$$\Delta h_{\text{sub}} \stackrel{\Delta}{=} h_f - h_{\text{in}} \quad .$$

Thus, to match the quality profile, we must match the dimensionless parameters, q''/Gh_{fg} and $\Delta h_{\text{sub}}/h_{fg}$ in the two fluids. In particular, for critical conditions, these parameters must be matched; however, other system and thermodynamic parameters also must be considered. Over the years, experiments have indicated that critical quality is a function of the following key parameters,

$$\langle x_e \rangle_c = F_1\left(\frac{\overline{q''_c}}{Gh_{fg}}, \frac{\Delta h_{\text{sub}}}{h_{fg}}, L_B, D_H, \rho_f, \rho_g, \sigma, G, g_c \right) \quad . \tag{4.164}$$

Other parameters, not considered, lead to an imperfect modeling relation. However, for most practical applications, the parameters given in Eq. (4.164) can be used to establish similarity for quasi-steady-state conditions.

Note that there are seven dimensional parameters in Eq. (4.164) and four characteristic dimensions: length, L, mass, M, force, F, and time, t. Thus, the Buckingham Pi theorem (Murphy, 1950) states that we can expect four independent dimensionless groups. By applying this method to Eq. (4.164),

$$\pi \sim L^a L^b (ML^{-3})^c (ML^{-3})^d (FL^{-1})^e (Mt^{-1}L^{-2})^f (MF^{-1}Lt^{-2})^g \quad . \tag{4.165}$$

Since quality is nondimensional, the sum of the exponents for each characteristic dimension must separately add up to zero:

$$\begin{aligned} &(\text{For } L) \quad a+b-3c-3d-e-2f+g=0 \\ &(\text{For } M) \quad c+d+f+g=0 \\ &(\text{For } F) \quad e-g=0 \\ &(\text{For } t) \quad -f-2g=0 \ . \end{aligned}$$

There are seven unknown exponents and only four equations; thus, any three exponents may be chosen arbitrarily to determine the desired dimensionless parameters.

First we choose $g=0$, $b=0$, and $c=1$. Then, the system of equations above implies,

$$\pi_1 = \rho_f / \rho_g \quad . \tag{4.166}$$

This makes sense physically since the slip ratio (i.e., the ratio of vapor phase velocity to the liquid phase velocity) is closely related to the density ratio (see Chapter 5). We expect that it is important to model the slip ratio in the two fluids accurately since slip is strongly related to entrainment and entrainment is a very important mechanism in BT.

Next we choose $g=0$, $c=0$, and $a=1$. This choice implies that,

$$\pi_2 = L_B / D_H \ , \tag{4.167}$$

which is a statement of geometric similarity. Finally, we choose $b=1$, $d=1$, and $e=-1$ to obtain,

$$\pi_3 = \left(\frac{G^2 D_H}{g_c \rho_g \sigma} \right) \quad . \tag{4.168}$$

This can be recognized as the Weber number, which is known to be of importance in the entrainment process.

The three nondimensional groups now have been established and, thus, Eq. (4.164) can be rewritten as,

$$\langle x_e \rangle_c = F_2\left(\frac{\overline{q_c''}}{Gh_{fg}}, \frac{\Delta h_{\text{sub}}}{h_{fg}}, \frac{L_B}{D_H}, \frac{\rho_f}{\rho_g}, \frac{G^2 D_H}{g_c \rho_g \sigma}\right) . \tag{4.169}$$

When the five dimensionless parameters given in Eq. (4.169) have been matched in the two fluids, the thermal performance should be scaled appropriately. Note that equating the so-called steaming rate, $\overline{q_c''}/Gh_{fg}$, implies that much lower power (i.e., heat flux) is required since the latent heat of Freon is much less than that of water. In addition, the density ratio of Freon is matched to that of water at a much lower pressure. For instance, at 123 psia, the density ratio of Freon-114 matches that of water at 1000 psia. Finally, equating the Weber number for the two fluids implies a mass flux modeling law that is the key to the accurate scaling of BT,

$$G_{\text{Freom}} = G_{\text{water}}\left[\frac{(\sigma\rho_g)_{\text{Freon}}}{(\sigma\rho_g)_{\text{water}}}\right]^{1/2} . \tag{4.170}$$

For example, conditions of interest in BWR technology (1000 psia) imply that water BT results can be achieved in simple geometry test sections when the Freon-114 mass flux is ~62% that of water. It has been found (Dix, 1970) that a small empirical correction to Eq. (4.170) can improve the scaling. This minor correction is apparently a reflection of system parameters that were not considered in Eq. (4.164), and essentially validates our choice of the key parameters.

Although the modeling relations implicitly given in Eq. (4.169) are fairly accurate for simple geometry experiments, they normally are not adequate when applied on a global basis to full-scale rod bundles. That is, in general, we cannot use these modeling laws for bundle-average scaling between Freon and water. This is indeed unfortunate since reactor designers really are interested only in rod bundle scaling. The problem is fundamental (Torbeck, 1975) and basically is due to differences in fluid properties that, as shown in Fig. 4-61, cause the variation of the total two-phase pressure drop with quality to be different. That is, as the quality increases at constant mass flux and system pressure, the total two-phase pressure drop initially decreases because the density head drops off faster than friction increases. As quality continues to increase, the friction term dominates and the trend reverses; however, this reversal occurs at a lower quality in water than in Freon due to the differences in fluid density. This implies that the diversion cross flows in the various subchannels of a rod bundle can be quite different in water and Freon, particularly when nonuniform local (rod-to-rod) peaking patterns are involved.

The situation that might exist in a typical subchannel can be seen in Fig. 4-62. It is found that for typical conditions of interest, we might have net flow diversion into a particular subchannel in a Freon experiment, while

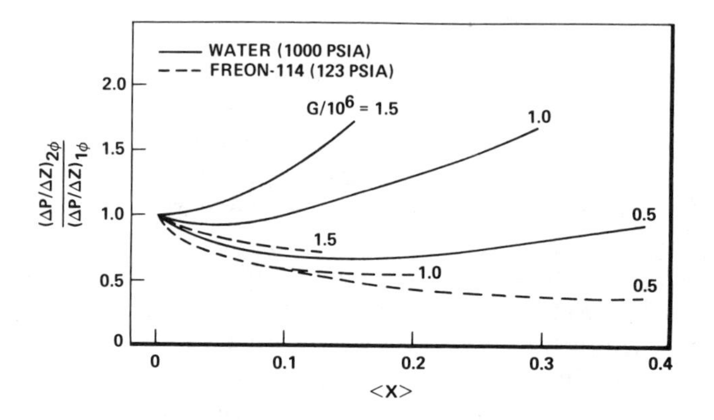

Fig. 4-61 Typical total pressure gradients.

in a corresponding water experiment, flow would be diverted out. The situation is even more aggravated when the bundle mass fluxes are scaled in accordance with Eq. (4.170). This implies that bundle-averaging scaling for BT does not produce hydrodynamic scaling in the various subchannels. It also explains why Freon tends to give higher thermal performance than an equivalent water bundle. This underscores the importance of accurate subchannel techniques, since it should be obvious that effective fluid-to-fluid modeling in rod bundles must be done on a subchannel basis.

When more understanding of subchannel phenomena is generally available, the cost of full-scale BT experiments can be reduced greatly through fluid-to-fluid modeling techniques such as those just discussed. The current state-of-the-art is that it is not yet possible to accomplish fluid independent subchannel modeling and, thus, only water results such as discussed in Sec. 4.4.2 are currently used in the design and licensing of modern BWRs.

References

Ahmad, S. Y., and D. C. Groeneveld, "Fluid Modeling of Critical Heat Flux in Uniformly Heated Annuli," *Progress in Heat and Mass Transfer*, Vol. VI, Pergamon Press, New York (1972).

Bailey, N.A., J. G. Collier, and J. C. Ralph, "Post Dryout Heat Transfer in Nuclear and Cryogenic Equipment," ASME preprint 73-HT-16, American Society of Mechanical Engineers (1973).

Barnett, P. G., "The Scaling of Forced Convection Boiling Heat Transfer," AEEW-R134, United Kingdom Atomic Energy Authority (1963).

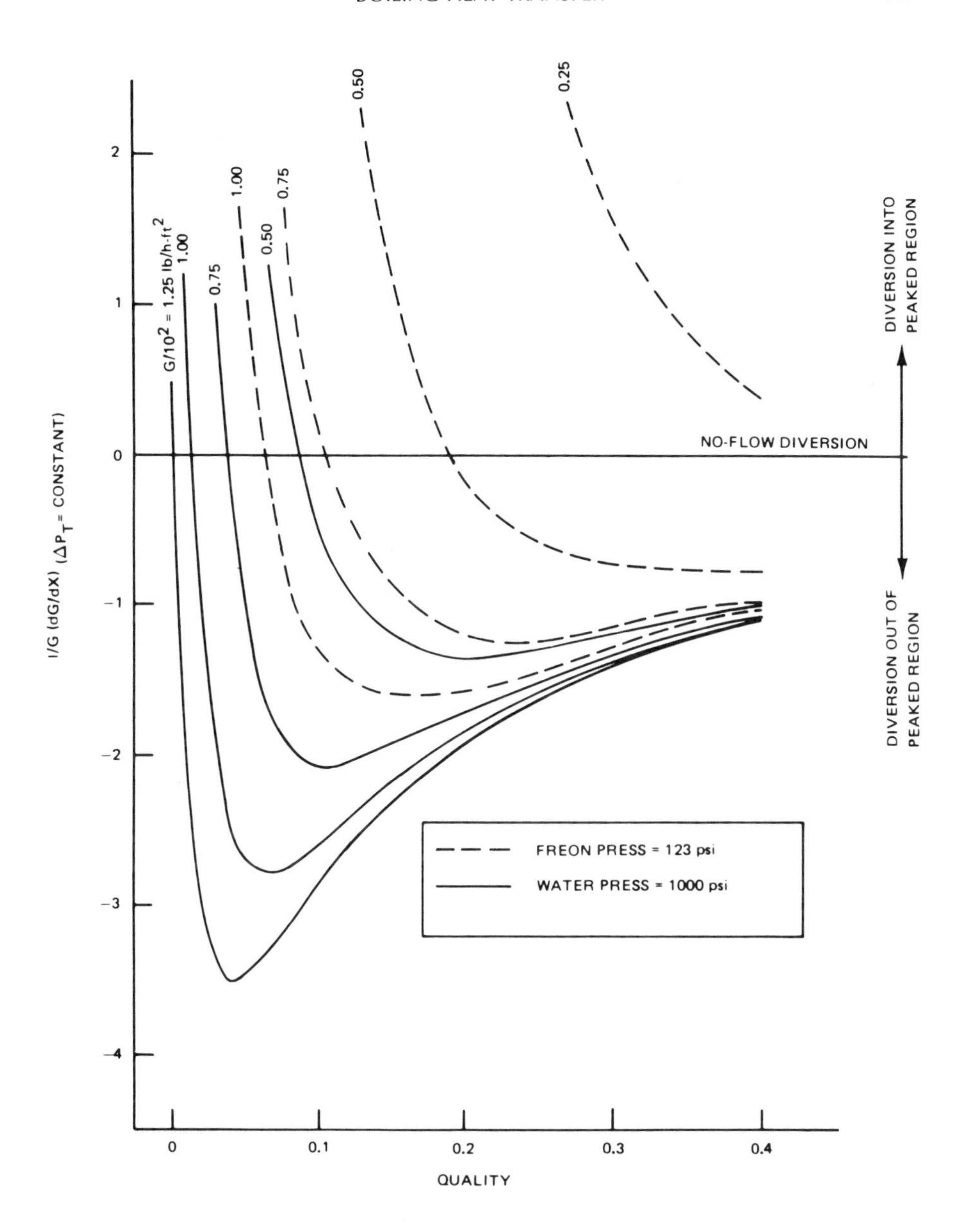

Fig. 4-62 Comparison of flow diversion parameter for water and Freon.

Barnett, P. G., "An Experimental Investigation to Determine the Scaling Law of Forced Convection Boiling Heat Transfer, Part I: The Preliminary Examination of Burnout Data for Water and Arcton-12," AEEW-R363, United Kingdom Atomic Energy Authority (1964).

Barnett, P. G., and R. W. Wood, "An Experimental Investigation to Determine the Scaling Laws of Forced Convection Boiling Heat Transfer, Part II: An Examination of Burnout Data for Water, Freon-12 and Freon-21 in Uniformly Heated Round Tubes," AEEW-R433, United Kingdom Atomic Energy Authority (1965).

Bennet, J. A. R., J. G. Collier, A. R. C. Pratt, and J. D. Thornton, "Heat Transfer in Two-Phase Gas-Liquid Systems, Part I: Steam-Water Mixtures in the Liquid-Dispersed Region in an Annulus," *Trans. Inst. Chem. Eng. (London),* **39** (1961).

Bennett, A. W., G. F. Hewitt, H. A. Kearsey, and R. K. F. Keeys, and D. J. Pulling, "Studies of Burnout in Boiling Heat Transfer to Water in Round Tubes with Non-Uniform Heating," AERE-R5076, Atomic Energy Research Establishment, Harwell (1966).

Bennett, A. W., G. F. Hewitt, H. A. Kearsey, and R. K. F. Keeys, "Heat Transfer to Steam-Water Mixtures Flowing in Uniformly Heated Tubes in Which the Critical Heat Flux has been Exceeded," AERE-R5375, Atomic Energy Research Establishment, Great Britain (1967).

Berenson, P. J., "Transition Boiling Heat Transfer from a Horizontal Surface," *J. Heat Trans.,* **83C**(3), 351–358 (1961).

Bertoletti, S., G. P. Gaspari, C. Lombardi, G. Peterlongo, M. Silvestri, and F. A. Tacconi, "Heat Transfer Crisis with Steam-Water Mixtures," *Energ. Nucl.,* **12,** *3* (1965).

Bishop, A. A., R. O. Sandberg, and L. S. Tong, "Forced Convection Heat Transfer at High Pressure After the Critical Heat Flux," ASME preprint 65-HT-31, American Society of Mechanical Engineers (1965).

Bouré, J. A., "A Method to Develop Similarity Laws for Two-Phase Flows," ASME paper 70-HT-25, American Society of Mechanical Engineers (1970).

Bowring, R. W., "HAMBO, A Computer Programme for the Subchannel Analysis of the Hydraulic and Burnout Characteristics of Rod Clusters—Part 2, The Equations," AEEW-R-582, United Kingdom Atomic Energy Authority (1968).

Chen, J. C., "A Correlation for Boiling Heat Transfer to Saturated Fluids in Convective Flow," ASME preprint 63-HT-34, American Society of Mechanical Engineers (1963).

Collier, J. G., *Convective Boiling and Condensation,* McGraw-Hill Book Company, New York (1972).

Davis, E. J., and G. H. Anderson, "The Incipience of Nucleate Boiling in Forced Convection," *AIChE J.,* **12** (1966).

De Bortoli, R. A., et al., "Forced Convection Heat Transfer Burnout Studies for Water in Rectangular Channels and Round Tables at Pressures Above 500 PSIA," WAPD-188, Westinghouse Electric Corporation (1958).

Dengler, C. E., and J. N. Addoms, *Chem. Eng. Prog. Ser.,* **52,** *18* (1956).

Dix, G. E., "Freon-Water Modeling of Critical Heat Flux in Round Tubes," NEDO-13091, General Electric Company (1970).

Dougall, R. S., and W. M. Rohsenow, "Flow Boiling on the Inside of Vertical Tubes with Upward Flow of the Fluid at Low Qualities," MIT Report 9079-26, Massachusetts Institute of Technology (1963).

Drew, D. A., and R. T. Lahey, Jr., "An Analytic Derivation of a Subchannel Drift-Flux Model," *ANS Trans.,* **53** (1979).

Forti, G., and J. M. Gonzalez-Santalo, "A Model for Subchannel Analysis of BWR Rod Bundles in Steady-State and Transient," *Proc. Int. Heat Transfer Conf.*, Karlsruhe, Germany (1973).

Gaspari, G. P., R. Granzini, and A. Hassid, "Dryout Onset in Flow Stoppage, Depressurization and Power Surge Transients," *Proc. CREST Specialist Mtg. Emergency Core Cooling for Light Water Reactors*, Munich, Germany (1972).

Gaspari, G. P., A. Hassid, and F. Lucchini, "A Rod-Centered Subchannel Analysis with Turbulent (Enthalpy) Mixing for Critical Heat Flux Prediction in Rod Clusters Cooled by Boiling Water," *Proc. Fifth Int. Heat Transfer Conf.*, paper B6.12 (1974).

"General Electric BWR Thermal Analysis Basis (GETAB): Data, Correlation and Design Application," NEDO-10958, General Electric Company (1973).

Gonzalez-Santalo, J. M., "Two-Phase Flow Mixing in Rod Bundle Subchannels," Ph.D. Thesis, Massachusetts Institute of Technology, Mechanical Engineering Department (1971).

Grace, T. M., "The Mechanism of Burnout in Initially Subcooled Forced Convective Systems," Ph.D. Thesis, University of Minnesota (1963).

Groeneveld, D. C., "The Thermal Behavior of a Heated Surface at and Beyond Dryout," AECL-4309, Atomic Energy of Canada Limited (1972).

Groeneveld, D. C., "Post-Dryout Heat Transfer at Operating Conditions," *Proc. Topl. Mtg. Water-Reactor Safety*, CONF-730304, U.S. Atomic Energy Commission (1973).

Hahne, E., and U. Grigull, *Heat Transfer in Boiling*, Hemisphere Publishing Corporation (1977).

Hatsopoulas, G. N., and J. H. Keenan, *Principles of General Thermodynamics*, John Wiley & Sons, New York (1965).

Healzer, J. M., J. E. Hench, E. Jannsen, and S. Levy, "Design Basis for Critical Heat Flux Condition in Boiling Water Reactors," APED-5286, General Electric Company (1966).

Heineman, J. B., "An Experimental Investigation of Heat Transfer to Superheated Steam in Round and Rectangular Channels," ANL-6213, Argonne National Laboratory (1960).

Hetsroni, G., *Handbook of Multiphase Systems*, Hemisphere Publishing Corporation (1982).

Hewitt, G. F., "Experimental Studies on the Mechanisms of Burnout in Heat Transfer to Steam-Water Mixtures," *4th IHTC*, Versailles, France, B6.6 (1970).

Hewitt, G. F., and N. S. Hall-Taylor, *Annular Two-Phase Flow*, Pergamon Press, New York (1970).

Hsu, Y-Y., and R. W. Graham, *Transport Processes in Boiling and Two-Phase Systems*, Hemisphere Publishing Corporation (1976).

Iloeje, O. C., D. N. Plummer, W. M. Rohsenow, and P. Griffith, "A Study of Wall Rewet and Heat Transfer in Dispersed Vertical Flow," MIT Report 72718-92, Massachusetts Institute of Technology (1974).

Janssen, E., and S. Levy, "Burnout Limit Curves for Boiling Water Reactors," APED-3892, General Electric Company (1962).

Janssen, E., "Two-Phase Flow and Heat Transfer in Multirod Geometries—Final Report," GEAP-10347, General Electric Company (1971).

Jens, W. H., and P. A. Lottes, "Analysis of Heat Transfer, Burnout, Pressure Drop and Density Data for High Pressure Water," USAEC Report ANL-4627, U.S. Atomic Energy Commission (1951).

Judd, R. L., and K. S. Hwang, "A Comprehensive Model for Nucleate Pool Boiling Heat Transfer Including Microlayer Evaporation," *J. Heat Transfer*, **98**(4), 623–629 (1976).

Kalinin, E. K., S. A. Yarkho, I. I. Berlin, Kochelaev, S. Yu, and V. V. Kostyuk, "Investigation of the Crisis of Film Boiling in Channels," *Two-Phase Flow and Heat Transfer in Rod Bundles*, ASME Booklet, American Society of Mechanical Engineers (1969).

Kays, W. M., *Convective Heat and Mass Transfer*, McGraw-Hill Book Company, New York (1966).

Keeys, R. K. F., J. C. Ralph, and D. N. Roberts, "Post Burnout Heat Transfer in High Pressure Steam-Water Mixtures in a Tube with Cosine Heat Flux Distribution," AERE-R6411, Atomic Energy Research Establishment, Great Britain (1971).

Kutateladze, S. S., "A Hydrodynamic Theory of Changes in the Boiling Process under Free Convection," *Izv. Akad. Nauk, S.S.S.R. Otd. Tekh. Nauk.*, **4** (1951).

Lahey, R. T., Jr., "Two-Phase Flow in Boiling Water Nuclear Reactors," NEDO-13388, General Electric Company (1974).

Lahey, R. T., Jr., "Subchannel Measurements of the Equilibrium Quality and Mass Flux Distribution in a Rod Bundle," *Proceedings of the 8th IHTC* (1986).

Lahey, R. T., E. E. Polomik, and G. E. Dix, "The Effect of Reduced Clearance and Rod Bow on Critical Power in Simulated Nuclear Reactor Rod Bundles," *Int. Mtg. Reactor Heat Transfer*, Karlsruhe, Germany, paper #5; see also NEDM-10991, General Electric Company (1973).

Lahey, R. T., and F. A. Schraub, "Mixing, Flow Regimes and Void Fraction for Two-Phase Flow in Rod Bundles," *Two-Phase Flow and Heat Transfer in Rod Bundles*, ASME Booklet, American Society of Mechanical Engineers (1969).

Lahey, R. T., Jr., B. S. Shiralkar, and D. W. Radcliffe, "Two-Phase Flow and Heat Transfer in Multirod Bundles: Subchannel and Pressure Drop Measurements for Diabatic and Adiabatic Conditions," GEAP-13049, General Electric Company (1970).

Lahey, R. T., Jr., B. S. Shiralkar, and D. W. Radcliffe, "Mass Flux and Enthalpy Distribution in a Rod Bundle for Single- and Two-Phase Flow Conditions," *J. Heat Transfer* (1971).

Lahey, R. T., Jr., B. S. Shiralkar, D. W. Radcliffe, and E. E. Polomik, "Out-of-Pile Subchannel Measurements in a Nine-Rod Bundle for Water at 1000 psia," *Progress in Heat and Mass Transfer*, Vol. VI, Pergamon Press, New York (1972).

Levy, S., "Prediction of Two-Phase Pressure Drop and Density Distribution from Mixing Length Theory," *J. Heat Transfer* (1963).

Levy, S., and A. P. Bray, "Reliability of Burnout Calculations in Nuclear Reactors," *Nucl. News* (1963).

Lienhard, J. H., and V. K. Dhir, "Hydrodynamic Prediction of Peak Pool-Boiling Heat Fluxes from Finite Bodies," ASME paper 72-WA/HT-10, American Society of Mechanical Engineers (1972).

McAdams, W. H., *Heat Transmission*, 3rd ed., McGraw-Hill Book Company, New York (1954).

Murphy, G., *Similitude in Engineering*, Ronald Press Company, New York (1950).

Plummer, D. N., O. C. Iloeje, P. Griffith, and W. M. Rohsenow, "A Study of Post Critical Heat Flux Transfer in a Forced Convective System," MIT Report DSR 73645-80, Massachusetts Institute of Technology (1973).

Plummer, D. N., O. C. Iloeje, W. M. Rohsenow, P. Griffith, and E. Ganic, "Post Critical Heat Transfer in Flowing Liquid in a Vertical Tube," MIT Report 72718-91, Massachusetts Institute of Technology (1974).

Polomik, E. E., et al., "Deficient Cooling Program, 12th Quarterly Progress Report (April 1–June 30, 1972)," GEAP-10221-12, General Electric Company (1972).

Quinn, E. P., "Forced Flow Heat Transfer Beyond the Critical Heat Flux," ASME paper 66-WA/HT-36, American Society of Mechanical Engineers (1966).

Reichardt, M., "Complete Representation of Turbulent Velocity Distribution in Smooth Pipe," *ZAMM.*, **31** (1951).

Rogers, J. T., and R. G. Rosehart, "Mixing by Turbulent Interchange in Fuel Bundles, Correlations and Inferences," ASME paper 72-HT-53, American Society of Mechanical Engineers (1972).

Rogers, J. T., and N. E. Todreas, "Coolant Interchange Mixing in Reactor Fuel Rod Bundles Single-Phase Coolants," *Heat Transfer in Rod Bundles,* ASME Booklet, American Society of Mechanical Engineers (1968).

Rohsenow, W. M., "A Method of Correlating Heat Transfer Data for Surface Boiling Liquids," *Trans. ASME,* **74,** 969–976 (1952).

Rohsenow, W., and P. Griffith, "Correlation of Maximum Heat Transfer Data for Boiling of Saturated Liquids," *Chem. Eng. Prog. Symp. Ser.*, **52** (1956).

Rohsenow, W. M., and J. P. Hartnett, *Handbook of Heat Transfer,* McGraw-Hill Book Company, New York (1973).

"The Role of Nucleation in Boiling and Cavitation," ASME Monograph, American Society of Mechanical Engineers (1970).

Rowe, D. S., "COBRA IIIC: A Digital Computer Program for Steady-State and Transient Thermal-Hydraulic Analysis of Rod Bundle Nuclear Fuel Elements," BNWL-1695, Battelle Pacific Northwest Laboratory (1973).

Rowe, D. S., and C. W. Angle, "Cross Flow Mixing Between Parallel Flow Channels During Boiling—Part III," BNWL-371, Pt-3, Battelle Pacific Northwest Laboratory (1969).

Saha, P., Personal Communication (1975).

Schraub, F. A., R. L. Simpson, and E. Janssen, "Two-Phase Flow and Heat Transfer in Multirod Geometries; Air-Water Flow Structure Data for a Round Tube, Concentric and Eccentric Annulus, and Nine-Rod Bundle," GEAP-5739, General Electric Company (1969).

Shiralkar, B. S., "Local Void Fraction Measurements in Freon-114 with a Hot-Wire Anemometer," NEDE-13158, General Electric Company (1970).

Shiralkar, B. S., "Analysis of Non-Uniform Flux CHF Data in Simple Geometries," NEDM-13279, General Electric Company (1972).

Shiralkar, B. S., and R. T. Lahey, "The Effect of Obstacles on a Liquid Film," *J. Heat Transfer,* **95** (1973).

Shiralkar, B. S., E. E. Polomik, R. T. Lahey, J. M. Gonzalez, D. W. Radcliffe, and L. E. Schnebly, "Transient Critical Heat Flux—Experimental Results," GEAP-13295, General Electric Company (1972).

Silvestri, M., "On Burnout Equation and Location of Burnout Points," *Energ. Nucl.,* **13,** 9 (1966).

Smith, O. G., L. S. Tong, and W. M. Rohrer, "Burnout in Steam-Water Flow with Axially Nonuniform Heat Flux," ASME preprint #65: WA/HT-33, American Society of Mechanical Engineers (1965).

Spiegler, P., J. Hopenfeld, M. Silberberg, C. F. Bumpus, and A. Norman, "Onset of Stable Film Boiling and the Foam Limit," *J. Heat Mass Transfer*, **6** (1963).

Stevens, G. F., and G. J. Kirby, "A Quantitative Comparison Between Burnout Data for Water at 1000 PSIA and Freon-12 at 155 PSIA. Uniformly Heated Round Tubes, Vertical Upflow," AEEW-R327, United Kingdom Atomic Energy Authority (1964).

Thom, J. R. S., W. M. Walker, T. A. Fallon, and G. F. S. Reising, "Boiling in Subcooled Water During Flow in Tubes and Annuli," *Proc. Inst. Mech. Eng.*, 3C180 (1966).

Tippets, F. E., "Analysis of the Critical Heat-Flux Condition in High Pressure Boiling Water Flows," *J. Heat Transfer* (1964).

Tong, L. S., *Boiling Heat Transfer and Two-Phase Flow*, John Wiley & Sons, New York (1965).

Tong, L. S., "Boiling Crisis and Critical Heat Flux," TID-25887, AEC Critical Review Series, U.S. Atomic Energy Commission (1972).

Torbeck, J., Personal Communication (1975).

Van Stralen, S., and R. Cole, *Boiling Phenomena*, Vols. I and II, McGraw-Hill Book Company, New York (1979).

Volmer, M., "Kinetic der Phasenbildung," *Die Chemische Reaktion*, **4** (1939).

Volyak, L. D., "Surface Tension of Water," *Dokl. Akad. Nauk. S.S.S.R.*, **74,** 2 (1950).

Walkush, J. P., and P. Griffith, "Counterflow Critical Heat Flux Related to Reactor Transient Analysis," EPRI-292-2, Interim Report, Electric Power Research Institute (1975).

Wallis, G. B., *One-Dimensional, Two-Phase Flow*, McGraw-Hill Book Company, New York (1969).

Whalley, P. B., *Boiling, Condensation and Gas-Liquid Flow*, Oxford Science Publications, Clarendon Press (1987).

Whalley, P. B., P. Hutchinson, and G. F. Hewitt, "The Calculation of Critical Heat Flux in Forced Convection Boiling," *Fifth Int. Heat Transfer Conf.*, Tokyo, Japan, paper B6.11 (1974).

Zuber, N., "Stability of Boiling Heat Transfer," *Trans. ASME*, **80,** 711–720 (1958).

Zuber, N., "Hydrodynamic Aspects of Boiling Heat Transfer," AECU-4439, U.S. Atomic Energy Commission (1959).

CHAPTER FIVE

Two-Phase Flow

In BWRs, the coolant is typically composed of both the liquid and vapor phases of water. The behavior of this one-component, two-phase mixture is usually more difficult to analyze than corresponding single-phase flows, and empirical techniques invariably must be employed. Nevertheless, in recent years, considerable progress has been made in the field of two-phase flow, such that reliable methods are currently available to the BWR designer.

The purpose of this chapter is to discuss and develop those relationships and techniques that are of importance in the design and development of modern BWRs. Hence, Chapter 5 is not an exhaustive treatment of two-phase flow phenomena in general but, rather, stresses those aspects that are of current practical interest in BWR technology.

5.1 Basic Principles and Definitions

It is useful to establish a consistent set of notation and to develop some fundamental concepts and relationships.

5.1.1 Concepts and Notations

5.1.1.1 *Void Fraction, α*

The local vapor, or void, fraction represents the time fraction of vapor in a two-phase mixture. Mathematically,

$$\alpha = \frac{1}{T} \int_{t-T}^{t} X_v(\underline{x},t')\, dt' \ , \tag{5.1}$$

where the vapor phase indicator fraction is given by:

$$X_v(\underline{x},t) = \begin{cases} 1.0, & \text{when vapor is at point } \underline{x} \text{ at time } t \\ 0.0, & \text{otherwise} \end{cases} . \tag{5.2}$$

It is often convenient to work with the volumetric average vapor fraction or holdup void fraction. This is given by,

$$\langle\!\langle \alpha(t) \rangle\!\rangle \stackrel{\Delta}{=} \frac{1}{V} \iiint_V \alpha(\underline{x},t)\, dv = \iiint_{V_g} dv \Big/ \iiint_V dv = V_g/(V_l + V_g) \ , \tag{5.3}$$

where the subscript, l, denotes the liquid phase and g denotes the vapor phase.

If the volume in question consists of the cross-sectional area of a flow tube times a differential length element, the void fraction can also be considered the time-averaged area fraction. That is, as shown in Fig. 5-1,

$$\langle \alpha \rangle = \Delta z \iint_{A_g} dA \Big/ \left(\Delta z \iint_A dA \right) = A_g/(A_l + A_g) \ . \tag{5.4}$$

The void fractions defined in Eqs. (5.1) and (5.3) are deterministic. However, the phase indicator function is a random variable in space and time.

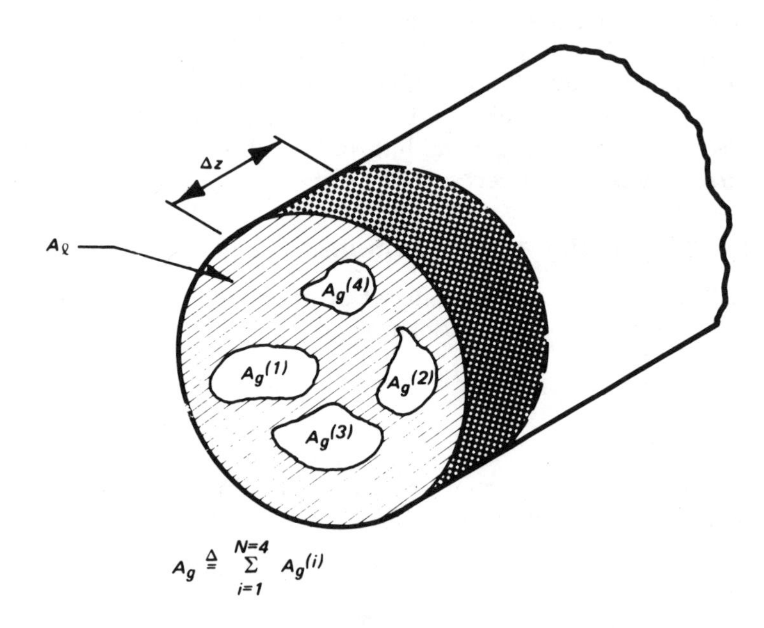

Fig. 5-1 Representation of void fraction.

If we perform N statistically independent measurements of the instantaneous phase content at location, $\underline{x}$, and time, t, then the ensemble-averaged local void fraction is given as,

$$\alpha(x, t) = N_g(\underline{x}, t)/N \ , \tag{5.5}$$

where $N_g(\underline{x}, t)$ is the number of experimental observations in which the location, $\underline{x}$, was occupied by vapor at time, t.

The link between Eqs. (5.5) and (5.1) is the Ergodic hypothesis, which essentially states that for a stationary random process, the ensemble average and the time average of a given member of the ensemble are identical. These concepts can be developed more fully (Bouré and Delhaye, 1981) for those readers interested in a rigorous statistical treatment of two-phase phenomena. However, for our present purpose, it is important only to realize that the void fraction that is used in subsequent analyses is a time-averaged deterministic quantity.

5.1.1.2 *Phase Velocity, u_l, u_g*

The true one-dimensional mean velocity of each phase is defined as the volumetric flow rate of that phase through its cross-sectional flow area. This velocity should be considered as averaged in the same way as classical single-phase turbulence analyses to eliminate the random fluctuations.

In accordance with Fig. 5-1,

$$\begin{aligned} u_l &\triangleq Q_l/A_l \\ u_g &\triangleq Q_g/A_g \ , \end{aligned} \tag{5.6}$$

where Q_i is the volumetric flow rate, ft^3/sec, of phase i.

5.1.1.3 *Volumetric Flux, j_l, j_g*

The mean volumetric flux, or so-called "superficial velocity," of each phase is defined as the volumetric flow rate of that phase divided by the total cross-sectional flow area in question. That is,

$$\begin{aligned} j_l &\triangleq Q_l/(A_l + A_g) \\ j_g &\triangleq Q_g/(A_l + A_g) \ . \end{aligned} \tag{5.7}$$

The total volumetric flux of the mixture, j, can be expressed as

$$j \triangleq j_l + j_g \ . \tag{5.8a}$$

Now, using the relations given in Eqs. (5.4) and (5.6), we can write,

$$\begin{aligned} j_l &= u_l(1-\alpha) \\ j_g &= u_g\alpha \ . \end{aligned} \tag{5.8b}$$

Thus we see that for two-phase flows, the phase velocities are larger than the corresponding volumetric fluxes of each phase.

The notation used here for volumetric flux is due to Zuber and Wallis (Wallis, 1969). In general this notation is used to provide consistency with the current two-phase flow literature.

5.1.1.4 *Volumetric Flow Fraction, β*

This quantity is defined as the volumetric flow rate of the vapor divided by the total volumetric flow rate,

$$\beta \triangleq Q_g/(Q_l + Q_g) = j_g/(j_l + j_g) \ . \tag{5.9}$$

The global volumetric flow fraction, $\langle\beta\rangle$, is a particularly convenient quantity for the experimentalist since the volumetric flow rates are readily calculated or measured.

5.1.1.5 *Relative Velocity, u_R*

It has been observed experimentally that in flowing two-phase systems, the one-dimensional velocity of the vapor phase is often greater than the one-dimensional phase velocity of the liquid. Thus, a parameter that quantifies this velocity difference has been defined and is called the relative velocity,

$$u_R \triangleq u_g - u_l \ . \tag{5.10}$$

For the special case of homogeneous flow, we have "no slip" between the phases and, thus, $u_R = 0$. In this case, Eqs. (5.10) and (5.8b) can be combined to yield,

$$j_g/\alpha = j_l(1-\alpha) \ .$$

Or solving for the void fraction,

$$\alpha = j_g/(j_l + j_g) \ . \tag{5.11}$$

A comparison between Eqs. (5.11) and (5.9) shows that for the homogeneous flow case (only), $\alpha = \beta$.

5.1.1.6 *Slip Ratio, S*

The slip ratio is defined as the ratio of the phase velocity of the vapor to that of the liquid. That is,

$$S \triangleq u_g/u_l \ . \tag{5.12}$$

For homogeneous flow, $S = 1$ by definition.

5.1.1.7 *Quality, x_e, x, x_s*

There are several qualities that are of importance in the analysis of two-phase flow. The first is the so-called "mixing cup" or thermodynamic equilibrium quality, x_e, which normally is calculated from the energy equation and is defined as,

$$x_e \triangleq (h - h_f)/h_{fg} \ , \tag{5.13}$$

where h is the enthalpy of the two-phase mixture and h_f and h_{fg} are functions only of the local static pressure. Here, x_e can be positive or negative, and can exceed unity. This is the quality that we would obtain if the flowing mixture were removed adiabatically, thoroughly mixed, and allowed to reach a condition of thermodynamic equilibrium. Thus, it is the flow fraction of vapor, when thermodynamic equilibrium prevails.

The next quality that must be considered is x, the so-called flow quality. This is the true flow fraction of vapor regardless of whether thermodynamic equilibrium exists or not. That is,

$$x \triangleq \rho_g u_g A_g/(\rho_l u_l A_l + \rho_g u_g A_g) \ . \tag{5.14}$$

The flow quality is always in the range $0.0 \leq x \leq 1.0$. It is of most interest during subcooled and post-dryout boiling conditions where thermodynamic equilibrium does not exist. As discussed in Sec. 5.3.1, x_e and x become equivalent for bulk boiling situations.

The final quality of interest is x_s, the so-called "static quality." It is defined as the mass fraction of vapor and is the quantity that is well known to students of classical thermodynamics. It is given by,

$$x_s \triangleq \rho_g A_g/(\rho_l A_l + \rho_g A_g) \ . \tag{5.15}$$

From the definitions given in Eqs. (5.14) and (5.15), x and x_s can be related through the following identity,

$$x/(1-x) = (u_g/u_l)x_s/(1-x_s) = Sx_s/(1-x_s) \ . \tag{5.16}$$

Hence, it is seen that for homogeneous flow (only), $x = x_s$.

5.1.1.8 *Two-Phase Density, $\bar{\rho}$*

There are numerous "densities" that can be defined in two-phase flow; however, following the standard definition of density as the average mass per unit volume, the density of the two-phase mixture is the volume-weighted density defined as,

$$\langle\!\langle \bar{\rho} \rangle\!\rangle \triangleq \iiint_{(V_l + V_g)} \bar{\rho}\, dv \Big/ (V_l + V_g)$$

$$= \left(\iiint_{V_l} \rho_l dv + \iiint_{V_g} \rho_g dv \right) \Big/ (V_l + V_g) \ .$$

By assuming constant density of each phase and employing Eq. (5.3),

$$\lessdot \bar{\rho} \gtrdot = (1 - \lessdot \alpha \gtrdot)\rho_l + \lessdot \alpha \gtrdot \rho_g = \rho_l - (\rho_l - \rho_g)\lessdot \alpha \gtrdot \ . \tag{5.17}$$

Similarly the local density is given by,

$$\bar{\rho} = (1 - \alpha)\rho_l + \alpha l_g \ , \tag{5.18}$$

and the cross-sectional average density by,

$$\langle \bar{\rho} \rangle = (1 - \langle \alpha \rangle)\rho_l + \langle \alpha \rangle \rho_g \ . \tag{5.19}$$

5.1.1.9 *Cross-Sectional Average Notation, ⟨ ⟩*

This is the notation used by Zuber to denote the cross-sectional average of an arbitrary variable, ζ. For instance,

$$\langle \zeta \rangle \triangleq \frac{1}{A_{x-s}} \iint_{A_{x-s}} \zeta dA$$

$$\langle \zeta_l \rangle_l \triangleq \frac{1}{A_{x-s}(1 - \langle \alpha \rangle)} \iint_{A_{x-s}} \zeta_l (1 - \alpha) dA = \frac{\langle (1 - \alpha)\zeta_l \rangle}{\langle 1 - \alpha \rangle} \tag{5.20}$$

$$\langle \zeta_g \rangle_g \triangleq \frac{1}{A_{x-s}\langle \alpha \rangle} \iint_{A_{x-s}} \zeta_g \alpha dA = \frac{\langle \alpha \zeta_g \rangle}{\langle \alpha \rangle} \ .$$

5.1.1.10 *Mass Flux, G*

Mass flux is a parameter that is frequently used in the analysis of flow systems. It is defined as the total mass flow rate divided by the cross-sectional flow area. That is,

$$G \triangleq w/A_{x-s} = \rho_l \langle u_l \rangle_l (1 - \langle \alpha \rangle)/(1 - \langle x \rangle) = \rho_g \langle u_g \rangle_g \langle \alpha \rangle / \langle x \rangle \ , \tag{5.21}$$

where the last two equalities come from the identities,

$$w_l = G(1 - \langle x \rangle)A_{x-s} = \rho_l \langle u_l \rangle_l (1 - \langle \alpha \rangle)A_{x-s}$$

and,

$$w_g = G\langle x \rangle A_{x-s} = \rho_g \langle u_g \rangle_g \langle \alpha \rangle A_{x-s} \ ,$$

which also imply,

$$\langle u_l \rangle_l = G(1 - \langle x \rangle)/[\rho_l(1 - \langle \alpha \rangle)] \tag{5.22}$$

$$\langle u_g \rangle_g = G\langle x \rangle/(\rho_g \langle \alpha \rangle) \ . \tag{5.23}$$

5.1.1.11 *The Fundamental Void-Quality Relation*

The equivalence of the last two quantities in Eq. (5.21) implies,

$$\langle x\rangle/(1-\langle x\rangle)=(\rho_g/\rho_l)S[\langle\alpha\rangle/(1-\langle\alpha\rangle)] \ , \tag{5.24}$$

which can be solved for void fraction to yield the fundamental void-quality relation,

$$\langle\alpha\rangle=\langle x\rangle/[\langle x\rangle+S(\rho_g/\rho_l)(1-\langle x\rangle)] \ . \tag{5.25}$$

This equation can be used to evaluate the void fraction once the flow quality, $\langle x\rangle$, is known and once the slip ratio, S, is known or estimated. Also note that Eqs. (5.24) and (5.16) imply,

$$\langle x_s\rangle/(1-\langle x_s\rangle)=(\rho_g/\rho_l)[\langle\alpha\rangle/(1-\langle\alpha\rangle)] \ , \tag{5.26}$$

which can be solved for void fraction,

$$\langle\alpha\rangle=\langle x_s\rangle\rho_l/[\langle x_s\rangle(\rho_l-\rho_g)+\rho_g] \ . \tag{5.27}$$

Equations (5.19) and (5.27) can be combined to eliminate void fraction in favor of static quality and, thus, the two-phase density can be written as,

$$\langle\bar{\rho}\rangle=\rho_l\rho_g/[\langle x_s\rangle\rho_l+(1-\langle x_s\rangle)\rho_g]=1/(v_l+\langle x_s\rangle v_{lg}) \ , \tag{5.28}$$

where the specific volumes are defined as,

$$v_l \triangleq 1/\rho_l$$

$$v_{lg} \triangleq (1/\rho_g-1/\rho_l)=\frac{(\rho_l-\rho_g)}{\rho_l\rho_g} \ .$$

Equation (5.28) has the same functional form as the classical thermodynamic density of a static multiphase system. Note that Eq. (5.16) implies that static quality is not equal to flow quality except for the special case of homogeneous flow. Thus, for slip flow conditions,

$$\langle\bar{\rho}\rangle\neq 1/(v_l+\langle x\rangle v_{lg}) \ .$$

The same result also can be obtained by inserting Eq. (5.25) into Eq. (5.19) to yield,

$$\langle\bar{\rho}\rangle=[(1-\langle x\rangle)S+\langle x\rangle]/[v_lS(1-\langle x\rangle)+v_g\langle x\rangle] \ . \tag{5.29}$$

Thus, only for the no slip, $S=1$, case do we have $\langle\bar{\rho}\rangle=\langle\rho_h\rangle$, where we have defined the so-called homogeneous density as,

$$\langle\rho_h\rangle\triangleq\frac{G}{\langle j\rangle}=\frac{w}{Q}=1/(v_l+\langle x\rangle v_{lg}) \ . \tag{5.30}$$

5.1.1.12 *Drift Velocity, V_{gj}*

This concept is originally due to Zuber. It is defined as the void-weighted average velocity of the vapor phase with respect to the velocity of the center-of-volume of the flowing mixture. That is,

$$V_{gj} \triangleq \langle \alpha(u_g - j) \rangle / \langle \alpha \rangle \ . \tag{5.31}$$

This parameter will be considered in more detail later when drift-flux relations are discussed.

5.1.1.13 *Heat Transfer and Fluid Flow Notation*

In general, the heat transfer notation that we employ is due to Jakob (1957). Specifically,

$$\begin{aligned} q &\triangleq \text{heat transferred (Btu/h)} \\ q' &\triangleq q/L_H = \text{linear heat generation rate (Btu/h-ft)} \\ q'' &\triangleq q/A_{HT} = \text{heat flux (Btu/h-ft}^2\text{)} \\ q''' &\triangleq q/V = \text{volumetric heat generation rate (Btu/h-ft}^3\text{)} \ , \end{aligned}$$

where

$$\begin{aligned} L_H &\triangleq \text{heated length, ft} \\ P_H &\triangleq \ = \text{heated perimeter, ft} \\ A_{HT} &\triangleq P_H L_H = \text{heat transfer area, ft}^2 \\ V &\triangleq \text{volume in which the heat is generated, ft}^3 \end{aligned}$$

Some of the notations used in the hydraulic analysis of BWRs are:

$$\begin{aligned} A_{x-s} &\triangleq \text{cross-sectional flow area, ft}^2 \\ p &\triangleq \text{system pressure, lb}_f\text{/ft}^2 \\ P_f &\triangleq \text{friction perimeter, ft} \\ \tau_w &\triangleq \text{wall shear stress, lb}_f\text{/ft}^2 \\ g_c &\triangleq \text{gravitational conversion factor, 32.17 lb}_m\text{-ft/lb}_f\text{-sec}^2 \\ J &\triangleq \text{mechanical equivalent of heat, 778 ft-lb}_f\text{/Btu} \\ w &\triangleq \text{flow rate, lb}_m\text{/h} \\ G &\triangleq w/A_{x-s} = \text{mass flux, lb}_m\text{/h-ft}^2 \ . \end{aligned}$$

5.1.2 Conservation Equations

The basic conservation equations of two-phase flow can be derived in many different ways. One method is to start with the differential equations of each phase and to integrate them across the cross section to obtain the working equations (Delhaye, 1981; Ishii, 1975).

Since the expressed purpose of this monograph is to convey physical understanding rather than to present analyses of complete generality, the derivations presented here are phenomenologically based. The one-dimensional conservation equations are derived for the limiting cases of homogeneous flow and the fully separated slip flow of a subcooled two-phase mixture. The appropriate set of equations to use depends on the situation and apparatus to be analyzed. In all the derivations that follow, it is implicitly assumed that the product of the cross-sectional average of the various quantities is equal to the average of the product. Essentially, this is equivalent to assuming flat profiles in each phase, which although in agreement with standard practice, may be a rather poor assumption in some cases of practical concern. For a more rigorous derivation in which flat profiles are not assumed, the reader is referred to the work of Yadigaroglu and Lahey (1975).

5.1.2.1 *Continuity Equation of Homogeneous Flow*

The basic conservation principle of mass can be stated as,

$$\left\{\begin{matrix}\text{rate of}\\ \text{creation of}\\ \text{mass}\end{matrix}\right\} \triangleq \left\{\begin{matrix}\text{mass}\\ \text{outflow}\\ \text{rate}\end{matrix}\right\} - \left\{\begin{matrix}\text{mass}\\ \text{inflow}\\ \text{rate}\end{matrix}\right\} + \left\{\begin{matrix}\text{mass}\\ \text{storage}\\ \text{rate}\end{matrix}\right\} = 0 \ . \tag{5.32}$$

This is represented schematically in Fig. 5-2a and implies,

$$\frac{\partial}{\partial z}(\langle\rho_h\rangle\langle j\rangle A_{x-s}) + A_{x-s}\frac{\partial\langle\rho_h\rangle}{\partial t} = 0 \ . \tag{5.33}$$

Equation (5.33) can also be written as,

$$\frac{D_j(\langle\rho_h\rangle A_{x-s})}{Dt} = -\langle\rho_h\rangle A_{x-s}\frac{\partial\langle j\rangle}{\partial z} \ , \tag{5.34}$$

where the material derivative is defined as,

$$\frac{D_j(\ )}{Dt} \triangleq \frac{\partial(\ )}{\partial t} + \langle j\rangle\frac{\partial(\ )}{\partial z} \ . \tag{5.35}$$

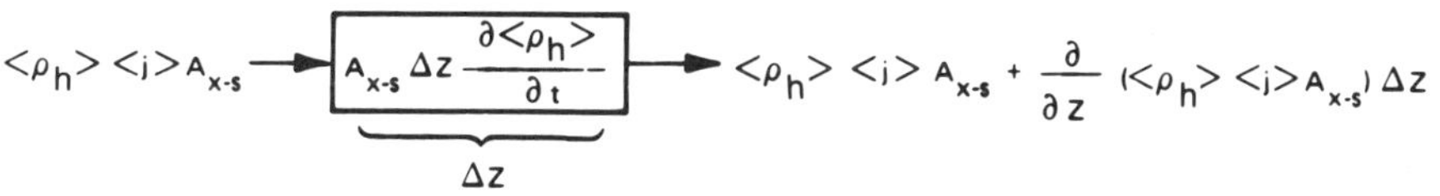

$\langle\rho_h\rangle\langle j\rangle A_{x-s}$ → $A_{x-s}\Delta z\dfrac{\partial\langle\rho_h\rangle}{\partial t}$ → $\langle\rho_h\rangle\langle j\rangle A_{x-s} + \dfrac{\partial}{\partial z}(\langle\rho_h\rangle\langle j\rangle A_{x-s})\Delta z$

Δz

Fig. 5-2a Continuity equation of homogeneous flow.

5.1.2.2 *Momentum Equation of Homogeneous Flow*

The basic conservation principle for momentum can be stated as,

$$\begin{Bmatrix}\text{rate of}\\ \text{creation of}\\ \text{momentum}\end{Bmatrix} \triangleq \begin{Bmatrix}\text{momentum}\\ \text{outflow}\\ \text{rate}\end{Bmatrix} - \begin{Bmatrix}\text{momentum}\\ \text{inflow}\\ \text{rate}\end{Bmatrix} + \begin{Bmatrix}\text{momentum}\\ \text{storage}\\ \text{rate}\end{Bmatrix}$$

$$= \begin{Bmatrix}\text{sum of the forces}\\ \text{acting on the}\\ \text{control volume}\end{Bmatrix} . \tag{5.36}$$

This is represented schematically in Fig. 5-2b and implies,

$$\frac{1}{g_c}\left[\frac{\partial}{\partial z}(GA_{x-s}\langle j\rangle)\Delta z + \frac{\partial}{\partial t}(G)A_{x-s}\Delta z\right]$$

$$= -\frac{\partial p}{\partial z}A_{x-s}\Delta z - \frac{g}{g_c}\langle \rho_h\rangle \sin\theta(A_{x-s}\Delta z) - \tau_w P_f \Delta z . \tag{5.37}$$

By rearranging, Eq. (5.37) becomes,

$$\frac{1}{g_c}\left[\frac{\partial G}{\partial t} + \frac{1}{A_{x-s}}\frac{\partial(G^2 A_{x-s}/\langle \rho_h\rangle)}{\partial z}\right] = -\frac{\partial p}{\partial z} - \frac{g}{g_c}\langle \rho_h\rangle \sin\theta - \frac{\tau_w P_f}{A_{x-s}} , \tag{5.38}$$

where the identity, $G = \langle \rho_h\rangle\langle j\rangle$, has been used.

Equation (5.38) is the standard Eulerian form of the one-dimensional momentum equation for homogeneous two-phase flow. This equation also can be written in its Lagrangian form by expanding the left side of Eq. (5.38),

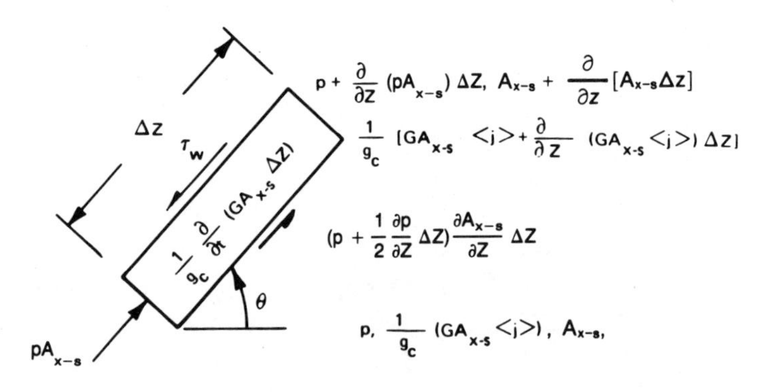

Fig. 5-2b Momentum equation of homogeneous flow.

$$\frac{1}{g_c}\left[\frac{\partial(\langle\rho_h\rangle\langle j\rangle)}{\partial t}+\frac{1}{A_{x-s}}\,\frac{\partial(\langle\rho_h\rangle\langle j\rangle^2 A_{x-s})}{\partial z}\right]$$

$$=\frac{1}{g_c}\left\{\langle\rho_h\rangle\left(\frac{\partial\langle j\rangle}{\partial t}+\langle j\rangle\frac{\partial\langle j\rangle}{\partial z}\right)+\langle j\rangle\left[\frac{\partial\langle\rho_h\rangle}{\partial t}+\frac{1}{A_{x-s}}\,\frac{\partial(\langle\rho_h\rangle\langle j\rangle A_{x-s})}{\partial z}\right]\right\} \;. \tag{5.39}$$

The first term of this expanded form can be recognized as the material derivative defined in Eq. (5.35), while the next term vanishes by continuity, Eq. (5.33). Hence, Eq. (5.38) can be rewritten in Lagrangian form as,

$$\frac{\langle\rho_h\rangle}{g_c}\,\frac{D_j\langle j\rangle}{Dt}=\frac{-\partial p}{\partial z}-\frac{g}{g_c}\,\langle\rho_h\rangle\,\sin\theta-\frac{\tau_w P_f}{A_{x-s}} \;. \tag{5.40}$$

5.1.2.3 *Energy Equation of Homogeneous Flow*

The final one-dimensional conservation equation required is the energy equation. The basic conservation principle for energy can be stated as,

$$\left\{\begin{matrix}\text{rate of}\\ \text{creation of}\\ \text{energy}\end{matrix}\right\}\triangleq\left\{\begin{matrix}\text{energy}\\ \text{outflow}\\ \text{rate}\end{matrix}\right\}-\left\{\begin{matrix}\text{energy}\\ \text{inflow}\\ \text{rate}\end{matrix}\right\}+\left\{\begin{matrix}\text{energy}\\ \text{storage}\\ \text{rate}\end{matrix}\right\}=0 \;. \tag{5.41}$$

This is represented schematically in Fig. 5-2c and implies,

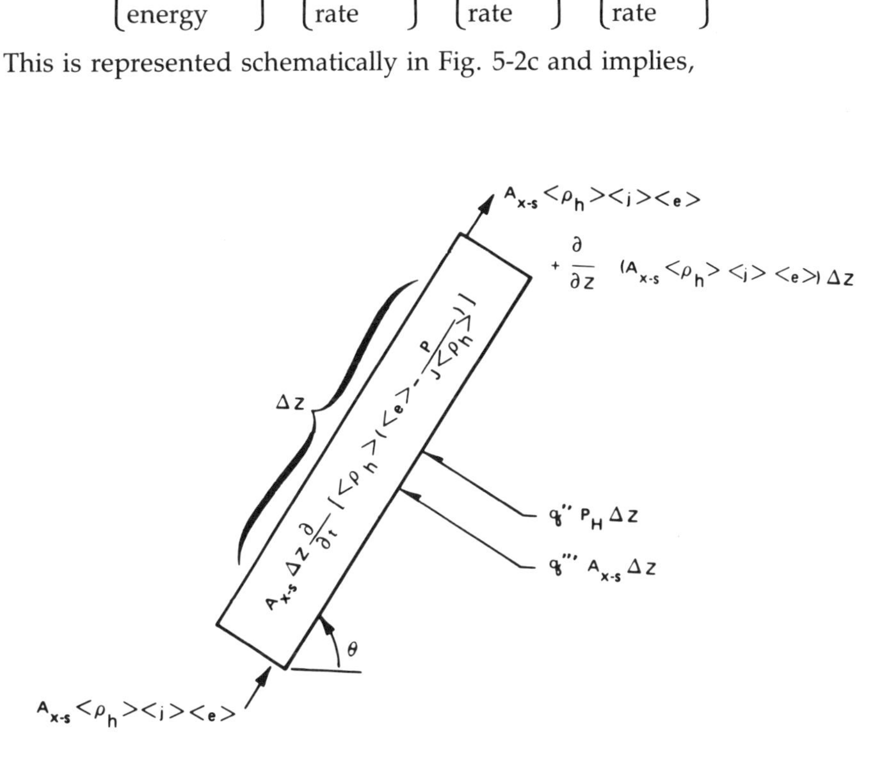

Fig. 5-2c Energy equation of homogeneous flow.

$$\frac{\partial}{\partial z}(\langle\rho_h\rangle\langle j\rangle A_{x-s}\langle e\rangle)\Delta z - q''P_H\Delta z - q'''A_{x-s}\Delta z + A_{x-s}\Delta z\frac{\partial}{\partial t}(\langle\rho_h\rangle\langle e\rangle - p/J) = 0 \ , \tag{5.42}$$

where,

$$\langle e\rangle \triangleq \left(\langle h\rangle + \frac{\langle j\rangle^2}{2g_cJ} + \frac{g}{g_c}\frac{z\sin\theta}{J}\right) \ . \tag{5.43}$$

Equation (5.42) can be recast into various different forms. The form that is used in subsequent analyses is the so-called "Lagrangian formulation." By expanding Eq. (5.42), we obtain,

$$\begin{aligned} &A_{x-s}\langle\rho_h\rangle\langle j\rangle\frac{\partial\langle e\rangle}{\partial z} + \langle e\rangle\left[\frac{\partial}{\partial z}(\langle\rho_h\rangle\langle j\rangle A_{x-s})\right] - q''P_H - q'''A_{x-s} \\ &\quad + A_{x-s}\langle\rho_h\rangle\frac{\partial\langle e\rangle}{\partial t} + A_{x-s}\langle e\rangle\frac{\partial\langle\rho_h\rangle}{\partial t} - \frac{A_{x-s}}{J}\frac{\partial p}{\partial t} = 0 \ . \end{aligned} \tag{5.44}$$

By regrouping Eq. (5.44),

$$\begin{aligned} &\langle\rho_h\rangle\left(\frac{\partial\langle e\rangle}{\partial t} + \langle j\rangle\frac{\partial\langle e\rangle}{\partial z}\right) + \frac{\langle e\rangle}{A_{x-s}}\left[\frac{\partial}{\partial z}(\langle\rho_h\rangle\langle j\rangle A_{x-s}) + A_{x-s}\frac{\partial\langle\rho_h\rangle}{\partial t}\right] \\ &\quad = q''\left(\frac{P_H}{A_{x-s}}\right) + q''' + \frac{1}{J}\frac{\partial p}{\partial t} \ . \end{aligned} \tag{5.45}$$

The first term on the left side can be recognized as the material derivative defined in Eq. (5.35) and the second term vanishes due to continuity, Eq. (5.33). Hence, Eq. (5.45) can be rewritten as,

$$\langle\rho_h\rangle\frac{D_j\langle e\rangle}{Dt} = q''\left(\frac{P_H}{A_{x-s}}\right) + q''' + \frac{1}{J}\frac{\partial p}{\partial t} \ . \tag{5.46}$$

Equation (5.46) is the thermodynamic energy equation of the homogeneous system. Another equation of interest is the thermal energy equation of the system. This is obtained by subtracting the mechanical energy equation from the thermodynamic energy equation. The mechanical energy equation can be derived by multiplying the momentum equation, Eq. (5.40), by the homogeneous velocity, $\langle j\rangle$, to yield,

$$\langle\rho_h\rangle\frac{D_j(\langle j\rangle^2/2g_cJ)}{Dt} = -(\langle j\rangle/J)\frac{\partial p}{\partial z} - \frac{g}{Jg_c}\langle j\rangle\langle\rho_h\rangle\sin\theta - \frac{\langle j\rangle\tau_w P_f}{JA_{x-s}} \ . \tag{5.47}$$

By using the definition of the total energy, $\langle e\rangle$, given in Eq. (5.43), Eq. (5.46) can be written as,

$$\langle\rho_h\rangle\frac{D_j\langle h\rangle}{Dt} + \langle\rho_h\rangle\frac{D_j(\langle j\rangle^2/2g_cJ)}{Dt} + \frac{g}{Jg_c}\langle j\rangle\langle\rho_h\rangle\sin\theta = q''\left(\frac{P_H}{A_{x-s}}\right) + q''' + \frac{1}{J}\frac{\partial p}{\partial t} \ . \tag{5.48}$$

Equation (5.47) then can be subtracted from Eq. (5.48) to yield the thermal energy equation,

$$\langle \rho_h \rangle \frac{D_j \langle h \rangle}{Dt} = q'' \left(\frac{P_H}{A_{x-s}} \right) + q''' + \frac{1}{J} \frac{D_j p}{Dt} + \langle j \rangle \left(\frac{\tau_w P_f}{J A_{x-s}} \right) \quad . \tag{5.49}$$

The last term on the right side of Eq. (5.49) represents the rate of increase in thermal energy per unit volume due to viscous dissipation. For many cases of practical significance it can be neglected.

The equations to solve for the transient response of a homogeneous two-phase system have now been formulated. Before leaving the case of homogeneous flow, it is useful to cast the energy equation, given as Eq. (5.49), into the form of a quality propagation equation. This is readily done by inserting the "state" equations of the homogeneous mixture into Eq. (5.49). That is, for saturated, homogeneous flow,

$$\begin{aligned} \langle \rho_h \rangle &= 1/(v_f + \langle x \rangle v_{fg}) \\ \langle h \rangle &= h_f + \langle x \rangle h_{fg} \quad . \end{aligned} \tag{5.50}$$

By neglecting internal generation, q''', and viscous dissipation effects, Eqs. (5.49) and (5.50) yield,

$$\begin{aligned} &\frac{D_j \langle x \rangle}{Dt} + \left\{ \frac{1}{h_{fg}} \left[\frac{D_j h_{fg}}{Dt} - \left(q'' P_H / A_{x-s} + \frac{D_j p}{Dt} \right) v_{fg} \right] \right\} \langle x \rangle \\ &\quad = \left\{ \frac{1}{h_{fg}} \left[\left(q'' P_H / A_{x-s} + \frac{D_j p}{Dt} \right) v_f - \frac{D_j h_f}{Dt} \right] \right\} \quad . \end{aligned} \tag{5.51}$$

Furthermore, if we consider the case of constant system pressure and neglect the axial pressure drop compared to the constant system pressure, Eq. (4.51) becomes,

$$\frac{D_j \langle x \rangle}{Dt} - \Omega \langle x \rangle = \Omega \frac{v_f}{v_{fg}} \quad , \tag{5.52}$$

where,

$$\Omega \triangleq q'' \frac{P_H}{A_{x-s}} \frac{v_{fg}}{h_{fg}} \quad . \tag{5.53}$$

The parameter, Ω, which has the units of reciprocal time, is frequently referred to as the characteristic frequency of phase change. It physically represents the speed at which phase change takes place. Equations (5.34) and (5.52) are used extensively in Chapters 7 and 9 to analyze the occurrence of density-wave instability and transient boiling transition, respectively.

Now that the one-dimensional conservation equations for the case of homogeneous flow are in hand, we can turn our attention to the case of separated flow. The conservation equations for the case of subcooled one-dimensional slip flow can be derived in terms of each phase separately or in terms of the two-phase mixture. In this chapter the phasic conservation equations will be derived and the mixture conservation equations will be deduced from them. Note, however, that the same results can be obtained in other ways; the technique is largely a matter of choice.

The derivations that follow are for an idealized annular flow situation; however, the resultant conservation equations are of more general applicability. The reader interested in a more general treatment is referred to the work of Meyer (1960).

5.1.2.4 *Continuity Equations of Separated Flow*

Consider the control volume shown in Fig. 5-3 for the case of idealized annular flow with no liquid entrainment. The conservation of mass principle for each phase is given by Eq. (5.32). For the liquid phase, this implies,

$$A_{x-s}\Gamma\Delta z+\frac{\partial}{\partial z}[\rho_l(1-\langle\alpha\rangle)\langle u_l\rangle_l A_{x-s}]\Delta z+\frac{\partial}{\partial t}[\rho_l A_{x-s}(1-\langle\alpha\rangle)\Delta z]=0 \ ,$$

or

$$\frac{\partial}{\partial t}[\rho_l(1-\langle\alpha\rangle)A_{x-s}]+\frac{\partial}{\partial z}[\rho_l(1-\langle\alpha\rangle)\langle u_l\rangle_l A_{x-s}]=-A_{x-s}\Gamma \ , \qquad (5.54)$$

where Γ is the amount of liquid phase evaporated per unit volume in the differential control volume. The magnitude of this evaporation term depends on the thermodynamic conditions that exist. For the special case of bulk boiling, thermodynamic equilibrium is assumed to exist and, neglecting flashing, a simple heat balance yields the following approximation[a] for Γ,

$$\Gamma=q''_w P_H/(h_{fg}A_{x-s}) \ , \qquad (5.55)$$

in which the subscript, l, has been changed to f to denote liquid phase saturation. Similarly, for the gas phase, Eq. (5.32) implies,

$$\frac{\partial}{\partial t}(\rho_g\langle\alpha\rangle A_{x-s})+\frac{\partial}{\partial z}(\rho_g\langle\alpha\rangle\langle u_g\rangle_g A_{x-s})=A_{x-s}\Gamma \ . \qquad (5.56)$$

Equations (5.54) and (5.56) represent the continuity equations for the liquid and gas phases, respectively. They can be added to obtain the continuity equation for the two-phase mixture,

$$\frac{\partial}{\partial t}(\langle\bar{\rho}\rangle A_{x-s})+\frac{\partial}{\partial z}(GA_{x-s})=0 \ , \qquad (5.57)$$

---

[a]See Eqs. (5.95), (5.107), and (5.232) for more general expressions.

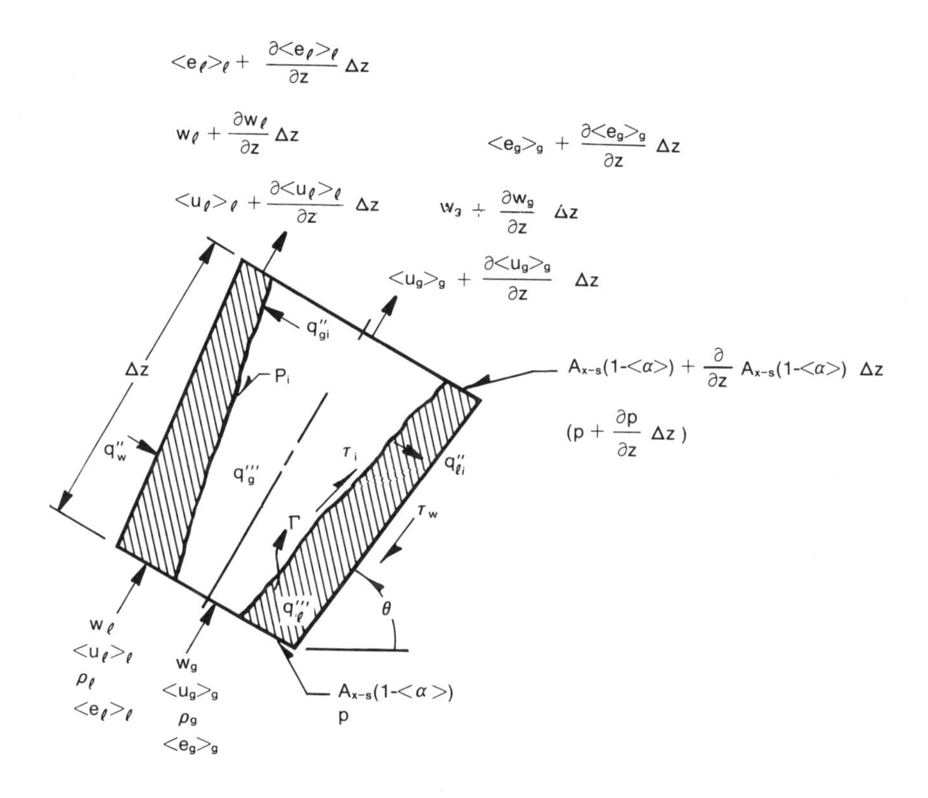

Fig. 5-3 Control volume for separated flow (each phase).

where the identities in Eqs. (5.22) and (5.23) have been employed to yield the mass flux,

$$G = \rho_l(1 - \langle\alpha\rangle)\langle u_l\rangle_l + \rho_g\langle\alpha\rangle\langle u_g\rangle_g \ . \tag{5.58}$$

5.1.2.5 *Momentum Equations of Separated Flow*

Next, the momentum equations for each phase are derived. In this derivation, surface tension, Reynolds stress, and virtual mass effects are neglected, and the pressure of each phase is assumed to be equal. Considering the control volume shown in Fig. 5-3, Eq. (5.36) applied to the liquid phase yields,

$$
-\frac{\partial}{\partial z}[p(1-\langle\alpha\rangle)A_{x-s}]\Delta z - p_i \frac{\partial[\langle\alpha\rangle A_{x-s}]}{\partial z}\Delta z + p_w \frac{\partial A_{x-s}}{\partial z}\Delta z
$$

$$
-\frac{g}{g_c}\rho_l(1-\langle\alpha\rangle)A_{x-s}\Delta z \sin\theta - \tau_w P_f \Delta z + \tau_i P_i \Delta z
$$

$$
= \frac{\partial}{\partial t}\left[\frac{\rho_l}{g_c}(1-\langle\alpha\rangle)A_{x-s}\langle u_l\rangle_l \Delta z\right] + \frac{\partial}{\partial z}\left[\frac{\rho_l}{g_c}A_{x-s}(1-\langle\alpha\rangle)\langle u_l\rangle_l^2\right]\Delta z + \frac{A_{x-s}\Gamma}{g_c}U_i\Delta z \ , \tag{5.59}
$$

where:

1. The quantity, τ_i, is the shear stress (lb_f/ft^2) at the liquid-vapor interface. Neglecting any liquid entrainment,

$$
\tau_i = (c_f)_i \frac{\rho_g}{2g_c}(\langle u_g\rangle_g - \langle u_l\rangle_l)^2 \ .
$$

Wallis (1969) has proposed an expression for the interfacial Fanning friction factor of annular flow as

$$
(c_f)_i \cong 0.005[1 + 75(1-\langle\alpha\rangle)] \ .
$$

For other flow regimes different expressions are valid. For example, for distorted bubbly two-phase flow Harmathy (1960) recommends,

$$
(c_f)_i = \rho_l/(4\rho_g)\left[\frac{R_b^2 g(\rho_l - \rho_g)}{g_c\sigma(1-\langle\alpha\rangle)}\right]^{1/2} \ .
$$

More generally, Anderson and Chu (1981) have proposed a relation between the interfacial friction factor and the drift-flux parameters.

2. As can be seen in Fig. 5-4, the interfacial velocity, U_i, is in general different from the phase velocities, $\langle u_k\rangle_k$. To maintain generality, we may assume:

$$
U_i = \eta_m\langle u_l\rangle_l + (1-\eta_m)\langle u_g\rangle_g \ , \tag{5.60}
$$

where η_m is a momentum transfer weighting parameter ($0.0 \leq \eta_m \leq 1.0$). The classical assumption is to choose $\eta_m = 1$ and, thus, assume that the liquid phase velocity alone is important in evaporative momentum transfer. Alternately, based on entropy production considerations, $\eta_m = 1/2$ is recommended (Wallis, 1969). This implies that the effective velocity of the evaporating interfacial liquid is $1/2\ (\langle u_l\rangle_l + \langle u_g\rangle_g)$. Since evaporation is assumed to occur at the interface, this appears to be a physically realistic choice within the framework of the present one-dimensional analysis. Moreover, choosing $\eta_m = 1/2$ leads to momentum equations for each phase that are symmetrical for evaporation and condensation.

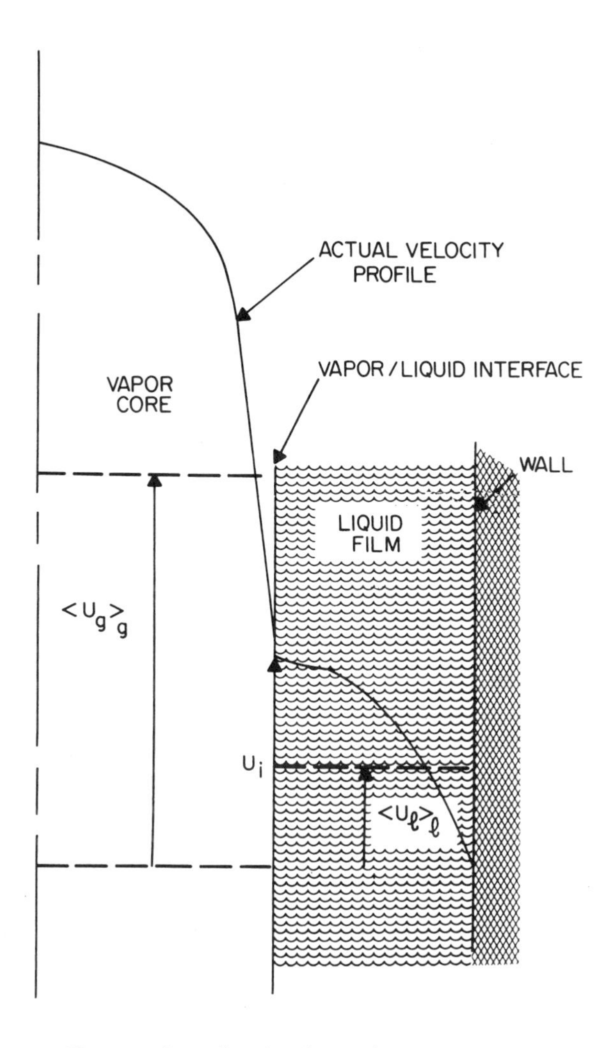

Fig. 5-4 Interfacial velocity for annular flow.

3. We have assumed that the radial pressure gradient is zero. Thus, in accordance with standard practice, we can also assume that the interfacial pressure, p_i, and wall pressure, p_w, are equal to the static pressure, p. Thus the first three terms on the left side of Eq. (5.59) can be expanded and simplified to yield, to first order,

$$-(1-\langle\alpha\rangle)A_{x-s}\frac{\partial p}{\partial z}\Delta z \ . \tag{5.61}$$

This is an interesting result since it is the same functional relationship we would obtain if we had neglected axial flow area changes and area changes due to evaporation in the duct.

Thus, for a nonconstant area duct, Eq. (5.59) yields, for the liquid phase momentum equation,

$$-(1-\langle\alpha\rangle)\frac{\partial p}{\partial z}-\frac{g}{g_c}\rho_l(1-\langle\alpha\rangle)\sin\theta-\frac{\tau_w P_f}{A_{x-s}}+\tau_i\frac{P_i}{A_{x-s}}$$
$$=\frac{1}{g_c}\left\{\frac{\partial}{\partial t}[\rho_l(1-\langle\alpha\rangle)\langle u_l\rangle_l]+\frac{1}{A_{x-s}}\frac{\partial}{\partial z}[\rho_l A_{x-s}(1-\langle\alpha\rangle)\langle u_l\rangle_l^2]\right\}+\frac{\Gamma}{g_c}U_i\ . \quad (5.62)$$

In a similar manner, Eq. (5.36) can be applied to the vapor phase in the control volume shown in Fig. 5-3 to yield,

$$-\frac{\partial}{\partial z}(p\langle\alpha\rangle A_{x-s})\Delta z+p_i\frac{\partial(\langle\alpha\rangle A_{x-s})}{\partial z}\Delta z-\frac{g}{g_c}\rho_g\langle\alpha\rangle A_{x-s}\Delta z\sin\theta-\tau_i P_i\Delta z$$
$$=\frac{\partial}{\partial t}\left(\frac{\rho_g}{g_c}\langle\alpha\rangle A_{x-s}\langle u_g\rangle_g\Delta z\right)+\frac{\partial}{\partial z}\left(\frac{\rho_g}{g_c}\langle\alpha\rangle A_{x-s}\langle u_g\rangle_g^2\right)\Delta z-\frac{\Gamma A_{x-s}}{g_c}U_i\ , \quad (5.63)$$

where we have made the same assumptions and approximations as in the liquid phase momentum equation.

By expanding the pressure drop term and neglecting higher order differentials as before, we obtain for the vapor phase momentum equation,

$$-\langle\alpha\rangle\frac{\partial p}{\partial z}-\frac{g}{g_c}\rho_g\langle\alpha\rangle\sin\theta-\frac{\tau_i P_i}{A_{x-s}}$$
$$=\frac{1}{g_c}\left[\frac{\partial}{\partial t}(\rho_g\langle\alpha\rangle\langle u_g\rangle_g)+\frac{1}{A_{x-s}}\frac{\partial}{\partial z}(\rho_g A_{x-s}\langle\alpha\rangle\langle u_g\rangle_g^2)\right]-\frac{\Gamma}{g_c}U_i\ . \quad (5.64)$$

It should be noted that because of the assumptions made there is no jump in momentum across the interface. Equations (5.62) and (5.64) can be added to yield the momentum equation of the two-phase mixture as,

$$-\frac{\partial p}{\partial z}-\frac{g}{g_c}\langle\bar{\rho}\rangle\sin\theta-\frac{\tau_w P_f}{A_{x-s}}=\frac{1}{g_c}\left\{\frac{\partial}{\partial t}[\rho_l(1-\langle\alpha\rangle)\langle u_l\rangle_l+\rho_g\langle\alpha\rangle\langle u_g\rangle_g]\right.$$
$$\left.+\frac{1}{A_{x-s}}\frac{\partial}{\partial z}[\rho_l A_{x-s}(1-\langle\alpha\rangle)\langle u_l\rangle_l^2+\rho_g A_{x-s}\langle\alpha\rangle\langle u_g\rangle_g^2]\right\}\ . \quad (5.65)$$

This equation also can be written in terms of mass flux through the use of the identities given in Eqs. (5.22), (5.23), and (5.58),

$$-\frac{\partial p}{\partial z}-\frac{g}{g_c}\langle\bar{\rho}\rangle\sin\theta-\frac{\tau_w P_f}{A_{x-s}}=\frac{1}{g_c}\left[\frac{\partial G}{\partial t}+\frac{1}{A_{x-s}}\frac{\partial}{\partial z}\left(\frac{G^2 A_{x-s}}{\langle\rho'\rangle}\right)\right]\ , \quad (5.66)$$

where,

$$\frac{1}{\langle\rho'\rangle} \triangleq \left[\frac{(1-\langle x\rangle)^2}{\rho_l(1-\langle\alpha\rangle)} + \frac{\langle x\rangle^2}{\rho_g\langle\alpha\rangle}\right] . \tag{5.67}$$

The parameter, $\langle\rho'\rangle$, is commonly referred to as the momentum density. It is not a "true" density, but rather it is defined so that Eq. (5.66) has the same functional form as Eq. (5.38).

Equation (5.66) is the form of the one-dimensional two-phase momentum equation originally derived by Meyer (1960). It is a widely used form of the mixture momentum equation; however, other forms are possible. First we consider a drift-flux form presented by Zuber (1967) and then a Lagrangian form from the Russian literature is considered.

To derive Zuber's form of the momentum equation, it is convenient to use the following identities,

$$\langle u_l\rangle_l = \frac{G}{\langle\bar{\rho}\rangle} - \frac{\langle\alpha\rangle}{(1-\langle\alpha\rangle)}\frac{\rho_g}{\langle\bar{\rho}\rangle}[V_{gj} + (C_0 - 1)\langle j\rangle] \tag{5.68}$$

$$\langle u_g\rangle_g = \frac{G}{\langle\bar{\rho}\rangle} + \frac{\rho_l}{\langle\bar{\rho}\rangle}[V_{gj} + (C_0 - 1)\langle j\rangle] , \tag{5.69}$$

where,

$$(C_0 - 1)\langle j\rangle + V_{gj} = \langle u_g\rangle_g - \langle j\rangle .$$

These relationships are easily verified by using the definition of V_{gj} given in Eq. (5.31) and using Eqs. (5.17) and (5.58). If Eq. (5.68) is multiplied through by $(1-\langle\alpha\rangle)$ and added to $\langle\alpha\rangle$ times Eq. (5.69), and if for convenience we define,

$$V'_{gj} \triangleq V_{gj} + (C_0 - 1)\langle j\rangle , \tag{5.70}$$

we obtain,

$$\langle j\rangle = \frac{G}{\langle\bar{\rho}\rangle} + \frac{(\rho_l - \rho_g)\langle\alpha\rangle}{\langle\bar{\rho}\rangle}V'_{gj} = \frac{\left[G/\langle\bar{\rho}\rangle + \frac{(\rho_l - \rho_g)}{\langle\bar{\rho}\rangle}\langle\alpha\rangle V_{gj}\right]}{\left[1 - (C_0 - 1)\frac{(\rho_l - \rho_g)}{\langle\bar{\rho}\rangle}\langle\alpha\rangle\right]} , \tag{5.71}$$

where the relationships given in Eq. (5.8b) have been used. Equation (5.71) is a fundamental identity that relates the velocity of the center-of-volume, $\langle j\rangle$, to the velocity of the center-of-mass, U_m. As will be shown later in this chapter, this latter velocity is given by,

$$U_m = G/\langle\bar{\rho}\rangle .$$

If the identities in Eqs. (5.68) and (5.69) are introduced into the spatial acceleration term of Eq. (5.65), we obtain,

$$\frac{\partial}{\partial z}[\rho_l A_{x-s}(1-\langle\alpha\rangle)\langle u_l\rangle_l^2 + \rho_g A_{x-s}\langle\alpha\rangle\langle u_g\rangle_g^2]$$

$$= \frac{\partial}{\partial z}\left(\frac{G^2}{\langle\bar{\rho}\rangle}A_{x-s}\right) + \frac{\partial}{\partial z}\left[A_{x-s}\left(\frac{\rho_l - \langle\bar{\rho}\rangle}{\langle\bar{\rho}\rangle - \rho_g}\right)\frac{\rho_l\rho_g}{\langle\bar{\rho}\rangle}(V'_{gj})^2\right] . \quad (5.72)$$

Thus, the one-dimensional two-phase momentum equation can be rewritten in Eulerian form as,

$$-\frac{\partial p}{\partial z} - \frac{g}{g_c}\langle\bar{\rho}\rangle\sin\theta - \frac{\tau_w P_f}{A_{x-s}} - \frac{1}{g_c A_{x-s}}\frac{\partial}{\partial z}\left[A_{x-s}\left(\frac{\rho_l - \langle\bar{\rho}\rangle}{\langle\bar{\rho}\rangle - \rho_g}\right)\frac{\rho_l\rho_g}{\langle\bar{\rho}\rangle}(V'_{gj})^2\right]$$

$$= \frac{1}{g_c}\left[\frac{\partial G}{\partial t} + \frac{1}{A_{x-s}}\frac{\partial}{\partial z}\left(\frac{G^2 A_{x-s}}{\langle\bar{\rho}\rangle}\right)\right] . \quad (5.73)$$

Several things should be noted about Eq. (5.73). First, since it was derived directly from Eq. (5.65), it contains no more information about momentum conservation than Eq. (5.66) does, which also was obtained from Eq. (5.65). That is, integration and numerical evaluation of Eqs. (5.66) and (5.73) produce the same results. However, Eq. (5.73) is written in terms of the center-of-mass and, thus, uses the correct two-phase density, $\langle\bar{\rho}\rangle$, in the spatial acceleration term, rather than a defined quantity, $\langle\rho'\rangle$. The spatial drift gradient [i.e., the last term on the left side of Eq. (5.73)] is due to the relative slip between the phases and can be considered as an additional volumetric force in the same sense that the classical Reynolds stress term of single-phase turbulence is considered a force. Physically it represents the net momentum flux with respect to the center-of-mass of the flowing two-phase system.

Thus, we find that the center-of-mass formulation, given by Eq. (5.73), lends itself to a readily understandable interpretation of the various terms and is consistent with the classical techniques used in the kinetic theory of gases. However, we must have information about the functional dependence of V_{gj} before it is a useful formulation, a requirement that can limit the practical application of this formulation.

Before leaving the center-of-mass formulation given in Eq. (5.73), note that the spatial drift gradient can also be written in terms of the relative velocity through the use of the identity,

$$V'_{gj} \triangleq \langle u_g\rangle_g - \langle j\rangle = \langle u_g\rangle_g - [(1-\langle\alpha\rangle)\langle u_l\rangle_l + \langle\alpha\rangle\langle u_g\rangle_g] = (1-\langle\alpha\rangle)U_r , \quad (5.74)$$

where,

$$U_r \triangleq \langle u_g\rangle_g - \langle u_l\rangle_l , \quad (5.75)$$

By using Eq. (5.74) and noting that,

$$\left(\frac{\rho_l - \langle\overline{\rho}\rangle}{\langle\overline{\rho}\rangle - \rho_g}\right) = \frac{\langle\alpha\rangle}{(1-\langle\alpha\rangle)} \ ,$$

we find,

$$\frac{\partial}{\partial z}\left[A_{x-s}\left(\frac{\rho_l - \langle\overline{\rho}\rangle}{\langle\overline{\rho}\rangle - \rho_g}\right)\frac{\rho_l\rho_g}{\langle\overline{\rho}\rangle}(V'_{gj})^2\right] = \frac{\partial}{\partial z}\left[A_{x-s}\frac{\langle\alpha\rangle(1-\langle\alpha\rangle)\rho_l\rho_g}{\langle\overline{\rho}\rangle}U_r^2\right] \ . \quad (5.76)$$

This may be a more useful form in certain applications.

Next we consider the Lagrangian form of the one-dimensional two-phase mixture momentum equation. Equation (5.73) can be recast into a Lagrangian formulation by expanding its right side and combining it with the continuity equation, Eq. (5.57), written in the form,

$$\frac{D_m}{Dt}(\langle\overline{\rho}\rangle A_{x-s}) + \langle\overline{\rho}\rangle A_{x-s}\frac{\partial U_m}{\partial z} = 0 \ , \quad (5.77)$$

where,

$$\frac{D_m(\ )}{Dt} \triangleq \frac{\partial(\ )}{\partial t} + U_m\frac{\partial(\ )}{\partial z} \ , \quad (5.78)$$

to obtain,

$$\frac{\langle\overline{\rho}\rangle}{g_c}A_{x-s}\frac{D_m U_m}{Dt} + \frac{1}{g_c}\frac{\partial}{\partial z}\left[A_{x-s}\left(\frac{\rho_l - \langle\overline{\rho}\rangle}{\langle\overline{\rho}\rangle - \rho_g}\right)\frac{\rho_l\rho_g}{\langle\overline{\rho}\rangle}(V'_{gj})^2\right]$$

$$= -A_{x-s}\frac{\partial p}{\partial z} - \tau_w P_f - \frac{g}{g_c}A_{x-s}\langle\overline{\rho}\rangle \sin\theta \ . \quad (5.79)$$

This is a form of the momentum equation that has been used by Ishii and Zuber (1970) to investigate two-phase hydrodynamic stability phenomena.

Another Lagrangian formulation of the one-dimensional two-phase mixture momentum equation frequently appears in the literature. This formulation can readily be derived from Eqs. (5.62) and (5.64). Consider the right side of the liquid phase momentum equation, Eq. (5.62), minus $\langle u_l\rangle_l/(g_c A_{x-s})$ times the liquid phase continuity equation, Eq. (5.54), to obtain,

$$\sum F_l = \frac{1}{g_c}\left\{\frac{\partial}{\partial t}[\rho_l(1-\langle\alpha\rangle)\langle u_l\rangle_l] + \frac{1}{A_{x-s}}\frac{\partial}{\partial z}[\rho_l A_{x-s}(1-\langle\alpha\rangle)\langle u_l\rangle_l^2]\right\}$$

$$+ \frac{\delta w'}{g_c A_{x-s}}[\eta_m\langle u_l\rangle_l + (1-\eta_m)\langle u_g\rangle_g] - \frac{\langle u_l\rangle_l}{g_c A_{x-s}}\left\{\frac{\partial}{\partial t}[\rho_l(1-\langle\alpha\rangle)A_{x-s}]\right.$$

$$\left. + \frac{\partial}{\partial z}[\rho_l(1-\langle\alpha\rangle)\langle u_l\rangle_l A_{x-s}] + \Gamma A_{x-s}\right\} \ . \quad (5.80)$$

By expanding these terms and simplifying,

$$\sum F_l = (1-\langle\alpha\rangle)\frac{\rho_l}{g_c}\left(\frac{\partial\langle u_l\rangle_l}{\partial t} + \langle u_l\rangle_l\frac{\partial\langle u_l\rangle_l}{\partial z}\right) + (1-\eta_m)\frac{\Gamma}{g_c}(\langle u_g\rangle_g - \langle u_l\rangle_l) \ . \tag{5.81}$$

For the normal case of interest in BWRs, the last term in Eq. (5.81) can be regarded as a negative volumetric force due to "vapor thrust." That is, it can be interpreted as a reaction force on the liquid due to evaporation. Only for the special case of $\eta_m = 1$ does this term vanish. In a similar fashion, the vapor phase momentum equation can be recast into a Lagrangian formulation by subtracting $\langle u_g\rangle_g/(g_c A_{x-s})$ times the vapor continuity equation, Eq. (5.56), from the right side of the vapor momentum equation, Eq. (5.64),

$$\sum F_g = \frac{1}{g_c}\left\{\frac{\partial}{\partial t}(\rho_g\langle\alpha\rangle\langle u_g\rangle_g) + \frac{1}{A_{x-s}}\frac{\partial}{\partial z}(\rho_g A_{x-s}\langle\alpha\rangle\langle u_g\rangle_g^2)\right.$$
$$\left. - \frac{\Gamma}{g_c}[\eta_m\langle u_l\rangle_l + (1-\eta_m)\langle u_g\rangle_g]\right\} - \frac{\langle u_g\rangle_g}{g_c A_{x-s}}\left[\frac{\partial}{\partial t}(\rho_g\langle\alpha\rangle A_{x-s})\right.$$
$$\left. + \frac{\partial}{\partial z}(\rho_g\langle\alpha\rangle\langle u_g\rangle_g A_{x-s}) - \Gamma A_{x-s}\right] \ . \tag{5.82}$$

By expanding these terms and simplifying,

$$\sum F_g = \langle\alpha\rangle\frac{\rho_g}{g_c}\left(\frac{\partial\langle u_g\rangle_g}{\partial t} + \langle u_g\rangle_g\frac{\partial\langle u_g\rangle_g}{\partial z}\right) + \frac{\eta_m\Gamma}{g_c}(\langle u_g\rangle_g - \langle u_l\rangle_l) \ . \tag{5.83}$$

For the evaporative case, the last term in Eq. (5.83) can be regarded as a negative volumetric force due to the slower moving evaporated liquid retarding the vapor phase due to momentum transfer.

Now we are in a position to write down the momentum equation for the liquid and vapor in terms of the material derivative of each phase. Equations (5.62) and (5.81) yield, for the liquid phase,

$$-(1-\langle\alpha\rangle)\frac{\partial p}{\partial z} - \frac{g}{g_c}\rho_l(1-\langle\alpha\rangle)\sin\theta - \frac{\tau_w P_f}{A_{x-s}} + \frac{\tau_i P_i}{A_{x-s}}$$
$$-(1-\eta_m)\frac{\Gamma}{g_c}(\langle u_g\rangle_g - \langle u_l\rangle_l) = (1-\langle\alpha\rangle)\frac{\rho_l}{g_c}\frac{D_l\langle u_l\rangle_l}{Dt} \ , \tag{5.84}$$

where,

$$\frac{D_l\langle u_l\rangle_l}{Dt} \triangleq \frac{\partial\langle u_l\rangle_l}{\partial t} + \langle u_l\rangle_l\frac{\partial\langle u_l\rangle_l}{\partial z} \ . \tag{5.85}$$

Similarly, for the vapor phase, Eqs. (5.64) and (5.83) yield,

$$-\langle\alpha\rangle\frac{\partial p}{\partial z} - \frac{g}{g_c}\rho_g\langle\alpha\rangle\sin\theta - \frac{\tau_i P_i}{A_{x-s}} - \frac{\eta_m\Gamma}{g_c}(\langle u_g\rangle_g - \langle u_l\rangle_l) = \langle\alpha\rangle\frac{\rho_g}{g_c}\frac{D_g\langle u_g\rangle_g}{Dt} \ , \tag{5.86}$$

where,

$$\frac{D_g\langle u_g\rangle_g}{Dt} \triangleq \frac{\partial\langle u_g\rangle_g}{\partial t} + \langle u_g\rangle_g \frac{\partial\langle u_g\rangle_g}{\partial z} \quad . \tag{5.87}$$

Equations (5.84) and (5.86) can be added to obtain the momentum equation for the two-phase mixture,

$$-\frac{\partial p}{\partial z} - \frac{g}{g_c}\langle\overline{\rho}\rangle \sin\theta - \frac{\tau_w P_f}{A_{x-s}} - \frac{\Gamma}{g_c}(\langle u_g\rangle_g - \langle u_l\rangle_l)$$
$$= \frac{1}{g_c}\left[(1-\langle\alpha\rangle)\rho_l \frac{D_l\langle u_l\rangle_l}{Dt} + \langle\alpha\rangle\rho_g \frac{D_g\langle u_g\rangle_g}{Dt}\right] \quad . \tag{5.88}$$

This is a form of the two-phase momentum equation that has appeared in the European and Russian literature. The last term on the left side of Eq. (5.88) is the net volumetric force due to evaporation and is frequently referred to as the "Meshcherskiy force" in the Russian literature (Kutateladze and Styrikovich, 1960). By comparing Eq. (5.79) with Eq. (5.88), it can be seen that a fundamental relationship between the acceleration of the center-of-mass of the system and the acceleration of the individual phases is given by,

$$\frac{\langle\overline{\rho}\rangle}{g_c}\frac{D_m U_m}{Dt} = \frac{1}{g_c}\left[(1-\langle\alpha\rangle)\rho_l \frac{D_l\langle u_l\rangle_l}{Dt} + \langle\alpha\rangle\rho_g \frac{D_g\langle u_g\rangle_g}{Dt}\right] + \frac{\Gamma}{g_c}(\langle u_g\rangle_g - \langle u_l\rangle_l)$$
$$- \frac{1}{g_c A_{x-s}}\frac{\partial}{\partial z}\left[A_{x-s}\left(\frac{\rho_l - \langle\overline{\rho}\rangle}{\langle\overline{\rho}\rangle - \rho_g}\right)\frac{\rho_l\rho_g}{\langle\overline{\rho}\rangle}(V'_{gj})^2\right] \quad . \tag{5.89}$$

It should be obvious to the reader that the continuity and momentum equations of two-phase flow can be manipulated and recast into many different forms. We have not tried to tabulate all the possible forms, but rather, to discuss and relate several popular forms that frequently appear in the literature of two-phase flow. For some applications, one form may have a definite advantage over another; however, all forms correctly describe the physics involved and yield the same quantitative information about the flow field.

5.1.2.6 *Energy Equation of Separated Flow*

In Secs. 5.1.2.4 and 5.1.2.5, the continuity and momentum equations of the two-phase mixture have been derived from consideration of the conservation principles for each phase.

We now consider the derivation of the phasic energy equations. As before let us write down the liquid phase conservation law for the control volume shown in Fig. 5-3. Neglecting surface tension, turbulence, shear-induced dissipation, conduction, and gravitational work terms, we have:

$$\frac{\partial}{\partial z}(\rho_l\langle u_l\rangle_l A_{x-s}\langle e_l\rangle_l)\Delta z+\Gamma e_{l_i}A_{x-s}\Delta z+p_i\frac{\partial(1-\langle\alpha\rangle)}{\partial t}A_{x-s}\Delta z-q''_w P_H\Delta z$$

$$-q'''_l(1-\langle\alpha\rangle)A_{x-s}\Delta z-q''_{l_i}P_i\Delta z+\frac{\partial}{\partial t}\left[\rho_l(1-\langle\alpha\rangle)\left(\langle e_l\rangle_l-\frac{p}{J\rho_l}\right)A_{x-s}\right]\Delta z=0\ , \tag{5.90}$$

where,

$$\langle e_k\rangle_k=h_k+\frac{\langle u_k\rangle_k^2}{2g_cJ}+\frac{gz\sin\theta}{g_cJ}\ ; \tag{5.91}$$

thus we have,

$$\frac{\partial}{\partial z}(\rho_l\langle u_l\rangle_l A_{x-s}\langle e_l\rangle_l)+\Gamma A_{x-s}e_{l_i}-p_iA_{x-s}\frac{\partial\langle\alpha\rangle}{\partial t}-q''_w P_H$$

$$-q'''_l(1-\langle\alpha\rangle)A_{x-s}-q''_{l_i}P_i+\frac{\partial}{\partial t}\left[\rho_l(1-\langle\alpha\rangle)\left(\langle e_l\rangle_l-\frac{p}{J\rho_l}\right)A_{x-s}\right]=0\ . \tag{5.92}$$

Similarly for the control volume shown in Fig. 5-3 we have for the vapor phase,

$$\frac{\partial}{\partial z}(\rho_g\langle u_g\rangle_g A_{x-s}\langle e_g\rangle_g)-\Gamma A_{x-s}e_{g_i}+p_iA_{x-s}\frac{\partial\langle\alpha\rangle}{\partial t}$$

$$-q'''_g\langle\alpha\rangle A_{x-s}+q''_{g_i}P_i+\frac{\partial}{\partial t}\left[\rho_g\langle\alpha\rangle\left(\langle e_g\rangle_g-\frac{p}{J\rho_g}\right)A_{x-s}\right]=0\ . \tag{5.93}$$

Equations (5.92) and (5.93) represent the first law of thermodynamics for each phase. The mixture energy equation can be derived by adding these equations and applying the interfacial jump condition for energy transfers. For the assumptions made, the energy jump condition is:

$$\Gamma A_{x-s}(e_{g_i}-e_{l_i})=(q''_{g_i}-q''_{l_i})P_i\ . \tag{5.94}$$

This is just a mathematical statement of the postulate that the difference in the heat fluxes at the interface results in phase change. If we assume that the interface is at saturation temperature and neglect kinetic and potential energy in Eq. (5.91), then Eq. (5.94) implies,

$$\Gamma=(q''_{g_i}-q''_{l_i})P_i/(A_{x-s}h_{fg})\ . \tag{5.95}$$

Once the interfacial heat fluxes (q''_{k_i}) are constituted, Eq. (5.95) can be used to evaluate Γ.

Adding Eqs. (5.92) and (5.93) and using Eq. (5.94) we obtain the mixture energy equation in the form:

$$\frac{\partial}{\partial z}(w_l\langle e_l\rangle_l)+\frac{\partial}{\partial z}(w_g\langle e_g\rangle_g)-q_w''P_H-q'''A_{x-s}$$
$$+\frac{\partial}{\partial t}\left[\rho_l(1-\langle\alpha\rangle)\left(\langle e_l\rangle_l-\frac{p}{J\rho_l}\right)+\rho_g\langle\alpha\rangle\left(\langle e_g\rangle_g-\frac{p}{J\rho_g}\right)\right]A_{x-s}=0 \;, \qquad (5.96)$$

where,

$$q'''\triangleq q_l'''(1-\langle\alpha\rangle)+q_g'''\langle\alpha\rangle \;. \qquad (5.97)$$

By combining Eqs. (5.91) and (5.21) with Eq. (5.96), we obtain

$$\frac{\partial}{\partial z}[GA_{x-s}(1-\langle x\rangle)h_l]+\frac{\partial}{\partial z}\left[A_{x-s}\frac{G^3(1-\langle x\rangle)^3}{2g_cJ\rho_l^2(1-\langle\alpha\rangle)^2}\right]$$
$$+\frac{\partial}{\partial z}\left[GA_{x-s}(1-\langle x\rangle)\frac{gz\sin\theta}{g_cJ}\right]+\frac{\partial}{\partial z}(GA_{x-s}\langle x\rangle h_g)$$
$$+\frac{\partial}{\partial z}\left(A_{x-s}\frac{G^3\langle x\rangle^3}{2g_cJ\rho_g^2\langle\alpha\rangle^2}\right)+\frac{\partial}{\partial z}\left(GA_{x-s}\langle x\rangle\frac{g}{g_c}\frac{z\sin\theta}{J}\right)$$
$$-q''P_H-q'''A_{x-s}+\frac{\partial}{\partial t}[\rho_l(1-\langle\alpha\rangle)h_l+\rho_g\langle\alpha\rangle h_g]A_{x-s}$$
$$+\frac{\partial}{\partial t}\left\{\frac{G^2}{2g_cJ}\left[\frac{(1-\langle x\rangle)^2}{\rho_l(1-\langle\alpha\rangle)}+\frac{\langle x\rangle^2}{\rho_g\langle\alpha\rangle}\right]\right\}A_{x-s}$$
$$+\frac{\partial}{\partial t}\left(\langle\overline{\rho}\rangle\frac{g}{g_c}\frac{z\sin\theta}{J}\right)A_{x-s}-\frac{1}{J}\frac{\partial p}{\partial t}A_{x-s}=0 \;. \qquad (5.98)$$

Simplifying, Eq. (5.98) becomes,

$$\frac{\partial}{\partial t}(\langle\overline{\rho}\rangle\overline{h})A_{x-s}+\frac{\partial}{\partial z}(GA_{x-s}\langle h\rangle)=q''P_H+q'''A_{x-s}-\frac{\partial}{\partial z}\left(\frac{G^3A_{x-s}}{2g_cJ\langle\rho'''\rangle^2}\right)$$
$$-\frac{g}{g_c}\frac{\sin\theta}{J}GA_{x-s}-\frac{\partial}{\partial t}\left(\frac{G^2}{2g_cJ\langle\rho'\rangle}\right)A_{x-s}+\frac{1}{J}\frac{\partial p}{\partial t}A_{x-s} \;, \qquad (5.99)$$

where Eq. (5.57) has been used to simplify the potential energy term, and we have defined,

$$\overline{h}\triangleq\frac{1}{\langle\overline{\rho}\rangle}[\rho_l(1-\langle\alpha\rangle)h_l+\rho_g\langle\alpha\rangle h_g]\equiv h_m \qquad (5.100)$$

$$\frac{1}{\langle\rho'''\rangle^2} \triangleq \left[\frac{(1-\langle x\rangle)^3}{\rho_l^2(1-\langle\alpha\rangle)^2} + \frac{\langle x\rangle^3}{\rho_g^2\langle\alpha\rangle^2}\right] \tag{5.101}$$

$$\langle h\rangle \triangleq h_l + \langle x\rangle h_{lg} \ ,$$

and $\langle\rho'\rangle$ was defined previously in Eq. (5.67).

In some cases, it may be more convenient to expand the left side of Eq. (5.99) and to combine it with Eq. (5.57) to yield (Meyer, 1960),

$$\langle\rho''\rangle\frac{\partial\langle h\rangle}{\partial t} + G\frac{\partial\langle h\rangle}{\partial z} = \frac{q''P_H}{A_{x-s}} + q''' - \frac{1}{A_{x-s}}\frac{\partial}{\partial z}\left(\frac{G^3A_{x-s}}{2g_cJ\langle\rho'''\rangle^2}\right)$$

$$-\frac{g}{g_c}\frac{\sin\theta}{J}G - \frac{\partial}{\partial t}\left(\frac{G^2}{2g_cJ\langle\rho'\rangle}\right) + \frac{1}{J}\frac{\partial p}{\partial t} \ , \tag{5.102}$$

where for the special case of a saturated two-phase mixture, in which the properties are only a function of $\langle h\rangle$,

$$\langle\rho''\rangle \triangleq \left[\langle\bar{\rho}\rangle\frac{d\bar{h}}{d\langle h\rangle} + (\bar{h} - \langle h\rangle)\frac{d\langle\bar{\rho}\rangle}{d\langle h\rangle}\right] = [\rho_g + (\rho_f - \rho_g)\langle x\rangle]\frac{d\langle\alpha\rangle}{d\langle x\rangle} \ . \tag{5.103}$$

Equation (5.102) has been widely used throughout the nuclear industry; however, when it is used, we are required to deal with several pseudo densities, $\langle\rho'\rangle$, $\langle\rho''\rangle$, $\langle\rho'''\rangle$. Moreover, Eq. (5.102) is not a true Lagrangian formulation and, thus, it has a limited physical interpretation.

Zuber has addressed himself to these shortcomings and has introduced an alternate drift-flux form of the energy equation (Zuber, 1967), which more readily lends itself to phenomenological interpretation. Equation (5.96) can be rewritten as,

$$\frac{1}{J}\frac{\partial p}{\partial t}A_{x-s} + q''P_H + q'''A_{x-s}$$

$$= \frac{\partial}{\partial t}(\langle\bar{\rho}\rangle\bar{e}A_{x-s}) + \frac{\partial}{\partial z}[\langle e_g\rangle_g\rho_g\langle\alpha\rangle A_{x-s}\langle u_g\rangle_g + \langle e_l\rangle_l\rho_l(1-\langle\alpha\rangle)A_{x-s}\langle u_l\rangle_l] \ , \tag{5.104}$$

where,

$$\bar{e} \triangleq \frac{1}{\langle\bar{\rho}\rangle}[\rho_l(1-\langle\alpha\rangle)\langle e_l\rangle_l + \rho_g\langle\alpha\rangle\langle e_g\rangle_g] \equiv e_m \ . \tag{5.105}$$

The phase velocities can be eliminated from Eq. (5.104) by combining it with Eqs. (5.68), (5.69), and (5.70) to yield,

$$\frac{1}{J}\frac{\partial p}{\partial t}A_{x-s} + q''P_H + q'''A_{x-s} - \frac{\partial}{\partial z}\left[A_{x-s}\frac{\langle\alpha\rangle\rho_l\rho_g}{\langle\bar{\rho}\rangle}V'_{gj}(\langle e_g\rangle_g - \langle e_l\rangle_l)\right] = \langle\bar{\rho}\rangle A_{x-s}\frac{D_m\bar{e}}{Dt} \ , \tag{5.106}$$

where Eqs. (5.57) and (5.78) have been used in conjunction with the identity, $U_m = G/\langle\overline{\rho}\rangle$.

Equation (5.106) is a generalization of the Lagrangian form of the two-phase energy equation (Zuber, 1967). The last term on the left side of Eq. (5.106) represents the drift of energy through the center-of-mass plane.

Another form of the two-phase energy equation, which has appeared in European literature, can be obtained by combining Eqs. (5.54) and (5.56) with Eq. (5.96) to yield,

$$\rho_l(1-\langle\alpha\rangle)\frac{D_l\langle e_l\rangle_l}{Dt} + \rho_g\langle\alpha\rangle\frac{D_g\langle e_g\rangle_g}{Dt} = \frac{q''P_H}{A_{x-s}} + q''' + \frac{1}{J}\frac{\partial p}{\partial t} - \Gamma(\langle e_g\rangle_g - \langle e_l\rangle_l) \quad . \tag{5.107}$$

This form is compatible with the form of the two-phase momentum equation given in Eq. (5.88). The last term on the right side of Eq. (5.107) represents the volumetric rate of energy transfer due to the evaporation process. Equation (5.107) is frequently used as the general equation for evaluating Γ. For the special case of saturated equilibrium, constant pressure, constant thermodynamic properties, and negligible internal generation (q'''), kinetic energy and potential energy, Eq. (5.107) yields Eq. (5.55).

It should be obvious that the same information about mixture energy conservation is contained in Eqs. (5.99), (5.102), (5.106), and (5.107). Furthermore, numerous other forms can be derived from Eq. (5.96).

Before discussing the application of these equations, it is interesting to consider a compact form of the conservation equations (Yadigaroglu and Lahey, 1975). To derive these equations, we consider the velocities shown in Figs. 5-5a, b, and c. In Fig. 5-5a, U_m is the velocity of the center-of-mass of the two-phase system. That is, it is the velocity of propagation of the plane through which no net mass flux passes. By equating the mass flux terms in Fig. 5-5a,

$$\rho_g\langle\alpha\rangle(\langle u_g\rangle_g - U_m) = \rho_l(1-\langle\alpha\rangle)(U_m - \langle u_l\rangle_l) \quad ,$$

which yields,

$$U_m \triangleq \frac{[\rho_l\langle u_l\rangle_l(1-\langle\alpha\rangle) + \rho_g\langle u_g\rangle_g\langle\alpha\rangle]}{[(1-\langle\alpha\rangle)\rho_l + \langle\alpha\rangle\rho_g]} = G/\langle\overline{\rho}\rangle \quad , \tag{5.108}$$

in which the definitions in Eqs. (5.19) and (5.58) have been used.

Similarly, by equating the momentum flux terms in Fig. 5-5b to define the plane through which no net momentum flux passes,

$$\frac{\langle u_g\rangle_g}{g_c}[\rho_g\langle\alpha\rangle(\langle u_g\rangle_g - U_p)] = \frac{\langle u_l\rangle_l}{g_c}[\rho_l(1-\langle\alpha\rangle)(U_p - \langle u_l\rangle_l)] \quad ,$$

which yields the center-of-momentum velocity as,

$$U_p \triangleq \frac{[\rho_l\langle u_l\rangle_l^2(1-\langle\alpha\rangle) + \rho_g\langle u_g\rangle_g^2\langle\alpha\rangle]}{[\rho_l\langle u_l\rangle_l(1-\langle\alpha\rangle) + \rho_g\langle u_g\rangle_g\langle\alpha\rangle]} = G/\langle\rho'\rangle \quad , \tag{5.109}$$

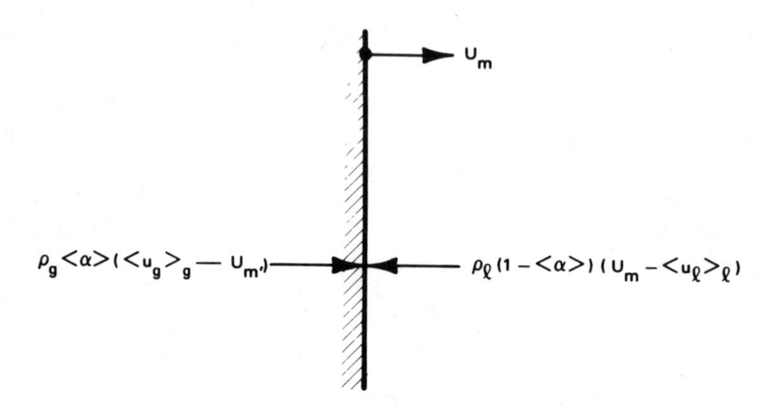

a. THE VELOCITY OF THE CENTER-OF-MASS (U_m)

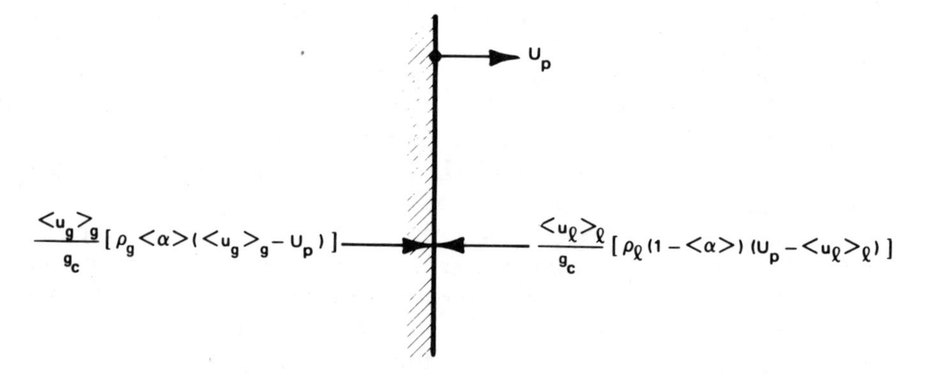

b. THE VELOCITY OF THE CENTER-OF-MOMENTUM (U_p)

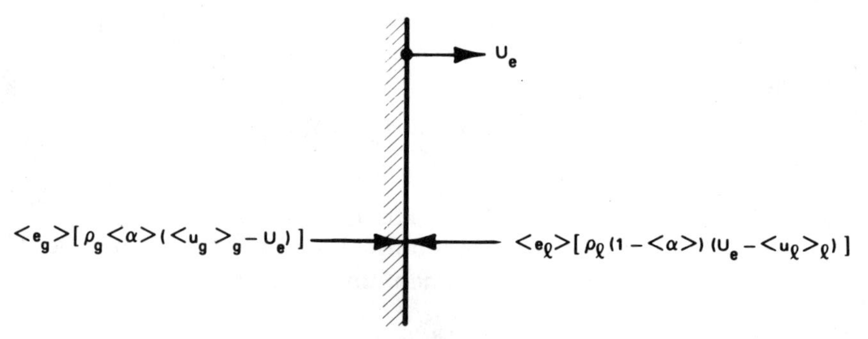

c. THE VELOCITY OF THE CENTER-OF-ENERGY (U_e)

Fig. 5-5 Two-phase velocities.

in which the definitions in Eqs. (5.22), (5.23), and (5.67) have been used.

Finally, defining the plane through which no net energy flux passes, by equating the energy flux terms in Fig. 5-5c,

$$\langle e_g\rangle_g[\rho_g\langle\alpha\rangle(\langle u_g\rangle_g - U_e)] = \langle e_l\rangle_l[\rho_l(1-\langle\alpha\rangle)(U_e - \langle u_l\rangle_l] \ ,$$

which yields the center-of-energy velocity,

$$U_e \triangleq \frac{[\rho_g\langle e_g\rangle_g\langle\alpha\rangle\langle u_g\rangle_g + \rho_l\langle e_l\rangle_l(1-\langle\alpha\rangle)\langle u_l\rangle_l]}{[\rho_l(1-\langle\alpha\rangle)\langle e_l\rangle_l + \rho_g\langle\alpha\rangle\langle e_g\rangle_g]} \ . \tag{5.110}$$

By using the identities given in Eqs. (5.22), (5.23), and (5.105), and by noting that $\langle e\rangle \triangleq \langle e_l\rangle + \langle x\rangle\langle e_{lg}\rangle$,

$$U_e \triangleq \frac{G\langle e\rangle}{\bar{e}\langle\bar{\rho}\rangle} = U_m \frac{\langle e\rangle}{\bar{e}} \ . \tag{5.111}$$

The continuity equation in terms of the velocity of the center-of-mass has already been given as,

$$\frac{D_m}{Dt}(\langle\bar{\rho}\rangle A_{x-s}) + \langle\bar{\rho}\rangle A_{x-s}\frac{\partial U_m}{\partial z} = 0 \ . \tag{5.112}$$

The momentum equation can be written in terms of the velocity of the center-of-momentum by expanding the right side of Eq. (5.66) and combining it with Eq. (5.109) to yield,

$$-\frac{\partial p}{\partial z}A_{x-s} - \frac{g}{g_c}\langle\bar{\rho}\rangle A_{x-s}\sin\theta - \tau_w P_f = \frac{1}{g_c}\left[\frac{D_p(GA_{x-s})}{Dt} + (GA_{x-s})\frac{\partial U_p}{\partial z}\right] \ , \tag{5.113}$$

where,

$$\frac{D_p(\square)}{Dt} \triangleq \frac{\partial(\square)}{\partial t} + U_p\frac{\partial(\square)}{\partial z} \ . \tag{5.114}$$

Finally, the energy equation can be written in terms of the velocity of the center-of-energy by introducing the identities in Eqs. (5.22) and (5.23) into Eq. (5.96) and regrouping the terms to obtain,

$$\frac{D_e}{Dt}(\langle\bar{\rho}\rangle\bar{e}A_{x-s}) + \langle\bar{\rho}\rangle\bar{e}A_{x-s}\frac{\partial U_e}{\partial z} = q''P_H + q'''A_{x-s} + \frac{\partial p}{\partial t}A_{x-s} \ , \tag{5.115}$$

where,

$$\frac{D_e(\square)}{Dt} \triangleq \frac{\partial(\square)}{\partial t} + U_e\frac{\partial(\square)}{\partial z} \ . \tag{5.116}$$

Clearly the conservation equations given by Eqs. (5.112), (5.113), and (5.115) are quite compact and phenomenologically appealing. Nevertheless, we must deal with three different velocities and, thus, they yield no fundamental simplification. Moreover, these equations contain no more or less information about the flow field than any of the other forms presented so far; hence, their use is largely a matter of choice.

Again, let us note that this has not been an exhaustive treatment of all possible forms of the two-phase conservation equations. However, several important forms, which appear in the two-phase literature, have been derived from basic principles by using the same assumptions and approximations. The reader should use the set of equations that he or she is most familiar with and that best satisfies the particular need. In the subsequent analyses performed in this monograph, Meyer's form of the conservation equations (Meyer, 1960) is frequently used, primarily because this form is so widely accepted in the nuclear engineering literature. Naturally, any of the relationships that are derived can be obtained from the other forms; however, this exercise is left to the interested reader.

5.2 Flow Regimes

When discussing two-phase flow phenomena, it is necessary to discuss flow regime configurations and analysis. The unique feature of multiphase flow is its ability to take on different spatial distributions. Indeed, this feature is one of the prime obstacles to an exact analytical treatment. That is, it may be difficult to specify with any degree of accuracy when a given flow regime exists and where transitions occur.

For many practical applications, the conservation equations, developed in Sec. 5.1, are solved independently of any explicit consideration as to what flow regime exists. This global point of view requires correlations for τ_w, $\langle\alpha\rangle$, V_{gj}, etc., which, in effect, account for variations with flow regime.

An alternate approach is to use the appropriate analysis for each flow regime and to couple these analyses at transition points. The present state-of-the-art is such that this is quite difficult to accomplish; however, some knowledge of flow regimes is available for possible use in BWR technology.

The various flow regimes that can occur in BWR thermal-hydraulic analyses are indicated in Fig. 5-6. It is customary to display these flow regimes in a so-called "flow regime map." A typical flow regime map is shown in Fig. 5-7. It can be seen that the coordinates of this map are the superficial velocities $\langle j_g\rangle$ and $\langle j_l\rangle$. Since,

$$\langle j_g\rangle = G\langle x\rangle/\rho_g \tag{5.117a}$$

and,

$$\langle j_l\rangle = G(1-\langle x\rangle)/\rho_l \ , \tag{5.117b}$$

we see that the coordinates could have equally well have been G and $\langle x\rangle$. Indeed, other flow regime maps (e.g., Bergles and Suo, 1966) have been displayed in this fashion.

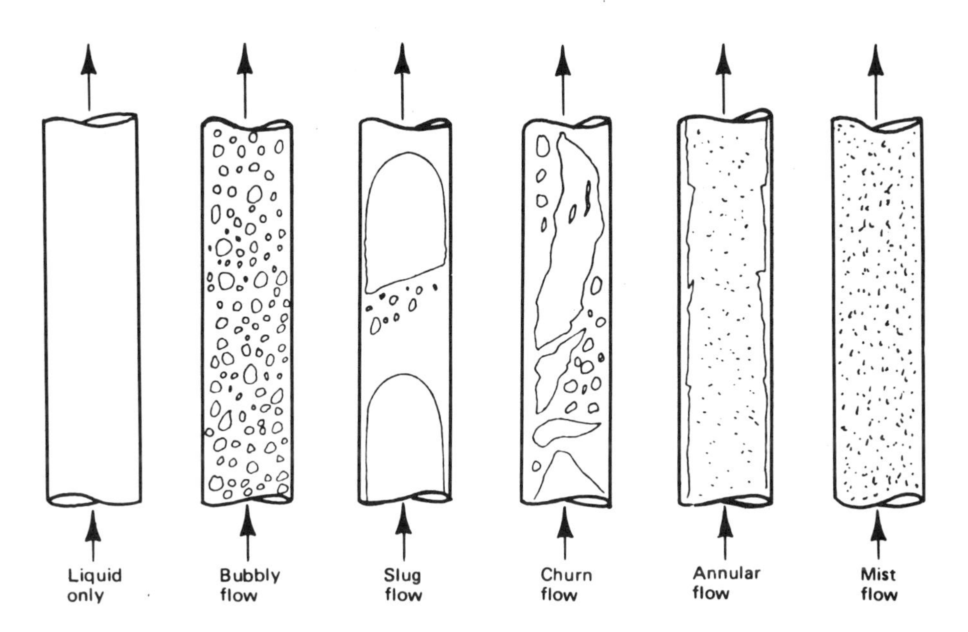

Fig. 5-6 Typical flow regime patterns in vertical flow.

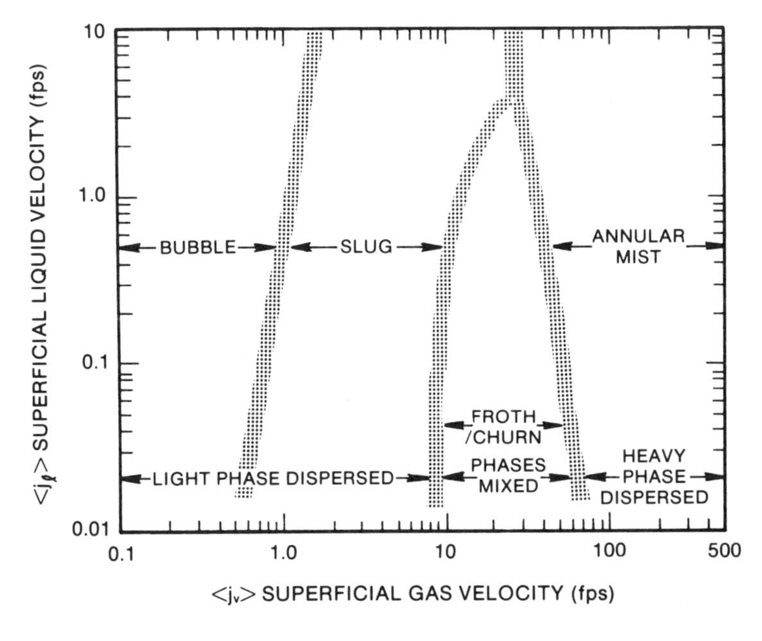

Fig. 5-7 Govier and Aziz flow regime map of vertical air/water at standard temperature and pressure.

It should be stressed that a flow regime map is nothing more than an empirical display of the experimentally observed flow regimes. Moreover, the coordinates, $\langle j_k \rangle$, do not account for variations in fluids or operating conditions (e.g., pressure). As a consequence, the flow regime map shown in Fig. 5-7 is, at best, only valid for steady air/water flow at standard temperatures and pressure (STP).

Let us now consider the flow regime boundaries. Radovich and Moissis (1962) have demonstrated that a large increase in the amount of bubble collision and coalescence occurs as the global void fraction, $\langle \alpha \rangle$, approaches 30%. Thus, a frequently used "rule of thumb" is to assume that the bubbly/slug flow regime transition occurs at,

$$\langle \alpha \rangle_c = 0.30 \ . \tag{5.118a}$$

Similarly, it is frequently assumed that the slug/froth[b] flow regime transition occurs in the range of global void fractions from 65 to 80%. That is,

$$\langle \alpha \rangle_c = 0.65 - 0.80 \ . \tag{5.118b}$$

One must use caution in applying these criteria. In particular, it is well known that water purity and the addition of surface-active agents can dramatically affect the bubbly/slug transition criterion. A good example is seen for foaming fluids (e.g., in fire extinguishers) in which bubbly flow still occurs for void fractions much greater than 30%. Nevertheless, for relatively pure water, such as that used in BWRs, Eq. (5.118a) is often a good first approximation.

The so-called Zuber/Findlay drift-flux model will be discussed in detail in Sec. 5.3. This relationship can be written as,

$$\langle j_g \rangle = \frac{\langle \alpha \rangle C_0}{(1 - \langle \alpha \rangle C_0)} \langle j_l \rangle + \frac{\langle \alpha \rangle V_{gj}}{(1 - \langle \alpha \rangle C_0)} \ . \tag{5.119}$$

It is interesting to note that if the transition void fractions, $\langle \alpha \rangle_c$, given in Eqs. (5.118) are used in Eq. (5.119) the bubbly/slug and the slug/froth flow regime boundaries shown in Fig. 5-7 are predicted fairly well.

5.2.1 Flow Regime Analysis

The analysis of two-phase flow systems can be quite accurate for situations in which the flow regime is known. In these cases the conservation and constitutive equations that apply to that particular flow regime are used. For example, many cases of two-component two-phase flow, such as air/water systems, have been solved using this method. The general techniques of one-dimensional flow regime analysis have been well developed (Wallis, 1969) and are not repeated here.

---

[b]The froth flow regime is often called the churn-turbulent flow regime.

In this section some of the analytical features of those flow regimes of particular importance to BWR technology will be discussed, although the treatment will be necessarily quite abbreviated.

5.2.1.1 *Bubbly Flow*

There is no such thing as fully developed bubbly flow, since the bubbles tend to grow due to static pressure changes and may coalesce and agglomerate as they flow through a conduit. The speed with which this process occurs depends on many factors, including impurities and surface-active agents. Nevertheless, for void fractions less than about 50%, an idealized bubbly flow regime can be discussed. In this sense bubbly flow can be considered as the case in which the bubbles flow with negligible interaction (i.e., the ideal bubbly regime).

In the bubbly flow regime, the momentum equations should be modified to account for the effect of virtual mass. This effect is the result of the relative acceleration of the vapor phase with respect to the liquid phase. As the bubbles accelerate through the liquid, they increase the kinetic energy of the liquid phase that drains around them. It is convenient to model this inviscid effect as an increase in the effective mass of the accelerating bubble by some fraction of the mass of liquid displaced. That is, the virtual mass of the bubble is the true mass plus an additional hydrodynamic mass. For our purposes, it is convenient to regard this additional mass, times the relative acceleration, as an additional force on the bubble. This "virtual mass force" will be zero if there is no relative acceleration. In general, the importance of this term will depend on the magnitude of the relative acceleration, the interaction between bubbles, and the magnitude of the other forces on the bubble.

The virtual mass effect can be accounted for by adding a virtual mass force to the left-hand side of the vapor phase momentum equation, Eq. (5.64). This volumetric force is given by:

$$M_g^{(vm)} = \frac{-\rho_l C_{vm} \langle\alpha\rangle \Delta z A_{x-s}}{g_c} \left(\frac{D_g \langle u_g \rangle_g}{Dt} - \frac{D_l \langle u_l \rangle}{Dt} \right) . \tag{5.120a}$$

Similarly, we must add,

$$M_l^{(vm)} = \frac{\rho_l C_{vm} \langle\alpha\rangle}{g_c} \left(\frac{D_g \langle u_g \rangle_g}{Dt} - \frac{D_l \langle u_l \rangle}{Dt} \right) \tag{5.120b}$$

to the left-hand side of the liquid phase momentum equation, Eq. (5.62). It can be seen that these are equal and opposite "forces" such that they will cancel when the momentum equations for each phase are added to yield the momentum equation of the mixture. The virtual volume parameter, C_{vm}, depends on the geometric configuration of, and interaction between, the bubbles. For a spherical, noninteracting bubble, it can be shown that $C_{vm} = ½$ (Milne-Thompson, 1961).

It should be stressed that the force due to virtual mass is not a viscous drag term due to relative slip between the phases. This latter effect is given by the interfacial shear term,

$$\frac{\tau_i P_i}{A_{x-s}} = \frac{3}{4}\left(\frac{C_d}{\langle D_b \rangle}\right) \rho_l \langle\alpha\rangle(\langle\langle u_g\rangle_g - \langle u_l\rangle_l)\ |\langle u_g\rangle_g - \langle u_l\rangle_l| \ . \tag{5.121}$$

The magnitude of the term, $C_d/\langle D_b\rangle$, depends on the hydrodynamic conditions of the bubbly flow regime. For example, for distorted bubbly flow Harmathy (1960) recommends,

$$\left(\frac{C_d}{\langle D_b\rangle}\right) = \frac{2}{3}\left[\frac{g(\rho_l - \rho_g)}{g_c \sigma(1-\alpha)}\right]^{1/2} \ . \tag{5.122}$$

Quite a bit of experimental and analytical work has also been done on the kinematics of bubbly flow. The discussion presented below is based largely on the work of Zuber and Hench (1962).

Equations (5.8) and (5.10) yield,

$$\langle u_R\rangle \triangleq \langle u_g\rangle_g - \langle u_l\rangle_l = \frac{\langle j_g\rangle}{\langle\alpha\rangle} - \frac{\langle j_l\rangle}{(1-\langle\alpha\rangle)} \ . \tag{5.123}$$

Equation (5.123) can be combined with the identity,

$$\langle j\rangle = \langle j_g\rangle + \langle j_l\rangle \tag{5.124}$$

to yield,

$$\langle u_R\rangle = \frac{\langle j_g\rangle}{\langle\alpha\rangle} + \frac{\langle j_g\rangle}{(1-\langle\alpha\rangle)} - \frac{\langle j\rangle}{(1-\langle\alpha\rangle)} \ . \tag{5.125}$$

Equation (5.125) is a general kinematic result for relative velocity.

Zuber and Hench (1962) determined the following empirical relationship between the relative velocity and void fraction:

$$\langle u_R\rangle = U_t(1-\langle\alpha\rangle)^m \ , \tag{5.126}$$

where U_t is the terminal rise velocity of a single bubble in a still tank. Equation (5.126) is related to an equivalent drift-flux relationship synthesized by Wallis (1969):

$$\langle j_{gl}\rangle = U_t\langle\alpha\rangle(1-\langle\alpha\rangle)^n \tag{5.127}$$

through the identity,

$$\langle j_{gl}\rangle = \langle\alpha\rangle(1-\langle\alpha\rangle)\langle u_R\rangle \ . \tag{5.128}$$

Combining Eqs. (5.127) and (5.128),

$$\langle u_R\rangle = U_t(1-\langle\alpha\rangle)^{n-1} \ . \tag{5.129}$$

Thus, from Eqs. (5.126) and (5.129) we see that, $m=n-1$.

It is interesting to note that Eq. (5.70) implies,

$$V'_{gj}=\langle u_g\rangle_g-\langle j\rangle=(1-\langle\alpha\rangle)\langle u_R\rangle \ . \tag{5.130}$$

Hence, Eqs. (5.130) and (5.126) imply,

$$V'_{gj}=U_t(1-\langle\alpha\rangle)^{m+1} \ . \tag{5.131}$$

The numerical value of the parameter, m, depends on a number of parameters, including the size of the bubble. The data (Haberman and Morton, 1953) shown in Fig. 5-8 indicate this dependence and the range of typical values of the parameter $m(m\leqslant 1)$. For large-diameter bubbles, wake interactions become quite strong, and it is recommended (Zuber and Hench, 1962) that for churn turbulent bubbly flows $m=-1$. This is an interesting case, since, in accordance with Eq. (5.131) it implies,

$$V'_{gj}=U_t \ . \tag{5.132}$$

Equations (5.123) and (5.126) can be combined to yield,

$$\langle\alpha\rangle V'_{gj}=U_t\langle\alpha\rangle(1-\langle\alpha\rangle)^{m+1}=\langle j_g\rangle(1-\langle\alpha\rangle)-\langle j_l\rangle\langle\alpha\rangle \ . \tag{5.133}$$

This is the general kinematic relationship deduced by Zuber and Hench (1962). To acquaint the reader with its use, several special cases will be discussed.

Let us first consider a "batch process." This is the case in which a gas is bubbled through a still pool of liquid. In this case, $\langle j_l\rangle=0$, and Eq. (5.133) yields,

$$\langle j_g\rangle/U_t=\langle\alpha\rangle(1-\langle\alpha\rangle)^m \ . \tag{5.134}$$

Several interesting properties of this equation are worth noting. The maximum possible void fraction is obtained by differentiating $\langle j_g\rangle/U_t$ with respect to $\langle\alpha\rangle$ and setting the result equal to zero. This procedure yields $\langle\alpha\rangle_{max}=1/(m+1)$, which implies that if,

$$\langle j_g\rangle/U_t > \frac{m^m}{(m+1)^{m+1}} \ , \tag{5.135}$$

then a flow regime change must take place. For example, if $m=1$ (a typical value for small bubbles), then if $\langle j_g\rangle/U_t > 1/4$, a transition from ideal bubbly flow occurs.

It is interesting to note that at a given value of $\langle j_g\rangle/U_t$, Eq. (5.134) implies that several different $\langle\alpha\rangle$ may satisfy it. In particular, for $m=1$, this model shows that we may have a low void fraction region and a high void fraction region (i.e., a foam). This type of analysis is particularly useful in the case of foam drainage problems (Wallis, 1969).

Finally, we may also note that since $\langle j_l\rangle=0$,

$$\langle u_R\rangle=\langle j_g\rangle/\langle\alpha\rangle=\langle u_g\rangle \ . \tag{5.136}$$

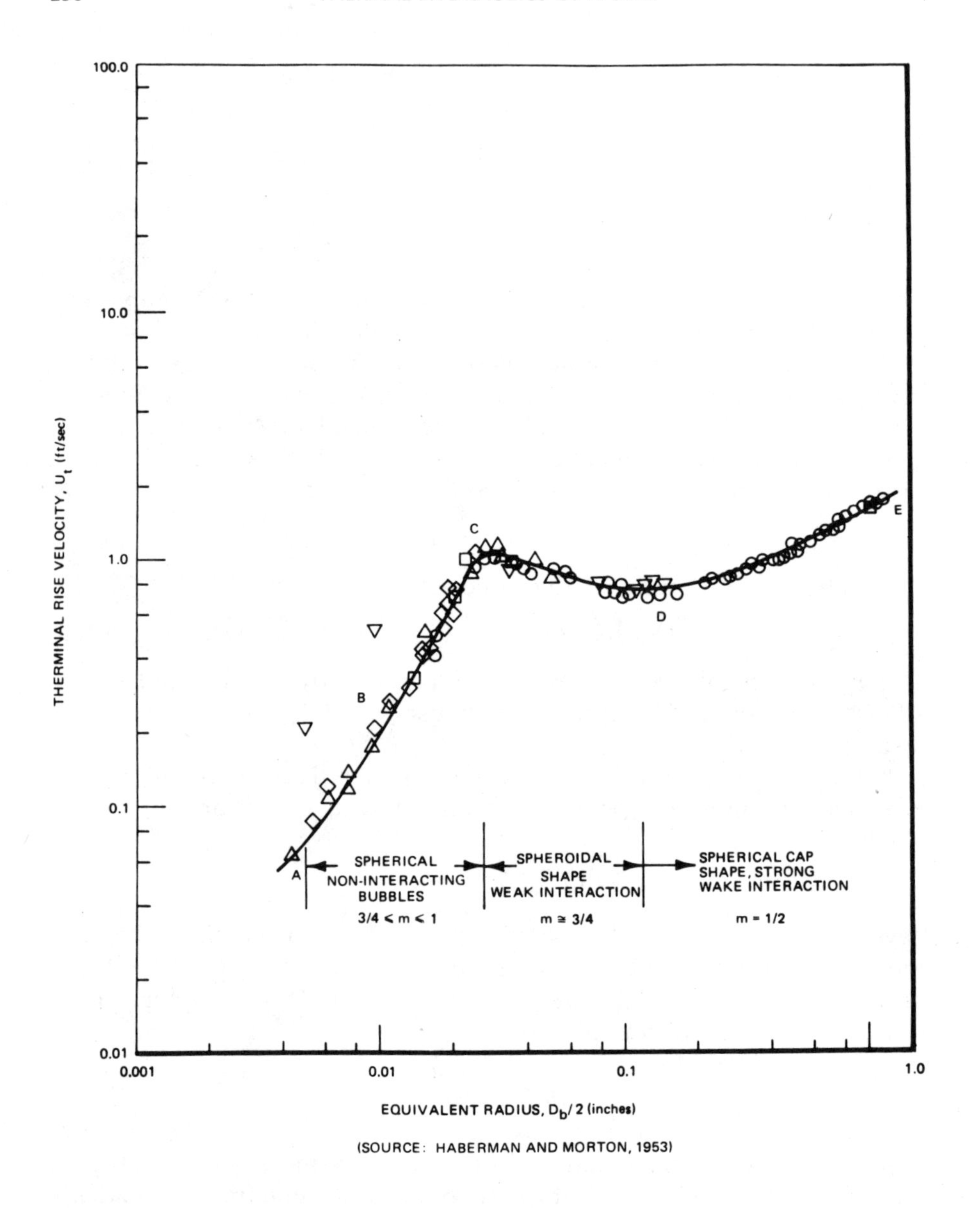

Fig. 5-8 Terminal rise velocity of air bubbles in filtered or distilled water as a function of bubble size.

Next, let us consider the case of countercurrent flow, in which the gas flows upward and the liquid flows downward. As an example, this is the case that occurs when a flask of liquid is drained. The unique feature of this case is that we have a volume-for-volume exchange of each phase. Hence, $\langle j_l \rangle = -\langle j_g \rangle$. Thus, Eq. (5.133) becomes,

$$\langle j_g \rangle / U_t = \langle \alpha \rangle (1 - \langle \alpha \rangle)^{m+1} \ . \tag{5.137}$$

By combining Eqs. (5.137) and (5.126) to eliminate the terminal rise velocity, U_t, we obtain:

$$\langle u_R \rangle (1 - \langle \alpha \rangle) = \frac{\langle j_g \rangle}{\langle \alpha \rangle} = \langle u_g \rangle \ . \tag{5.138}$$

This relationship is of interest, since it implies that the relative velocity, $\langle u_R \rangle$, is greater than the velocity of the vapor phase, $\langle u_g \rangle$. This is simply a consequence of the fact that the vapor velocity is taken relative to the liquid phase, which is flowing in a negative direction.

The case of most interest in BWR technology is co-current upward flow, a situation described by Eq. (5.133). Bubbly flow regimes of this type are common in the lower part of BWR fuel rod bundles.

5.2.1.2 *Slug Flow*

Unlike bubbly flow, slug flow is not random. That is, the void fraction fluctuates periodically. Although clearly discernible slug flow is rare in a diabatic system, it is very often seen in adiabatic systems. Fortunately, due in large part to the work of Griffith at MIT (e.g., Griffith and Wallis, 1961), the analysis of the slug flow is in fairly good shape.

It has been found that the terminal rise velocity can be taken to be the drift velocity,

$$U_t = V_{gj} \ . \tag{5.139}$$

Fractional analysis can be applied to determine the appropriate functional form of the terminal rise velocity. The most important group has been found to be the ratio of the inertia to the buoyancy force, given in Eq. (5.170). Solving for the terminal rise velocity yields

$$U_t = k_1 \left[\frac{g D_b (\rho_l - \rho_g)}{\rho_l} \right]^{1/2} \ . \tag{5.140}$$

The empirical parameter k_1 has been determined experimentally (White and Beardmore, 1962) and analytically (Davies and Taylor, 1950) to be,

$$k_1 = 0.345 \ . \tag{5.141}$$

For the more general case in which viscous and/or surface tension effects are important, one may work with the pertinent force terms in Eqs. (5.169) or modify the parameter k_1 in Eq. (5.140) to reflect the relative importance of these other effects.

A modified parameter has been proposed by Wallis (1969):

$$k_1 = 0.345[1 - \exp(-0.01N_l/0.345)]\{1 - \exp[(3.37 - N_{EÖ}/M]\} , \quad (5.142)$$

in which,

$$N_l \triangleq \frac{[D_b^3 g(\rho_l - \rho_g)\rho_l]^{1/2}}{g_c \mu_l} , \text{ dimensionless inverse viscosity} \quad (5.143)$$

$$N_{EÖ} \triangleq \frac{g D_b^2(\rho_l - \rho_g)}{g_c \sigma} , \text{ Eötvös number} \quad (5.144)$$

$$M \triangleq \begin{cases} 10, & N_l > 250 \\ 69 N_l^{-0.35}, & 18 < N_l < 250 \\ 25, & N_l < 18 \end{cases} . \quad (5.145)$$

Thus Eqs. (5.140) through (5.145) allow one to calculate the terminal rise velocity for slug flow for many cases of practical interest. As in bubbly flow, the values of k_1 may need to be modified due to channel inclination and geometric effects, thus design values of this parameter should be experimentally verified if at all possible.

The slug flow regime is of interest in BWR technology, since it may occur in the steam separator stand pipes. However, the flow regime of most interest in BWRs is the annular flow regime. Annular flow, in which the conduit boundaries are wet by the liquid while the vapor phase primarily collects in interior regions, occurs when slug flow breaks down. That is, when the liquid slugs between the large vapor bubbles disappear. There is no clearly defined transition between the slug and annular flow regimes. Moreover, as shown in Fig. 5-7, for lower liquid flow rates there is a region between the slug/annular transition that is frequently referred to as the froth (or churn-turbulent) flow regime.

A criterion has been proposed (Wallis, 1969) by which the slug/annular transition can be predicted. This criterion is given in terms of the normalized superficial velocities, $\langle j_g^* \rangle$ and $\langle j_l^* \rangle$. These parameters are the square root of the phasic Froude numbers and represent the ratio of the momentum flux of each phase to the buoyancy force. They are given by,

$$\langle j_g^* \rangle \triangleq \langle j_g \rangle \rho_g^{1/2} / [g D_H(\rho_l - \rho_g)]^{1/2} \quad (5.146a)$$

$$\langle j_l^* \rangle \triangleq \langle j_l \rangle \rho_l^{1/2} / [g D_H(\rho_l - \rho_g)]^{1/2} , \quad (5.146b)$$

where the bubble diameter, D_b, has been replaced by the hydraulic diameter, D_H, since the large, spherical cap bubbles of the slug flow regime generally fill the conduit. The transition criterion is given by,

$$\langle j_l^* \rangle = 0.4 + 0.6 \langle j_l^* \rangle . \quad (5.147)$$

If, $\langle j_g^* \rangle \geqslant (0.4 + 0.6\langle j_l^* \rangle)$, one should expect annular flow. Otherwise bubbly or slug slow conditions would be expected.

Equation (5.147) is compared with data (Bergles and Suo, 1966) in Fig. 5-9. It can be seen that for the lower mass fluxes ($G/10^6 \leqslant 1.0$ lb$_m$/h-ft^2) the trends are generally correct, and due to the subjective nature of determining the transition boundaries, it is reasonably accurate. Thus, Eq. (5.147) can be used for estimating the slug/annular transition boundary for the lower mass fluxes. Also the rule of thumb, $\langle \alpha \rangle_c \cong 0.65$, is shown in Fig. 5-9. We note that this simple criterion and Eq. (5.147) are reasonably consistent.

5.2.1.3 *Annular Flow*

The annular flow regime is quite important in BWR technology. This is primarily because modern BWRs normally operate so that annular flow conditions exist in the upper portion of the fuel rod bundles. This flow regime has been extensively studied and an excellent treatise has been written by Hewitt and Hall-Taylor (1970). Due to the existence of this reference, the treatment that will be given to the annular flow regime in this section will be rather brief.

The unique feature of annular flow is that the liquid phase collects on the walls of the conduit, while the vapor phase and any entrained liquid tend to collect in the open, higher velocity regions. There are various arbitrary subdivisions of annular flow that may exist. Ideal annular flow is the situation in which there is no entrainment of the liquid phase and there is a smooth, symmetric liquid-vapor interface. This situation is highly idealized, but is quite useful for analytical purposes. As indicated in Fig. 5-9, wispy annular flow is the situation frequently observed at higher mass fluxes ($G > 1.0 \times 10^6$ lb$_m$/h-ft^2). It is characterized by the entrained liquid flowing in large agglomerates somewhat resembling ectoplasm. Spray annular is the annular flow regime most frequently found in BWR fuel rod bundles. It is characterized by the entrainment of relatively small liquid droplets. The size of these droplets is determined by the stability of the liquid droplets to various shattering mechanisms. It has been shown both experimentally and analytically (Hinze, 1948) that the appropriate stability criterion can be given in terms of a critical droplet Weber number,

$$W_{e_{\text{crit}}} \triangleq \left[\frac{\rho_g(\langle u_g \rangle_g - \langle u_l \rangle_l)^2 D_d}{g_c^{\sigma}} \right]_{\text{crit}} = C^* \, , \tag{5.148}$$

where D_d is the droplet diameter and, for relatively nonviscous fluids, the constant C^* has a value between 6.5 and 22, depending on the process of droplet formation. A value in the range of 7.5 to 13 is recommended for application to problems in BWR technology.

In both the spray annular and wispy annular flow regimes, the liquid-vapor interface is normally quite irregular due to the presence of surface

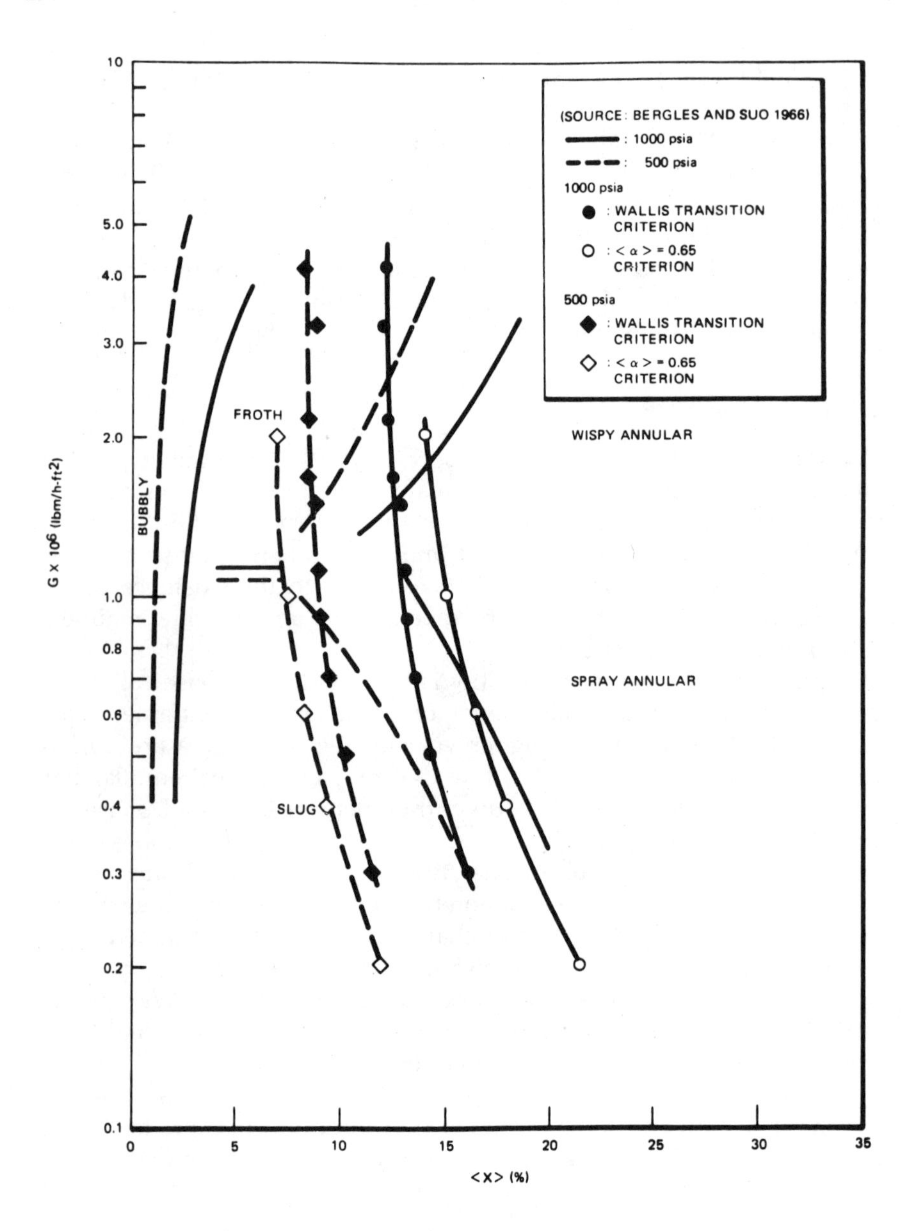

Fig. 5-9 Flow regime map.

waves. In some situations, the thin, almost stationary, liquid film is periodically washed by roll waves having amplitudes an order of magnitude greater than the thickness of the liquid film. The instability mechanisms causing surface waves are closely related to the process of liquid entrainment (Hewitt and Hall-Taylor, 1970), although in the diabatic case, nucleate boiling may also cause substantial entrainment (Newitt et al., 1954). The process of entrainment and the behavior of the entrained liquid is very complicated. The present state-of-the-art is such that our qualitative understanding far outweighs our quantitative knowledge. For instance, as shown in Fig. 5-10, the distribution of entrained liquid builds up slowly to some "fully developed" profile in which droplet deposition and entrainment are balanced. Thus, although there are approximate empirical relationships for the onset of entrainment and the fully developed entrainment fraction, very little is known about the nonequilibrium process of developing entrainment and, hence, our ability to calculate accurately the entrainment mass flux, m_e'', is limited. Progress continues to be made in this area as evidenced by the success of entrainment correlations in terms of the parameter $\tau_i\delta/\sigma$ (Hutchinson and Whalley, 1972). Nevertheless, much work remains to be done, particularly for complex geometries.

The mechanisms involved in liquid droplet deposition are also not completely understood. It is commonly supposed that equilibrium conditions in this mass transfer process are fairly rapidly achieved, and that for small droplets ($D_d < 10$ μm) the deposition mass flux (m_d'') can be given by a diffusion model,

$$m_d'' = kC \ , \tag{5.149}$$

where C is the mean concentration (lb_m/ft^3) of liquid droplets in the vapor stream and k is an empirically determined mass transfer coefficient, having the units of velocity. It is now known (James et al., 1980) that the larger droplets follow ballistic trajectories. Clearly, this feature of annular flow greatly complicates the analysis that must be performed.

In a diabatic situation, such as that occurring in a BWR rod bundle, fully developed conditions are never achieved, primarily because the quality increases monotonically in the flow direction. Nevertheless, it has been shown (Bennett et al., 1966) that in the diabatic situation there is a strong trend toward the "hydrodynamic equilibrium" state, where the rate of entrainment of liquid balances the deposition rate. Quantitative relationships for this equilibrium state are not well known nor is the process controlling the dynamics of the developing entrained fraction well established. A consequence of this state of understanding on BWR technology is that it is currently virtually impossible to keep an accurate inventory of the liquid films in rod bundles (having spacer grids) using a physically based analytical model. That is, first principle film dry-out models for critical power determination in rod bundles are beyond the present state-

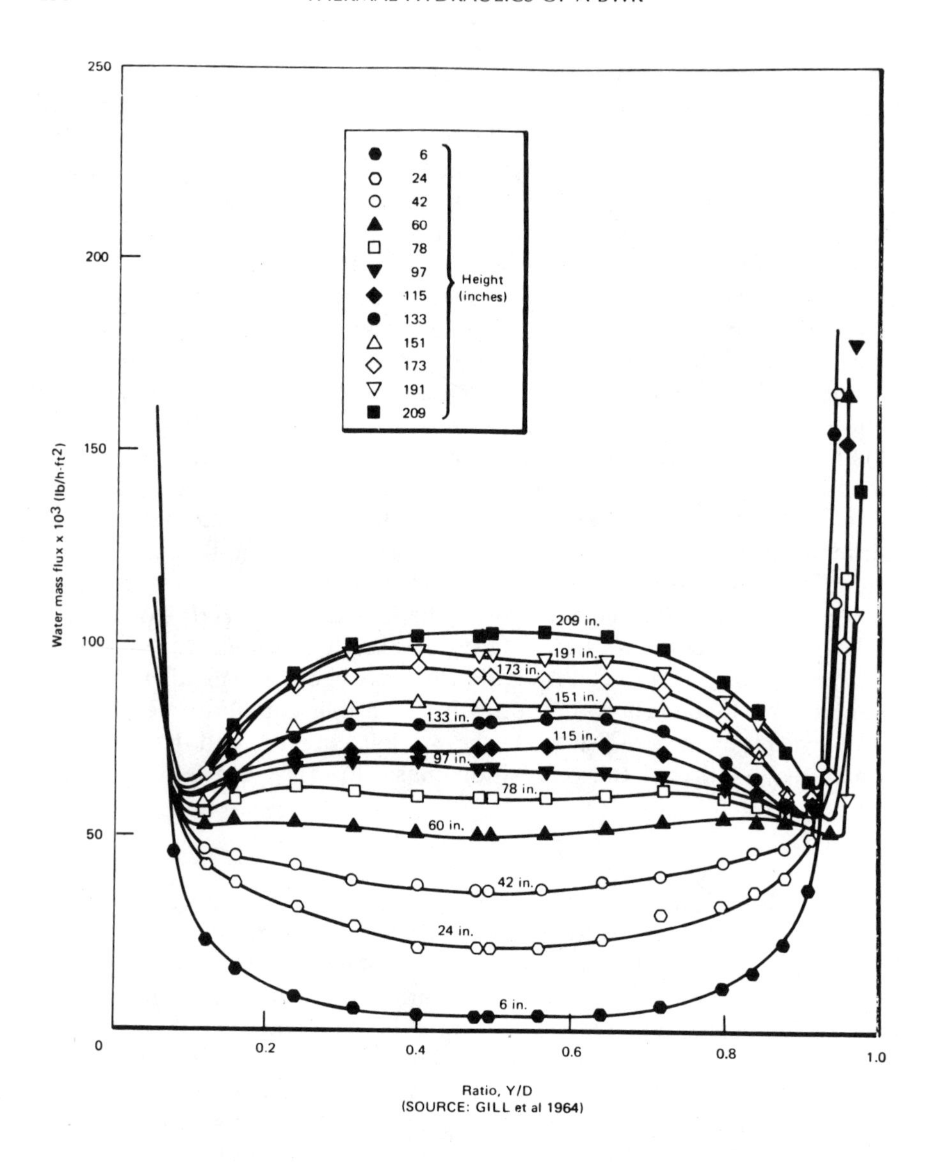

Fig. 5-10 Distribution of water mass flux in air/water flow as a function of distance from injection sinter.

of-the-art. Nevertheless, some progress is being made toward this goal. Whalley (1987) gives a good review of film flow modeling techniques for those readers interested in further information on this subject.

There are other cases, however, where our current understanding is sufficient to improve the analysis of two-phase flow phenomena. For instance, it has been shown that the momentum equations of separated two-phase flow can be modified to account for the effect of the entrained liquid (Hewitt and Hall-Taylor, 1970). If we define E as the fraction of the liquid phase that is entrained, and $\langle\Lambda\rangle$ as the volumetric fraction of the liquid film, then the various velocities can be written as,

$$\langle u_g\rangle_g = \frac{G\langle x\rangle}{\langle\alpha\rangle\rho_g} \tag{5.150a}$$

$$\langle u_l\rangle_{l_F} = \frac{G(1-\langle x\rangle)(1-E)}{\langle\Lambda\rangle\rho_l} \tag{5.150b}$$

$$\langle u_l\rangle_{l_e} = \frac{G(1-\langle x\rangle)E}{(1-\langle\alpha\rangle-\langle\Lambda\rangle)\rho_l}\ , \tag{5.150c}$$

where the subscripts F and e in Eqs. (5.150b) and (5.150c) refer to the average velocity of the liquid film and the entrained liquid, respectively. Now, the right-hand side of Eq. (5.65) represents the acceleration of the two-phase mixture. Taking into account entrainment it can be rewritten as,

$$\begin{aligned}\mathrm{RHS}_{(5.65)} = \frac{1}{g_c}\Bigg\{ &\frac{\partial}{\partial_t}\left[(\rho_l(1-\langle\alpha\rangle-\langle\Lambda\rangle)\langle u_l\rangle_{l_e} + \rho_l\langle\Lambda\rangle\langle u_l\rangle_{l_F} + \rho_g\langle\alpha\rangle\langle u_g\rangle_g\right] \\ &+ \frac{1}{A_{x-s}}\frac{\partial}{\partial z}\left[\rho_l A_{x-s}(1-\langle\alpha\rangle-\langle\Lambda\rangle)\langle u_l\rangle_{l_e}^2 + \rho_l A_{x-s}\langle\Lambda\rangle\langle u_l\rangle_{l_F}^2\right. \\ &\left. + \rho_g A_{x-s}\langle\alpha\rangle\langle u_g\rangle_g^2\right]\Bigg\}\ . \end{aligned} \tag{5.151}$$

Introducing Eqs. (5.150a), (5.150b), and (5.150c) into Eq. (5.151), one obtains,

$$\begin{aligned}\mathrm{RHS}_{(5.65)} = \frac{1}{g_c}\Bigg[\frac{\partial G}{\partial t} + \frac{1}{A_{x-s}}\frac{\partial}{\partial z}\Bigg\{ A_{x-s}G^2\Bigg[&\frac{E^2(1-\langle x\rangle)^2}{(1-\langle\alpha\rangle-\langle\Lambda\rangle)\rho_l} \\ &+ \frac{(1-E)^2(1-\langle x\rangle)^2}{\langle\Lambda\rangle\rho_l} + \frac{\langle x\rangle^2}{\langle\alpha\rangle\rho_g}\Bigg]\Bigg\}\Bigg]\ . \end{aligned} \tag{5.152}$$

Hence, Eqs. (5.152) and (5.65) show that, aside from small changes in void fractions, the only effect of liquid entrainment on the momentum equation of the two-phase mixture is to modify the spatial acceleration term. The resultant momentum equation is thus:

$$-\frac{\partial p}{\partial z}-\frac{g}{g_c}\langle\bar{\rho}\rangle\sin\theta-\frac{\tau_w P_f}{A_{x-s}}=\frac{1}{g_c}\left(\frac{\partial G}{\partial t}+\frac{1}{A_{x-s}}\right.$$

$$\left.\times\frac{\partial}{\partial z}\left\{A_{x-s}G^2\left[\frac{E^2(1-x)^2}{(1-\langle\alpha\rangle-\langle\Lambda\rangle)\rho_l}+\frac{(1-E)^2(1-\langle x\rangle)^2}{\langle\Lambda\rangle\rho_l}+\frac{\langle x\rangle^2}{\langle\alpha\rangle\rho_g}\right]\right\}\right) \quad . \tag{5.153}$$

While there are currently no reliable correlations for developing flows, Ishii and Mishima (1984) recommend the following entrainment correlation for fully developed annular flows:

$$E=\tanh(7.25\times10^{-7}\mathrm{We}^{1.25}\mathrm{Re}_l^{0.25}) \ , \tag{5.154}$$

where,

$$\mathrm{We}=\frac{\rho_g\langle j_g\rangle^2 D_H}{\sigma}\left(\frac{\rho_l-\rho_g}{\rho_g}\right)^{1/3}$$

$$\mathrm{Re}=\rho_l\langle j_l\rangle/\mu_l \ ,$$

and the inception of entrainment is given by the criterion:

$$\frac{\mu_l\langle j_g\rangle}{\sigma}\sqrt{\rho_g/\rho_l}\geqslant\begin{cases}11.78N_\mu^{0.8}\mathrm{Re}_l^{-1/3}, & \mathrm{Re}_l<1635\\ N_\mu^{0.8}, & \mathrm{Re}_l\geqslant1635\end{cases} \tag{5.155}$$

where,

$$N_\mu=\mu_l\Bigg/\left\{\rho_l\frac{\sigma}{g_c}\sqrt{\sigma/[g/g_c(\rho_l-\rho_g)]}\right\}^{1/2} \ .$$

Mathematically, the effect of including entrainment has been to introduce two new dependent variables, E and $\langle\Lambda\rangle$. Normally, a correlation for the entrainment fraction, E, is used. To quantify $\langle\Lambda\rangle$, one must either make experimental measurements or make some assumption about the velocity of the entrained liquid. One popular assumption, which is reasonably accurate in some cases, is to assume "no slip" between the vapor phase and the entrained liquid (i.e., $\langle u_g\rangle_g=\langle u_l\rangle_{le}$). Thus, by equating Eqs. (5.150a) and (5.150b),

$$\langle\Lambda\rangle=1-\langle\alpha\rangle-\frac{\rho_g\langle\alpha\rangle(1-\langle x\rangle)E}{\rho_l\langle x\rangle} \ . \tag{5.156}$$

We note in Eq. (5.156) that we can eliminate $\langle\Lambda\rangle$ in favor of the known variables $\langle\alpha\rangle$ and $\langle x\rangle$ and the parameter E.

Although the discussion concerning the effect of entrainment has been directed toward the momentum equation, the energy equation of the two-phase mixture would be modified as well, as can be seen by combining Eqs. (5.91), (5.96), and (5.150).

In actual practice, entrainment effects are either not considered, are lumped into the void correlations, or are treated approximately using correlations (i.e., in the advanced "two-fluid" models). Unfortunately, accurate correlations for the nonequilibrium entrainment fraction, E, are not yet available for detailed analysis of annular flows.

Some interesting features of annular two-phase flow can be appreciated by doing a simple force balance for the case of idealized annular flow in a tube, as shown schematically in Fig. 5-11. By neglecting any momentum contribution due to acceleration of the phases, the force balance becomes,

$$(2\pi r\Delta z)\tau(r) = \pi r^2[p-(p+\frac{dp}{dz}\Delta z)] - \frac{g\Delta z}{g_c}\sin\theta\int_0^r [\rho_l(1-\alpha)+\rho_g\alpha]2\pi r'dr' \ ,$$

which simplifies to,

$$\tau(r) = \frac{r}{2}\left(-\frac{dp}{dz}\right) - \frac{g}{g_c}\frac{\sin\theta}{r}\int_0^r [\rho_l(1-\alpha)+\rho_g\alpha]r'dr' \ . \tag{5.157}$$

For the special case of no liquid entrainment and a smooth interface, Eq. (5.157) gives the interfacial shear stress as,

$$\tau(R-\delta) \triangleq \tau_i = \frac{(R-\delta)}{2}\left(-\frac{dp}{dz} - \frac{g}{g_c}\rho_g \sin\theta\right) \ , \tag{5.158}$$

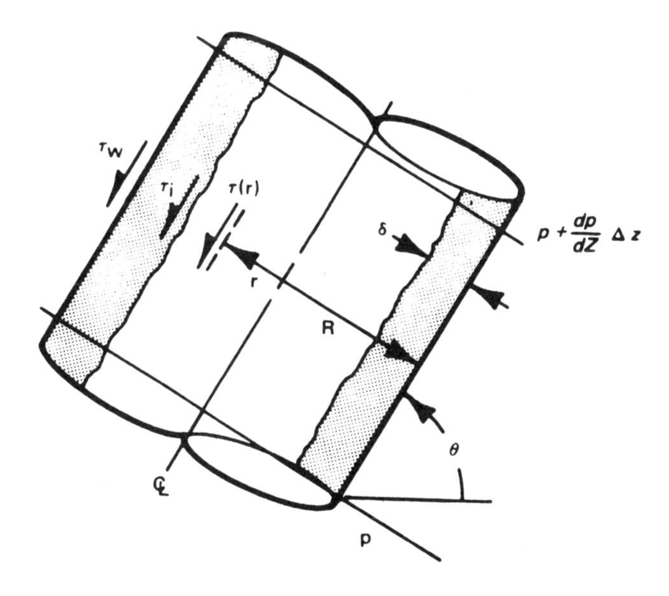

Fig. 5-11 Idealized annular flow.

where δ is the mean film thickness.

If it is also assumed that the liquid film contains no voids and it is noted that $R-\delta=R\sqrt{\langle\alpha\rangle}$, Eq. (5.157) can be used to obtain the shear stress at the wall, τ_w,

$$\tau(R)\triangleq\tau_w=\frac{R}{2}\left(-\frac{dp}{dz}\right)-\frac{g}{g_c}\frac{\sin\theta}{R}\left[\int_0^{R\sqrt{\langle\alpha\rangle}}\rho_g r'dr'+\int_{R\sqrt{\langle\alpha\rangle}}^{R}\rho_l r'dr'\right] ,$$

which simplifies to,

$$\tau_w=\frac{R}{2}\left(-\frac{dp}{dz}-\frac{g}{g_c}\langle\bar{\rho}\rangle\sin\theta\right) . \tag{5.159}$$

The pressure gradient, which has been assumed to be the same in both the liquid and vapor phase, can be eliminated between Eqs. (5.158) and (5.159) to obtain,

$$\tau_w=\frac{R}{2}\left[\frac{2\tau_i}{(R-\delta)}-\frac{g}{g_c}(\rho_l-\rho_g)(1-\langle\alpha\rangle)\sin\theta\right] . \tag{5.160}$$

Several things can be noted from these simple relationships. First of all, if,

$$-\frac{dp}{dz}=\frac{g}{g_c}\rho_g\sin\theta , \tag{5.161}$$

then Eq. (5.160) implies that $\tau_i=0$, and Eq. (5.161) implies that for 0 deg $<\theta<$ 180 deg, $\tau_w<0$. That is, there is down-flow of the liquid film. When,

$$\tau_i=\frac{(R-\delta)}{2}\left[\frac{g}{g_c}(\rho_l-\rho_g)(1-\langle\alpha\rangle)\sin\theta\right] , \tag{5.162}$$

then Eq. (5.160) implies $\tau_w=0$ and Eq. (5.159) yields,

$$-\frac{dp}{dz}=\frac{g}{g_c}\langle\bar{\rho}\rangle\sin\theta . \tag{5.163}$$

Physically this means that for 0 deg $<\theta<$ 180 deg, the pressure gradient is just equal to the value it would have for a static two-phase system. This is essentially the minimum value it can obtain (Hewitt and Hall-Taylor, 1970). For the case under consideration, the system flow increases the interfacial shear stress in order to maintain the annular flow configuration; however, it has no pressure loss due to wall friction. In contrast, for horizontal flow ($\theta=0$), this flow rate will produce a frictional pressure drop and thus a positive wall shear stress. This simple example should warn the reader about the limitations inherent in pressure drop models, such as those discussed in Sec. 5.4, which are based on the work of Martinelli

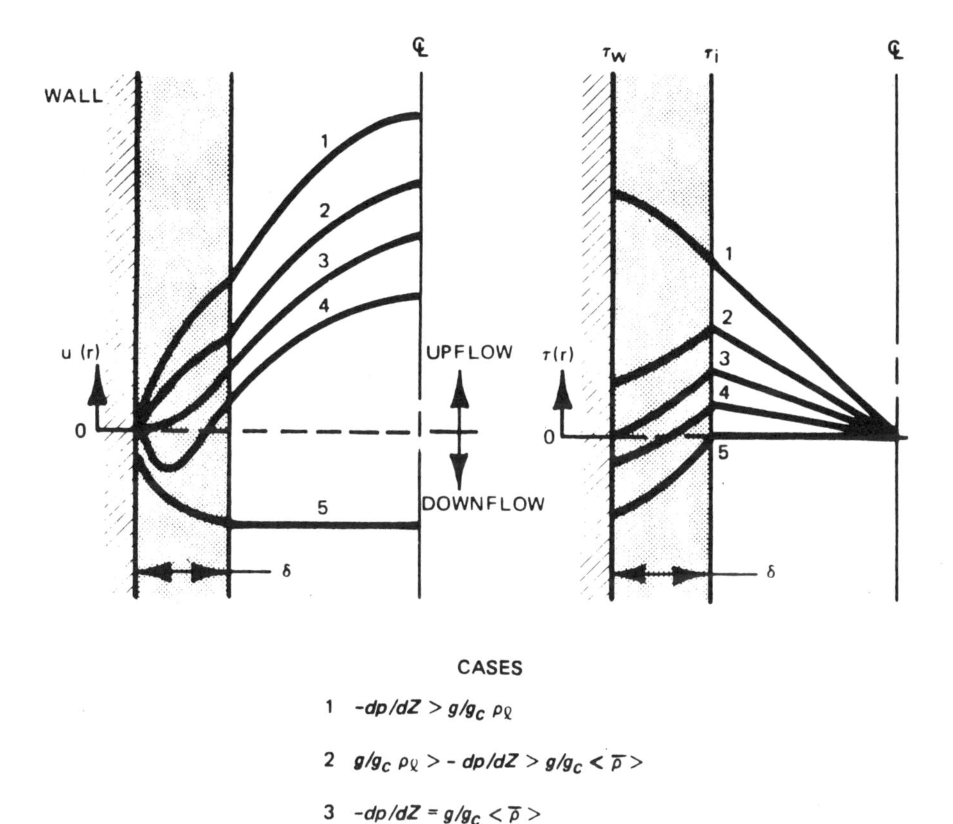

CASES

1 $-dp/dZ > g/g_c\ \rho_\ell$

2 $g/g_c\ \rho_\ell > -dp/dZ > g/g_c < \bar{\rho} >$

3 $-dp/dZ = g/g_c < \bar{\rho} >$

4 $(g/g_c < \bar{\rho} >) > -dp/dZ > g/g_c\ \rho_g$

5 $-dp/dZ = g/g_c\ \rho_g$

Fig. 5-12 Typical velocity and shear stress profiles for vertical, idealized annular two-phase flow.

(Martinelli and Nelson, 1948) and in which it is implicitly assumed the wall shear stress is constant regardless of orientation. Figure 5.12 is a schematic representation of some of the shear stress and velocity profiles that can occur in vertical, annular two-phase flow. It can be seen that many interesting situations are possible.

The consideration of flow regime analysis just presented has been necessarily brief and has been slanted toward items having possible applications in BWR technology. The interested reader is referred to the references cited for a more thorough treatment. In the next section we will be concerned with void-quality relations, which are of use in the thermal-hydraulic analysis of light water nuclear reactors.

5.3 Void-Quality Analysis

To have the number of independent equations equal the number of dependent variables, the analysis of nonhomogeneous two-phase systems requires that we relate the flowing quality to the holdup void fraction. In this sense, the appropriate void-quality relation can be thought of as the two-phase equation-of-state. We discuss the specification of this important relationship, first for diabatic conditions of saturated equilibrium (bulk boiling) and then for nonequilibrium (subcooled boiling) conditions.

5.3.1 Bulk Boiling

The situation of diabatic one-component two-phase flow in saturated equilibrium is commonly known as bulk boiling. It is characterized by the fact that both the liquid and vapor phases are saturated and is by far the most important case of interest in BWR technology. In this subsection, we discuss various void-quality models that have been used successfully in the analysis of two-phase flow phenomena, in particular for bulk boiling situations.

We now consider the derivation of a generalized void-quality model. Following the work of Zuber (Zuber and Findlay, 1965), we write the identity,

$$u_g = j + (u_g - j) \; . \tag{5.164}$$

By using the relations given in Eq. (5.6), Eq. (5.164) yields,

$$j_g = \alpha_j + \alpha(u_g - j) \; . \tag{5.165}$$

By averaging across the cross-sectional flow area,

$$\langle j_g \rangle = C_0 \langle j \rangle \langle \alpha \rangle + V_{gj} \langle \alpha \rangle \tag{5.166a}$$

or

$$\langle u_g \rangle_g = \langle j_g \rangle / \langle \alpha \rangle = C_0 \langle j \rangle + V_{gj} \; , \tag{5.166b}$$

where,

$$C_0 \triangleq \langle j\alpha \rangle / (\langle j \rangle \langle \alpha \rangle) \tag{5.167a}$$

$$V_{gj} \triangleq \langle (u_g - j)\alpha \rangle / \langle \alpha \rangle \; . \tag{5.167b}$$

Equation (5.166a) can be solved for a fundamental relationship for determining void fraction,

$$\langle \alpha \rangle = \frac{\langle j_g \rangle}{(C_0 \langle j \rangle + V_{gj})} \; . \tag{5.168}$$

The concentration parameter, C_0, quantifies the effect of the radial void and volumetric flux distribution. This parameter can be regarded as a measure of the global slip due to cross-sectional averaging. The drift velocity, V_{gj}, is a measure of the local slip and is closely related to the terminal rise velocity, U_t, of the vapor phase through the liquid.

To gain some appreciation for the various functional forms of V_{gj}, which appear in the literature, we can perform fractional analysis (Kline, 1965) on a flowing two-phase system. In fractional analysis, we write down the important forces and then form the various ratios of these forces. For instance, in bubbly/slug flow, the important forces are given by,

$$\begin{Bmatrix} \text{net buoyancy} \\ \text{force on a bubble} \end{Bmatrix} = \frac{1}{6}\pi D_b^3(\rho_l - \rho_g)\frac{g}{g_c} \tag{5.169a}$$

$$\begin{Bmatrix} \text{inertia force} \\ \text{on a bubble} \end{Bmatrix} = \frac{\rho_l}{g_c}U_t^2\left(\frac{1}{4}\pi D_b^2\right) \tag{5.169b}$$

$$\begin{Bmatrix} \text{surface tension} \\ \text{force on a bubble} \end{Bmatrix} = \pi D_b \sigma \tag{5.169c}$$

$$\begin{Bmatrix} \text{viscous force}^c \\ \text{on a bubble} \end{Bmatrix} = 3\pi U_t \nu_l D_b \rho_l \ . \tag{5.169d}$$

In the important case of churn-turbulent bubbly flow, the most important force groups have been found to be the ratio of the inertia force to the buoyancy force and the ratio of the inertia force to the surface tension force. In accordance with the rules of classical fractional analysis, we assume that the ratio between these forces is constant. Equations (5.169a), (5.169b), and (5.169c) then yield,

$$\frac{\rho_l U_t^2}{D_b g(\rho_l - \rho_g)} \triangleq k_1^2 \tag{5.170}$$

and,

$$\frac{\rho_l U_t^2 D_b}{g_c \sigma} = k_2^2 \ . \tag{5.171}$$

The product of these dimensionless groups yields,

$$\frac{\rho_l^2 U_t^4}{g g_c \sigma(\rho_l - \rho_g)} = k_1^2 k_2^2 \triangleq k_3^4 \ . \tag{5.172}$$

By solving Eq. (5.172) for the terminal rise velocity,

$$U_t = k_3\left[\frac{(\rho_l - \rho_g)\sigma g g_c}{\rho_l^2}\right]^{1/4} \ . \tag{5.173}$$

---

[c]Stokes' drag law.

As noted in Eq. (5.132), for churn-turbulent flows, $V'_{gj} = U_t$ and, since $C_0 \simeq 1.0$ for this flow regime, Eq. (5.70) implies $V_{gj} \simeq U_t$. Here Eq. (5.173) is a frequently used expression for the drift velocity, V_{gj}, in two-phase flow.

Similar expressions for V_{gj} can be derived for other flow regimes (Lahey, 1974); however, when generalized void-quality correlations are constructed, Eq. (5.173) is frequently assumed to be valid over all flow regimes.

For many practical applications, Eq. (5.168) is frequently rewritten in terms of flow quality. That is, by combining Eqs. (5.8b), (5.22), and (5.23), we obtain,

$$\langle j_g \rangle = \frac{G\langle x \rangle}{\rho_g} \tag{5.174}$$

$$\langle j_l \rangle = \frac{G(1-\langle x \rangle)}{\rho_l} \tag{5.175}$$

$$\langle j \rangle = \langle j_g \rangle + \langle j_l \rangle = G\left[\frac{\langle x \rangle}{\rho_g} + \frac{(1-\langle x \rangle)}{\rho_l}\right] . \tag{5.176}$$

For bulk boiling, $\rho_l = \rho_f$; however, the subscript, l, is retained for generality.

Equation (5.168) can be combined with Eqs. (5.174), (5.175), and (5.176) to yield,

$$\langle \alpha \rangle = \frac{\langle x \rangle}{\left\{ C_0\left[\langle x \rangle + \frac{\rho_g}{\rho_l}(1-\langle x \rangle)\right] + \frac{\rho_g V_{gj}}{G} \right\}} . \tag{5.177}$$

Comparison of Eqs. (5.25) and (5.177) indicates that the slip ratio implicit in the Zuber-Findlay void-quality model is,

$$S = C_0 + \frac{\langle x \rangle (C_0 - 1)\rho_l}{\rho_g (1-\langle x \rangle)} + \frac{\rho_l V_{gj}}{G(1-\langle x \rangle)} \tag{5.178a}$$

or

$$S = \frac{(1 - C_0\langle \alpha \rangle)}{(1-\langle \alpha \rangle)} + \frac{V_{gj}(1-\langle \alpha \rangle)}{(1-C_0\langle \alpha \rangle)\langle j_l \rangle} . \tag{5.178b}$$

The first two terms in Eq. (5.178a) represent the slip due to cross-sectional averaging of a nonuniform void fraction profile. This is frequently referred to as "integral slip." The last term represents the contribution due to local slip between the phases. If we neglect the local slip between the phases (i.e., set $V_{gj} = 0$), Eqs. (5.166a) and (5.9) yield the so-called "Bankoff void model" (Bankoff, 1960),

$$\frac{\langle \alpha \rangle}{\langle \beta \rangle} = \frac{1}{C_0} \triangleq \kappa . \tag{5.179}$$

The variation with system pressure of the Armand (1959) parameter, κ, has been given by Bankoff (1960) for water as,

$$\kappa = 0.71 + 0.0001p \text{ (psia)} \ . \tag{5.180}$$

Thus, in general, $\kappa \leq 1.0$; for instance, at a system pressure of $p = 1000$ psia, $\kappa = 0.81$.

Bankoff's void-quality relation only contains integral slip and can be obtained easily from Eq. (5.177) by formally setting $V_{gj} = 0$ to obtain,

$$\langle \alpha \rangle = \frac{\langle x \rangle}{C_0\left[\langle x \rangle + \frac{\rho_g}{\rho_l}(1 - \langle x \rangle)\right]} = \langle \beta \rangle / C_0 \ . \tag{5.181}$$

It is obvious that if we let $C_0 = 1$ in Eq. (5.179), then Eq. (5.11) implies that we have no slip. That is, the homogeneous void-quality relation can be obtained from Eq. (5.181) by setting $C_0 = 1$ to obtain,

$$\langle \alpha \rangle = \frac{\langle x \rangle}{\left[\langle x \rangle + \frac{\rho_g}{\rho_l}(1 - \langle x \rangle)\right]} = \langle \beta \rangle \ . \tag{5.182}$$

Note that Eq. (5.182) is identical to Eq. (5.25) for the special case of no slip, $S = 1$.

Thus, comparison of these various void-quality relations indicates that Bankoff's model, Eq. (5.181), and the homogeneous model, Eq. (5.182), are both special cases of the Zuber-Findlay model, Eq. (5.177).

It is of interest to consider typical values that the concentration parameter, C_0, could be expected to obtain in a diabatic system. Some insight can be obtained by considering the definition of C_0 in Eq. (5.167a),

$$C_0 \stackrel{\Delta}{=} \frac{1}{\langle j \rangle}\left[\frac{1}{\langle \alpha \rangle A_{x-s}} \iint_{A_{x-s}} (j\alpha) \, dA\right] \ . \tag{5.183}$$

It is reasonable to assume that the local volumetric flux of the two-phase mixture can be expected to follow a power law expression. For example, in a tube this expression is,

$$j/j_{\mathbb{C}} = \left[1 - \left(\frac{r}{R}\right)\right]^{1/n} , \tag{5.184}$$

and, thus, j becomes very small as the heated surface is approached (i.e., $r \to R$).

Figure 5-13 is a representation of various typical void profiles in a once-through evaporator. Case ⑦ is the situation in which there is single-phase vapor flow ($\langle \alpha \rangle = 1.0$) and, thus, Eq. (5.183) clearly gives $C_0 = 1.0$. Case ⑥ is the situation of annular flow and is of great interest in BWR technology.

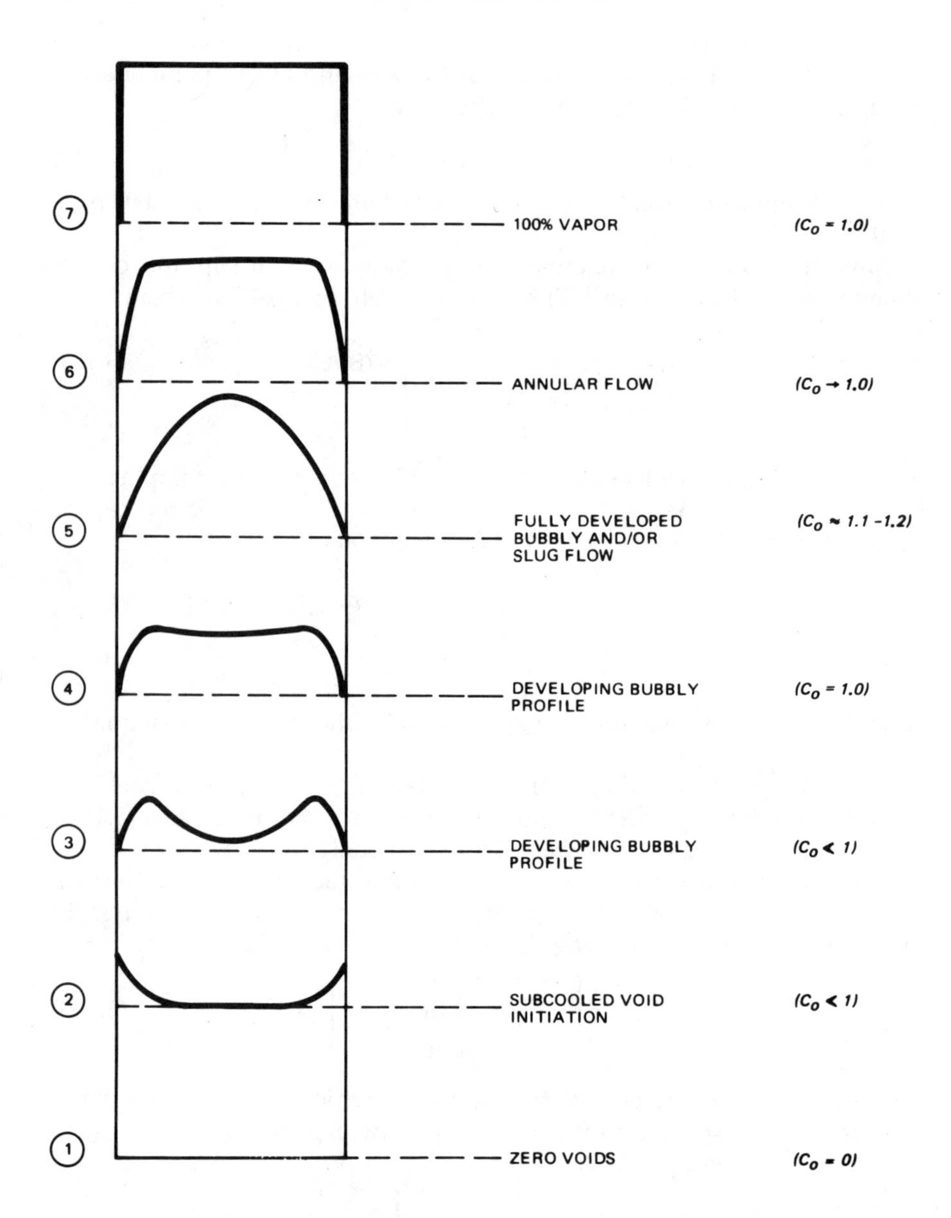

Fig. 5-13 Diabatic void concentration profiles and flow regimes.

Typically, the vapor phase collects in the high-velocity core and the liquid annulus normally is rather thin, such that Eq. (5.183) implies $C_0 \simeq 1.0$. Developing and fully developed bubbly/slug flow is shown in Cases ④ and ⑤, respectively. In these situations the local void fraction has been found to have a radial profile of the power law type, similar in form to Eq. (5.184). It is well known that values of C_0 around 1.2 are not unusual in adiabatic experiments. In fact, slug flow typically is characterized by a value of $C_0 = 1.2$. In a diabatic system, fully developed conditions and slug flow are rarely obtained and, thus, C_0 can be expected to be somewhat smaller than its corresponding adiabatic value. Furthermore, a typical BWR rod bundle has a rather complex geometry (grid spacers, etc.), which could be expected to further limit the upper value that C_0 can obtain. Nevertheless, for diabatic bubbly/slug flow, C_0 can be expected to be larger than unity.

Case ③ is again a developing profile. However, in this case, subcooled boiling is presumed to occur and, thus, due to condensation and other effects discussed in Sec 5.3.2, the local void fraction is highest in the lower velocity region. The implication of this type of distribution is that Eq. (5.183) yields $C_0 < 1.0$. Case ② is an even more pronounced case of subcooled boiling in that virtually all the voids are in the very low velocity region and, thus, Eq. (5.183) implies $C_0 \ll 1.0$. In the limit, as we approach zero diabatic voids, we obtain Case ①, in which $C_0 \simeq 0.0$.

A functional form for the variation of C_0 with pressure and flow quality has been deduced by Dix (1971),

$$C_0 = \langle\beta\rangle[1 + (1/\langle\beta\rangle - 1)^b] \ , \tag{5.185}$$

where,

$$b \triangleq (\rho_g/\rho_l)^{0.1} \ . \tag{5.186}$$

Equation (5.182) implies that,

$$\langle\beta\rangle = \frac{\langle x\rangle}{\left[\langle x\rangle + \frac{\rho_g}{\rho_l}(1 - \langle x\rangle)\right]} \ . \tag{5.187}$$

Note that Eq. (5.185) satisfies the necessary constraints,

$$C_0 \to 0 \text{ as } \langle\beta\rangle \to 0$$

$$C_0 \to 1 \text{ as } \langle\beta\rangle \to 1 \ ,$$

and the behavior of C_0 for intermediate flow qualities can be seen in Fig. 5-14.

The final parameter in Eq. (5.177) that must be specified is the drift velocity, V_{gj}. Note that the parameters, C_0 and V_{gj}, are not independent. That is, if the variation of one of these parameters is specified, the other necessarily must be optimized to ensure that Eq. (5.177) agrees with the available data. Equation (5.185) gives the functional form of the concen-

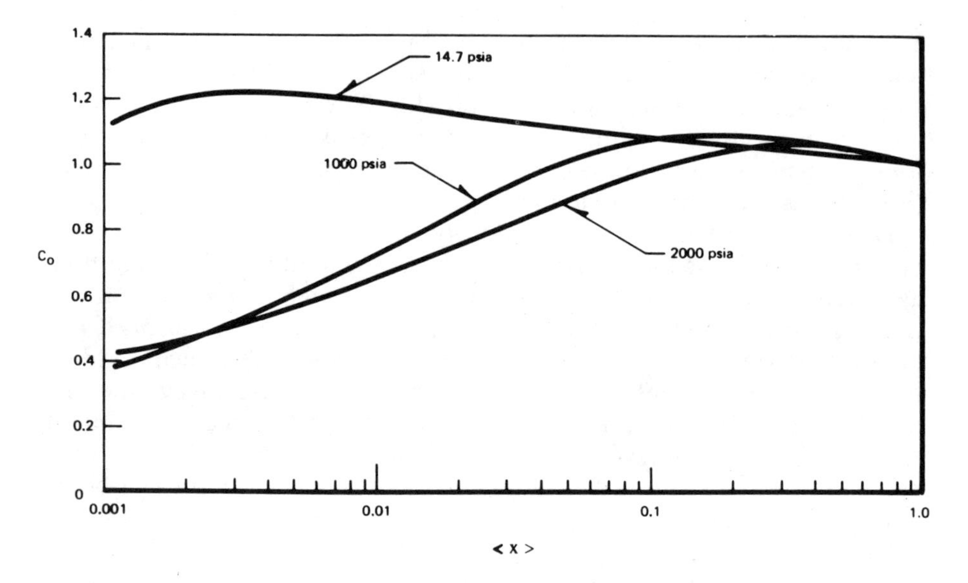

Fig. 5-14 Variation of concentration parameter, C_0, with pressure and flow quality.

tration parameter, C_0, obtained from a diabatic experiment (Dix, 1971). It is evident from Figs. 5-13 and 5-14 that C_0 is a function of flow regime. In general, V_{gj} is also a function of flow regime. However, as discussed previously, Eq. (5.173) frequently is assumed to be flow regime independent and, thus, the appropriate drift velocity is taken as,

$$V_{gj} = k_3\left[\frac{(\rho_l - \rho_g)\sigma g g_c}{\rho_l^2}\right]^{1/4} . \tag{5.188}$$

The value of V_{gj} compatible with Eq. (5.185) is obtained by setting $k_3 = \pm 2.9$ for upflow and downflow, respectively. Thus, Eq. (5.188) becomes,

$$V_{gj} = 2.9\left[\frac{(\rho_l - \rho_g)\sigma g g_c}{\rho_l^2}\right]^{1/4} \sin\theta . \tag{5.189}$$

Equations (5.177), (5.185), (5.186), (5.187), and (5.189) yield a void-quality relation that is reasonably accurate for many practical applications.

Another technique that has been employed in the past is to use the empirical correlation for void fraction developed by Lockhart and Martinelli (1949). This correlation is shown in Fig. 5-15. It can be shown (Wallis, 1969) that the mean curve through the data is well represented by,

$$\langle\alpha\rangle = (1 + X_{tt}^{0.8})^{-0.378} , \tag{5.190}$$

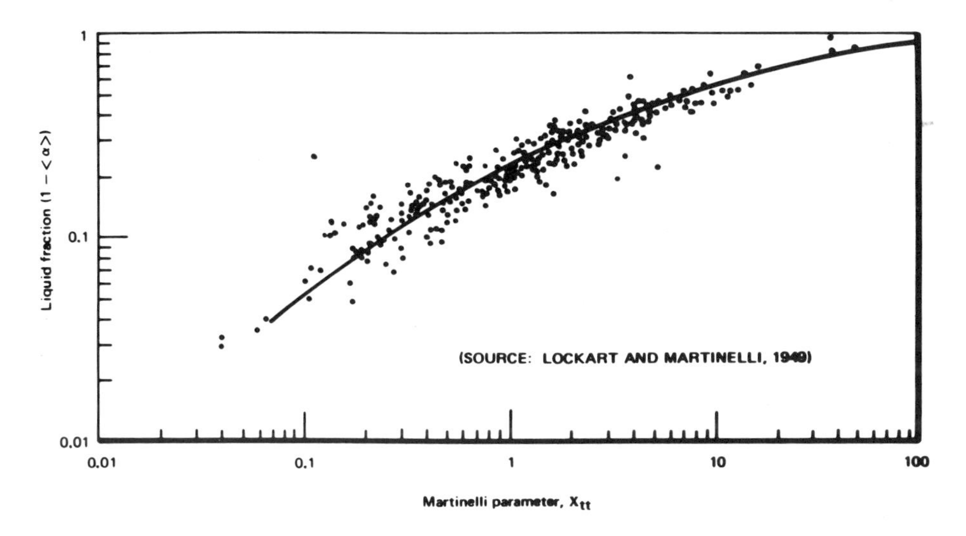

Fig. 5-15 Martinelli's correlation for void fraction.

where, as discussed in more detail in Sec. 5.4, Martinelli's parameter, X_{tt}, is given by,

$$X_{tt} \triangleq \left[\frac{\left(\frac{dp}{dz}\right)_l}{\left(\frac{dp}{dz}\right)_g}\right]^{1/2} = \left(\frac{\rho_g}{\rho_l}\right)^{1/2}\left(\frac{\mu_l}{\mu_g}\right)^{n/2}\left(\frac{1-\langle x\rangle}{\langle x\rangle}\right)^{\frac{(2-n)}{2}} \tag{5.191}$$

Here, n is the exponent of the Reynold's number in $f = C/R_e^n$, and normally has a value in the range of 0.2 to 0.25. The empirical approach taken by Martinelli (Lockhart and Martinelli, 1949) and the phenomenological approach taken by Zuber (Zuber and Findlay, 1965) are excellent examples of the state of void-quality models. Neither model is based strictly on first principles. However, both have been useful to workers in the field of two-phase flow.

A more basic approach has been taken by Levy (1960) in that his void-quality model, although only valid for a special case, is derived directly from first principles. To appreciate Levy's model, consider the case of highly accelerating "flashing flow." In this case, we can neglect the wall and interfacial shear and body force terms in comparison with the spatial acceleration and momentum exchange terms. Thus, for steady state, Eq. (5.84) yields,

$$-\frac{dp}{dz} = \frac{(1-\eta_m)}{g_c(1-\langle\alpha\rangle)}\Gamma(\langle u_g\rangle_g - \langle u_l\rangle_l) + \frac{\rho_l}{g_c}\langle u_l\rangle_l\frac{d\langle u_l\rangle_l}{dz}\ , \tag{5.192}$$

and Eq. (5.83) yields,

$$-\frac{dp}{dz}=\frac{\eta_m}{g_c\langle\alpha\rangle}\Gamma(\langle u_g\rangle_g-\langle u_l\rangle_l)+\frac{\rho_g}{g_c}\langle u_g\rangle_g\frac{d\langle u_g\rangle_g}{dz}\ . \tag{5.193}$$

Equations (5.192) and (5.193) are the appropriate steady-state momentum equations for the liquid and vapor phases, respectively.

For steady flow in a constant area duct, the equation of vapor continuity, Eq. (5.56), yields,

$$\Gamma=G\frac{d\langle x\rangle}{dz}\ . \tag{5.194}$$

If we subtract Eq. (5.193) from Eq. (5.192) to eliminate $-dp/dz$, and combine the resultant equation[d] with Eq. (5.194), we obtain Levy's "momentum exchange" model in the form,

$$\frac{\rho_g}{g_c}\langle u_g\rangle_g\frac{d\langle u_g\rangle_g}{dz}-\left[\frac{(1-\eta_m)}{g_c(1-\langle\alpha\rangle)}-\frac{\eta_m}{g_c\langle\alpha\rangle}\right](\langle u_g\rangle_g-\langle u_l\rangle_l)G\frac{d\langle x\rangle}{dz}-\frac{\rho_l}{g_c}\langle u_l\rangle_l\frac{d\langle u_l\rangle_l}{dz}=0\ . \tag{5.195}$$

To reduce this to Levy's (1960) original notation, it is significant to note,

$$G\frac{d\langle x\rangle}{dz}=\frac{d}{dz}(\rho_g\langle\alpha\rangle\langle u_g\rangle_g)=-\frac{d}{dz}[\rho_l(1-\langle\alpha\rangle)\langle u_l\rangle_l]\ . \tag{5.196}$$

For the high evaporation rates associated with rapidly accelerating flashing flow, it is often assumed (Wallis, 1969) that $\eta_m=1$. Thus, Eqs. (5.195) and (5.196) yield,

$$\frac{1}{g_c}\left\{\frac{d}{dz}(\rho_g\langle\alpha\rangle\langle u_g\rangle_g^2)+\frac{d}{dz}[\rho_l(1-\langle\alpha\rangle)\langle u_l\rangle_l^2]-\rho_l\langle u_l\rangle_l\frac{d\langle u_l\rangle_l}{dz}\right\}=0\ . \tag{5.197}$$

Equation (5.197) can also be written as an exact differential,

$$\frac{1}{g_c}\frac{d}{dz}\left[\rho_g\langle\alpha\rangle\langle u_g\rangle_g^2+\rho_l(1-\langle\alpha\rangle)\langle u_l\rangle_l^2-\frac{\rho_l}{2}\langle u_l\rangle_l^2\right]=0. \tag{5.198}$$

By combining Eqs. (5.198), (5.22), and (5.23), we obtain,

$$\frac{G^2}{g_c\rho_l}\frac{d}{dz}\left[\frac{\rho_l}{\rho_g}\frac{\langle x\rangle^2}{\langle\alpha\rangle}+\frac{(1-\langle x\rangle)^2}{(1-\langle\alpha\rangle)}-\frac{1}{2}\frac{(1-\langle x\rangle)^2}{(1-\langle\alpha\rangle)^2}\right]=0\ , \tag{5.199}$$

Integration of Eq. (5.199), with the initial conditions $\langle\alpha\rangle=0$ at $\langle x\rangle=0$, yields,

$$\frac{\rho_l}{\rho_g}\frac{\langle x\rangle^2}{\langle\alpha\rangle}+\frac{(1-\langle x\rangle)^2}{(1-\langle\alpha\rangle)}-\frac{1}{2}\frac{(1-\langle x\rangle)^2}{(1-\langle\alpha\rangle)^2}=\frac{1}{2}\ . \tag{5.200}$$

[d] It is implicitly assumed that there is no radial pressure gradient in the two-phase mixture.

By solving Eq. (5.200) for flow quality, $\langle x \rangle$, we obtain Levy's (1960) void-quality model,

$$\langle x \rangle = \frac{\langle \alpha \rangle (1 - 2\langle \alpha \rangle)}{2(1 - \langle \alpha \rangle)^2 \frac{\rho_l}{\rho_g} + \langle \alpha \rangle (1 - 2\langle \alpha \rangle)} + \frac{\langle \alpha \rangle \left\{ (1 - 2\langle \alpha \rangle)^2 + \langle \alpha \rangle \left[2(1 - \langle \alpha \rangle)^2 \frac{\rho_l}{\rho_g} + \langle \alpha \rangle (1 - 2\langle \alpha \rangle) \right] \right\}^{1/2}}{2(1 - \langle \alpha \rangle)^2 \frac{\rho_l}{\rho_g} + \langle \alpha \rangle (1 - 2\langle \alpha \rangle)} . \quad (5.201)$$

Although not of general applicability, Eq. (5.201) has been found to correlate two-phase critical flow data reasonably well.

Several void-quality models now have been discussed. The most accurate and generally useful of these void-quality models is the modified Zuber-Findlay model given by Eqs. (5.177), (5.185), (5.186), (5.187), and (5.189). Although the present discussion has been oriented toward bulk boiling situations, the void-quality relationships derived have been in terms of flow quality and thus are also applicable to subcooled boiling situations, assuming that C_0 accurately reflects subcooled boiling characteristics.

5.3.2 Subcooled Boiling

Where there is local boiling from the heated surface, even though the mean enthalpy of the liquid phase is less than saturation (i.e., $\langle h_l \rangle < h_f$), a situation of subcooled boiling exists. That is, subcooled boiling is characterized by the fact that thermodynamic equilibrium does not exist.

Bulk boiling tends to dominate the nuclear and thermal-hydraulic performance of a BWR, nevertheless, any sophisticated analysis of a modern BWR requires the use of a reasonably accurate subcooled void-quality model. In this subsection, we attempt to summarize and unify the most important analytical and experimental work done to date and recommend those techniques considered to be the most accurate.

Figure 5-16 is a schematic of a typical subcooled void-fraction profile in a heated tube. In accordance with a suggestion by Griffith (Griffith et al., 1958), the subcooled boiling process can be subdivided into two regions. Region I is commonly referred to as the region of wall voidage. Originally it was suggested that the voids in this region adhered to the heated surface. However, experimental observations (Dix, 1971) have shown that the voids actually travel in a narrow bubble layer close to the wall. This bubble boundary layer continues to grow under the competing effects of bubble coalescence and condensation until the void departure point, z_d, is reached. From this point on, voids are ejected into the subcooled core and significant void fraction begins to appear.

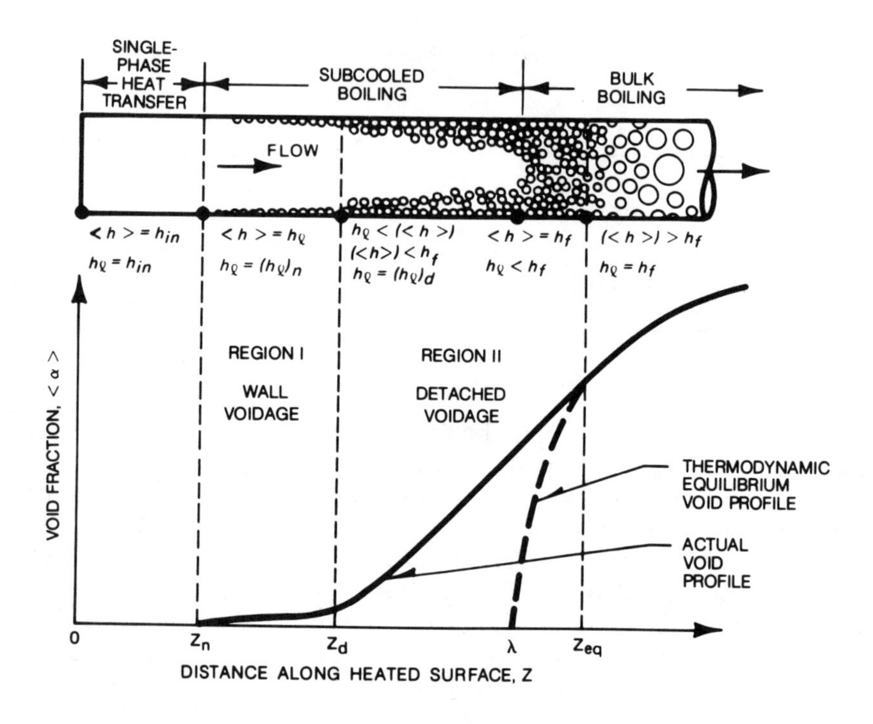

Fig. 5-16 Void fraction during forced convection subcooled boiling.

For many practical applications, the void fraction existing in Region I can be neglected. In contrast, the void fraction in Region II can be appreciable and normally must be considered. In Region II, voids that are ejected from the bubble layer condense in the subcooled core and raise the mean liquid enthalpy, h_l. At some point in the channel, $z_{\mathrm{bulk}} = \lambda$, an equilibrium heat balance would give the onset of bulk boiling. Note in Fig. 5-16 that at $z = z_{\mathrm{bulk}}$, there already are significant subcooled voids and, although the bulk enthalpy is saturated (i.e., $\langle h \rangle = h_f$), the mean liquid enthalpy is still subcooled (i.e., $\langle h_l \rangle < h_f$). At some point, z_{eq}, further up the heated channel, true thermodynamic equilibrium is achieved and both phases are saturated. From this point on, the subcooled and bulk boiling void profiles merge.

The ability to predict accurately the boundaries of Regions I and II is essential to the prediction of accurate void fraction profiles. The location of the point of initial void nucleation is rather difficult since it is dependent on surface finish and chemistry, and the subjective interpretation of the experimentalist. For analyses that require considerations of Region I, the nucleation point, z_n, can be defined in terms of a critical subcooling by equating the single-phase forced convection heat flux at $z = z_n$,

$$q'' = H_{1\phi}(T_{w_n} - T_l) \ , \tag{5.202}$$

to an appropriate nucleation criterion of the form,

$$q'' = C_2(T_{w_n} - T_{sat})^n \quad . \tag{5.203}$$

By equating Eqs. (5.202) and (5.203),

$$C_2(T_{w_n} - T_{sat})^n = H_{1\phi}[(T_{w_n} - T_{sat}) + (T_{sat} - T_l)] \quad , \tag{5.204}$$

which can be solved numerically or analytically for T_{w_n}.

The recommended values of the nucleation parameters, C_2 and n, have been obtained from the nucleation model discussed in Chapter 4, either analytically as,

$$\left.\begin{aligned} C_2 &= \frac{\kappa_f}{\left(\dfrac{8\sigma T_{sat} v_{fg}}{778 h_{fg}}\right)} \\ n &= 2.0 \end{aligned}\right\} \text{(Davis and Anderson, 1966)} \quad , \tag{5.205}$$

or graphically as,

$$\left.\begin{aligned} C_2 &= 15.6p^{1.156} \\ n &= 2.3/p^{0.0234} \end{aligned}\right\} \text{(Bergles and Rohsenow, 1963)} \quad , \tag{5.206}$$

where p is in psia, T_{sat} is in °R, and q'' is in Btu/h-ft^2.

The parameters given in Eqs. (5.205) and (5.206) are approximately equivalent over a wide pressure range and, thus, either set can be used. It may turn out that the actual wall temperature, T_w, is greater at every axial position than the value calculated by Eq. (5.204). In this case, it should be assumed that nucleation begins at the inlet of the heated section.

The most important element in any effective subcooled boiling model is to be able to calculate accurately where significant void fraction appears, that is, the location of the void departure point, z_d, in Fig. 5-16. The location of the initial void ejection into the subcooled liquid core can be determined fairly well experimentally and normally is given in terms of a critical subcooling. Table 5-1 tabulates several of the most widely used void departure criteria. The most accurate of these criteria has been found to be those of Levy (1966) and Saha (Saha and Zuber, 1974).

Note that all of these criteria predict a similar dependence on the heat and mass flux; that is, the critical subcooling is proportional to the heat flux and inversely proportional to the mass flux to some power. Thus, based on the departure criterion alone, we cannot determine whether initial void ejection is controlled by hydrodynamic or heat transfer processes.

To better understand subcooled boiling, let us consider a control volume at the heated surface of our system. Assuming that subcooled boiling occurs, the energy flux at the surface can be arbitrarily partitioned into that required to form vapor, q''_{evap}, that associated with single-phase convection,

TABLE 5-1
Summary of Void Departure Criterion

Criterion (Critical Subcooling, Btu/lb$_m$)	Source	Principle
$[h_f-(h_l)_d]=\dfrac{c_{p_l}q''}{5.0H_{1\phi}}$	(Griffith et al., 1958)	Heat Transfer Model
$[h_f-(h_l)_d]=c_{p_l}\eta\,\dfrac{q''}{(G/\rho_f)}$, where $\eta \triangleq 0.94+0.00046p(156\leqslant p\leqslant 2000,\ \text{psia})$.	(Bowring, 1962)	Empirical
If: $0\leqslant y_b^+\leqslant 5.0$ $[h_f-(h_l)_d]=c_{p_l}\dfrac{q''}{H_{1\phi}}-\dfrac{q''}{G(f/8)^{1/2}}\mathrm{Pr}\,y_b^+$ If: $5.0\leqslant y_b^+\leqslant 30.0$ $[h_f-(h_l)_d]=c_{p_l}\dfrac{q''}{H_{1\phi}}-\dfrac{5.0q''}{G(f/8)^{1/2}}\times\{\mathrm{Pr}+\ln[1+\mathrm{Pr}(y_b^+/5.0-1.0)]\}$ If: $y_b^+\geqslant 30.0$ $[h_f-(h_l)_d]=c_{p_l}\dfrac{q''}{H_{1\phi}}-\dfrac{5.0q''}{G(f/8)^{1/2}}\times[\mathrm{Pr}+\ln(1.0+5.0\,\mathrm{Pr})+0.5\ln(y_b^+/30.0)]$, where $y_b^+=0.015(\sigma g_c D_H\rho_f)^{1/2}/\mu_f$.	(Levy, 1966)	Force Balance
If: $\mathrm{Pe}\triangleq\dfrac{GD_Hc_{pl}}{\kappa_l}<70\,000$ $[h_f-(h_l)_d]=0.0022\dfrac{q''D_hc_{pl}}{\kappa_l}$. If: Pe>70,000 $[h_f-(h_l)_d]=154q''/G$.	(Saha and Zuber, 1974)	Empirical

$q''_{1\phi}$, and that due to liquid agitation or pumping, q''_{pump}. The total energy flux from the heated surface, q'', is then given as,

$$q''\triangleq q''_{1\phi}+q''_b=q''_{1\phi}+q''_{\text{evap}}+q''_{\text{pump}}\ . \tag{5.207}$$

The single-phase convection term normally is given as,

$$q''_{1\phi}=\frac{A_{\text{eff}}}{A_{\text{HT}}}H_{1\phi}(T_w-T_l)\ , \tag{5.208}$$

where A_{eff} is the effective area for single-phase heat transfer. For most cases of practical interest (Dix, 1971), $A_{\text{eff}} \approx A_{\text{HT}}$, thus,

$$q''_{\text{evap}} = j_{\text{B}} \rho_g h_{fg} \ , \tag{5.209}$$

where j_{B} has velocity units (ft/h) and represents the volumetric rate of vapor formation per unit heated area.

The pumping term is made up of the energy flux required to raise, h_{cv}, the enthalpy of liquid leaving the control volume (due to displacement by bubble formation) from the mean core enthalpy, $\langle h_l \rangle$ to h_{cv}, and the energy flux required to raise the mass of liquid that evaporates to saturation temperature. Thus,

$$q''_{\text{pump}} \triangleq j_{\text{B}}[\rho_l(h_{cv} - \langle h_l \rangle) + \rho_g(h_f - h_{cv})] \ . \tag{5.210}$$

By following Bowring's notation (Bowring, 1962), we define the ratio of the heat flux due to pumping to that causing vapor formation as,

$$\varepsilon \triangleq \frac{q''_{\text{pump}}}{q''_{\text{evap}}} = \frac{\rho_l(h_{cv} - \langle h_l \rangle) + \rho_g(h_f - h_{cv})}{\rho_g h_{fg}} \ . \tag{5.211}$$

If we assume that the enthalpy of the liquid leaving the control volume is saturated, then $h_{cv} = h_f$ and Eq. (5.211) reduces to Rouhani's (Rouhani and Axelsson, 1970) result,

$$\varepsilon = \frac{\rho_l(h_f - h_l)}{\rho_g h_{fg}} \ . \tag{5.212}$$

Two distinctly different approaches have been taken to quantify the prediction of forced convection subcooled void fraction. In the first approach, a phenomenological description of the boiling heat transfer process is postulated and, thus, the subcooled flow quality and void fraction are calculated from a mechanistic model. The other approach is to postulate a convenient mathematical fit for the flow quality or liquid enthalpy profile between the void departure point, z_d, and the point at which thermodynamic equilibrium is achieved, z_{eq}. Some of the more widely used references for each method are

1. *Mechanistic Models:* Griffith et al. (1958), Bowring (1962), Rouhani and Axelsson (1970), Larsen and Tong (1969), and Hancox and Nicoll (1971).
2. *Profile-Fit Models:* Zuber et al. (1966), Staub (1968), Levy (1966), and Saha and Zuber (1974).

The procedure that we follow is to first discuss a mechanistic model that is representative of previous models, modified as necessary to reflect our most recent understanding. As will be discussed, in light of the present state-of-the-art, the mechanistic model presented herein is a far from perfect description of the observed phenomenon, but it does afford an opportunity to discuss the basic physics involved and to focus on those areas that need further experimental investigation. Next, we discuss a modified version of Levy's (1966) profile-fit model and point out the merit and shortcomings of such an approach. Finally, recommendations are made for both steady-state and transient calculational techniques.

5.3.2.1 *A Mechanistic Model*

There are various ways to derive the appropriate flow quality relationship. The method that is presented here comes directly from the conservation equations derived in Sec. 5.1. Consider the energy equation of the two-phase mixture, Eq. (5.107),

$$\rho_l(1-\langle\alpha\rangle)\left(\frac{\partial\langle e_l\rangle_l}{\partial t}+\langle u_l\rangle_l\frac{\partial\langle e_l\rangle_l}{\partial z}\right)+\rho_g\langle\alpha\rangle\left(\frac{\partial\langle e_g\rangle_g}{\partial t}+\langle u_g\rangle_g\frac{\partial\langle e_g\rangle_g}{\partial z}\right)$$
$$=\frac{q''P_H}{A_{x-s}}+q'''+\frac{1}{J}\frac{\partial p}{\partial t}-\Gamma(\langle e_g\rangle_g-\langle e_l\rangle_l) \ , \tag{5.213}$$

where the volumetric vapor generation term, Γ, has been defined previously and is related to other notation in Table 5-2. From the definition of the total energy of each phase given in Eq. (5.97), if the kinetic and potential energy terms are neglected we obtain,

$$\langle e_g\rangle_g = h_g \tag{5.214}$$

$$\langle e_l\rangle_l = \langle h_l\rangle \ . \tag{5.215}$$

For most cases of practical interest, we can neglect internal generation, q''', and the kinetic and potential energy terms without introducing any serious error. Under these simplifying assumptions, Eq. (5.213) becomes,

$$\rho_l(1-\langle\alpha\rangle)\left(\frac{\partial\langle h_l\rangle}{\partial t}+\langle u_l\rangle_l\frac{\partial\langle h_l\rangle}{\partial z}\right)+\rho_g\langle\alpha\rangle\left(\frac{\partial h_g}{\partial t}+\langle u_g\rangle_g\frac{\partial h_g}{\partial z}\right)$$
$$=\frac{q''P_H}{A_{x-s}}+\frac{1}{J}\frac{\partial p}{\partial t}-\Gamma(h_g-\langle h_l\rangle) \ . \tag{5.216}$$

TABLE 5-2
Relationship Between Present Notation and that of Previous Subcooled Vapor Generation Models

Notation	Source	Relationship	Units
Γ	(Zuber and Staub, 1966)	Present Notation	lb_m/ft^3-h
ψ	(Forti, 1968)	$\psi_{net}=\Gamma/\rho_g$	1/h
Ω	(Wallis, 1969)	$\Omega=v_{lg}\Gamma$	1/h

For steady state, this becomes,

$$G(1-\langle x\rangle)\frac{d\langle h_l\rangle}{dz}+G\langle x\rangle\frac{dh_g}{dz}=\frac{q''P_H}{A_{x-s}}-\Gamma(h_g-\langle h_l\rangle)\ , \tag{5.217}$$

where the identities given in Eqs. (5.22) and (5.23) have been used. The next step is to consider Eq. (5.56), the steady-state version of the steam phase continuity equation. By using Eq. (5.23), we obtain for a constant area duct,

$$G\frac{d(\langle x\rangle)}{dz}=\Gamma\ , \tag{5.218}$$

Equations (5.217) and (5.218) can be combined to eliminate the volumetric vapor generation term, Γ, yielding,

$$G(1-\langle x\rangle)\frac{d\langle h_l\rangle}{dz}+G\langle x\rangle\frac{dh_g}{dz}+G(h_g-\langle h_l\rangle)\frac{d\langle x\rangle}{dz}=\frac{q''P_H}{A_{x-s}}\ . \tag{5.219}$$

This equation can be used to yield several important results. First, we can regroup Eq. (5.219) into an exact differential,

$$\frac{d}{dz}[(1-\langle x\rangle)\langle h_l\rangle+h_g\langle x\rangle]=\frac{q''P_H}{GA_{x-s}}\ , \tag{5.220}$$

which is the same result that we would obtain from a steady-state heat balance. Equation (5.220) can be integrated to yield,

$$\langle h_l\rangle(1-\langle x\rangle)+h_g\langle x\rangle=h_{\text{in}}+\frac{1}{GA_{x-s}}\int_0^z P_H q''dz\ .$$

Noting that,

$$\langle h\rangle\triangleq h_{\text{in}}+\frac{1}{GA_{x-s}}\int_0^z P_H q''dz\ ,$$

we obtain the interesting result,

$$\langle x(z)\rangle=\frac{[\langle h(z)\rangle-\langle h_l(z)\rangle]}{[h_g-\langle h_l(z)\rangle]}\ . \tag{5.221}$$

From the definition of thermodynamic equilibrium quality, $\langle x_e\rangle$, given in Eq. (5.13), it is obvious that $\langle x(z)\rangle$ becomes equal to $\langle x_e(z)\rangle$ as $\langle h_l(z)\rangle$ approaches the saturation enthalpy, h_f.

Equation (5.219) can be rewritten under the reasonable assumption that dh_g/dz is negligible,

$$\frac{q''P_H}{GA_{x-s}} = (1-\langle x\rangle)\frac{d\langle h_l\rangle}{dz} + (h_f - \langle h_l\rangle)\frac{d\langle x\rangle}{dz} + h_{fg}\frac{d\langle x\rangle}{dz} \ . \tag{5.222}$$

During subcooled boiling, part of the heat flux goes into raising the mean liquid temperature and part goes into vapor formation. This latter term is a balance between the vapor generated at the wall and that condensed by the subcooled fluid. Thus, we can write,

$$\frac{q''P_H}{AG_{x-s}} = \frac{P_H}{GA_{x-s}}[(q''_{\text{evap}} - q''_{\text{cond}}) + (q''_{1\phi} + q''_{\text{pump}} + q''_{\text{cond}})] \ . \tag{5.223}$$

By comparing Eqs. (5.222) and (5.223), we can arbitrarily partition that portion of energy that goes into the net formation of vapor as,

$$\frac{P_H}{GA_{x-s}}(q''_{\text{evap}} - q''_{\text{cond}}) = h_{fg}\frac{d\langle x\rangle}{dz} \ . \tag{5.224}$$

Now, by using the definition of ε given in Eq. (5.211), we can write,

$$\frac{d\langle x(z)\rangle}{dz} = \frac{P_H}{GA_{x-s}h_{fg}}\left\{\frac{q''_b(z)}{[1+\varepsilon(z)]} - q''_{\text{cond}}(z)\right\} , \tag{5.225}$$

where the boiling heat flux, q''_b, is defined as $q''_b \triangleq q''_{\text{evap}} + q''_{\text{pump}}$.

Equation (5.225) can be readily integrated from some location, z_0, to any axial position downstream,

$$\langle x(z)\rangle = \frac{1}{GA_{x-s}h_{fg}}\left\{\int_{z_0}^{z} \frac{P_H q''_b(z)dz}{[1+\varepsilon(z)]} - \int_{z_0}^{z} P_H q''_{\text{cond}}(z)dz\right\} . \tag{5.226}$$

Bowring (1962) derived a similar expression for flow quality in a constant area duct, such that the heated perimeter, P_H, was constant. He assumed that the void departure point, z_d, was the appropriate value for z_0, and performed analyses which indicated that for water the magnitude of the condensation heat flux, q''_{cond}, should be negligible. Moreover, he found that by comparing his model to the available data, the parameter ε could be assumed to be piecewise constant, thus, Eq. (5.226) yields,

$$\langle x(z)\rangle = \frac{P_H}{GA_{x-s}h_{fg}(1+\varepsilon)}\int_{z_d}^{z} q''_b(z)dz \ . \tag{5.227}$$

A more careful consideration of the problem exposes the fact that Bowring's analysis of the effect of the condensation term did not take into account the effect of turbulence and the resultant eddy diffusivity transport of thermal energy. In general, it has been found that one cannot neglect the condensation heat flux term, and comparisons with a wide range of subcooled void fraction data have shown that rather than being constant, ε

varies axially. Thus, in general, Eq. (5.226) is the appropriate expression to use for steady-state flow quality calculations.

If one is to use Eq. (5.226) for calculations of the steady-state flow quality, then we must have explicit expressions for z_0, $q_b''(z)$, $\varepsilon(z)$, and $q_{\text{cond}}''(z)$. We will now consider the specification of these parameters one at a time.

First, we consider the appropriate expression for z_0. In Fig. 5-17, it is seen that the effect of boiling on the heat transfer process is felt from the nucleation point on. Thus, one possible choice for z_0 is the nucleation point, z_n. However, it has been found experimentally that the net void production below the void departure point, z_d, is negligible. That is, we normally can neglect Region I voids. This observation would suggest that the appropriate choice for z_0 is z_d, and this is the recommended choice. If we choose $z_0 = z_d$ for evaluation of Eq. (5.226), then we must somehow modify $q_b''(z)$ or $q_{\text{cond}}''(z)$ to reflect the fact that Region I voids have been neglected. The easiest way to accomplish this is to consider the heat flux profile between the nucleation point and the void departure point and to modify $q_b''(z)$ to avoid any discontinuity in our calculations.

As shown in Fig. 5-17, at the void departure point, z_d, $q_b''>0$; thus, $q''>q_{1\phi}''$. Indeed, it has been found (Griffith et al., 1958) that the heat flux

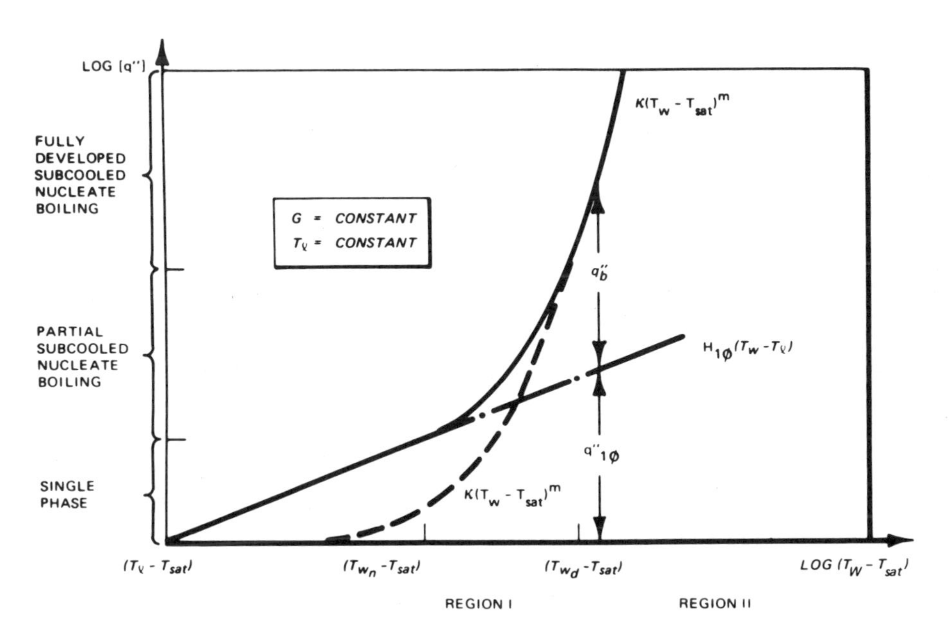

Fig. 5-17 A forced convection subcooled boiling curve.

is approximately five times that predicted with single-phase flow correlations. That is,

$$q''(z_d)=\kappa(T_{wd}-T_{sat})^m \cong 5.0H_{1\phi}(T_w-T_1)_d \ . \tag{5.228}$$

However, to be consistent with our assumption of neglecting Region I voids, the boiling heat flux, q_b'', must be zero below the departure point. To ensure this, we can postulate a pseudo-single-phase contribution for $(\langle h_l\rangle)\geqslant h_{l_d}$,

$$q_{1\phi}''(z)_{\text{eff}} \triangleq q''(z)\left[\frac{h_f-\langle h_l(z)\rangle}{h_f-h_{l_d}(z)}\right] , \tag{5.229}$$

such that for steady-state calculations, the net boiling heat flux is given by

$$q_b''(z)=\begin{cases}0.0, \ h_l\leqslant h_{l_d}\\ q''(z)-q_{1\phi}''(z)_{\text{eff}}=q''(z)\left\{1-\left[\dfrac{h_f-\langle h_l(z)\rangle}{h_f-h_{l_d}(z)}\right]\right\} , \quad (\langle h_l\rangle)\geqslant h_{l_d}\end{cases} . \tag{5.230}$$

The next term, which must be considered before Eq. (5.226) can be integrated, is the condensation heat flux, q_{cond}''. Levenspiel (1959) experimentally determined a functional relationship for the condensation heat transfer process in the form,

$$P_H q_{\text{cond}}''=H_0\frac{h_{fg}}{v_{fg}}A_{x-s}\langle\alpha\rangle(T_{sat}-T_l) \ . \tag{5.231}$$

Since Levenspiel's experiment was run in a still tank, the arbitrary parameter, H_0, which he determined from his measurements, is not representative and must be determined from forced convection subcooled boiling data.

The last item to be specified, so that Eq. (5.226) can be integrated, is the appropriate form of Eq. (5.211). At this point, we have no better evidence to recommend anything other than the simple expression given in Eq. (5.212). The basic problem inherent in all mechanistic subcooled boiling models is that there are insufficient data to specify accurately the basic mechanisms involved and there are too many degrees of freedom. Specifically, in the present model, which is representative of most previous attempts, we have phenomenologically based functional forms of $\varepsilon(z)$, z_d, $q_b''(z)$, and $q_{\text{cond}}''(z)$. Any, or all, of these models could be incorrect, which would naturally bias the choice of the arbitrary condensation constant, H_0. Hence, considering the current state-of-the-art in subcooled boiling models and the present data base, we cannot currently synthesize a mechanistic model of complete generality. However, assuming the validity of the functional forms described above, we can optimize the choice of the arbitrary condensation parameter, H_0, such that the void data of interest to BWR technology are predicted accurately. Preliminary comparisons with representative subcooling boiling data indicated that the appropriate value for H_0 was,

$$H_0 = 150 \quad (\text{h-°F})^{-1} \ .$$

More recent work (Park et al., 1983) has indicated that a better choice is,

$$H_0 = [1.0 + 4.0(\langle j \rangle / j_{\text{in}})^{2.0}]^{-1} \ .$$

Note that if we were to consider Region I voids, thus setting $z_0 = z_n$ in Eq. (5.226), the optimized value of H_0 would need to be much larger in Region I and vary axially to provide agreement with the data.

It can be seen in Figs. 5-18 and 5-19 that the agreement with steady-state data is quite good for the mechanistic model and the profile-fit model, to be discussed in Sec. 5.3.2.2.

Until now, we have been considering the prediction of subcooled void fraction for steady-state conditions only. However, many important cases exist for which we must be able to predict the subcooled void fraction during transients. For such predictions, we can generalize the mechanistic method just discussed. Equation (5.216) can be partitioned [in the same manner that led to Eq. (5.224)] to yield[e],

$$\Gamma = \frac{P_H q_b''(z,t)}{A_{x-s} h_{fg}[1+\varepsilon(z,t)]} - \frac{P_H q_{\text{cond}}''(z,t)}{A_{x-s} h_{fg}} + \frac{1}{h_{fg}} \left[\frac{1}{J} \frac{\partial p}{\partial t} - \rho_f (1 - \langle \alpha \rangle) \frac{D_f h_f}{Dt} - \rho_g \langle \alpha \rangle \frac{D_g h_g}{Dt} \right] \ . \tag{5.232}$$

The remainder of the energy goes directly into the liquid phase and,thus, Eq. (5.216) yields,

$$\rho_l (1 - \langle \alpha \rangle) \frac{D_l \langle h_l \rangle}{Dt} = \frac{P_H}{A_{x-s}} q''(z,t) - \Gamma (h_g - \langle h_l \rangle) - \rho_g \langle \alpha \rangle \frac{D_g h_g}{Dt} + \frac{1}{J} \frac{\partial p}{\partial t} \ . \tag{5.233}$$

Equations (5.232) and (5.233) are coupled through the former's dependence of q_b'', ε, and q_{cond}'' on $\langle h_l \rangle$. Basically, Eq. (5.232) defines the vapor generation term and Eq. (5.233) yields the mean liquid enthalpy, $\langle h_l(z,t) \rangle$.

In transient calculations where Region I voids are neglected, it is still important to consider the effects of boiling in this region so that representative heat transfer coefficients and accurate cladding temperatures can be evaluated. Various methods are available for approximating the heat transfer in Region I, the partial subcooled nucleate boiling region (Lahey, 1974). The next section is concerned with approximate profile-fit methods and their application to BWR design-type problems.

---

[e]Note that for steady, saturated, two-phase flow, Eq. (5.232) reduces to Eq. (5.55).

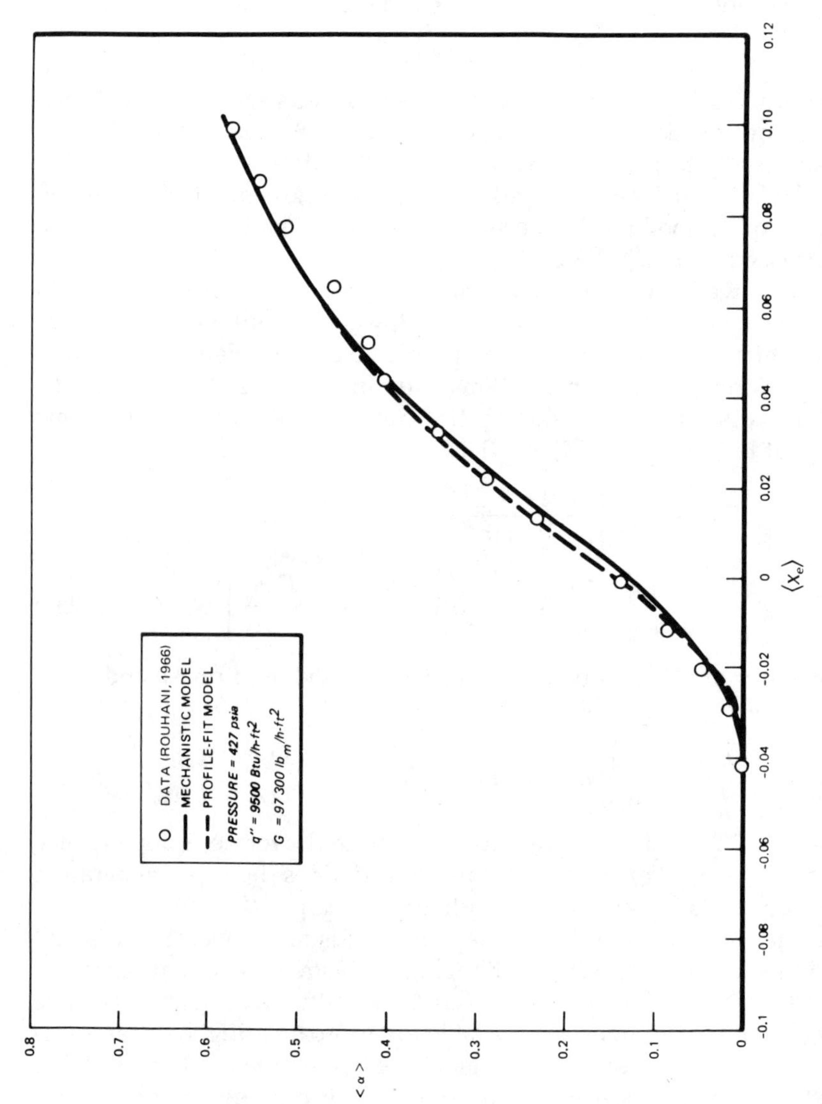

Fig. 5-18 Comparison of mechanistic and profile-fit model with void fraction data, $P = 427$ psia.

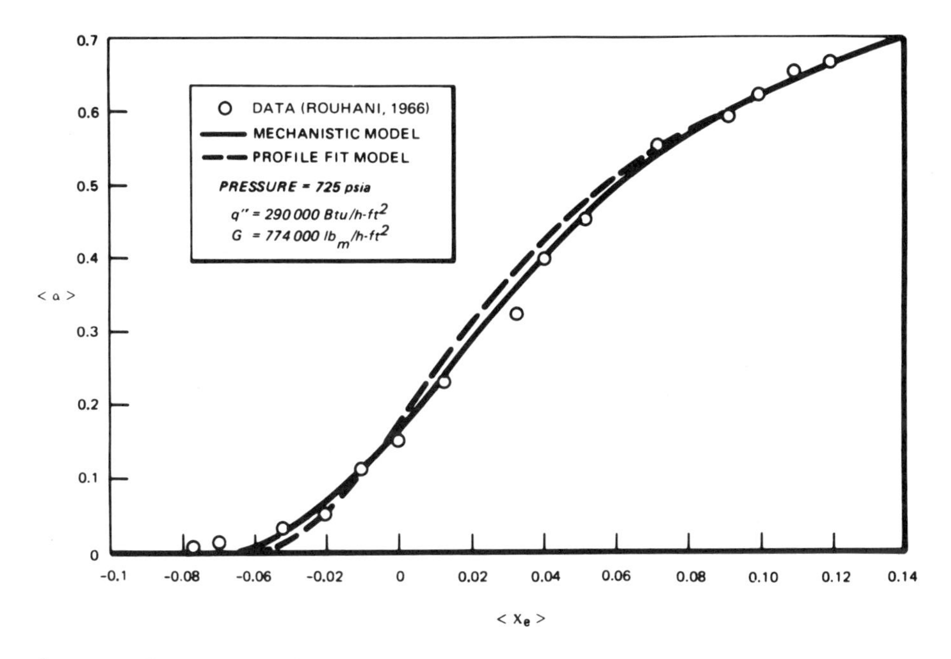

Fig. 5-19 Comparison of mechanistic and profile-fit models with void fraction data.

5.3.2.2 *A Profile-Fit Model*

As mentioned earlier in this section, a number of subcooled void fraction models are available in the literature that do not attempt to specify the mechanism of subcooled boiling, but rather, directly assume a profile between z_d and z_{eq}. The assumed profile may be either that of the mean liquid enthalpy, $\langle h_l(z)\rangle$, or that of the flow quality, $\langle x(z)\rangle$. In this section, we present a version of this technique that hopefully combines the best features of previous models. Following Zuber et al. (1966), we assume that the mean liquid enthalpy, $\langle h_l\rangle$, is a function of the bulk enthalpy, $\langle h\rangle$, such that,

$$\frac{(h_f-\langle h_l\rangle)}{[h_f-(h_l)_d]}=\mathcal{F}\left[\frac{\langle h\rangle-(h_l)_d}{h_f-(h_l)_d}\right] . \tag{5.234}$$

In accordance with Fig. 5-16, if we assume no voids in Region I, the function $\mathcal{F}$ must satisfy,

$$\mathcal{F}=1, \text{ at } \langle h\rangle=\langle h_l\rangle=(h_l)_d \tag{5.235}$$

$$\mathcal{F}\rightarrow 0, \text{ for } \langle h\rangle>>h_f . \tag{5.236}$$

Also, since $d\langle h\rangle/dz = d\langle h_l\rangle/dz$ for $z \leqslant z_d$,

$$\frac{d\mathcal{F}}{d\langle h\rangle} = \frac{-1}{h_f - (h_l)_d}\ , \text{ at } z = z_d\ . \tag{5.237}$$

Two functions that satisfy these requirements are,

$$\mathcal{F} = \exp\left\{-\left[\frac{\langle h\rangle - (h_l)_d}{h_f - (h_l)_d}\right]\right\} \tag{5.238}$$

$$\mathcal{F} = 1 - \tanh\left[\frac{\langle h\rangle - (h_l)_d}{h_f - (h_l)_d}\right]\ . \tag{5.239}$$

Data comparisons have indicated that Eq. (5.239) gives a slightly better fit and, thus, we will use it as the basis for our profile-fit model. Before we formulate this model, it is instructive to combine Eqs. (5.234) and (5.221) to obtain,

$$\langle x\rangle = \frac{(\langle h\rangle - h_f) + [h_f - (h_l)_d]\mathcal{F}}{\{h_{fg} + [h_f - (h_l)_d]\mathcal{F}\}}\ . \tag{5.240}$$

For virtually all cases of practical concern, $h_{fg} >> [h_f - (h_l)_d]\mathcal{F}$. Thus, Eq. (5.240) reduces to,

$$\langle x\rangle = \frac{(\langle h\rangle - h_f)}{h_{fg}} + \frac{[h_f - (h_l)_d]}{h_{fg}}\mathcal{F}\ . \tag{5.241}$$

Equation (5.13) yields,

$$\langle x_e\rangle = \frac{(\langle h\rangle - h_f)}{h_{fg}}\ , \tag{5.242}$$

and by analogy,

$$\langle x_e\rangle_d = -\frac{[h_f - (h_l)_d]}{h_{fg}}\ . \tag{5.243}$$

Thus, Eq. (5.241) can be written as,

$$\langle x\rangle = \langle x_e\rangle - \langle x_e\rangle_d\mathcal{F}\ . \tag{5.244}$$

If we assume the validity of Eq. (5.238) and use the definitions in Eqs. (5.242) and (5.243), Eq. (5.244) can be written as,

$$\langle x\rangle = \langle x_e\rangle - \langle x_e\rangle_d \exp\left(\frac{\langle x_e\rangle}{\langle x_e\rangle_d} - 1\right)\ , \tag{5.245}$$

which is Levy's (1966) assumed expression for the functional relationship between flow quality, $\langle x\rangle$, and thermodynamic equilibrium quality, $\langle x_e\rangle$. Thus, it is obvious that $\langle x(z)\rangle$ and $\langle h_l(z)\rangle$ are not independent. That is, we can either directly assume the form of the flow quality variation or the liquid enthalpy variation. However, if the available data are to be correlated, such assumptions necessarily must have the same basic functional dependence on $\langle h \rangle$ and $(h_l)_d$. Moreover, once the $\langle h_l \rangle$ or $\langle x \rangle$ profile has been assumed, the volumetric vapor generation term, Γ, is uniquely defined. For example, Eqs. (5.218) and (5.224) yield,

$$\Gamma = \frac{P_H}{A_{x-s}h_{fg}}(q''_{\text{evap}} - q''_{\text{cond}}) = G\frac{d\langle x \rangle}{dz} \ . \tag{5.246}$$

Assuming that Eq. (5.245) is valid, we obtain,

$$\Gamma = G\frac{d\langle x_e \rangle}{dz}\left[1 - \exp\left(\frac{\langle x_e \rangle}{\langle x_e \rangle_d} - 1\right)\right] \ , \tag{5.247}$$

which implies that,

$$\Gamma = \begin{cases} 0, \text{ at } z \leqslant z_d \\ G\dfrac{d\langle x_e \rangle}{dz} = \dfrac{q'' P_H}{A_{x-s}h_{fg}} \ , \qquad \text{for } z >> z_d \ . \end{cases}$$

For steady-state calculations, a profile-fit method is normally easier to use than a mechanistic method and is as accurate. The main disadvantage of this simple method is that it is based on a fit to uniform axial heat flux data and, thus, is uncomfirmed for the prediction of subcooled void fraction in cases of nonuniform axial heat flux. In particular, "cold patch" experiments indicate that the subcooled void fraction should decrease in the adiabatic section, however, only mechanistic models follow this trend. Nevertheless, for most cases of interest in BWR technology, the profile-fit technique is quite adequate for steady-state situations and is recommended. For transients calculations, the mechanistic model is required, although in some instances profile-fit techniques have been used. In Figs. 5-18 and 5-19 it is seen that the recommended profile-fit model agrees with the data and the proposed mechanistic model.

In summary, several techniques have been discussed for the prediction of subcooled voids during steady-state and/or transient conditions typical of BWR technology. Basically, these models consist of either postulating a phenomenological description of the subcooled boiling process (mechanistic model) or assuming a convenient mathematical fit to the data (profile-fit model) to determine flow quality. Once flow quality is known, Eq. (5.177) can be used as the appropriate void-quality relationship to calculate subcooled void fraction. That is, the same void-quality relation is used for both subcooled and bulk boiling conditions. Since these methods have been found to be in reasonably good agreement with available data, they are recommended for application to problems in BWR technology.

The next section is concerned with the evaluation of pressure drop in BWR fuel and components. This subject logically follows void fraction models since adequate void models are essential to the accurate evaluation of the elevation component of total pressure drop. The accurate evaluation of irreversible pressure loss is quite important to the reactor designer to ensure adequate performance of the reactor system.

5.4 Pressure Drop

The field of two-phase flow is empirically based. That is, we can rarely make calculations solely based on first principles, but instead must rely heavily on correlations synthesized from experimental data. The prediction of pressure drop is a particularly good example of the empirical basis of these calculations.

To focus attention on the various types of pressure drop that can occur in a two-phase system, we rewrite Eq. (5.66) in the form,

$$-\frac{\partial p}{\partial z}=\frac{1}{g_c}\left[\frac{\partial G}{\partial t}+\frac{1}{A_{x-s}}\frac{\partial}{\partial z}\left(\frac{G^2 A_{x-s}}{\langle\rho'\rangle}\right)\right]+\frac{\tau_w P_f}{A_{x-s}}+\frac{g}{g_c}\langle\bar{\rho}\rangle\sin\theta \ . \tag{5.248}$$

The first term on the right side is the pressure gradient due to the effect of acceleration. The $\partial G/\partial t$ term is frequently called the "temporal acceleration," while the other term is called the "spatial acceleration." The next term on the right side is the irreversible pressure drop due to frictional effects. The last term is the gravitational, or elevation, pressure gradient. Once the empirical void-quality relation has been specified, the evaluation of the acceleration and the elevation pressure gradient is straightforward and is not discussed further here.

Empirical methods are also employed for the evaluation of friction pressure gradients and local losses; e.g., sudden expansions, spacers, etc. The next section is devoted to discussions of the techniques commonly employed to evaluate these components.

5.4.1 Frictional Pressure Drop

In single-phase flow, it is well known that frictional pressure losses are commonly correlated in terms of the dynamic head. That is, they are normally written as,

$$\Delta p_{1\phi}=K\frac{1}{2g_c}\rho_l\langle j_l\rangle^2=K\frac{G^2}{2g_c\rho_l} \ , \tag{5.249}$$

where $G^2/2g_c\rho_l$ is the dynamic head and K is an empirical irreversible loss coefficient, which is written for frictional losses as,

$$K=f\left(\frac{\Delta z}{D_H}\right) \ , \tag{5.250}$$

and f is the Darcy-Weisbach friction factor (Vennard, 1959).

In flowing two-phase systems, it has been observed experimentally that for a given mass flux, the pressure drop can be much greater than for a corresponding single-phase system. The classical approach, which has been taken to correlate two-phase frictional losses, is to multiply the equivalent saturated single-phase pressure loss by an empirical multiplier, ϕ_{l0}^2, which is a function of (at least) flow quality and system pressure. That is,

$$\Delta p_{2\phi} = K \frac{G^2}{2 g_c \rho_f} \phi_{l0}^2(\langle x \rangle, p, \ldots) \ . \tag{5.251}$$

Before we present some commonly accepted correlations for ϕ_{l0}^2, let us examine the expected behavior of ϕ_{l0}^2. For this purpose, we note that for two-phase flow we can write,

$$\Delta p_{2\phi} = K \frac{G^2}{2 g_c \langle \rho_{2\phi} \rangle} \ , \tag{5.252}$$

in which $\langle \rho_{2\phi} \rangle$ is the appropriate two-phase density. For the limiting case of saturated, homogeneous two-phase flow, Eq. (5.30) yields,

$$\Delta p_{2\phi} = K \frac{G^2}{2 g_c \rho_f} \left(\frac{\rho_f}{\langle \rho_h \rangle} \right) \ . \tag{5.253}$$

Thus, by comparing Eqs. (5.253) and (5.251), it is found that,

$$\phi_{l0}^2 = \left(1 + \frac{v_{fg}}{v_f} \langle x \rangle \right) \ . \tag{5.254}$$

Note that $v_{fg}/v_f = 19.6$ at 1000 psia, thus it is obvious that even for elevated pressures, ϕ_{l0}^2 increases rapidly with flow quality.

The friction pressure-drop multiplier, ϕ_{l0}^2, is sometimes modified to account for the effect of the vapor phase on apparent viscosity. Using the definition of loss coefficient given in Eq. (5.250), and recalling that over a wide range of turbulent Reynolds numbers we can approximate the friction factor by,

$$f = \frac{C}{\left(\dfrac{G D_H}{\mu_{2\phi}} \right)^n} \ , \tag{5.255}$$

we can choose an appropriate correlation for two-phase viscosity such as that due to McAdams et al. (1942),

$$\mu_{2\phi}/\mu_f = \frac{1}{\left[\dfrac{\mu_f}{\mu_g} \langle x \rangle + (1 - \langle x \rangle) \right]} \ , \tag{5.256}$$

such that Eq. (5.252) can be written as,

$$\Delta p_{2\phi} = \frac{C\Delta z G^2}{\left(\frac{GD_H}{\mu_f}\right)^n 2g_c D_H \rho_f} \left(\frac{\rho_f}{\langle\rho_{2\phi}\rangle}\right)\left(\frac{\mu_{2\phi}}{\mu_f}\right)^n . \tag{5.257}$$

Thus, for homogeneous flow,

$$\phi_{l0}^2 = \left(1 + \frac{v_{fg}}{v_f}\langle x\rangle\right)\left[\frac{\mu_f}{\mu_g}\langle x\rangle + (1 - \langle x\rangle)\right]^{-n} . \tag{5.258}$$

Comparison of Eqs. (5.254) and (5.258) indicates the appropriate viscosity correction. Due to uncertainty of the correct two-phase viscosity correlation and the fact that flow qualities of practical significance normally are much less than unity, viscosity corrections are normally not employed in two-phase analyses.

The appropriate two-phase density of separated flow is not as well defined. It has been suggested (Chisholm, 1973) that the momentum density should be used. Thus, Eq. (5.252) becomes,

$$\Delta p_{2\phi} = K\frac{G^2}{2g_c\langle\rho'\rangle} = K\frac{G^2}{2g_c\rho_f}\left(\frac{\rho_f}{\langle\rho'\rangle}\right) . \tag{5.259}$$

Thus, Eqs. (5.251), (5.259), and (5.67) yield,

$$\phi_{l0}^2 = \left[\frac{(1-\langle x\rangle)^2}{(1-\langle\alpha\rangle)} + \frac{\rho_f}{\rho_g}\frac{\langle x\rangle^2}{\langle\alpha\rangle}\right] . \tag{5.260}$$

As discussed in Chapter 9, we can rewrite Eq. (5.260) as,

$$\phi_{l0}^2 = \rho_f/\langle\rho'\rangle = \rho_f\left[\frac{\langle x\rangle}{\rho_g} + \frac{S(1-\langle x\rangle)}{\rho_f}\right]\left[\langle x\rangle + \frac{(1-\langle x\rangle)}{S}\right] . \tag{5.261}$$

Thus,

$$\phi_{l0}^2 = (1-\langle x\rangle)^2\left[\frac{\langle x\rangle}{(1-\langle x\rangle)}\left(\frac{\rho_f}{\rho_g}\right) + S\right]\left[\frac{\langle x\rangle}{(1-\langle x\rangle)} + \frac{1}{S}\right] . \tag{5.262}$$

It is interesting to note that for high Reynolds number flows ($n=0$), Eq. (5.191) yields,

$$X_{tt} = \frac{(1-\langle x\rangle)}{\langle x\rangle}\sqrt{\rho_g/\rho_l} . \tag{5.263}$$

Thus we can rewrite Eq. (5.262) as,

$$\phi_{l0}^2 = (1-\langle x\rangle)^2\left(\frac{1}{X_{tt}\sqrt{\rho_g/\rho_l}} + S\right)\left(\frac{1}{X_{tt}\sqrt{\rho_f/\rho_g}} + \frac{1}{S}\right) . \tag{5.264}$$

Equation (5.264) can also be rewritten in the form proposed by Chisholm and Sutherland (1969),

$$\phi_{l0}^2 = (1-\langle x\rangle)^2\left[1+\frac{C}{X_{tt}}+\frac{1}{X_{tt}^2}\right] , \tag{5.265a}$$

where,

$$C = \frac{1}{S}\sqrt{\rho_f/\rho_g} + S\sqrt{\rho_g/\rho_f} \ . \tag{5.265b}$$

Equation (5.265a) is a popular form of the friction loss multiplier for two-phase flows.

Levy (1960) has provided further insight into the nature of the increase in dissipation in annular two-phase flow compared to single-phase flow in terms of the resultant speedup of the liquid phase. Consider the single-phase frictional pressure gradient,

$$-\left(\frac{dp}{dz}\right)_{l0} = f_{1\phi}\frac{G^2}{2g_c D_H \rho_f} = \frac{C}{\left(\dfrac{GD_H}{\mu_f}\right)^n}\frac{G^2}{2g_c D_H \rho_f} \ . \tag{5.266}$$

In ideal two-phase annular flow, we can write similarly,

$$-\left(\frac{dp}{dz}\right)_{2\phi} = \frac{4\tau_w}{D_H} = 4\left(\frac{1}{4}f_{1\phi}\right)\frac{\rho_f\langle u_f\rangle_f^2}{2g_c D_H} = f_{1\phi}\frac{\rho_f\langle u_f\rangle_f^2}{2g_c D_H} \ . \tag{5.267}$$

By combining Eqs. (5.267), (5.22), and (5.255), we obtain,

$$-\left(\frac{dp}{dz}\right)_{2\phi} = \frac{C}{\left[\dfrac{G(1-\langle x\rangle)D_H}{(1-\langle\alpha\rangle)\mu_f}\right]^n}\frac{G^2(1-\langle x\rangle)^2}{2g_c D_H \rho_f (1-\langle\alpha\rangle)^2} \ . \tag{5.268}$$

By definition,

$$\phi_{l0}^2 \triangleq \frac{-\left(\dfrac{dp}{dz}\right)_{2\phi}}{-\left(\dfrac{dp}{dz}\right)_{l0}} \ . \tag{5.269}$$

Thus, Eqs. (5.266) and (5.268) yield,

$$\phi_{l0}^2 = \left[\frac{(1-\langle x\rangle)}{(1-\langle\alpha\rangle)}\right]^{2-n} , \tag{5.270}$$

where a typical value of the Reynolds number coefficient is $n = 0.25$.

We see that ϕ_{l0}^2 increases rapidly with quality, which is a reflection of the fact that the average velocity of the liquid phase, $\langle u_f \rangle_f = G(1-\langle x \rangle)/\rho_f(1-\langle\alpha\rangle)$, which is presumed to flow as a film on the wall, increases rapidly with quality.

Equations (5.260), (5.270), and (5.177) imply that for a given system pressure and flow quality, ϕ_{l0}^2 tends to increase as mass flux increases. Unfortunately, this is the opposite of observed experimental trends (Isbin et al., 1959). Nevertheless, these models are often good first approximations and serve to help explain why two-phase pressure drops are normally much higher than for single-phase flow.

Now that the need for, and the significance of, an appropriate two-phase multiplier, ϕ_{l0}^2, has been motivated, we turn our attention to several frequently used empirical correlations. By far the most famous and widely used of these correlations is due to Martinelli (Martinelli and Nelson, 1948). This correlation is shown in Fig. 5-20. Note that the two-phase dissipation goes up markedly with decreasing pressure level and that no flow effect has been included. This correlation was developed for horizontal flow conditions under the assumption that one has turbulent flow in both phases. Nevertheless, it has been found to be reasonably accurate for vertical two-phase flows in which the flow rates are high enough to ensure co-current flow.

Another popular empirical correlation is due to Thom (1964). Thom's correlation for ϕ_{l0}^2 is shown in Fig. 5-21. Note that like the Martinelli-Nelson correlation, no flow effect has been included in the Thom correlation.

A convenient empirical approximation to ϕ_{l0}^2 for water, which is valid for low qualities and includes the flow effect synthesized by Jones (1961), is given by,

$$\phi_{l0}^2 = \P(G,p)\{1.2[(\rho_f/\rho_g)-1]\langle x \rangle^{0.824}\} + 1.0 \; , \qquad (5.271)$$

where,

$$\P(G,p) \triangleq \begin{cases} 1.36 + 0.0005p + 0.1(G/10^6) - 0.000714p(G/10^6); & (G/10^6) < 0.7 \\ 1.26 - 0.0004p + 0.119\left(\dfrac{10^6}{G}\right) + 0.00028p\left(\dfrac{10^6}{G}\right); & (G/10^6) > 0.7 \end{cases}$$

$\langle x \rangle$ = decimal fraction

p = psia

$G = \mathrm{lb}_m/\mathrm{h\text{-}ft}^2$.

Equation (5.271) is shown plotted for $p = 1000$ psia in Fig. 5-22. Note that the flow effect is in the same direction as observed experimentally (Isbin

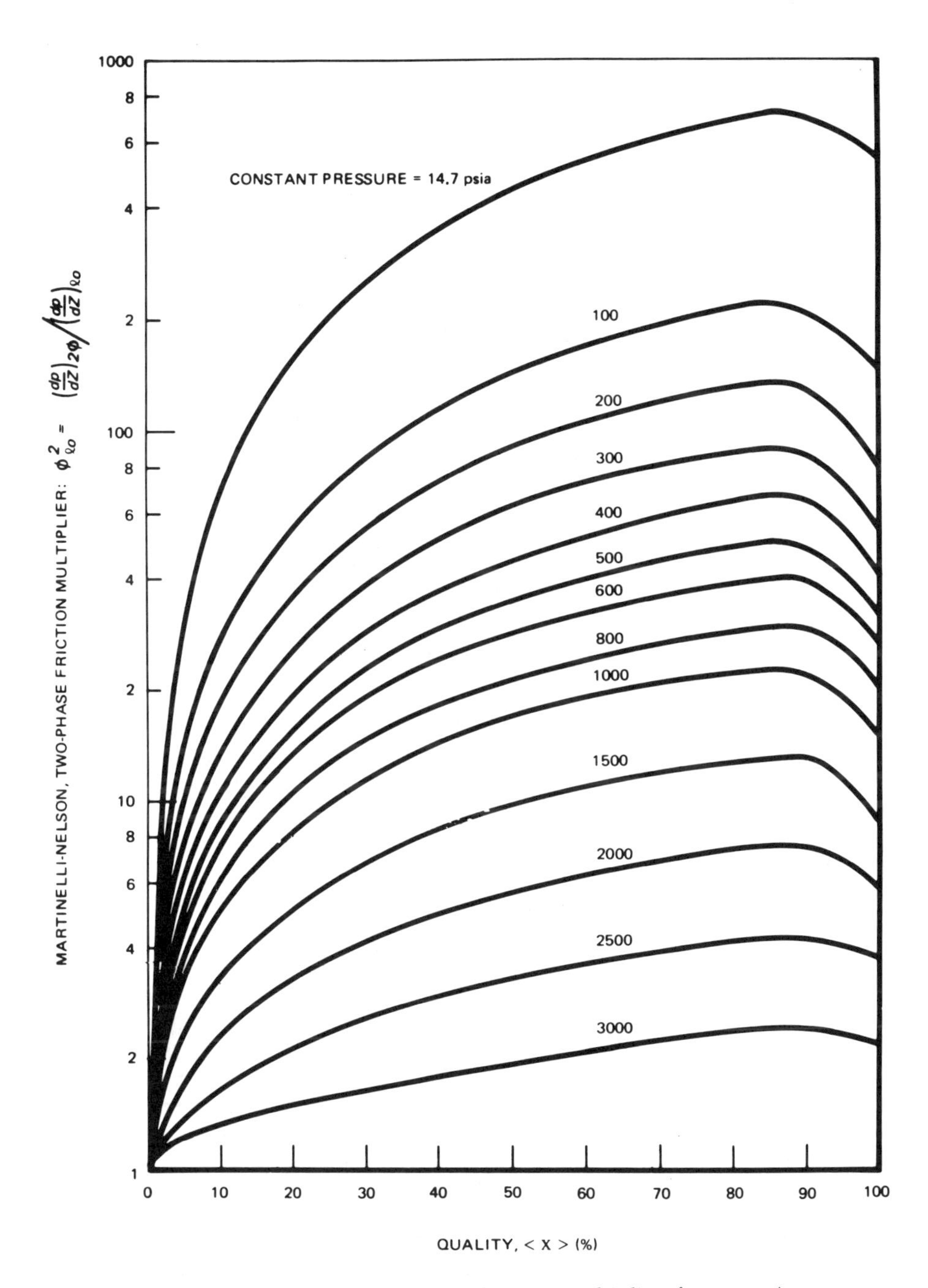

Fig. 5-20 Martinelli-Nelson, two-phase friction multiplier for steam/water as a function of quality and pressure.

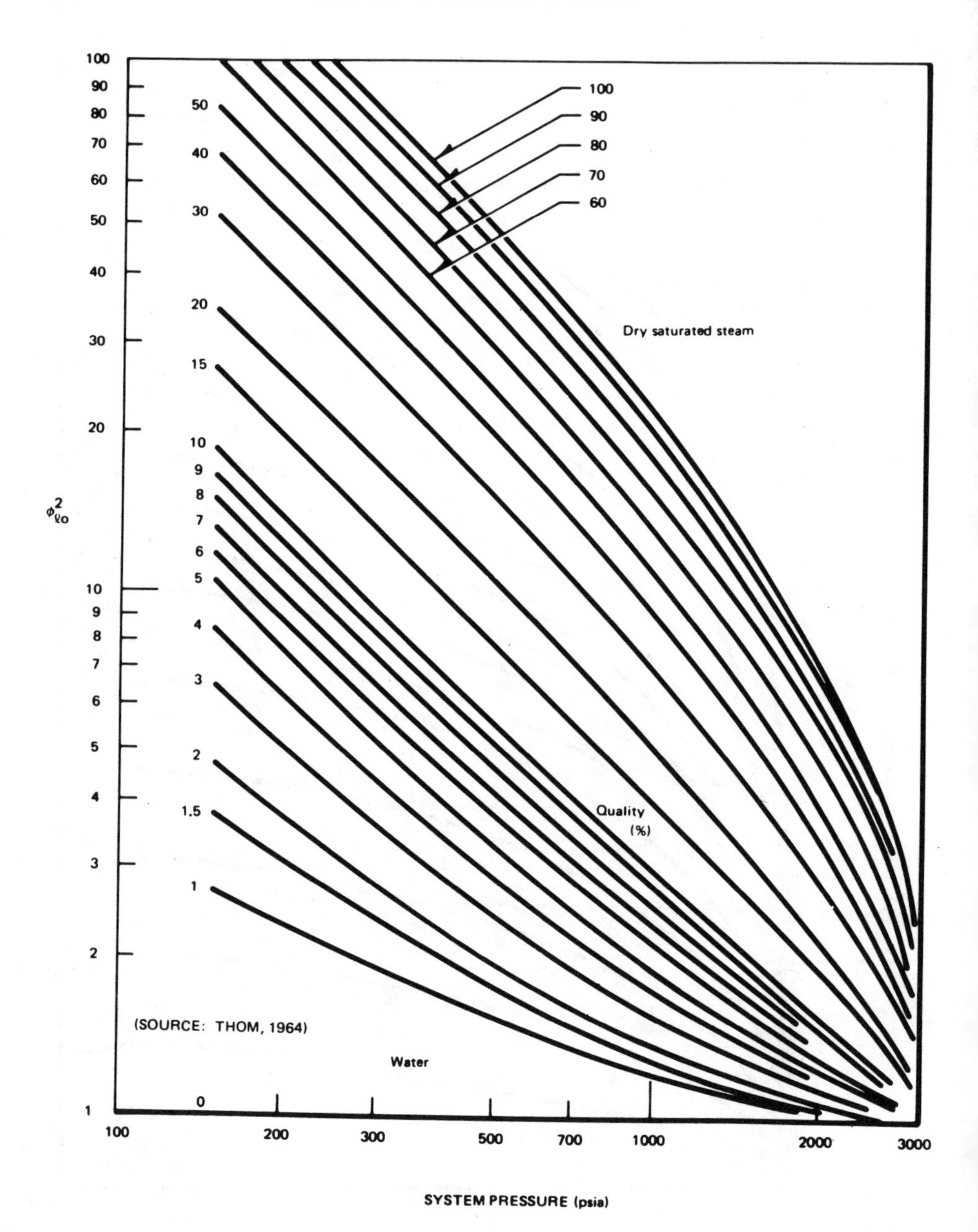

Fig. 5-21 Thom's two-phase friction multiplier, ϕ^2_{l0}.

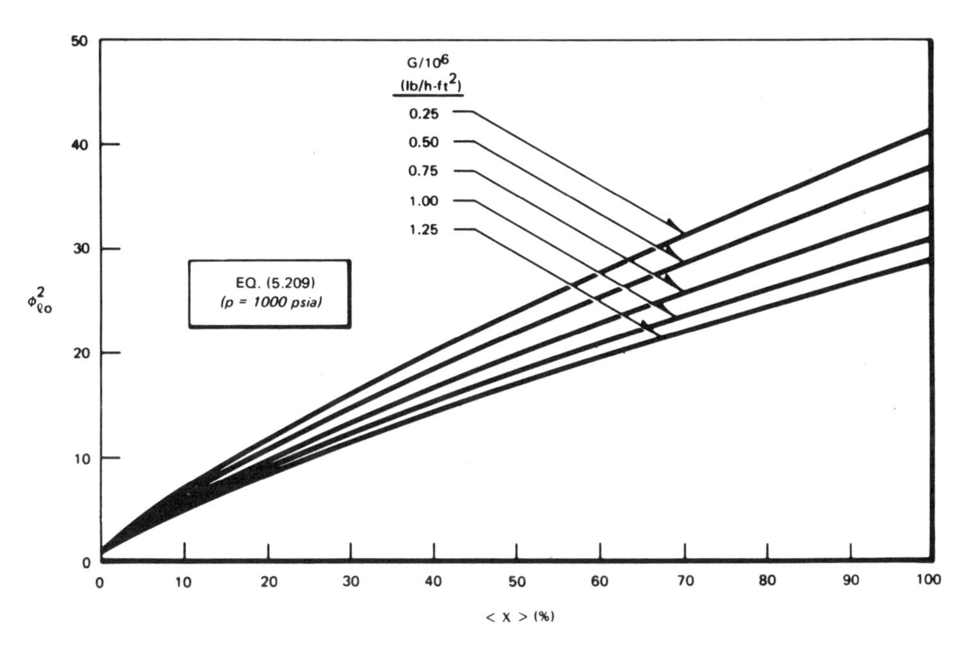

Fig. 5-22 Martinelli-Nelson approximation with Jones' flow effect.

et al., 1959). However, this correlation does not go to the correct (all vapor) limit.

The final correlation we consider for ϕ_{l0}^2 is due to Baroczy (1965). To better appreciate this correlation, we recall that the Martinelli parameter, given in Eq. (5.191), has been found to do a good job of correlating the pressure drop characteristics of a wide range of fluids. If we square both sides of Eq. (5.191), we obtain,

$$X_{tt}^2 = \left(\frac{\rho_g}{\rho_l}\right)\left(\frac{\mu_l}{\mu_g}\right)^n \left(\frac{1-\langle x\rangle}{\langle x\rangle}\right)^{(2-n)} . \tag{5.272}$$

When we compare the pressure drop characteristics of different fluids, it normally is done at the same flow rate and quality. The term $(1-\langle x\rangle/\langle x\rangle)^{2-n}$ can be considered a constant and we can write Eq. (5.272) as,

$$X_{tt}^2\left(\frac{\langle x\rangle}{1-\langle x\rangle}\right)^{2-n} = \frac{\rho_g}{\rho_l}\left(\frac{\mu_l}{\mu_g}\right)^n . \tag{5.273}$$

The nondimensional parameter on the right of Eq. (5.273) has been called the property index by Baroczy, who postulated a value of $n=0.2$ as the appropriate Reynolds number exponent.

Figure 5-23 is Baroczy's ϕ_{l0}^2 multiplier plotted versus the property index for various fluids. The values for water at various pressures of interest are

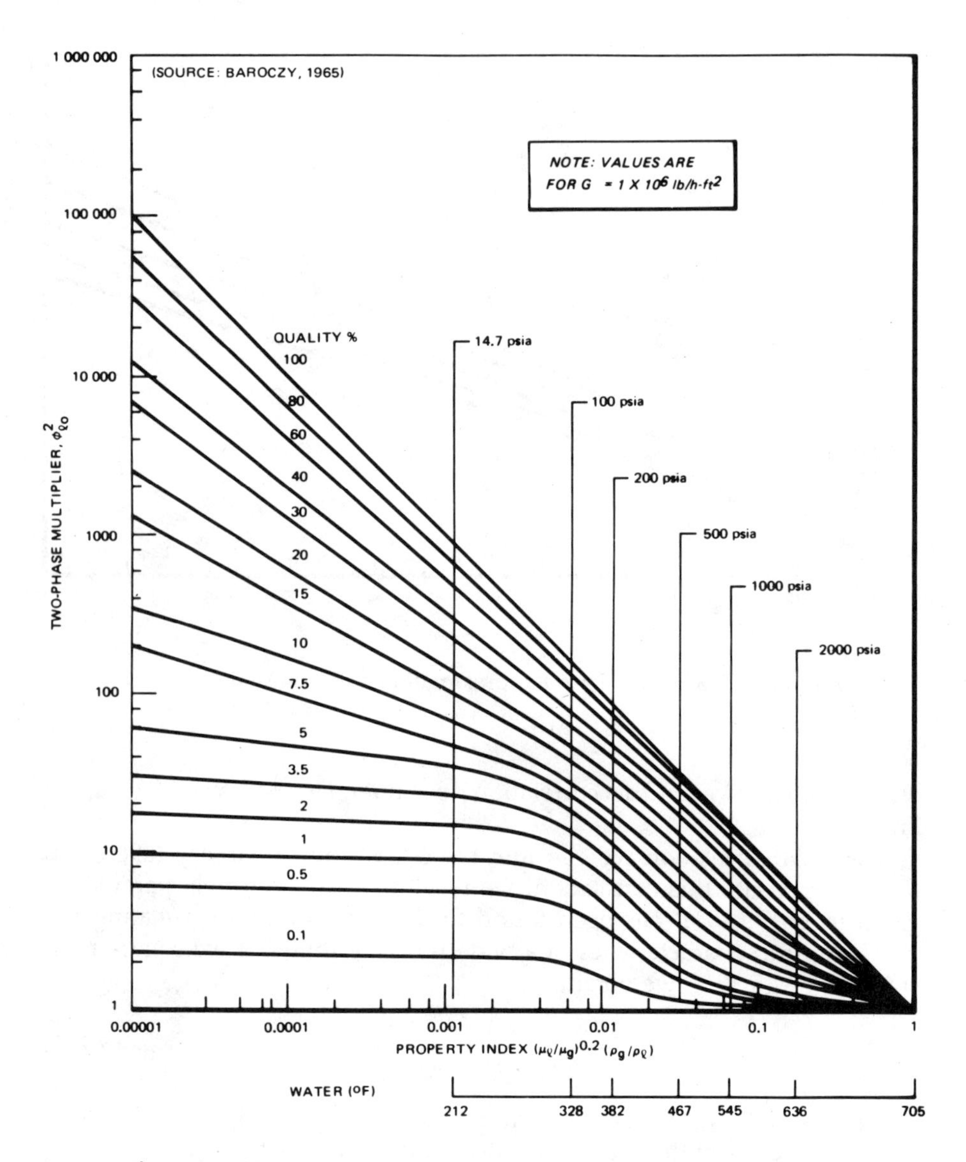

Fig. 5-23 Baroczy's two-phase friction pressure drop correlation.

explicitly shown. Figure 5-24 is the mass flux correction ratio that should multiply the appropriate value from Fig. 5-23 for flows other than $G = 1.0 \times 10^6$ $\text{lb}_m/\text{h-ft}^2$.

Figure 5-25 is a composite of the various ϕ_{l0}^2 correlations discussed above for the commercially interesting case of 1000 psia steam/water at $G = 1.0 \times 10^6$ $\text{lb}_m/\text{h-ft}^2$. Note that the Thom multiplier is very close to that for homoge-

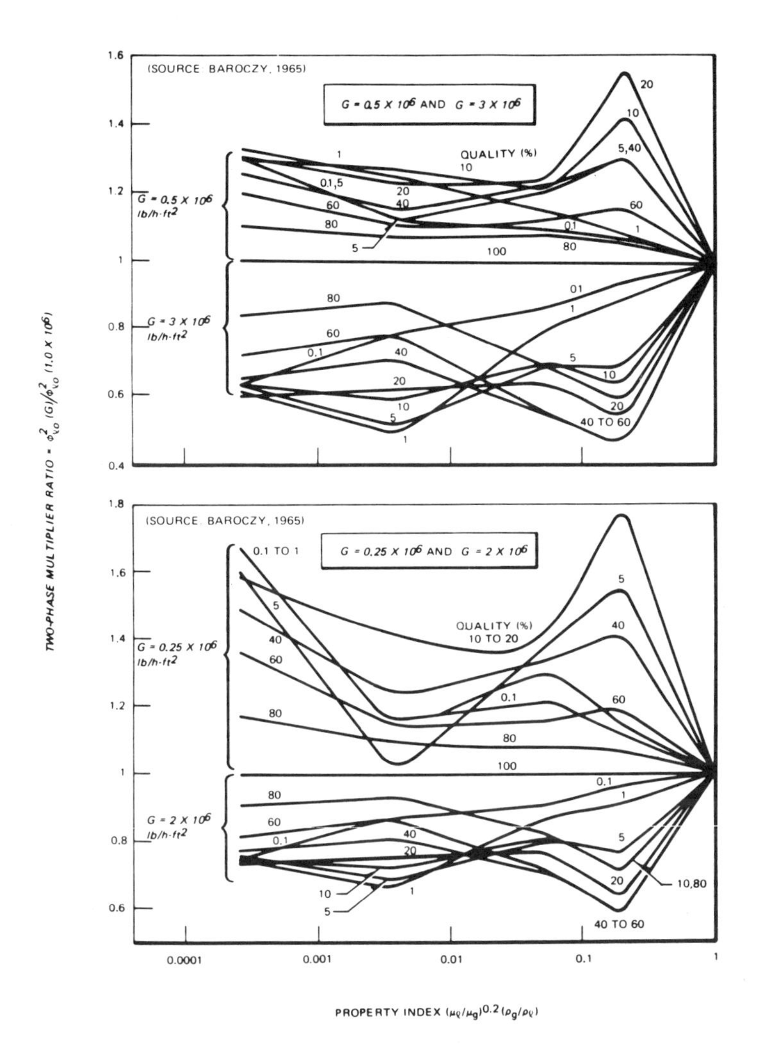

Fig. 5-24 Mass flux correction versus property index.

neous (no slip) conditions. Also note that the Martinelli-Nelson multiplier and the ϕ_{l0}^2 approximation given by Eq. (5.271) are the highest in the quality range of interest to BWR technology.

5.4.2 Local Pressure Drop

The two-phase pressure loss due to local flow obstructions, such as grid spacers or sudden contractions and enlargements, is treated in much the same manner as frictional pressure loss. That is, as in Eq. (5.251), the

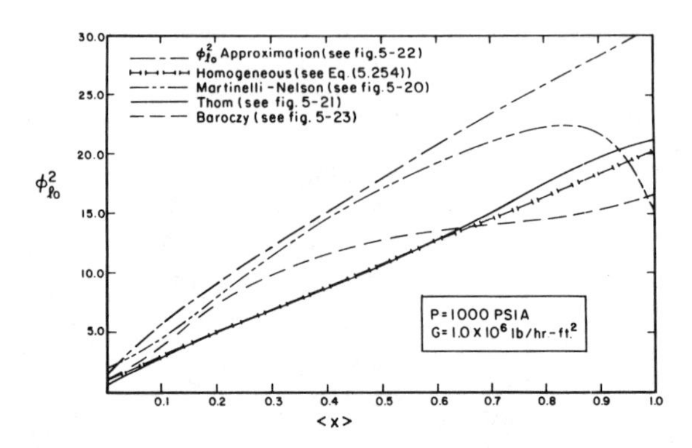

Fig. 5-25 A comparison of ϕ_{l0}^2 correlations.

corresponding single-phase pressure drop is multiplied by an appropriate two-phase multiplier to yield the local two-phase pressure drop. Historical precedent requires us to reserve the notation, ϕ_{l0}^2, for frictional pressure drop. Thus, we define the two-phase local loss multiplier as Φ. Thus,

$$\Delta p_{2\phi\mathrm{local}} \triangleq K\frac{G^2}{2g_c\langle\rho_{2\phi}\rangle} = K\frac{G^2}{2g_c\rho_f}\Phi \quad . \tag{5.274}$$

The functional form of Φ can sometimes be derived from the two-phase conservation equations, but it is normally synthesized empirically.

To provide some appreciation of the analysis required in the derivation of two-phase local loss multipliers, we deduce Φ for the special case of a sudden expansion. This is the type of flow situation that might be expected at the outlet of a BWR fuel bundle.

Before we launch into a detailed derivation of the two-phase pressure change caused by area changes, we recall the equivalent situation in single-phase flow. Normally, the designer is given pressure loss coefficients such as K_{exp} and K_{cont}, for sudden expansion and sudden contraction losses, respectively. The static pressure change is then calculated as,

$$(p_2 - p_1)_{\mathrm{exp}} = [(1-\sigma^2) - K_{\mathrm{exp}}]\frac{G_1^2}{2g_c\rho_l} \tag{5.275}$$

and,

$$(p_2 - p_1)_{\mathrm{cont}} = [(1-\sigma^2) - \sigma^2 K_{\mathrm{cont}}]\frac{G_1^2}{2g_c\rho_l} \ , \tag{5.276}$$

where σ is the flow area ratio, $\sigma \triangleq (A_1/A_2)$, and subscript 1 refers to conditions upstream of the area change and subscript 2 refers to downstream conditions. The term, $(1-\sigma^2)G_1^2/2g_c\rho_l$, is the single-phase Bernoulli equation (with no elevation and irreversible head-loss term) and is associated with the reversible pressure change. Thus, Eqs. (5.275) and (5.276) indicate,

$$\Delta p_{\text{static}} = \Delta p_{\text{rev}} - \Delta p_{\text{irrev}} \ . \tag{5.277}$$

That is, the static pressure change is equal to the reversible pressure change minus the irreversible loss.

By focusing attention on the case of a sudden expansion, it can be shown (Vennard, 1959) that the irreversible loss coefficient can be approximated by,

$$K_{\text{exp}} = (1-\sigma)^2 \ . \tag{5.278}$$

By combining Eqs. (5.275) and (5.278), for the static pressure change due to a sudden expansion we obtain,

$$(p_2 - p_1) = 2\sigma(1-\sigma)\frac{G_1^2}{2g_c\rho_l} \ . \tag{5.279}$$

Equation (5.279) can also be derived directly from the single-phase momentum equation. However, it is normally better to use Eq. (5.275) in conjunction with tabulated values of K_{exp} since Eq. (5.278) is only an approximation for a certain type of geometric configuration; that is, it does not account for ragged edges, nonflat velocity profiles, etc.

In two-phase flow, the situation is different. As we will soon show, the equivalent two-phase irreversible loss coefficients are rather complex and are not normally tabulated. Thus, it is usually more convenient for the analyst to work out the two-phase equivalent of Eq. (5.279) directly from the momentum equation.

Consider the pressure rise associated with a sudden expansion as shown in Fig. 5-26. The engineering approximation to the actual pressure rise is

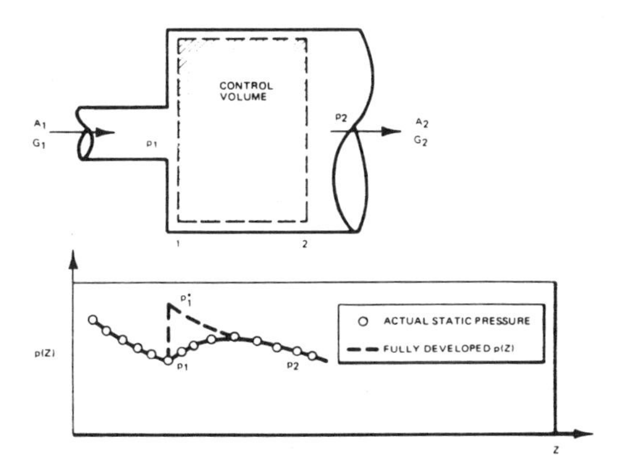

Fig. 5-26 Sudden expansion.

the "sawtooth"-shaped profile labeled fully developed $p(z)$. Thus, to do an accurate analysis, we must be able to calculate the ideal pressure rise, $(p_1^* - p_1)$. For steady conditions, the two-phase momentum equation for separated flow, Eq. (5.66), becomes,

$$-\frac{dp}{dz} = \frac{1}{g_c A_{x-s}} \frac{d}{dz}\left(\frac{G^2 A_{x-s}}{\langle \rho' \rangle}\right) + \frac{g}{g_c}\langle \bar{\rho} \rangle \sin\theta + \frac{\tau_w P_f}{A_{x-s}} \quad . \tag{5.280}$$

For a sudden expansion, the pressure rise is dominated by spatial acceleration effects, thus we may neglect the wall friction and gravity head terms in Eq. (5.280). If we note that,

$$A_{x-s}(z) = A_{xs_1} + (A_{xs_2} - A_{xs_1})U(z - z^*) \quad , \tag{5.281}$$

where z^* is the axial position of the sudden expansion and U is the so-called Heaviside step operator (which is discussed in Chapter 7), then Eq. (5.280) can be easily integrated.

Making the assumptions noted above, Eq. (5.280) can be rewritten as,

$$A_{x-s}\frac{dp}{dz} = -\frac{1}{g_c}\frac{d}{dz}\left(\frac{G^2 A_{x-s}}{\langle \rho' \rangle}\right) \quad . \tag{5.282}$$

The left-hand side of this equation can be integrated by parts:

$$\int_1^2 A_{x-s}\frac{dp}{dz}dz = \int_1^2 d[A_{x-s}p] - \int_1^2 p\frac{dA_{x-s}}{dz}dz \quad , \tag{5.283}$$

since,

$$\frac{dU(z-z^*)}{dz} = \delta(z - z^*) \quad , \tag{5.284}$$

where δ is a Dirac delta function, Eqs. (5.281) and (5.283) yield,

$$\int_1^2 A_{x-s}\frac{\partial p}{\partial z}dz = p_1^* A_{x-s_2} - p_1 A_{x-s_1} - p(z^*)[A_{x-s_2} - A_{x-s_1}] \quad . \tag{5.285}$$

If we assume (based on experimental observations) that,

$$p(z^*) = p_1 \quad ,$$

then Eq. (5.285) yields,

$$\int_1^2 A_{x-s}\frac{\partial p}{\partial z}dz = (p_1^* - p_1)A_{x-s_2} \quad . \tag{5.286}$$

Similarly, the right-hand side of Eq. (5.282) can be integrated, yielding,

$$-\frac{1}{g_c}\int_1^2 \frac{d}{dz}\left(\frac{G^2A_{x-s}}{\langle\rho'\rangle}\right)dz=\frac{1}{g_c}\left(\frac{G_1^2A_{x-s_1}}{\langle\rho'_1\rangle}-\frac{G_2^2A_{x-s_2}}{\langle\rho'_2\rangle}\right) . \tag{5.287}$$

Thus, Eqs. (5.286) and (5.287) yield,

$$(p_1^*-p_1)A_{x-s_2}=\frac{1}{g_c}\left(\frac{G_1^2A_{x-s_1}}{\langle\rho'_1\rangle}-\frac{G_2^2A_{x-s_2}}{\langle\rho'_2\rangle}\right) , \tag{5.288}$$

or, equivalently,

$$(p_1^*-p_1)=\frac{G_1^2}{2g_c\rho_l}\left(\frac{2\sigma\rho_l}{\langle\rho'_1\rangle}-\frac{2\sigma^2\rho_l}{\langle\rho'_2\rangle}\right) . \tag{5.289}$$

Thus, to a good first approximation, Eq. (5.280) yields the following general expression for a sudden expansion,

$$(p_2-p_1)\stackrel{\Delta}{=}\Delta p_{\text{static}}\cong(p_1^*-p_1)-\frac{g}{g_c}\langle\bar{\rho}\rangle\sin\theta(z_2-z_1)-\tau_w P_f(z_2-z_1)/A_{x-s_2} , \tag{5.290}$$

where, as discussed previously, the last two terms on the right side of Eq. (5.290) are normally negligible.

For the special case of no phase change across the sudden expansion, $\langle\rho'_1\rangle=\langle\rho'_2\rangle$, and Eq. (5.289) becomes,

$$(p_1^*-p_1)=2\sigma(1-\sigma)\frac{G_1^2}{2g_c\rho_l}\left(\frac{\rho_l}{\langle\rho'\rangle}\right)$$

$$=2\sigma(1-\sigma)\frac{G_1^2}{2g_c\rho_l}\left[\frac{(1-\langle x\rangle)^2}{1-\langle\alpha\rangle}+\frac{\rho_l}{\rho_g}\frac{\langle x\rangle^2}{\langle\alpha\rangle}\right] . \tag{5.291}$$

Comparison of Eqs. (5.291), (5.279), and (5.274) indicates that the appropriate two-phase multiplier for sudden expansion in which no phase change is assumed is,

$$\Phi=\left[\frac{(1-\langle x\rangle)^2}{1-\langle\alpha\rangle}+\frac{\rho_l}{\rho_g}\frac{\langle x\rangle^2}{\langle\alpha\rangle}\right]=\frac{\rho_l}{\langle\rho'\rangle} . \tag{5.292}$$

This multiplier is frequently called the "Romie multiplier" and has been found to correlate the available data rather well (Lottes, 1961).

At this point, the analyst has sufficient information to predict the static pressure rise accompanying a sudden expansion. Nevertheless, it is of interest to pursue the derivation of the irreversible component of the pressure change so that the reader can better appreciate the complexity involved. The steady, two-phase energy equation can be deduced from Eq. (5.99) as,

$$\frac{d}{dz}(GA_{x-s}\langle h\rangle)=q''P_H+q'''A_{x-s}-\frac{d}{dz}\left(\frac{G^3A_{x-s}}{2g_cJ\langle\rho'''\rangle^2}\right)-\frac{g}{g_c}\frac{\sin\theta}{J}GA_{x-s}\ . \tag{5.293}$$

For steady flow conditions, the flow rate (GA_{x-s}) is constant, by definition. Thus, Eq. (5.293) can be divided through by GA_{x-s} to obtain,

$$\frac{d\langle h\rangle}{dz}=\frac{q''P_H}{GA_{x-s}}+\frac{q'''}{G}-\frac{d}{dz}\left(\frac{G^2}{2g_cJ\langle\rho'''\rangle^2}\right)-\frac{g}{g_c}\frac{\sin\theta}{J}\ . \tag{5.294}$$

In accordance with the thermodynamic definition of enthalpy, $h\triangleq\mu+p/J\rho$, and the definition of $\langle h\rangle$ given in Eq. (5.101), we can write,

$$\langle h\rangle=\langle\mu\rangle+\frac{p}{J\langle\rho_h\rangle}\ , \tag{5.295}$$

where,

$$\langle\mu\rangle=\mu_l+\langle x\rangle\mu_{lg} \tag{5.296}$$

and $\langle\rho_h\rangle$ previously has been defined in Eq. (5.30).

By combining Eqs. (5.294) and (5.295), we obtain, after some algebra,

$$-\frac{dp}{dz}=\psi+\frac{g}{g_c}\sin\theta\langle\rho_h\rangle+\frac{\langle\rho_h\rangle}{2g_c}\frac{d}{dz}\left(\frac{G^2}{\langle\rho'''\rangle^2}\right)\ , \tag{5.297}$$

where the generalized two-phase dissipation function, ψ, is given by,

$$\psi\triangleq J\langle\rho_h\rangle\frac{d\langle\mu\rangle}{dz}+\langle\rho_h\rangle p\frac{d}{dz}\left(\frac{1}{\langle\rho_h\rangle}\right)-J\frac{\langle\rho_h\rangle}{G}\left(\frac{q''P_H}{A_{x-s}}\right)-J\left(\frac{q'''\langle\rho_h\rangle}{G}\right);\ (\mathrm{lb}_f/\mathrm{ft}^3)\ . \tag{5.298}$$

This dissipation function can be easily related to entropy production. For example, consider the adiabatic case in which,

$$\psi=J\langle\rho_h\rangle\left[\frac{d\langle\mu\rangle}{dz}+\frac{p}{J}\frac{d}{dz}\left(\frac{1}{\langle\rho_h\rangle}\right)\right]\ . \tag{5.299}$$

It can be shown (Wallis, 1969) that for a general two-phase mixture,

$$T\frac{d\langle s\rangle}{dz}=\frac{d\langle\mu\rangle}{dz}+\frac{p}{J}\frac{d}{dz}\left(\frac{1}{\langle\rho_h\rangle}\right)\ . \tag{5.300}$$

Thus, by comparing Eqs. (5.299) and (5.300),

$$\psi=J\langle\rho_h\rangle T\frac{d\langle s\rangle}{dz}\ . \tag{5.301}$$

The last two terms on the right side of Eq. (5.298) involve heat transfer and represent the entropy production due to irreversible transfers of thermal energy.

The reversible pressure rise between points 1 and 2 in Fig. 5-26 can be derived by considering the case of no dissipation; that is, $\psi = 0$. For this case, Eq. (5.297) integrates to,

$$\Delta p_{\text{rev}} = (p_1 - p_2) = \frac{g}{g_c}\,\sin\theta \int_1^2 \frac{\langle\rho_h\rangle}{2g_c} dz + \int_1^2 \frac{\langle\rho_h\rangle}{2g_c} \frac{d}{dz}\left(\frac{G^2}{\langle\rho'''\rangle^2}\right) dz \ . \tag{5.302}$$

Note that for situations in which there is no quality change across the expansion, $\langle\rho_h\rangle_1 = \langle\rho_h\rangle_2$ and Eq. (5.302) becomes,

$$\frac{p_1}{\langle\rho_h\rangle} + \frac{1}{2g_c}\frac{G_1^2}{\langle\rho_1'''\rangle^2} + \frac{g}{g_c}\,\sin\theta z_1 = \frac{p_2}{\langle\rho_h\rangle} + \frac{1}{2g_c}\frac{G_2^2}{\langle\rho_2'''\rangle^2} + \frac{g}{g_c}\,\sin\theta z_2 \ , \tag{5.303}$$

which can be recognized as the two-phase Bernoulli equation.

To obtain the ideal reversible pressure rise, $p_1^* - p_1$, Eq. (5.302) yields for the special case of a horizontal conduit,

$$p_1^* - p_1 \stackrel{\Delta}{=} \Delta p_{\text{rev}} = -\int_1^2 \frac{\langle\rho_h\rangle}{2g_c} \frac{d}{dz}\left(\frac{G^2}{\langle\rho'''\rangle^2}\right) dz \ . \tag{5.304}$$

For adiabatic flow, or diabatic flow in which no phase change occurs across the expansion, Eq. (5.304) integrates to

$$\Delta p_{\text{rev}} = \frac{\langle\rho_h\rangle G_1^2}{2g_c}(1-\sigma^2)\left[\frac{(1-\langle x\rangle)^3}{\rho_l^2(1-\langle\alpha\rangle)^2} + \frac{\langle x\rangle^3}{\rho_g\langle\alpha\rangle^2}\right] , \tag{5.305}$$

where we can recognize $(1-\sigma^2)$ as the single-phase reversible pressure rise coefficient discussed previously, and we note that the two-phase multiplier is $\Phi = \langle\rho_h\rangle\rho_l/\langle\rho'''\rangle^2$.

The most general expression for the irreversible pressure loss due to dissipative mechanisms occurring during the sudden expansion of a two-phase mixture can be obtained by combining Eqs. (5.277), (5.290), and (5.302),

$$\Delta p_{\text{irrev}} = -\int_1^2 \frac{\langle\rho_h\rangle}{2g_c} \frac{d}{dz}\left(\frac{G^2}{\langle\rho'''\rangle^2}\right) dz + \frac{G_1^2}{g_c}\left(\frac{\sigma^2}{\langle\rho_2'\rangle} - \frac{\sigma}{\langle\rho_1'\rangle}\right) + \tau_w P_f (z_2 - z_1)/A_{x-s_2} \ , \tag{5.306}$$

where,

$$\tau_w = \frac{1}{4} f \frac{G^2}{2g_c\rho_l} \phi_{l0}^2 \ . \tag{5.307}$$

Neglecting the residual elevation head, this can often be approximated by

$$\Delta p_{\text{irrev}} \cong \frac{G_1^2}{2g_c\rho_l}\left[\frac{(\langle\rho_h\rangle_1+\langle\rho_h\rangle_2)}{2}\left(\frac{\rho_l}{\langle\rho_1'''\rangle^2}-\frac{\rho_l\sigma^2}{\langle\rho_2'''\rangle^2}\right)+2\rho_l\left(\frac{\sigma^2}{\langle\rho_2'\rangle}-\frac{\sigma}{\langle\rho_1'\rangle}\right)\right] \quad . \tag{5.308}$$

It should be apparent from Eq. (5.308) that the two-phase multiplier for the irreversible pressure losses during a sudden expansion is quite complex and, thus, unlike single-phase flow, two-phase loss multipliers are rarely tabulated.

A completely analogous derivation can be performed for the case of a sudden contraction. This analysis is not reproduced here since it has been found experimentally (Geiger, 1964) that a homogeneous multiplier does a good job of correlating the data. That is, Eqs. (5.274) and (5.276) yield,

$$(p_2-p_1)_{\text{cont}}=[(1-\sigma^2)-\sigma^2K_{\text{cont}}]\frac{G_1^2}{2g_c\rho_l}\Phi \ , \tag{5.309}$$

where,

$$\Phi=\left(1+\frac{v_{l_g}}{v_l}\langle x\rangle\right) \tag{5.310}$$

and values for the irreversible loss coefficient (K_{cont}) are normally taken as (Vennard, 1959):

$\sigma \triangleq A_1/A_2$:	0.0	0.2	0.4	0.5	0.6	0.8	1.0
$K_{\text{cont}} \triangleq \left(\frac{1}{C_c}-1\right)^2$:	0.385	0.340	0.266	0.219	0.164	0.053	0.0.

The situation shown in Fig. 5-27 is typical of various hydraulic components such as pipe bends, tees, elbows, and valves. These components have not been investigated thoroughly. However, there is evidence (Fitzsimmons, 1964) indicating that the homogeneous multiplier, given by Eq. (5.310), does a reasonably good job of correlating two-phase component pressure drop data and is recommended for design purposes for all local losses.

One of the most important local losses that must be considered in BWR technology is that due to grid-type spacers that maintain the proper geometric configurations of the fuel rod bundles. Normal design practice is to superimpose the local loss due to spacers on the fully developed friction pressure gradient. Two situations can exist: as shown in Fig. 5-28, the spacers can be located axially such that the two-phase mixture never gets redeveloped before it encounters the next spacer (Case A), or the situation can be shown in Case B, in which the flow becomes fully developed before the next spacer. In Case A, the designer must account for the pressure recovery that occurs in the outlet region of the fuel bundle.

In contrast, Case B, which is typical of the $L/D_H \cong 40$ spacing in BWRs, the normal design approximation yields the correct static pressure drop

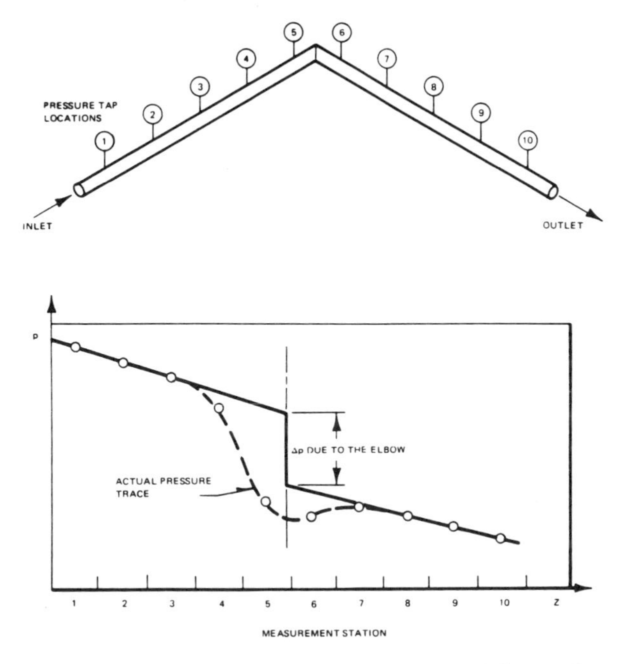

Fig. 5-27 Typical static pressure profile due to a typical fitting (i.e., an elbow).

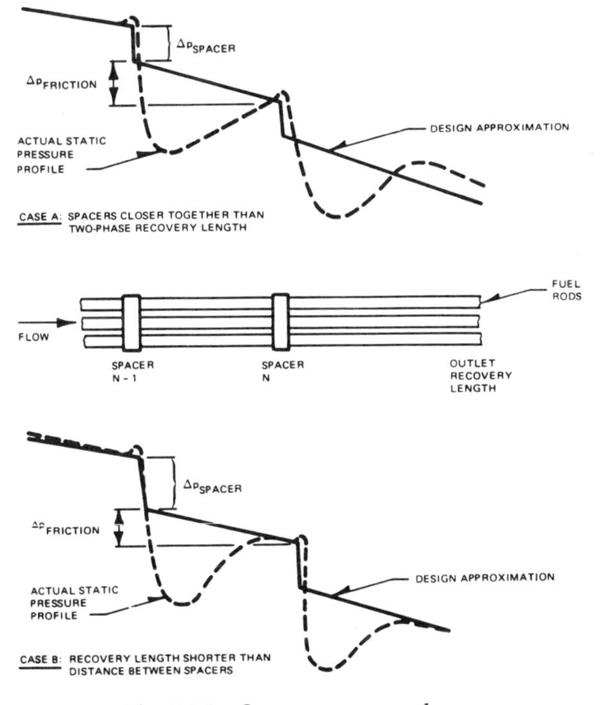

Fig. 5-28 Spacer pressure drop.

across the fuel rod bundle. These considerations become quite important when one is trying to measure the single-phase loss coefficient, K_{SP}. The only reliable approach is to measure the gradient downstream of the spacer and, as shown in Fig. 5-27, to extrapolate the fully developed gradients back to the spacer.

Extensive experimental data taken on a wide range of grid-type spacers have indicated that the homogeneous multiplier does a fairly good job of correlating the two-phase pressure drop data. That is, Eqs. (5.310) and (5.274) yield

$$\Delta p_{\text{spacer}} = K_{SP}\frac{G^2}{2g_c\rho_f}\left(1+\frac{v_{fg}}{v_f}\langle x\rangle\right) , \tag{5.311}$$

where the single-phase spacer loss coefficient, K_{SP}, is normally determined experimentally.

5.4.3 Alternate Schemes

Previously it has been suggested (Isbin, 1962) that there may be some advantage to correlate friction pressure drop data in terms of the dissipation function, ψ, in Eq. (5.297), rather than the usual procedure of trying to correlate the wall shear, τ_w, in Eq. (5.248). Since the effect of elevation is different in these equations, this suggestion has caused some confusion among workers in the field as to what the appropriate density head term should be. More specifically, Eq. (5.299) uses the homogeneous density, $\langle\rho_h\rangle$, and Eq. (5.248) uses the slip density, $\langle\bar{\rho}\rangle$, to account for the effect of elevation on the static pressure gradient. Our purpose here is to show that both approaches are correct and that the difference in density head terms is just a reflection of the fact that the dissipation function, ψ, contains interphase dissipation due to slip as well as wall shear.

To gain some appreciation for the relationship between these results, we eliminate the static pressure gradient between Eq. (5.297) and the steady version of Eq. (5.248),

$$\psi = \frac{\tau_w P_f}{A_{x-s}} + \frac{g}{g_c}\sin\theta(\langle\bar{\rho}\rangle - \langle\rho_h\rangle) + G\frac{d}{dz}\left(\frac{G}{\langle\rho'\rangle}\right) - \frac{\langle\rho_h\rangle}{2g_c}\frac{d}{dz}\left(\frac{G^2}{\langle\rho'''\rangle^2}\right) . \tag{5.312}$$

If we multiply both sides of Eq. (5.312) by $\langle j\rangle/J$, we can recognize[f] the first term on the right as the volumetric dissipation due to viscous effects at the wall. The next term,[g]

$$\frac{g}{g_c}\frac{\langle j\rangle}{J}\sin\theta(\langle\bar{\rho}\rangle - \langle\rho_h\rangle) \tag{5.313}$$

represents the volumetric dissipation due to slip between the phases. That is, interfacial shear increases with elevation (θ) and the resultant increase in dissipation is reflected in Eq. (5.313).

The final two terms in Eq. (5.312) are the volumetric dissipation due to vaporization effects; i.e., due to axial acceleration and kinetic energy changes of the two-phase mixture.

It is evident from Eq. (5.312) that ψ and τ_w are uniquely related. Thus, there is no particular advantage in correlating ψ if we are required to perform calculations involving the momentum equation, since we must know, or assume, an appropriate void-quality relation to determine the corresponding correlation for two-phase wall shear (τ_w).

In summary, either approach of data correlation is possible, but since the two-phase momentum equation is normally required in BWR analysis (e.g., transient evaluations), the classical method of correlating two-phase τ_w (i.e., $\phi_{l0}^2\tau_{w_{l0}}$) is preferred.

This section on two-phase pressure drop completes our treatment of two-phase flow phenomena. By no means has it been an exhaustive study of the subject, since we have only attempted to cover those topics of particular significance to BWR technology. In the following chapters, we use the analytical results presented here and apply them to evaluating the performance of a modern BWR nuclear steam supply system.

References

Anderson, J. G. M., and K. H. Chu, "BWR Refill-Reflood Program, Task 4.7, Constitutive Correlations for Shear and Heat Transfer for BWR Version of TRAC," GEAP-24940, General Electric Company (1981).

Armand, A., AERE Translation from Russian, 828, United Kingdom Atomic Energy Authority (1959).

Bankoff, S. G., "A Variable Density Single-Fluid Model for Two-Phase Flow with Particular Reference to Steam-Water Flow," *J. Heat Transfer*, **82** (1960).

Baroczy, C. J., "A Systematic Correlation for Two-Phase Pressure Drop," preprint #37, Eighth Nat'l Heat Transfer Conf., Los Angeles, American Institute of Chemical Engineers (1965).

Bennett, A. W., et al., "Studies of Burnout in Boiling Heat Transfer to Water in Round Tubes with Non-Uniform Heating," AERE-R 5076, Atomic Energy Research Establishment (1966).

Bergles, A. E., and W. M. Rohsenow, "The Determination of Forced Convection Surface Boiling Heat Transfer," paper 63-HT-22, American Society of Mechanical Engineers (1963).

Bergles, A. E., and M. Suo, "Investigation of Boiling Water Flow Regimes at High Pressure," *Proc. 1966 Heat Transfer and Fluid Mechanics Institute*, Stanford University Press, pp. 79–99 (1966).

---

[f] See Eq. (5.49).

[g] This term is always positive since $(\langle\overline{\rho}\rangle - \langle\rho_h\rangle) > 0.0$.

Bouré, J., and J. M. Delhaye, *Two-Phase Flow and Heat Transfer in the Power and Process Industries,* Chapter 1.2, Hemisphere Publishing Corporation (1981).

Bowring, R. W., "Physical Model Based on Bubble Detachment and Calculations of Steam Voidage in the Subcooled Region of a Heated Channel," Report HPR-10, OECD Halden Reactor Project (1962).

Chisholm, D., "Pressure Gradients due to Friction during the Flow of Evaporating Two-Phase Mixtures in Smooth Tubes and Channels," *Int. J. Heat Mass Transfer,* **16** (1973).

Chisholm, D., and L. A. Sutherland, "Prediction of Pressure Gradients in Pipeline Systems During Two-Phase Flow," *Proc. Inst. Mech. Engs.,* **184,** P7–3C (1969).

Davies, R. M., and G. I. Taylor, "The Mechanics of Large Bubbles Rising through Extended Liquids in Tubes," *Proc. Roy Soc. London 200,* Series A (1950).

Davis, E. J., and G. H. Anderson, "The Incipience of Nucleate Boiling in Forced Convection," *AIChE J.,* **12** (1966).

Dix, G. E., "Vapor Void Fractions for Forced Convection with Subcooled Boiling at Low Flow Rates," NEDO-10491, General Electric Company (1971).

Fitzsimmons, D. E., "Two Phase Pressure Drop in Piping and Components," HW-08970, Rev. 1, General Electric Company (1964).

Forti, G., "A Dynamic Model for the Cooling Channels of a Boiling Nuclear Reactor with Forced Convection and High Pressure Level," EUR-4052e, European Atomic Energy Community (1968).

Geiger, G. E., "Sudden Contraction Losses in Single- and Two-Phase Flow," Ph.D. Thesis, University of Pittsburgh (1964).

Gill, L. E., G. F. Hewitt, and P. M. C. Lacey, "Sampling Probe Studies of the Gas Core in Annular Two-Phase Flow: II, Studies of the Effect of Phase Flow Rates on Phase and Velocity Distribution," *Chem. Eng. Sci.,* **19** (1964).

Govier, G. W., and K. Aziz, *The Flow of Complex Mixtures in Pipes,* Van Nostrand Reinhold Company, New York (1972).

Griffith, P., J. A. Clark, and W. M. Rohsenow, "Void Volumes in Subcooled Boiling Systems," paper 58-HT-19, American Society of Mechanical Engineers (1958).

Griffith, P., and G. B. Wallis, "Two-Phase Slug Flow," *Trans. ASME. Ser. C.,* **83** (1961).

Haberman, W. L., and R. K. Morton, "An Experimental Investigation of the Drag and Shape of Air Bubbles Rising in Various Liquids," David Taylor Model Basin Report DTMB802 (1953).

Hancox, W. T., and W. B. Nicoll, "A General Technique for the Prediction of Void Distributions in Non-Steady Two-Phase Forced Convection," *Int. J. Heat Mass Transfer,* **14** (1971).

Harmathy, T. Z., "Velocity of Large Drops and Bubbles in Media of Infinite or Restricted Extent," *AIChE J.,* **6**(2) (1960).

Hewitt, G. F., "Analysis of Annular Two-Phase Flow: Application of the Dukler Analysis to Vertical Upward Flow in a Tube," AERE-R-3680, Atomic Energy Research Establishment (1961).

Hewitt, G. F., and Hall-Taylor, N. S., *Annular Two-Phase Flow,* Pergamon Press, New York (1970).

Hinze, J. O., "Critical Speed and Sizes of Liquid Globules," *Appl. Sci. Res.,* **A1** (1948).

Hutchinson, P., and P. B. Whalley, "A Possible Characterization of Entrainment in Annular Flow," AERE-R-7126, Atomic Energy Research Establishment (1972).

Isbin, H. S., "Reply," *AIChE J.*, **8** (1962).

Isbin, H. S., R. H. Moen, R. O. Wickey, D. R. Mosher, and H. C. Larson, "Two-Phase Steam-Water Pressure Drops," *Chem. Eng. Progr. Symp. Ser.*, **55**(23) (1959).

Ishii, M., "Thermo-Fluid Dynamic Theory of Two-Phase Flow," *Eyrolles* (1975).

Ishii, M., and K. Mishima, "Two-Fluid Model and Hydrodynamic Constitutive Relations," *Nucl. Eng. Des.*, **86**(2&3) (1984).

Ishii, M., and N. Zuber, "Thermally Induced Flow Instabilities in Two-Phase Mixtures," *Proc. Fourth Int. Heat Transfer Conf. (Paris)*, Vol. V (1970).

Jakob, M., *Heat Transfer*, Vol. I, John Wiley & Sons, New York (1957).

James, P. W., G. F. Hewitt, and P. B. Whalley, "Droplet Motion in Two-Phase Flow," *ANS/ASME/NRC Topl. Mtg. Nuclear Reactor Thermal-Hydraulics*, NUREG/CP-0014, Vol. 3, pp. 1484–1503 (1980).

Jens, W. H., and P. A. Lottes, "Analysis of Heat Transfer, Burnout, Pressure Drop and Density Data for High Pressure Water," ANL-4627, Argonne National Laboratory (1951).

Jones, A. B., "Hydrodynamic Stability of a Boiling Channel," KAPL-2170, Knolls Atomic Power Laboratory (1961).

Kline, S. J., *Similitude and Approximation Theory*, McGraw-Hill Book Company, New York (1965).

Kutateladze, S. S., and M. A. Styrikovich, "Hydraulics of Gas-Liquid System," p. 24, Wright-Patterson Air Force Base translation (1960).

Lahey, R. T., "Two-Phase Flow in Boiling Water Nuclear Reactors," NEDO-13388, General Electric Company (1974).

Larsen, P. S., and L. S. Tong, "Void Fractions in Subcooled Flow Boiling," *Trans. ASME*, **91** (1969).

Levenspiel, O., "Collapse of Steam Bubbles in Water," *Ind. Eng. Chem.*, **51** (1959).

Levy, S., "Forced Convection Subcooled Boiling—Prediction of Vapor Volumetric Fraction," GEAP-5157, General Electric Company (1966).

Levy, S., "Analysis of Various Types of Two-Phase Annular Flow, Part II—Prediction of Two-Phase Annular Flow with Liquid Entrainment," GEAP-4615, General Electric Company (1964).

Levy, S., "Steam Slip-Theoretical Prediction from Momentum Model," *J. Heat Mass Transfer*, **82** (1960).

Lockhart, R. W., and R. C. Martinelli, "Proposed Correlation of Data for Isothermal Two-Phase Two-Component Flow in Pipes," *Chem. Eng. Progr.*, **45** (1949).

Lottes, P. A., "Expansion Losses in Two-Phase Flow," *Nucl. Sci. Eng.*, **9** (1961).

Martinelli, R. C., and D. B. Nelson, "Prediction of Pressure Drop During Forced-Circulation Boiling of Water," *Trans. ASME*, **70** (1948).

McAdams, W. H., et al., "Vaporization Inside Horizontal Tubes—II. Benzene-Oil Mixtures," *Trans. ASME*, **64** (1942).

Meyer, J. E., "Conservation Laws in One-Dimensional Hydrodynamics," Bettis Technical Review, WAPD-BT-20, Westinghouse Electric Company (1960).

Milne-Thomson, L. M., *Theoretical Hydrodynamics*, 4th ed., Macmillan Publishing Company (1961).

Newitt, D. M., N. Dombrowski, and F. H. Knelman, "Liquid Entrainment: 1, The Mechanism of Drop Formation from Gas or Vapor Bubbles," *Trans. Inst. Chem. Eng.*, **32** (1954).

Park, G.-C., M. Popowski, M. Becker, and R. T. Lahey, Jr., "The Development of NURREQ-N, an Analytical Model for the Stability Analysis of Nuclear-Coupled Density-Wave Oscillations in Boiling Water Nuclear Reactors," NUREG/CR-3375, U.S. Nuclear Regulatory Commission (1983).

Quandt, E. R., "Measurements of Some Basic Parameters in Two-Phase Annular Flow," *AIChE J.*, **11** (1965).

Radovich, N. A., and R. Moissis, "The Transition from Two-Phase Bubble Flow to Slug Flow," Dept. of Mech. Eng. Report 7-7632-22, MIT (1962).

Rouhani, S. Z., "Void Measurements in the Region of Subcooled and Low-Quality Boiling, Part II," AE-RTL-788, Aktiebolaget Atomenergi, Studsvik, Sweden (Apr. 1966).

Rouhani, S. Z., and E. Axelsson, "Calculation of Void Volume Fraction in Subcooled and Quality Boiling Regions," *Int. J. Heat Mass Transfer*, **13** (1970).

Saha, P., and N. Zuber, "Point of Net Vapor Generation and Vapor Void Fraction in Subcooled Boiling," *Proc. Fifth Int. Heat Transfer Conf.*, Vol. IV (1974).

Staub, F. W., "The Void Fraction in Subcooled Boiling—Prediction of the Initial Point of Net Vapor Generation," *Trans. ASME*, **90** (1968).

Thom, J. R. S., "Prediction of Pressure Drop During Forced Circulation Boiling of Water," *Int. J. Heat Mass Transfer*, **7** (1964).

van Deemter, J. J., and van der Laan, "Momentum and Energy Balances for Dispersed Two-Phase Flow," *Appl. Sci. Res., Sec. A*, **10** (1961).

Vennard, J. K., *Elementary Fluid Mechanics*, 3rd ed., John Wiley & Sons, New York (1959).

Wallis, G. B., *One-Dimensional Two-Phase Flow*, McGraw-Hill Book Company, New York (1969).

Whalley, P. B., "Boiling, Condensation and Gas-Liquid Flow," Oxford Engineering Science Series 21, Clarendon Press, Oxford (1987).

White, E. T., and R. H. Beardmore, "Velocity of the Rise of Single Cylindrical Air Bubbles through Liquids Contained in Vertical Tubes," *Chem. Eng. Sci.*, **17** (1962).

Yadigaroglu, G., and R. T. Lahey, "On the Various Forms of the Conservation Equations in Two-Phase Flow," *Int. J. Multiphase Flow* (1975).

Zuber, N., "Flow Excursions and Oscillations in Boiling, Two-Phase Flow Systems with Heat Addition," *Proc. Symp. Two-Phase Flow Dynamics*, EUR-4288, University of Eindhoven (1967).

Zuber, N., and J. A. Findlay, "Average Volumetric Concentration in Two-Phase Flow Systems," *J. Heat Transfer* (1965).

Zuber, N., and J. Hench, "Steady-State and Transient Void Fraction of Bubbling Systems and Their Operating Limits. Parts I and II," GEL Reports 62GL100 and 62GL111 (1962).

Zuber, N., and F. W. Staub, "The Propagation and the Wave Form of the Vapor Volumetric Concentration in Boiling, Forced Convection Systems Under Oscillatory Conditions," *Int. J. Heat Mass Transfer*, **9** (1966).

Zuber, N., F. W. Staub, and G. Bijwaard, "Vapor Void Fractions in Subcooled Boiling and Saturated Boiling Systems," *Proc. 3rd Int. Heat Transfer Conf.*, Chicago (1966).

PART THREE

Performance of the Nuclear Steam Supply System

CHAPTER SIX

Thermal-Hydraulic Performance

This chapter deals with several of the most important conductive and convective heat transfer modes in BWR technology.

6.1 Heat Conduction in Nuclear Fuel Elements

A detailed mechanical description of the fuel elements used in modern BWRs was given in Chapter 2. It was shown that they are typically composed of a series of cylindrical UO_2 fuel pellets of ~95% theoretical density. These pellets are enclosed in hollow Zircaloy-2 tubes, which serve to contain the radioactive fission products and to protect the UO_2 pellets from the steam-water environment in the rod bundles.

The properties and physical characteristics of UO_2 are well known (Belle, 1961) and are not discussed in detail here. However, a few of the salient physical characteristics, such as thermal conductivity, heat capacity, coefficient of thermal expansion, and density are tabulated in Table 6-1 for handy reference. In general, sintered UO_2 fuel pellets are an economical and durable fuel material, which have excellent stability in a steam/water environment. One feature of these fuel pellets, which is of importance in subsequent discussions of the so-called "gap conductance" between the fuel pellets and the Zircaloy-2 cladding, is that because they are ceramic, they readily crack and fragment due to thermal stress and irradiation effects.

Of particular interest is the temperature dependence of UO_2 thermal conductivity. Figure 6-1 is a plot of Eq. (A.1) given in Table 6-1. This correlation is based on General Electric (GE) data reported by Lyons (1966) and an International Panel (1965). It can be seen that this correlation in-

TABLE 6-1
Some Selected Thermal Properties of UO_2 and Zircaloy-2

A. Thermal conductivity

1. $\kappa_{(UO_2)}(T) = \dfrac{3978.1}{(692.61+T)} + (6.02366 \times 10^{-12})(T+460)^3, \ \dfrac{\text{Btu}}{\text{h-ft-°F}},$ [a]

where

$$T = T(°\text{F}) \ .$$

2. $\kappa_{(Zr\text{-}2)}(T) = 7.151 + (2.472 \times 10^{-3})T + (1.674 \times 10^{-6})T^2 - (3.334 \times 10^{-10})T^3, \ \dfrac{\text{Btu}}{\text{h-ft-°F}},$ [b]

where

$$T = T(°F) \ .$$

B. Coefficient of thermal expansion $\alpha_T \triangleq \dfrac{\Delta L}{L[T(°\text{F}) - 72°\text{F}]}$

1. $\alpha_{T(UO_2)} \cong 6.6 \times 10^{-6}, \ \dfrac{\text{ft}}{\text{ft-°F}},$ [c]

2. $\alpha_{T(Zr\text{-}2)} \cong 3.65 \times 10^{-6}, \ \dfrac{\text{ft}}{\text{ft-°F}},$ [d]

C. Density

$$\rho_{(UO_2)} = \begin{cases} 684.76, \ \text{lb/ft}^3 \ (100\% \ \text{TD})^{e,f} \\ 650.52, \ \text{lb/ft}^3 \ (95\% \ \text{TD})^{e,f} \end{cases}$$

$$\rho_{(Zr\text{-}2)} = 410.17, \ \text{lb/ft}^3 \ ^{g}$$

D. Heat capacity, c_p

1. $c_{p(UO_2)}(T) = 0.07622 + 1.16 \times 10^{-6}(5/9T + 255.22) + \dfrac{X}{(1+X)^2}$

$$\times \{6.76 \times 10^6/[R(5/9T + 255.22)^2]\} \frac{\text{Btu}}{\text{lb-°F}},^{h}$$

where

$X \triangleq \exp\{6.25 - 42659/[R(5/9T + 255.22)]\}$

$R = 1.987$

$T = T(°\text{F})$

2. $c_{p(Zr\text{-}2)}(T) = \begin{cases} 0.06805 + 2.3872 \times 10^{-5}[(T-32)/1.8], \ \dfrac{\text{Btu}}{\text{lb-°F}} (32°\text{F} < T < 1175°\text{F}) \\ 0.085, \ \dfrac{\text{Btu}}{\text{lb-°F}} (1175°\text{F} < T < 1922°\text{F})^{i} \end{cases}$

[a]See FSAR Report (1971).
[b]See Scott (1965).
[c]See Belle (1961).
[d]See Mehon and Wiesinger (1961).
[e]TD = theoretical density.
[f]See Belle (1961).
[g]See Seddon (1962).
[h]See Hein and Flagella (1968).
[i]See Eldridge and Deem (1967).

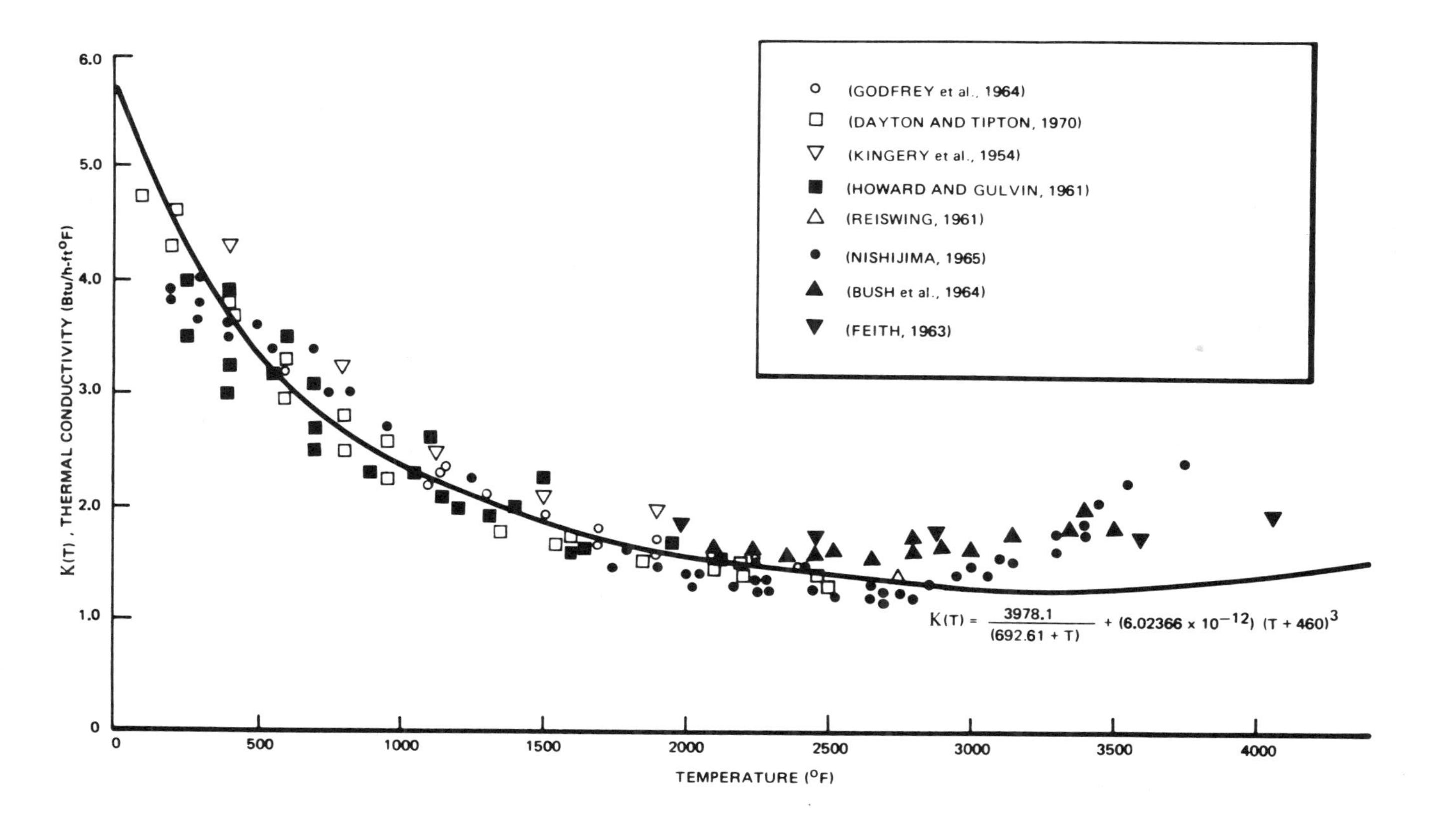

Fig. 6-1 The thermal conductivity of UO_2.

dicates some increase in thermal conductivity due to densification at the higher temperatures, but that this increase is somewhat less than shown in the data plotted, thus, Eq. (A.1) is obviously conservative for BWR design calculations.

The Zircaloy-2 cladding material is also a well-known reactor material and, hence, only a limited discussion of its properties is given here. It is chosen for use in BWR fuel elements because it has superior properties of strength, corrosion resistance, and neutron economy; i.e., Zircaloy-2 has a relatively small neutron capture cross section. Table 6-1 contains information on Zircaloy-2 thermal conductivity, heat capacity, coefficient of thermal expansion, and density that proves useful in subsequent discussions.

On initial loading of the fuel pellets, there is a small diametric gap between the surface of the fuel pellets and the inside surface of the Zircaloy-2 cladding. Before initial operation, this gap is filled with helium gas. Helium is inert and thus is nonreactive during seal welding of the end plugs. Furthermore, helium has a relatively high thermal conductivity so that the operating temperature drop across the gap is minimized. Subsequent to initial operation, small quantities of fission gases such as xenon and krypton are released. These gases have fairly low thermal conductivity compared to the helium fill gas (i.e., $\kappa_{Xe}/\kappa_{He} = 1/23$ and $\kappa_{Kr}/\kappa_{He} = 1/15$); hence, in the absence of any other effects, we might expect that the gap, ΔT, would increase with irradiation. However, shortly after initial operation, the gap essentially is closed. This gap closure occurs due to thermal stress-induced cracking and relocation of the UO_2 pellets, and due to the fact that the fuel pellets are at a higher temperature than the cladding and have a coefficient of thermal expansion about twice that of the Zircaloy-2 cladding [see Eqs. (B.1) and (B.2)].

A detailed analytical model known as GEGAP-III (1973) is used in the evaluation of the fuel/clad gap conductance, H_g. This detailed technique accounts for the effects of clad and fuel thermal expansion, cladding creep-down, fuel densification kinetics, fuel relocation, and irradiation-induced swelling of the fuel. Figures 6-2 and 6-3 indicate how gap conductance varies with burnup in 7×7 and 8×8 fuel, respectively. It is seen that for most cases of practical significance, it is conservative to use a constant value of $H_g = 1000$ Btu/h-ft^2-°F in design evaluations. A more thorough discussion of the actual heat transfer process at the fuel gap interface is given later in this chapter, when the thermal "contact problem" is considered.

Now that the fuel element geometry, thermal properties, and physical characteristics have been described, we can consider the thermal analysis of typical BWR fuel elements. This analysis yields the transient and steady-state temperature distributions required in the design/safeguards analysis of a BWR. The general procedure is to solve the classical conduction equation,

$$\rho c_p \frac{\partial T}{\partial t} = \nabla \cdot \kappa \nabla T + q''' \quad . \tag{6.1}$$

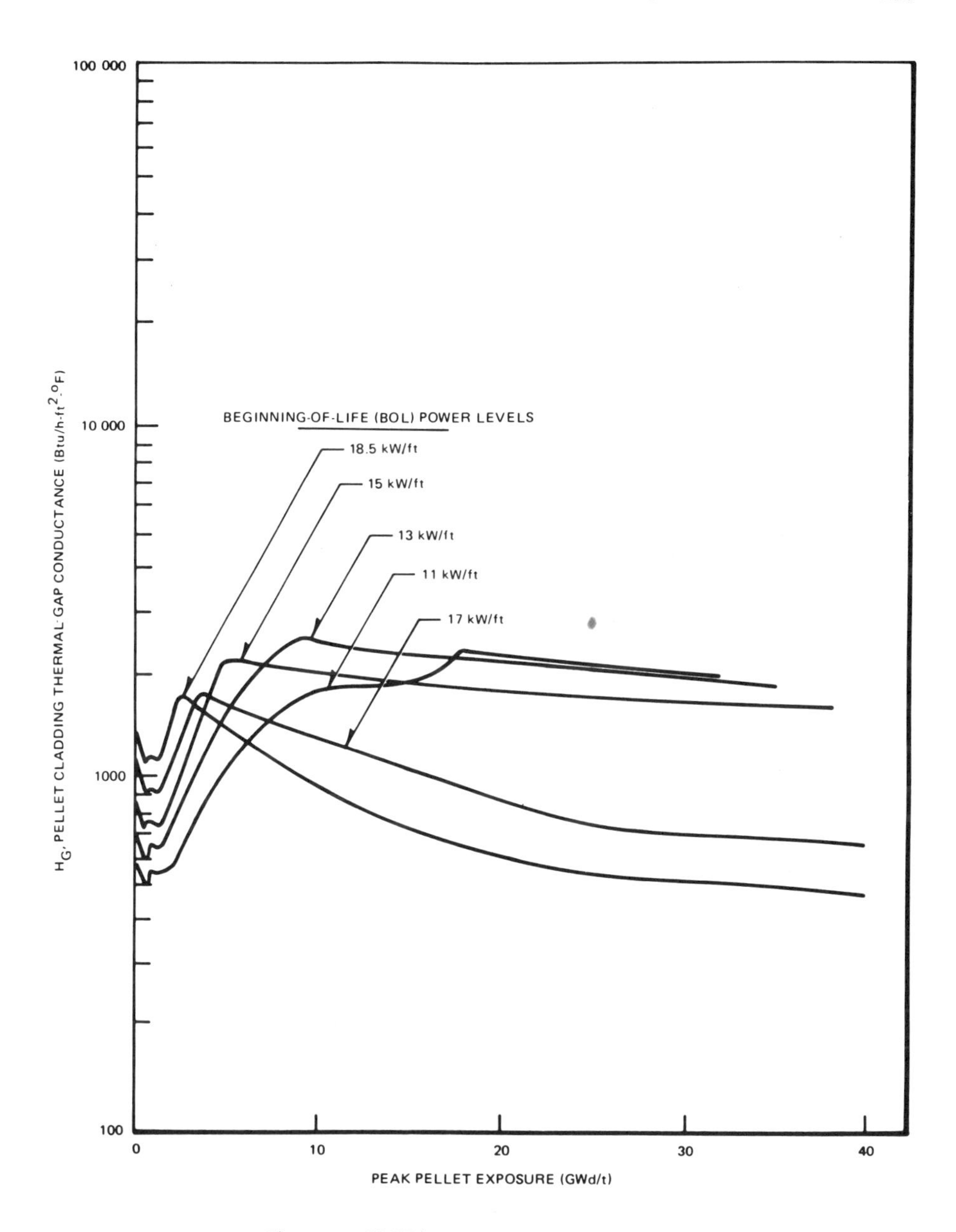

Fig. 6-2 BWR/5, 7×7 gap conductance.

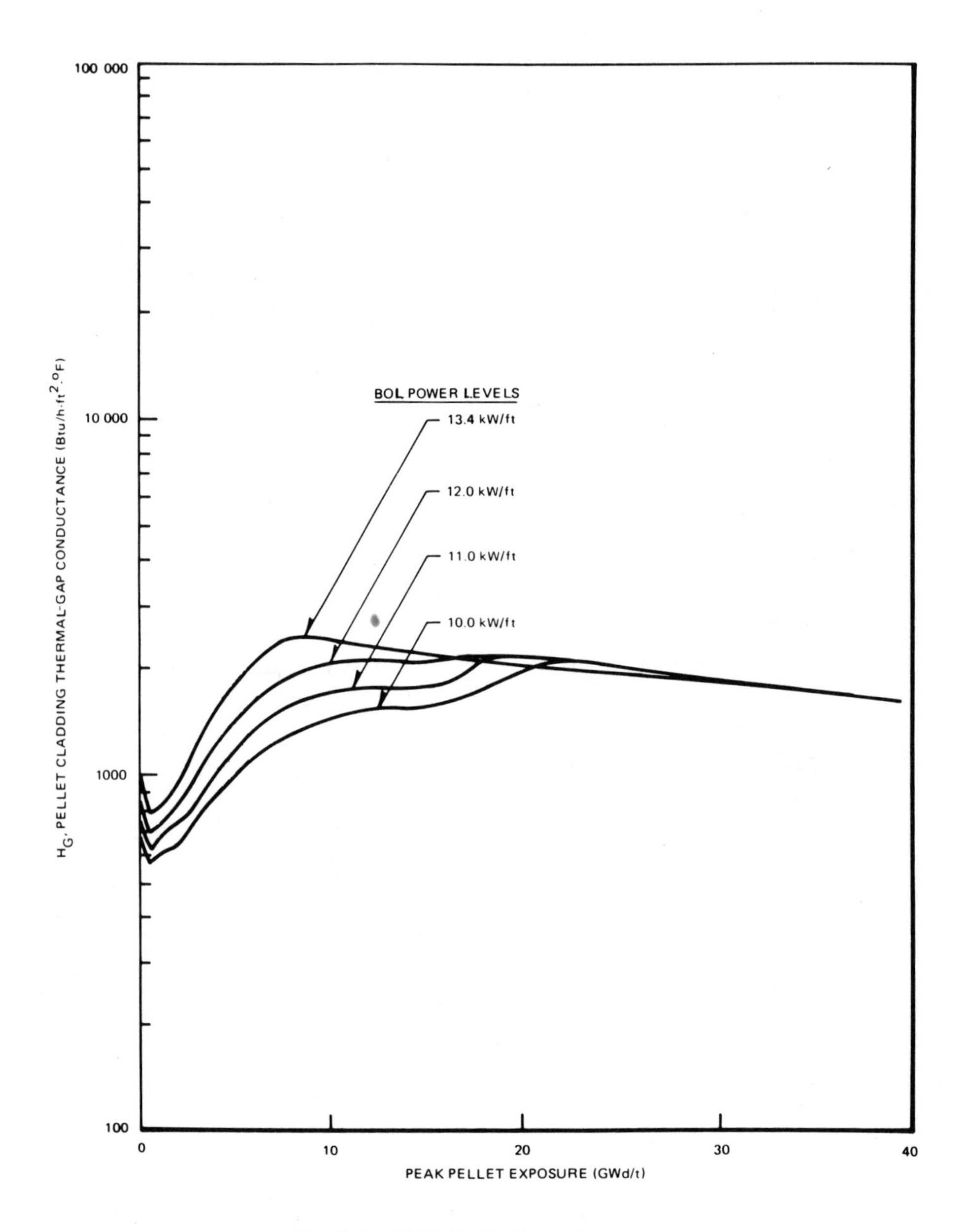

Fig. 6-3 BWR/6, 8×8 conductance.

One technique is to write this parabolic partial differential in finite difference form and to solve it numerically with the appropriate initial and boundary conditions. This is the approach that is taken in the subsequent discussion of thermal transients.

Another approach is to attempt to obtain exact solutions such as those tabulated by Carslaw and Jaeger (1959). Typically, these are quite complex, involving Bessel functions of various types and order. Also, many simplifying assumptions invariably must be made. A convenient discussion of solutions of this type is given by Tong and Weisman (1970).

The technique that we employ in the discussion of steady-state temperature distributions is to obtain an exact solution for the case of variable internal generation and conductivity. The significance and relative importance of the conductivity integral concept and the linear heat generation (kW/ft) parameter are clearly indicated by this form of the solution.

6.1.1 Steady-State Temperature Distributions

For steady-state conditions in which angular symmetry is assumed and axial effects are neglected, Eq. (6.1) becomes,

$$\frac{1}{r}\frac{d}{dr}\left[\kappa(T)r\frac{dT}{dr}\right] = -q'''(r) \ . \tag{6.2}$$

It is convenient to express the internal generation, $q'''(r)$, in terms of its average value, $\overline{q'''}$, and a normalized power distribution function, $F(r)$, such that,

$$\int_0^R F(r)r dr = R^2/2 \ .$$

By using these parameters, Eq. (6.2) becomes,

$$\frac{d}{dr}\left[\kappa(T)r\frac{dT}{dr}\right] = -\overline{q'''}rF(r) \ . \tag{6.3}$$

By integrating Eq. (6.3),

$$\kappa(T)\frac{dT}{dr} = -\frac{\overline{q'''}}{r}\int_0^r r'F(r')dr' \ .$$

Integrating, once again,

$$\int_{T(R)}^{T(r)} \kappa(T)dT = \overline{q'''}\int_r^R \left[\frac{1}{r''}\int_0^{r''} r'F(r')dr'\right]dr'' \ , \tag{6.4}$$

where R is the pellet radius. The left side of Eq. (6.4) is called the conductivity integral, which is an important parameter in the thermal analysis of reactor fuel elements (Belle, 1961).

It is interesting to note from the first integral of Eq. (6.3) that heat flux is given by,

$$q''(R) = \frac{\overline{q'''}}{R} \int_0^R F(r) r dr = R\overline{q'''}/2 \ , \tag{6.5}$$

and the linear heat generation rate (i.e., kW/ft) is given by,

$$q'(R) \triangleq 2\pi R q''(R) = \pi R^2 \overline{q'''} \ . \tag{6.6}$$

For the special case of constant thermal conductivity and uniform internal generation, Eqs. (6.4) and (6.6) yield,

$$\kappa[T(0) - T(R)] = q'(R)/4\pi \ , \tag{6.7}$$

where the integration has been carried out from the centerline, $r=0$, to the pellet surface, $r=R$. Equation (6.7) implies that the temperature difference across the fuel pellet is a function only of the linear heat generation of the fuel and thus is independent of the radius of the pellet. Since the avoidance of fuel melting (i.e., center melt) is normally a thermal design criterion, we can readily appreciate why the kW/ft rating of reactor fuel is such an important parameter in the thermal design of nuclear fuel elements.

Returning to the more general case, Eq. (6.4) can be written as,

$$M_2 = \overline{q'''} R^2 M_1 = 2Rq''(R) M_1 \ , \tag{6.8}$$

where,

$$M_1 \triangleq - \int_1^{\alpha} \left[\frac{1}{\alpha''} \int_0^{\alpha''} \alpha' F(\alpha') d\alpha' \right] d\alpha'' \tag{6.9}$$

$$M_2 \triangleq \int_{T(R)}^{T(r)} \kappa(T) dT \ , \tag{6.10}$$

and,

$$\alpha \triangleq r/R$$
$$\alpha' \triangleq r'/R$$
$$\alpha'' \triangleq r''/R \ .$$

The integral, M_1, given by Eq. (6.9), can be obtained once the normalized power distribution function, $F(\alpha')$, has been specified.

There are several cases of practical interest that are considered. The first, and simplest, case is that of uniform power density such that $F(\alpha') \triangleq 1.0$. The next case of interest is a parabolic power distribution such that $F(\alpha') \triangleq 1 + [(\alpha')^2 - ½]f$, which is normally a good approximation to the actual power distribution in fuel pellets. A better approximation is given by neutron diffusion theory as,

TABLE 6-2

Type of Power Distribution	Form of the Power Distribution Function, $F(\alpha')$	Internal Generation Integral, M_1
Uniform	1.0	$\frac{1}{4}(1-\alpha^2)$
Parabolic	$1+[(\alpha')^2-\frac{1}{2}]f$	$(1-\alpha^2)/4-[(1-\alpha^2)/4]^2 f$
Diffusion theory approximation	$\beta I_0(\beta\alpha')/2I_1(\beta)$	$\dfrac{I_0(\beta)-I_0(\beta\alpha)}{2\beta I_1(\beta)}$

$$F(\alpha')=\frac{\beta}{2I_1(\beta)}I_0(\beta\alpha') \ .$$

Equation (6.9) can be integrated using these assumed forms of the power distribution function. The results are summarized in Table 6-2.

The internal generation integrals, M_1, can be displayed conveniently. Figure 6-4 is a parametric plot of the internal generation integral for an assumed parabolic power distribution function. Note that the $f=0.0$ case implies uniform internal generation. A plot of the internal generation integral for the diffusion theory approximation is given in Fig. 6-5. As discussed later, these figures can be used to estimate rapidly the temperature profile in the fuel pellets.

A parameter frequently used in nuclear reactor fuel design is the "flux depression factor." It is defined as the ratio of the power at the centerline of the fuel pellet to the value at the surface of the pellet. For a parabolic power distribution, the flux depression factor is simply,

$$F(0)/F(1.0)=(1-0.5f)/(1+0.5f) \ ,$$

which is displayed in Fig. 6-6. Note that for all finite f, internal heat generation due to nuclear fission is normally less at the centerline of the fuel pellet than at the surface. This is due to the self-shielding effect of the uranium atoms and the higher thermal neutron flux in the moderator. If the flux depression factor is specified, Fig. 6-6 can be used to determine f in the parabolic distribution.

To complete the thermal analysis of a fuel pellet, we now consider the conductivity integral, M_2, given by Eq. (6.10). The temperature dependence of the thermal conductivity of UO_2 has been tabulated as Formula (A.1) in Table 6-1. This is readily integrated such that Eq. (6.10) yields,

$$M_2=3978.1\ \ln\left[\frac{692.61+T(r)}{692.61+T(R)}\right]+\frac{6.02366\times10^{-12}}{4.0}$$
$$\times\{[T(r)+460]^4-[T(R)+460]^4\} \ . \tag{6.11}$$

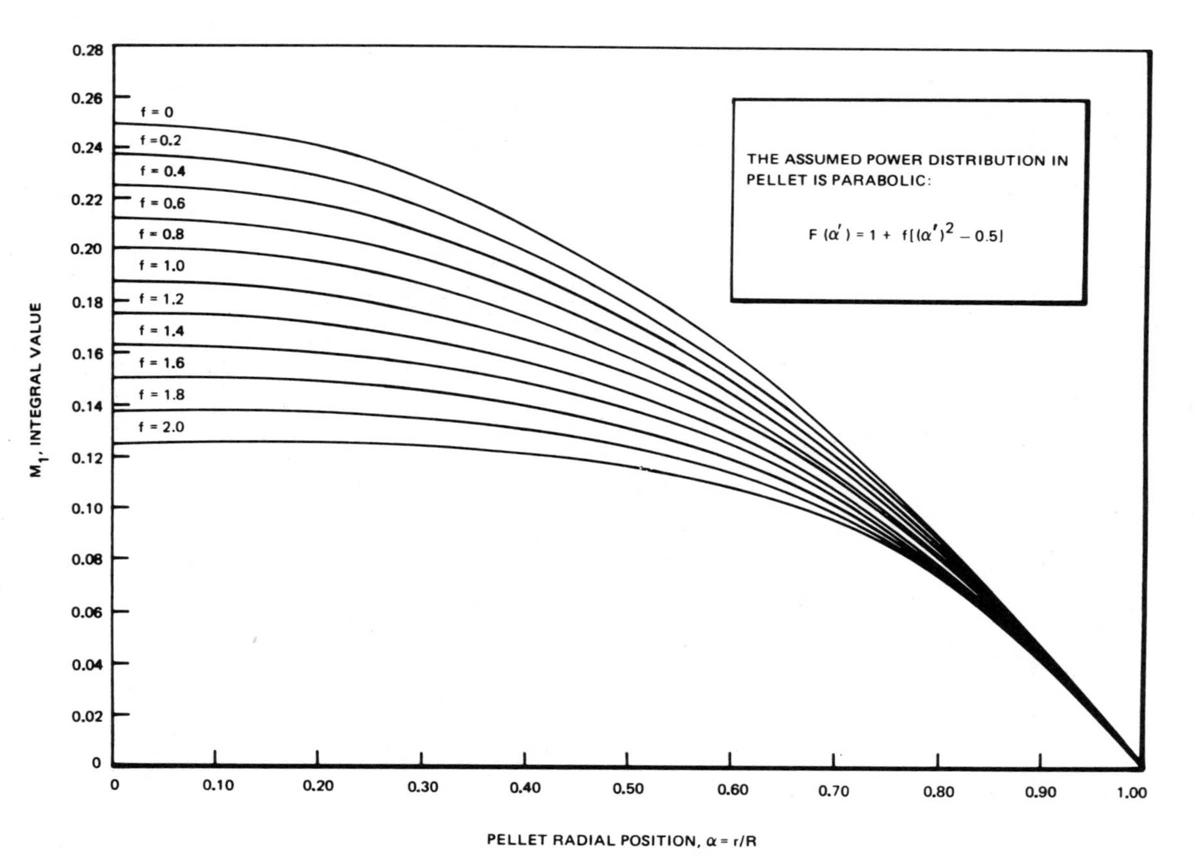

Fig. 6-4 Graphic presentation of M_1 integral for parabolic power distribution.

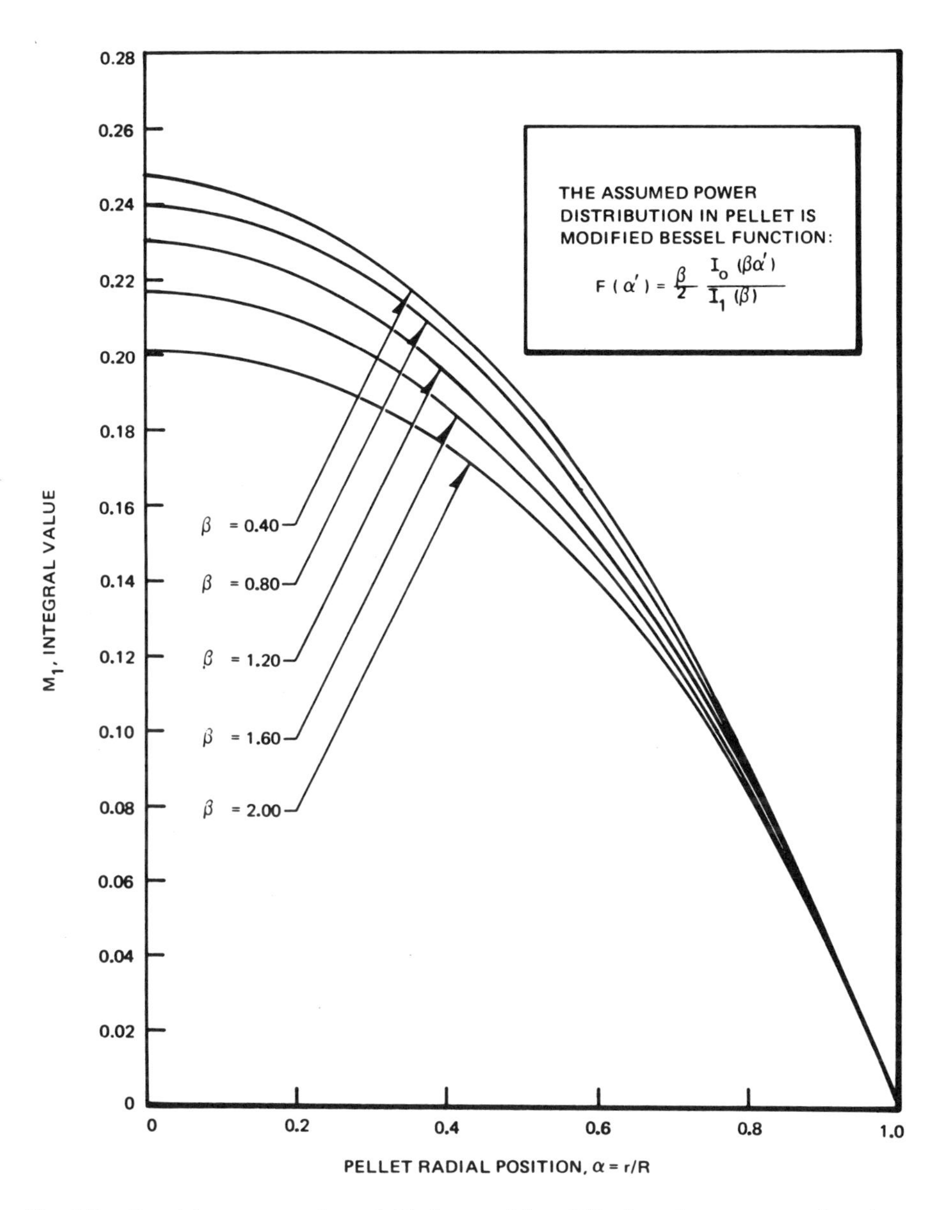

Fig. 6-5 Graphic presentation of M_1 integral for diffusion theory approximation to power distribution.

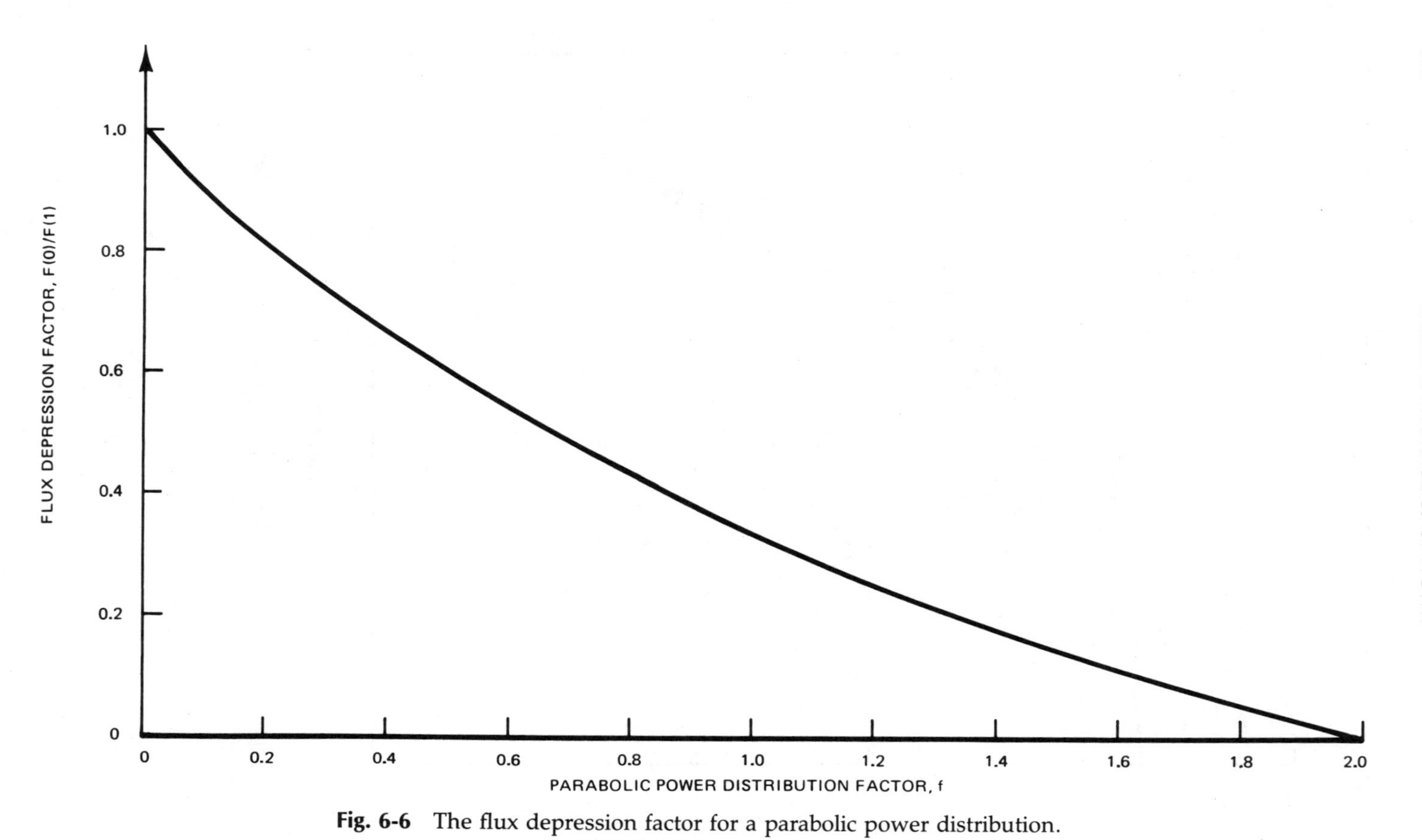

Fig. 6-6 The flux depression factor for a parabolic power distribution.

This expression can be used to evaluate $T(r)$, the temperature in degrees Fahrenheit at any radius, in degrees Fahrenheit r, when M_2 and the pellet surface temperature, $T(R)$, are known. If graphic solutions are to be obtained, it is convenient to evaluate the related conductivity integral,

$$N_2(T) \triangleq \int_{T_0}^{T} \kappa(T')dT' \ ,$$

in which T_0 is an arbitrary constant temperature. By integrating as before, we obtain,

$$N_2(T) \triangleq \int_{T_0}^{T} \kappa(T')dT' = 3978.1 \ln\left(\frac{692.61+T}{692.61+T_0}\right) + \frac{6.02366\times 10^{-12}}{4.0}[(T+460)^4-(T_0+460)^4] \ . \quad (6.12)$$

By comparing Eq. (6.12) with Eqs. (6.10) and (6.11), it is seen that,

$$M_2 = N_2[T(r)] - N_2[T(R)] \ . \quad (6.13)$$

As discussed later, a graphic display of the conductivity integral, $N_2(T)$, allows us to estimate rapidly the radial temperature distribution in a fuel pellet.

The choice of reference temperature, T_0, is arbitrary. Some authors (International Panel, 1965) prefer to use $T_0 = 500°C$, since it has been observed that below this temperature the thermal conductivity of UO_2 decreases with irradiation, while above 500°C irradiation effects apparently anneal out. Nevertheless, for our purposes here, the conventional value of $T_0 = 32°F$ is used. Equation (6.12) can be evaluated up to the melting temperature of UO_2, which is nominally 5080°F. Between these limits, Eq. (6.12) yields,

$$\int_{32°F}^{5080°F} \kappa(T')dT' = 9674 \frac{\text{Btu}}{\text{h-ft}} = 93 \text{ W/cm} \ . \quad (6.14)$$

For $T \leq 5080°F$, Eq. (6.12) is displayed in Fig. 6-7.

Now it is possible to determine the temperature distribution in the fuel pellets by using Eqs. (6.8), (6.12), and (6.13) once the fuel pellet radius, surface heat flux, surface temperature, and internal generation profile (Table 6-2) have been specified. To determine the surface temperature of the fuel pellet, we normally must add the temperature drop between the pellet surface and the coolant to the bulk temperature, T_∞, of the coolant. That is, for a given size pellet operating at a heat flux, $q''(R)$, we write,

$$T(R) = \Delta T_{\text{gap}} + \Delta T_{\text{Zr-2}} + \Delta T_{\text{oxide}} + \Delta T_{\text{crud}} + \Delta T_{\text{film}} + T_\infty \ , \quad (6.15)$$

where the physical significance of these temperature drops is shown in Fig. 6-8. Next we must direct our attention to the determination of these temperature drops.

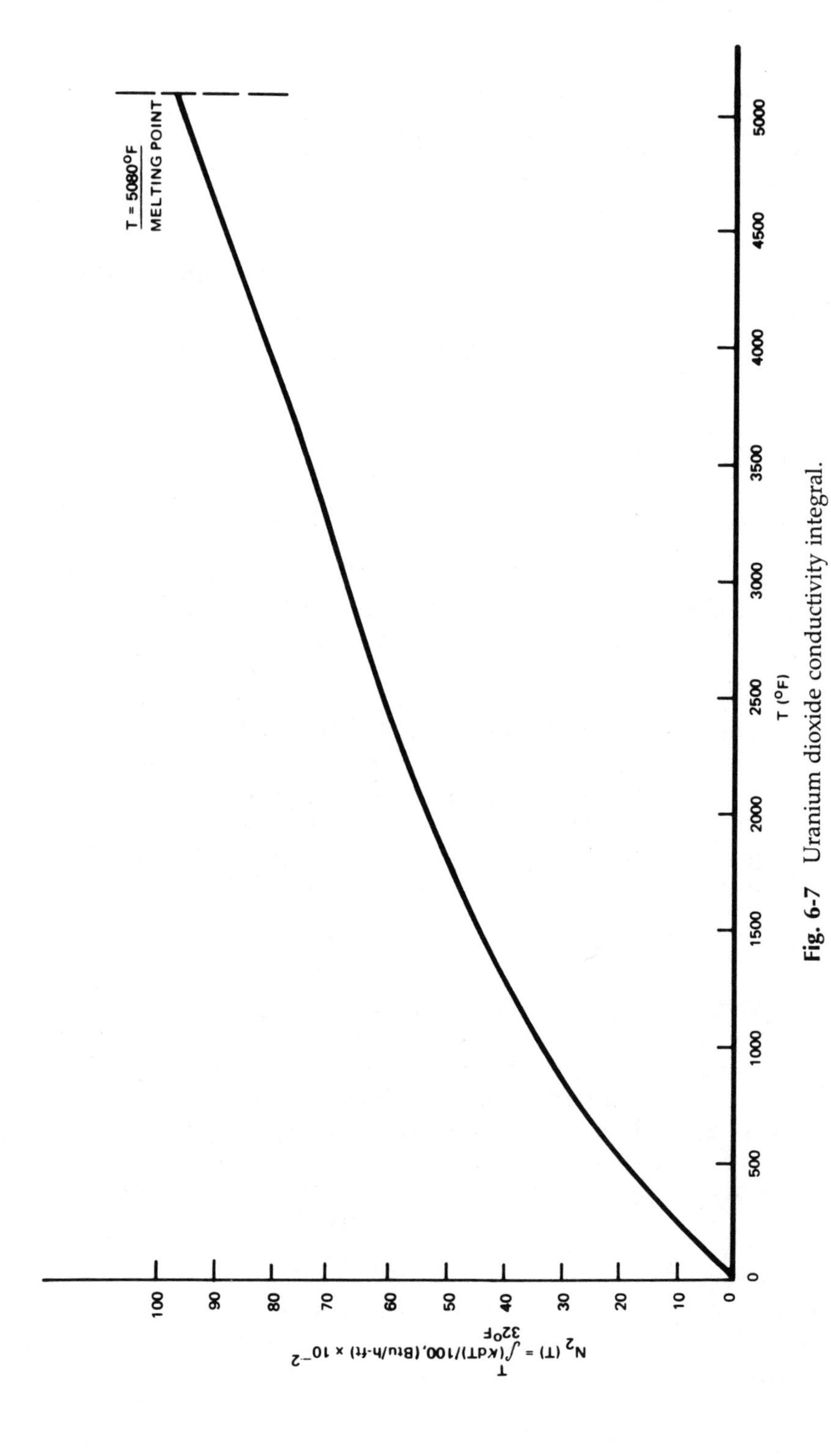

Fig. 6-7 Uranium dioxide conductivity integral.

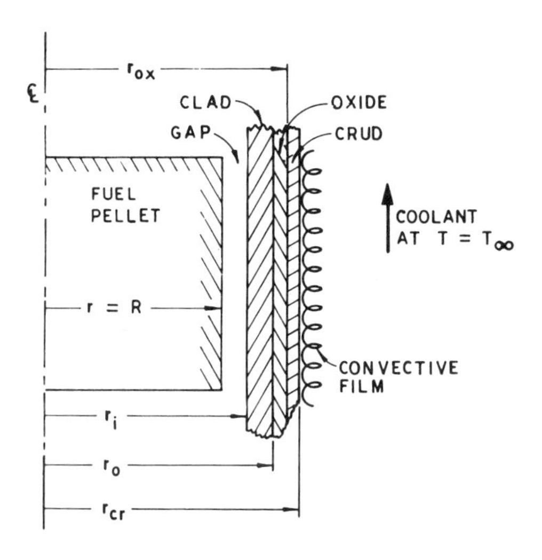

Fig. 6-8 Representation of thermal resistances of nonfuel components in a typical fuel element.

Figure 6-8 is a representation of the various resistances in the thermal circuit. Since the Zircaloy-2 cladding is relatively thin, 32 to 37 mils, and since its thermal conductivity is fairly large and not a strong function of temperature, it is usually a good approximation to assume constant thermal conductivity. For this case, Eq. (6.2) reduces to,

$$\frac{d}{dr}\left(r\frac{dT}{dr}\right)=0 \ .$$

This can be integrated to yield,

$$\Delta T_{\text{clad}}=q''(R)R\ \ln(r_0/r_i)/\kappa_{\text{Zr-2}} \ , \tag{6.16}$$

where the identity, $2\pi q''(R)R=2\pi q''(r_i)r_i$, has been used to reference the heat flux to the pellet surface. For more detailed design calculations, the temperature dependence of the cladding thermal conductivity given in Table 6-1 can be considered and Eq. (6.2) can be evaluated numerically. This procedure is discussed in Sec. 6.1.3.

The temperature drops across the zirconium oxide zone and the crud layer are of the same form as Eq. (6.16) since the constant thermal conductivity assumption is usually valid. Specifically, we can write,

$$\Delta T_{\text{crud}} = q''(R)R \ln(r_{cr}/r_0)/\kappa_{cr} \tag{6.17}$$

$$\Delta T_{\text{oxide}} = q''(R)R \ln(r_{ox}/r_0)/\kappa_{ox} \ . \tag{6.18}$$

The next component of the thermal circuit is the so-called "gap conductance," the specification of which is a very complex problem of determining the appropriate contact resistance. The contact resistance problem subsequently is considered in Sec. 6.1.2; however, it has been experimentally verified that an equivalent gap conductance is a useful concept and, as previously discussed, a constant, normally conservative, value of $H_g = 1000$ Btu/h-ft^2-°F can be used in scoping calculations.

Gap conductance normally is defined as,

$$H_g \triangleq q''(r_i)/\Delta T_{\text{gap}} = q''(R)(R/r_i)/\Delta T_{\text{gap}} \ . \tag{6.19}$$

Thus,

$$\Delta T_{\text{gap}} = q''(R)\left(\frac{R}{r_i}\right)/H_g \ . \tag{6.20}$$

Also, we can integrate as before to obtain the gap temperature drop in terms of an equivalent conductivity. This results in,

$$\Delta T_{\text{gap}} = q''(R)R \ln(r_i/R)/\kappa_g \ . \tag{6.21}$$

Equating Eqs. (6.20) and (6.21), we obtain,

$$H_g = \frac{\kappa_g}{r_i \ln(r_i/R)} \cong \frac{\kappa_g}{(r_i - R)} \ , \tag{6.22}$$

which implies that gap conductance is approximately equal to the equivalent thermal conductivity of the gap divided by the gap thickness. This expression is frequently modified to more accurately account for the fact that $H_g << \infty$ when the gap is zero. Previous investigators (Dean, 1963) have proposed that a pseudo-gap distance, l_0, be employed and that an accommodation factor, C, be used to approximate the effect of temperature discontinuities. Dean's expression takes the form,

$$H_g = \frac{\kappa_g}{C[(r_i - R) + l_0]} \ . \tag{6.23}$$

For most BWR design purposes, H_g is taken as an empirically determined parameter. Further discussion of the details of the contact problem is deferred to Sec. 6.1.2.

The final temperature drop in Eq. (6.15) that must be considered is that across the coolant convective film. The appropriate heat transfer coefficient depends on whether the coolant is single phase or two phase (i.e., boiling) at the axial location of interest. A discussion of the appropriate empirical heat transfer coefficients was given in Chapter 4. Once the coolant boundary condition has been specified, the film drop can be written as,

$$\Delta T_{film} = q''(R)(R/r_{cr})/H_{film} \ . \tag{6.24}$$

Equations (6.15) through (6.18), (6.20), and (6.24) can be combined to give,

$$T(R) - T_\infty = q''(R)[(R/r_i)H_g + R\ \ln(r_0/r_i)/\kappa_{Zr\text{-}2} + R\ \ln(r_{ox}/r_0)/\kappa_{ox} + R\ \ln(r_{cr}/r_{ox}/\kappa_{cr} + (R/r_{cr})H_{film}] \ . \tag{6.25}$$

This frequently is written as,

$$T(R) - T_\infty = q''(r_0)/U \ , \tag{6.26}$$

where U is the overall heat transfer coefficient based on $q''(r_0)$, the surface heat flux of the cladding, and is given by,

$$1/U = \left[\frac{r_0}{r_i H_g} + r_0\ \ln(r_0/r_i)/\kappa_{Zr\text{-}2} + r_0\ \ln(r_{ox}/r_0)/\kappa_{ox} + r_0\ \ln(r_{cr}/r_{ox})/\kappa_{cr} + \frac{r_0}{r_{cr}H_{film}} \right] \ . \tag{6.27}$$

Hence, when the bulk coolant temperature, T_∞, and the heat flux at the pellet surface, $q''(R) = (r_0/R)q''(r_0)$, have been specified, Eqs. (6.26) and (6.27) allow us to calculate the fuel pellet surface temperature, $T(R)$. In addition, the individual temperature drops through the convective film, crud layer, oxide layer, cladding, and fuel-cladding gap can be calculated using Eqs. (6.24), (6.17), (6.18), (6.16), and (6.20), respectively.

Equations (6.8) and (6.11), combined with the appropriate internal heat generation integral, M_1, given in Table 6-2 can be used to evaluate numerically the radial temperature distribution, $T(r)$. If computer facilities are not readily available, then a graphic solution scheme can be employed. One procedure is to choose a radial position, r/R, in Figs. 6-4 or 6-5, depending on which internal generation profile is assumed, to yield the internal generation integral, $M_1(r)$. Equation (6.8) then assigns a unique value to $M_2(r)$. Figure 6-7, in conjunction with Eq. (6.13), then is used to evaluate the temperature, $T(r)$, at the radial position originally assumed. This procedure can be repeated until the radial temperature profile through the fuel pellet has been evaluated completely. A typical temperature profile generated in this manner is given in Fig. 6-9.

Note that the calculation just discussed includes many assumptions and simplifications. For instance, we have assumed no circumferential variations in the internal generation distribution factor (i.e., no flux tilt), and have neglected axial effects. In addition, we have taken a macroscopic point

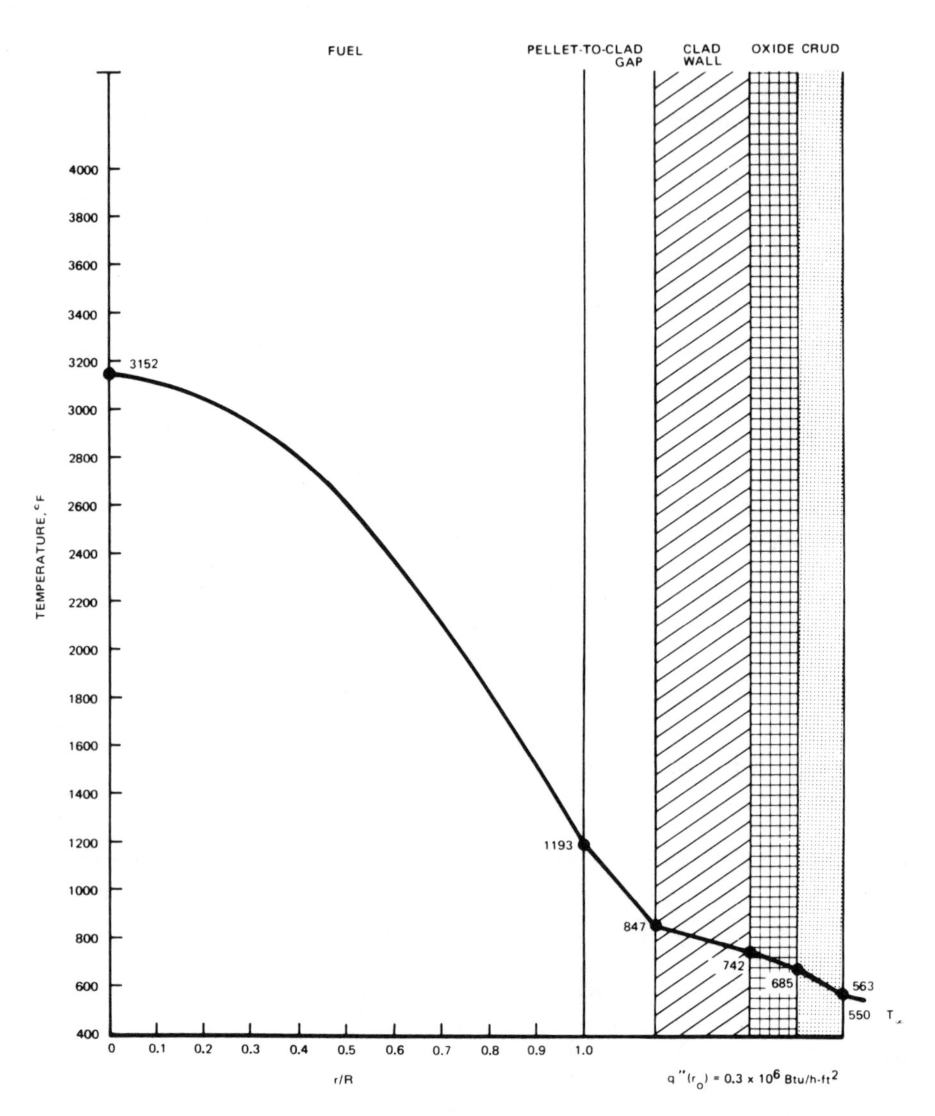

Fig. 6-9 Typical end-of-life (EOL) temperature distribution in a BWR/4 fuel rod.

of view in analyzing the fuel pellet in that we have employed an equivalent thermal conductivity rather than trying to consider the details of the heat flow paths in a fragmented fuel pellet. This approach is motivated largely by the complexity of a microscopic approach and the fact that an empirically determined equivalent thermal conductivity is normally adequate for design purposes. In some instances, conservative assumptions were made to purposely simplify the analysis. Typical of these assumptions is that we have neglected the "coring" of fuel pellets, which has been observed in high-performance fuel elements as the fabricated porosities and fission gases migrate to the center of the pellets. By assuming a solid pellet throughout its in-core life, the calculated EOL centerline temperatures can be higher than actual.

Another conservative assumption that has been made to simplify the analysis is the use of a constant gap conductance, which has a value of 1000 Btu/h-ft^2-°F. As discussed previously, the actual physical situation at the fuel-cladding interface can be considered a thermal contact problem. The details of this type of analysis and its role in BWR fuel design are considered next.

6.1.2 Gap Conductance—The Thermal Contact Problem

So far, the discussion concerning the heat transfer modes at the fuel-cladding interface has indicated that the physical mechanisms involved are very complex and that an exact, analytic solution is quite difficult. Indeed, this is the case. Nevertheless, there are many semiempirical models available that can be used to estimate the effect of interface parameters on the effective gap conductance. A concise summary of many of the models currently in use was given previously by Tong and Weisman (1970). The most widely used models are probably those of Ross and Stoute (1962) and Rapier et al. (1963). Both of these models are based on the earlier work of Centinkale and Fishenden (1951) and both involve several key assumptions concerning the dependence of gap conductance on contact pressure. In addition, these models assume that there is a uniform distribution of surface contact points (i.e., protrusions), and that when contact is made at the fuel-cladding interface, the contact pressure will be sufficient to cause plastic deformation on at least one of the surfaces.

The assumptions made in the Centinkale and Fishenden-type models can be realistic if the surface finish of the fuel pellets and cladding is controlled carefully. However, in general we expect a nonuniform distribution of surface protrusions. These protrusions or contact points are normally of various heights and their deformation can be plastic or elastic depending on the contact pressure at the interface. A more physically realistic model for the gap conductance has been put forward by Jacobs and Todreas (1973). This model is based on the work of Mikic (Cooper et al., 1969) and, as indicated in Fig. 6-10, it is assumed that the surface protrusions are of variable size with a Gaussian height distribution. Neglecting heat transfer through the entrapped gas pockets between contact points, mathematical expressions for gap conductance are given in Table 6-3 for the three models just discussed.

Jacobs and Todreas (1973) have compared the three models given in Table 6-3 with some selected gap conductance data. As seen in Fig. 6-11,

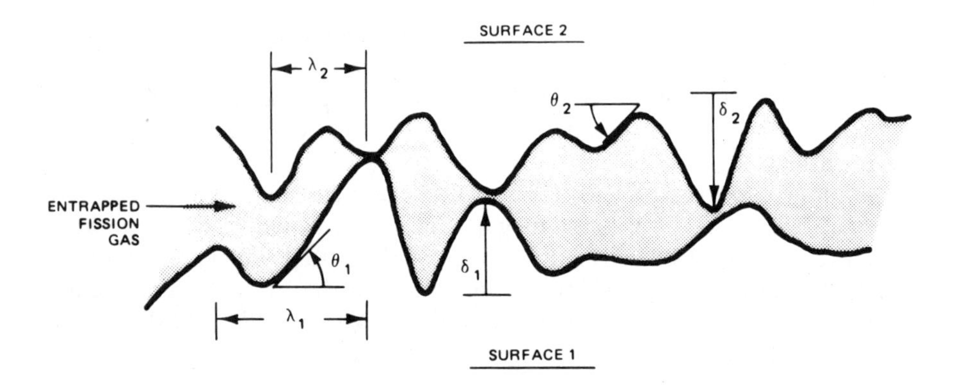

Fig. 6-10 Typical surface profile (Jacobs and Todreas, 1972).

Mikic's model does a somewhat better job of predicting the gap conductance, although the data scatter is quite large. When no information is available about the surface finish or contact pressure, Mikic's model is recommended. However, if the surface finish is known and the contact pressure is relatively high, O (1000 psi), Ross and Stoute's model can be used to estimate the effect of varying the height of the surface protrusions on gap conductance.

6.1.3 Transient Conduction Analysis

During the course of its in-core life, a nuclear reactor fuel element normally experiences many thermal transients. These operational transients are associated with such things as reactor startup or shutdown and changes in power level. In addition, we must be able to analyze the thermal response to various hypothetical accident-type transients. The time history of the fuel element temperatures can be obtained by numerically evaluating the conduction equation, Eq. (6.1).

In the interest of simplicity, we retain the previous assumptions of angular symmetry and neglect axial effects. Under these assumptions, Eq. (6.1) can be written for the fuel element shown in Fig. 6-12 as,

$$\rho c_p \frac{\partial T}{\partial t} = \frac{1}{r}\frac{\partial}{\partial r}\left(r\kappa\frac{\partial T}{\partial r}\right) + q''' \ . \tag{6.28}$$

To solve for the transient temperature distribution, Eq. (6.28) must be finite differenced and numerically evaluated, subject to the following boundary conditions:

$$-\kappa\frac{\partial T}{\partial r}\bigg|_{r=r_n} = q''(r_n, t) = H_\infty[T_n(t) - T_\infty(t)] + q''_R \tag{6.29a}$$

TABLE 6-3
Expressions for Gap Conductance, H_g

A. The Ross and Stoute model[a]

$$H_g = \frac{k_m}{0.0905 R^{1/2}}\left(\frac{p_c}{H}\right), \text{ (Btu/h-ft}^2\text{-°F)},$$

where

$k_m \triangleq 2k_1k_2/(k_1+k_2)$, harmonic mean of the thermal conductivities of the contacting surfaces 1 and 2, (Btu/h-ft-°F)

δ_i = height of protrusion on surface, i, ft

$R \triangleq [(\delta_1^2+\delta_2^2)/2]^{1/2}$, root mean square (rms) of surface protrusions,[b] ft

$p_c \triangleq$ contact pressure, psi

$H \triangleq$ hardness (Meyer) of softer solid, psi.

This model is valid strictly for plastic deformation at the fuel-clad interface.

B. The Rapier et al. model[c]

$$H_g = \frac{k_m}{\lambda_m}\left(\frac{p_c}{H}\right)^{1/2}, \text{ (Btu/h-ft}^2\text{-°F)},$$

where

$\lambda_m \triangleq \left(\frac{\lambda_1^2+\lambda_2^2}{2}\right)$, rms protrusion-to-protrusion wave length,[b] ft

k_m, p_c and H : defined in A above.

This model is valid strictly for plastic deformation at the fuel-clad interface.

C. The Mikic model[d]

$$H_g = \frac{C_1 k_m \overline{|\tan(\theta)|}}{\sigma}\left(\frac{p_c}{H}\right)^{C_2},$$

where

$\overline{|\tan(\theta)|} \triangleq [\tan^2(\theta_1)+\tan^2(\theta_2)]^{1/2}$, mean absolute slope of the surface protrusions of the contacting surfaces

$\tan(\theta_i) \triangleq \delta_i/(\lambda_i/2)$, the slope of the surface protrusion on surface i

$\sigma_i \triangleq$ rms surface roughness of surface i, ft

$\sigma \triangleq (\sigma_1^2+\sigma_2^2)^{1/2}$, ft

$$C_1 \triangleq \begin{cases} 1.45, & \text{fresh fuel} \\ 0.29, & \text{irradiated fuel}^{e} \end{cases}$$

$$C_2 \triangleq \begin{cases} 1.0, & \text{plastic deformation at fuel-clad interface} \\ 0.50, & \text{elastic deformation at fuel-clad interface.} \end{cases}$$

[a]See Ross and Stoute (1962).

[b]See Fig. 6-10 for further definition.

[c]See Rapier et al. (1963).

[d]See Jacobs and Todreas (1972).

[e]The gap conductance of irradiated fuel has been found (Skipper and Woolton, 1958) to decrease by approximately a factor of 5 due to the presence of fission products and the buildup of oxide layers on the inner surface of the cladding.

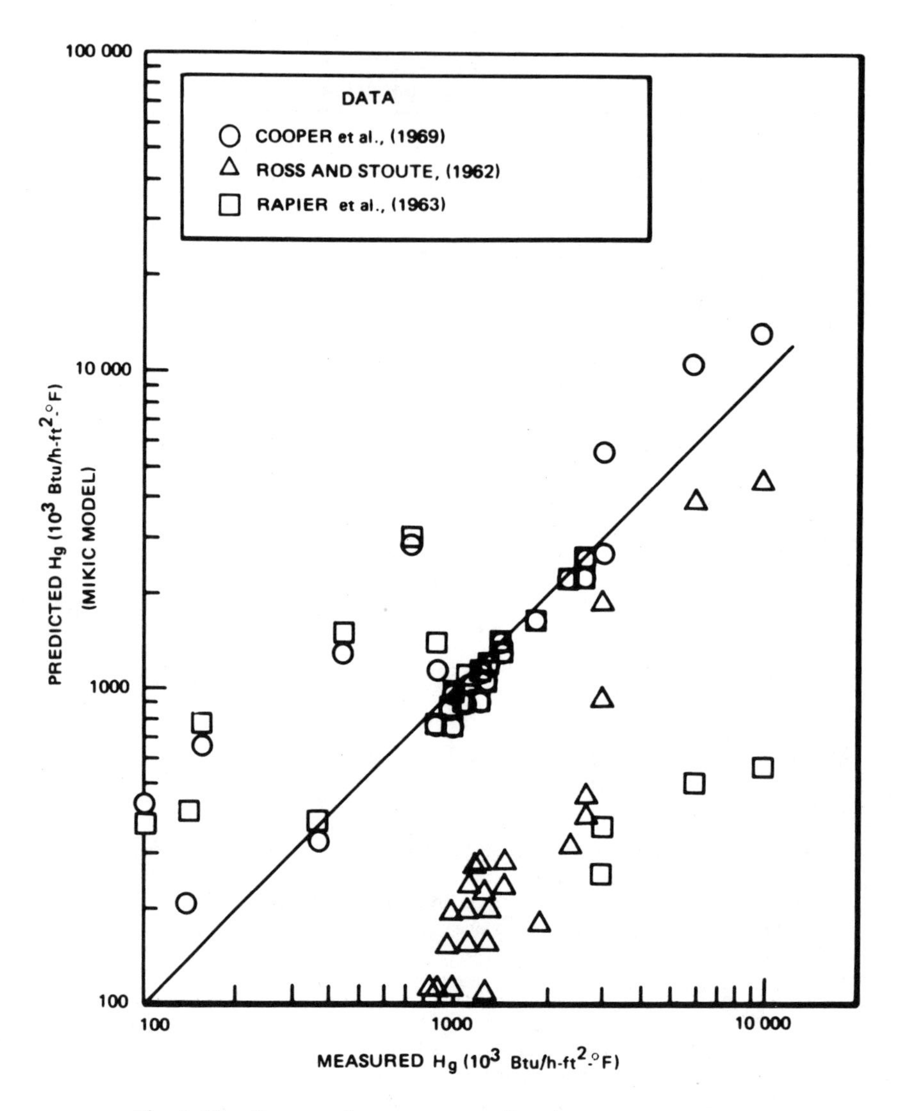

Fig. 6-11 Gap conductance—predicted versus measured.

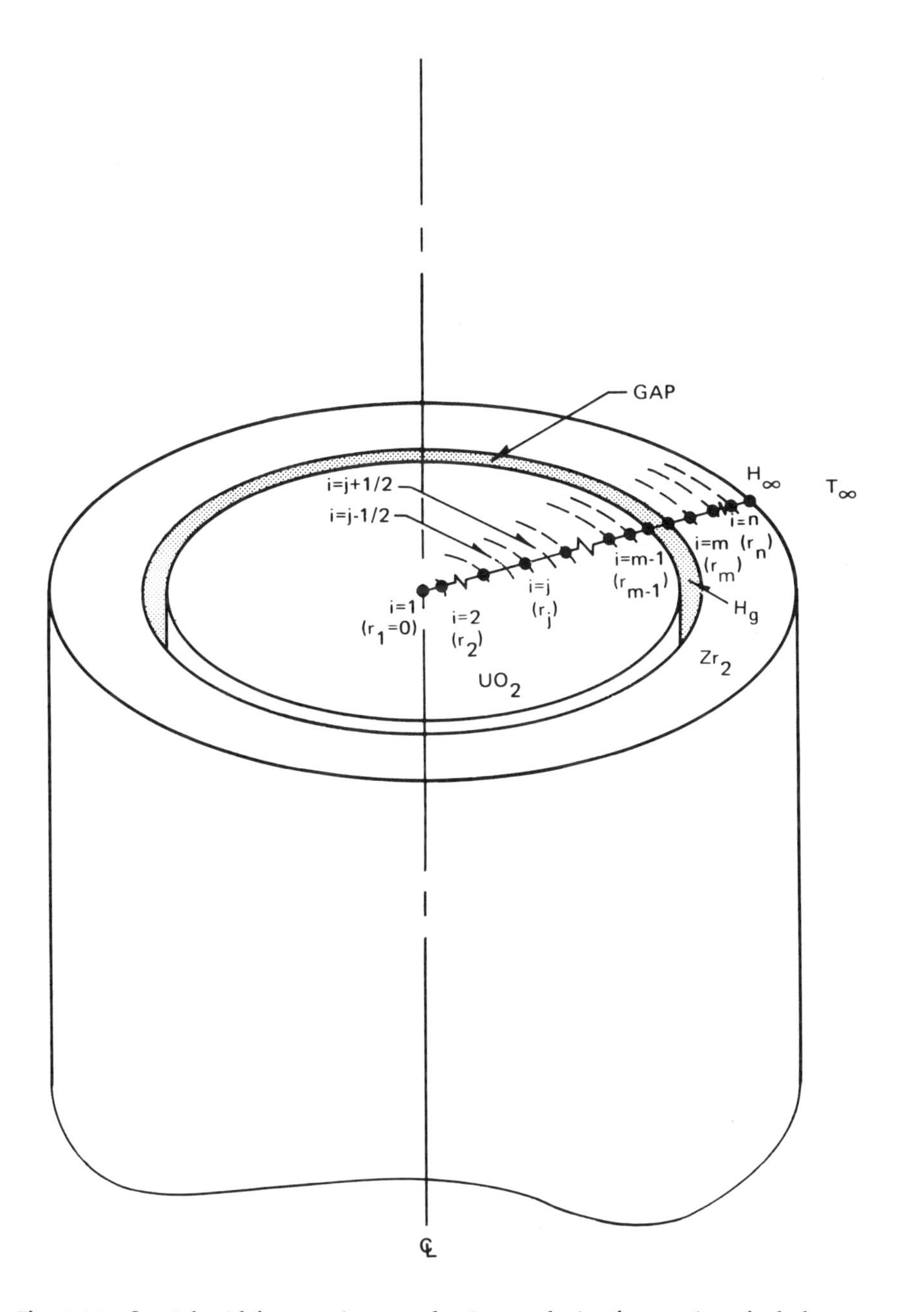

Fig. 6-12 Spatial grid for transient conduction analysis of a uranium fuel element.

$$-\kappa\frac{\partial T}{\partial r}\bigg|_{r=r_m} = q''(r_m,\ t) = H_g[T_{m-1}(t) - T_m(t)] \tag{6.29b}$$

$$-\kappa\frac{\partial T}{\partial r}\bigg|_{r=r_{m-1}} = q''(r_{m-1},\ t) = \left(\frac{r_m}{r_{m-1}}\right) q''(r_m,\ t) \tag{6.29c}$$

$$\frac{\partial T}{\partial r}\bigg|_{r=r_1=0} = 0\ , \tag{6.29d}$$

where H_∞ is the "convective" heat transfer coefficient, H_g is the gap conductance, and q''_R is the heat flux due to thermal radiation. Equation (6.29c) implies negligible energy storage in the fuel-cladding gap and Eq. (6.29d) implies radial symmetry. Both of these boundary conditions are rather good for most cases of practical concern. However, there are special cases in which different boundary conditions may need to be imposed. For our purpose here, the system given by Eqs. (6.28) and (6.29) is well posed and can be used to demonstrate the solution scheme of interest. To set up the solution scheme, the spatial derivative in Eq. (6.28) must be finite differenced. There are three cases that must be considered: the interior nodes, the boundary nodes, and the node on the centerline of the UO_2 fuel pellet. First, we consider the interior nodes in both the fuel pellet and the Zircaloy cladding.

By using standard central difference techniques, the term, $r\kappa(\partial T/\partial r)$, can be approximated at,

$$r_{i+1/2} \triangleq \tfrac{1}{2}(r_i + r_{i+1})\ ,$$

and time, t, by,

$$r\kappa\frac{\partial T}{\partial r}\bigg|_{r_{i+1/2}} = r_{i+1/2}\kappa_{i+1/2}\left(\frac{T_{i+1} - T_i}{r_{i+1} - r_i}\right) + O(\Delta r^2)\ . \tag{6.30}$$

Similarly,

$$r\kappa\frac{\partial T}{\partial r}\bigg|_{r_{i-1/2}} = r_{i-1/2}\kappa_{i-1/2}\left(\frac{T_i - T_{i-1}}{r_i - r_{i-1}}\right) + O(\Delta r^2)\ . \tag{6.31}$$

Next, at interior mesh point, r_i, we write,

$$\frac{\partial}{\partial r}\left(r\kappa\frac{\partial T}{\partial r}\right)\bigg|_{r_i} = \frac{1}{(r_{i+1/2} - r_{i-1/2})}\left[\left(r\kappa\frac{\partial T}{\partial r}\right)_{r_{i+1/2}} - \left(r\kappa\frac{\partial T}{\partial r}\right)_{r_{i-1/2}}\right] + O(\Delta r)\ . \tag{6.32}$$

Hence, by combining Eqs. (6.30), (6.31), and (6.32), and by dropping $O(\Delta r)$ terms, we obtain, after regrouping and dividing through by r_i,

$$\begin{aligned}\frac{1}{r}\frac{\partial}{\partial r}\left(r\kappa\frac{\partial T}{\partial r}\right)_{r_i} = &\left[\left(1+\frac{r_{i-1}}{r_i}\right)\frac{\kappa_{i-1/2}}{(r_{i+1}-r_{i-1})(r_i-r_{i-1})}\right]T_{i-1} \\ &-\left[\left(1+\frac{r_{i-1}}{r_i}\right)\frac{\kappa_{i-1/2}}{(r_{i+1}-r_{i-1})(r_i-r_{i-1})}\right. \\ &\left.+\left(1+\frac{r_{i+1}}{r_i}\right)\frac{\kappa_{i+1/2}}{(r_{i+1}-r_{i-1})(r_{i+1}-r_i)}\right]T_i \\ &+\left[\left(1+\frac{r_{i+1}}{r_i}\right)\frac{\kappa_{i+1/2}}{(r_{i+1}-r_{i-1})(r_{i+1}-r_i)}\right]T_{i+1} \ . \end{aligned} \tag{6.33}$$

Equation (6.33) is the appropriate finite difference expression of the spatial derivative in Eq. (6.28) for all interior nodes. Note that the thermal conductivity, κ, can vary spatially in this expression and that we do not need to have a uniform spatial mesh; i.e., Δr can vary. We now turn our attention to the node on the centerline of the fuel pellet.

Note that on the centerline, $r=0$, the spatial derivative in Eq. (6.28) is undefined. The easiest way to proceed is to expand the spatial derivative as,

$$\frac{1}{r}\frac{\partial}{\partial r}\left(r\kappa\frac{\partial T}{\partial r}\right) = \kappa\frac{\partial^2 T}{\partial r^2}+\frac{\partial \kappa}{\partial r}\frac{\partial T}{\partial r}+\frac{1}{r}\kappa\frac{\partial T}{\partial r} \ . \tag{6.34}$$

On the centerline, the symmetry condition given by Eq. (6.29d) implies that the second term on the right side of Eq. (6.34) vanishes and the last term is undefined. By using L'Hopital's rule on this latter term, we find,

$$\lim_{r\to 0}\frac{1}{r}\kappa\frac{\partial T}{\partial r} = \kappa\frac{\partial^2 T}{\partial r^2}\bigg|_{r=0} \ . \tag{6.35}$$

Thus, Eqs. (6.34) and (6.35) imply,

$$\frac{1}{r}\frac{\partial}{\partial r}\left(r\kappa\frac{\partial T}{\partial r}\right)\bigg|_{r=0} = 2\kappa\frac{\partial^2 T}{\partial r^2}\bigg|_{r=0} \ . \tag{6.36}$$

Equation (6.36) can be central differenced at time, t, to yield,

$$\frac{1}{r}\frac{\partial}{\partial r}\left(r\kappa\frac{\partial T}{\partial r}\right)\bigg|_{r=0} = 2\kappa_1\frac{(T_2-2T_1+T_0)}{r_2^2}+O(\Delta r^2) \ , \tag{6.37}$$

where, due to symmetry, the fictitious temperature, T_0, is given by $T_0=T_2$. Thus, Eq. (6.37) yields,

$$\frac{1}{r}\frac{\partial}{\partial r}\left(r\kappa\frac{\partial T}{\partial r}\right)\bigg|_{r=0} = 4\kappa_1\frac{(T_2-T_1)}{r_2^2} \ , \tag{6.38}$$

or, by regrouping and neglecting $O(\Delta r^2)$ terms,

$$\frac{1}{r}\frac{\partial}{\partial r}\left(r\kappa\frac{\partial T}{\partial r}\right)\bigg|_{r=0} = -\frac{4\kappa_1}{r_2^2}T_1 + \frac{4\kappa_1}{r_2^2}T_2 \ . \tag{6.39}$$

Now we must consider the form of the spatial derivative on the various boundary nodes. First we consider the node at r_{m-1}, the surface of the fuel pellet. Using Taylor's theorem,

$$r\kappa\frac{\partial T}{\partial r}\bigg|_{r_{m-3/2}} = r\kappa\frac{\partial T}{\partial r}\bigg|_{r_{m-1}} - \frac{1}{2}(r_{m-1}-r_{m-2})\frac{\partial}{\partial r}\left(r\kappa\frac{\partial T}{\partial r}\right)_{r_{m-1}} + O(\Delta r^2) \ . \tag{6.40}$$

By using Eqs. (6.29c) and (6.29b), Eq. (6.40) can be written as,

$$\frac{1}{2}(r_{m-1}+r_{m-2})\kappa_{m-3/2}\left(\frac{T_{m-1}-T_{m-2}}{r_{m-1}-r_{m-2}}\right) = -r_m H_g(T_{m-1}-T_m)$$

$$-\frac{1}{2}(r_{m-1}-r_{m-2})\frac{\partial}{\partial r}\left(r\kappa\frac{\partial T}{\partial r}\right)\bigg|_{r_{m-1}} + O(\Delta r^2) \ . \tag{6.41}$$

By rearranging, dividing through by r_{m-1}, and neglecting terms of $O(\Delta r)$, we obtain,

$$\frac{1}{r}\frac{\partial}{\partial r}\left(r\kappa\frac{\partial T}{\partial r}\right)\bigg|_{r_{m-1}} = \left[\left(1+\frac{r_{m-2}}{r_{m-1}}\right)\frac{\kappa_{m-3/2}}{(r_{m-1}-r_{m-2})^2}\right]T_{m-2}$$

$$-\left[\left(1+\frac{r_{m-2}}{r_{m-1}}\right)\frac{\kappa_{m-3/2}}{(r_{m-1}-r_{m-2})^2} + \left(\frac{r_m}{r_{m-1}}\right)\left(\frac{2H_g}{r_{m-1}-r_{m-2}}\right)\right]T_{m-1}$$

$$+\left[\left(\frac{r_m}{r_{m-1}}\right)\left(\frac{2H_g}{r_{m-1}-r_{m-2}}\right)\right]T_m \ . \tag{6.42}$$

At the inner surface of the cladding, position r_m, a Taylor's expansion yields,

$$r\kappa\frac{\partial T}{\partial r}\bigg|_{r_{m+1/2}} = r\kappa\frac{\partial T}{\partial r}\bigg|_{r_m} + \frac{1}{2}(r_{m+1}-r_m)\frac{\partial}{\partial r}\left(r\kappa\frac{\partial T}{\partial r}\right)\bigg|_{r_m} + O(\Delta r^2). \tag{6.43}$$

As outlined above, Eqs. (6.43) and (6.29b) can be combined to yield,

$$\frac{1}{r}\frac{\partial}{\partial r}\left(r\kappa\frac{\partial T}{\partial r}\right)\bigg|_{r_m} = \left[\frac{2H_g}{(r_{m+1}-r_m)}\right]T_{m-1}$$

$$-\left[\frac{2H_g}{(r_{m+1}-r_m)} + \left(1+\frac{r_{m+1}}{r_m}\right)\frac{\kappa_{m+1/2}}{(r_{m+1}-r_m)^2}\right]T_m$$

$$+\left[\left(1+\frac{r_{m+1}}{r_m}\right)\frac{\kappa_{m+1/2}}{(r_{m+1}-r_m)^2}\right]T_{m+1} \ . \tag{6.44}$$

The final node that must be considered is the surface node of the Zircaloy cladding at r_n. The appropriate finite difference form can be derived using exactly the same method as employed to obtain Eq. (6.42). The corresponding result is,

$$\frac{1}{r}\frac{\partial}{\partial r}\left(r\kappa\frac{\partial T}{\partial r}\right)\bigg|_{r_n} = \left[\left(1+\frac{r_{n-1}}{r_n}\right)\frac{\kappa_{n-½}}{(r_n-r_{n-1})^2}\right]T_{n-1}$$

$$-\left[\left(1+\frac{r_{n-1}}{r_n}\right)\frac{\kappa_{n-½}}{(r_n-r_{n-1})^2}+\frac{2H_\infty}{(r_n-r_{n-1})}\right]T_n$$

$$+\left[\frac{2H_\infty}{(r_n-r_{n-1})}\right]T_\infty-\frac{2q_R''}{(r_n-r_{n-1})} \ . \tag{6.45}$$

The various finite difference approximations for the spatial derivative in Eq. (6.28) are given by Eqs. (6.33), (6.39), (6.42), (6.44), and (6.45). The temporal term can be approximated as,

$$\frac{\partial T}{\partial t}=\frac{(T_{i,k+1}-T_{i,k})}{\Delta t}+O(\Delta t) \ , \tag{6.46}$$

where,

$$T_{i,k+1}\stackrel{\Delta}{=}T(r_i,t+\Delta t) \ .$$

Thus, Eqs. (6.46), (6.33), (6.39), (6.42), (6.44), and (6.45) comprise the *explicit* finite difference formulation of the transient conduction equation. It is well known that the stability of explicit numerical schemes limits the size of the time step, Δt, relative to the spatial step, Δr. Crank and Nicolson (1947) have suggested a mathematical substitution that essentially eliminates the stability problems experienced in explicit schemes. The basis for this method is to substitute the arithmetic average of the temperature at the present and future time step into the spatial derivatives. That is, in Eqs. (6.33), (6.39), (6.42), (6.44), and (6.45), terms of the form $T_{i,k}$ would be replaced by $(T_{i,k}+T_{i,k+1})/2$. The resultant spatial derivatives can be combined with the temporal derivative, given by Eq. (6.46), and the transient conduction equation, Eq. (6.28), to yield a *semi-implicit* finite difference formulation of the conduction equation. Approximating the conductivity, heat capacity, heat transfer coefficients, and surface radiation heat flux at time $t+\Delta t$ by their value at time t, this system of equations can be written conveniently in matrix form as,

$$C_k(\mathbf{T}'-\mathbf{T})=L_k(\mathbf{T}'+\mathbf{T})+\mathbf{Q}_k \ , \tag{6.47}$$

where, using standard indicial notation,

$$\mathbf{T}\stackrel{\Delta}{=}(T_{1,k},\ T_{2,k},\ \dots\ T_{n,k})^T,\ \text{a vector} \tag{6.48}$$

$$\mathbf{T}'\stackrel{\Delta}{=}(T_{1,k+1},\ T_{2,k+1},\ \dots\ T_{n,k+1})^T,\ \text{a vector} \tag{6.49}$$

$$C_k \stackrel{\Delta}{=} [(\rho c_p)_{i,k}\delta_{ij}], \text{ a diagonal matrix,}^{a}\ 1 \leqslant i,\ j \leqslant n \tag{6.50}$$

$$\mathbf{Q}_k \stackrel{\Delta}{=} \frac{\Delta t}{2}\left[q'''_{1,k} + q'''_{1,k+1},\ \ldots\ q'''_{n-1,\ k} + q'''_{n-1,k+1}\ ,\right.$$

$$\left. \frac{2H_{\infty,k}}{(r_n - r_{n-1})}(T_{\infty,k} + T_{\infty,k+1}) - \frac{4}{(r_n - r_{n-1})} q''_{R,k} + q'''_{n,k} + q'''_{n,k+1} \right]^T \text{, a vector} \tag{6.51}$$

$$L_k \stackrel{\Delta}{=} l_{i,k}^{(D-1)}\delta_{ij+1} + l_{i,k}^{(D)}\delta_{ij} + l_{i,k}^{(D+1)}\delta_{ij-1}, \text{ a tridiagonal matrix, } 1 \leqslant i,j \leqslant n\ . \tag{6.52}$$

The elements, $l_{i,k}$, of the tridiagonal matrix, **L**, are tabulated in Table 6-4.
Equation (6.47) can be rewritten as,

$$(C-L)_k\mathbf{T}' = (C+L)_k\mathbf{T} + \mathbf{Q}_k\ . \tag{6.53}$$

In this form, we can readily solve for $\mathbf{T}'$, the temperature vector at time $t + \Delta t$ using standard matrix inversion techniques such as a modified Gaussian elimination scheme (Varga, 1962). Thus, the required solution is,

$$\mathbf{T}' = (C-L)_k^{-1}(C+L)_k\mathbf{T} + (C-L)_k^{-1}\mathbf{Q}_k\ . \tag{6.54}$$

To start the calculation, the initial temperature distribution is given in the **T** vector and the internal generation vector, $\mathbf{Q}_k$, is specified. The elements of the C_k and L_k matrices are evaluated for the given initial conditions and the new temperature distribution is obtained, as indicated in Eq. (6.54), by matrix multiplication. For the special case of constant properties, the matrix $(C-L)_k$ is constant and only needs to be inverted once. For the more general case of variable thermal conductivity, it must be inverted at each time step. Fortunately, $(C-L)_k$ is tridiagonal and, thus, the linear system given in Eq. (6.53) is solved readily on modern digital computers.

The procedure outlined above is a powerful method to evaluate the transient temperature distribution in a nuclear reactor fuel pin. A very similar analysis can be performed on the steady-state version of Eq. (6.28) to yield the finite difference formulation of the steady conduction equation as,

$$\mathbf{T} = B^{-1}\mathbf{G}\ . \tag{6.55}$$

Thus, the steady-state temperature distribution can be easily evaluated by numerical integration in which the effect of spatially varying properties can be included. This can be used to initialize a transient calculation, for direct design analysis, or to check the accuracy of the approximate analytical methods discussed in Sec. 6.1.1.

---

[a] $\delta_{ij} \stackrel{\Delta}{=} \begin{cases} 1, & i=j \\ 0, & i \neq j \end{cases}\ .$

TABLE 6-4
The Elements of the Matrix, $\mathbf{L}_k$*

A. Interior nodes, $\begin{cases}(2 \leqslant i \leqslant m-2),\ UO_2 \\ (m+1 \leqslant i \leqslant n-1),\ Zr_2\end{cases}$

$$l_{i,k}^{(D-1)} \triangleq \frac{\Delta t}{2}\left(1+\frac{r_{i-1}}{r_i}\right)\frac{\kappa_{i-1/2,k}}{(r_{i+1}-r_{i-1})(r_i-r_{i-1})}$$

$$l_{i,k}^{(D+1)} \triangleq \frac{\Delta t}{2}\left(1+\frac{r_{i+1}}{r_i}\right)\frac{\kappa_{i+1/2,k}}{(r_{i+1}-r_{i-1})(r_{i+1}-r_i)}$$

$$l_{i,k}^{(D)} \triangleq -[l_{i,k}^{(D-1)}+l_{i,k}^{(D+1)}] \ .$$

B. Node on centerline, $i=1$

$$l_{i,k}^{(D)} - \triangleq \frac{2\Delta t \kappa_{1,k}}{(r_2)^2}$$

$$l_{1,k}^{(D+1)} \triangleq -l_{1,k}^{(D)} \ .$$

C. Node on fuel pellet surface, $i=m-1$

$$l_{m-1,k}^{(D-1)} \triangleq \frac{\Delta t}{2}\left(1+\frac{r_{m-2}}{r_{m-1}}\right)\frac{\kappa_{m-3/2,k}}{(r_{m-1}-r_{m-2})^2}$$

$$l_{m-1,k}^{(D+1)} \triangleq \frac{\Delta t}{2}\left(\frac{r_m}{r_{m-1}}\right)\left(\frac{2H_{g,k}}{r_{m-1}-r_{m-2}}\right)$$

$$l_{m-1,k}^{(D)} \triangleq -[l_{m-1,k}^{(D-1)}+l_{m-1,k}^{(D+1)}] \ .$$

D. Node on the inner clad surface, $i=m$

$$l_{m,k}^{(D-1)} \triangleq \frac{\Delta t H_{g,k}}{(r_{m+1}-r_m)}$$

$$l_{m,k}^{(D+1)} \triangleq \frac{\Delta t}{2}\left[\left(1+\frac{r_{m+1}}{r_m}\right)\frac{\kappa_{m+1/2,k}}{(r_{m+1}-r_m)^2}\right]$$

$$l_{m,k}^{(D)} \triangleq -[l_{m,k}^{(D-1)}+l_{m,k}^{(D+1)}] \ .$$

E. Node on outer clad surface, $i=n$

$$l_{n,k}^{(D-1)} \triangleq \frac{\Delta t}{2}\left(1+\frac{r_{n-1}}{r_n}\right)\frac{\kappa_{n-1/2,k}}{(r_n-r_{n-1})^2}$$

$$l_{n,k}^{(D)} \triangleq -\left[l_{n,k}^{(D-1)}+\frac{\Delta t H_{\infty,k}}{(r_n-r_{n-1})}\right] \ .$$

*Defined quantities: $\kappa_{i\pm 1/2} \triangleq \frac{1}{2}(\kappa_i + + \kappa_{i\pm 1})$.

Although the Crank-Nicolson substitution is not used in time invariant problems, we obtain here the functional relationship between the matrices in Eqs. (6.54) and (6.55) by formally setting $\mathbf{T}'=\mathbf{T}$ in Eq. (6.54) to yield,[b]

$$\mathbf{T}=[I-(C-L)^{-1}(C+L)]^{-1}(C-L)^{-1}\mathbf{Q} \ . \tag{6.56}$$

[b]I is the identity matrix, δ_{ij}.

Thus, by comparing Eqs. (6.55) and (6.56),

$$B = [I - (C - L)^{-1}(C + L)] \tag{6.57}$$

$$\mathbf{G} = (C - L)^{-1}\mathbf{Q} \ . \tag{6.58}$$

This completes our discussion of both steady-state and transient heat conduction in nuclear fuel elements. The intent has been to make this discussion representative of BWR fuel rod analysis. There are many other areas of BWR technology in which the analysis of heat conduction is necessary. Nevertheless, the basic approach and considerations are similar to those discussed in this section. Next, we consider the analysis of convective heat transfer in a BWR core.

6.2 Core Convection Heat Transfer

In this section, analytical models for the thermal-hydraulic analysis of a BWR core are summarized. The model consists of parallel but noncommunicating flow channels. Core geometry, operating power level, and distribution in the core, operating pressure, and core inlet enthalpy are presumed known. Techniques are summarized for determining either core pressure drop for a specified core flow or the core flow split for a specified core pressure drop.

6.2.1 Core Hydraulics

Flow entering the core from the lower plenum divides into two parts, the active coolant that passes through the fuel bundles, and the bypass (i.e., leakage) flow. Figure 6-13 shows the various leakage flow paths in a typical BWR fuel assembly. Normally, the leakage flow, w_L, is specified as a known fraction, F_L, of total core flow rate, w_T, such that,

$$w_L = F_L w_T \ . \tag{6.59}$$

The active core coolant distributes itself among the bundles in such a way that the total static pressure drop across each fuel assembly is the same.

The fuel channels are normally grouped into discrete types, which provide a number of parallel flow paths. Each type is characterized by similar hydraulic characteristics, power distribution, and power level.

The total static pressure drop of a BWR fuel channel is composed of the sum of frictional, local, spatial acceleration, and elevation effects,

$$\Delta p_{\text{total}} = \Delta p_{\text{friction}} + \Delta p_{\text{local}} + \Delta p_{\text{acceleration}} + \Delta p_{\text{elevation}} \ . \tag{6.60}$$

Analytical formulations for determining these pressure drops, including reversible pressure effects caused by flow area changes, were given in Chapter 5.

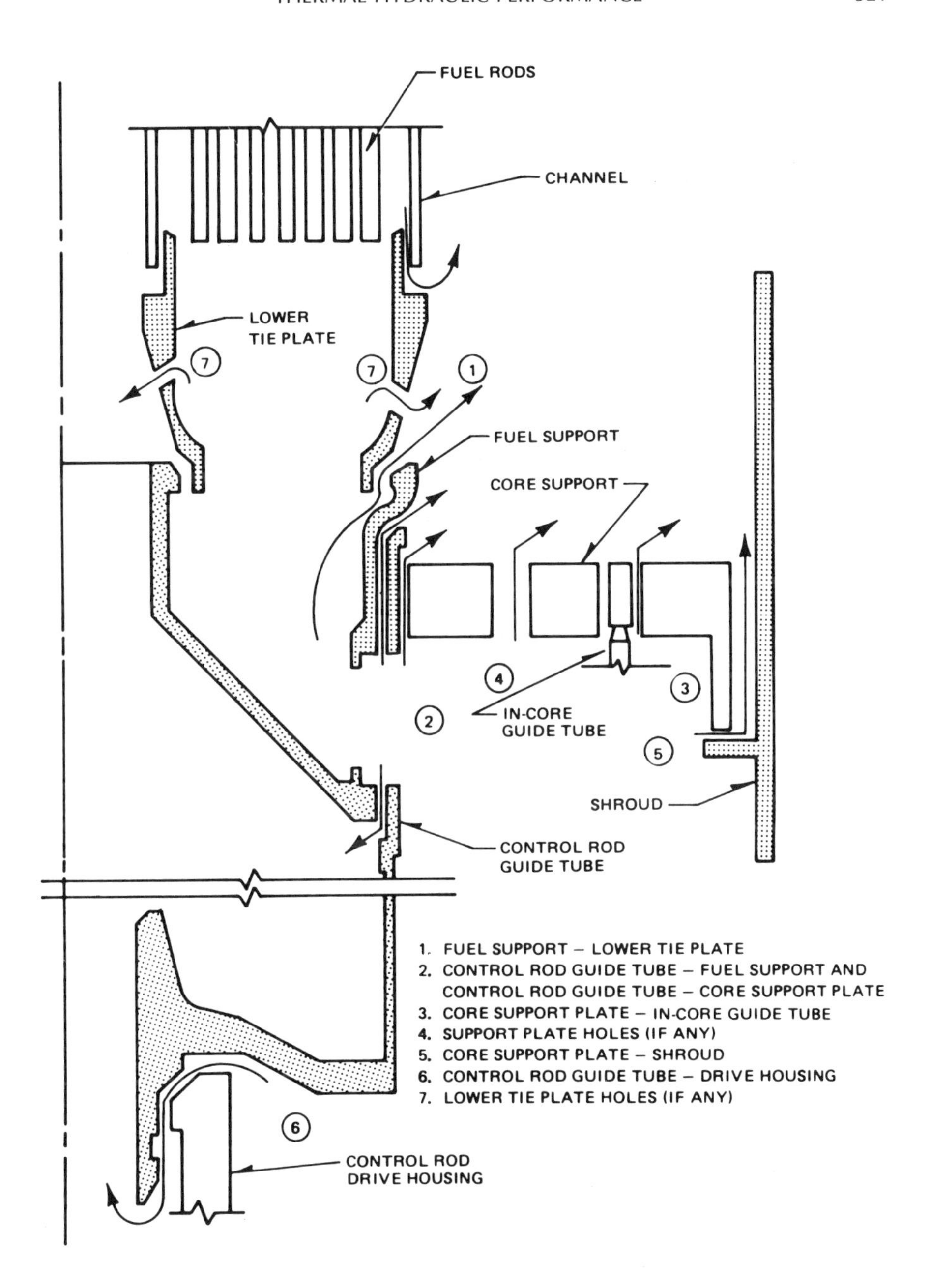

Fig. 6-13 Schematic of a typical BWR fuel assembly showing the leakage flow paths.

6.2.2 Core Enthalpy Rise

The actual core power distribution and hydraulic characteristics are closely coupled. That is, the flow split between the bundles influences the core power distribution, which in turn influences the flow split. Hence, the thermal-hydraulic and physics models are normally solved iteratively.

Core power is absorbed in the coolant flowing through the channels and also in the leakage flow. The energy absorbed by the leakage flow is due to neutron thermalization, gamma heating, and the heat flux from the channel walls. The power absorbed by the active coolant, Q_{active}, is used to calculate the active core coolant enthalpy rise, and is normally expressed in terms of total power, Q_{total}, less that fraction of power associated with leakage flow, F_{QL},

$$Q_{\text{active}} = (1 - F_{QL})Q_{\text{total}} \ . \tag{6.61}$$

The active core coolant energy gain normally is allocated to the N individual fuel bundles, based on the associated normalized radial peaking factor, F_R. The energy absorbed by the active coolant in a given bundle is, for steady-state operation, given by a normalized axial peaking factor, F_A. It follows that the energy absorbed by the coolant at the k'th axial node of the N'th fuel bundle is,

$$q_{\text{active}}(N,k) = F_R(N)F_A(k)Q_{\text{active}}/N \ . \tag{6.62}$$

By applying the First Law, as discussed in Chapter 3, to axial node k of fuel channel N, the enthalpy rise is given by

$$\Delta h(N,k) = \frac{q_{\text{active}}(N,k)}{w(N)} \ . \tag{6.63}$$

If h_{in} is fluid enthalpy entering fuel channel N, then the enthalpy at the M'th axial node is given by,

$$h(N,M) = \sum_{k=1}^{M} \Delta h(N,k) + h_{\text{in}} \ . \tag{6.64}$$

The enthalpy rise through the leakage path is also obtained from the First Law of Thermodynamics as,

$$h_{\text{leak}} - h_{\text{in}} = \frac{Q_{\text{total}}F_{QL}}{w_L} \ . \tag{6.65}$$

This section dealt with the steady-state analysis of BWR thermal hydraulics. The next three chapters are concerned with transient thermal-hydraulic models and analyses.

References

Belle, J., "Uranium Dioxide: Properties and Nuclear Applications," *Naval Reactor Handbook* (1961).

Bush, A. J., J. A. Christensen, H. M. Ferrari, and R. J. Allio, "Uranium Dioxide Thermal Conductivity," *Trans. Am. Nucl. Soc.*, **7**, 392 (1964).

Carslaw, H. S., and J. C. Jaeger, *Conduction of Heat in Solids*, 2nd ed., Oxford University Press, New York (1959).

Cetinkale, T. N., and M. Fishenden, "Thermal Conductance of Metal Surfaces in Contact," *Proc. General Discussion on Heat Transfer*, Institute of Mechanical Engineers, London (Sep. 1951).

Cooper, M. G., B. B. Mikic, and M. M. Yovanovich, "Thermal Contact Conductance," *Int. J. Heat Mass Transfer*, **12** (1969).

Crank, J., and P. Nicolson, "A Practical Method for Numerical Evaluations of Solutions of Partial Differential Equations of the Heat Conduction Type," *Proc. Cambridge Phil. Soc.*, **43** (1947).

Dayton, R., and C. Tipton, "Progress on the Development of Uranium Carbide-Type Fuels. Phase II: Report on the AEC Fuel-Cycle Program," Frank A. Rough and Walston Chubb, Eds., U.S. Atomic Energy Commission (Dec. 27, 1970).

Dean, R. A., "Thermal Contact Conductance," M.S. Thesis, University of Pittsburgh (1963).

Eldridge, E. A., and H. W. Deem, "Specific Heats and Heats of Transformation of Zr-2 and Low Nickel Zr-2," BMI-1803, Battelle Memorial Institute (1967).

Feith, A. D., "Thermal Conductivity of UO_2 by Radial Heat Flow Method," USAEC Report TM-63-9-5, General Electric Company, Cincinnati, Ohio (1963).

"FSAR—Browns Ferry Nuclear Plant," Final Safety Analysis Report, Vol. I (1971).

"GEGAP-III: A Model for the Prediction of Pellet Cladding Thermal Conductance in BWR Fuel Rods," NEDO-20181, General Electric Company (1973).

Godfrey, T. G., W. Fulkerson, T. G. Kollie, J. P. Moore, and D. L. McElroy, "Thermal Conductivity of UO_2 and Armco Iron by an Improved Radial Flow Technique," ORNL-3556, Oak Ridge National Laboratory (1964).

Hein, R. A., and P. N. Flagella, "Enthalpy Measurements of UO_2 and Tungsten to 3260°K," GEMP-578, General Electric Company (Feb. 1968).

Howard, V. C., and T. F. Gulvin, "Thermal Conductivity Determination on Uranium Dioxide by a Radial Flow Method," 1G Report 51 (RD/C) (1961).

International Panel of Experts, "Thermal Conductivity of Uranium Dioxide," Technical Report Series 59, International Atomic Energy Agency, Vienna (1965).

Jacobs, G., and N. Todreas, "Thermal Contact Conductance in Reactor Fuel Elements," *Nucl. Sci. Eng.*, **50** (1973).

Kingery, W. D., J. Francl, R. L. Coble, and T. Vasilas, "Thermal Conductivity: X, Data for Several Pure Oxide Materials Corrected to Zero Porosity," *J. Am. Ceram. Soc.*, **37**, 107 (1954).

Lyons, M. F., D. H. Coplin, H. Hausner, B. Weidenbaum, and T. J. Pashos, "UO_2 Powder and Pellet Thermal Conductivity During Irradiation," GEAP-5100-1, General Electric Company (Mar. 1966).

Mehon, R. L., and F. W. Wiesinger, "Mechanical Properties of Zircaloy-2," KAPL-2110, Knolls Atomic Power Laboratory (Feb. 1961).

Nishijima, T. J., "Thermal Conductivity of Sintered UO_2 and Al_2O_3 at High Temperatures," *J. Am. Ceram. Soc.*, **48,** 31 (1965).

Rapier, A. C., T. M. Jones, and J. E. McIntosh, "The Thermal Conductance of UO_2/Stainless Steel Interface," *Int. J. Heat Mass Transfer*, **6** (Mar. 1963).

Reiswing, R. D., "Thermal Conductivity of UO_2 to 2100°C," *J. Am. Ceram. Soc.*, **44,** 48 (1961).

Ross, A. M. and R. L. Stoute, "Heat Transfer Coefficient Between UO_2 and Zircaloy-2," AECL-1152, Atomic Energy of Canada Limited (1962).

Scott, D. B., "Physical and Mechanical Properties of Zircaloy-2 and Zircaloy-4," WCAP-3269-41, Westinghouse Electric Corporation (May 1965).

Seddon, B. J., "Zirconium Data Manual—Properties of Interest in Reactor Design," TRG-108(R), United Kingdom Atomic Energy Authority (1962).

Skipper, R. G. S., and K. J. Woolton, "Thermal Resistance Between Uranium and Can," *Proc. 2nd U.N. Int. Conf. Peaceful Uses of At. Energy*, **7,** United Nations, Geneva, Switzerland (1958).

Slifer, B., "Loss-of-Coolant Accident and Emergency Core Cooling Models for General Electric BWR's," NEDO-10329, General Electric Company (1971).

Tong, L. S., and J. Weisman, *Thermal Analysis of Pressurized Water Reactors*, American Nuclear Society, La Grange Park, Illinois (1970).

Varga, R. S., *Matrix Iterative Analysis*, Prentice-Hall, Englewood Cliffs, New Jersey (1962).

CHAPTER SEVEN

Boiling Water Reactor Stability Analysis

During the early days of BWR technology, there was considerable concern about nuclear-coupled instability; that is, the interaction between the random boiling process and the void-reactivity feedback modes. This concern caused Argonne National Laboratory (ANL) to conduct a rather extensive series of experiments, which indicated that, while instability was observed at lower pressures, it was not expected to be a problem at the higher system pressures typical of modern BWRs (Kramer, 1958). Indeed, this has proven to be the case in the many operating BWRs in commercial use today. The lack of instability problems due to void-reactivity feedback mechanisms is because BWR void reactivity coefficients ($\partial k/\partial\langle\alpha\rangle$) are several orders of magnitude smaller at 1000 psia than at atmospheric pressure and, thus, only relatively small changes in reactivity are experienced due to void fluctuations. Moreover, modern BWRs use Zircaloy-clad UO_2 fuel pins that have a thermal time constant in the range of 6 to 8 sec and, thus, the change in internal heat generation (i.e., reactivity) due to changes in voids tends to be strongly damped.

7.1 Classification of Thermal-Hydraulic Instabilities

In addition to nuclear-coupled instability, the reactor designer must consider a number of static and dynamic thermal-hydraulic instabilities. The so-called "static" instabilities are explainable in terms of steady-state laws, while explanation of the dynamic instabilities requires the use of the dynamic conservation equations and servo (i.e., feedback control) analysis. Examples of the static instabilities that are normally considered are:

1. flow excursion (Ledinegg) instability
2. flow regime "relaxation" instability
3. geysering or chugging.

Excursive instability was discussed by Ledinegg (1938) and is concerned with the interaction between the pump's head-flow characteristics and the hydraulic characteristics of the boiling channel. This instability mode is discussed in more detail in Sec. 7.2.1.

Flow regime relaxation instability is due to flow regime changes. For instance, if a churn-turbulent flow pattern changes to annular flow due to a small perturbation, a lower pressure drop is experienced, which may tend to increase the system flow rate and, thus, cause the flow regime to return to its previous state. This periodic phenomenon can have an undesirable effect on system performance. In BWR technology, one must guard against the occurrence of this instability mechanism in components such as rod bundles and steam separator standpipes.

Geysering or chugging instability is frequently observed during the reflood phase of BWR emergency core cooling system (ECCS) experiments. That is, when the flooding water rises into the lower part of hot fuel rod assemblies, rapid vapor formation occurs and a two-phase mixture surges up through the fuel assembly. After a short dwell period, this expulsion process is repeated. In this case, the instability mode in question is desirable, because it enhances the cooling characteristics of the flooding mode of the ECCS.

Examples of dynamic instabilities of interest in BWR technology are:

1. density-wave oscillations
2. pressure drop oscillations
3. flow regime-induced instability
4. acoustic instabilities.

In the past decade, the phenomenon of density-wave oscillations has received rather thorough experimental and analytical investigation. This instability mode is due to the feedback and interaction between the various pressure drop components and is caused by the lag introduced due to the finite speed of propagation of kinematic density waves. In BWR technology nuclear coupling occurs, and the reactor should be designed so that it is stable from both the standpoint of parallel channel and system (loop) oscillations. This instability mode obviously is quite important and is examined at length in Sec. 7.2.2.

Pressure drop oscillations (Maulbetsch and Griffith, 1965) is the name given to the instability mode in which Ledinegg-type instability and a compressible volume in the boiling system interact to produce a fairly low-frequency ($\sim$0.1-Hz) oscillation. Although this instability is normally not a problem in modern BWRs, care frequently must be exercised to avoid its occurrence in low-pressure out-of-core thermal-hydraulic test loops.

Another instability mode of interest is due to the flow regime itself. For example, the slug flow regime is periodic and its occurrence in an adiabatic riser can drive a dynamic oscillation (Wallis and Heasley, 1961). In a BWR system, we must guard against this type of instability in components such as steam separation standpipes. The design of the BWR steam separator complex is normally given a full-scale out-of-core proof test to demonstrate both static and dynamic performance and, thus, flow regime related instability is not normally a problem of practical concern.

The final dynamic instability to be considered herein is acoustic instability. This instability occurs when standing waves are excited in a single or two-phase system. For example, steam line resonance and acoustic instabilities in the upper-plenum/steam-separator-complex have been observed in BWRs. The former is a single-phase flow and the latter a two-phase flow. Both can be analyzed using standard "organ pipe" methods. The only difference is that, as discussed in Section 9.2.1.3, in two-phase systems the sonic velocity is a strong function of void fraction.

The static and dynamic instability modes discussed above are of interest in BWR design and operation. However, other instability modes can occur in two-phase systems. Since this monograph is concerned exclusively with BWR technology, we do not consider them here. The interested reader can pursue these topics more fully by reference to an excellent review paper on two-phase instability (Bouré et al., 1971).

7.2 Analysis of Thermal-Hydraulic Instabilities

From the point of view of the reactor designer, the most important instabilities that must be considered are Ledinegg-type flow excursions and density-wave oscillations. In this section, we are concerned with the analysis of these instability modes.

7.2.1 The Analysis of Excursive (Ledinegg) Instability

First we examine the case of nonperiodic flow excursions due to the Ledinegg instability mechanism. Consider a system, having a static pressure drop of Δp_{system}, which has an externally imposed pressure gain, Δp_{ext} (e.g., a pump). A dynamic force balance on the boiling channel yields,

$$\Delta p_{\text{ext}} = \Delta p_{\text{system}} + I\frac{dw}{dt}\ , \tag{7.1}$$

where,

$$I \triangleq \left(\frac{L}{g_c A_{x-s}}\right)\ . \tag{7.2}$$

Equation (7.1) can be linearized for small flow perturbations, δw, to yield,

$$I\frac{d(\delta w)}{dt}+\left[\frac{\partial(\Delta p_{\text{system}})}{\partial w}-\frac{\partial(\Delta p_{\text{ext}})}{\partial w}\right]\delta w=0 \ . \tag{7.3}$$

The general solution of this first-order differential equation is,

$$\delta w(t)=\delta w(0)\left(\exp\left\{-\left[\frac{\partial(\Delta p_{\text{system}})}{\partial w}-\frac{\partial(\Delta p_{\text{ext}})}{\partial w}\right]\frac{t}{I}\right\}\right) \ . \tag{7.4}$$

Thus, this first-order system is stable from the point of view of excursive instabilities if,

$$\frac{\partial(\Delta p_{\text{system}})}{\partial w}>\frac{\partial(\Delta p_{\text{ext}})}{\partial w} \ . \tag{7.5}$$

Various situations of interest are shown in Fig. 7-1. Case 1 is an approximation to the situation that would exist with a positive displacement pump in the external loop. In this case,

$$\frac{\partial(\Delta p_{\text{ext}})}{\partial w}=-\infty \ .$$

Thus, Eq. (7.5) is always satisfied and the system is stable.

Case 2 is an approximation to a parallel channel situation in which a fixed external pressure drop is imposed on the boiling channel. In this case, operation at state 1 is impossible since a small negative flow perturbation drives the channel to operating state 2, while a small positive perturbation drives it to state 3. Clearly, the designer must guard against this phenomenon since flow excursions from state 1 to 2 could cause physical burnout of the heated channel.

Case 3 is typical of the situation that can occur when a centrifugal or jet pump is used in the external circuit of a low-pressure boiling loop. As in Case 2, small negative and positive flow perturbations cause flow excursions from state 1 to states 2′ or 3′, respectively. Again, this is an undesirable situation that must be prevented through appropriate design.

Case 4 is typical of the situation that exists in a low-pressure boiling loop, in which the pump in the external circuit is well throttled. In this case, state 1 is stable since Eq. (7.5) is satisfied and, thus, we would not experience any Ledinegg-type instability.

Note that neither Case 3 or 4 is typical of a modern BWR operating at high pressure. For these operating conditions the hysteresis in the pressure drop/flow curve does not occur.

7.2.2 The Analysis of Density-Wave Oscillations

We now turn our attention to the analysis of so-called "density-wave oscillations." The physical mechanism leading to density-wave oscillations is

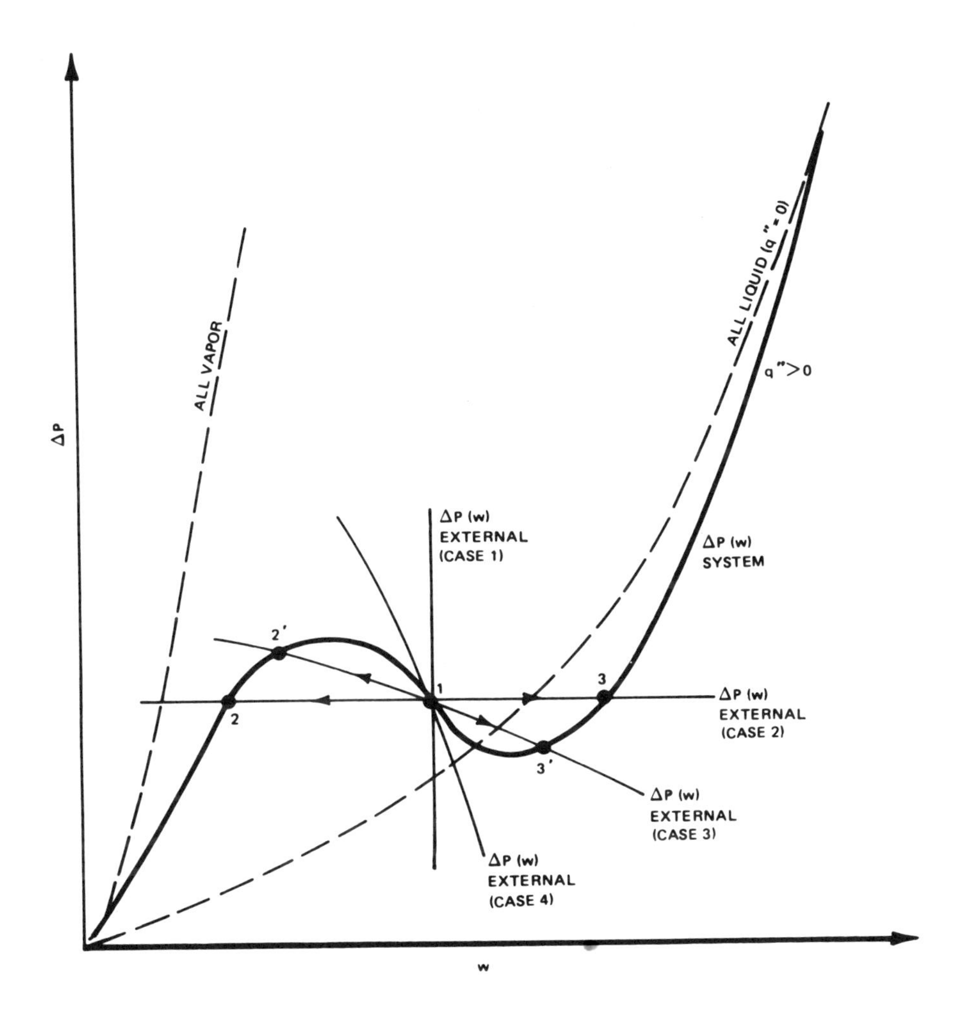

Fig. 7-1 Excursive instability.

now clearly understood and can be described in a number of equivalent ways. The following description closely reflects the essence of the physical phenomena.

Consider an oscillatory subcooled flow entering a heated channel. The inlet flow fluctuations create propagating enthalpy perturbations in the single-phase region. The boiling boundary, $\lambda(t)$, defined as the instantaneous location of the point at which the bulk fluid temperature reaches saturation, oscillates due to these enthalpy perturbations. At the boiling boundary, the enthalpy perturbations are transformed into quality (or void fraction) perturbations that travel up the heated channel with the flow. The combined effects of the flow and void fraction perturbations and variations of the two-phase length create a two-phase pressure drop perturbation. Since the total pressure drop across the boiling channel is imposed externally by the characteristics of the system feeding the channel, the two-phase pressure drop perturbation produces a feedback perturbation of the opposite sign in the single-phase region, which can either reinforce or attenuate the imposed perturbation.

Those familiar with the techniques of linear systems analysis recognize that the phenomenon described above readily lends itself to frequency-domain interpretation. For the benefit of those readers not familiar with classical servo techniques, a brief review of frequency-domain stability criteria is in order. An excellent monograph is available (Weaver, 1963) on this subject; our remarks here are largely motivational.

Linear system instability occurs when the system becomes self-excited. In terms of the block diagram given in Fig. 7-2, self-excitation occurs when the external forcing function, $\delta\hat{\upsilon}_{ext}$, can be removed and the system continues to oscillate in an undamped fashion. This situation can be expressed in analytical form as,

$$\delta\hat{\upsilon}_{ext}[G(s)H(s)] = -\delta\hat{\upsilon}_{ext} \; , \tag{7.6}$$

where $G(s)$ is the so-called "forward loop transfer function" and $H(s)$ is the transfer function of the feedback loop.

Equation (7.6) can be rewritten as,

$$[1+G(s)H(s)]\delta\hat{\upsilon}_{ext}=0 \; . \tag{7.7}$$

Since by definition $\delta\hat{\upsilon}_{ext} \neq 0$, the appropriate frequency-domain stability criterion is,

$$1+G(s)H(s)=0 \; . \tag{7.8}$$

Equation (7.8) gives the zeroes of the characteristic equation, which, as seen in Fig. 7-2, are the poles of the closed-loop transfer function. The

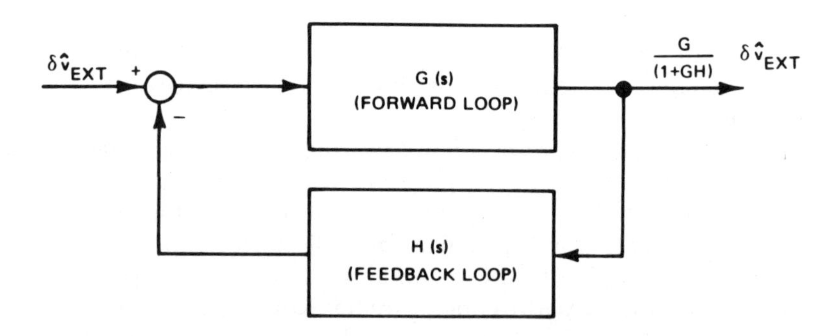

Fig. 7-2 Schematic of a linear negative feedback system.

criterion given in Eq. (7.8) is frequently stated in terms of the Nyquist theorem (Cheng, 1961), which, for the transfer functions associated with BWR stability analysis, says, "A necessary condition for a linear system to be unstable is that the complex locus of the open-loop transfer function (GH) passes through or encircles (in a clockwise fashion) the unity point on the negative real axis."

Before we become too engrossed in the mathematics involved in frequency-domain analysis of density-wave oscillations, note that we can always appraise the stability margin of a boiling system in the time domain through the use of a transient computer code and the application of an appropriate damping criterion. Indeed, this is frequently done in practice. Time-domain solution schemes use the nonlinear conservation equations developed in Chapter 5 and, thus, are theoretically able to predict limit cycle phenomena as well as being of use for general transient analysis. The real motivation for linear frequency-domain stability analysis is that exact solutions are possible, the computer costs are far less, and the threshold of instability can be predicted at least as accurately as with a nonlinear time-domain analysis. Moreover, for those familiar with servo techniques, instability mechanisms are much easier to understand in the frequency domain. Hence, we now investigate the stability of a boiling channel by constructing the appropriate transfer functions, as shown in Fig. 7-3. In this diagram, the subscript, w, refers to flow effects and λ refers to boiling boundary effects.

To simplify the analysis and yet retain the essential physics involved, we consider the case of diabatic homogeneous two-phase flow in a heated

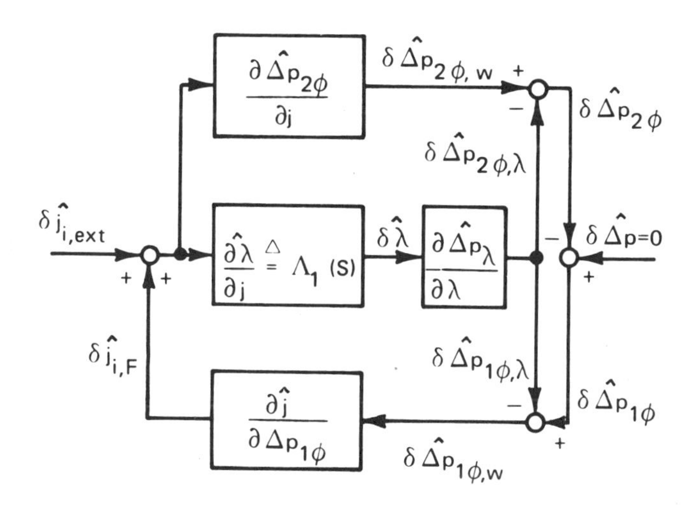

Fig. 7-3 Block diagram of a boiling channel.

channel with constant system pressure, uniform axial heat flux, and no subcooled boiling.

The basic solution scheme consists of coupling analytical models for the single- and two-phase regions with a model for the dynamic thermal response of the heated surface, such that the imposed boundary conditions are satisfied.

In this section, we derive each of these dynamic models. They are then linearized, Laplace-transformed, and combined such that the boundary condition imposed on the momentum equation is satisfied. The resultant system of equations then readily lends itself to a frequency-domain interpretation of the stability margin.

First we consider the model for the heated wall (i.e., fuel rod) dynamics. As discussed previously by Yadigaroglu and Bergles (1971), various levels of sophistication can be incorporated into this model. The simplest model, which still captures the basic physics involved, is that of a single-node lumped heat capacity model. This model can be written as,

$$Mc_{pH}\frac{dT_H}{dt} = V_H q''' - q'' A_{HT} \ . \tag{7.9}$$

Equation (7.9) can be linearized using standard perturbation techniques and Laplace-transformed to convert from the time domain to the frequency domain. The resultant equation is,

$$Mc_{pH} s \delta\hat{T}_H = V_H q_0''' \left(\frac{\delta\hat{q}'''}{q_0'''}\right) - A_{HT} q_0'' \left(\frac{\delta\hat{q}''}{q_0''}\right) \ , \tag{7.10}$$

where s is the Laplace-transform parameter and the transformed variables are denoted by $\hat{}$.

The perturbation in heater wall temperature, $\delta\hat{T}_H$, can be expressed in terms of the heat flux perturbation, $\delta\hat{q}''$. In the single-phase region,

$$q'' = H_{1\phi}(T_H - T_\infty) \ . \tag{7.11}$$

By perturbing and Laplace-transforming,

$$\delta\hat{q}'' = \left(\frac{\delta\hat{H}_{1\phi}}{H_{1\phi 0}}\right) q_0'' + H_{1\phi 0}(\delta\hat{T}_H - \delta\hat{T}_\infty) \ . \tag{7.12}$$

Since the single-phase continuity equation implies $\partial j/\partial z = 0$, the velocity of the single-phase fluid is $j_i(t)$, independent of axial location. Thus, the perturbation in the single-phase heat transfer coefficient, $\delta\hat{H}_{1\phi}$, can be related to the inlet velocity perturbation, $\delta\hat{j}_i$, as,

$$\frac{\delta\hat{H}_{1\phi}}{H_{1\phi 0}} = a\frac{\delta\hat{j}_i}{j_{i0}} \ , \tag{7.13}$$

where the Reynolds number exponent, a, is typically around 0.8. Equations (7.12) and (7.13) can be combined to yield,

$$\left(\frac{\delta\hat{q}''}{q_0''}\right)=a\frac{\delta\hat{j}_i}{j_{i0}}+\frac{H_{1\phi0}}{q_0''}(\delta\hat{T}_H-\delta\hat{T}_\infty) \ . \tag{7.14}$$

It is now convenient to eliminate the heated wall temperature perturbation, $\delta\hat{T}_H$, between Eqs. (7.10) and (7.14) to obtain,

$$\left[\frac{\delta\hat{q}''(s,z)}{q_0''}\right]=\frac{Mc_{pH}s}{(Mc_{pH}s+H_{1\phi0}A_{HT})}\left[a\frac{\delta\hat{j}_i}{j_{i0}}+\frac{H_{1\phi0}A_{HT}}{Mc_{pH}s}\left(\frac{\delta\hat{q}'''}{q_0'''}\right)-\left(\frac{H_{1\phi0}}{c_{pf}q_0''}\right)\delta\hat{h}(s,z)\right] \ , \tag{7.15}$$

which gives the response of the heated wall in the single-phase region to perturbations in inlet velocity, internal heat generation, and fluid enthalpy. In the two-phase region, the relationship is similar. However, it is simpler since it is well known (Jens and Lottes, 1951) that the surface heat flux in the boiling region is not affected by flow and quality perturbations. For out-of-core stability experiments, constant heat flux may be assumed in the two-phase region. In contrast, for nuclear-coupled stability evaluations (in which $\delta q'''$ is nonzero), the assumption of constant heat flux in the boiling region is not valid and may introduce appreciable error.

Let us first consider the analysis of the single-phase region of the heater. The single-phase energy equation can be written,

$$\rho_f\frac{\partial h}{\partial t}+\rho_f j_i\frac{\partial h}{\partial z}=\frac{q''P_H}{A_{x-s}} \ . \tag{7.16}$$

By perturbing and Laplace transforming,

$$\frac{d(\delta\hat{h})}{dz}+\frac{s\delta\hat{h}}{j_{i0}}=\frac{q_0''}{\rho_f j_{i0}}\left(\frac{P_H}{A_{x-s}}\right)\left(\frac{\delta\hat{q}''}{q_0''}-\frac{\delta\hat{j}_i}{j_{i0}}\right) \ , \tag{7.17}$$

where the identity, $(\partial h/\partial z)_0=q_0''P_H/\rho_f j_{i0}A_{x-s}$, has been used. The normalized heat flux variation, $\delta\hat{q}''/q_0''$, can be eliminated by combining Eqs. (7.15) and (7.17) to yield,

$$\frac{d(\delta\hat{h})}{dz}\phi(s)\delta\hat{h}=\theta(s)\left(\frac{\delta\hat{j}_i}{j_{i0}}\right)+\beta(s)\left(\frac{\delta\hat{q}'''}{q_0''}\right) \ , \tag{7.18}$$

where,

$$\phi(s)\stackrel{\Delta}{=}\frac{s}{j_{i0}}+\left[\frac{H_{1\phi0}P_H Mc_{pH}s}{c_{pf}\rho_f j_{i0}A_{x-s}(Mc_{pH}s+H_{1\phi0}A_{HT})}\right] \tag{7.19}$$

$$\theta(s) \triangleq \frac{q_0'' P_H}{\rho_f j_{i_0} A_{x-s}} \left[\frac{a M c_{pH} s}{(M c_{pH} s + H_{1\phi 0} A_{HT})} - 1 \right] \tag{7.20}$$

$$\beta(s) \triangleq \frac{P_H H_{1\phi 0} q_0'' A_{HT}}{\rho_f j_{i_0} A_{x-s}(M c_{pH} s + H_{1\phi 0} A_{HT})} = \frac{P_H H_{1\phi 0} q_0''' V_H}{\rho_f j_{i_0} A_{x-s}(M c_{pH} s + H_{1\phi 0} A_{HT})} \ . \tag{7.21}$$

Equation (7.18) is a Bernoulli-type ordinary differential equation that can be readily integrated from the inlet to some arbitrary point, z. For uniform axial heat flux, we obtain,

$$\delta\hat{h}(s,z) = \exp[-\phi(s)z] \left[\theta(s)\left(\frac{\delta\hat{j}_i}{j_{i_0}}\right) + \beta(s)\left(\frac{\delta\hat{q}'''}{q_0'''}\right) \right] z \left\{ \frac{\exp[\phi(s)z] - 1}{\phi(s)z} \right\}$$

$$+ \exp[-\phi(s)z] h_{i_0} \left(\frac{\delta\hat{h}_i}{h_{i_0}} \right) \ . \tag{7.22}$$

The axial position of interest is the location of the steady-state boiling boundary, $z = \lambda_0$, shown in Fig. 7-4. This location is unambiguously defined

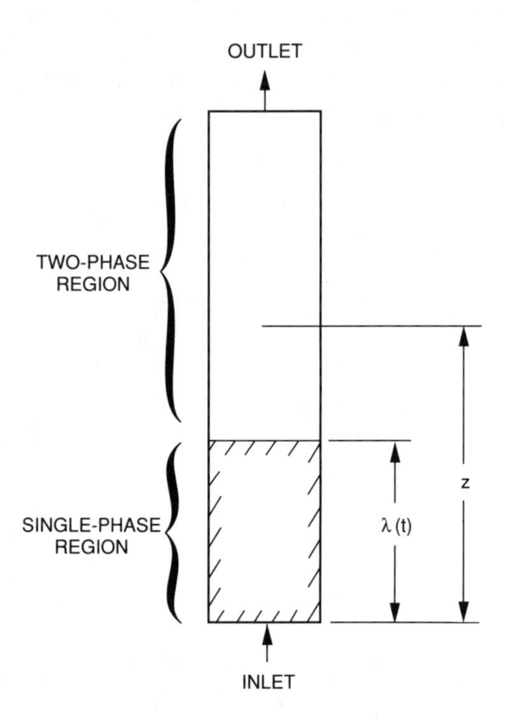

Fig. 7-4 Schematic of a heated section.

as the position at which saturation conditions are achieved, assuming thermodynamic equilibrium. At this location, Eq. (7.22) becomes,

$$\delta\hat{h}(s,\lambda_0)=\left[\theta(s)\frac{\delta\hat{j}_i}{j_{i_0}}+\beta(s)\frac{\delta\hat{q}'''}{q_0'''}\right]\lambda_0\left\{\frac{1-\exp[-\phi(s)\lambda_0]}{\phi(s)\lambda_0}\right\}$$

$$+h_{i_0}\exp[-\phi(s)\lambda_0]\left(\frac{\delta\hat{h}_i}{h_{i_0}}\right) \quad . \tag{7.23}$$

The term, $\{1-\exp[-\phi(s)\lambda_0]\}/\phi(s)\lambda_0$, is a generalized form of the so-called "zero-hold" operator of servo theory. It has the property of delaying and "smearing out" the response of any input perturbation imposed on it. The term, $\exp[-\phi(s)\lambda_0]$, is a generalized form of pure transport delay. For the special case of constant heat flux (i.e., no heater wall dynamics), we can set $M=V_H=0$ to obtain,

$$\delta\hat{h}(s,\lambda_0)=h_{i_0}\exp(-s\lambda_0/j_{i_0})\left(\frac{\delta\hat{h}_i}{h_{i_0}}\right)-\frac{q_0''\lambda_0P_H}{\rho_f j_{i_0}A_{x-s}}\left[\frac{1-\exp(-s\lambda_0/j_{i_0})}{s\lambda_0/j_{i_0}}\right]\left(\frac{\delta\hat{j}_i}{j_{i_0}}\right) \quad . \tag{7.24}$$

Equation (7.24) readily lends itself to phenomenological interpretation. That is, the enthalpy perturbation at the boiling boundary, $\delta\hat{h}(s,\lambda_0)$, is comprised of the delayed inlet enthalpy perturbation, $\delta\hat{h}_i$, and the inlet velocity perturbation, $\delta\hat{j}_i$, times the standard zero-hold operator.

The next step in our derivation is to determine the dynamics of the boiling boundary. For the assumption of uniform axial heat flux, the enthalpy increases linearly with axial distance up the heater. As shown in Fig. 7-5, a positive perturbation in enthalpy at the boiling boundary causes a negative perturbation in the boiling boundary. Hence,

$$\delta\lambda(t)=\frac{h(t,\lambda)-h(t,\lambda_0)}{\left[\dfrac{dh(\lambda_0)}{dz}\right]_0} \quad .$$

Thus,

$$\delta\hat{\lambda}=\frac{-\delta\hat{h}(s,\lambda_0)}{\left[\dfrac{dh(\lambda_0)}{dz}\right]_0}=-\left(\frac{\rho_f A_{x-s}j_{i_0}}{q_0''P_H}\right)\delta\hat{h}(s,\lambda_0) \quad . \tag{7.25}$$

Equations (7.23) and (7.25) can be combined to yield,

$$\delta\hat{\lambda}(s)=\Lambda_1(s)\delta\hat{j}_i+\Lambda_2(s)\delta q'''+\Lambda_3(s)\delta\hat{h}_i \quad , \tag{7.26}$$

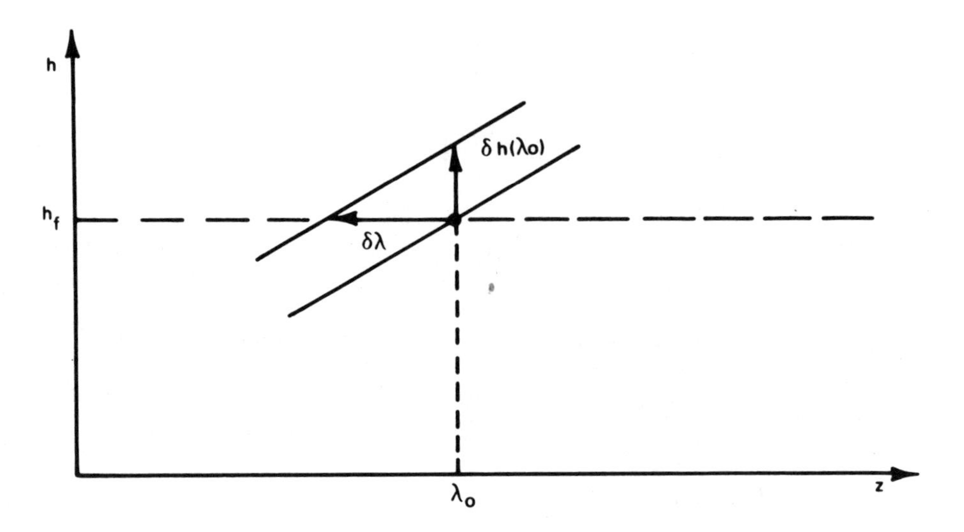

Fig. 7-5 Boiling boundary perturbations.

where the appropriate transfer functions are given as,

$$\Lambda_1(s) \triangleq -\left(\frac{\rho_f A_{x-s}}{q_0'' P_H}\right)\left(\lambda_0 \theta(s)\left\{\frac{1-\exp[-\phi(s)\lambda_0]}{\phi(s)\lambda_0}\right\}\right) \tag{7.27}$$

$$\Lambda_2(s) \triangleq -\left(\frac{\rho_f A_{x-s} j_{i_0}}{q_0'' P_H}\right)\left(\frac{\lambda_0}{q_0'''}\beta(s)\left\{\frac{1-\exp[-\phi(s)\lambda_0]}{\phi(s)\lambda_0}\right\}\right) \tag{7.28}$$

$$\Lambda_3(s) \triangleq -\left(\frac{\rho_f A_{x-s} j_{i_0}}{q_0'' P_H}\right) \exp[-\phi(s)\lambda_0] \ . \tag{7.29}$$

The main purpose of the single-phase analysis has been to derive the expression for the dynamics of the boiling boundary. Equation (7.26) is the required result. We now turn our attention to the two-phase portion of the heater.

For constant system pressure, the continuity equation for a homogeneous two-phase mixture can be obtained from Eqs. (9.19) and (9.23) in the form,

$$\frac{\partial \langle j \rangle}{\partial z} = \Omega(t) \ , \tag{7.30}$$

where,

$$\Omega(t) \triangleq \frac{q''(t) P_H v_{fg}}{A_{x-s} h_{fg}} \ . \tag{7.31}$$

The corresponding energy equation for a homogeneous two-phase system is, from Eq. (9.23),

$$\frac{D_j\langle x\rangle}{Dt} - \Omega(t)\langle x\rangle = \Omega(t)\frac{v_f}{v_{fg}} \quad . \tag{7.32}$$

Gonzalez-Santalo and Lahey (1973) have shown that by neglecting acoustic phenomena, Eqs. (7.30) and (7.32) form a hyperbolic system having two characteristics given by,

$$\frac{dz}{dt} = \langle j(z,t)\rangle \quad , \tag{7.33}$$

and

$$\frac{dt}{dz} = 0 \quad . \tag{7.34}$$

The first characteristic, Eq. (7.33), can be recognized as the equation of the particle paths. The second characteristic, Eq. (7.34), is a family of constant time lines in the space-time plane. This characteristic implies an infinite speed of sound and is a consequence of the assumption of incompressibility. It is well known that a system of hyperbolic partial differential equations can be integrated along their characteristics as if they were ordinary differential equations.

Equation (7.30) can be integrated along a constant time characteristic to yield,

$$\langle j(z,t)\rangle = j_i(t) + \int_{\lambda(t)}^{z} \Omega \, dz' = j_i(t) + \Omega(t)[z - \lambda(t)] \quad , \tag{7.35}$$

where $\lambda(t)$ is the instantaneous position of the boiling boundary (Wallis and Heasley, 1961) and we have allowed for the time-varying heat flux (and thus Ω).

The next step in the analysis is to consider the boundary conditions that must be satisfied by the conservation equations. Typically, the boundary conditions are in terms of pressure drops. Since we have implicitly neglected acoustic effects (thus restricting the analysis to transients of longer duration than the time it takes an acoustic wave to propagate through the heated section), we are able to obtain information from the continuity and energy equations without having to consider the momentum equation. The momentum equation, however, must be considered to complete the analysis, since it is through the momentum equation that the imposed boundary conditions are satisfied.

To simplify the analysis, we first consider the single-phase and then the two-phase momentum equation. These results are then coupled such that the required boundary condition is satisfied.

The appropriate momentum equation for the single-phase portion of the heater can be written,

$$-\frac{\partial p}{\partial z}=\frac{\rho_f}{g_c}\frac{dj_i}{dt}+\frac{f\rho_f j_i^2}{2g_c D_H}+\frac{g}{g_c}\rho_f \ . \tag{7.36}$$

By integrating and perturbing,

$$\delta[\Delta p_{1\phi,H}]=\int_0^{\lambda_0}\left[\frac{\rho_f}{g_c}\frac{d(\delta j_i)}{dt}+\frac{f\rho_f j_{i_0}\delta j_i}{g_c D_H}\right]dz+K_{\text{inlet}}\frac{\rho_f j_{i_0}}{g_c}\delta j_i+\left(\frac{f\rho_f j_{i_0}^2}{2g_c D_H}+\frac{g}{g_c}\rho_f\right)\delta\lambda \ . \tag{7.37}$$

After Laplace-transforming, this equation becomes,

$$\delta[\Delta\hat{p}_{1\phi,H}]=\frac{\rho_f}{g_c}\lambda_0 S\delta\hat{j}_i+\frac{f\rho_f j_{i_0}\lambda_0}{g_c D_H}\delta\hat{j}_i+\sum_{i\in 1\phi}^{N_{1\phi}}K_i\frac{\rho_f j_{i_0}}{g_c}\delta\hat{j}_i+\left(\frac{f\rho_f j_{i_0}^2}{2g_c D_H}+\frac{g}{g_c}\rho_f\right)\delta\hat{\lambda} \ . \tag{7.38}$$

where $\delta\hat{\lambda}$ can be eliminated in favor of $\delta\hat{j}_i$, $\delta\hat{q}'''$, and $\delta\hat{h}_i$ by using Eq. (7.26). Physically, the first term on the right side of Eq. (7.38) represents the inertia term, which has a 90-deg phase lead. The next two terms represent in-phase friction and local single-phase losses, respectively, while the last term represents the dynamic boiling boundary effect. The resultant single-phase pressure drop perturbation transfer function for the heater can be rewritten as,

$$\delta[\Delta\hat{p}_{1\phi,H}]=\Gamma_{1,H}(s)\delta\hat{j}_i+\Gamma_{2,H}(s)\delta q'''+\Gamma_{3,H}(s)\delta\hat{h}_i \ , \tag{7.39}$$

where,

$$\Gamma_{1,H}(s)\triangleq\left[\frac{\rho_f}{g_c}\lambda_0 s+\frac{f\rho_f j_{i_0}\lambda_0}{g_c D_H}+\sum_{k\in 1\phi}^{N_{1\phi}}K_k\frac{\rho_f j_{i_0}}{g_c}+\left(\frac{f\rho_f j_{i_0}^2}{2g_c D_H}+\frac{g}{g_c}\rho_f\right)\Lambda_1(s)\right] \tag{7.40}$$

$$\Gamma_{2,H}(s)\triangleq\left(\frac{f\rho_f j_{i_0}^2}{2g_c D_H}+\frac{g}{g_c}\rho_f\right)\Lambda_2(s) \tag{7.41}$$

$$\Gamma_{3,H}(S)\triangleq\left(\frac{f\rho_f j_{i_0}^2}{2g_c D_H}+\frac{g}{g_c}\rho_f\right)\Lambda_3(s) \ . \tag{7.42}$$

Now we are ready to deal with the two-phase momentum equation. For the case of vertical, homogeneous two-phase flow in a heated channel having constant cross-sectional area, the momentum equation in the two-phase portion of the loop is given by Eq. (5.33) as,

$$-\frac{\partial p}{\partial z}=\frac{\partial G}{\partial t}+\frac{\partial}{\partial z}(G^2/\langle\rho_h\rangle)+\frac{f}{D_H}\frac{G^2}{2\langle\rho_h\rangle}+g\langle\rho_h\rangle \ . \tag{7.43}$$

By integrating, perturbating, and Laplace-transforming,

$$\delta(\Delta\hat{p}_{2\phi,H}) = \int_{\lambda_0}^{L_H} \left\{ \left[(s+\Omega_0)\langle\rho_{h_0}(z)\rangle + \frac{f}{D_H}G_0 \right] \delta\langle\hat{j}(s,z)\rangle \right.$$

$$\left. + \left[\Omega_0\langle j_0(z)\rangle + \frac{f}{2D_H}\langle j_0(z)\rangle^2 + g \right] \delta\langle\hat{\rho}_h(s,z)\rangle + G_0\delta\hat{\Omega}(s) \right\} dz$$

$$- \left(G_0\Omega_0 + \frac{f}{D_H}G_0 j_{i_0} + g\rho_l \right) \delta\hat{\lambda}$$

$$+ \sum_{k\in L_{2\phi}}^{N_{2\phi}} K_k \left[G_0\delta\langle\hat{j}(s,z_k)\rangle + \frac{1}{2}\langle j_0(z_k)\rangle^2 \delta\langle\hat{\rho}_h(s,z_k)\rangle \right] . \tag{7.44}$$

The next step in the analysis is to specify the various terms in the integrand of Eq. (7.44) so that spatial integration can be performed. The first term to be considered is the two-phase velocity perturbation.

Perturbating Eq. (7.35) at a fixed z and Laplace-transforming yields,

$$\delta\langle\hat{j}(s,z)\rangle = (z-\lambda_0)\delta\hat{\Omega}(s) - \Omega_0\delta\hat{\lambda}(s) + \delta j_i(s) . \tag{7.45}$$

Thus, we find that the two-phase velocity perturbation is a linear function of axial position.

The next terms in the integrand of Eq. (7.44), which must be considered, are the axial varying steady-state terms. Equation (7.35) implies,

$$\langle j_0(z)\rangle = \Omega_0(z-\lambda_0) + j_{i_0} . \tag{7.46}$$

Finally, since,

$$\langle\rho_{h_0}(z)\rangle = G_0/\langle j_0(z)\rangle , \tag{7.47}$$

then,

$$\langle\rho_{h_0}(z)\rangle = \frac{\rho_f}{[1+\Omega_0(z-\lambda_0)/j_{i_0}]} . \tag{7.48}$$

Another term of interest is the perturbation in the time constant for phase change, $\delta\hat{\Omega}(s)$. Although this parameter is spatially independent it must be related to one of the state variables of the system, in particular, to $\delta\hat{q}'''$.

Let us begin by taking the transformed perturbation of the boiling heat transfer correlation given in Eq. (4.134),

$$\delta\hat{q}'' = m\kappa'(T_{H_0} - T_{sat})^{m-1}\,\delta\hat{T}_H = m[q_0''/(T_{H_0} - T_{sat})]\delta\hat{T}_H , \tag{7.49}$$

where,

$$\kappa' = \kappa \times 10^6 .$$

Now we note from Eq. (7.10), that for a lumped parameter model of the heated wall,

$$\delta\hat{T}_H = \left[\frac{1}{\rho_H c_{p_H} s}\right]\delta\hat{q}''' - \left[\frac{A_{HT}}{M_H c_{p_H} s}\right]\delta\hat{q}'' \ . \tag{7.50}$$

Thus combining Eqs. (7.49) and (7.50), and using the definition of $\Omega(t)$,

$$\Omega(t) = \left[\frac{q''(t)P_H}{A_{x-s}}\right]\frac{v_{fg}}{h_{fg}} \tag{7.51}$$

to deduce,

$$\delta\hat{\Omega}(s) = \left(\frac{P_H v_{fg}}{A_{x-s} h_{fg}}\right)\delta\hat{q}''(s) \ , \tag{7.52}$$

we obtain,

$$\delta\hat{\Omega} = Z_4(s)\delta\hat{q}''' \ , \tag{7.53}$$

where,

$$Z_4(s) \triangleq \frac{P_H v_{fg}}{A_{x-s} h_{fg}}\left[\frac{1/(\rho_H c_{p_H} s)}{(T_{H_0} - T_{\text{sat}})/(m q_0'') + A_{HT}/(M_H c_{p_H} s)}\right] \ . \tag{7.54}$$

The last term in the integrand of Eq. (7.44) to be determined is $\delta\langle\hat{\rho}_h(s,z)\rangle$. To do so, we can consider the perturbation in flow quality. The perturbed form of the quality propagation equation, Eq. (7.32), is given as,

$$\frac{\partial\delta\langle x\rangle}{\partial t} + \left(\frac{d\langle x\rangle}{dz}\right)_0 \delta\langle j\rangle + \langle j_0\rangle\frac{\partial\delta\langle x\rangle}{\partial z} - \langle x_0\rangle\delta\Omega - \Omega_0\delta\langle x\rangle = \delta\Omega v_f/v_{fg} \ . \tag{7.55}$$

By Laplace-transforming we obtain,

$$\frac{d\delta\langle\hat{x}\rangle}{dz} + \frac{(s-\Omega_0)\delta\langle\hat{x}\rangle}{[j_{L_0} + \Omega_0(z-\lambda_0)]} = \frac{\left(\frac{v_f}{v_{fg}}\delta\hat{\Omega} + \frac{\Omega_0^2}{G_0 v_{fg}}\delta\hat{\lambda} - \frac{\Omega_0}{G_0 v_{fg}}\delta\hat{j}_i\right)}{[j_{i_0} + \Omega_0(z-\lambda_0)]} \ , \tag{7.56}$$

where we note,

$$\frac{\Omega_0}{G_0 v_{fg}} = \frac{q_0'' P_H}{G_0 A_{x-s} h_{fg}} = \left(\frac{d\langle x\rangle}{dz}\right)_0 \ . \tag{7.57}$$

Equation (7.56) is a Bernoulli-type differential equation and has an integrating factor (μ) given by,

$$\mu = [j_{i_0} + \Omega_0(z-\lambda_0)]^{(s/\Omega_0 - 1)} \ . \tag{7.58}$$

The required boundary condition comes from Eq. (7.25),

$$\delta\hat{h} = h_{fg}\delta\langle\hat{x}\rangle = -\frac{\Omega_0}{G_0 v_{fg}}\delta\hat{\lambda} \ . \tag{7.59}$$

Thus, Eq. (7.56) can be integrated subject to Eq. (7.59), yielding,

$$\delta\langle\hat{x}(s,z)\rangle = \frac{\left(\dfrac{v_f}{v_{fg}}\delta\hat{\Omega} + \dfrac{\Omega_0^2}{G_0 v_{fg}}\delta\hat{\lambda} - \dfrac{\Omega_0}{G_0 v_{fg}}\delta\hat{j}_i\right)}{(s-\Omega_0)}[1 - \langle j_0(z)\rangle/j_{i_0}]^{(1-s/\Omega_0)}$$

$$-\frac{\Omega_0}{G_0 v_{fg}}\langle j_0(z)\rangle/j_{i_0}]^{(1-s/\Omega_0)}\delta\hat{\lambda} \ . \tag{7.60}$$

Now we can perturb and Laplace transform Eq. (5.25) to obtain,

$$\delta\langle\hat{\rho}_h\rangle = -\frac{G_0^2 v_{fg}\delta\langle\hat{x}\rangle}{[j_{i_0} + \Omega_0(z-\lambda_0)]^2} = -\frac{G_0^2 v_{fg}\delta\langle\hat{x}\rangle}{\langle j_0(z)\rangle^2} \ . \tag{7.61}$$

Combining Eqs. (7.60) and (7.61) we obtain the result we sought,

$$\delta\langle\hat{\rho}_h(s,z)\rangle = \{s[\langle j_0(z)\rangle/j_{i_0}]^{(1-s/\Omega_0)} - \Omega_0\}\left[\frac{\Omega_0}{(s-\Omega_0)}\right]\frac{G_0}{\langle j_0(z)\rangle^2}\delta\hat{\lambda}$$

$$+\{1 - [\langle j_0(z)\rangle/j_{i_0}]^{1-s/\Omega_0}\}\left[\frac{\Omega_0}{(s-\Omega_0)}\right]\frac{G_0}{\langle j_0(z)\rangle^2}\delta\hat{j}_i$$

$$-\{1 - [\langle j_0(z)\rangle/j_{i_0}]^{1-s/\Omega_0}\}\frac{1}{(s-\Omega_0)}\frac{Gj_{i_0}}{\langle j_0(z)\rangle^2}\delta\hat{\Omega} \ . \tag{7.62}$$

Equation (7.44) can now be integrated. The result is the two-phase pressure drop perturbation in the heater. It is given by:

$$\delta[\hat{p}_{2\phi,H}] = \Pi_{1,H}(s)\delta\hat{j}_i + \Pi_{2,H}(s)\delta\hat{q}''' + \Pi_{3,H}(s)\delta\hat{h}_i \ , \tag{7.63}$$

where,

$$\Pi_{1,H}(s) = G_0[F_1(s) - F_2(s)\Lambda_1(s)] \tag{7.64}$$

$$\Pi_{2,H}(s) = -G_0[F_2(s)\Lambda_2(s) - F_3(s)] \tag{7.65}$$

$$\Pi_{3,H}(s) = -G_0 F_s(s)\Lambda_3(s) \tag{7.66}$$

$$F_1(s) = \frac{s^2\tau_{ex}}{(s-\Omega_0)} + \frac{f(L_H-\lambda_0)}{2D_H}\frac{(2s-\Omega_0)}{(s-\Omega_0)} + \frac{\Omega_0^2}{(s-\Omega_0)^2}\{\exp[(\Omega_0-s)\tau_{ex}]-1\}$$

$$+\frac{fj_{i_{0,0}}}{2D_H}\frac{\Omega_0}{(s-\Omega_0)(s-2\Omega_0)}\{\exp[(2\Omega_0-s)\tau_{ex}]-1\}$$

$$+\frac{g}{j_{i0,0}}\frac{1}{(s-\Omega_0)}\left\{1-\exp(-\Omega_0\tau_{ex})+\frac{\Omega_0}{s}[\exp(-s\tau_{ex})-1]\right\}$$

$$+K_{ex}\left\{1+\frac{\Omega_0}{2(s-\Omega_0)}(1-\exp[(\Omega_0-s)\tau_{ex}])\right\} \tag{7.67}$$

$$F_2(s)=\frac{\Omega_0 s^2\tau_{ex}}{(s-\Omega_0)}+\frac{f(L_H-\lambda_0)}{2D_H}\frac{\Omega_0(2s-\Omega_0)}{s-\Omega_0}+\frac{\Omega_0^2 s}{(s-\Omega_0)^2}\{\exp[(\Omega_0-s)\tau_{ex}]-1\}$$

$$+\frac{fj_{i0,0}}{2D_H}\frac{\Omega_0 s}{(s-\Omega_0)(s-2\Omega_0)}\{\exp[(2\Omega_0-s)\tau_{ex}]-1\}$$

$$+\frac{g}{j_{i0,0}}\frac{\Omega_0}{(s-\Omega_0)}[\exp(-s\tau_{ex})-\exp(-\Omega_0\tau_{ex})]$$

$$+\Omega_0+\frac{fj_{i0,0}}{2D_H}+\frac{g}{j_{i0,0}}-K_{ex}\Omega_0\left\{\frac{1}{2}\exp[(\Omega_0-s)\tau_{ex}]-1\right.$$

$$\left.-\frac{\Omega_0}{2(s-\Omega_0)}[1-\exp[(\Omega_0-s)\tau_{ex}]]\right\} \tag{7.68}$$

$$F_3(s)=\left[\frac{(s+2\Omega_0)}{\Omega_0}(L_H-\lambda_0)-\frac{s^2}{(s-\Omega_0)\Omega_0}j_{i0,0}\tau_{ex}-\frac{\Omega_0 j_{i0,0}}{(s-\Omega_0)^2}[\exp(\Omega_0-s)\tau_{ex}-1]\right.$$

$$+\frac{f}{2D_H}\left((L_H-\lambda_0)^2-\frac{j_{i0,0}}{(s-\Omega_0)}(L_H-\lambda_0)-\frac{j_{i0,0}^2\{\exp[(2\Omega_0-s)\tau_{ex}]-1\}}{(s-\Omega_0)(s-2\Omega_0)}\right.$$

$$\left.-\frac{g}{(s-\Omega_0)}\left\{\frac{1}{\Omega_0}[1-\exp(-\Omega_0\tau_{ex})]-\frac{1}{s}[1-\exp(-s\tau_{ex})]\right\}\right)$$

$$\left.+\left((L_H-\lambda_0)-\frac{j_{i0,0}}{2(s-\Omega_0)}\{1-\exp[\Omega_0-s)\tau_{ex}]\}\right)K_{ex}\right]Z_4(s) \tag{7.69}$$

$$\tau_{ex}=\frac{1}{\Omega_0}\ln\left[\frac{\langle j_0(L_H)\rangle}{j_{i0,0}}\right]. \tag{7.70}$$

Note that the perturbations in density and volumetric flux at the exit of the heated channel can be evaluated from Eqs. (7.45) and (7.62) by substituting $z=L_H$. The result for $\delta\langle\hat{j}(s)\rangle_{H,ex}\overset{\Delta}{=}\delta\langle\hat{j}(s,L_H)\rangle$ is,

$$\delta\langle\hat{j}\rangle_{H,ex}=J_{H,1}(s)\delta\hat{j}_i+J_{H,2}(s)\delta\hat{q}'''+J_{H,3}(s)\delta\hat{h}_i\ , \tag{7.71}$$

where

$$J_{H,1}(s)=1-\Omega_0\Lambda_1(s) \tag{7.72}$$

$$J_{H,2}(s)=L_B Z_4(s)-\Omega_0\Lambda_2(s) \tag{7.73}$$

$$J_{H,3}(s) = -\Omega_0 \Lambda_3(s) \ , \tag{7.74}$$

where L_B is the steady-state length of the boiling region,

$$L_B \triangleq (L_H - \lambda_0) \ . \tag{7.75}$$

Evaluating Eq. (7.62) at $z = L_H$, we obtain,

$$\delta\langle\hat{\rho}_h\rangle_{H,\text{ex}} = R_{H,1}(s)\delta\hat{j}_i + R_{H,2}(s)\delta\hat{q}''' + R_{H,3}(s)\delta\hat{h}_i \ , \tag{7.76}$$

where

$$R_{H,1}(s) = \frac{G_0}{j_{i_0,0}^2}\frac{\Omega_0}{(s-\Omega_0)}\{\exp[-(s-\Omega_0)\tau_{\text{ex}}][s\Lambda_1(s) - 1]$$

$$+ [1 - \Omega_0\Lambda_1(s)]\} \exp(-2\Omega_0\tau_{\text{ex}}) \tag{7.77}$$

$$R_{H,2}(s) = \frac{G_0}{j_{i_0,0}^2}\frac{\Omega_0}{(s-\Omega_0)}\{\exp[(\Omega_0 - s)\tau_{\text{ex}}][s\Lambda_2(s) + (j_{i_0,0}/\Omega_0)Z_4(s)]$$

$$- [\Omega_0\Lambda_2(s) + (j_{i_0,0}/\Omega_0)Z_4(s)]\} \exp(-2\Omega_0\tau_{\text{ex}}) \tag{7.78}$$

$$R_{H,3}(s) = \frac{G_0}{j_{i_0,0}^2}\frac{\Omega_0}{(s-\Omega_0)}\Lambda_3(s)\{s \exp[(\Omega_0 - s)\tau_{\text{ex}} - \Omega_0] \exp(-2\Omega_0\tau_{\text{ex}})\} \ . \tag{7.79}$$

To evaluate the enthalpy perturbation at the exit of the heater, we note that for a homogeneous two-phase flow in thermodynamic equilibrium [see Eq. (7.87)]:

$$\delta\langle\hat{\rho}_h\rangle = -\langle\rho_{h_0}\rangle^2 \frac{v_{fg}}{h_{fg}}\delta\langle\hat{h}\rangle \ . \tag{7.80}$$

Applying Eq. (7.80) to the heater exit, with $\delta\langle\hat{\rho}_h\rangle_{H,\text{ex}}$ given by Eq. (7.76), we obtain,

$$\delta\langle\hat{h}\rangle_{H,\text{ex}} = H_{H,1}(s)\delta\hat{j}_i + H_{H,2}(s)\delta\hat{q}''' + H_{H,3}(s)\delta\hat{h}_i \ , \tag{7.81}$$

where,

$$H_{H,i}(s) = -\left[\langle\rho_{h_0}\rangle_{H,\text{ex}}^2 \frac{v_{fg}}{h_{fg}}\right]^{-1} R_{H,i}(s) \ , \quad \text{for } i = 1, 2, 3 \ . \tag{7.82}$$

Let us now extend the analysis of density-wave oscillations in a heated channel to density-wave oscillations in a recirculation loop. In particular, let us consider the analysis of the recirculation loop of a boiling water nuclear reactor. Loops in which nuclear feedback does not occur (e.g., fossil boilers or electrically heated test loops) will then be a special case of this analysis. To analyze loop oscillations, the model of the heated channel must be combined with similar models for the remaining loop components. For a typical BWR loop, shown schematically in Fig. 7-6, these components

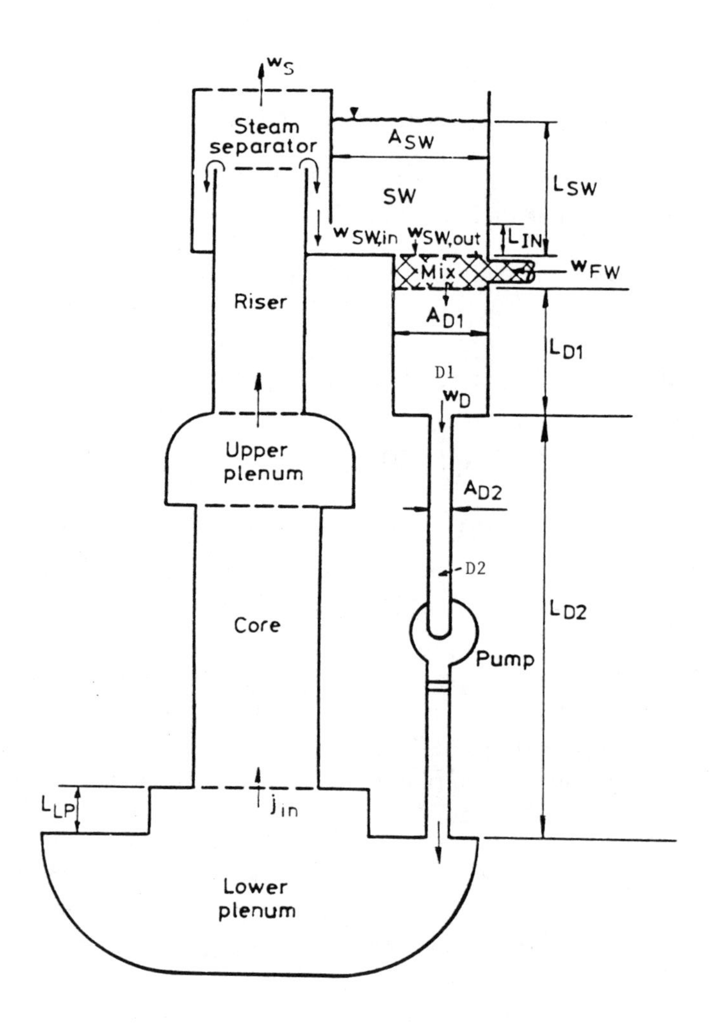

Fig. 7-6 BWR loop schematic.

are the upper plenum, riser and steam separator complex, downcomer, and lower plenum.

The upper plenum has a fairly large cross-sectional area as compared to its height, thus a lumped parameter model can be used. Assuming adiabatic conditions, the mass and energy conservation equations are,

$$V_{\mathrm{UP}}\frac{d}{dt}\langle\rho_h\rangle_{\mathrm{UP}} = w_{\mathrm{UP,in}} - w_{\mathrm{UP,ex}} \tag{7.83}$$

$$V_{\mathrm{UP}}\frac{d}{dt}(\langle\rho_h\rangle_{\mathrm{UP}}\langle h\rangle_{\mathrm{UP}} - p) = w_{\mathrm{UP,in}}\langle h\rangle_{\mathrm{UP,in}} - w_{\mathrm{UP,ex}}\langle h\rangle_{\mathrm{UP,ex}}\ , \tag{7.84}$$

where $\langle\rho_h\rangle_{UP}$ and $\langle h\rangle_{UP}$ are volume-averaged parameters. For a simple homogeneous two-phase flow we have,

$$\langle\rho_h\rangle = \rho_f/(1+\frac{v_{fg}}{v_f}\langle x\rangle) \tag{7.85}$$

$$\langle h\rangle = \langle x\rangle h_{fg} + h_f \ . \tag{7.86}$$

Combining Eqs. (7.85) and (7.86) yields,

$$\langle h\rangle = \frac{h_{fg}}{v_{fg}}\left(\frac{1}{\langle\rho_h\rangle} - \frac{1}{\rho_f}\right) + h_f \ . \tag{7.87}$$

Using a perfect mixing assumption for the upper plenum, we obtain,

$$\langle h\rangle_{UP,ex} = \langle h\rangle_{UP} \ . \tag{7.88}$$

Consequently,

$$\langle x\rangle_{UP,ex} = \langle x\rangle_{UP} \ . \tag{7.89}$$

Thus, Eq. (7.85) yields,

$$\langle\rho_h\rangle_{UP,ex} = \langle\rho_h\rangle_{UP} \ . \tag{7.90}$$

Assuming a constant system pressure, p, and substituting Eqs. (7.88), (7.89) and (7.90) into Eqs. (7.83) and (7.84), perturbing and Laplace-transforming the resultant expression and rearranging yields,

$$(sV_{UP}\langle\rho_{h_0}\rangle_{UP} + w_0)\frac{\delta\langle\hat{\rho}_h\rangle_{UP,ex}}{\langle\rho_{h_0}\rangle_{UP}} + w_0\frac{\delta\langle\hat{j}\rangle_{UP,ex}}{\langle j_0\rangle_{UP,ex}} = w_0\left(\frac{\delta\langle\hat{\rho}_h\rangle_{UP,in}}{\langle\rho_{h_0}\rangle_{UP}} + \frac{\delta\langle\hat{j}\rangle_{UP,in}}{\langle j_0\rangle_{UP,in}}\right) \tag{7.91}$$

$$(sV_{UP}\langle\rho_{h_0}\rangle_{UP} + w_0)\left(1 - \frac{h_{fg}}{\langle h_0\rangle_{UP}\ v_{fg}\langle\rho_{h_0}\rangle_{UP}}\right)\frac{\delta\langle\hat{\rho}_h\rangle_{UP,ex}}{\langle\rho_{h_0}\rangle_{UP}} + w_0\frac{\delta\langle\hat{j}\rangle_{UP,ex}}{\langle j_0\rangle_{UP,ex}} =$$

$$w_0\left[\left(1 - \frac{h_{fg}}{\langle h_0\rangle_{UP}\ v_{fg}\langle\rho_{h_0}\rangle_{UP}}\right)\frac{\delta\langle\hat{\rho}_h\rangle_{UP,in}}{\langle\rho_{h_0}\rangle_{UP}} + \frac{\delta\langle\hat{j}\rangle_{UP,in}}{\langle j_0\rangle_{UP,in}}\right] \ . \tag{7.92}$$

Solving Eqs. (7.91) and (7.92) for $\delta\langle\hat{\rho}_h\rangle_{UP,ex}$ and $\delta\langle\hat{j}\rangle_{UP,ex}$, we obtain,

$$\delta\langle\hat{\rho}_h\rangle_{UP,ex} = (1+\tau_{UP}\ s)^{-1}\delta\langle\hat{\rho}_h\rangle_{UP,in} \tag{7.93}$$

$$\delta\langle\hat{j}\rangle_{UP,ex} = \frac{A_{UP,in}}{A_{UP,ex}}\delta\langle\hat{j}\rangle_{UP,in} \ , \tag{7.94}$$

where τ_{UP} is the (perfect) mixing time constant for the two-phase mixture in the upper plenum, and is given by,

$$\tau_{UP} \triangleq (V_{UP}\langle\rho_{h_0}\rangle_{UP})/w_0 = (M_0)_{UP}/w_0 \ . \tag{7.95}$$

We note that both the density and flow rate at the inlet to the upper plenum are equal to those at the core exit. Thus Eqs. (7.93) and (7.94) can be rewritten as,

$$\delta\langle\hat{\rho}_h\rangle_{UP,ex} = R_{UP}(s)\delta\langle\hat{\rho}_h\rangle_{H,ex} \tag{7.96}$$

$$\delta\langle\hat{j}\rangle_{UP,ex} = J_{UP}\,\delta\langle\hat{j}\rangle_{H,ex}\ , \tag{7.97}$$

where,

$$R_{UP}(s) \triangleq (1+\tau_{UP}\,s)^{-1} \tag{7.98}$$

$$J_{UP} \triangleq A_{x-s}/A_{UP,ex}\ . \tag{7.99}$$

Similarly, the enthalpy perturbation at the exit of the upper plenum can be expressed as,

$$\delta\langle\hat{h}\rangle_{UP,ex} = U_{UP}(s)\delta\langle\hat{h}\rangle_{H,ex}\ , \tag{7.100}$$

where,

$$H_{UP}(s) = R_{UP}(s) = (1+\tau_{UP}\,s)^{-1}\ . \tag{7.101}$$

The perturbation in pressure drop across the upper plenum can be evaluated from Eq. (5.33). Lumping any frictional pressure losses into the pressure drops at the heater exit and riser inlet, we can integrate Eq. (5.33) to obtain,

$$\Delta p_{UP} \triangleq p_{UP,in} - p_{UP,ex} = \left[\frac{d}{dt}\int_0^{L_{UP}} G_{UP}\,dz + \int_0^{L_{UP}} \frac{1}{A_{UP}} \times \frac{\partial}{\partial z}\left(\frac{G_{UP}^2 A_{UP}}{\langle\rho_h\rangle_{UP}}\right) dz + g\int_0^{L_{UP}} \langle\rho_h\rangle_{UP}\,dz\right]\ . \tag{7.102}$$

It is convenient to assume,

$$G_{UP}A_{UP} = w_{UP} = w_{UP,ex} = G_{UP,ex}A_{UP,ex}\ . \tag{7.103}$$

Substituting Eqs. (7.90) and (7.103) into Eq. (7.102), perturbing and Laplace-transforming the resultant expression and rearranging, we obtain,

$$\delta(\Delta\hat{p}_{UP}) = \left[A_{UP,ex}I_{UP}\frac{(G_0)_{UP,ex}}{\langle j_0\rangle_{UP,ex}}s + (1 - A_{UP,ex}^2/A_{UP,in}^2)(G_0)_{UP,ex}\right]\delta\langle\hat{j}\rangle_{UP,ex} + \left[sA_{UP,ex}I_{UP}\langle j_0\rangle_{UP,ex} + \frac{1}{2}\langle j_0\rangle_{UP,ex}^2(1 - A_{UP,ex}^2/A_{UP,in}^2) + gL_{UP}\right]\delta\langle\hat{\rho}_h\rangle_{UP,ex}\ , \tag{7.104}$$

where L_{UP} is the hydraulic inertia of the upper plenum, given by,

$$I_{UP}=\int_0^{L_{UP}}[A_{UP}(z)]^{-1}\,dz\ . \tag{7.105}$$

Using Eqs. (7.87), (7.97), and (7.100), we can rewrite Eq. (7.104) as,

$$\delta(\Delta\hat{p}_{UP})=\Pi_{1,UP}(s)\delta\langle\hat{j}\rangle_{H,ex}+\Pi_{3,UP}(s)\delta\langle\hat{h}\rangle_{H,ex}\ , \tag{7.106}$$

where,

$$\Pi_{1,UP}=G_0\left[(A_{x-s}I_{UP}/\langle j_0\rangle_{UP,ex})s+(1-A^2_{UP,ex}/A^2_{UP,in})\right](A_{x-s}/A_{UP,ex}) \tag{7.107}$$

$$\Pi_{3,UP}(s)=-G_0^2\left[A_{UP,ex}I_{UP}\frac{s}{\langle j_0\rangle_{UP,ex}}+\frac{1}{2}(1-A^2_{UP,ex}/A^2_{UP,in})\right.$$
$$\left.+\frac{gL_{UP}}{\langle j_0\rangle^2_{UP,ex}}\right]\frac{A^2_{x-s}}{A^2_{UP,ex}}\frac{v_{fg}}{h_{fg}}(1+\tau_{UP}s)^{-1}\ . \tag{7.108}$$

Assuming constant system pressure, p, and an adiabatic model for the riser (i.e., setting $\Omega=0$), we obtain from Eq. (7.30),

$$\delta\langle j(\hat{s})\rangle_R=\delta\langle\hat{j}\rangle_{R,in}\ . \tag{7.109}$$

Taking into account that $\langle\rho_{h,0}\rangle_R=\langle\rho_{h,0}\rangle_{R,in}$ is a constant, substituting Eq. (7.109) into Eq. (7.62), and integrating the resultant expression, yields,

$$\delta\langle\hat{\rho}_h(s,z)\rangle_R=\exp[-(s/\langle j_0\rangle_R)z]\delta\langle\hat{\rho}_h(s)\rangle_{R,in}\ . \tag{7.110}$$

Using the following boundary conditions at the upper plenum/riser interface:

$$\delta\langle\hat{\rho}_h(s)\rangle_{R,in}=\delta\langle\hat{\rho}_h(s)\rangle_{UP,ex} \tag{7.111}$$

$$\delta\langle\hat{j}(s)\rangle_{R,in}=(A_{UP,ex}/A_R)\delta\langle\hat{j}(s)\rangle_{UP,ex}\ , \tag{7.112}$$

we have,

$$\delta\langle\hat{\rho}_h\rangle_{R,ex}=R_R(s)\delta\langle\hat{\rho}_h\rangle_{UP,ex} \tag{7.113}$$

$$\delta\langle\hat{j}\rangle_{R,ex}=J_R\delta\langle\hat{j}\rangle_{UP,ex}\ , \tag{7.114}$$

where,

$$R_R(s)=\exp(-s\tau_R) \tag{7.115}$$

$$J_R=A_{UP,ex}/A_R \tag{7.116}$$

$$\tau_R=L_R/\langle j_0\rangle_R\ . \tag{7.117}$$

Also,

$$\delta\langle\hat{h}\rangle_{R,ex}=H_R(s)\delta\langle\hat{h}\rangle_{UP,ex}\ , \tag{7.118}$$

where,

$$H_R(s) = R_R(s) = \exp(-s\tau_R) \ . \tag{7.119}$$

The pressure drop perturbation in the riser, $\delta(\hat{\Delta p})_R$, can be obtained by integrating Eq. (5.33) up the length of the riser and then perturbing it and Laplace-transforming. Including both the inlet and exit losses, and taking into account Eqs. (7.90) and (7.103), we obtain,

$$\delta(\hat{\Delta p})_R = \Pi_{1,R}(s)\delta\langle\hat{j}\rangle_{UP,ex} + \Pi_{3,R}(s)\delta\langle\hat{h}\rangle_{UP,ex} \ , \tag{7.120}$$

where,

$$\Pi_{1,R}(s) \triangleq G_0\left(s\tau_R + f\frac{L_R}{D_{H,R}} + K_{R,in} + K_{R,ex}\right)\frac{A_{x-s}A_{UP,ex}}{A_R^2} \tag{7.121}$$

$$\Pi_{3,R}(s) \triangleq -G_0\left\{\left(\frac{f}{2D_{H,R}}\langle j_0\rangle_R^2 + g\right)\frac{G_0}{\langle j_0\rangle_R s}[1-\exp(-s\tau_R)] + \frac{1}{2}G_0[K_{R,in} + K_{R,ex}\exp(-s\tau_R)]\right\}\frac{v_{fg}}{h_{fg}}\frac{A_{x-s}^2}{A_R^2} \ . \tag{7.122}$$

Note that the hydraulic loss characteristics of the cyclone-type steam separator are contained in $K_{R,ex}$.

As can be seen in Fig. 7-6, the downcomer model can be derived by dividing the total length of the downcomer into four zones: the saturated separated water region, (SW), the mixing region (MIX), the upper downcomer (D1) region, and the lower downcomer (D2) region.

In region SW we can assume an adiabatic model, so that both the inlet and outlet enthalpy are equal to the saturation enthalpy, h_f. In addition, the outlet flow rate from this region, $w_{SW,out}$, will be constant.

In the mixing region (MIX) we can ignore the storage terms in both the mass and energy conservation equations. Thus,

$$w_D = w_{SW,ex} + w_{FW} \tag{7.123}$$

$$w_D h_{D1,in} = w_{SW,ex}h_f + w_{FW}h_{FW} \ . \tag{7.124}$$

Assuming that the inlet feedwater parameters, w_{FW} and h_{FW}, are time independent, Eqs. (7.123) and (7.124) can be perturbed, to yield,

$$\delta\hat{w}_D = \delta\hat{w}_{SW,ex} \tag{7.125}$$

$$\delta\hat{h}_{D1,in} = \frac{(h_f - h_{FW})w_{FW}}{w_{D,0}^2}\delta\hat{w}_{SW,ex} \ . \tag{7.126}$$

Because of the assumed incompressibility of the water, the downcomer flow rate, w_D, does not depend on position in the downcomer. Consequently, the energy equation for the upper regions of the downcomer, D1, can be written as,

$$\rho_f A_{D1}\frac{\partial(h_{D1}-p/\rho_f)}{\partial t}+w_D\frac{\partial h_{D1}}{\partial z}=q''_V P_{H,D1} \;, \tag{7.127}$$

where the heat flux between the vessel wall and liquid is given by Newton's law of cooling as,

$$q''_V = H_{D1}(T_V - T_{D1}) \;. \tag{7.128}$$

The heat flux, q''_V, and vessel wall temperature, T_V, are related through the vessel wall dynamics model. Assuming that the vessel wall's internal heat generation rate is negligible, and ignoring any heat loss to the ambient, Eq. (7.50) reduces to,

$$\delta\hat{T}_V(s,z) = Z_{1V}(s)\delta\hat{q}''_V(s,z) \;, \tag{7.129}$$

where, for a lumped parameter model,

$$Z_{1V}(s) = -A_{HT}/(M_V c_{P_V} s) \;. \tag{7.130}$$

After perturbing and Laplace-transforming Eqs. (7.127) and (7.128), and combining the result with Eq. (7.129), we obtain,

$$\delta\hat{h}_{D1,ex} = \exp[-\phi_{D1}(s)L_{D1}]\delta\hat{h}_{D1,in} \;, \tag{7.131}$$

where,

$$\phi_{D1}(s) \triangleq \left\{\frac{A_{D1}\rho_f}{w_{D,0}}s + \frac{P_{H,D1}H_{D1,0}}{w_{D,0}c_{pf}[1-H_{D1,0}Z_{1V}(s)]}\right\} \;. \tag{7.132}$$

Similarly, for the lower downcomer section, we can write,

$$\delta\hat{h}_{D2,ex} = \exp[-\phi_{D2}(s)L_{D2}]\delta\hat{h}_{D1,ex} \;, \tag{7.133}$$

where $\phi_{D2}(s)$ is given by an expression similar to Eq. (7.132). Substituting Eq. (7.131) into Eq. (7.133) with $\delta\hat{h}_{D1,in}$ given by Eq. (7.126), we obtain,

$$\delta\hat{h}_{D2,ex} = \frac{(h_f - h_{FW})w_{FW}}{w_{D,0}^2}\exp\{-[\phi_{D1}(s)L_{D1} + \phi_{D2}(s)L_{D2}]\}\delta\hat{w}_{SW,ex} \;. \tag{7.134}$$

To evaluate the pressure drop perturbation along the downcomer, we assume that for the steady-state there is no net flow in region SW and integrate the single-phase momentum equation [i.e., Eq. (5.33) with $\langle\rho_h\rangle = \rho_f$] over the component-volumes, SW, D1, and D2, to obtain,

$$\delta(\Delta\hat{p}_{DC}) = \left[I_{DC}s + f\frac{1}{\rho_f}\left(\frac{w_{SW,0}L_{in}}{D_{H,SW}A_{SW}^2} + \frac{w_{D,0}L_{D1}}{D_{H,D1}A_{D1}^2} + \frac{w_{D,0}L_{D2}}{D_{H,D2}A_{D2}^2}\right.\right.$$
$$\left.\left. + \frac{w_{D,0}K_{DC}}{fA_{D2}^2}\right) - \frac{\xi}{\rho_f}\right]\delta\hat{w}_{SW,ex} \;, \tag{7.135}$$

where ξ is the slope of the pump-head curve of the (equivalent) recirculation pump at steady state.

The last element of the recirculation loop is the lower plenum. A lumped parameter model yields the following energy equation,

$$V_{LP}\rho_f \frac{d(h_{LP} - p/\rho_f)}{dt} = w_{LP}(h_{LP,in} - h_{LP,ex}) + q''_{LP} A_{HT,LP} \ , \tag{7.136}$$

where,

$$w_{LP} = w_D \equiv w = A_{x-s} \rho_f j_{in} \tag{7.137}$$

$$q''_{LP} = H_{LP}(T_{W,LP} - T_{LP}) \ . \tag{7.138}$$

As with the upper plenum, we shall assume constant system pressure, p, perfect mixing, and adiabatic steady-state flow. This implies no internal heat generation in the head of the lower plenum, and no heat loss to the ambient. Using these assumptions we have,

$$h_{LP} = h_{LP,ex} = h_{in} \tag{7.139}$$

$$q''_{LP,0} = 0 \ . \tag{7.140}$$

Perturbing and Laplace-transforming Eq. (7.136), combining it with a transfer function (i.e., $Z_{1,LP}$), which is analogous to that in Eq. (7.129), and rearranging yields,

$$\delta \hat{h}_{LP,ex} = H_{LP}(s) \delta \hat{h}_{D2,ex} \ , \tag{7.141}$$

where,

$$H_{LP}(s) \triangleq \left(1 + \tau_{LP} s + \frac{A_{HT,LP} H_{LP,0}}{w_0 c_{pf}} \{1/[1 - H_{LP,0} Z_{1,LP}(s)]\}\right)^{-1} \tag{7.142}$$

and τ_{LP} is the mixing time of the single-phase water in the lower plenum,

$$\tau_{LP} \triangleq \frac{V_{LP}\rho_f}{w_0} = \frac{M_{LP}}{w_0} \ . \tag{7.143}$$

Ignoring the frictional pressure drop, the momentum equation for the lower plenum reduces to,

$$\Delta p_{LP} = g \rho_f L_{LP} \ . \tag{7.144}$$

After perturbing, Eq. (7.144) yields,

$$\delta(\Delta \hat{p}_{LP}) = 0 \ . \tag{7.145}$$

Perturbing Eqs. (7.137) and (7.139), combining them with Eqs. (7.134) and (7.141), and rearranging, we can relate the enthalpy perturbation at the core inlet, $\delta \hat{h}_{in}$, to the inlet velocity perturbation, $\delta \hat{j}_{in}$, as,

$$\delta \hat{h}_{in} = H_{in}(s) \delta \hat{j}_{in} \ , \tag{7.146}$$

where,

$$H_{in}(s) = \frac{(h_f - h_{FW})w_{FW}}{w_0^2} \exp\{-[\phi_{D1}(s)L_{D1} + \phi_{D2}(s)L_{D2}]\} H_{LP}(s) A_{x-s}\rho_f \ . \quad (7.147)$$

Also, Eq. (7.135) can be rewritten as,

$$\delta(\Delta\hat{p}_{DC}) = \Gamma_{1,DC}(s)\delta\hat{j}_{in} \ , \quad (7.148)$$

where,

$$\Gamma_{1,DC}(s) = \left[\rho_f I_{DC} s + f\left(\frac{w_{SW,0} L_{in}}{D_{H,SW} A_{SW}^2} + \frac{w_0 L_{D1}}{D_{H,D1} A_{D1}^2} + \frac{w_0 L_{D2}}{D_{H,D2} A_{D2}^2} + \frac{w_0 K_{DC}}{f A_{D2}^2} \right) - \xi \right] A_{x-s} \quad (7.149)$$

and,

$$I_{DC} = \sum_{i=1}^{N_{DC}} \frac{L_i}{A_i} \ . \quad (7.150)$$

All the necessary component pressure drop perturbations have now been derived, and the hydraulic loop shown in Fig. 7-6 can be closed.

We now have in hand the equations with which to appraise the hydrodynamic stability margins of a boiling system. We are normally interested in appraising the likelihood of density-wave oscillations in a boiling loop (loop oscillations) and/or in the "hot channel" of a heated array (parallel channel oscillations).

Both loop and parallel oscillations must satisfy essentially the same boundary condition. For the parallel channel case in which a constant core-average pressure drop is impressed on the hot channel,

$$\delta(\Delta\hat{p}_{1\phi,H}) + \delta(\Delta\hat{p}_{2\phi,H}) = 0.0 \ , \quad (7.151)$$

where Eqs. (7.39) and (7.63) are used for the single- and two-phase pressure drop perturbations, respectively.

For the loop oscillations case, the static pressure is continuous around the closed loop; thus,

$$\delta(\Delta\hat{p}_{1\phi,L}) + \delta(\Delta\hat{p}_{2\phi,L}) = 0.0 \ , \quad (7.152)$$

where, combining Eqs. (7.146), (7.120), (7.106), (7.81) and (7.71), we have,

$$\delta[\Delta\hat{p}_{1\phi,L}] = \Gamma_{1,L}(s)\delta\hat{j}_i + \Gamma_{2,L}(s)\delta\hat{q}''' \quad (7.153)$$

and,

$$\delta[\Delta\hat{p}_{2\phi,L}] = \Pi_{1,L}(s)\delta\hat{j}_i + \Pi_{2,L}(s)\delta\hat{q}''' \ , \quad (7.154)$$

where,

$$\Gamma_{1,L}(s) = \Gamma_{1,H}(s) + \Gamma_{1,DC}(s) + \Gamma_{3,H}(s)H_{in}(s) \quad (7.155a)$$

$$\Gamma_{2,L}(s)=\Gamma_{2,H}(s) \tag{7.155b}$$

$$\begin{aligned}\Pi_{1,L}(s)=\Pi_{1,H}(s)+\Pi_{3,H}(s)H_{\text{in}}(s)+[\Pi_{1,\text{UP}}(s)+\Pi_{1,\text{R}}(s)J_{\text{UP}}(s)]\\ \times[J_{H,1}(s)+J_{H,3}(s)H_{\text{in}}(s)]+[\Pi_{3,\text{UP}}(s)+\Pi_{3,\text{R}}(s)H_{\text{UP}}(s)]\\ \times[H_{H,1}(s)+H_{H,3}(s)H_{\text{in}}(s)]\end{aligned} \tag{7.156a}$$

$$\begin{aligned}\Pi_{2,L}(s)=\Pi_{2,H}(s)+[\Pi_{1,\text{UP}}(s)+\Pi_{1,\text{R}}(s)J_{\text{UP}}(s)]J_{H,2}(s)\\ +[\Pi_{3,\text{UP}}(s)+\Pi_{3,\text{R}}(s)H_{\text{UP}}(s)]H_{H,2}(s)\ .\end{aligned} \tag{7.156b}$$

The application of these boundary conditions to the equations just derived allows us to construct the characteristic equation of the system for frequency-domain evaluations of stability margins. In fact, Eqs. (7.151) and (7.152) are essentially the stability criterion, $1+GH=0$, for parallel channel and loop oscillations, respectively. Before discussing the block diagram of these coupled equations, let us consider the extension of the analysis required to accommodate nuclear feedback effects.

7.3 Analysis of Void-Reactivity Feedback

In a BWR, the nuclear feedback is largely through void-reactivity coupling. Thus, a parameter of interest, especially in boiling water nuclear reactor stability analysis, when the dynamics of the core thermal hydraulics is coupled with a point-kinetics model for neutronics, is the core-average void perturbation. This is because BWR reactivity feedback effects deal primarily with the perturbation in core-average void fraction, defined as,

$$\delta\langle\hat{\alpha}(s)\rangle_{\text{ave}}=\frac{1}{L_H}\int_0^{L_H}\delta\langle\hat{\alpha}(s,z)\rangle\,dz=\frac{1}{L_H}\int_{\lambda_0}^{L_H}\delta\langle\hat{\alpha}(s,z)\rangle\,dz \tag{7.157}$$

where, for homogeneous flow, $\delta\langle\hat{\alpha}(s,z)\rangle$ can be expressed in terms of $\delta\langle\hat{\rho}_h(s,z)\rangle$ as,

$$\delta\langle\hat{\alpha}(s,z)\rangle=-\frac{1}{\rho_{fg}}\delta\langle\hat{\rho}_h(s,z)\rangle\ . \tag{7.158}$$

Consequently, substituting Eq. (7.158) into Eq. (7.157), with $\delta\hat{\lambda}(s)$, $\delta\langle\hat{j}(s,z)\rangle$, and $\delta\hat{\Omega}(s)$ given by Eqs. (7.26), (7.45), and (7.53), respectively, performing the integration, and rearranging, we obtain,

$$\delta\langle\hat{\alpha}(s)\rangle_{\text{ave}}=T_1(s)\delta\hat{j}_i+T_2(s)\delta\hat{q}'''+T_3(s)\delta\hat{h}_i\ , \tag{7.159}$$

where,

$$\begin{aligned}T_1(s)=\frac{G_0\Omega_0}{(\rho_f-\rho_g)j_{\text{in},0}L_H(s-\Omega_0)}\Bigg\{\frac{1}{s}[1-\exp(-s\tau_{\text{ex}})][1-s\Lambda_1(s)]\\ -\frac{(L_H-\lambda_0)[1-\Omega_0\Lambda_1(s)]}{\Omega_0(L_H-\lambda_0)+j_{\text{in},0}}\Bigg\}\end{aligned} \tag{7.160a}$$

$$T_2(s) = \frac{G_0\Omega_0}{(\rho_f - \rho_g)j_{in,0}L_H(s - \Omega_0)}\left\{ -\frac{1}{s}[1 - \exp(-s\tau_{ex})][(j_{in,0}/\Omega_0)Z_4(s) + s\Lambda_2(s)] + \frac{(L_H - \lambda_0)}{\Omega_0(L_H - \lambda_0) + j_{in,0}}[(j_{in,0}/\Omega_0)Z_4(s) + \Omega_0\Lambda_2(s)]\right\} \quad (7.160b)$$

$$T_3(s) \triangleq \frac{G_0\Omega_0\Lambda_3(s)}{(\rho_f - \rho_g)j_{in,0}L_H(s - \Omega_0)}\left\{\frac{\Omega_0(L_H - \lambda_0)}{\Omega_0(L_H - \lambda_0) + j_{in,0}} - [1 - \exp(-s\tau_{ex})]\right\} . \quad (7.160c)$$

The equations are now in-hand with which to assess nuclear-coupled density-wave instabilities in BWRs.

The reactivity perturbation is obtained by multiplying the core-average void perturbation by a suitable void reactivity coefficient, C_α,

$$\frac{\delta\hat{k}}{k_0} = C_\alpha \frac{\delta\langle\hat{\alpha}\rangle_{ave}}{\langle\alpha_0\rangle_{ave}} , \quad (7.161)$$

and the reactivity is related to the nuclear heat generation through a classical zero-power kinetics transfer function (Weaver, 1963), Φ, such that

$$\delta\hat{q}''' = \Phi(s)\delta\hat{k}_t , \quad (7.162)$$

where,

$$\Phi(s) = q_0'''\left\{s\left[\Lambda + \sum_{i=1}^{6}\frac{\beta_i}{(s + \lambda_i)}\right]\right\}^{-1} \quad (7.163)$$

and

Λ = the neutron generation time
λ_i = the decay constant of delayed neutron precursor i
β_i = delayed neutron fraction of group i.

The transfer functions that we have constructed have been combined in Fig. 7-7 into a compact block diagram for a BWR. The blocks labeled $\underline{T}$, $\underline{\Pi}$, and $\underline{\Gamma}$ represent the matrix (vector) transfer functions tabulated here:

Driving Variable \ Driven Variable	Core-Average Voids [Eqs. (7.160)]	Pressure Drop In Single-Phase Region [Eq. (7.153)]	Pressure Drop In Two-Phase Region [Eq. (7.154)]
Inlet velocity	T_1	$\Gamma_{1,L}$	$\Pi_{1,L}$
Heat generation	T_2	$\Gamma_{2,L}$	$\Pi_{2,L}$

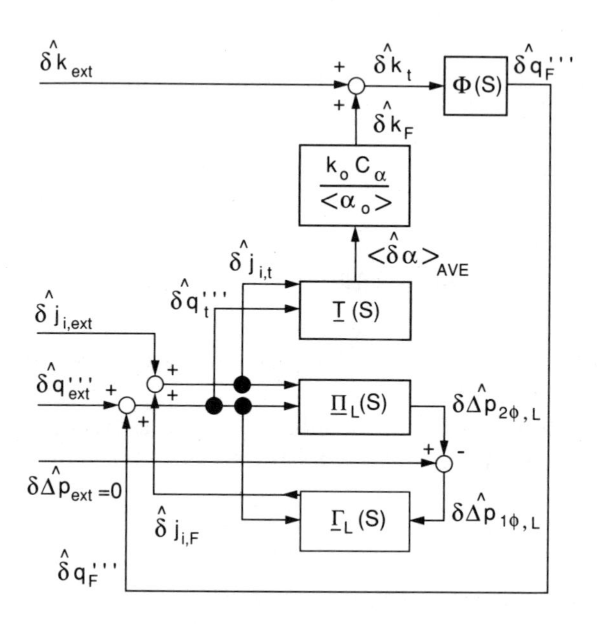

Fig. 7-7 Simplified BWR block diagram.

The stability of the system can be investigated by externally oscillating the inlet velocity, heat generation rate, total pressure drop, or the reactivity in Fig. 7-7, the external perturbation quantities have the subscript, ext; the subscript, F, denotes feedback quantities, and subscript, t, the net total perturbation. Note that in a BWR $\delta\hat{q}'''_{\text{ext}}=0$, while in an electrically heated test loop $\delta\hat{k}_{\text{ext}}=0$, $C_\alpha=0$, and $\Phi(s)=0$.

The system derived is linear and thus has a unique characteristic equation. That is, its stability is independent of the forcing function.

If we choose the state variable vector as,

$$\delta\hat{\underline{x}}_j=(\delta\hat{j}_{i,j}\delta\hat{q}'''_j,\delta\hat{k}_j)^T \ , \tag{7.164}$$

where,

$$j=t,\ F,\ \text{or ext}$$

then the block diagram in Fig. 7-7 can be written in matrix form as,

$$\underline{\underline{A}}(s)\delta\hat{\underline{x}}_t=\underline{\underline{B}}(s)\delta\hat{\underline{x}}_{\text{ext}} \ , \tag{7.165}$$

where,

$$\delta\hat{\underline{x}}_t=\delta\hat{\underline{x}}_F+\delta\hat{\underline{x}}_{\text{ext}} \tag{7.166}$$

$$\underline{\underline{A}}(s)=\begin{bmatrix} [\Gamma_{1,L}(s)+\Pi_{1,L}(s)] & [\Gamma_{2,L}(s)+\Pi_{2,L}(s)] & 0 \\ 0 & 1 & -\Phi(s) \\ T_1(s) & T_2(s) & 1/C_\alpha \end{bmatrix} \tag{7.167a}$$

$$\underline{\underline{B}}(s)=\begin{bmatrix} T_{1,L}(s) & 0 & 0 \\ 0 & 1 & 0 \\ 0 & 0 & 1/C_\alpha \end{bmatrix} . \tag{7.167b}$$

The characteristic equation of the system (i.e., a BWR) is given by,

$$\det[\underline{\underline{A}}(s)]=0 . \tag{7.168}$$

That is,

$$\Phi(s)\{[\Gamma_{1,L}(s)+\Pi_{1,L}(s)]T_2(s)-[\Gamma_{2,L}(s)+\Pi_{2,L}(s)]T_1(s)\} + 1/C_\alpha[\Gamma_{1,L}(s)+\Pi_{1,L}(s)]=0 . \tag{7.169}$$

Hence the stability of a BWR can be appraised by evaluating the roots of Eq. (7.169). That is, if we have roots, s, with positive real parts, then the system is linearly unstable. Alternatively, other servo techniques can be used to assess stability. In particular, the Nyquist locus can be plotted in terms of either:

$$-\frac{\delta\hat{j}_{i,F}}{\delta\hat{j}_{i,t}}, \quad -\frac{\delta\hat{q}'''_F}{\delta\hat{q}'''_t}, \quad -\frac{\delta\hat{k}_F}{\delta\hat{k}_t}, \quad \text{or} \quad -\frac{\delta\Delta\hat{p}_{2\phi}}{\delta\Delta\hat{p}_{1\phi}} .$$

It is also interesting to note that for the evaluation of thermal-hydraulic instabilities one need only set $C_\alpha=\Phi(s)=0$ in Eq. (7.169) to obtain the characteristic equation for a non-nuclear boiling loop or parallel channel array:

$$\Gamma_{1,j}(s)+\Pi_{1,j}(s)=0 , \tag{7.170}$$

where,

$$j=\begin{cases} H, \text{ parallel channels} \\ L, \text{ loop (with no nuclear coupling)} \end{cases} .$$

It should be noted that Eq. (7.170) is often used to evaluate parallel channel instability in BWRs since the nuclear coupling with an individual channel is weak.

For nuclear-coupled stability evaluations, the model given in Fig. 7-7 can be evaluated numerically. For hydrodynamic stability evaluations, nuclear feedback is suppressed by setting $C_\alpha=0$ and $\Phi(s)=0$ and either loop or parallel channel stability margins can be appraised.

The model discussed above can be easily programmed. The resultant digital computer code can be used to gain insight into the mechanism of hydrodynamic instability. To this end, let us investigate the stability of the hot channel of a typical BWR/4. For simplicity, a uniform axial power distribution was assumed. Since under normal operating conditions the reactor is quite stable, it became necessary to examine the pathological case of reduced inlet flow and unrealistically high power generation (50% flow, 355% power) to achieve the threshold of instability. Figure 7-8 shows the frequency variation of several important perturbation phasors in the complex plane. Note that $\delta\Delta\hat{p}_{1\phi,w}$ leads the inlet velocity perturbation because the inertia effects (having a 90-deg lead) are combined with the frictional effects, which are in phase. In contrast, $\delta\Delta\hat{p}_{2\phi,w}$ acquires large lags as the frequency is increased. This effect is due to the lags in the density perturbations caused by the finite speed of the void waves through the channel. At the highest frequency shown ($\omega = 0.956$ rad/sec), the threshold of instability is achieved, as evidenced by the fact that the single- and two-phase pressure drop perturbation phasors become equal and opposite.

This same information can be displayed more compactly in a Nyquist plot, such as that shown in Fig. 7-9. For channel stability the Nyquist locus is seen to cross the negative real axis at an absolute value of unity and, thus, for the pathological conditions investigated, the system is marginally stable.

This model has also been used to generate a stability map for the "hot channel" of a typical BWR/4. The results are given in Fig. 7-10, in which the abscissa is the phase change number (N_{pch}) and the ordinate the subcooling number (N_{sub}). For the higher subcoolings, the stability boundaries tend to follow lines of constant exit quality, $\langle x_e \rangle$. Also note that at lower subcoolings as the inlet subcooling is raised, the stability margins can decrease. Finally, note that for typical BWR/4 operating conditions ($N_{sub} = 0.56$, $N_{pch} = 5.2$, and $\langle x_e \rangle = 0.24$), we are quite far from the threshold of parallel channel density-wave oscillations. This stability margin is designed into BWRs, largely through the use of appropriate inlet orificing to increase the single-phase loss, which stabilizes the system since it is always in phase with inlet velocity perturbations.

The model developed in this chapter is, although simplified, quite capable of predicting the trends given by more exact solutions (Peng et al., 1986). The purpose of this simplified analysis has been to expose the reader to the phenomena and techniques involved without getting too involved in refinements. Those readers interested in a more detailed understanding of BWR stability and dynamic performance are referred to more complete works on these subjects (e.g., Lahey and Podowski, 1989).

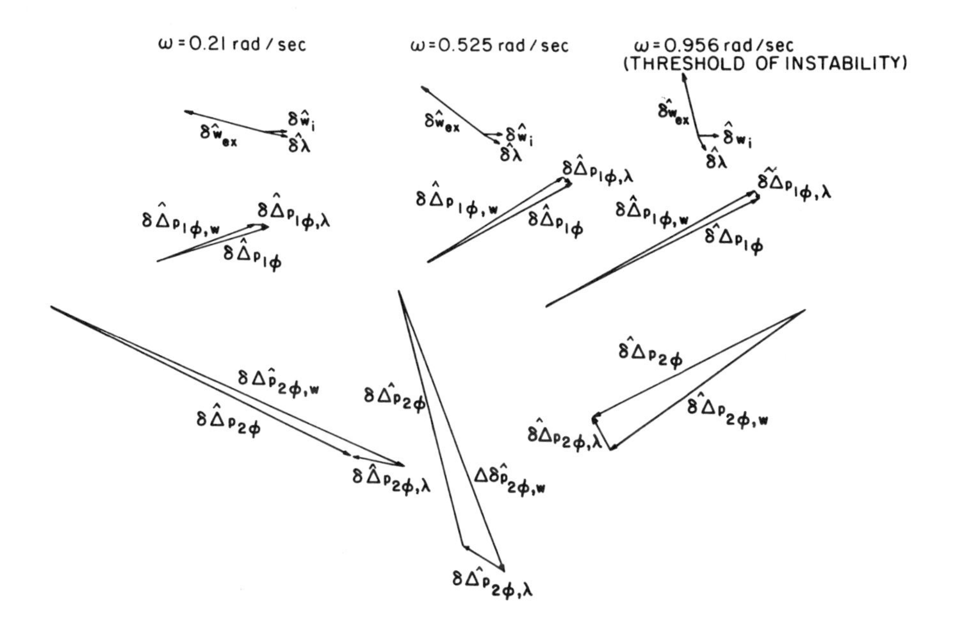

Fig. 7-8 Phasors for a limiting BWR/4 parallel channel.

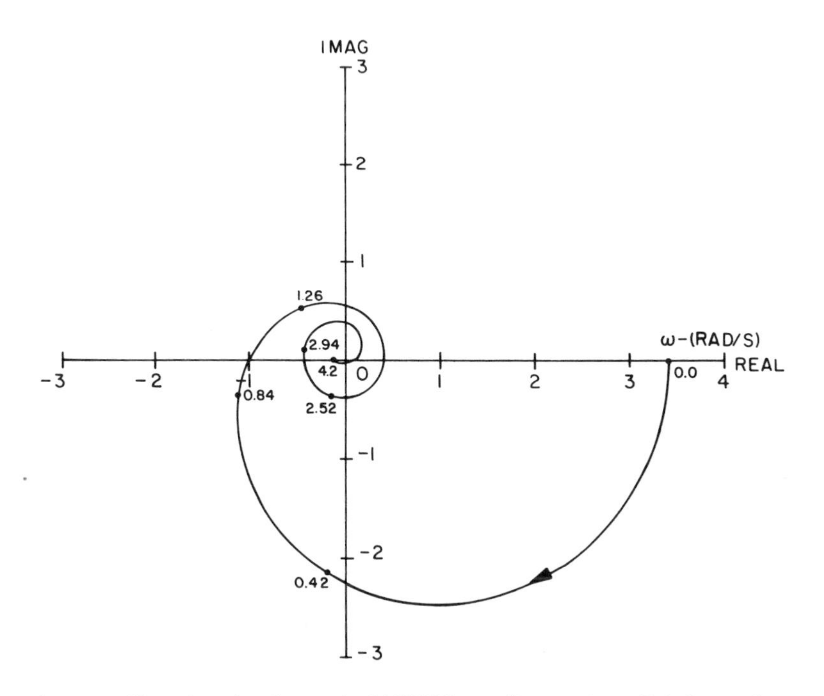

Fig. 7-9 Nyquist plot for typical BWR/4 conditions (parallel channels).

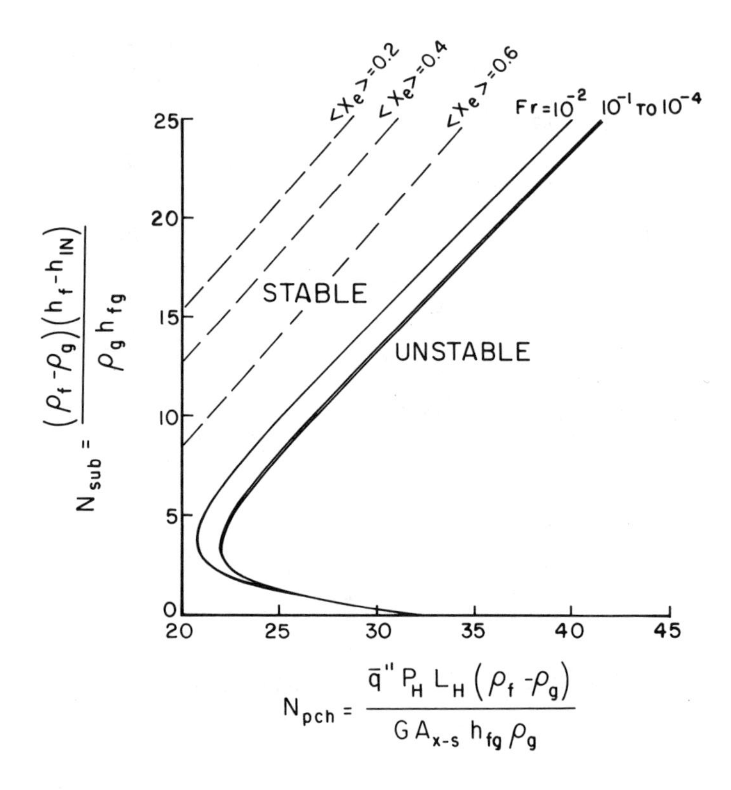

Fig. 7-10 Typical BWR/4 stability map ($K_{in}=27.8$, $K_{exit}=0.14$).

References

Bouré, J. A., A. E. Bergles, and L. S. Tong, "Review of Two-Phase Flow Instability," ASME preprint 71-HT-42, American Society of Mechanical Engineers (1971).

Carmichael, L. A., and G. J. Scatena, "Stability and Dynamic Performance of the General Electric Boiling Water Reactor," APED-5652, General Electric Company (1969).

Cheng, D. K., *Analysis of Linear Systems,* Addison-Wesley Publishing Company, New York (1961).

Gonzalez-Santalo, J. M., and R. T. Lahey, Jr., "An Exact Solution for Flow Transients in Two-Phase Systems by the Method of Characteristics," *J. Heat Transfer,* **95** (1973).

Jens, W. H., and P. A. Lottes, "Analysis of Heat Transfer, Burnout, Pressure Drop and Density Data for High Pressure Water," USAEC Report ANL-4627, U.S. Atomic Energy Commission (1951).

Kramer, A. W., *Boiling Water Reactors,* Addison-Wesley Publishing Company, New York (1958).

Lahey, R. T., Jr., and M. Z. Podowski, "On the Analysis of Various Instabilities in Two-Phase Flows," *Multiphase Science and Technology*, Vol. 4, Hemisphere Publishing Corporation (1989).

Ledinegg, M., "Instability of Flow During Natural and Forced Circulation," *Die Wärme*, **61,** 8 (1938).

Maulbetsch, J. S., and P. Griffith, "A Study of System-Induced Instabilities in Forced-Convection Flows with Subcooled Boiling," MIT Engineering Projects Lab Report 5382-35, Massachusetts Institute of Technology (1965).

Peng, S. J., M. Z. Podowski, and R. T. Lahey, Jr., "BWR Linear Stability Analysis," *J. Nucl. Eng. Des.*, **93** (1986).

Wallis, G. B., and J. H. Heasley, "Oscillations in Two-Phase Flow Systems," *J. Heat Transfer*, **83C** (1961).

Weaver, L. E., *A System Analysis of Nuclear Reactor Dynamics*, American Nuclear Society, La Grange Park, Illinois (1963).

Yadigaroglu, G., and A. E. Bergles, "Fundamental and Higher-Mode Density-Wave Oscillations in Two-Phase Flow: The Importance of the Single-Phase Region," ASME paper 71-HT-13, American Society of Mechanical Engineers (1971).

PART FOUR

Performance of BWR Safety Systems

CHAPTER EIGHT

Postulated Abnormal Operating Conditions, Accidents, and Engineered Safeguards

A nuclear plant is designed not only to operate in a prescribed range of steady-state conditions, but also to undergo successfully changes between different operating conditions without exceeding established design limits. Furthermore, the design must accommodate various expected abnormal conditions. Such abnormal conditions are associated with any deviation from normal conditions anticipated to occur often enough that the design should include a capability to withstand the conditions without operational impairment. Included are transients that may result from any single operator error or control malfunction, transients caused by a fault in a system component requiring its isolation from the system, transients caused by loss of load or power, and any system upset not resulting in a forced outage.

An operator error can result from a deviation from written operating procedures or nuclear plant standard operating practices. A single operator error is the set of actions that is a direct consequence of a single reasonably expected erroneous decision by the reactor plant operator.

In addition to expected abnormal conditions, the BWR design must withstand various postulated accidents. An accident is defined as a single event, not reasonably expected during the course of plant operation, that has been hypothesized for analysis purposes or postulated from unlikely but possible situations and that has the potential to cause a release of an unacceptable amount of radioactive material. Thus, for example, a reactor coolant pressure boundary rupture would be considered an accident, while a fuel cladding defect would not. Another interesting hypothetical accident is the anticipated transient without scram (ATWS). Such accidents have received particular attention since they have a significant potential to produce core damage.

General safety features of the BWR and its containment are designed to handle anticipated and abnormal transients as well as postulated accidents. A safety function is identified for each event and is incorporated into the design of one or more engineered safety systems. An engineered safety feature is a safety system that provides a safety function to mitigate the consequences of accidents that may cause major fuel damage. The purpose of these engineered safety features is to prevent the release of radioactive material in excess of the requirements of the U.S. Nuclear Regulatory Commission (NRC) and to ensure that the radiation exposure to plant personnel does not exceed allowable limits.

Since the accidents at Three Mile Island 2 (TMI-2) and Chernobyl, interest has increased in the evaluation of severe degraded core events in which significant core damage may result. Analyses of this type are normally performed using digital computer models having various levels of sophistication (e.g., MARCH, MAAP, and APRIL).

Due to the complexity of the phenomena involved, and the fact that the field of degraded core analysis is still evolving, none of the models currently available is completely satisfactory. Nevertheless, they can provide useful guidance as to likely scenarios and appropriate steps for accident mitigation. The current state-of-the-art in degraded BWR core analysis will be summarized in Sec. 8.6.

8.1 Single Operator Error or Equipment Malfunction

To assure compliance with the established limits, specific abnormal conditions that could result from a possible operator error or equipment malfunction must be analyzed. The following describes some abnormal conditions that are considered in the design of a BWR:

1. Nuclear reactor system pressure increases caused by
   a. generator trip
   b. turbine trip (with bypass)
   c. loss of condenser vacuum (turbine trip without bypass)
   d. isolation of one main steam line
   e. isolation of all main steam lines.
2. Positive reactivity insertion caused by moderator temperature decrease due to
   a. loss of feedwater heating
   b. inadvertent recirculation pump start.

3. Positive reactivity insertion caused by
   a. control rod withdrawal error
   b. improper fuel assembly insertion or drop
   c. control rod removal or dropout.
4. Loss-of-coolant inventory caused by
   a. pressure regulator failure
   b. inadvertent opening of relief or safety valve
   c. loss of feedwater flow
   d. total loss of off-site power (i.e., loss of pumping power).
5. Core coolant flow decrease caused by
   a. recirculation flow control failure
   b. trip of one recirculation pump
   c. trip of two recirculation pumps
   d. recirculation pump seizure.
6. Positive reactivity insertion caused by core coolant flow increase due to
   a. recirculation flow control failure
   b. inadvertent startup of idle recirculation pump.
7. Core coolant temperature increase caused by feedwater controller failure.
8. Excess of coolant inventory caused by feedwater controller failure.
9. Various ATWS events.

In addition to human (operator) error and equipment malfunctions, other potential causes of these conditions include natural phenomena such as earthquakes, tornadoes, tropical storms and/or hurricanes, floods, drought, excessive rain, ice, and snow. Other potential hazards that must be considered include the plant proximity to airports, ordnance plants, chemical plants, and transportation routes for shipping explosive or corrosive materials.

8.2 Design Basis Loss-of-Coolant Accident

The BWR design basis accident (DBA) of a modern (jet pump) BWR is defined as the instantaneous, double-ended circumferential break on the suction side of a recirculation line, resulting in discharge from both ends of the broken line. This postulated pipe break results in the largest coolant loss rate from the reactor and imposes the most severe thermal transient on the core. The engineered safety systems are designed to provide adequate core cooling to keep radioactive releases and exposure within established limits for the DBA.

8.3 Emergency Core Cooling Systems

Protection against a highly unlikely loss-of-coolant accident (LOCA) is an essential safety feature of all nuclear reactors. The main purpose of the emergency core cooling system (ECCS) is to provide sufficient cooling of the core to prevent gross core meltdown and/or fuel cladding fragmentation, thereby limiting release of radioactive materials and ensuring that the core maintains a coolable geometry. To assure sufficient safety margin and to avoid exceeding cladding fragmentation limits, a maximum cladding temperature criterion of 2200°F has been legislated by the NRC for acceptable ECC system performance. To ensure the adequacy of the ECCS design, it is necessary to determine the important thermal-hydraulic phenomena and to consider the overall effect of each accident on the maximum temperature in the core. We now summarize the important considerations and philosophy employed in the design of current generation BWR ECC systems.

8.3.1 Important Emergency Core Cooling System Design Considerations

The major considerations that contribute to the conservative design of a BWR ECCS are:

1. *Stored Energy:* The assumption of 102% of maximum power, the highest allowed peaking factors, and conservatively calculated thermal resistance between the UO_2 fuel pellets and Zircaloy cladding provides a conservatively high value of the stored thermal energy.
2. *Blowdown Heat Transfer:* Heat transfer during the blowdown phase of a LOCA is also calculated in a very conservative manner. There is strong evidence (Sutherland and Lahey, 1975) that more stored energy would actually be removed during blowdown than current design assumptions permit, resulting in conservatively high calculated cladding temperatures.
3. *Decay Heat:* It is conservatively assumed that the heat generation rate from the radioactive decay of fission products is 20% higher than the proposed American Nuclear Society (ANS) standard (ANS-5 Subcommittee, 1971), and that the fuel has been irradiated at full power for infinite time.
4. *The Peak Clad Temperature Criterion:* The limitation of the calculated peak clad temperature to 2200°F on the hottest fuel rod provides substantial conservatism (e.g., the average temperature rise would be much less than the peak rod's temperature rise) to ensure that the core would suffer little damage in the event of a LOCA.
5. *Blowdown Flow Rate:* There is considerable evidence (Sozzi and Sutherland, 1975) that the critical flow rates assumed are conservative and, thus, the calculated blowdown transient is more rapid than would actually occur.

These conservatisms are incorporated into the following design bases for BWR ECC systems.

8.3.2 Design Bases for BWR Emergency Core Cooling Systems

The overall objective of the BWR ECCS, in conjunction with the containment, is to limit the release of radioactive material following a hypothetical LOCA so that resulting radiation exposures are within the established guidelines. Moreover, the ECCS must be designed to meet its objective even if certain associated equipment is damaged. Ground rules for equipment damage are summarized by the so-called "single-failure" criterion (10CFR50, Appendix-K, 1974) that states, "An analysis of possible failure modes of ECCS equipment and of their effects on ECCS performance must be made. In carrying out the accident evaluation the combination of ECCS subsystems assumed to be operating shall be those available after the worst damaging single failure of ECCS equipment has taken place."

To satisfy the single-failure criterion, evaluation of the BWR LOCA is performed assuming the single active component failure that results in the most severe consequences, and also assuming loss of normal auxiliary power. The combination of ECC subsystems assumed to be operating are those remaining after the component failure has occurred.

The term "active component" means a component in which moving parts must operate to accomplish its safety function. A single failure may involve only one individual component or it may include the failures of several components resulting from the cascading effect of an initial failure. For either a large main steam line or recirculation line break, the worst single failure that can be hypothesized for a BWR/6 is failure of the low-pressure coolant injection (LPCI) diesel-generator (diesel-B, see Fig. 8-1).

Therefore, following the hypothetical LOCA and assuming the most damaging single failure, the BWR ECCS must:

1. Provide adequate core cooling in the event of any size break or leak in any pipe up to and including a double-ended recirculation line break.
2. Remove both residual stored energy and radioactive decay heat from the reactor core at a rate that limits the maximum fuel cladding temperature to a value less than the established limit of 2200°F.
3. Have sufficient capacity, diversity, reliability, and redundancy to cool the reactor core under all accident conditions.
4. Be initiated automatically by conditions that sense the potential inadequacy of normal core cooling.
5. Be capable of startup and operation regardless of availability of off-site power and the normal generation systems of the plant.
6. Operate independently of containment back pressure.
7. Have mechanical components that are designed to withstand transient mechanical loads during LOCA.
8. Have essential components that can withstand such effects as mis-

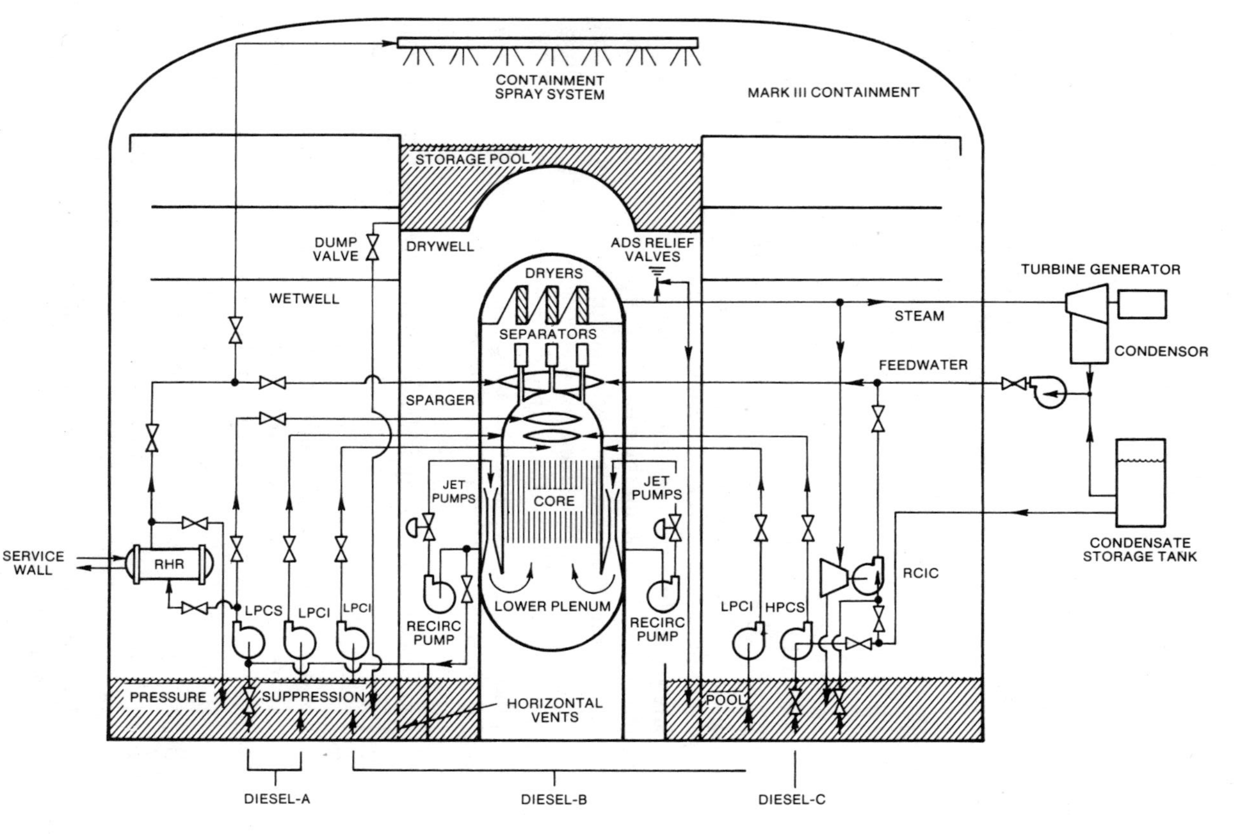

Fig. 8-1 BWR/6 ECCS schematic.

siles (i.e., hurled objects), fluid jets, pipe whip, high temperature, pressure, humidity, and seismic acceleration.

9. Be able to obtain cooling water from a stored source located within the containment barrier.
10. Have flow rate and sensing networks that are testable during normal reactor operation to ensure that all active components are operational.

The BWR/6 ECCS is composed of four separate subsystems: the high-pressure core spray (HPCS) system, the automatic depressurization system (ADS), the low-pressure core spray (LPCS) system, and the LPCI systems. In addition, the reactor core isolation cooling (RCIC) system and the residual heat removal (RHR) system can be used for emergency core cooling. These systems are shown schematically in Fig. 8-1 for a BWR/6 plant with a Mark-III containment.

The HPCS pump obtains suction from the condensate storage tank or the pressure suppression pool. Injection piping enters the vessel near the top of the shroud and feeds two semicircular spargers that are designed to spray water radially over the core into the fuel assemblies. The system functions over the full range of reactor pressure. For smaller breaks, the HPCS cools the core by resubmerging it. For large breaks, it cools the core by spray cooling and reflooding of the lower plenum. The HPCS system is activated by either a low reactor water level signal or a high drywell pressure.

If the HPCS cannot maintain water level or, if HPCS failure occurs, reactor pressure can be reduced by the independent actuation of the ADS so that the flow from the LPCI and LPCS can provide sufficient cooling. The ADS employs pressure relief valves for steam discharge to the pressure suppression pool.

The LPCS pump draws suction flow from the suppression pool and discharges from a circular spray sparger in the top of the reactor vessel above the core. Low water level or high drywell pressure activates this system, which begins injection when reactor pressure is low enough.

The LPCI is an operating mode of the RHR system. It is actuated by low water level or high drywell pressure and, in conjunction with other ECC subsystems, can reflood the core before cladding temperatures reach 2200°F, and, thereafter, maintain a sufficient water level in the core.

The HPCS, ADS, LPCS, and LPCI are designed to accommodate steam line and liquid line breaks of any size. Breaks that impose the most severe demands on the ECCS will be considered in the next section.

The RCIC system is comprised of a steam turbine driven centrifugal pump, which takes suction from the condensate storage tank or the pressure suppression pool. The RCIC system discharges coolant into the feedwater ring in the upper part of the downcomer. The spent steam from the

RCIC turbine is routed into the pressure suppression pool (PSP), where it is condensed.

The LPCI mode is the dominant operating mode of the RHR system. All other modes are submissive to the LPCI mode and will be overridden when ECCS initiation conditions are sensed.

As shown schematically in Fig. 8-1, the RHR may also be manually aligned to accomplish shutdown cooling. In this mode the centrifugal pump takes suction from a recirculation line, passes the coolant through the RHR heat exchangers, and discharges into the reactor pressure vessel (RPV) via the feedwater ring. It is also possible to discharge the coolant through the containment spray system to accomplish cooling and scrubbing. While not shown in Fig. 8-1, the RHR system can also be operated in a steam condensing mode. In this mode, steam from the steam line is passed through the RHR heat exchangers. The resultant condensate is then routed to the suction of the RCIC centrifugal pump and discharged, via the feedwater ring, into the RPV.

It should be clear that there are many ways in which core cooling can be accomplished during normal and emergency conditions.

8.3.2.1 *Steam Line Breaks*

An instantaneous, guillotine severance of the main steam pipe upstream of the steam line flow restrictors is the worst steam line break that can be postulated. Vessel depressurization causes sufficient in-core voids to shut down the reactor, although an automatic control rod scram also occurs to ensure shutdown. About 20 sec after the postulated break, the emergency diesels are running at rated conditions and the HPCS is activated. About 20 sec later, the LPCS and LPCI also start to inject coolant into the vessel. The worst single failure to be assumed, in conjunction with a large steam line break, is failure of the LPCI diesel-generator. Even in this severe case, the core is always submerged and the minimum critical power ratio (MCPR) is always greater than unity. Thus, boiling transition would not be expected and fuel cladding temperatures would not rise.

8.3.2.2 *Liquid Line Breaks*

The double-ended recirculation line break is the DBA for the ECCS. That is, it is the largest pipe break that can be postulated in a BWR. The reactor is assumed to be operating at 102% of rated design steam flow when, as shown in Fig. 8-2, a double-ended circumferential rupture instantly occurs in one of the two recirculation pump suction lines simultaneously with the loss of off-site power. Pump coastdown in the intact loop and natural circulation continue to provide relatively high core flow until the falling water level in the downcomer reaches the elevation of the jet pump suction, as shown in Fig. 8-3. Shortly thereafter, the break discharge flow changes

Fig. 8-2 Hypothetical BWR LOCA event—time of initiation.

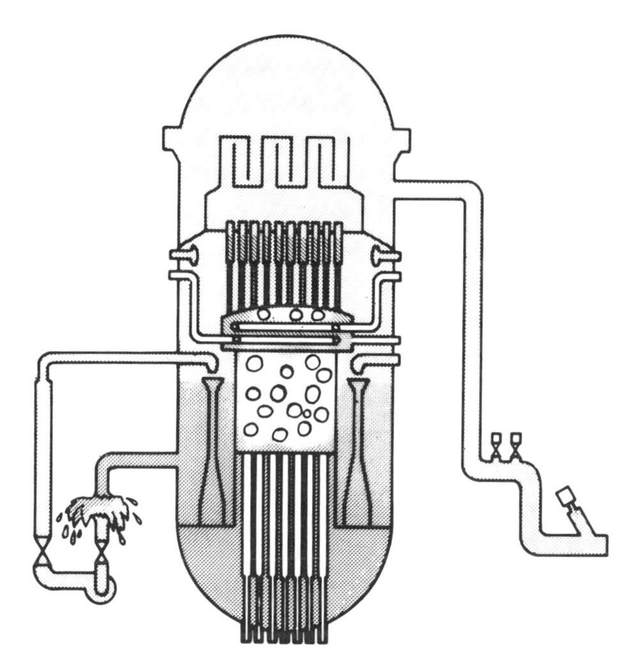

Fig. 8-3 Hypothetical BWR LOCA event—time of jet pump suction uncovery.

to steam and an increased vessel depressurization rate causes vigorous flashing of the residual water in the lower plenum, which in turn forces a two-phase mixture up through the jet pump diffusers and core, resulting in enhanced core cooling.

The HPCS is automatically initiated by low RPV water level or high drywell pressure within ~30 sec after the DBA break occurs. We know (Dix, 1983) that flashing of the liquid in the lower plenum may create a countercurrent flow limitation (CCFL) at the inlet orifices of the fuel assemblies, preventing the two-phase mixture in the assemblies from draining out. In contrast, the interstitial region between the fuel assemblies rapidly fills with HPCS liquid since the CCFL in that region breaks down almost immediately after HPCS activation.

These conditions continue until the two-phase level in the lower plenum falls to the bottom of the jet pumps, and vapor venting through the jet pumps occurs. This reduces the CCFL at the inlet orifices and allows some of the liquid in the fuel assemblies to drain into the lower plenum, uncovering the fuel rods.

Significant parallel channel effects have been found to occur during the subsequent portion of the transient (Dix, 1983; Fakory and Lahey, 1984). As shown schematically in Fig. 8-4, a quasiequilibrium condition can be

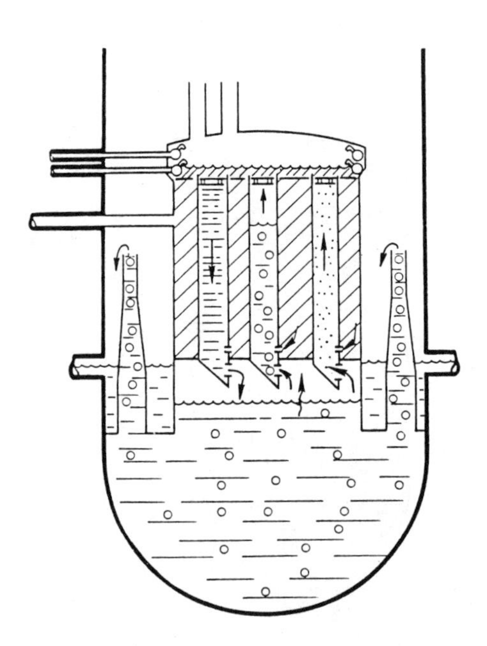

Fig. 8-4 Typical multichannel conditions.

established due to the CCFL at some of the inlet orifices and upper tie plates and the difference in power levels of the various fuel assemblies. In particular, CCFL breakdown at the upper tie plate and inlet orifice of the lower powered peripheral fuel assemblies may occur, leading to a rapid draining of the ECC from the upper plenum to the lower plenum.

In contrast, the average and high-powered assemblies do not normally experience a CCFL breakdown at the upper tie plate or the inlet orifice. Due to differences in the hydrostatic pressure, these fuel assemblies are fed ECC from the interstitial region via the leakage paths in the lower tie plate of the fuel rod bundles (see Fig. 6-13). As shown in Fig. 8-4, the assemblies having average power are cooled by a relatively low-quality two-phase mixture, while the high-powered assemblies are cooled by a high-quality spray flow. As core power delays, CCFL breakdown will occur in all fuel assemblies and they will eventually fill with coolant.

It is significant to note in Fig. 8-4 that reflood from the interstitial region occurs well before bottom reflood from the lower plenum. Moreover, it has been observed (Dix, 1983) that, unless the ECC is near its saturation temperature, the liquid level in the upper plenum does not rise much above the level of the spray header.

While the LOCA reflood scenario just described is quite different from that envisioned in the original design, it does lead to core cooling that is superior (Dix, 1983) to that obtained using Appendix-K licensing assumptions (10CFR50, Appendix-K, 1974).

It is also interesting to note that while break size changes the timing of the events discussed above, a small-break LOCA (SBLOCA) in a BWR responds similarly to a large-break LOCA (Dix, 1983). Indeed, a SBLOCA is normally converted into an equivalent large-break LOCA depressurization through actuation of the ADS.

It is currently conservatively assumed that in licensing calculations only very limited heat transfer occurs subsequent to boiling transition and that no rewetting occurs during lower plenum flashing. However, in spite of the conservative assumptions employed in the licensing analysis of ECCS performance, the calculated peak clad temperature for a BWR/6 is substantially less than 2200°F; so low, in fact, that no perforation of the fuel cladding is calculated to occur.

The analysis used to predict DBA core heatup is discussed in Sec. 8.5, which is concerned with the evaluation of hypothetical BWR accidents.

8.4 Anticipated Transients Without Scram

The category of postulated accidents known as anticipated transients without scram (ATWS) has been extensively studied since the late 1970s. These low-probability accidents are concerned with the occurrence of various anticipated transients coupled with the failure of automatic control rod insertion (i.e., SCRAM). Examples that may lead to severe conditions include a turbine trip without bypass or an event involving the closure of the main steam isolation valves (MSIVs). Both of these events lead to pressurization of the reactor and, due to the resultant void collapse in the core, a power excursion.

While the recommended operator actions for ATWS events have not yet been finalized, current mitigation procedures are based on interim emergency procedure guidelines (EPGs), which are symptom oriented. In particular, various operator actions are recommended to reduce core power and to prevent overpressurization of the containment due to loss of condensing capability (i.e., heatup) of the pressure suppression pool. Typical actions include recirculation pump trip (automatic), manual insertion of the individual control rods, and, if necessary, injection of a sodium pentaborate solution into the lower plenum with the standby liquid control system (SLCS), the reduction of the water level in the downcomer to the top of the active fuel (TAF), and cooling of the pressure suppression pool with the RHR system. Most of these actions are intended to reduce the reactivity in the core through either flow control, which results in a void fraction increase in the core, or (providing mixing and convective transport are adequate) the introduction of a neutron-absorbing solution into the core region via SLCS injection into the lower plenum.

The ATWS rule established under 10CFR50.62 requires hardware criteria for ATWS compliance. BWRs are required to have boron injection equivalent in control capacity to 86 gal/min of 13% sodium pentaborate solution. The statement of considerations for the rule is based on analysis demonstrating that all design criteria can be met with this boron injection capacity. Specifically, this means that the suppression pool temperature remains below the quencher condensation stability temperature limit with the high-pressure makeup systems (RCIC/HPCI) available and without operator action to reduce the downcomer water level.

The operator is instructed to activate boron injection into the reactor vessel if the suppression pool temperature reaches 110°F and there is evidence of no scram based on APRM power and SRV opening. The operator is also instructed to drop the downcomer water level to the top of the active fuel to reduce the power level. Reducing the water level is not normally a desirable action because it could challenge core coverage and may increase the potential for core flux oscillations. Thus, the issue is whether this step could be avoided by taking credit for faster boron delivery to the core and a lower integrated energy release to the containment. The operator's actions should be viewed in the context of the primary requirement to maintain containment integrity. The philosophy of the EPGs requires the operator to have a series of contingency actions in the event of equipment failure. In the extremely unlikely event that the boron injection fails, the operator has the option of depressurizing the vessel to reduce power. The heat capacity temperature limit (HCTL) is calculated for the pool such that the vessel blowdown energy can be accommodated without exceeding the suppression pool quencher condensation stability temperature limit. If the HCTL is reached, the operator is expected to depressurize the system. When boron is available, the hot shutdown condition should be achieved before the pool reaches HCTL to avoid an unnecessary blowdown. The energy required to heat the pool from 110°F (boron initiation) to the HCTL (typically, about 150°F) is of the order of 1.5 full-power minutes. Even if perfect mixing of boron with the vessel contents is assumed, there would not be sufficient time to reach the hot shutdown condition before HCTL is reached unless the downcomer water level is reduced. This action reduces the power level, from about 30% to about 20% of rated, before the boron starts taking effect. The shutdown rate is currently limited by the boron injection rate rather than mixing efficiency.

The question of pressure control is also a complicated one. As an example, during the most limiting ATWS events, in which the SLCS is inoperative, system pressure is predicted to increase, and safety relief valves (SRV) open. The SRV discharge is routed to the pressure suppression pool (PSP), which will, in turn, heat up due to the discharge of the steam from the reactor pressure vessel. As the discharge continues to the PSP the mass of liquid in the reactor may become depleted until the lo-lo level, shown in Fig. 8-5, is reached. At this point the HPCI and RCIC systems are activated. If unthrottled, these systems can lead to an operating state more severe than the previous one (i.e., higher core power). Moreover, due to intermittent SRV actuation, the water level in the RPV may continue to fall and the PSP may continue to heat up until condensation capability is lost, thus leading to excessive hydraulic loads and containment overpressurization. If the lo-lo-lo level, shown in Fig. 8-5, is reached and there is a high drywell pressure, the automatic depressurization system (ADS) will be actuated. This system will blow the reactor down to the PSP. Once the shutoff heads of the low-pressure ECCS and the condensate pumps are reached, they will begin to inject water into the core, resulting in a large positive reactivity insertion.

Clearly the above scenario is quite undesirable since it may lead to containment overpressurization. The current EPGs recommend either early manual ADS actuation (if the SLCS is inoperative) and thus blowdown, coupled with well-throttled ECCS, to increase voids and thus reduce power level, or, a manual reduction of the water level in the downcomer to the top of the active fuel level, coupled with ECCS throttling to control core power and RHR system cooling of the PSP. While there are a number of potential problems with both of these techniques (e.g., core stability), they appear to be capable of mitigating ATWS events if prompt and proper operator action is taken (Dallman et al., 1987; Anderson et al., 1987).

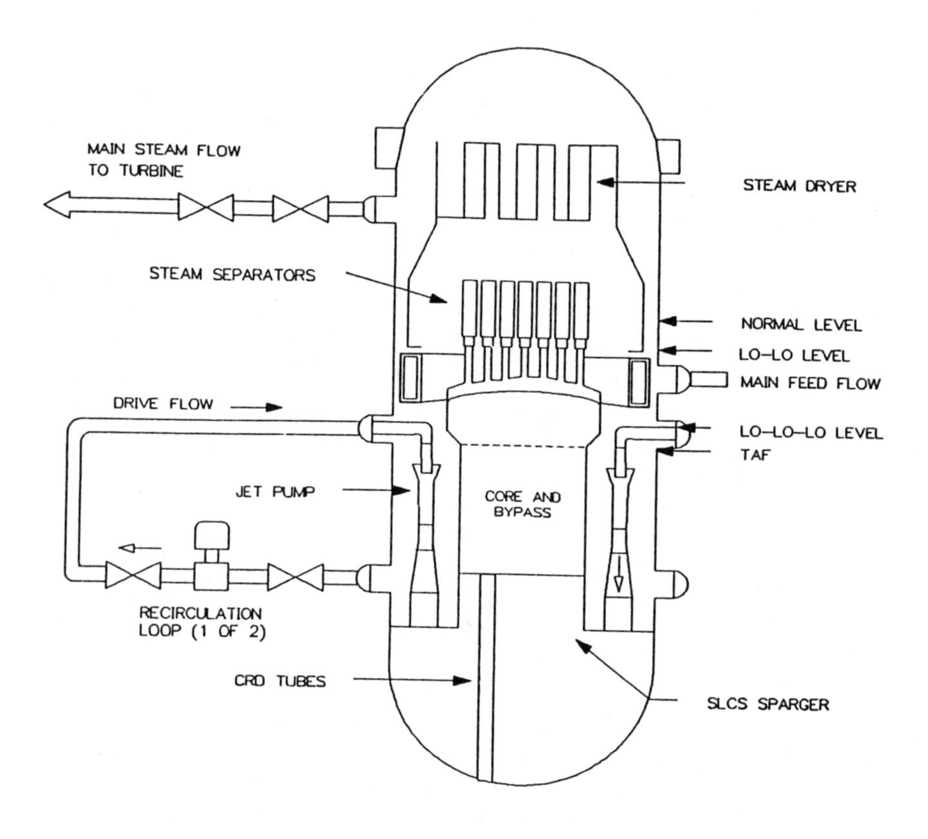

Fig. 8-5 Reactor pressure vessel.

8.5 Core Heatup

The analysis of a hypothetical BWR LOCA requires the consideration of heat transfer mechanisms that exist during the core heatup period. Of particular interest are those heat transfer mechanisms that are dominant during ECCS operation. These mechanisms are characterized by radiant heat transfer phenomena, the rewetting of hot surfaces, and liquid droplet evaporation in the superheated steam.

In this section, we try to summarize the essence of the phenomena that may occur during core heatup. A more in-depth treatment is found in the work of Andersen (1973).

8.5.1 Radiation Heat Transfer in Rod Bundles

Several definitions and identities are used frequently in the radiation heat transfer analysis of gray surfaces and medium. It is assumed that the reader is familiar with classical radiation heat transfer and, thus, these relations are considered only briefly here. The reader unfamiliar with these relations should consult a standard text on radiation heat transfer for clarification; e.g., Sparrow and Cess (1967).

The fraction of radiation in a transparent medium that leaves a uniformly distributed, diffusely emitting surface, i, and is intercepted by surface j, is normally denoted by F_{i-j} and frequently is called the view (or angle) factor. Mathematically, the view factor, which is really just the line-of-sight between the two surfaces, is given by,

$$F_{i-j}=\frac{1}{A_i}\int_{A_i}\int_{A_j}\frac{\cos\beta_i\ \cos\beta_j dA_i dA_j}{\pi r^2} \ . \tag{8.1}$$

Here, F_{i-j} has two important properties. One property is,

$$\sum_{j=1}^{N} F_{i-j}=1 \ , \tag{8.2}$$

where N denotes the number of surfaces in the enclosure. The other is known as reciprocity,

$$A_i F_{i-j}=A_j F_{j-i} \ . \tag{8.3}$$

Another useful relation in radiation heat transfer is that the sum of the reflectivity, ρ_k, the absorptivity, α_k, and the transmissivity, τ_k, of body, k, must equal unity. That is,

$$\rho_k+\alpha_k+\tau_k=1 \ . \tag{8.4}$$

In addition, it is well known that Kirchhoff's law implies that the absorptivity, α_k, and the emissivity, ε_k, are equal,

$$\alpha_k=\varepsilon_k \ . \tag{8.5}$$

We now consider the components of the radiant energy flux from surface i. Figure 8-6 is the diagram of surface i, which is taken to be a gray, opaque ($\tau_i=0$) body. Here, H_i denotes the radiant energy flux incident on surface i. The radiant flux emitted from surface i is given by $\varepsilon_i\sigma T_i^4$, and that portion of the radiant energy that is diffusively[a] reflected is given by $\rho_i H_i$ or, by using Eqs. (8.4) and (8.5), $(1-\varepsilon_i)H_i$.

The radiosity, B_i, at surface i is defined as the net radiant energy flux leaving surface i,

$$B_i \triangleq \varepsilon_i\sigma T_i^4+(1-\varepsilon_i)H_i \ . \tag{8.6}$$

[a]The inclusion of specular reflectance improves the modeling, but greatly complicates the resultant analysis and is not considered here.

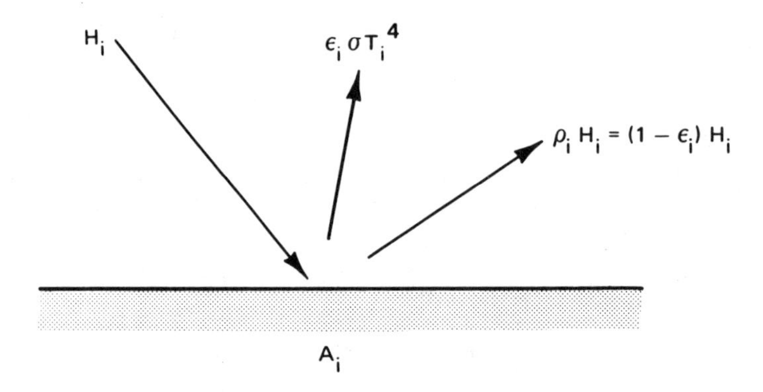

Fig. 8-6 The components of radiation at surface, i.

Now consider the case of a homogeneous mixture of a gray gas and liquid droplets in a gray enclosure consisting of the fuel rods and the channel. If we employ Eq. (8.3), assume that the radiosity is uniform over the individual surfaces, and that all reflections are diffuse, the incident radiation flux is given by,

$$H_i = \sum_{j=1}^{N} B_j F_{i-j} \tau_M + \varepsilon_v \sigma \langle T_v \rangle^4 \tau_l F_{i-v} + \varepsilon_l \sigma \langle T_l \rangle^4 \tau_v F_{i-l} \ . \tag{8.7}$$

The first term on the right side of Eq. (8.7) represents the sum of the radiant energy flux from all surfaces of the enclosure, attenuated by the average transmissivity of the medium, τ_M, due to absorption in the medium. The second and third terms on the right side of Eq. (8.7) represent reemission by the vapor and liquid phases, respectively. Since the situation of interest in BWR LOCA analysis is optically thin (Sun et al., 1975), to a good first-order approximation,

$$F_{i-v} = F_{i-l} = 1.0 \tag{8.8}$$

$$\tau_l = 1 - \varepsilon_l \tag{8.9}$$

$$\tau_v = 1 - \varepsilon_v \tag{8.10}$$

$$\tau_M = 1 - \varepsilon_v \tau_l - \varepsilon_l \tau_v \ . \tag{8.11}$$

The term $\varepsilon_v \tau_l$ can be thought of as the net absorption by the vapor phase and $\varepsilon_l \tau_v$ can be thought of as the net absorption by the liquid phase. Note that the absorption coefficient for a single-phase medium (only) has been reduced by the transmissivity of the other phase. In a more exact analysis, the absorption coefficient would be a function of the beam length between surfaces i and j. Here, for simplicity, we use equivalent average coefficients based on a mean beam length, given by (Siegel and Howell, 1972),

$$L_M \triangleq \frac{4V}{A_{HT}} = D_H \ . \tag{8.12}$$

The emissivity for the vapor and liquid are given by

$$\varepsilon_v = 1 - \exp(-a_v L_M) \ , \tag{8.13}$$

where a_v is the Plank mean absorption coefficient (Sparrow and Cess, 1967) and,

$$\varepsilon_l = 1 - \exp(-a_l L_M) \ , \tag{8.14}$$

where a_l can be expressed in terms of the droplet size and number density (Sun et al., 1975).

It is informative to combine Eqs. (8.6) and (8.7) to yield,

$$B_i = \varepsilon_i \sigma T_i^4 + (1 - \varepsilon_i)\left(\sum_{j=1}^{N} B_j F_{i-j}\tau_M + \varepsilon_v \sigma \langle T_v \rangle^4 \tau_l + \varepsilon_l \sigma \langle T_l \rangle^4 \tau_v\right) \ . \tag{8.15}$$

Now, by noting the identity,

$$B_i = \varepsilon_i B_i + (1 - \varepsilon_i) B_i \ ,$$

and by using Eq. (8.2), Eq. (8.15) can be rewritten as,

$$(\sigma T_i^4 - B_i)\varepsilon_i/(1 - \varepsilon_i) = \sum_{j=1}^{N} (B_i - B_j) F_{i-j}\tau_M + \varepsilon_v \tau_l (B_i - \sigma\langle T_v \rangle^4) + \varepsilon_l \tau_v (B_i - \sigma \langle T_l \rangle^4) \ , \tag{8.16}$$

which can be recognized as a statement of Kirchhoff's "current" law for node i.

Equation (8.16) can be thought of as an equivalent electrical network. Before drawing the network, however, we need to complete the circuit between the liquid and vapor phases. This can be accomplished by writing down an expression for the net radiant heat flux from the vapor to the liquid phase,

$$q''_{v-il} = \varepsilon_v \sigma \langle T_v \rangle^4 \varepsilon_l - \varepsilon_l \sigma \langle T_l \rangle^4 \varepsilon_v \ . \tag{8.17}$$

The first term on the right side of Eq. (8.17) is the product of the radiant energy emitted by the vapor phase times the fraction absorbed by the liquid phase. The second term is just the converse statement for the liquid phase.

Equation (8.17) can be rewritten in terms of an equivalent electrical resistance as,

$$q''_{v-l} = (\sigma \langle T_v \rangle^4 - \sigma \langle T_l \rangle^4)/(\varepsilon_v \varepsilon_l)^{-1} \ . \tag{8.18}$$

Equations (8.16) and (8.18) can now be used to construct Fig. 8-7, the electrical analog for radiation heat transfer in a fuel rod bundle.

Normally it is convenient to recast the equations of radiative heat transfer into matrix form. To do so, we consider the net radiant heat loss from surface i,

$$q''_{Ri} = B_i - H_i = \varepsilon_i \sigma T_i^4 - \varepsilon_i H_i \ . \tag{8.19}$$

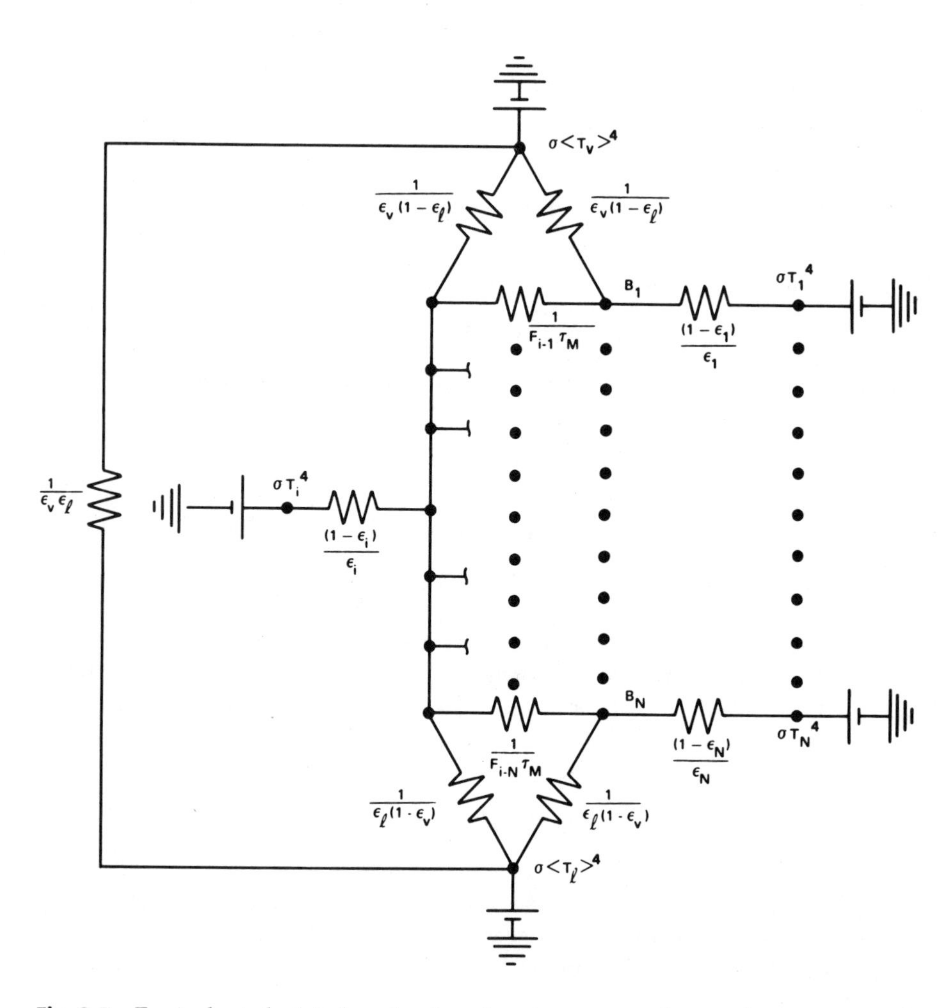

Fig. 8-7 Equivalent electrical analog for a homogeneous mixture of a gray gas and liquid droplets in a gray enclosure (rod bundle).

By combining this equation with Eq. (8.6) to eliminate H_i,

$$q''_{Ri} = \frac{\varepsilon_i}{(1-\varepsilon_i)}(\sigma T_i^4 - B_i) \ . \tag{8.20}$$

We now explore expressions for the vector, $\mathbf{B}_i$.

Equation (8.15) can be rewritten as,

$$\sigma T_i^4 = \frac{1}{\varepsilon_i}\left[\mathbf{B}_i - (1-\varepsilon_i)\left(\sum_{j=1}^{N}\mathbf{B}_j F_{i-j}\tau_M + \varepsilon_v \sigma \langle T_v\rangle^4 \tau_l + \varepsilon_l \sigma \langle T_l\rangle^4 \tau_v\right)\right] \ , \tag{8.21}$$

or in matrix form as,

$$X\mathbf{B} = \mathbf{\Omega} + \mathbf{S} \ , \tag{8.22}$$

where, in terms of the Kronecker delta function, δ_{ij},

$$X_{ij} \triangleq [\delta_{ij} - (1-\varepsilon_i)F_{i-j}\tau_M]/\varepsilon_i$$
$$\mathbf{\Omega}_i \triangleq \sigma T_i^4$$
$$S_i \triangleq \frac{(1-\varepsilon_i)}{\varepsilon_i}(\varepsilon_v \sigma \langle T_v\rangle^4 \tau_l + \varepsilon_l \sigma \langle T_l\rangle^4 \tau_v) \ .$$

Equation (8.22) now can be solved in terms of the radiosity vector $\mathbf{B}$,

$$\mathbf{B} = \mathbf{\Psi}\mathbf{\Omega} + \mathbf{\Psi}\mathbf{S} \ , \tag{8.23}$$

where,

$$\Psi \triangleq X^{-1} \ .$$

Normally, we are not interested in the radiosity, $\mathbf{B}$, but rather the relationship between the $\mathbf{q}''_R$ and $\mathbf{\Omega}$ vectors. By combining Eqs. (8.20) and (8.23), we obtain,

$$\mathbf{q}''_R = \mathbf{\Lambda}\mathbf{\Omega} - \mathbf{Q}\mathbf{S} \ , \tag{8.24}$$

where,

$$\Lambda_{ij} \triangleq \frac{\varepsilon_i}{(1-\varepsilon_i)}(\delta_{ij} - \Psi_{ij})$$
$$Q_{ij} \triangleq \frac{\varepsilon_i}{(1-\varepsilon_i)}\Psi_{ij} \ . \tag{8.25}$$

For steady-state conditions, Eq. (8.24) represents N linear algebraic equations, $1 < j < N$ and $N+2$ unknowns; i.e., N surface temperatures T_i and the two medium temperatures, $\langle T_v\rangle$ and $\langle T_l\rangle$. For transient conditions, Eq. (8.24) is normally used to determine $\mathbf{q}''_R$ in conjunction with an appropriate conduction model. For all cases of practical significance, the liquid droplets are saturated. Thus, $\langle T_l\rangle = T_{sat}$. To close the system of equations, the vapor temperature, which is frequently superheated, must be calculated separately through the use of the appropriate conservation equations for mass, momentum, and energy (Andersen, 1973). The radiation interaction with the vapor temperature, $\langle T_v \rangle$, is through the evaporation rate of the liquid droplets,[b]

$$\Gamma A_{x-s} = \frac{[\beta_d A_{x-s}(\langle T_v \rangle - T_{sat}) + q'_{R_l}]}{h_{fg}} , \tag{8.26}$$

and the net radiant energy absorbed by the vapor and liquid phases,

$$q'_{R_v} = \sum_{j=1}^{N} \sum_{i=1}^{N} P_{H_i} F_{i-j} \varepsilon_v \tau_l (B_i - \sigma \langle T_v \rangle^4) + \sum_{i=1}^{N} P_{H_i} \varepsilon_v \varepsilon_l (\sigma T_{sat}^4 - \sigma \langle T_v \rangle^4) \tag{8.27}$$

$$q'_{R_l} = \sum_{j=1}^{N} \sum_{i=1}^{N} P_{H_i} F_{i-j} \varepsilon_l \tau_v (B_i - \sigma T_{sat}^4) + \sum_{i=1}^{N} P_{H_i} \varepsilon_v \varepsilon_l (\sigma \langle T_v \rangle^4 - \sigma T_{sat}^4) . \tag{8.28}$$

Equations (8.27) and (8.28) can be interpreted as being the amount of radiant energy absorbed by the phase in question, less the amount reemitted, plus the radiant energy emitted by the other phase and absorbed by the phase under consideration. That is, q'_{Rv} and q'_{Rl} are the net rate of radiant energy absorbed per unit axial length by the vapor and liquid phases, respectively.

The radiative heat transfer process, just described, acts in conjunction with convective and conductive heat transfer processes during ECCS operation. Any effective model for core heatup must also consider these mechanisms as well as radiation effects. One of the most important heat transfer processes is the rewetting of the hot surfaces (e.g., the channel wall) with ECCS spray water. Indeed, the rewetting of the Zircaloy channel changes the temperature and the emissivity of this surface, $\varepsilon_{Z_rO_2} \cong 0.67$ and $\varepsilon_{H_2O} \cong 0.96$, and thus greatly enhances the ability of the channel wall to act as a heat sink for radiative energy transfer.

8.5.2 Falling Film Rewetting

The mechanism for rewetting the channel wall with ECCS spray water is through the conduction-controlled advance of a liquid film. To gain some appreciation of the physics involved, we consider the falling liquid film shown in Fig. 8-8, assuming that the bundle's CCFL flooding characteristics, discussed in Sec. 4.2, are such that sufficient liquid is available to support a falling film.

A one-dimensional energy balance on the Eulerian control volume shown in Fig. 8-8 yields the classical "fin" equation

[b]Here, β_d represents the heat transfer rate between the droplets and the superheated vapor, per unit temperature, per unit volume (Sun et al., 1975).

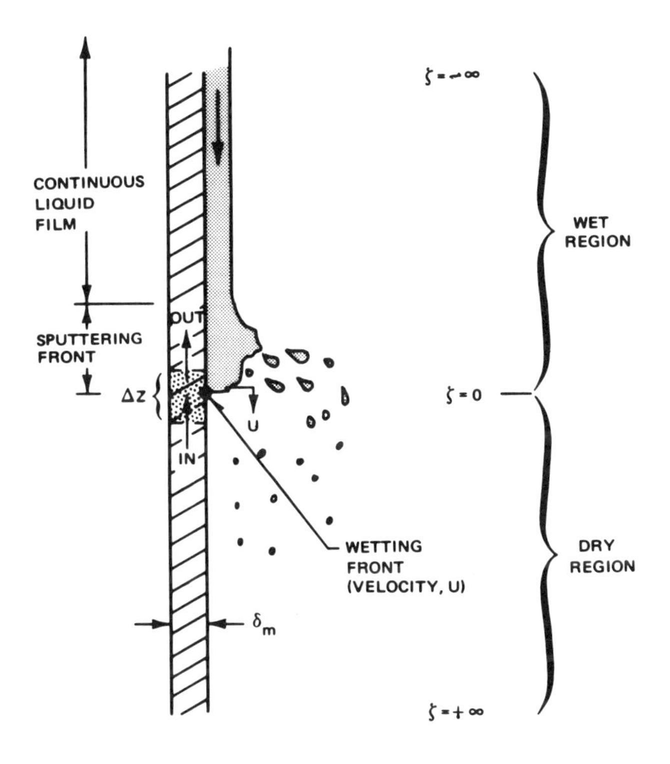

Fig. 8-8 Falling film conduction-controlled rewet.

$$\rho_m c_{p_m} \frac{\partial T}{\partial t} + H(T - T_{\text{sat}}) = k_m \delta_m \frac{\partial^2 T}{\partial z^2} \ . \tag{8.29}$$

It is convenient to make a transformation to a coordinate system moving with the assumed constant[c] velocity, U, of the advancing film front. If we let $\zeta = z - Ut$, then,

$$\frac{\partial T}{\partial t} = \frac{dT}{d\zeta}\frac{\partial \zeta}{\partial t} = -U\frac{dT}{d\zeta}$$

$$\frac{\partial^2 T}{\partial z^2} = \frac{d^2 T}{d\zeta^2}\left(\frac{\partial \zeta}{\partial z}\right)^2 = \frac{d^2 T}{d\zeta^2} \ .$$

[c]For situations in which $U = U(z)$, the heated surface can be divided into nodes of length, ΔL, such that the velocity of the advancing film front can be considered constant within each node.

Thus, Eq. (8.29) becomes,

$$\rho_m c_{p_m} \delta_m U \frac{dT}{d\zeta} + k_m \delta_m \frac{d^2T}{d\zeta^2} = H(T - T_{sat}) \ . \tag{8.30}$$

In essence Eq. (8.30) is a statement of the relationship between the rewetting front velocity, U, and the ease with which axial heat conduction can occur. Hence, the name "conduction-controlled" rewet. Equation (8.30) can be transformed further into a more convenient nondimensional form,

$$\frac{d^2\theta(\eta)}{d\eta^2} + P\frac{d\theta(\eta)}{\partial\eta} = B(\eta)\theta(\eta) \ , \tag{8.31}$$

where,

$$P \triangleq \rho_m c_{p_m} \delta_m U/k_m; \text{ the Peclet number} \tag{8.32}$$

$$B(\eta) \triangleq H(\eta)\delta_m/k_m; \text{ the Biot number} \tag{8.33}$$

$$\theta \triangleq (T - T_{sat})/(T_0 - T_{sat}) \tag{8.34}$$

$$T_0 \triangleq \text{rewetting temperature} \tag{8.35}$$

$$\eta \triangleq \zeta/\delta_m \ . \tag{8.36}$$

Once the axial dependence of the convective heat transfer coefficient, $H(\eta)$, is specified, Eq. (8.31) can be solved with the appropriate boundary conditions,

$$\theta^-(-\infty) = 0 \tag{8.37a}$$

$$\theta^-(0) = \theta^+(0) = 1 \tag{8.37b}$$

$$\theta^+(\infty) \triangleq 1 + \theta_1 = 1 + (T_w - T_0)/(T_0 - T_{sat}) \ , \tag{8.37c}$$

$$\left.\frac{d\theta^+}{d\eta}\right|_{\eta=0} = \left.\frac{d\theta^-}{d\eta}\right|_{\eta=0} \tag{8.37d}$$

where T_w is the initial upstream temperature of the channel wall.

If we assume that $H(\eta) = H_b$ for $\eta \leqslant 0$ and $H(\eta) = 0$ for $\eta > 0$, then on integration we obtain the well-known result of Yamanouchi (1968),

$$P/\sqrt{B} = [\theta_1(\theta_1 + 1)]^{-1/2} \ . \tag{8.38}$$

Equation (8.38) frequently is written in terms of the velocity of the falling film, U, as,

$$U = \frac{1}{\rho_m c_{p_m}} \left(\frac{H_b k_m}{\delta_m}\right)^{1/2} \left\{\frac{(T_0 - T_{sat})}{[(T_w - T_0)(T_w - T_{sat})]^{1/2}}\right\} \ . \tag{8.39}$$

For high spray rates, the analysis can be improved by considering the precursory cooling of the liquid droplets through the assumption that for $\eta > 0$,

$$B(\eta)=\left(\frac{H_b\delta_m}{k_m}\right)\exp(-a\delta_m\eta)\ . \tag{8.40}$$

Solution of Eqs. (8.31), (8.35), (8.36), (8.37), and (8.40) yields an implicit expression for the dimensionless wet front velocity, P (Sun et al., 1974).

The one-dimensional analysis resulting in Eq. (8.29) can be extended to a two-dimensional analysis for those cases in which the simple one-dimensional model is inadequate. For conditions in which two-dimensional conduction effects in the channel wall are important, but upstream precursory cooling is negligible (e.g., low spray rates), the appropriate expression for the dimensionless wet front propagation velocity, P, is given by (Anderson, 1974),

$$P/B=2^{-\sqrt{\pi/4}}[\theta_1(\theta_1+1)]^{-(1/2+\sqrt{\pi/4})}\ . \tag{8.41}$$

In general, Eq. (8.38) is valid for $B \ll 1$ and Eq. (8.41) for $B \gg 1$. These equations are shown in Fig. 8-9. Note that intersection occurs at,

$$B=2^{\sqrt{\pi/2}}[\theta_1(\theta_1+1)]^{\sqrt{\pi/2}} \tag{8.42}$$

$$P=2^{\sqrt{\pi/4}}[\theta_1(\theta_1+1)]^{(\sqrt{\pi/4}-1/2)}\ . \tag{8.43}$$

Equation (8.42) represents the threshold for two-dimensional effects. For Biot numbers, B, larger than those given by Eq. (8.42), two-dimensional effects are important and, thus, Eq. (8.41) should be used to evaluate the rewetting front velocity. For Biot numbers less than those given by Eq. (8.42), a one-dimensional analysis is adequate and Eq. (8.39) can be used. For most cases of practical significance in BWR technology, Eq. (8.41) should be used.

If precursory cooling is important, the resultant expressions are somewhat different; however, it is always conservative (i.e., the wet front velocity is always smaller) if precursory cooling is neglected.

Once the velocity of the advancing liquid front has been calculated, the rewetting time, t_{RW} (i.e., the time for the wet front to advance a distance, L), can be approximated by,

$$t_{RW}=\sum_{i=1}^{N=L/\Delta L}\frac{\Delta L}{U_i}\ . \tag{8.44}$$

It is important to determine when the liquid film arrives at a given axial position since, as mentioned previously, a wetted surface is a much more effective radiation heat sink than a dry one. That is, as shown in Fig. 8-10, the fuel rod temperatures in the region for $z \leq L$ have a more effective radiation heat sink than those for $z > L$.

To complete the analysis of core heatup, we must consider the effects of decay heat and the exothermic metal-water (Zircaloy-steam) reaction that can occur at elevated ($T_w > 1800°F$) cladding temperatures.

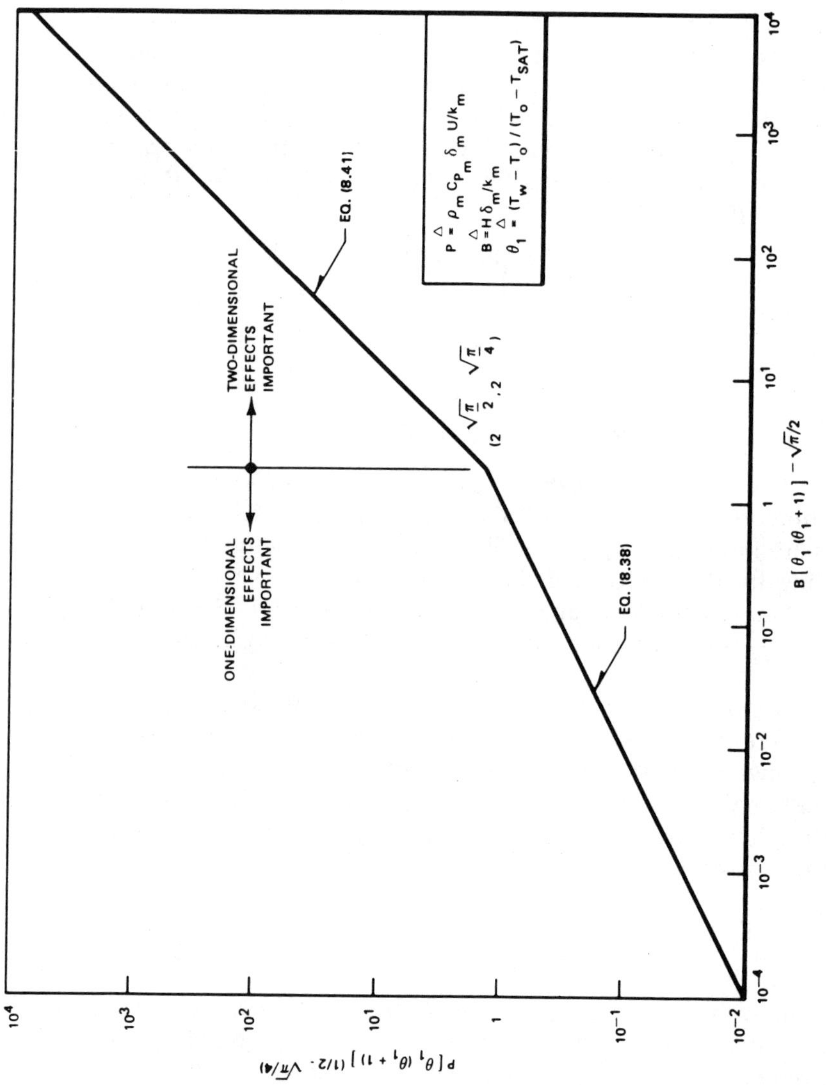

Fig. 8-9 Modified Peclet number, P, as a function of modified Biot number, B.

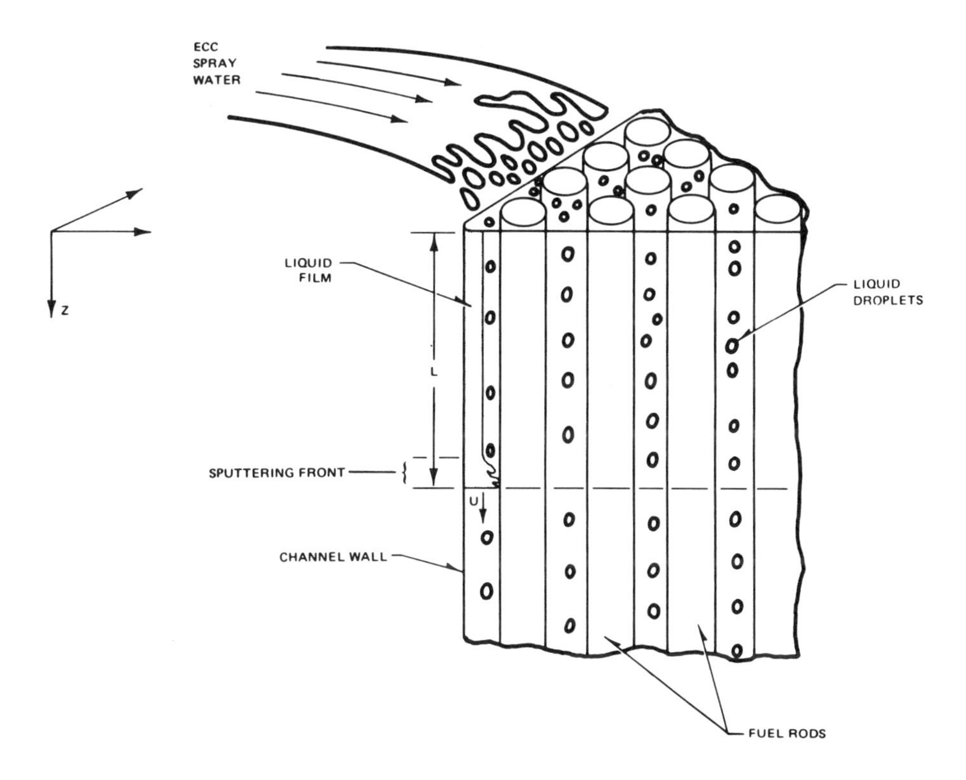

Fig. 8-10 A segment of a BWR fuel rod bundle during ECCS spray cooling.

8.5.3 Heat Generation Effects

The exothermic reaction that can occur between the hot Zircaloy cladding and the steam environment is given by

$$Zr + 2H_2O \rightarrow ZrO_2 + 2H_2 + \Delta \ . \tag{8.45}$$

The heat of reaction, Δ, is dependent on the reaction rate, which is, in turn, dependent on the temperature and the amount of steam available at the reacting surface. In general, the reaction rate is limited by the gaseous diffusion of steam from the coolant through a hydrogen boundary layer near the oxidizing surface and/or the solid-state diffusion process through the ZrO_2 layer to the reacting Zr-ZrO_2 interface. The actual reaction rate would be the most limiting of these two processes. Normally, however, it is conservatively assumed that there is an infinite amount of steam available at the reacting surface and that the parabolic rate law of Baker and Just (1962) is valid. Thus, for current BWR ECCS calculations, the heat of reaction, Δ, is conservatively based on the well-known Baker-Just model.

Another extremely important energy input to the calculation of peak clad temperature is due to the effect of stored sensible heat, delayed neutron-induced fission, and decay heat. Figure 8-11 is a curve of the energy input due to delayed neutrons and decay heat. This curve is based on the conservative 10CFR50 Appendix-K guidelines established by the U.S. Nuclear Regulatory Commission. It frequently is known as the "ANS + 20%" curve.

Finally, internal heat generation due to neutron and gamma-ray absorption in the fuel cladding can be considered in the rigorous calculation of core heatup, but these effects generally are quite small and normally are neglected.

8.5.4 Analytical Modeling

We are now in position to synthesize an analytical model for the calculation of core heatup. It is obvious that a transient calculation is required to determine the thermal response of the fuel rod cladding to the postulated accident.

The essence of the radiation model required has been described in Sec. 8.5.1, the conduction model for film rewetting in Sec. 8.5.2, and for cladding temperature response in Sec. 6.1.3, and the convection model is given by,

$$q''_{\text{conv}} = H_{\text{conv}}(z, t)[T(z, t) - \langle T_g(z, t) \rangle] , \tag{8.46}$$

where the appropriate convective heat transfer coefficient, H_{conv}, to the droplet-ladened flow stream can be calculated in a manner similar to that presented previously by Sun et al. (1975).

Figure 8-12 shows how the various essential pieces fit together into an analytical core heatup model. Note that the calculation is iterative in nature and invariably takes the form of a large digital computer code. An example of the transient peak clad temperatures calculated with such a technique is shown in Fig. 8-13 for a typical BWR/6 LOCA analysis in which Appendix-K assumptions have been used. Licensing calculations of this type have shown that the ultimate peak clad temperature (PCT) is controlled by the maximum axial planar linear heat generation rate (MAPLHGR). Thus, limiting the MAPLHGR of 8×8 BWR/6 fuel to 13.4 kW/ft inherently limits the thermal excursion during a hypothetical LOCA. Moreover, best estimate calculations (Dix, 1983) indicate a much lower PCT than that shown in Fig. 8-13.

8.6 Severe Accidents

Subsequent to the accident at TMI-2, a lot of attention has been given to postulated accidents that are outside of the original design basis for the ECCS and containment design. In particular, accidents have been studied that may cause a degraded core due to inadequate core cooling. Such

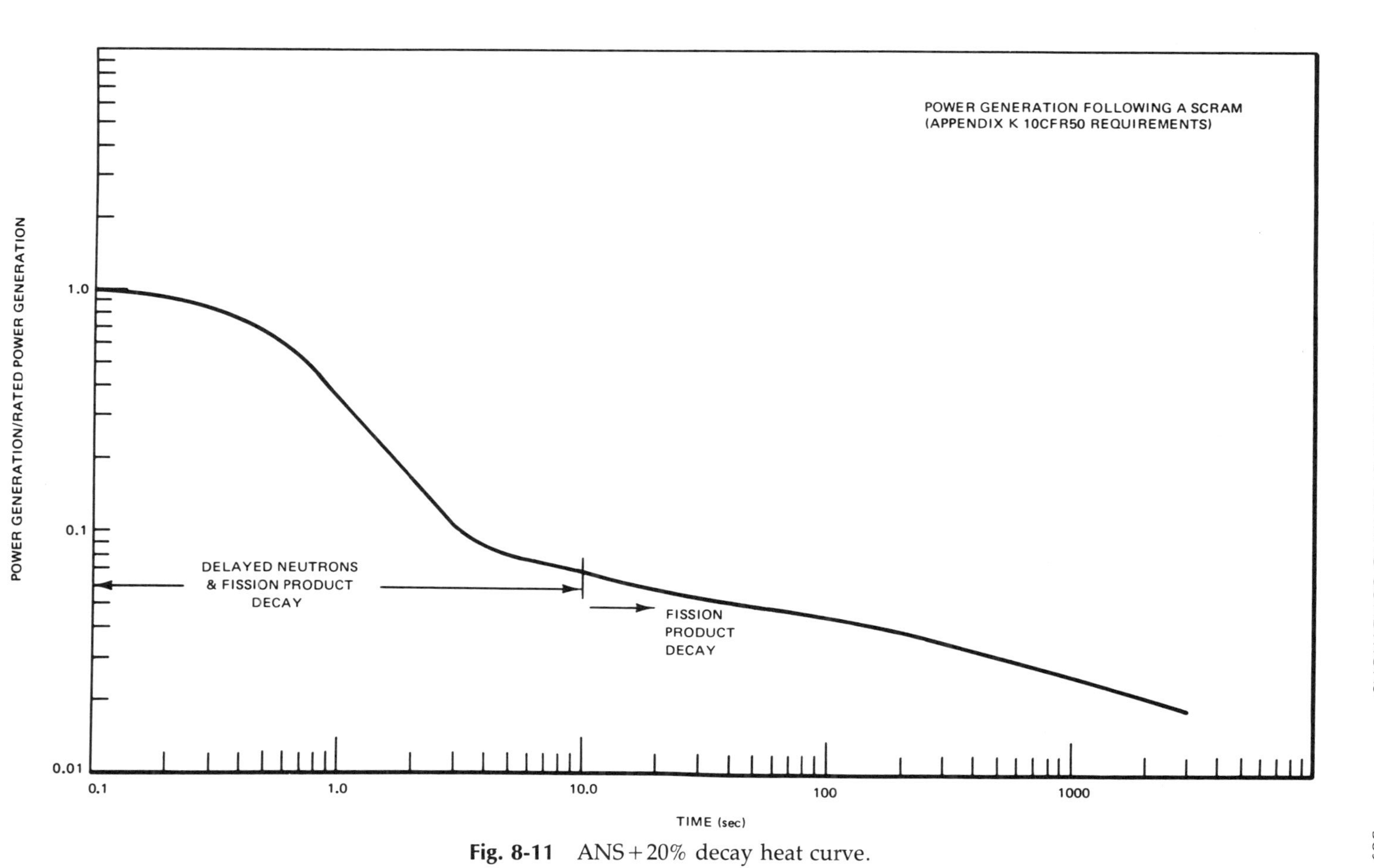

Fig. 8-11 ANS + 20% decay heat curve.

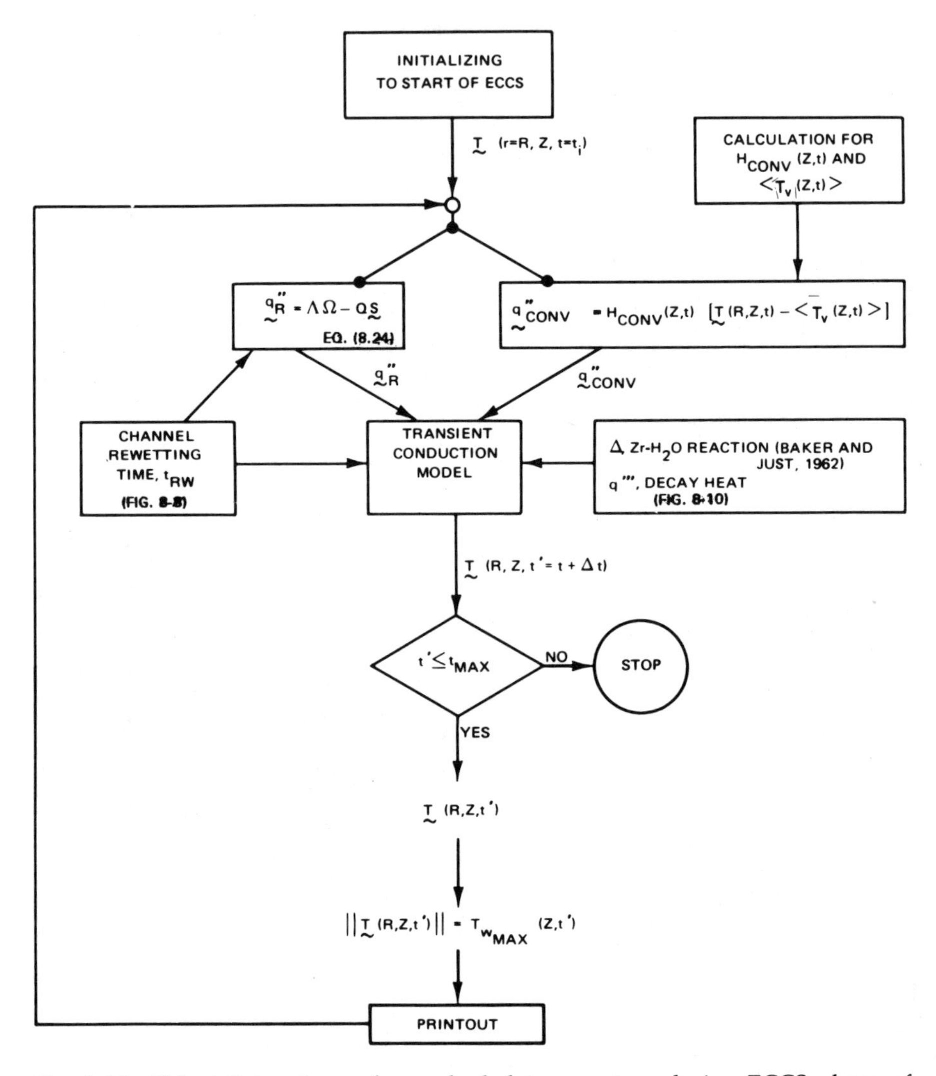

Fig. 8-12 Calculation scheme for peak clad temperature during ECCS phase of core heatup.

hypothetical accidents may result in the release of molten core and structural materials (i.e., corium) to the containment and may also result in a loss of containment integrity. Clearly these are undesirable events since there may be an associated release of radioactive materials to the environment.

The purpose of this section is to present an overview of the status of BWR severe accident sequence analysis (SASA). Since these issues are still under study no final conclusions can be made at this time. Nevertheless,

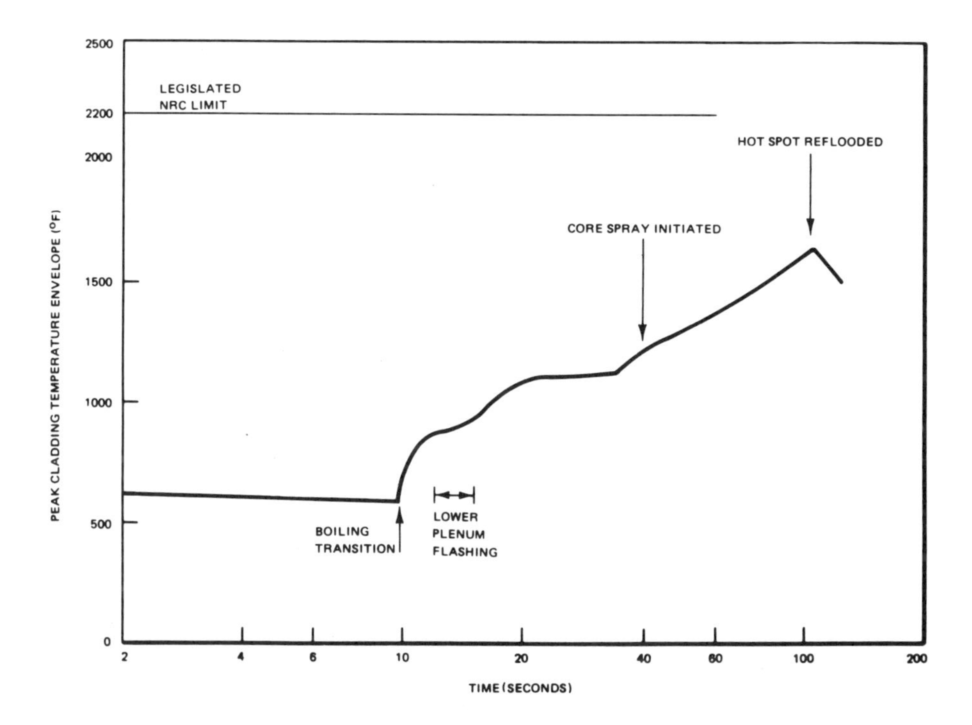

Fig. 8-13 Typical BWR/6 peak cladding temperature following a design basis accident calculated with Appendix-K licensing assumptions (LPCS diesel-generator failure systems available—2 LPCI + HPCS + ADS + RCIC).

a qualitative understanding has emerged as to how BWRs might respond during severe accidents. It is the purpose of this section to document some of these insights.

8.6.1 Severe Accident Sequence Analysis

The overall goal of severe accident analysis is the determination of the probability, timing, mode, and magnitude of the releases of fission products to the environment. The dominant accident sequences that may lead to core uncovering and damage are normally evaluated through a probabilistic risk assessment (PRA) of a particular plant. Once the dominant sequences have been identified, a SASA study can be conducted to explore the details of the postulated accident's scenario. The results of the SASA study may then be fed back into a revised PRA study if the original assumptions concerning accident scenario and consequences were invalid.

The final, and most important, result of any severe accident study is an evaluation of the amount and type of radioactive material released to the environment. Once this is known, the possible environmental and health consequences can be estimated using standard aerosol plume dispersal methodology such as that embodied in the CRAC2 code (Ritchie et al., 1981).

A number of detailed severe accident analyses have been performed, starting with the well-known WASH-1400 study. The results of these studies are generally consistent and indicate that there are four sequences that give BWR accidents having dominant risks:

1. transients with SCRAM coupled with a failure to provide adequate makeup water to the core
2. transients with SCRAM, which result in a loss of containment (i.e., suppression pool) heat removal capability
3. loss-of-coolant accidents (e.g., a SBLOCA, such as a stuck-open SRV)
4. transients in which the reactor remains critical (i.e., ATWS events).

Sequences 3 and 4 have been previously discussed in this chapter. Sequences 1 and 2 are situations in which there is an inability to remove the decay heat generated during the accident. Such situations can arise from postulated accidents such as loss of all off-site power, coupled with a loss of the ECCS diesel generators.

To specify unambiguously the type of severe accident under consideration, it has become customary to identify a particular accident with the symbols given in Table 8-1. Thus, for example, a TQUV event represents an accident in which an inadequate amount of makeup water is supplied to the core. Interestingly, an important source of makeup water for such accidents may be the CRD cooling water that enters the reactor. This flow increases to about 100 gal/min subsequent to SCRAM when the reactor is pressurized, and to about 170 gal/min when the reactor has been depressurized. Although this cooling water does not come from a safety grade system, it can successfully mitigate many accidents of the TQUV type.

8.6.2 Degraded Core Cooling Phenomena

There are too many BWR accident scenarios to discuss them all in detail. Fortunately, most of them are not "risk dominant" sequences. Nevertheless, to gain some appreciation of the current state-of-the-art in degraded core analysis, consideration will be given to some typical accident scenarios in which severe core damage is predicted to occur.

To this end, let us consider a TQUV sequence in which CRD cooling water is assumed to be inoperative. In such an accident the decay heat of the fuel will boil off the water in the core and uncover it. As the core uncovers, it will begin the core heatup. At some point (about 1600°F) a vigorous exothermic reaction will take place between steam and the Zircaloy of the fuel cladding and channel walls, producing zirc oxide (ZrO_2) and hydrogen (H_2). This exothermic reaction will greatly accelerate the core

TABLE 8-1
BWR Severe Accident Sequence Symbols

A – Rupture of reactor coolant boundary with an equivalent diameter of greater than 6 in.
B – Failure of electric power to engineered safety features.
C – Failure of the reactor protection system.
D – Failure of vapor suppression system.
E – Failure of emergency core cooling injection.
F – Failure of emergency core cooling functionability.
G – Failure of containment isolation to limit leakage to less than 100% volume/day.
H – Failure of core spray recirculation system.
I – Failure of low-pressure recirculation system.
J – Failure of high-pressure service water system.
M – Failure of safety/relief valves to open.
P – Failure of safety/relief valves to reclose after opening.
Q – Failure of normal feedwater system to provide core makeup water.
S_1 – Small pipe break with an equivalent diameter of ~2 to 6 in.
S_2 – Small pipe break with an equivalent diameter of ~0.5 to 2 in.
T – Transient event.
U – Failure of HPCI, HPS, or RCIC to provide core makeup water.
V – Failure of low-pressure ECCS to provide core makeup water.
W – Failure to remove residual core heat.
α – Containment failure due to steam explosion in vessel.
β – Containment failure due to steam explosion in containment.
γ – Containment failure due to overpressure—releases through the reactor building.
γ' – Containment failure due to overpressure—releases direct to the atmosphere.
δ – Containment isolation failure in the drywell.
ε – Containment isolation failure in the wetwell.
ζ – Containment leakage greater than 2400% volume/day.
η – Reactor building isolation failure.
θ – Standby gas treatment system failure.

heatup process. As the uncovered portion of the core heats up, energy will be radiated to stainless steel control rod blades. As a consequence, local melting and relocation of control rod material is predicted to occur before melting of the Zircaloy cladding.

When the cladding reaches high enough temperatures, the internal fission gas pressure in the fuel rods may cause them to balloon and rupture, leading to a release of the fission gas. Subsequently, a molten Zircaloy/uranium eutectic mixture may form. The molten materials (i.e., corium) may oxidize and relocate to a lower portion of the core where freezing and blockage may occur. If blockage occurs, the steam flow from the lower regions of the reactor will be reduced (or terminated) and thus the production of hydrogen will be correspondingly reduced.

It can be expected that the local blockages may reopen as the frozen corium heats up, remelts, and again relocates. However, due to fuel slumping and rubble bed formation, a cohesive global blockage (i.e., crust) may occur leading to the formation of a corium pool. As shown schematically in Fig. 8-14, at some point the corium pool may be relocated into the lower plenum, due to either crust failure or pool overflow. In any event, the corium that collects on the lower head will attack the lower head and its penetrations. It has been found that the 2-in.-diam lower head drain plug in the center of the lower head of the reactor pressure vessel (RPV) should fail first, followed by the CRD mechanism stub tubes.

Once molten corium begins to discharge into the drywell, it will ablate the original opening in the lower head and thus enlarge it. The amount of dispersal, throughout the drywell and wetwell regions of the containment, of the corium being discharged will depend on the containment type and if the reactor is at pressure at the time of lower head failure.

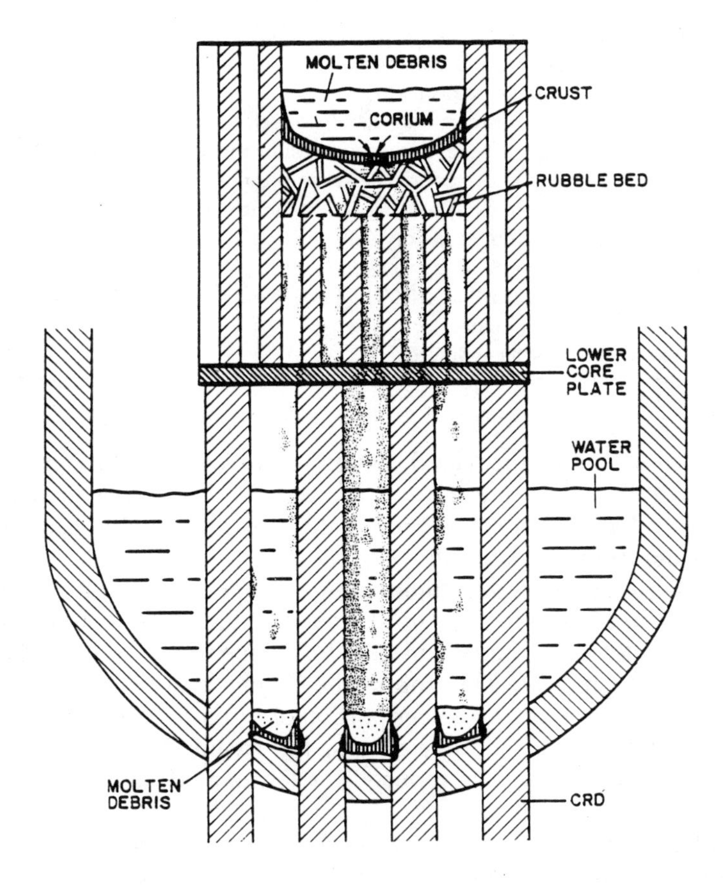

Fig. 8-14 A schematic of melt propagation.

The molten corium that collects on the floor of the drywell may attack and ablate it. As the concrete decomposes, it will release gases that may react with the unoxidized materials in the corium pool on the drywell floor. The gases released from the core and the corium pool on the drywell floor, as well as the steam that is present, will lead to aerosol formation and to further containment pressurization. Naturally, if the temperature and/or pressure in the containment is high enough, failure may occur. In addition, the molten corium may spread (particularly in MARK-I containments) and attack the drywell liner, causing it to fail. Such phenomena are considered in Sec. 11.4.

Large digital computer-based models have been developed to predict the thermal-hydraulic scenarios just described. Typical mechanistic models include APRIL.MOD2 (Kim et al., 1988), APRIL.MOD3 (Cho et al., 1990), and MAAP-3.0 (Henry et al., 1987). Other nonmechanistic models include MARCH (Wooton and Avci, 1980), a code that was used in the WASH-1400 study and has been employed in the NRC's Source Term Code Package (STCP).

To appreciate better the state-of-the-art, we can compare the thermal-hydraulic predictions of various codes for the same accident. APRIL is more mechanistic than MAAP, while MARCH/STCP is rule-based, rather than being mechanistic; thus, significant differences in the predicted results can occur.

An interesting accident that can be used for code comparison is a hypothetical BWR/4 ATWS initiated by a MSIV closure (Abramson and Komoriya, 1983). For ATWS events the Chexal-Layman correlation (Chexal and Layman, 1986) has been used to approximate the ATWS core power.

Figure 8-15 shows that APRIL.MOD2 predictions of the maximum core temperature are higher than those of the other codes because, unlike the other codes, arbitrary assumptions concerning the eutectic melting temperature of the fuel are not made in APRIL.MOD2. Moreover, we note MAAP-3.0 predicts a period of corium superheating starting at about 2 h into the accident.

Figure 8-16 shows that the fraction of the core relocated by APRIL.MOD2 is similar to that relocated by MAAP-3.0 (however, the composition of the corium melt relocated is quite different). In contrast, Fig. 8-17 shows that, because of differences in material relocation and blockage models, APRIL.MOD2 predictions of the fraction of cladding reacted are somewhat higher than MAAP-3.0 (which assumes irreversible blockage) but much lower than MARCH (which assumes no blockage). As a consequence, as can be seen in Fig. 8-18, APRIL.MOD2 predicts that more hydrogen (H_2) is evolved from oxidation processes than does MAAP-3.0, but less than MARCH/STCP.

It is important to note that, because of the mechanistic in-core blockage model used in APRIL.MOD2, this code predicts that sustained blockage does not occur. Thus, as shown in Fig. 8-19, the core outlet gas temper-

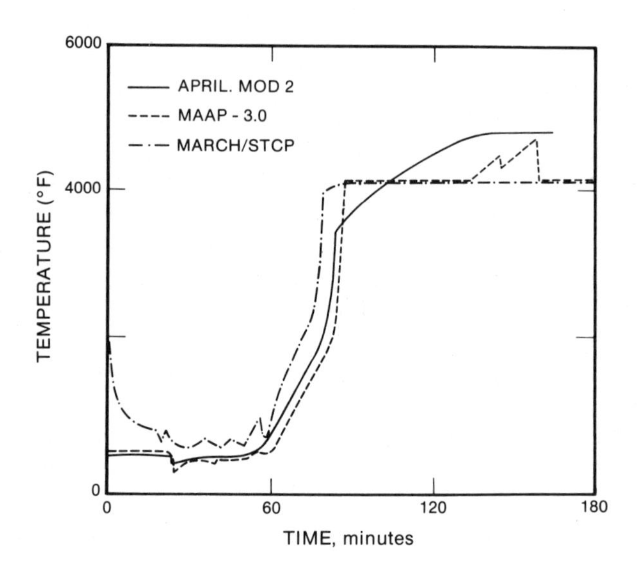

Fig. 8-15 Maximum core temperature.

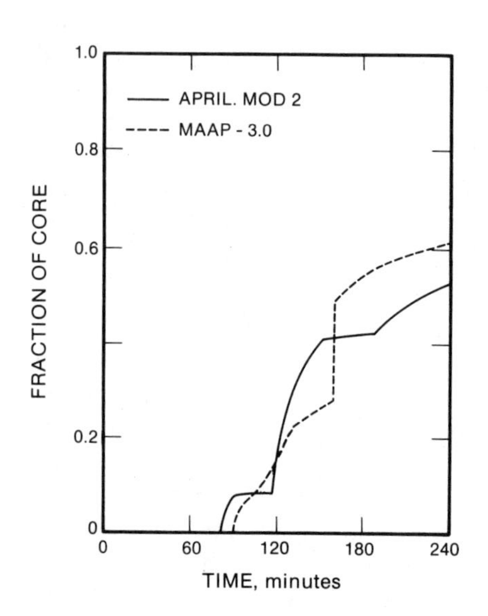

Fig. 8-16 Fraction of core relocated.

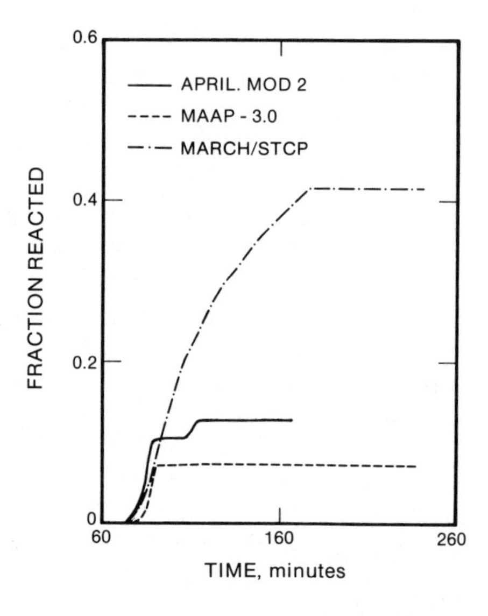

Fig. 8-17 Fraction of cladding reacted.

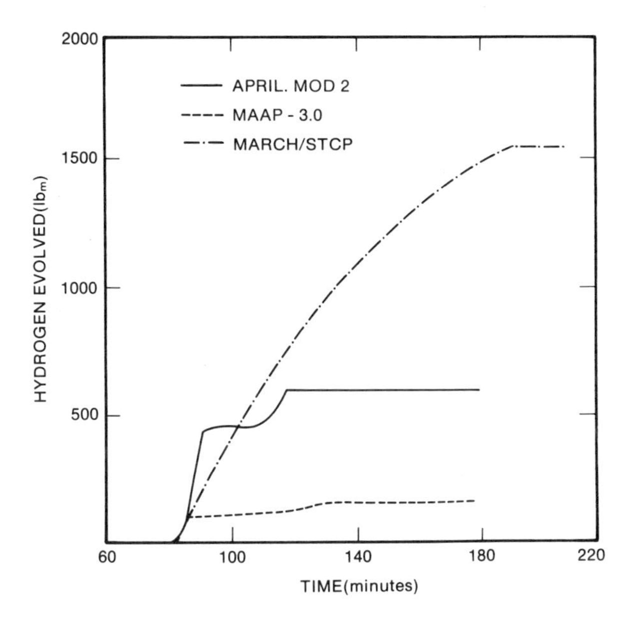

Fig. 8-18 Total hydrogen (H_2) mass evolved.

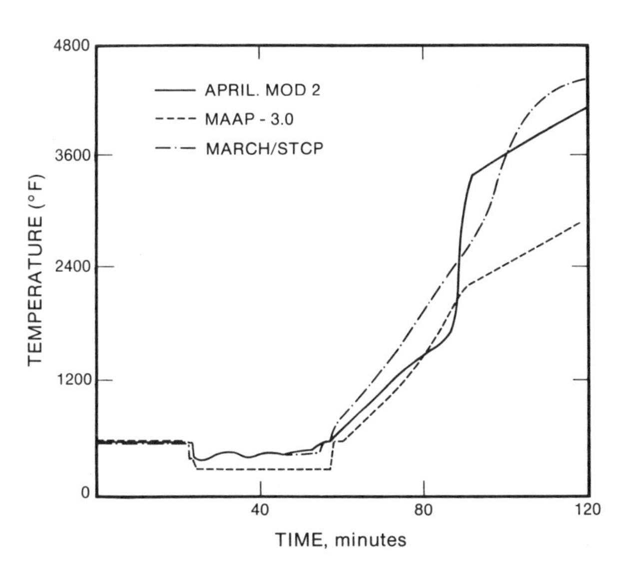

Fig. 8-19 Core outlet gas temperature.

atures predicted by APRIL.MOD2 are similar to those predicted by MARCH (which assumes that blockage does not occur), but MAAP-3.0, which assumes that irreversible core blockage occurs, predicts much lower core exit gas temperatures. As a consequence, unlike MAAP-3.0 predictions, both APRIL.MOD2 and MARCH predict melting of the shroud head. Since this structure supports the steam separator assembly, failure of the steam separator complex can be expected. Such a failure may have a significant effect on the subsequent accident scenario; thus, it is clear that the implications of the blockage and other assumptions made in these codes may have a significant effect on the predicted accident scenario.

While it is not clear at this time which, if any, of these core predictions is correct, it appears that the variance between these predictions is reflective of the current uncertainty in the state of the art in degraded core analysis.

Since the main purpose of all degraded core analyses is to determine the amount of radioactivity that may be released to the environment, let us now turn our attention to the evaluation of the radiation "source term."

8.6.3 The Source Term

The evaluation of the so-called "radioactive source term" is concerned with the hypothetical release and transport of fission products from a degraded core to the environment. During a postulated BWR accident, in which the core dries out and becomes severely overheated and damaged, many of the radioactive fission products may be released. In particular, the fission gases, for example, xenon (Xe) and krypton (Kr), will be released when the fuel cladding fails and many of the other radioactive fission products, such as cesium (Cs) and iodine (I), may subsequently vaporize and be released. The vaporized fission products may chemically react to form other compounds and the resultant mixture of condensible vapors and relatively noncondensible gases, such as hydrogen (H_2), may flow out of the reactor pressure vessel (RPV) into the containment building.

When the superheated vapors flow out of the core into relatively cool regions, fog formation is expected. That is, aerosol droplets will be formed, which are then convected with the bulk flow of the gas/vapor mixture. It is expected that a significant fraction of the aerosol droplets may be deposited on the relatively cool structures of the nuclear steam supply system (NSSS) and the containment building. As a consequence, the radioactive aerosol material deposited on these structures, for example, cesium iodine (CsI), cesium hydroxide (CsOH), and tellurium (Te), may be unable to escape to the environment through any leakage paths from the containment building. Thus aerosol deposition may significantly reduce the radioactive source term.

It should be obvious that aerosol transport and deposition mechanisms must be carefully evaluated if one is to predict realistic source terms. In this section models for the aerosol technology relevant to nuclear reactor safety analysis will be summarized.

The release of vaporized radioactive material from the core can be approximated from the burnup history of the core and a phenomenological model recommended by the NRC (NUREG-0772, 1981):

$$\frac{dM_j}{dt} = -K_j(T_f)M_j \ , \tag{8.47}$$

where M_j is the mass of isotope j in the fuel and the empirical release rate parameter, K_j, is assumed to be only a function of fuel temperature, T_f.

8.6.3.1 *Aerosol Formation Mechanisms*

Let us next consider aerosol formation mechanisms. A force balance on an aerosol droplet of radius r_d yields:

$$p_v \pi r_d^2 + 2\pi r_d \sigma = p_l \pi r_d^2 \ .$$

Thus,

$$p_l - p_v = \frac{2\sigma}{r_d} \ . \tag{8.48}$$

To understand better the heterogeneous droplet nucleation rates that may occur in the vapor stream, let us examine the differential of the specific Gibbs free energy function, g. As noted in Chapter 4 this is given by,

$$dg = vdp - sdT \ .$$

For constant temperature conditions, we have for a perfect gas,

$$\int_{g_{sat}(T)}^{g_v} dg = RT \int_{p_{sat}(T)}^{p_v} \frac{dp}{p}$$

$$g_v = g_{sat} + RT \ln(p_v/p_{sat}) \ . \tag{8.49}$$

Similarly, for an incompressible liquid,

$$\int_{g_{sat}(T)}^{g_l} dg = v_l \int_{p_{sat}(T)}^{p_l} dp$$

$$g_l = g_{sat} + v_l(p_l - p_{sat}) \ . \tag{8.50}$$

As discussed in Chapter 4, for phasic equilibrium:

$$g_l = g_v \ .$$

Thus Eqs. (8.49) and (8.50) yield,

$$p_l - p_{sat} = \frac{RT}{v_l} \ln(p_v/p_{sat}) \simeq \frac{RT}{v_l p_v}(p_v - p_{sat}) \ . \tag{8.51}$$

Hence, using Eqs. (8.48), (8.51), and the perfect gas law,

$$p_l - p_v = \frac{2\sigma}{r_d} = \frac{v_v}{v_l}(p_v - p_{sat}) - (p_v - p_{sat}) \ . \tag{8.52}$$

Thus, assuming $v_l \cong v_f$,

$$p_v - p_{sat} = \frac{v_f}{v_{fv}} \frac{2\sigma}{r_d} \ . \tag{8.53}$$

To compute the required vapor subcooling (ΔT_{sub}) for droplet nucleation, let us use the Clasius-Clapeyron equation,

$$\left(\frac{dp}{dT}\right)_{sat} = \frac{h_{fg}}{T_{sat} v_{fg}} \simeq \frac{[p_v - p_{sat}(T_v)]}{T_{sat}(p_v) - T_v} = \frac{[p_v - p_{sat}(T_v)]}{\Delta T_{sub}} \ . \tag{8.54}$$

Combining Eqs. (8.53) and (8.54), we have for the case where $v_v \simeq v_g$,

$$\Delta T_{sub} = \frac{v_f T_{sat}}{h_{fg}} \frac{2\sigma}{r_d} \ . \tag{8.55}$$

To appreciate better the implications of Eq. (8.55), let us consider a simple example. For atmospheric pressure steam, Eq. (8.55) yields a vapor subcooling of 0.2 K for a water droplet of diameter 0.1 mm. This is a low subcooling, thus it is clear that if small nucleation sites are present in the vapor stream, fogging can occur. It should be noted, however, that Eq. (8.55) implies that an infinite vapor subcooling is needed if no nucleation sites are present. This is incorrect since, due to molecular effects, the probability always exists that liquid molecules present in the vapor stream will collide and form a liquid droplet. This process is called homogeneous nucleation. The homogeneous nucleation rate is given by (Hill et al., 1963),

$$\frac{dn_d'''}{dt} = \left(\frac{2\sigma}{\pi m}\right)^{1/2} \frac{v_f}{v_v} N_d''' \exp(-\Delta A/kT_v) \ , \tag{8.56}$$

where,

n_d''' = droplets/m^3

m = molecular mass, (M/N_a)

M = molecular weight, (gm/mol)

N_a = Avogadro's number, (6.023×10^{23} mol^{-1})

N_d''' = molecules/m^3, $(N_a/M)\rho_g$

k = Boltzmann constant, (R/N_a)

and the so-called "activation energy" is given by:

$$\Delta A = \frac{16\pi\sigma^3}{3(p_l - p_v)^2} = \frac{4}{3}\pi(r_d^*)^2\sigma \ , \tag{8.57}$$

where r_d^* can be calculated from Eq. (8.55).

It is interesting to note that Eq. (8.56) only differs in form from the corresponding expression for the homogeneous nucleation of bubbles by the factor v_f/v_v. Evaluating Eq. (8.56) for atmospheric steam we find that at $\Delta T_{sub} = 25$ K, $dn_d'''/dt = 3.6 \times 10^{20}$ droplets/m^3-s, and $r_d^* = 8 \times 10^{-10}$ m.

Subsequent to fog formation one can determine what happens to the droplets formed by either computing the individual droplet trajectories (see, for example, Bunz et al., 1981) or by performing control volume analysis (Epstein and Ellison, 1988). Both approaches have been used (e.g., in NAVA and MAAP) in light water nuclear reactor safety analysis.

A complete discussion of aerosol mechanics is beyond the scope of this chapter. However, the aerosol phenomena of most interest in nuclear reactor safety will be discussed. They are aerosol coagulation and the associated gravitational settling, hydroscopic effects, inertial impaction, diffusiophoresis, thermophoresis, and the decontamination factor (DF) associated with pool scrubbing. The "bible" for aerosol phenomena is by Davies (1966), nevertheless, many other related works are also of interest (e.g., Brock, 1962; Dunbar, 1983; Epstein et al., 1986; Epstein et al., 1985; Friedlander and Johnstone, 1957; Loyalka, 1983; Owcarski et al., 1985; Ritchie et al., 1981; Von Smolnchowski, 1917; Walker et al., 1976).

The next section will summarize the aerosol mechanics of most interest for the evaluation of severe accidents in the BWRs. This description is based on the models used in APRIL.MOD3 (Cho et al., 1990).

8.6.3.2 *Aerosol Transport Equations*

The time rate of change of the suspended aerosol content in an arbitrary volume, V (see Fig. 8-20), can be derived by balancing the various mechanisms through which aerosols can be gained or lost from the volume (Duderstadt and Martin, 1979). By taking into account convective transport, events that change the particle's size, and aerosol sources/sinks, the conservation of aerosol mass in the control volume V is given by:

$$\begin{bmatrix}\text{outflow rate of} \\ \text{aerosol mass from} \\ \text{the control volume}\end{bmatrix} - \begin{bmatrix}\text{inflow rate of} \\ \text{aerosol mass to} \\ \text{the control volume}\end{bmatrix} + \begin{bmatrix}\text{storage rate of} \\ \text{aerosol mass in} \\ \text{the control volume}\end{bmatrix} = 0 \ .$$

This balance can be written mathematically as,

$$\int_S \int_v \mathbf{ds} \cdot \mathbf{u}_a(\mathbf{r}, v, t) n(\mathbf{r}, v, t) \rho_a v \, dv - \int_S \int_v \mathbf{ds} \cdot D \nabla n(\mathbf{r}, v, t) \rho_a v \, dv$$

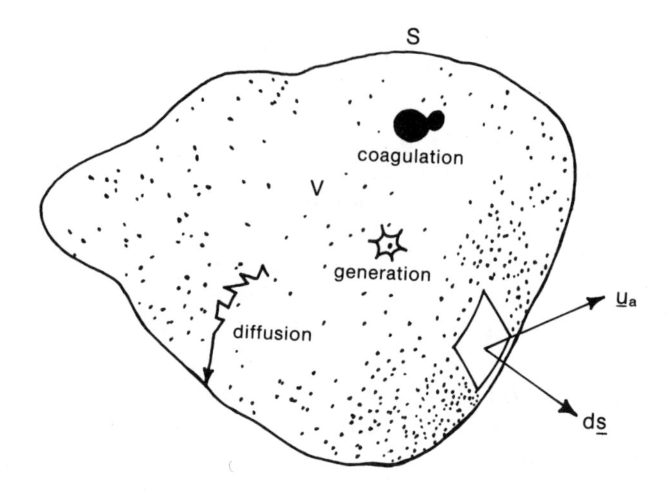

Fig. 8-20 An arbitrary control volume V with surface area S.

$$
-\int_V\int_v \dot{N}(\mathbf{r},\ v,\ t)\rho_a v\ dv\ dV - \int_V\int_v \dot{G}(\mathbf{r},\ v,\ t)\rho_a v\ dv\ dV
$$
$$
+\frac{\partial}{\partial t}\int_V\int_v n(\mathbf{r},\ v,\ t)\rho_a v\ dv\ dV = 0\ , \tag{8.58}
$$

where $n(\mathbf{r},\ v,\ t)$ is the aerosol size distribution density function such that $n(\mathbf{r},v,t)dv$ is the number density of aerosol particles within size range $(v,\ v+dv)$ at time t and location $\mathbf{r}$, v is the volume of an aerosol particle, D is the diffusion coefficient for aerosol, and $\mathbf{u}_a$ is the velocity of the aerosol particles. This velocity can be rewritten as the sum of the fluid velocity, $\mathbf{u}_f$, and the relative velocity of the aerosol particle, $\mathbf{u}_r$ (i.e., $\mathbf{u}_a = \mathbf{u}_f + \mathbf{u}_r$). Also, in Eq. (8.58), S is the surface area of the control volume (see Fig. 8-20), V is the size of the control volume, $\dot{G}(\mathbf{r},\ v,\ t)$ is the source rate density defined in such a way that $\dot{G}(\mathbf{r},\ v,\ t)\ dv$ is the number density of particles generated/lost within size range $(v,\ v+dv)$ at time t and location $\mathbf{r}$, ρ_a is the density of the aerosol particles, and $\dot{N}(\mathbf{r},\ v,\ t)$ is the rate of aerosol particle size change per unit volume such that $\dot{N}(\mathbf{r},\ v,\ t)\ dv$ is the rate at which the number density of aerosol particles change in the size range $(v,\ v+dv)$ at time t and location $\mathbf{r}$; this term accounts for coagulation and hygroscopic effects. It is interesting to note that coagulation may either increase or decrease the number density of aerosol particles in size range $(v,\ v+dv)$.

The term $\dot{N}$ can be expressed as,

$$
\dot{N} = 1/2\int_0^v \beta(\tilde{v},\ v-\tilde{v})n(\tilde{v})n(v-\tilde{v})\ d\tilde{v} - \int_0^\infty \beta(v,\ \tilde{v})n(v)n(\tilde{v})\ d\tilde{v}\ , \tag{8.59}
$$

where $\beta(v, \tilde{v})$ is the so-called coagulation kernel (Friedlander, 1977). The surface integral in Eq. (8.58) can be rewritten as a volume integral by using the divergence theorem. That is,

$$\int_S\int_v \mathbf{ds}\cdot\mathbf{u}_a(\mathbf{r}, v, t)n\rho_a v\, dv = \int_S\int_v \mathbf{ds}\cdot(\mathbf{u}_f + \mathbf{u}_r)n\rho_a v\, dv$$

$$= \int_V\int_v \nabla\cdot(\mathbf{u}_f + \mathbf{u}_r)n\rho_a v\, dv\, dV\ . \tag{8.60}$$

Similarly,

$$\int_S\int_v \mathbf{ds}\cdot D\nabla n\rho_a v\, dv = \int_V\int_v \nabla\cdot D\nabla n\rho_a v\, dv\, dV\ . \tag{8.61}$$

Thus, the aerosol mass conservation equation, Eq. (8.58), can be rewritten as,

$$\int_V\int_v \rho_a v\left(\frac{\partial n}{\partial t} + \nabla\cdot\mathbf{u}_f n + \nabla\cdot\mathbf{u}_r n - \nabla\cdot D\nabla n - \dot{N} - \dot{G}\right) dV\, dv = 0\ . \tag{8.62}$$

Since this equation must be satisfied for any control volume, V, its integrand must be equal to zero, thus,

$$\frac{\partial n}{\partial t} + \nabla\cdot\mathbf{u}_f n = \nabla\cdot D\nabla n - \nabla\cdot\mathbf{u}_r n + \dot{N} + \dot{G}\ . \tag{8.63}$$

As can be seen, Eqs. (8.59) and (8.63) comprise a nonlinear, partial integrodifferential equation. A significant amount of computing is required to solve such equations numerically. A more tractable expression for the aerosol mass balance can be obtained using a lumped parameter approach. In particular, the total aerosol mass, M, in volume V is given by,

$$M(t) = \int_V\int_v n\rho_a v\, dv\, dV\ . \tag{8.64}$$

If we assume that the relative velocity of the aerosol particles ($\mathbf{u}_r$) is also the deposition velocity ($\mathbf{u}_d$) of those particles, the divergence terms in Eq. (8.58) can be written separately for the inflows, outflows, and deposition as:

$$\int_S\int_v \mathbf{ds}\cdot\mathbf{u}_a n\rho_a v\, dv - \int_S\int_v \mathbf{ds}\cdot D\nabla n(\mathbf{r}, v, t)\rho_a v\, dv =$$

$$\int_S\int_v \mathbf{ds}\cdot\mathbf{u}_f n\rho_a v\, dv + \int_S\int_v \mathbf{ds}\cdot\mathbf{u}_r n\rho_a v\, dv - \int_S\int_v \mathbf{ds}\cdot D\nabla n(\mathbf{r}, v, t)\rho_a v\, dv =$$

$$\int_{S_{\text{out}}}\int_v \mathbf{ds}\cdot\mathbf{u}_f n\rho_a v\, dv + \int_{S_{\text{in}}}\int_v \mathbf{ds}\cdot\mathbf{u}_f n\rho_a v\, dv$$

$$+\frac{\int_S\int_v \mathbf{ds}\cdot(\mathbf{u}_r n - D\nabla n)\rho_a v\,dv}{\int_V\int_v n\rho_a v\,dv\,dV}\int_V\int_v n\rho_a v\,dv\,dV \triangleq w_{out} - w_{in} + \lambda M(t)\ , \quad (8.65)$$

where S_{out} is the surface area where the angle between the positive outward normal vector of the surface and the flow direction is less than 90 deg, S_{in} is the surface area where the angle between the positive outward normal vector of the surface and the flow direction is greater than 90 deg, $w_{out} = \int_{S_{out}}\int_v \mathbf{ds}\cdot\mathbf{u}_f n\rho_a v\,dv$ is the aerosol outflow rate, $w_{in} = \int_{S_{in}}\int_v \mathbf{ds}\cdot\mathbf{u}_f n\rho_a v\,dv$ is the aerosol inflow rate, and,

$$\lambda \triangleq \frac{\int_S\int_v \mathbf{ds}\cdot(\mathbf{u}_r n - D\nabla n)\rho_a v\,dv}{\int_V\int_v n\rho_a v\,dv\,dV}$$

is the fraction of deposited mass to the total mass in the control volume per unit time. This parameter, λ, represents a rate constant for aerosol loss.

The terms implicit in $\dot{N}$ imply a gain or loss of aerosol in $(v,\ v+dv)$ due to aerosol size change. Since the total aerosol mass in the control volume does not change due to coagulation, we find that,

$$\int_V\int_v \dot{N}\rho_a v\,dv\,dV = 0\ . \quad (8.66)$$

Finally, integrating the source/sink term, yields:

$$\int_V\int_v \dot{G}(\mathbf{r},\ v,\ t)\rho_a v\,dv\,dV \triangleq \dot{\mathcal{G}}(t)\ . \quad (8.67)$$

Combining Eqs. (8.65), (8.66) and (8.67) with Eq. (8.62), we finally obtain the following global aerosol mass conservation equation,

$$\frac{dM(t)}{dt} + w_{out} - w_{in} + \lambda M(t) = \dot{\mathcal{G}}(t)\ . \quad (8.68)$$

It is interesting to note that Eq. (8.68) is a generalization of the result obtained previously by Epstein et al. (1986, 1988).

8.6.3.3 *Aerosol Deposition Mechanisms*

Aerosols suspended in the primary system may deposit on solid surfaces because of various mechanisms, depending on the flow conditions and geometry. The net deposition rate of aerosol is usually written as a sum of a number of individual deposition rates representing the different removal processes, such as gravitational settling, inertial impaction, diffusiophoresis, and thermophoresis. That is,

$$\lambda_{tot} = \lambda_{sed} + \lambda_{imp} + \lambda_{dif} + \lambda_{th} \ , \tag{8.69}$$

where λ_{sed} is the deposition rate due to sedimentation, λ_{imp} is the deposition rate due to inertial impaction, λ_{dif} is the deposition rate due to diffusiophoresis, and λ_{th} is the deposition rate due to thermophoresis.

The deposition rates for sedimentation and inertial impaction can be obtained using the functional forms of the correlations (Epstein and Ellison, 1988) developed by numerically evaluating (Epstein et al., 1986) the full integrodifferential equation, Eqs. (8.63) and (8.59). The deposition rates for diffusiophoresis and thermophoresis may be expressed as,

$$\lambda_d = \frac{u_d}{h_{eff}} \ , \tag{8.70}$$

where u_d is a deposition velocity, and h_{eff} is the effective height (i.e., the projected-area/volume ratio). This expression will be used subsequently to define the various deposition constants, λ_d.

Gravitational Settling. Gravitational settling is the dominant deposition process in most severe accidents since, due to the relatively high aerosol concentration expected during accident-induced releases, the aerosol size may grow very rapidly as a result of coalescence. The resultant larger aerosol particles, in turn, settle out more rapidly. The rate of deposition due to gravitational settling is proportional to the sum of the horizontal projections of the receiving surfaces times the settling velocity of the aerosol particles, and inversely proportional to the control volume for well-mixed conditions.

As noted previously, Epstein et al. (1986) developed the functional forms of the correlations for both source-reinforced aerosols and aging aerosols by numerically evaluating the integrodifferential equation for coagulation-dominated situations using a sectional method developed by Gelbard et al. (1980). In the present model convective outflows are explicitly accounted for, thus the deposition rate due to sedimentation is given by,

$$\lambda_{sed} = \lambda_{FAI} - \frac{u_f}{h_{eff}} \ , \tag{8.71}$$

where λ_{FAI} is the deposition rate using the appropriate FAI correlations (Epstein et al., 1986), u_f is the fluid velocity, and h_{eff} is the effective height (i.e., the projected area/volume).

Inertial Deposition. Aerosol particles in gas streams moving along solid surfaces may impact and be removed by these surfaces due to aerosol inertia. The Stokes number (Stk) is the basic quantity characterizing the intensity of inertial deposition,

$$\mathrm{Stk} = \frac{4C\rho_a r_a^2 u_f}{9\mu_f D_s} , \tag{8.72}$$

where C is the so-called Cunningham correction factor (Fuchs, 1964), r_a is the aerosol particle's radius, ρ_a is the aerosol density, u_f is the fluid velocity, μ_f is the fluid's viscosity, and D_s is the characteristic dimension of the surface.

The efficiency of inertial deposition, ε_i, is defined as the ratio of the number of captured particles to the number of aerosol particles that would be captured if the particles did not change direction. The parameter ε_i is a function of the Stokes number and is often represented by an empirical formula (Fuchs, 1964). Finally, the deposition velocity for inertial impaction, u_i, is given as,

$$u_i = \varepsilon_i u_f . \tag{8.73}$$

Diffusiophoresis. Diffusiophoresis accounts for the force acting on aerosol particles suspended in a nonuniform but isothermal gas mixture due to aerosol concentration gradients. Such phenomena may occur in the flow of a single condensible vapor (component 1) and a noncondensible gas (component 2). A flow of the vapor directed toward a condensing surface exists near the surface of the condensing body. The velocity of this flow (i.e., the Stefan velocity), u_D, is governed by the following diffusion equation:

$$D_{12}\frac{dC_2}{dx} = u_D C_2 , \tag{8.74}$$

where D_{12} is the diffusion coefficient of the vapor in the noncondensible gas, and C_2 is the mass concentration of the gas. Using the perfect gas law, Eq. (8.74) can be rewritten as,

$$u_D = \frac{D_{12}}{p_2}\frac{dp_2}{dx} = -\frac{D_{12}}{p_2}\frac{dp_1}{dx} . \tag{8.75}$$

The aerosol particles near a condensing surface might be expected to move toward it with velocity u_D. However, a diffusion-induced flow of noncondensible gas away from the condensing surface will reduce the velocity of the aerosol particles toward the surface. The reduction in particle velocity in a binary gas mixture in which two components are diffusing into each other is given by (Davies, 1966):

$$u'_{\mathrm{dif}} = \left(\frac{\sqrt{M_1} - \sqrt{M_2}}{\tilde{y}_1\sqrt{M_1} + \tilde{y}_2\sqrt{M_2}}\right) D_{12}\frac{d\tilde{y}_1}{dx} , \tag{8.76}$$

where M_1 is the molar weight of the vapor, M_2 is the molar weight of the gas, $\tilde{y}_1$ is the mole fraction of the vapor, and $\tilde{y}_2$ is the mole fraction of the gas.

Subtracting u'_{dif} from u_D, the Stefan velocity, the resultant deposition velocity due to diffusiophoresis, u'_{dif}, is:

$$u_{\mathrm{dif}} = u_D - u'_{\mathrm{dif}} = -\frac{D_{12}}{p_2}\frac{dp_1}{dx} - \frac{\sqrt{M_1}-\sqrt{M_2}}{(\tilde{y}_1\sqrt{M_1}+\tilde{y}_2\sqrt{M_2})}D_{12}\frac{d\tilde{y}_1}{dx}$$

$$= -\frac{\sqrt{M_1}}{(\tilde{y}_1\sqrt{M_1}+\tilde{y}_2\sqrt{M_2})}\frac{D_{12}}{p_2}\frac{dp_1}{dx} = -\mathcal{F}\frac{D_{12}}{p_2}\frac{dp_1}{dx} = \mathcal{F}u_D \ , \qquad (8.77)$$

where p_v is the mixture pressure, and $\mathcal{F} = \sqrt{M_1}/(\tilde{y}_1\sqrt{M_1}+\tilde{y}_2\sqrt{M_2})$. Noting that by Dalton's law,

$$p_v = p_1 + p_2 \ ,$$

Eq. (8.77) can be rewritten as,

$$\frac{dp_1}{dx} - \frac{u_{\mathrm{dif}}p_1}{\mathcal{F}D_{12}} + \frac{u_{\mathrm{dif}}p_v}{\mathcal{F}D_{12}} = 0 \ . \qquad (8.78)$$

Assuming that u_{dif} and D_{12} are constant, Eq. (8.78) can be integrated from the condensing surface, $x=0$, to the edge of the mass transfer "boundary layer," $x=\delta$, using an integrating factor. Thus,

$$\int_0^\delta d\left[p_1 \exp\left(-\frac{u_{\mathrm{dif}}}{\mathcal{F}D_{12}}x\right)\right] = -\frac{u_{\mathrm{dif}}p_v}{\mathcal{F}D_{12}}\int_0^\delta \exp\left(-\frac{u_{\mathrm{dif}}}{\mathcal{F}D_{12}}x\right)dx \ .$$

Hence,

$$[p_1(\delta)-p_v]\exp\left(-\frac{u_{\mathrm{dif}}}{\mathcal{F}D_{12}}\delta\right) = p_1(0)-p_v \ . \qquad (8.79)$$

Thus, the deposition velocity due to diffusiophoresis is,

$$u_{\mathrm{dif}} = \frac{\mathcal{F}D_{12}}{\delta}\ln\left[\frac{p_v-p_1(0)}{p_v-p_1(\delta)}\right] = \frac{\mathcal{F}\beta_{12}}{\tilde{\rho}_1}\ln\left[\frac{p_v-p_1(0)}{p_v-p_1(\delta)}\right] \ , \qquad (8.80)$$

where $\beta_{12} = D_{12}\tilde{\rho}_1/\delta$ is a mass transfer parameter that can be obtained using the Chilton-Colburn analogy.

Thermophoresis. Aerosol particles suspended in the gas move with a constant velocity toward cooler temperature regions under the influence of a temperature gradient in the gas. This phenomenon is called *thermophoresis* and was apparently first described by Tyndall (1870).

The unbalanced force is presumably due to differences in molecular agitation and collision frequency due to local differences in the internal energy of the gas. The thermophoretic deposition velocity is given by Epstein et al. (1985) as:

$$u_{\text{th}} = \frac{\mu_g \kappa}{\chi \rho_g L}\left(\frac{T_\infty}{T_w} - 1\right)\left[\frac{1-(\kappa \text{Pr})^{1.25}\frac{T_w}{T_\infty}}{1-(\kappa \text{Pr})^{1.25}}\right]\text{Nu} \ , \tag{8.81}$$

where Nu is the Nusselt number for natural convection heat transfer, Pr is the gas Prandtl number, T_w is the temperature of the surface, T_∞ is the temperature of gas, κ is the nondimensional deposition velocity coefficient, L is the length of thermophoretic surface, μ_g is the gas viscosity, χ is a Stokes law correction factor,and ρ_g is the gas density.

Condensation and Evaporation. Noncondensible gas mixed with the condensing vapor has a significant influence on the resistance to heat transfer in the region of the condensation interface. This occurs because noncondensible gas is carried with the vapor toward the interface where it accumulates. The partial pressure of gas at the interface increases above that in the bulk of the mixture, producing a driving force for gas diffusion away from the surface.

That is, as shown in Fig. 8-21, since the total pressure remains constant, a partial pressure gradient of the gas exists opposite that of the vapor

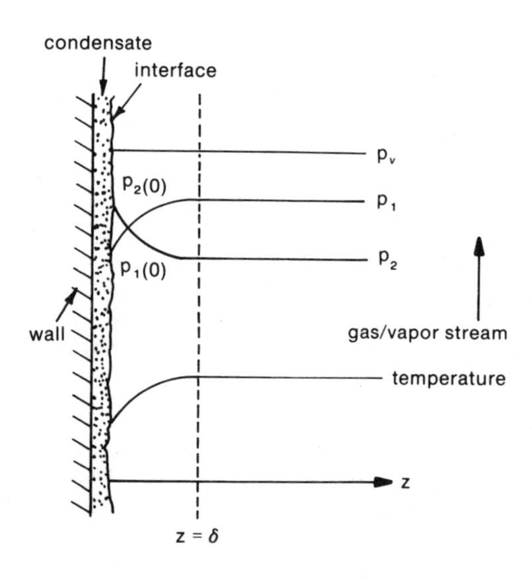

Fig. 8-21 Profiles of partial pressure during condensation.

(Collier, 1980). The velocity of this vapor toward the condensing surface is given by,

$$u_D = \frac{\beta_{12}}{\tilde{\rho}_1} \ln \left[\frac{p_v - p_1(0)}{p_v - p_1(\delta)} \right] , \tag{8.82}$$

where β_{12} is the mass transfer parameter, p_v is the mixture pressure, $p_1(0)$ is the partial pressure of vapor at the interface, $p_1(\delta)$ is the partial pressure of vapor at the edge of the diffusion "boundary layer" δ (i.e., bulk conditions), and $\tilde{\rho}_1$ is the molar density of vapor.

If u_D is defined to be positive for condensation, we see that it is just the Stefan velocity previously discussed.

Chemical Reaction with Surfaces. It has been widely assumed that, subsequent to the deposition of radioactive aerosols on a surface, they remain there unchanged and may heat up and revaporize due to decay heat and vapor superheat. Actually, depending on the chemical form of the surface and the deposited material, chemical reaction with the surface material may occur, dramatically changing the ability of the radioactive material to revolatilize.

The flux of fission products diffusing into a solid surface and reacting there is given by (Elrick et al., 1986):

$$n''_{ij} = K_{ij}[\rho_i(0) - \rho_i(y)] , \tag{8.83}$$

where n''_{ij} is the flux (g/cm^{2}sec) of fission product of species i reacting with material j, K_{ij} are the coefficients of chemical reaction (cm/sec), and $\rho_i(y)$ is the density of fission product i at distance y into the surface (g/cm^3), where $y = 0$ is the location of the surface of the solid structure.

At the chemical reaction front, $\rho_i(y)$ is assumed to be equal to zero. Thus, Eq. (8.83) reduces to,

$$n''_{ij} = K_{ij}\rho_i(0) . \tag{8.84}$$

Hence, the chemical reaction rate is given as,

$$\dot{R}_{\text{chem}} = n''_{ij} A_s , \tag{8.85}$$

where A_s is the surface area. A schematic of the various aerosol interactions, including surface chemistry, is shown in Fig. 8-22.

8.6.3.4 *Pool Scrubbing*

In BWR containment systems the radioactive material released from a degraded core may pass through a large (pressure suppression) pool of water before it is released into the so-called wetwell air space of the containment building. When an aerosol mixture of steam and condensibles is injected

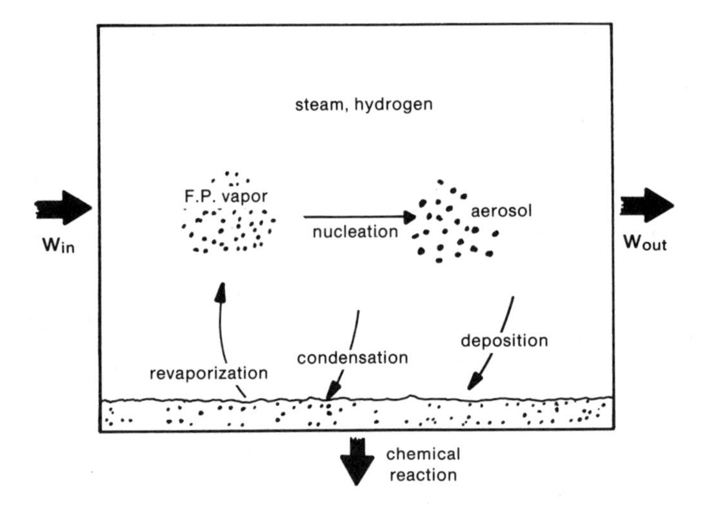

Fig. 8-22 Aerosol interaction in a control volume.

via a pipe or sparger into a pool of subcooled liquid, vapor bubbles are formed that rapidly condense until the partial pressure of the steam remaining in the bubbles reaches the saturation pressure corresponding to the local pool temperature. This condensation process causes significant diffusiophoresis to occur.

As the bubbles rise through the pool, many of the aerosol particles may be deposited on the liquid-vapor interface of the bubbles and are thus no longer entrained by the vapor-gas mixture in the bubbles. This process of aerosol removal is called *pool scrubbing*. A number of pool scrubbing mechanisms are involved, including sedimentation, diffusiophoresis, and centrifugal-induced aerosol deposition. Of these mechanisms, diffusiophoresis is normally the most important.

The efficiency with which pool scrubbing occurs is often given in terms of a decontamination factor (DF). This represents the ratio of the amount of radioactive material entering the liquid pool divided by the amount leaving. Typical DF values for BWR containment systems are around 100.

In addition to aerosol deposition due to diffusiophoresis, sedimentation must also be considered. This aerosol capture process occurs during the time the bubbles are rising through the pool. The sedimentation velocity, u_{sed}, is given by a force balance,

$$(\rho_a - \rho_v)\frac{g}{g_c}\frac{4}{3}\pi r_a^3 = 6\pi\mu_v r_a u_{sed}$$

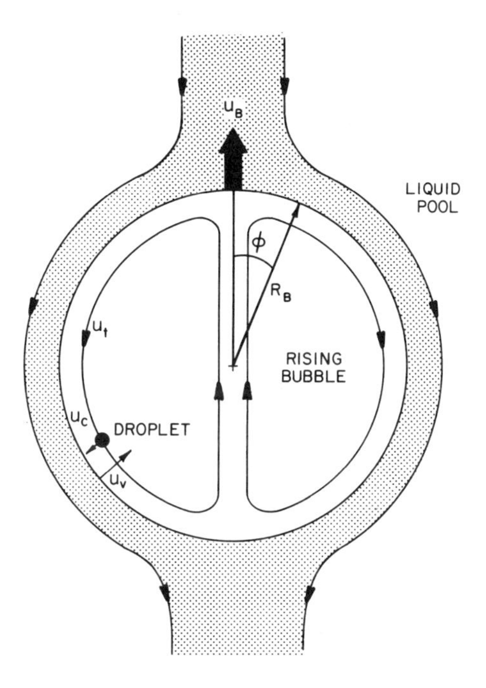

Fig. 8-23 The generation of centrifugal force on an aerosol droplet in a gas bubble.

or,

$$u_{sed} = \frac{2g(\rho_a - \rho_v)r_a^2}{9g_c\mu_v} \quad . \tag{8.86}$$

When applying Eq. (8.86) inside a bubble, the retarding effect of the vapor produced at the vapor-liquid interface of the rising bubble as the hydrostatic pressure is reduced should be considered. As can be seen in Fig. 8-23, the vertical upward component of the (radial) vapor velocity (u_v) is given by $-u_v \cos\phi$, or equivalently, $u_v \sin\theta$, where the angle θ is measured as positive when downward from the horizontal (i.e., $\theta = \phi - \pi/2$). Hence, the relative velocity in the vertical direction is,

$$u_r = u_{sed} - u_v \sin\theta \quad . \tag{8.87}$$

We note that no gravitational setting occurs when $u_r = 0$, and that this situation occurs at,

$$\theta_0 = \sin^{-1}(u_{sed}/u_v) \quad . \tag{8.88}$$

Hence, the aerosol sedimentation rate per unit volume (N'''_{sed}) is given by:

$$V_B N'''_{sed} = n'''_B \int_0^{\theta_0} u_r 2\pi R_B^2 \cos\theta \sin\theta \, d\theta$$

or,

$$N'''_{sed} = \pi R_B^2 (u_{sed}/u_v)^2 \left[u_{sed} - \frac{2u_v}{3}(u_{sed}/u_v) \right] (n'''_B/V_B) \; . \tag{8.89}$$

Another mechanism that should be considered is the deposition of aerosol particles due to the centrifugal force created by the gas circulation induced within the bubble due to the interfacial shear created by bubble motion through the pool. As can be seen in Fig. 8-23, when the water in the pressure suppression pool is clean (such that the bubble interface is mobil) the aerosol particles inside the bubbles will experience a centrifugal force, which is opposed by drag. The net drag on the aerosol particles must also include the effect of the radial vapor velocity just discussed.

A potential flow analysis indicates that the tangential velocity (u_t) inside the bubble at the vapor-liquid interface is given by,

$$u_t = \frac{3}{2} u_B \sin\phi \; , \tag{8.90}$$

where u_B is the bubble's rise velocity.

A force balance on an aerosol droplet within the bubble and adjacent to the interface yields,

$$\frac{\frac{4}{3} r_a^3 \rho_a \left(\frac{3}{2} u_B \sin\phi \right)^2}{g_c R_B} = \frac{1}{2g_c} \rho_v C_D \pi r_a^2 (u_c + u_v)^2 \tag{8.91}$$

or,

$$u_c = \left[\left(\frac{6 r_a \rho_a}{\rho_v C_D R_B} \right)^{1/2} u_B \sin\phi - u_v \right] . \tag{8.92}$$

We will only have droplet deposition due to centrifugal effects when $u_c > 0$. This starts at,

$$\phi_0 = \sin^{-1} \left[\frac{u_v}{u_B} \left(\frac{\rho_v C_D R_B}{6 r_a \rho_a} \right)^{1/2} \right] . \tag{8.93}$$

The net deposition rate due to centrifugal effects is thus,

$$V_B N'''_c = \int_{\phi_0}^{\pi - \phi_0} n'''_a u_c(\phi) 2\pi R_B^2 \sin\phi \, d\phi \tag{8.94}$$

or, integrating,

$$V_B N_c''' = 2\pi n_a''' \left(\frac{6 r_a R_B^3 \rho_a}{\rho_v C_D} \right)^{1/2} \left[\frac{(\pi - 2\phi_0)}{2} - \frac{1}{4} \sin 2(\pi - \phi_0) + \frac{1}{4} \sin(2\phi_0) \right]$$

$$- u_v n_a''' 2\pi R_B^2 [\cos(\pi - \phi_0) - \cos(\phi_0)] \ . \tag{8.95}$$

The material presented in this section briefly summarizes the current state-of-the-art in aerosol mechanics as it applies to BWR safety analysis. Needless to say, proper analysis of aerosol deposition mechanisms is essential to obtain realistic radiation source term evaluations for hypothetical nuclear reactor accidents.

References

Abramson, P. B., and H. Komoriya, "BWR Source Term Codes Comparison," Report EP86-19, ESEERCO (1983).

Anderson, J. G. M., Personal Communication, RISO, Denmark (1974).

Anderson, J. G. M., "REMI/HEAT COOL, A Model for Evaluation of Core Heatup and Emergency Core Spray Cooling System Performance for Light-Water-Cooled Nuclear Power Reactors," Report 296, RISO, Denmark (1973).

Anderson, J. G. M., L. B. Claassen, S. S. Dua, and J. K. Garrett, "Analysis of Anticipated Transients Without Scram in Severe BWR Accidents," EPRI NP-5562, Electric Power Research Institute (1987).

Baker, L., and L. C. Just, "Studies of Metal-Water Reactions at High Temperature, III. Experimental and Theoretical Studies of the Zircaloy-Water Reaction," ANL-6548, Argonne National Laboratory (1962).

Brock, J. R., "On the Theory of Thermal Forces Acting on Aerosol Particles," *J. Colloid Sci.*, **17**, 768 (1962).

Bunz, H., M. Koyro, and W. Schöck, "NAUA-Mod 3: Ein Computer Programm Zur Beschreiibung des Aerosolverhaltene in Kondensierender Atmosphäer," KfK-3514, Kernforschungzentrum Karlsruhe (1981).

Chexal, B., and W. Layman, "A Correlation for Predicting Reactor Power During a BWR ATWS," *Trans. Am. Nucl. Soc.*, **53** (1986).

Cho, C-S., H. Jia, D-H. Kim, S-W. Kim, R. T. Lahey, Jr., and M. Z. Podowski, "Degraded BWR Core Modeling—*APRIL. MOD3* Severe Accident Code," Final Report EP84-4, ESEERCO (1990).

Collier, J. G., *Convective Boiling and Condensation*, 2nd ed., McGraw-Hill, New York (1980).

Dallman, R. J., R. E. Gottula, E. E. Holcomb, W. C. Jouse, S. R. Wagoner, and P. D. Wheatley, "Severe Accident Sequence Analysis Program—Anticipated Transient Without Scram Simulations for Browns Ferry Nuclear Plant Unit-1," NUREG/CR-4165, U.S. Nuclear Regulatory Commission (1987).

Davies, C. N., *Aerosol Science*, Academic Press, New York (1966).

"Decay Energy Release Rates Following Shutdown of Uranium-Fueled Thermal Reactors," proposed ANS Standard, 5.1/N-18.6, American Nuclear Society (1971).

Dix, G. E., "BWR Loss of Coolant Technology Review," *Proc. ANS Symp. Thermal-Hydraulics of Nuclear Reactors*, Vol. 1 (1983).

Duderstadt, James J., and William R. Martin, *Transport Theory,* John Wiley and Sons, New York (1979).

Dunbar, I. H., "The Role of Diffusiophoresis in LWR Accidents," *Proc. Int. Mtg. Light Water Reactor Severe Accident Evaluation,* ANS-700085, American Nuclear Society, La Grange Park, Illinois (1983).

Elrick, R. M., R. A. Sallach, A. L. Ouellette, and S. C. Douglas, "Reaction Between Some Cesium-Iodine Compounds and the Reactor Materials 304 Stainless, Inconel 600 and Silver," NUREG/CR-3197, SAND 83-0395, Sandia National Laboratories (1986).

Epstein, M., and P. G. Ellison, "Correlations of the Rate of Removal of Coagulating and Depositing Aerosols for Application to Nuclear Reactor Safety Problems," *J. Nucl. Eng. Des.,* **107,** *3,* 327–344 (1988).

Epstein, M., P. G. Ellison, and R. E. Henry, "Correlation of Aerosol Sedimentation," *J. Colloid. Interface Sci.,* **113,** *2,* 342–355 (1986).

Epstein, M., G. M. Hauser, and R. E. Henry, "Thermophoretic Deposition of Particles in Natural Convection Flow from a Vertical Plate," *J. Heat Transfer,* **107,** 272–276 (1985).

Fakory, M. R., and R. T. Lahey, Jr., "An Investigation of BWR/4 Parallel Channel Effects During a Hypothetical LOCA for Both Intact and Broken Jet Pumps," *Nucl. Technol.,* **65,** *2* (1984).

Friedlander, S. K., *Smoke, Dust and Haze,* John Wiley and Sons, New York (1977).

Friedlander, S. K., and H. F. Johnstone, "Deposition of Suspended Particles from Turbulent Gas Streams," *Ind. Eng. Chem,* **49,** 1151 (1957).

Fuchs, N. A., *The Mechanics of Aerosols,* p. 162, Pergamon Press, New York (1964).

Gelbard, F., et al., "Sectional Representations for Simulating Aerosol Dynamics," *J. Colloid Interface Sci.,* **76,** 2 (1980).

Henry, R. E., et al., "MAAP User's Manual, Vol-I & II," Fauske & Associates (1987).

Hill, P. G., H. Witting, and E. P. Demetri, "Condensation of Metal Vapors During Rapid Expansion," *J. Heat Transfer,* **85,** 303–317 (1963).

Kim, S. H., D. H. Kim, B. R. Koh, J. Pessanha, El-K. SiAhmed, M. Z. Podowski, and R. T. Lahey, Jr., "The Development of APRIL.MOD2—A Computer Code for Core Meltdown Accident Analysis of Boiling Water Nuclear Reactors," NUREG/CR-5157, U.S. Nuclear Regulatory Commission (1988).

Loyalka, S. K., "Mechanics of Aerosols in Nuclear Reactory Safety: A Review," *Prog. Nucl. Energy,* **12,** *7,* 1–56 (1983).

Owczarski, P. C., R. I. Schreck, and A. K. Postma, "Technical Bases and User's Manual for the Prototype of a Suppression Pool Aerosol Removal Code (SPARC)," NUREG/CR-3317, PNL-4742, Pacific Northwest Laboratories (1985).

Ritchie, L. T., J. P. Johnson, and R. B. Bland, "Calculation of Reactor Accident Consequences, Version-2," NUREG/CR-2324, U.S. Nuclear Regulatory Commission (1981).

Siegel, R., and J. R. Howell, *Thermal Radiation Heat Transfer,* McGraw-Hill, New York (1972).

Sozzi, G. L., and W. A. Sutherland, "Critical Flow Measurements of Saturated and Subcooled Water at High Pressure," NEDO-13418, General Electric Company (1975).

Sparrow, E. M., and R. D. Cess, *Radiation Heat Transfer,* Brooks/Cole Publishing Company, Monterey, California (1967).

Sun, K. H., G. E. Dix, and C. L. Tien, "Effect of Precursory Cooling on Falling-Film Rewetting," ASME paper 74-WA/HT-52, American Society of Mechanical Engineers (1974).

Sun, K. H., J. M. Gonzalez, and C. L. Tien, "Calculations of Combined Radiation and Convection Heat Transfer in Rod Bundles Under Emergency Cooling Conditions," ASME paper 75-HT-64, American Society of Mechanical Engineers (1975).

Sutherland, W. A., and R. T. Lahey, "BWR Blowdown Heat Transfer Experiments," *Trans. Am. Nucl. Soc.*, **21,** 346 (1975).

"Technical Basis for Estimating Fission Product Behavior During LWR Accidents," NUREG/CR-0772, U.S. Nuclear Regulatory Commission (1981).

10CFR50, Appendix-K, *Federal Register,* **39,** *3* (1974).

Tyndall, J., *Proc. R. Inst. Gr. Br.*, **6,** *3* (1870).

Von Smolnchowski, M., "Mathematical Theory of the Kinetics of the Coagulation of Colloidal Solutions," *Z. Phys. Chem.*, **92,** 129 (1917).

Walker, B. C., C. R. Kirby, and R. J. Williams, "Discretisation and Integration of the Equation Governing Aerosol Behavior," J. K. Report, SRD-R-98 (1976).

Wooton, R. O., and H. I. Avci, "MARCH Code Description and User's Manual," NUREG/CR-1711, U.S. Nuclear Regulatory Commission (1980).

Yamanouchi, A., "Effect of Core Spray Cooling in Transient State After Loss-of-Coolant Accident," *Nucl. Appl.*, **5** (1968).

CHAPTER NINE

Boiling Water Reactor Accident Evaluations

All commercial nuclear reactors have as part of their design criteria the evaluations of certain postulated accidents. The designer must have available the analytical tools with which to appraise the consequence of these various accident conditions to determine whether relevant safety limits are satisfied. In Sec. 9.1, some analytical techniques, typical of those used in the evaluation of certain hypothetical single-operator error or equipment malfunction-type accidents, are discussed and in Sec. 9.2, analysis relevant to the design basis loss-of-coolant accident (LOCA) is presented.

As discussed in Chapter 8, there are a number of hypothetical accidents that can be classified as single-operator error or equipment malfunction-type accidents. All of these involve a transient evaluation of the thermal margin; that is, an evaluation of the transient critical power ratio (CPR). One of the best examples of the analysis of transient boiling transition (BT) in a BWR is that for a recirculation pump trip or seizure. Hence, this is the case examined in some detail in Sec. 9.1.

9.1 Transient Boiling Transition Evaluations

If a BT occurs at all in a nuclear reactor, it is most likely to occur during transient accident conditions of system pressure, power, and flow rate. Thus, the designer must be able to analyze these transients accurately to appraise realistically the potential for fuel rod damage.

The techniques required are quite similar for the various accidents that must be considered. Basically, we must have the analytical capability (typically in the form of a digital computer code) to evaluate the flow and enthalpy in the "hottest" fuel assembly (bundle) at each point in space and time. These flow parameters then can be used in conjunction with an appropriate bundle-average BT correlation to evaluate the transient CPR. If the minimum critical power ratio (MCPR) is always >1.0, then no BT would be expected. Conversely, if we predict an MCPR ≤ 1.0 at some point in space and time, then a BT may occur.

Rather than analyze each postulated accident individually, we will concentrate on the recirculation pump trip or seizure-type accident, since the analysis is somewhat less complex and, yet, it is typical of the whole class of transients under consideration.

Although the main purpose of this section is to describe the analysis involved, it is appropriate to summarize some salient experimental observations. Indeed, it has been the need to accurately predict transient experimental data of this type that has motivated the development of the analytical techniques to be discussed. For our purposes here, we consider a constant power, flow decay transient performed in a single-rod annulus with uniform axial heat flux (Shiralkar et al., 1972). This is typical of the separate effects-type experiments required to check the prediction capabilities of transient analytical methods.

Figure 9-1 gives the trajectories of several typical flow decay transients in the flux-quality plane. These trajectories are the loci of state points at the end of the heated length, where BT occurs, of a heater rod with a uniform axial heat flux distribution. The initial condition of these transients is a mass flux of $G=1.0\times10^6$ lb/h-ft^2 and an inlet subcooling of 50 Btu/lb. The flow is then decayed at some prescribed rate to $G=0.5\times10^6$ lb/h-ft^2. During this transient, the system pressure is constant at 1000 psia. Since typical flow transients are rather fast, O (0.5 to 3 sec), it is found that the inlet temperature, and thus the inlet subcooling, remains approximately constant during the transient. Hence, the end points of the transient are known and, to a good first approximation, can be connected by a trajectory of constant heat flux.

Trajectory number 1 is the locus of state points during the transient at constant heat flux, q_1''. Note that even at the end of the transient, BT would not be expected from steady-state considerations. Indeed, it has been observed experimentally that a BT does not occur for transients typical of trajectory number 1.

Trajectory number 2 is the locus of the same flow decay transient starting from a higher heat flux, q_2''. Note in Fig. 9-1 that a BT would be expected during the flow decay since the trajectory exceeds the steady-state thermal limits near the end of the transient. Indeed, it has been verified experimentally that the use of plots such as that given in Fig. 9-1 allows us to determine *a priori* whether a BT will occur, but does not give any information as to when in time it will occur. The fact that we desire to obtain accurate predictions of the time to BT has motivated much of the transient analysis that follows.

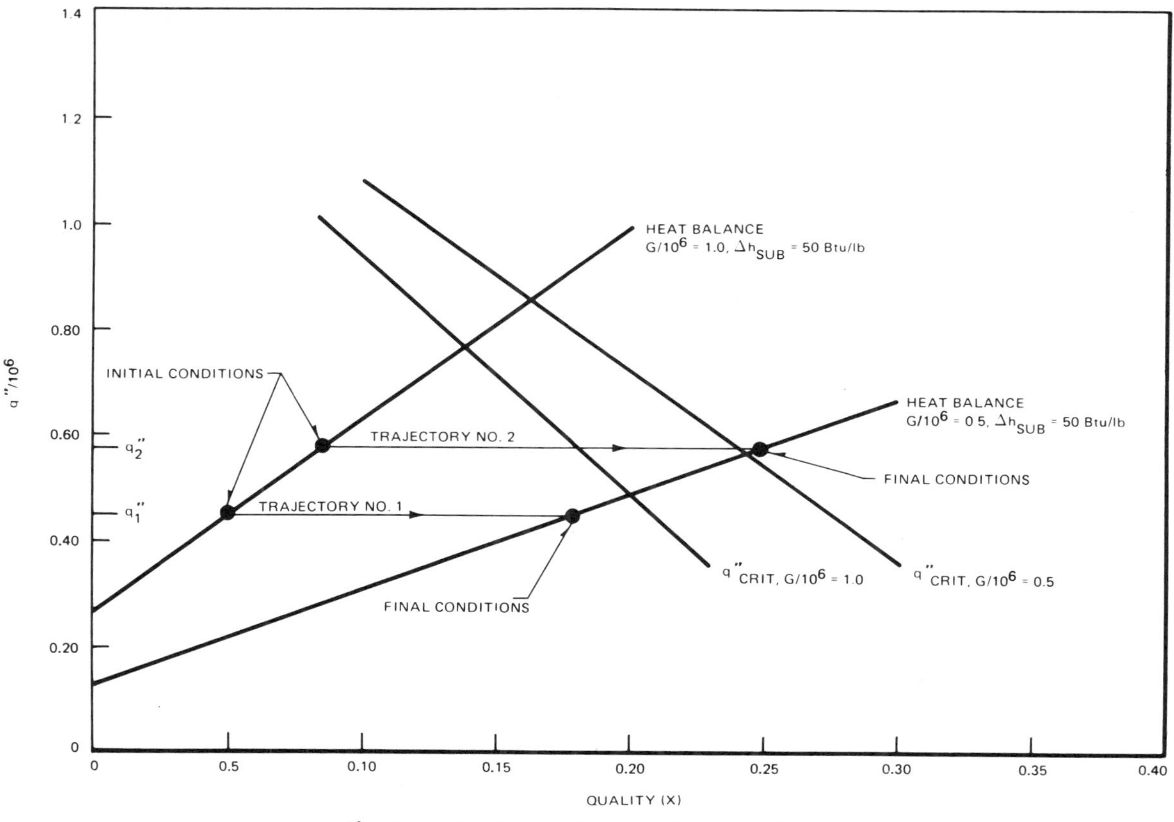

Fig. 9-1 The locus of typical flow transients.

We now consider the case of a BWR recirculation pump trip or seizure. To evaluate the transient thermal margin, we must be able to predict accurately the flow and enthalpy in the fuel assemblies at each point in space and time. In practice, this is normally accomplished by the simultaneous numerical solution of the appropriate two-phase conservation equations given in Sec. 5.1, in conjunction with void-quality relations such as those given in Sec. 5.2. Rather than detail the specific numerical solution scheme required to handle the most general problem, we consider a somewhat simpler problem for which an exact solution can be achieved. More specifically, we consider the case of an exponential flow decay in a homogeneous two-phase system.

The assumptions made in the subsequent analysis are:

1. The flow is homogeneous.[a]
2. The system pressure is constant.
3. The inlet conditions are specified functions of time.
4. The heat flux is constant in space and time.
5. Both phases are incompressible; i.e., constant thermodynamic properties.
6. The two phases are in thermodynamic equilibrium.
7. There is a constant cross-sectional flow area.
8. There is negligible internal generation, q''', and viscous dissipation in the coolant.

With these assumptions, the continuity and energy equations, as given in Eqs. (5.29) and (5.44), respectively, can be written,

$$\frac{D_j\langle\rho_h\rangle}{Dt}+\langle\rho_h\rangle\frac{\partial\langle j\rangle}{\partial z}=0 \tag{9.1}$$

$$\langle\rho_H\rangle\frac{D_j\langle h\rangle}{Dt}=q''\left(\frac{P_H}{A_{x-s}}\right) , \tag{9.2}$$

and the appropriate equation-of-state can be written as,

$$\langle\rho_h\rangle=\frac{1}{(v_f+\langle x\rangle v_{fg})} \tag{9.3}$$

$$\langle h\rangle=h_f+\langle x\rangle h_{fg} \ . \tag{9.4}$$

[a]This assumption can be relaxed at the expense of a somewhat more complicated analysis (Lahey et al., 1972).

It can be shown (Gonzalez-Santalo and Lahey, 1973) that this system of equations forms a degenerate hyperbolic system with characteristics given by,

$$\frac{dz}{dt} = \langle j(z, t) \rangle \ , \tag{9.5}$$

and,

$$t = \text{constant} \ . \tag{9.6}$$

Equation (9.5) can be recognized as the differential equation describing the trajectory of the particle paths.

Note that due to assumptions 2 and 5, we have been able to leave out the momentum equation [i.e., Eq. (5.38)], since the system of equations is decoupled. This implies that acoustic waves propagate through the test section instantaneously, an assumption that only introduces serious errors for very rapid flow transients where the duration of the transient is of the same order as the time it takes acoustic waves to propagate through the test section.

In the more general case, in which compressibility effects are included, it can be shown (Tong and Weisman, 1970) that there are three characteristic curves: the particle paths ($dz_1/dt = \langle j \rangle$) and two curves ($dz_{2,3}/dt = \langle j \rangle \pm C_{2\phi}$) representing (in the space-time plane) the trajectories of the acoustic waves propagating in the test section. In that solution, as the propagation velocity of the acoustic waves increases, the two acoustic characteristics collapse into a single trajectory that is identical to the constant time characteristic given in Eq. (9.6).

For all cases of practical significance in BWR technology, we can neglect acoustic effects and use the simplified analysis given by Eqs. (9.1) through (9.6) with the appropriate initial conditions. As we will see in the development that follows, this simplification allows us to construct exact, closed-form solutions rather than having to use numerical schemes. Although computer evaluations are normally used for actual design application, we frequently obtain more insight into the nature of the phenomena when exact solutions can be constructed.

The point of view that we adopt is the Lagrangian description in which the "observer" travels with a particular control volume. Physically, this is more meaningful than the Eulerian point of view, in which the observer is fixed in space, since one set of characteristics, given by Eq. (9.5), of this hyperbolic system is just the Langrangian trajectory (i.e., the particle path).

There are two initial conditions that we must consider for the control volumes of interest. For those control volumes that were in the two-phase portion of the heater at the start of the transient,

$$\langle x \rangle = \langle x_0 \rangle \text{ at } t = 0 \ . \tag{9.7}$$

For those subcooled control volumes that were either in the heater or had not yet entered the heater at the start of the transient,

$$\langle x\rangle = 0 \text{ at } t = t_0 \ , \tag{9.8}$$

where t_0 is the time at which the control volume under consideration first experiences bulk boiling; that is, the time at which that control volume crosses the so-called boiling boundary.

To establish the transient position of the boiling boundary, we must work out the dynamics of the single-phase portion of the heater. In the single-phase region, the continuity equation, Eq. (9.1), reduces to,

$$\frac{\partial\langle j\rangle}{\partial z} = 0 \ , \tag{9.9}$$

in which constant liquid density has been assumed. This equation implies that fluid motion is essentially that of a "rigid body." That is,

$$\langle j(z,\ t)\rangle = \langle j_i(t)\rangle \ , \tag{9.10}$$

in which the subscript, i, refers to inlet conditions. In the single-phase region, the energy equation, Eq. (9.2), reduces to,

$$\frac{D_j\langle h\rangle}{Dt} = \frac{q''P_H}{\rho_f A_{x-s}} \ . \tag{9.11}$$

In the Lagrangian frame of reference, this differential equation can be integrated as if it were an ordinary differential equation. The result of the integration gives us ν, the time required for a control volume that enters the heater to lose its subcooling. That is, for $t \geqslant \nu$,

$$\int_{h_i}^{h_f} d\langle h\rangle = \int_{t-\nu}^{t}\left(\frac{q''P_H}{\rho_f A_{x-s}}\right)dt' \ . \tag{9.12}$$

Thus, using assumption 4,

$$\nu = \left(\frac{\rho_f A_{x-s}}{q''P_H}\right)\Delta h_{\text{sub}} \ . \tag{9.13}$$

Knowing ν, the next step in the analysis is to evaluate the transient position of the boiling boundary, $\lambda(t)$, shown in Fig. 9-2. The required expression was formulated previously by Wallis (Wallis and Heasley, 1961) as,

$$\lambda(t) = \int_{t-\nu}^{t} \frac{G_i(t')}{\rho_f}dt' \ . \tag{9.14}$$

Once the time dependence of the inlet flow has been specified, Eq. (9.14) can readily be evaluated. For the case under consideration here (i.e., exponential flow decay),

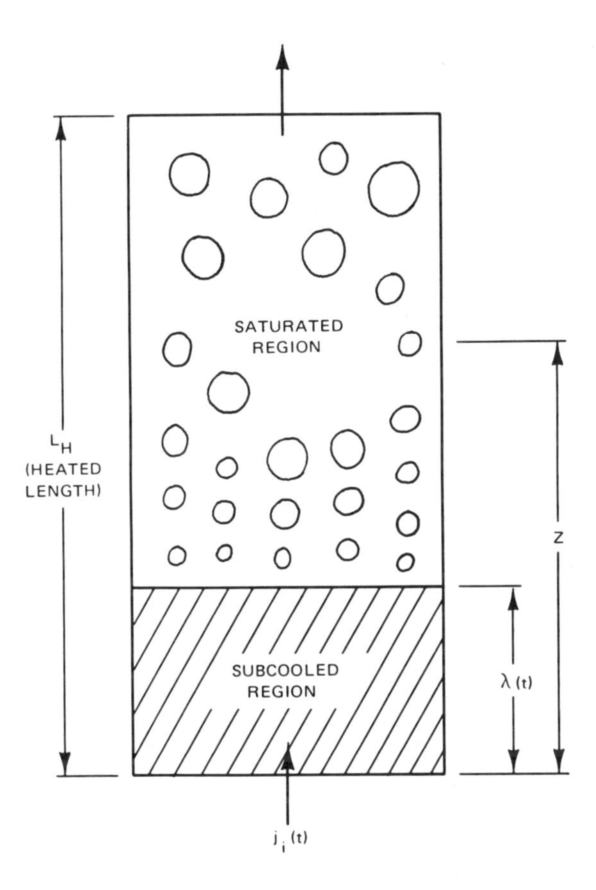

Fig. 9-2 Schematic of a heated channel undergoing convective boiling.

$$G_i(t) = \begin{cases} G_0, \; t \leqslant 0 \\ G_0 \exp(-K_2 t), \; t > 0 \end{cases} . \tag{9.15}$$

Equation (9.14) must be integrated in two parts, depending on how the time, t, compares to ν. That is, for $t \leqslant \nu$,

$$\lambda(t) = \int_{t-\nu}^{0} \frac{G_0}{\rho_f} dt' + \int_0^t \frac{G_0}{\rho_f} \exp(-K_2 t') dt' \; . \tag{9.16}$$

Thus,

$$\lambda(t) = \frac{G_0}{\rho_f}(\nu - t) + \frac{G_0}{\rho_f K_2}[1 - \exp(-K_2 t)] \; . \tag{9.17}$$

Similarly, for $t \geqslant \nu$,

$$\lambda(t)=\int_{t-\nu}^{t} \frac{G_0}{\rho_f} \exp(-K_2 t')dt' = \frac{G_0}{\rho_f K_2} \exp(-K_2 t)[\exp(K_2 \nu)-1] \ . \qquad (9.18)$$

Since there are two expressions for the boiling boundary, λ, depending on the value of the time considered, it should be anticipated that integrations in the two-phase region have to be done in two parts.

We now turn our attention to the two-phase region of the heater. To reduce the number of variables involved, it is convenient to write the continuity and energy equation in terms of quality, $\langle x \rangle$. In the case of the continuity equation, Eqs. (9.1) and (9.3) can be combined to yield,

$$\frac{\partial \langle j \rangle}{\partial z} = \frac{1}{(v_f + \langle x \rangle v_{fg})} \left(\frac{D_j v_f}{Dt} + \langle x \rangle \frac{D_j v_{fg}}{Dt} + v_{fg} \frac{D_j \langle x \rangle}{Dt} \right) \ . \qquad (9.19)$$

This equation now can be integrated from the heater inlet to some arbitrary position, z, in the two-phase region. Thus, for the heated channel shown in Fig. 9-2,

$$\frac{dz}{dt} = \frac{G_i(t)}{\rho_f} + \int_0^{\lambda(t)} \frac{1}{v_f} \frac{D_j v_f}{Dt} dz$$

$$+ \int_{\lambda(t)}^{z} \frac{1}{(v_f + \langle x \rangle v_{fg})} \left(\frac{D_j v_f}{Dt} + \langle x \rangle \frac{D_j v_{fg}}{Dt} + v_{fg} \frac{D_j \langle x \rangle}{Dt} \right) dz \ , \qquad (9.20)$$

where we have used Eq. (9.5) and the fact that $\langle j_i(t) \rangle = G_i(t)/\rho_f$ and, in the subcooled region, $\langle x \rangle = 0$, by definition.

Next we consider the appropriate energy equation. As shown in Chapter 5, we can combine Eqs. (5.49), (9.3), and (9.4) to eliminate $\langle \rho_h \rangle$ and $\langle h \rangle$ in favor of $\langle x \rangle$, thus obtaining Eq. (5.51),

$$\frac{D_j \langle x \rangle}{Dt} + \left\{ \frac{1}{h_{fg}} \left[\frac{D_j h_{fg}}{Dt} - \left(q'' P_H / A_{x-s} + \frac{D_j p}{Dt} \right) v_{fg} \right] \right\} \langle x \rangle$$

$$= \frac{1}{h_{fg}} \left[\left(q'' P_H / A_{x-s} + \frac{D_j p}{Dt} \right) v_f - \frac{D_j h_f}{Dt} \right] \ . \qquad (9.21)$$

Equations (9.20) and (9.21) are more general than required for solution of the present problem. That is, they include the "flashing" term, $D_j p / Dt$, due to depressurization and the effects of pressure on thermodynamic properties. For our particular problem, assumptions 2 and 5 can be applied to reduce these equations to,

$$\frac{dz}{dt} = \frac{G_i(t)}{\rho_f} + \int_{\lambda(t)}^{z} \frac{v_{fg}}{(v_f + \langle x \rangle v_{fg})} \left(\frac{D_j \langle x \rangle}{Dt} \right) dz \qquad (9.22)$$

$$\frac{D_j\langle x\rangle}{Dt}=\frac{\Omega}{v_{fg}}(v_f+\langle x\rangle v_{fg}) \ , \tag{9.23}$$

where Ω previously was defined in Eq. (5.48).

By taking the Lagrangian point of view, Eq. (9.23) can be regarded as an ordinary differential equation along the particle path and is readily integrated. For the initial conditions given in Eq. (9.7), the solution is,

$$\langle x(t,\ z_0)\rangle=\langle x_0(z_0)\rangle\ \exp(\Omega t)+\frac{v_f}{v_{fg}}[\exp(\Omega t)-1] \ , \tag{9.24}$$

where from a steady-state energy balance,

$$\langle x_0(z_0)\rangle=\frac{1}{h_{fg}}\left(\frac{q''P_H z_0}{G_0 A_{x-s}}-\Delta h_{\text{sub}}\right) \ . \tag{9.25}$$

For the initial conditions given in Eq. (9.8),

$$\langle x(t,\ t_0)\rangle=\frac{v_f}{v_{fg}}\{\exp[\Omega(t-t_0)]-1\} \ . \tag{9.26}$$

To integrate Eq. (9.22), Eq. (9.23) can be combined with the integrand to yield,

$$\frac{dz}{dt}=\frac{G_i(t)}{\rho_f}+\int_{\lambda(t)}^{z}\Omega dz \ . \tag{9.27}$$

For the case under consideration here, assumption 4 implies that Ω is constant. Thus, Eq. (9.27) integrates as,

$$\frac{dz}{dt}-\Omega[z-\lambda(t)]=\frac{G_i(t)}{\rho_f} \ . \tag{9.28}$$

Equation (9.28) can be integrated exactly for virtually any constant pressure flow transient of practical significance. For the specific case of an exponential flow decay, Eqs. (9.15), (9.17), and (9.18) can be combined with Eq. (9.28) and the various initial conditions of interest to yield the axial position of the control volume as a function of time; i.e., $z(t)$. Thus, by knowing dz/dt, $z(t)$, and $\langle x(t)\rangle$, we can easily evaluate the flow and enthalpy at each point in space and time.

We now turn our attention to the integration of Eq. (9.28). As discussed previously, there are three separate groups of fluid particles that must be considered. Each fluid particle can be identified by its state at the beginning of the transient. One group contains those fluid particles that were within the two-phase portion of the heater at $t=0$. As shown in Fig. 9-3, these particles are bounded by the "main particle path characteristic." That is, the trajectory of the control volume that was located at the boiling boundary

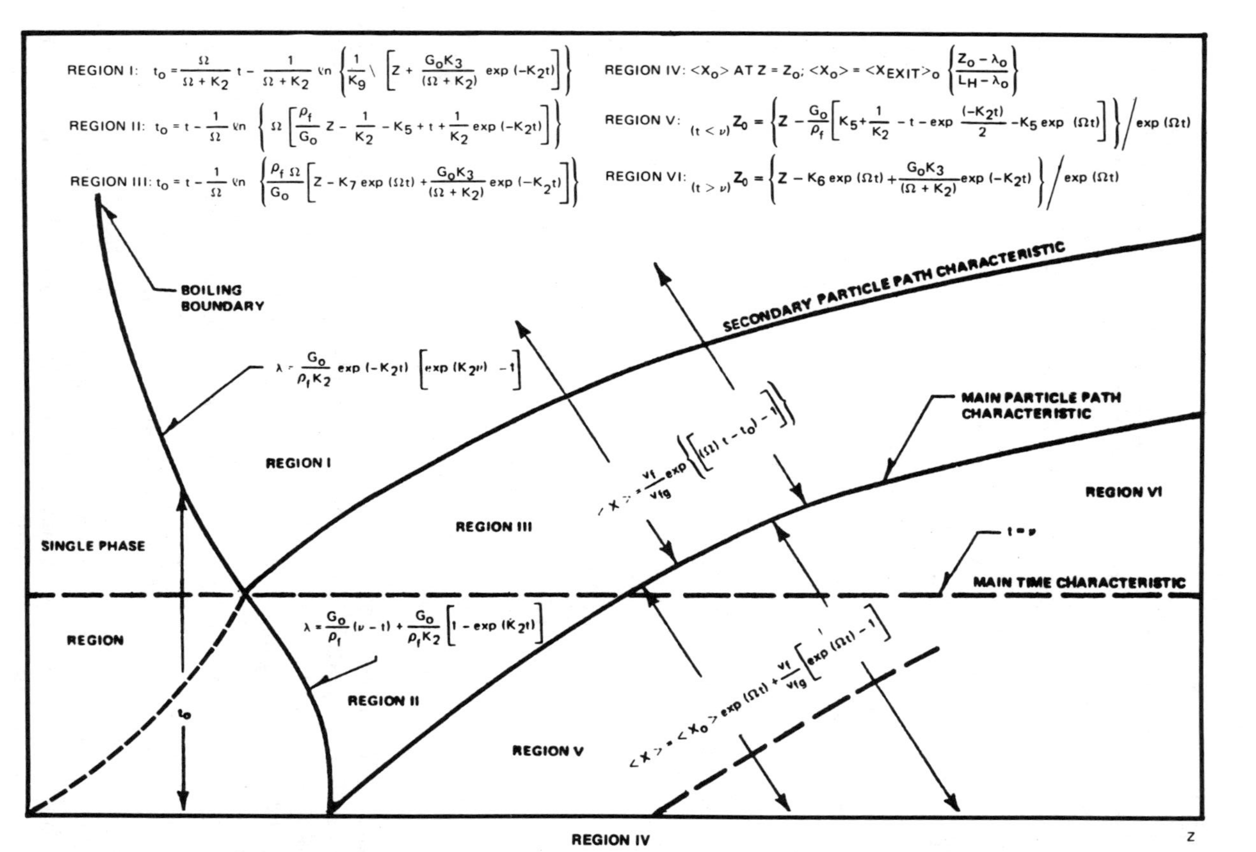

Fig. 9-3 Solution map for constant pressure exponential flow decay.

at $t=0$. The next group contains those particles that were in the single-phase portion of the heater at $t=0$. This group of particles is bounded by the "main" and "secondary particle path characteristic." This latter characteristic is defined as the trajectory of the control volume that was located at the inlet to the heater at the beginning of the transient. The final group contains those particles that had not yet entered the heater at $t=0$. These particles are bounded only by the secondary particle path characteristic.

In addition to the particle path characteristics forming natural boundaries for the various regions, the constant time characteristic at $t=\nu$ also subdivides the solution plane shown in Fig. 9-3. This characteristic has been defined as the "main time characteristic." Its importance is a mathematical consequence of the fact that $\lambda(t)$ has "branches" depending on whether $t \geq \nu$ or $t \leq \nu$.

As shown in Fig. 9-3, there are seven regions into which the solution naturally divides. The first region is for single-phase flow. For the assumptions made here, the solution in the single-phase region is trivial and we will not discuss it further. The physical significance of the other six regions is tabulated in Table 9-1. For each of these regions, the exact solution for quality, $\langle x(t) \rangle$, and axial position, $z(t)$, is shown in Fig. 9-3.

These exact solutions are obtained through integration of the continuity and energy equations. As a specific example, we consider the case of Region I. For this region, the initial conditions are given by Eq. (9.8) and the appropriate expression for $\lambda(t)$ is Eq. (9.18). The appropriate expression for quality is given by Eq. (9.26). To complete the solution, we must com-

TABLE 9-1
Tabulation of Various Space-Time Regions in the Two-Phase Portion of the Heater

Region	Expression for Boiling Boundary	Significance of Region
I	$\lambda(t)=\frac{G_0}{\rho_f K_2}\exp(-K_2 t)[\exp(K_2\nu-1)]$	The particles that had not yet entered the heater at $t=0$.
II	$\lambda(t)=\frac{G_0}{\rho_f}(\nu-t)+\frac{G_0}{\rho_f K_2}[1-\exp(K_2 t)]$	The particles that at $t=0$ were in the single-phase portion of the heater but only for $t \leq \nu$.
III	$\lambda(t)=\frac{G_0}{\rho_f K_2}\exp(-K_2 t)[\exp(K_2\nu)-1]$	The particles that at $t=0$ were in the single-phase region of the heater but only for $t \geq \nu$.
IV	$\lambda=\lambda_0$	The particles in the two-phase portion of the heater for $t \leq 0$.
V	$\lambda(t)=\frac{G_0}{\rho_f}(\nu-t)+\frac{G_0}{\rho_f K_2}[1-\exp(K_2 t)]$	The particles in the two-phase portion of the heater for $0 \leq t \leq \nu$.
VI	$\lambda(t)=\frac{G_0}{\rho_f K_2}\exp(-K_2 t)[\exp(K_2\nu)-1]$	The particles in the two-phase portion of the heater for $t \geq \nu$.

bine the continuity equation, Eq. (9.28), with Eqs. (9.15) and (9.18) and integrate. The resultant differential equation for Region I is,

$$\frac{dz}{dt} - \Omega z = \frac{G_0}{\rho_f} \exp(-K_2 t)\left\{1 - \frac{\Omega}{K_2}[\exp(K_2 \nu) - 1]\right\} . \tag{9.29}$$

This differential equation is of the Bernoulli type and is easily integrated,

$$z = K_9 \exp[-(\Omega + K_2)t_0] \exp(\Omega t) - \frac{G_0 K_3}{(\Omega + K_2)} \exp(-K_2 t) , \tag{9.30}$$

where,

$$K_3 \triangleq \frac{1}{\rho_f}\left\{1 - \frac{\Omega}{K_2}[\exp(K_2 \nu) - 1]\right\}$$

$$K_9 \triangleq \frac{G_0 \exp(K_2 \nu)}{\rho_f (K_2 + \Omega)} .$$

Equation (9.30) also can be written as,

$$t_0 = \frac{\Omega}{(\Omega + K_2)} t - \frac{1}{(\Omega + K_2)} \ln\left\{\frac{1}{K_9}\left[z + \frac{G_0 K_3}{(\Omega + K_2)} \exp(-K_2 t)\right]\right\} . \tag{9.31}$$

Equations (9.26) and (9.30) specify the solution for quality and axial location as a function of time. Region I is an interesting case, since if the parameter, K_2, is complex, this region represents the asymptotic response of the system, while Regions II, III, V, and VI represent the "startup" transient. In the special case of an imaginary K_2, only Region I exists and the results presented here are essentially the same as those previously used in the stability analysis of a boiling channel (Wallis and Heasley, 1961).

We can proceed in a similar fashion for the other regions. The results of these integrations are tabulated in Table 9-2. For real K_2 (i.e., noncomplex), these results represent the exact solution for an exponential flow decay in a homogeneous two-phase system. A typical flow chart for numerically evaluating this solution is given in Fig. 9-4. Note in this figure, the flow and quality are easily evaluated at each point in space and time. Thus, for this particular transient, the likelihood of transient BT can be readily appraised through the use of the exact solution for flow and quality and the appropriate BT correlation.

An example of typical prediction capability is given in Fig. 9-5. Note that for transients typical of BWR technology, predictions using a steady-state BT correlation (GETAB, 1973) are quite good. However, it has been found that for very rapid transients a steady-state BT correlation is generally conservative in that it underpredicts the transient critical power (Aoki et al., 1973).

To better appreciate the effect of transient speed on BT, let us consider some simple pool-boiling experiments. Figure 9-6 is typical of transient pool-boiling data (Sakurai and Shiotsu, 1974; Tachibana et al., 1968) taken for exponential power excursions. Sakurai (Sakurai and Shiotsu, 1974) found that due to time delays associated with activation of the boiling cavities, the stationary (i.e., quasi-steady) pool-boiling curve was normally not traversed during exponential power transients if the heated surface was not boiling before initiation of the transient. Curves 1 and 2 in Fig. 9-6 illustrate the observed phenomena for two different power excursions. Note that the more rapid power excursion (Curve 2) has a greater overshoot from the stationary pool-boiling curve. Also note that the critical heat flux (CHF) point associated with the stationary pool-boiling curve is exceeded for sufficiently rapid transients; i.e., for τ on the order of milliseconds. In addition, note from Curve 3 that if the heated surface were boiling at the time of initiation of the power excursions (i.e., if the cavities were already activated), the stationary pool-boiling curve is traversed during the excursion, although, as in Curve 2, the transient CHF can be much larger than the stationary CHF.

From these basic experiments, we can infer that the precise value of the transient critical power apparently involves history and surface effects as well as transient type and speed. Fortunately, as indicated in Fig. 9-5, for rather mild transients of interest in BWR technology, steady-state BT correlations do a very good job of predicting BT.

The idealized thermal-hydraulic analysis of transient BT, which resulted in Tables 9-1 and 9-2, was restricted to an exponential flow transient in which the heat flux and pressure were constant. Most transients of interest in BWR technology involve simultaneous variations of pressure, heat flux, and flow and, thus, do not readily lend themselves to exact, analytical solutions. Nevertheless, computerized numerical solution schemes based on the "method of characteristics" can be readily constructed (Lahey et al., 1972).

It is of interest to consider the results of the numerical evaluation of a constant heat flux and inlet flow rate, depressurization transient in a 0.5-in-diam heated tube. These results are typical of those that would be obtained from a transient computer code. In Fig. 9-7, several characteristics are shown in the space-time plane. In addition, lines of constant quality are shown. We consider the time varying quality at the end of the heated length, $z=6$ ft. Note that during the first part of the transient, the exit quality actually decreases even though significant flashing is occurring. This is best seen in Fig. 9-8, which is a plot of exit quality versus time. During the time it takes the control volume that initially was at the boiling boundary to exit the test section (i.e., one two-phase transit time), the exit quality decreases. This is due to the rapid expulsion of the two-phase fluid due to the flashing process. Once the main particle path characteristic has been crossed, a quasi-equilibrium situation is achieved and the exit quality monotonically increases. When the blowdown is terminated, in this particular case at $t=1$ sec, the transient quality rises rapidly to the new equi-

TABLE 9-2
Tabulation of Solution Scheme

Region	Time	Integration Procedure Eq. (9.28)	Elimination of Constant of Integration
I	$t \geqslant \nu$	Combine Eqs. (9.18) and (9.15) with Eq. (9.28) and integrate.	Set $z=\lambda$ at $t=t_0$.
II	$t \leqslant \nu$	Combine Eqs. (9.17) and (9.15) with Eq. (9.28) and integrate.	Set $z=\lambda$ at $t=t_0$.
III	$t \geqslant \nu$	Combine Eqs. (9.18) and (9.15) with Eq. (9.28) and integrate.	Equate result of integration with result of Region II at $t=\nu$; i.e., $z(\nu, t_0)\|_{II}=z(\nu, t_0)\|_{III}$.
IV	$t \leqslant 0$	Steady-State Initial Conditions	
V	$t \leqslant \nu$	Combine Eqs. (9.17) and (9.15) with Eq. (9.28) and integrate.	Use initial condition $z_0=z(t=0)$.
VI	$t \geqslant \nu$	Combine Eqs. (9.18) and (9.15) with Eq. (9.28) and integrate.	Equate result of integration with result of Region V at $t=\nu$; i.e., $z(\nu, z_0)\|_{V}=z(\nu, z_0)\|_{VI}$.

TABLE 9-2
Tabulation of Solution Scheme (continued)

Result: Explicitly for $z(t, z_0)$ or Implicitly for $z(t, t_0)$	Result: Explicitly for $\langle x(t, z_0)\rangle$ or $\langle x(t, t_0)\rangle$	Defined Quantities
$t_0=\frac{\Omega}{(\Omega+K_2)}t-\frac{1}{(\Omega+K_2)} \times \ln\left\{\frac{1}{K_9}\left[z(t, t_0)+\frac{G_0K_2}{(\Omega+K_2)}\exp(-K_2t)\right]\right\}$	Eq. (9.26)	$K_2 \triangleq$ Reciprocal of the time constant of the flow transient
$t_0=t-\frac{1}{\Omega}\ln\left\{\Omega\left[\frac{\rho_f}{G_0}z((t, t_0)-\frac{1}{K_2}-K_5+t+\frac{1}{K_2}\exp(-K_2t)\right]\right\}$	Eq. (9.26)	$K_3 \triangleq \frac{1}{\rho_f}\left\{1-\frac{\Omega}{K_2}[\exp(K_2\nu)-1]\right\}$
$t_0=t-\frac{1}{\Omega}\ln\left\{\frac{\rho_f}{G_0}\Omega\left[z-K_7\exp(\Omega t)+\frac{G_0K_3}{(\Omega+K_2)}\exp(-K_2t)\right]\right\}$	Eq. (9.26)	$K_5 \triangleq \left(\nu-\frac{1}{\Omega}\right)$
$z=z_0$	$\langle x_0\rangle=\langle x_{exit}\rangle_0\left[\frac{z_0-\lambda_0}{L_H-\lambda_0}\right]$	$K_6 \triangleq \exp(-\Omega\nu)\frac{G_0}{\rho_f}\left\{\frac{1}{K_2}-\frac{1}{\Omega}-\exp(-K_2\nu)\left[\frac{1}{K_2}-\frac{\rho_fK_3}{(\Omega+K_2)}\right]-K_5\exp(\Omega\nu)\right\}$
$z(t, z_0)=\frac{G_0}{\rho_f}\left[K_5+\frac{1}{K_2}-t-\frac{1}{K_2}\exp(-K_2t)-K_5\exp(\Omega t)\right]+Z_0\exp(\Omega t)$	Eq. (9.24)	$K_7 \triangleq \frac{G_0}{\rho_f}\exp-(\Omega\nu)\left[\frac{1}{K_2}-\frac{1}{\Omega}-\frac{1}{K_2}\exp(-K_2\nu)+\frac{\rho_fK_3}{(\nu+K_2)}\exp(-K_2\nu)\right]$
$z(t, z_0)=z_0\exp(\Omega t)+K_6\exp(\Omega t)-G_0\frac{K_3}{(\Omega+K_2)}\times\exp(-K_2t)$	Eq. (9.24)	$K_9 \triangleq \frac{G_0}{\rho_f(K_2+\Omega)}\exp(K_2\nu)$

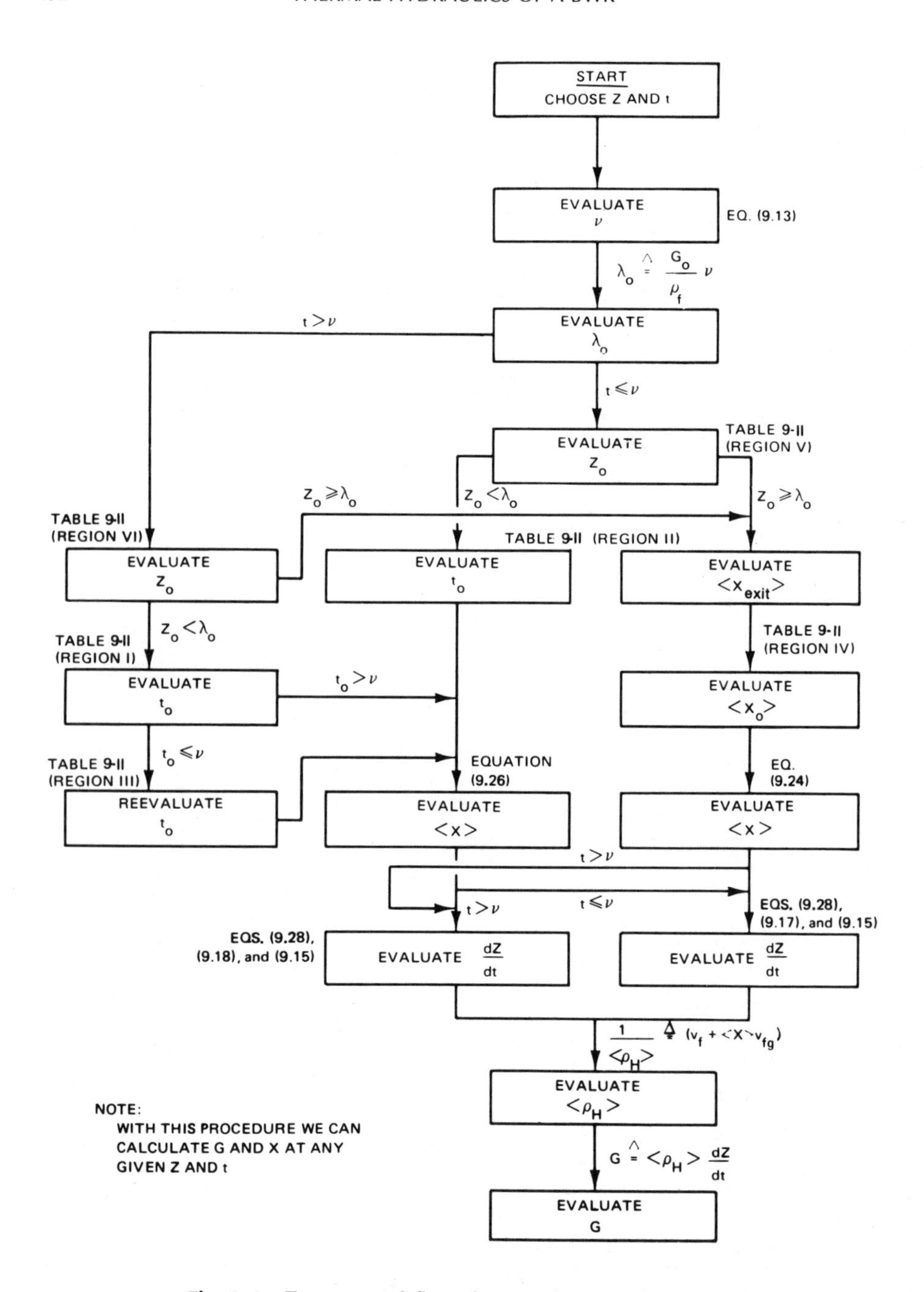

Fig. 9-4 Exponential flow decay solution scheme.

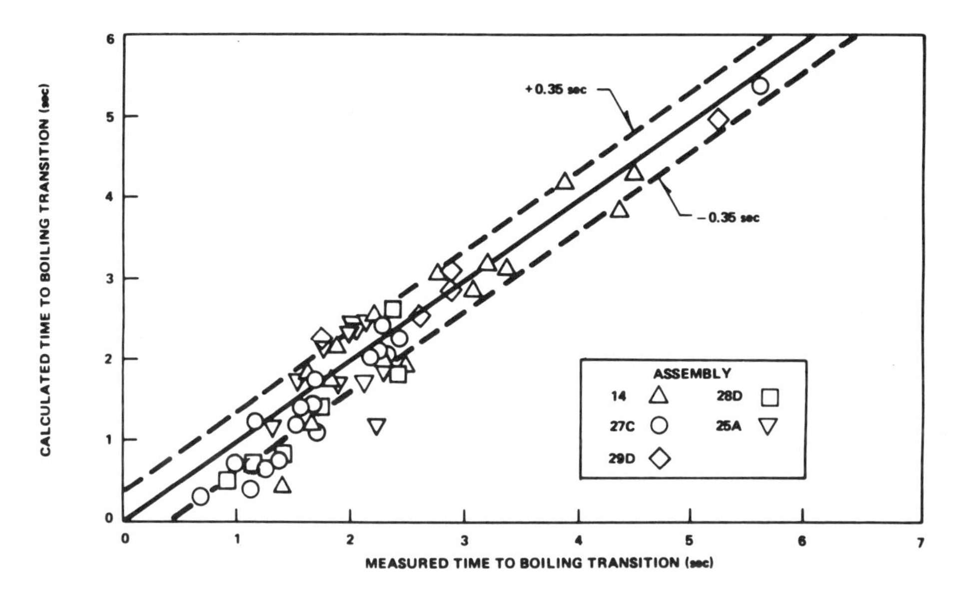

Fig. 9-5 Calculated versus measured time to initial BT (GETAB, 1973).

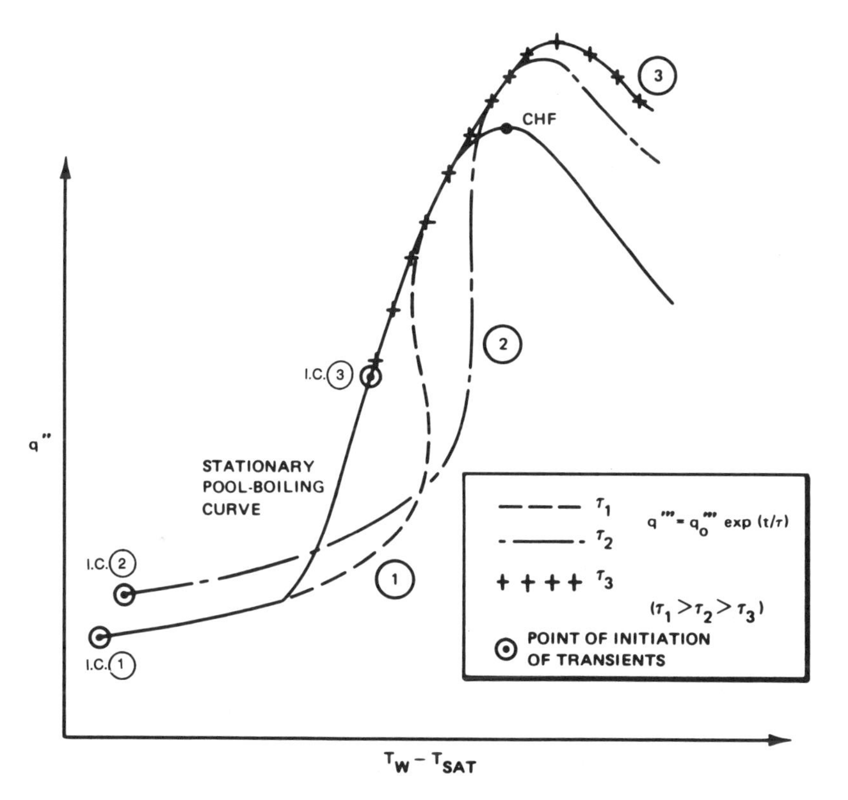

Fig. 9-6 The effect of exponential power transients on the pool-boiling curve.

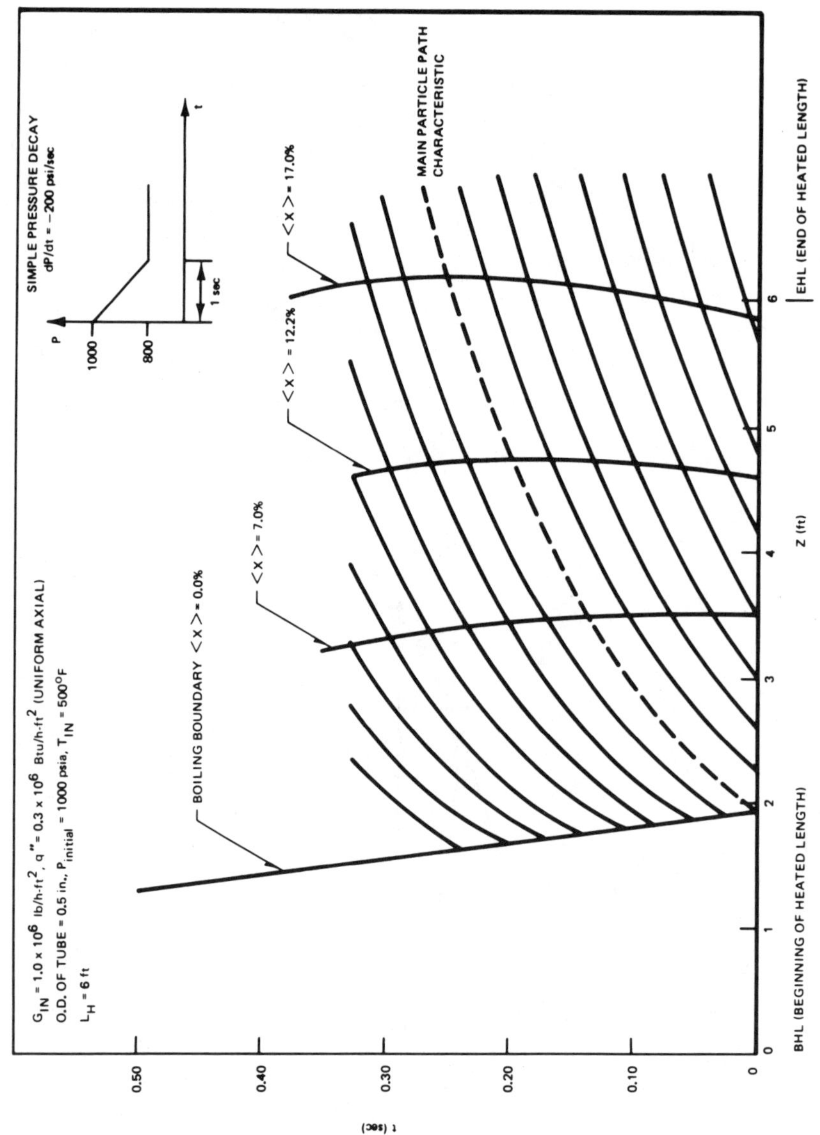

Fig. 9-7 Particle paths and constant quality lines for depressurization transient.

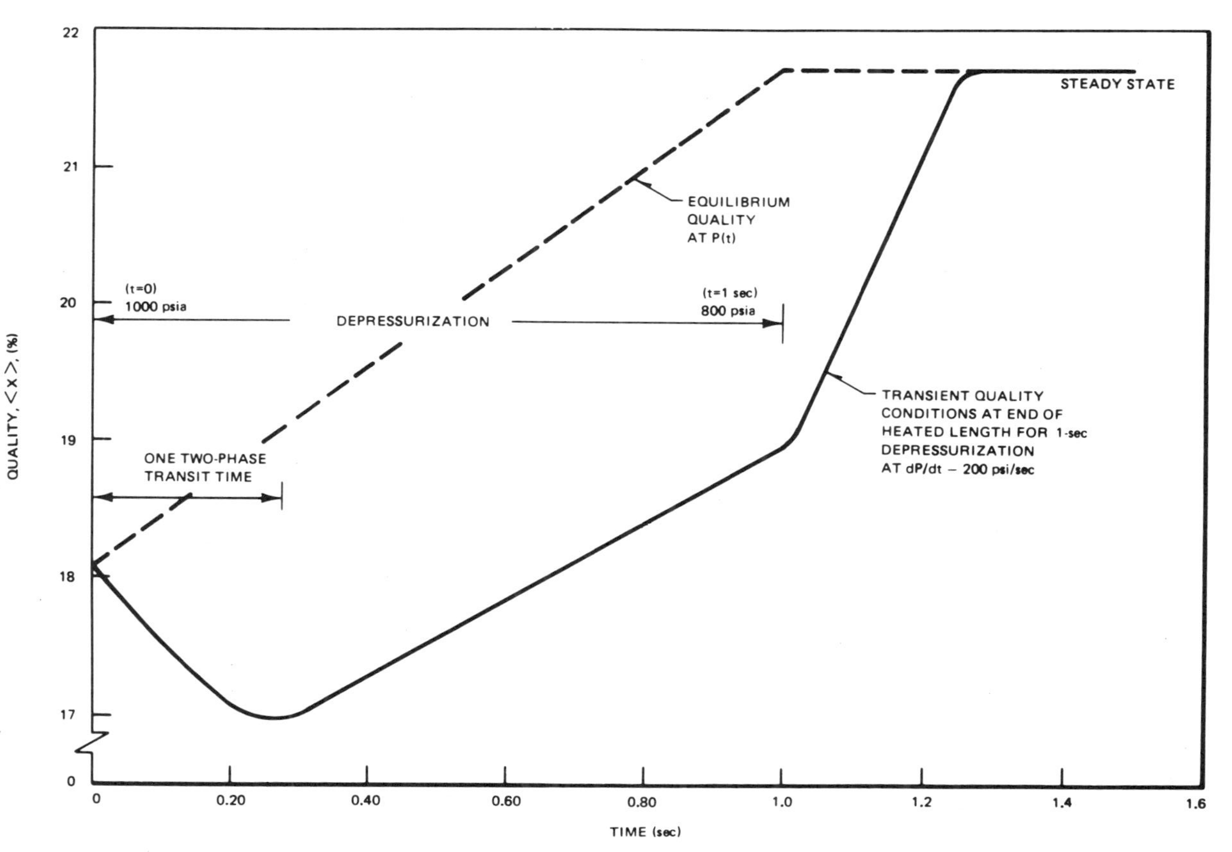

Fig. 9-8 Quality versus time during and after depressurization ($z = 6$ ft).

librium value at 800 psia. This type of analysis helps explain why there normally is no premature BT observed during depressurization experiments and why BT frequently occurs just after blowdown ceases.

In actual practice, similar analyses are carried out for the various hypothetical BWR accident-type transients of interest to identify which transients are the most limiting and what, if any, equipment modifications or additions may be necessary. The same type of analysis is performed for both single-operator error or equipment malfunction-type transients and the design-basis LOCA. Since the LOCA represents the worst-case BWR accident and is the most involved to analyze, it is considered separately in the next section.

9.2 Loss-of-Coolant Accident Evaluation

There are many aspects to the accurate evaluation of the hypothetical BWR LOCA. In the following sections, we concentrate on the thermal-hydraulic phenomena involved.

9.2.1 Critical Flow

When a flow passage opens between the reactor coolant system and its environment by either a pipe rupture or some other mechanism, fluid is expelled at a blowdown rate, w. If the blowdown passage transfers no energy to or from the fluid,

$$h_0 = \langle h \rangle + \frac{\langle u \rangle^2}{2 g_c J} = \text{constant} \;, \tag{9.32}$$

where h_0 is the stagnation enthalpy. Moreover, for ideal reversible flow, the second law of thermodynamics and the Gibbs equation yield,

$$d\langle h \rangle = \frac{dp}{J\langle \rho \rangle} \;. \tag{9.33}$$

By assuming a one-dimensional velocity profile and by introducing the definition of mass flux, $G = \langle \rho \rangle \langle u \rangle$, Eq. (9.33) can be integrated between the stagnation state in the vessel and properties at the throat of an ideal nozzle. By using Eqs. (9.32) and (9.33), the resulting isentropic mass flux (G_i) is,

$$G_i = \langle \rho \rangle [2 g_c J (h_0 - \langle h \rangle)]^{1/2} = \langle \rho \rangle \left(2 g_c \int_{p_t}^{p_0} \frac{dp}{\langle \rho \rangle} \right)^{1/2} \;. \tag{9.34}$$

For incompressible liquid flow, $\langle \rho \rangle = \rho_l$. Thus, Eq. (9.34) becomes,

$$G_{l_i} = [2 g_c \rho_l p_0 (1 - p_t/p_0)]^{1/2} \;, \tag{9.35}$$

where throat pressure, p_t, is equal to the downstream receiver pressure, p_R, and G_{l_i} is the so-called Bernoulli mass flux.

For compressible gas flow in an ideal nozzle, local pressure and density are normally characterized by the isentropic relationship,

$$p/\rho_g^K = p_0/\rho_{g0}^K \ . \tag{9.36}$$

By substituting $\langle\rho\rangle = \rho_g$ into Eq. (9.34), it follows that,

$$G_{g_i} = \left\{ 2g_c \left(\frac{K}{K-1}\right) p_0\rho_{g0} \left[\left(\frac{p_t}{p_0}\right)^{\frac{2}{K}} - \left(\frac{p_t}{p_0}\right)^{\frac{K+1}{K}} \right] \right\}^{1/2} \ . \tag{9.37}$$

The flow of incompressible liquid and an ideal gas have unique differences. Consider a nozzle discharging into a receiver at pressure, p_R. Before the flow rate can be predicted, it is necessary to obtain throat pressure, p_t. Whenever the discharge flow is subsonic, the receiver pressure readily propagates to the throat so that,

$$p_t = p_R \text{ (subsonic)} \ . \tag{9.38}$$

Equation (9.38) always applies for incompressible fluid flows and, as seen in Fig. 9-9, G_{l_i} continuously increases with decreasing p_t. However, as also shown in Fig. 9-9, if receiver pressure is decreased for the flow of an ideal compressible gas, a critical condition is reached at which the discharge rate reaches a maximum. For this condition, the corresponding throat velocity is sonic. Therefore, receiver pressure cannot propagate to the throat and further decrease of p_R does not change the value of flow rate or throat pressure. That is,

$$p_t \geqslant p_R \text{ (sonic)} \ . \tag{9.39}$$

The mathematical maximum value of G_{g_i} is determined by the condition at the throat,

$$\frac{dG_{g_i}}{dp_t} = 0 \ . \tag{9.40}$$

It can be easily shown from Eq. (9.37) that the condition of Eq. (9.40) leads to the so-called critical pressure ratio, p_c/p_0,

$$\frac{p_t}{p_0} \triangleq \frac{p_c}{p_0} = \left(\frac{2}{K+1}\right)^{\frac{K}{K-1}} \ . \tag{9.41}$$

By combining Eqs. (9.37) and (9.41), the corresponding critical mass flux for an ideal gas is,

$$G_{g_c} = \left[K g_c p_0 \rho_{g0} \left(\frac{2}{K+1}\right)^{\frac{K+1}{K-1}} \right]^{1/2} \ . \tag{9.42}$$

Equation (9.42) shows that the critical discharge rate of a given ideal gas

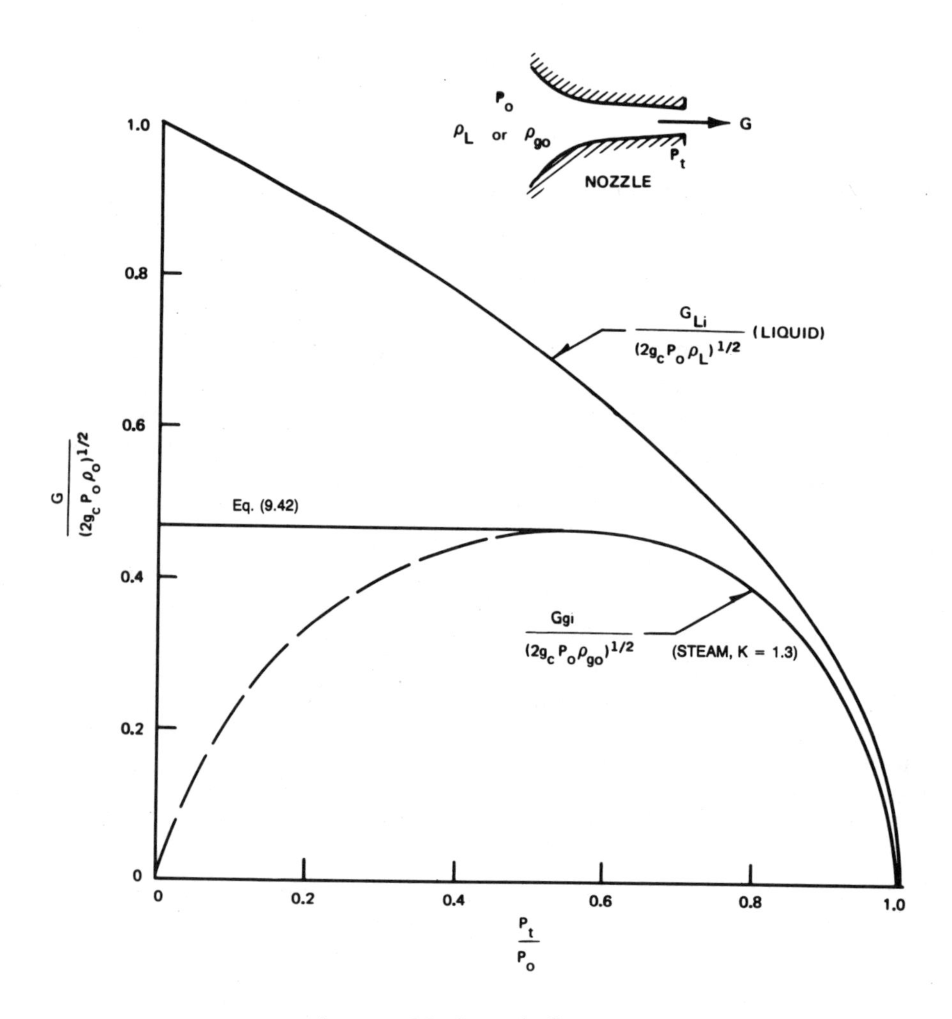

Fig. 9-9 Ideal nozzle flow rates.

from an isentropic nozzle is determined entirely by vessel stagnation properties.

When a particle of liquid travels from a high-pressure region through a nozzle, its pressure decreases until it reaches the throat. If a liquid particle is near the boiling point, its pressure may drop below the saturated value during its transit, resulting in vapor formation due to flashing. Therefore, two-phase flow occurs for most postulated liquid discharge processes in a BWR. Furthermore, vessel fluid may already be a liquid-vapor mixture near the point of discharge, such that the discharge of two-phase flow occurs. Therefore, it is important to consider the effects of two-phase discharge flow in the calculation of BWR blowdown rates.

There are at least two important additional degrees of freedom in single-component, two-phase flow systems that do not occur in single-phase flows; namely, nonequilibrium effects and flow regime effects. Later, it is demonstrated that nonequilibrium effects in two-phase critical flow are important only for certain geometric configurations of the flow passage. For most practical purposes, equilibrium flow assumptions apply to BWR LOCA analysis. Therefore, the equilibrium model is developed first with the effects of flow regime incorporated into a phase-averaged slip ratio.

9.2.1.1 *An Equilibrium Critical Flow Model*

The following is a derivation of a thermodynamic equilibrium critical flow model that has come to be known as the Moody model (Moody, 1965). Consider a two-phase system in which unequal phase velocities occur. For this situation, Eq. (9.32) becomes,

$$h_0 = \langle h\rangle + \langle x\rangle \frac{\langle u_g\rangle_g^2}{2g_cJ} + (1-\langle x\rangle)\frac{\langle u_f\rangle_f^2}{2g_cJ} , \tag{9.43}$$

where the two-phase enthalpy of the saturated mixture is given by,

$$\langle h\rangle \triangleq h_f + \langle x\rangle h_{fg} . \tag{9.44}$$

If the flow is approximated by an isentropic process,

$$s_f + \langle x\rangle s_{fg} = \text{constant} \triangleq s_0 . \tag{9.45}$$

By eliminating $\langle x\rangle$ between Eqs. (9.44) and (9.45), we obtain a saturation state equation in the form,

$$\langle h\rangle = h_f + \frac{h_{fg}}{s_{fg}}(s_0 - s_f) = \langle h(p, s_0)\rangle . \tag{9.46}$$

Furthermore, a functional relationship exists among stagnation properties in the form,

$$s_0 = s_0(p_0, h_0) . \tag{9.47}$$

Equations (9.43), (9.46), (9.47), (5.22), and (5.23) yield

$$G = \langle \rho'''\rangle\{2g_cJ[h_0 - \langle h(p, p_0, h_0)\rangle]\}^{1/2} , \tag{9.48}$$

where the so-called energy density, $\langle \rho'''\rangle$, defined in Eq. (5.101), can be rewritten in terms of the slip ratio, S, as,

$$\langle \rho'''\rangle = \left\{\left[\frac{\langle x\rangle}{\rho_g} + \frac{S(1-\langle x\rangle)}{\rho_f}\right]\left[\langle x\rangle + \frac{(1-\langle x\rangle)}{S^2}\right]^{1/2}\right\}^{-1} . \tag{9.49}$$

Note that Eq. (9.48) reduces to the Bernoulli mass flux of Eq. (9.35) when $\langle x\rangle \to 0$. It is readily seen from Eqs. (9.45) through (9.49) that for given stagnation properties, p_0 and h_0,

$$G = G(p, S; p_0, h_0) \ . \tag{9.50}$$

That is, for a known stagnation state, local static pressure, p, and slip ratio, S, the flow rate per unit area is uniquely determined.

Numerous experimental and theoretical studies have been made to determine the slip ratio (or equivalently, the void fraction $\langle\alpha\rangle$) in terms of other properties such as flow quality, $\langle x\rangle$, and the local static pressure, p; e.g., Levy (1960) and Zivi (1964). Various maps are available giving flow regimes in terms of G and $\langle x\rangle$ for various pressures. Published slip ratios cover the range from $S=1.0$ (homogeneous flow) to $S=(\rho_f/\rho_g)^{1/2}$. Here, however, we obtain the maximum two-phase equilibrium flow rate with respect to slip ratio by finding a particular value of S that maximizes G.

If an experiment similar to that which provides the critical flow rate of an ideal gas is considered, then we would reduce the downstream receiver pressure until a maximum discharge rate is achieved. Since throat and receiver pressures are equal until the discharge velocity is sonic, two-phase critical flow should correspond to the same condition expressed by Eq. (9.40). From Eq. (9.37), G_{g_i} is seen to be a function only of throat pressure for a given stagnation state, so that the total derivative of Eq. (9.40) is appropriate. However, for the two-phase system, Eq. (9.50) expresses G as a function of both local static pressure, p, and the slip ratio, S, for a given stagnation state. If both p and S are considered independent, a maximum G corresponds to the conditions,

$$\left(\frac{\partial G}{\partial S}\right)_p = 0; \ \left(\frac{\partial^2 G}{\partial S^2}\right)_p < 0 \ , \tag{9.51}$$

and,

$$\left(\frac{\partial G}{\partial p}\right)_S = 0; \ \left(\frac{\partial^2 G}{\partial p^2}\right)_S < 0 \ . \tag{9.52}$$

For the condition of Eq. (9.51), it follows from Eqs. (9.48) and (9.49) that for critical flow,

$$S = S(p) = \left(\frac{\rho_f}{\rho_g}\right)^{1/3} \ . \tag{9.53}$$

Equation (9.53) shows that at maximum G, the slip ratio is a function of pressure only. Employing Eq. (9.53) in Eq. (9.52) and using saturated steam-water properties, Fig. 9-10a can be obtained, which gives G_c in terms of p_0 and h_0. Moreover, Fig. 9-10b provides the nozzle throat pressure, p_c, at which critical two-phase steam-water flow occurs. Figure 9-10c gives G_c in terms of p_c and $\langle x_c\rangle$, where,

$$\langle x_c\rangle = \frac{s_0 - s_f(p_c)}{s_{fg}(p_c)} \ . \tag{9.54}$$

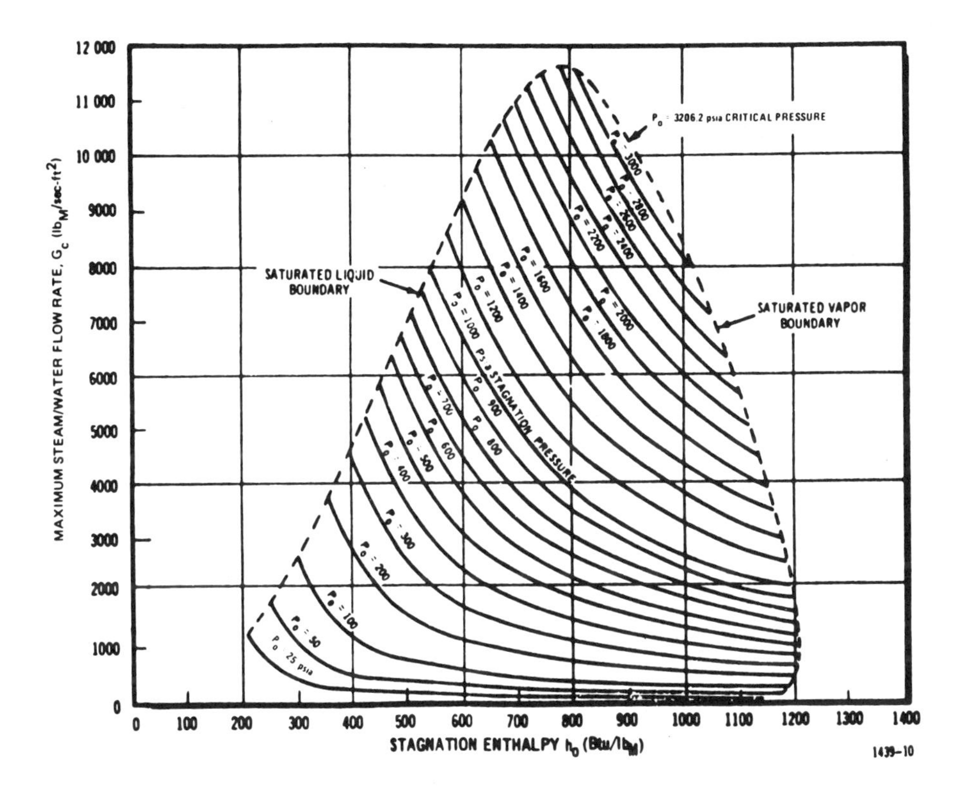

Fig. 9-10a Maximum steam/water flow rate and local stagnation properties (Moody model).

Note that the two-phase critical flow model just described minimizes the specific kinetic energy of the two-phase mixture with respect to slip ratio. That is, by using Eqs. (5.12), (5.22), and (5.23),

$$G\frac{\partial}{\partial S}\left[\langle x\rangle\frac{\langle u_g\rangle_g^2}{2g_cJ}+(1-\langle x\rangle)\frac{\langle u_f\rangle_f^2}{2g_cJ}\right]=\frac{G^3}{2g_cJ}\frac{\partial}{\partial S}\left\{\left[\frac{\langle x\rangle}{\rho_g}+\frac{S(1-\langle x\rangle)}{\rho_f}\right]^2\left(\langle x\rangle+\frac{1-\langle x\rangle}{S^2}\right)\right\}=0\ , \tag{9.55}$$

from which the result is again,

$$S=(\rho_f/\rho_g)^{1/3}\ .$$

9.2.1.2 *Other Equilibrium Slip Flow Models*

Several other workers in the field have formulated thermodynamic equilibrium models for two-phase critical flow. A formulation that minimizes

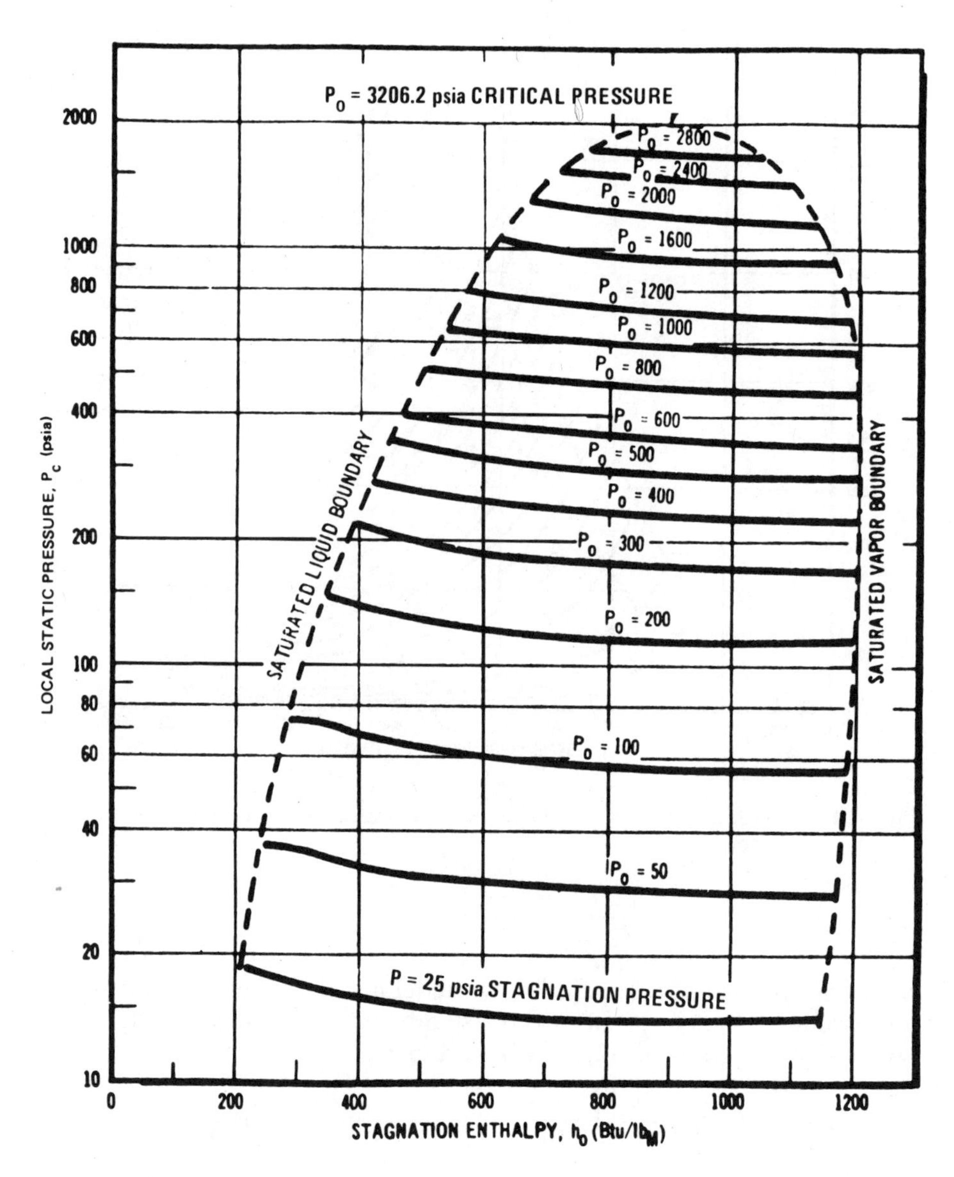

Fig. 9-10b Local static pressure and stagnation properties at maximum steam/water flow rate (Moody model).

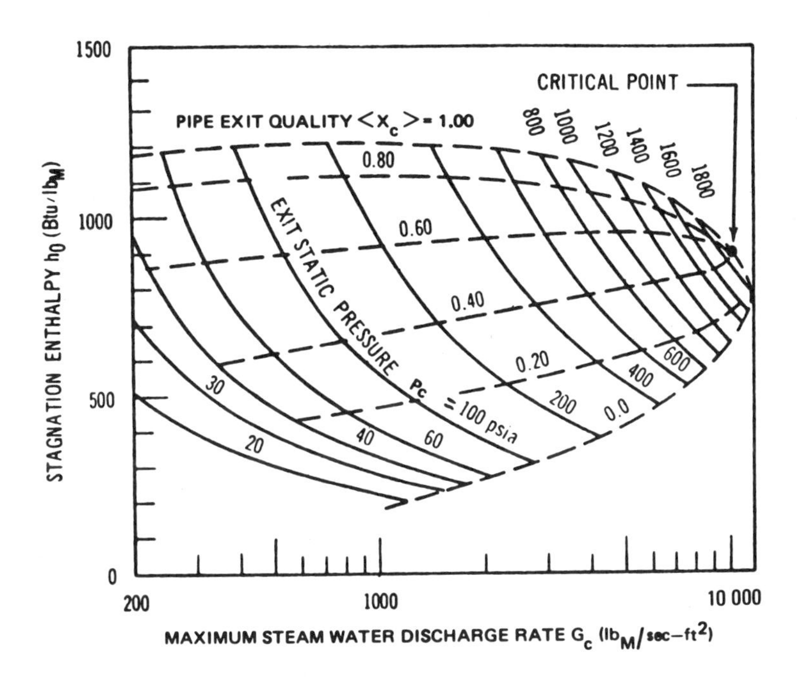

Fig. 9-10c Exit properties for maximum steam/water discharge (Moody model).

the specific two-phase momentum flux with respect to slip ratio was obtained (Fauske, 1962) by writing,

$$G\frac{\partial}{\partial S}[\langle x\rangle\langle u_g\rangle_g + (1-\langle x\rangle)\langle u_f\rangle_f] = G^2\frac{\partial}{\partial S}\left\{\left[\frac{\langle x\rangle}{\rho_g} + \frac{S(1-\langle x\rangle)}{\rho_f}\right]\left[\langle x\rangle + \frac{1-\langle x\rangle}{S}\right]\right\} = 0 \ , \tag{9.56}$$

which results in

$$S = (\rho_f/\rho_g)^{1/2} \ .$$

A theoretical formulation for slip ratio based on interphase momentum exchange was incorporated in another model for two-phase critical flow rate (Levy, 1965). In contrast to Moody's model, which is concerned only with upstream stagnation properties, both Levy's and Fauske's models require local properties at the nozzle throat to be known before the critical mass flux can be predicted.

Models employing either $(\rho_f/\rho_g)^{1/2}$, $(\rho_f/\rho_g)^{1/3}$, or Levy's slip ratio predict nearly the same critical blowdown flow rate for given values of local pressure and quality. The Moody model predicts equilibrium flow rates somewhat higher than data representative of BWR blowdowns and is, thus, conservative for accident analysis. Moreover, this model is readily presented in terms of known stagnation properties, rather than unknown local properties. Thus, it is much easier to use in design analysis.

Another important analytical critical flow model is the so-called homogeneous equilibrium model (HEM). When S is set equal to unity in Eq. (9.49), then for given values of p_0 and h_0, Eq. (9.50) gives G_c as a function of p.

To find the critical mass flux, the conditions of Eq. (9.52) were employed. Figures 9-11a and b give the homogeneous equilibrium G_c, and associated critical pressure p_c, in terms of p_0 and h_0.

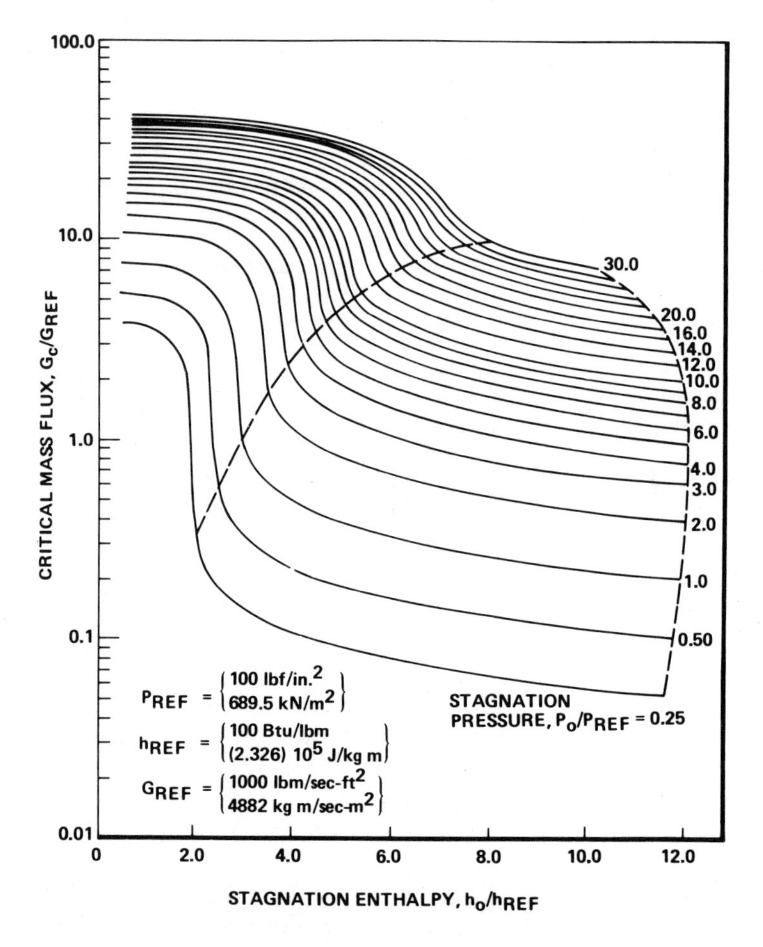

Fig. 9-11a Critical mass flux-homogeneous, equilibrium steam/water (Moody, 1975).

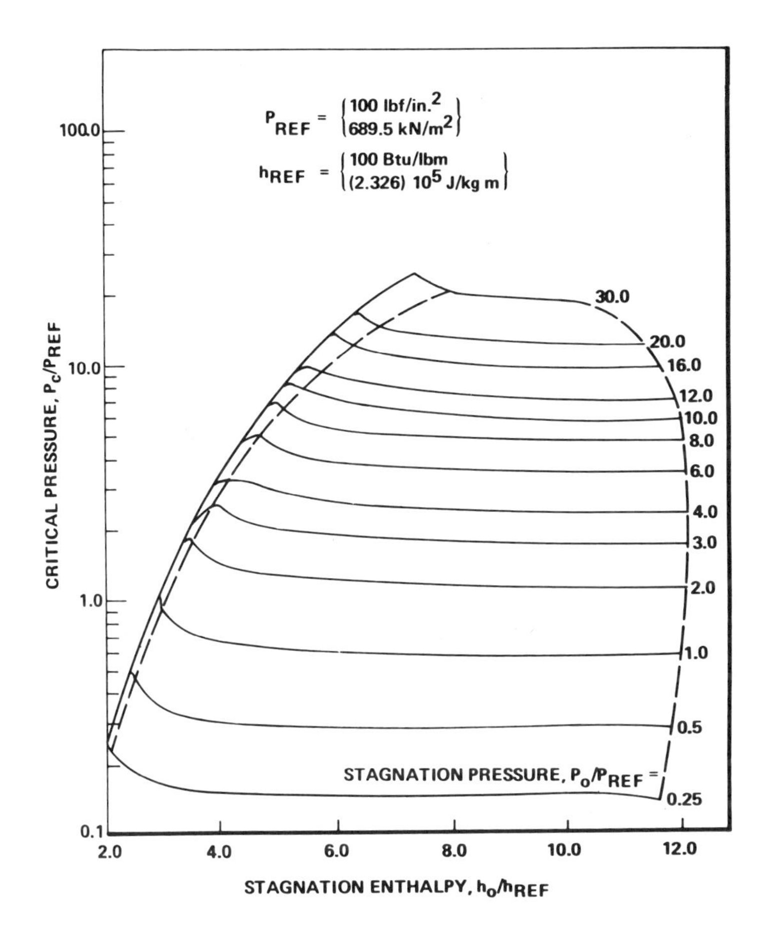

Fig. 9-11b Critical pressure-homogeneous, equilibrium steam/water (Moody, 1975).

When vessel fluid is saturated, the maximum mass flux, G_c, is readily identified. Figure 9-12 shows that for increasingly subcooled vessel liquid, the value of G_c occurs at the peak of a sharp spike. However, the critical pressure p_c is easily obtained, at which the quality is just zero. Therefore, incompressible liquid flow can be assumed between p_0 and p_c to obtain G_c for large subcooling.

A comparison of Figs. 9-10a and 9-11a shows that for $\langle x_c \rangle < 1.0$, values of G_c are greater for the Moody model than for the homogeneous equilibrium model. Experimental results (Sozzi and Sutherland, 1975) from tests on various nozzle types have shown that for conditions typical of BWR technology, the homogeneous equilibrium model agrees quite well with

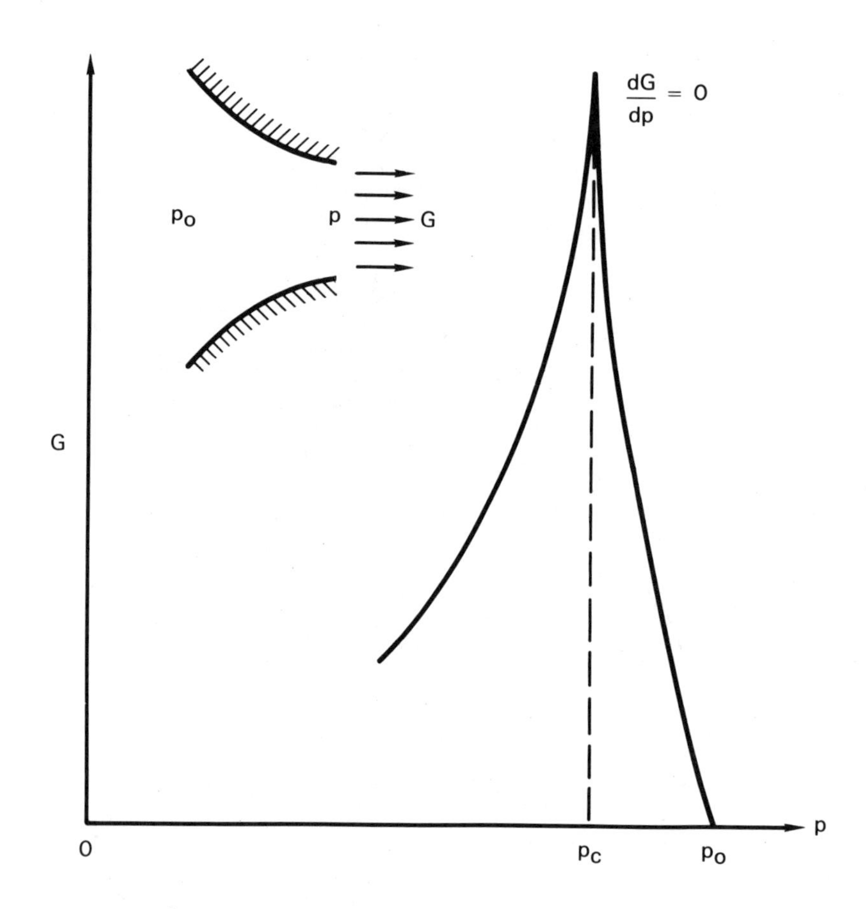

Fig. 9-12 Maximum G, subcooled liquid discharge.

the available data when vessel stagnation properties are used to evaluate G_c, as shown in Fig. 9-13.

Moreover, as seen in Fig. 9-14, a slip flow model (Moody, 1975) can also accurately predict the equilibrium discharge rate in terms of local properties at the nozzle throat or exit.

These observations have led to the apparent discrepancy indicated in the critical mass flux data shown in Fig. 9-15. Why does the slip flow model overpredict in terms of vessel properties, and yet accurately predict the flow based on pipe discharge properties? Moreover, why does the homogeneous model predict flows in terms of vessel properties, and yet underpredict flows in terms of pipe discharge properties?

It has been hypothesized (Moody, 1975) that bubbles form homogeneously in the decompression region near the pipe entrance, which leads to

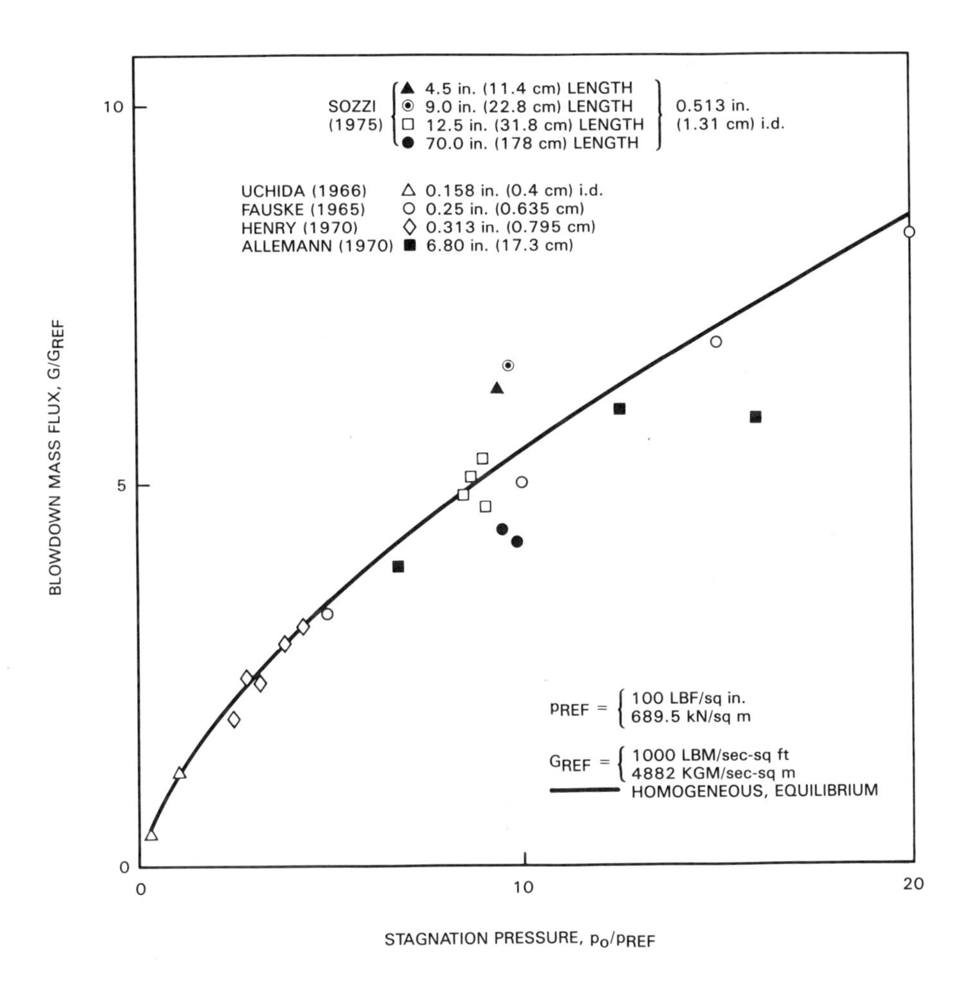

Fig. 9-13 Blowdown rate of saturated water.

a homogeneous critical flow condition, as shown schematically in Fig. 9-16. This condition is followed by phase separation with discharge in the annular or separated choked flow state. The mass flux is limited by upstream homogeneous critical flow, which is more restrictive. Comparisons with measurements, which include simultaneous sets of vessel and discharge properties and flow rates, verify the hypothesis for equilibrium stream/water flows, as seen in Fig. 9-17.

9.2.1.3 *An Alternate Approach*

The critical flow rate models just discussed have been based on thermodynamic principles. An alternate approach, which is frequently employed,

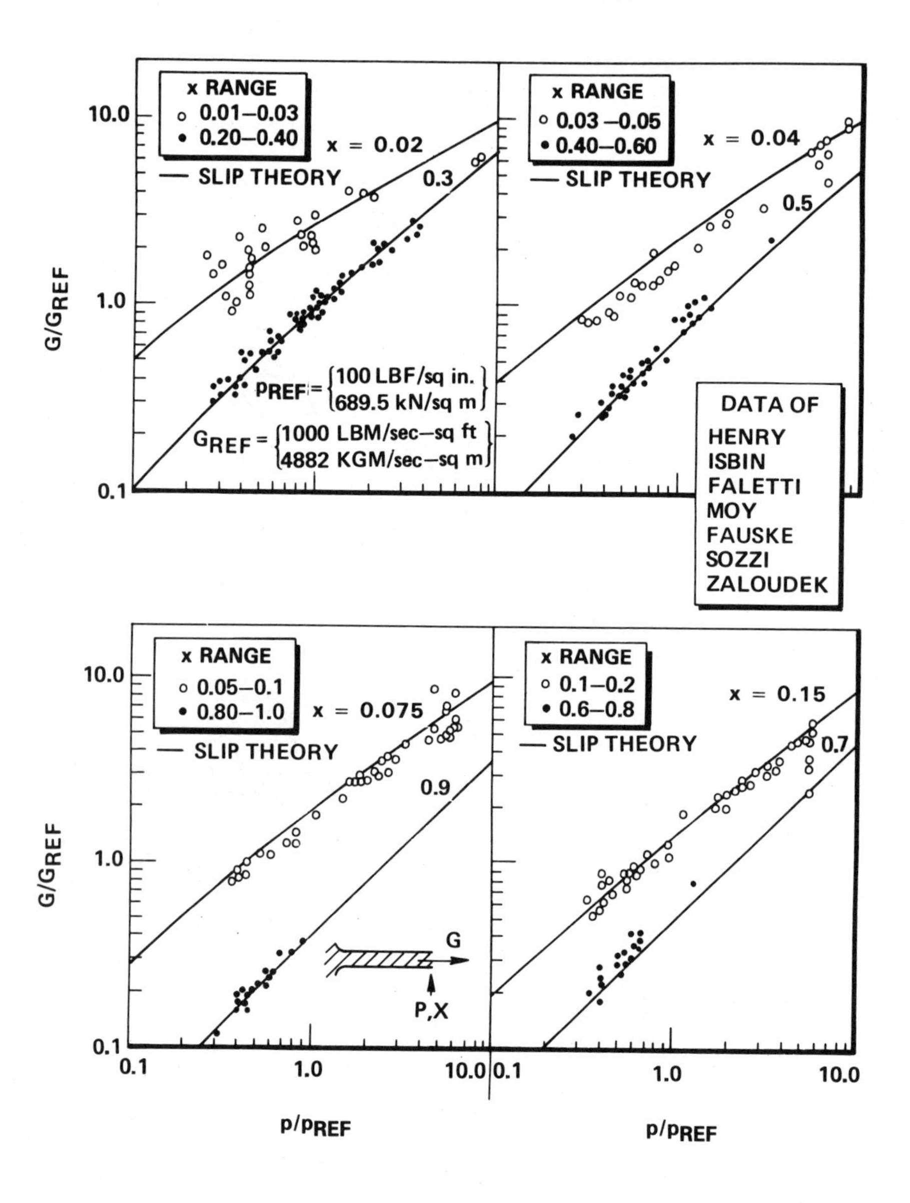

Fig. 9-14 Critical flow properties at pipe exit.

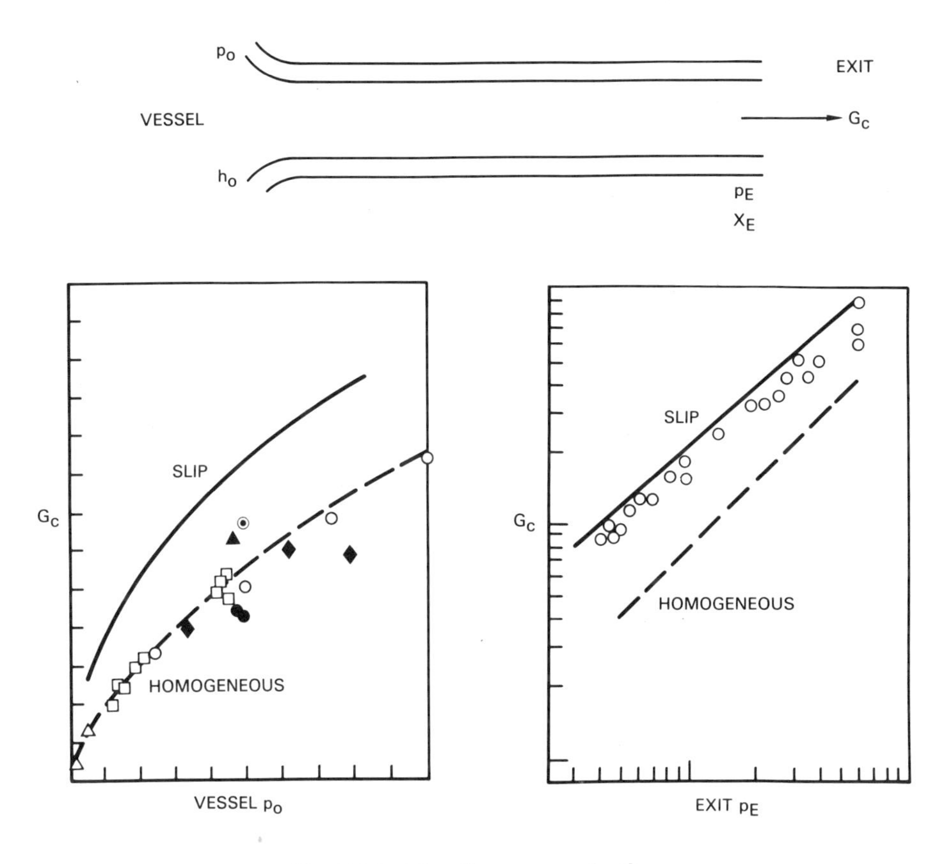

Fig. 9-15 The discrepancy in G_c.

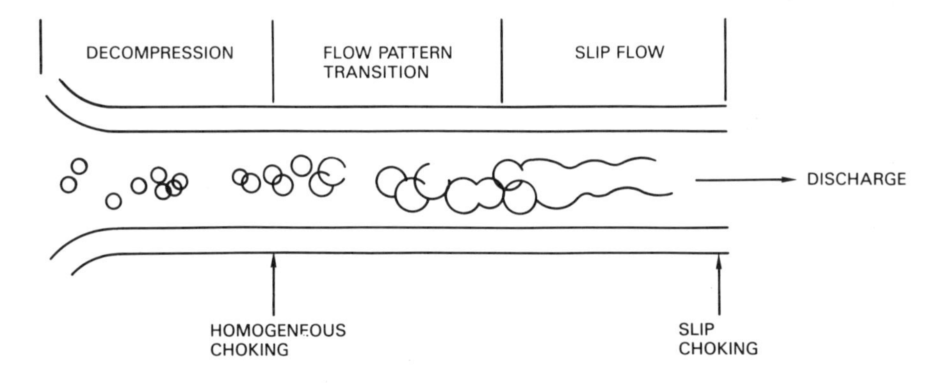

Fig. 9-16 Two choked conditions.

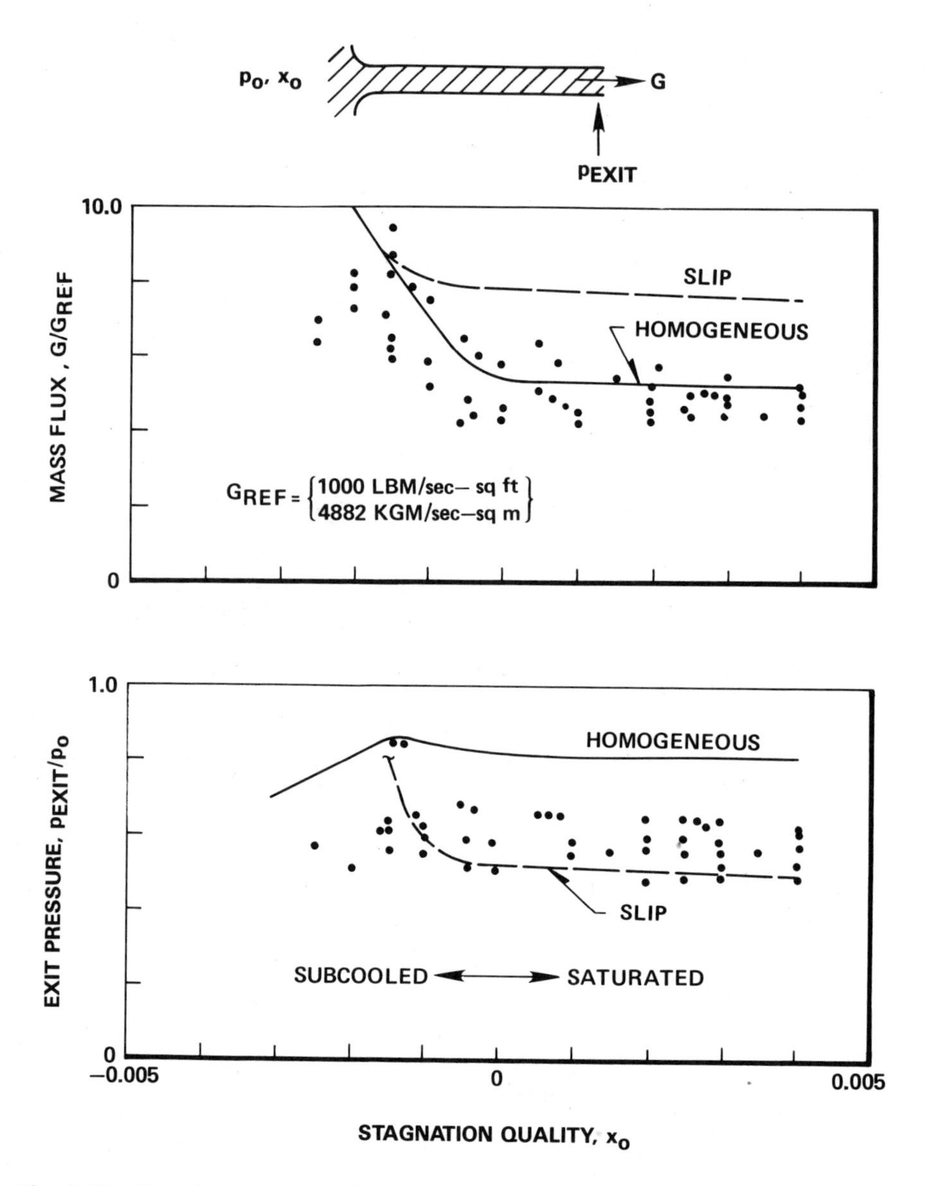

Fig. 9-17 Blowdown rates and exit pressures of Sozzi and Sutherland (1975) at 70 atm.

is to derive the critical condition directly from the conservation equations of two-phase flow. For illustration purposes, we derive the so-called "two-equation model" for two-phase critical flow. That is, we use only the momentum and continuity equations of the two-phase mixture in our derivations and, thus, the functional form for G_c is based on fluid mechanics rather than thermodynamics. Note, however, that to numerically evaluate the critical mass flux, G_c, the thermodynamic process must also be considered.

More complex models of this type are under development (Bouré and Reocreux, 1972). The most sophisticated of these is known as the "six-equation model," since it is based on the three conservation equations for each phase. The basic solution procedure is to solve the system of equations for the pressure gradient and to set the denominator of the system's matrix equal to zero to establish the choking conditions. Essentially, the same procedure is used with the two-equation model discussed below.

To derive the required result, we consider the steady-state version of the two-phase momentum equation given in Eq. (5.66),

$$\frac{1}{g_c A_{x-s}} \frac{d}{dz}\left(\frac{G^2 A_{x-s}}{\langle\rho'\rangle}\right) = -\frac{dp}{dz} - \frac{g}{g_c}\langle\bar{\rho}\rangle \sin\theta - \frac{\tau_w P_f}{A_{x-s}} \ . \tag{9.57}$$

The continuity equation implies that at the critical condition, the mass flow rate is constant at its maximum value (i.e., $G_c A_{x-s}$) and, thus, the spatial acceleration term on the left side of Eq. (9.57) can be rewritten as

$$\frac{G_c^2}{g_c}\left[\frac{-1}{A_{x-s}\langle\rho'\rangle}\frac{dA_{x-s}}{dz} + \frac{d\left(\dfrac{1}{\langle\rho'\rangle}\right)}{dp}\frac{dp}{dz}\right] \ . \tag{9.58}$$

By combining Eqs. (9.57) and (9.58), and solving for the pressure gradient,

$$-\frac{dp}{dz} = \frac{\left(\dfrac{-G_c^2}{g_c A_{x-s}\langle\rho'\rangle}\dfrac{dA_{x-s}}{dz} + \dfrac{g}{g_c}\langle\bar{\rho}\rangle \sin\theta + \dfrac{\tau_w P_f}{A_{x-s}}\right)}{\left[1 + \dfrac{G_c^2}{g_c}\dfrac{d}{dp}\left(\dfrac{1}{\langle\rho'\rangle}\right)\right]} \ . \tag{9.59}$$

Equation (9.59) can be recognized as a generalization of similar formulations deduced previously (Wallis, 1969). Hence, by using arguments established in this previous work, the denominator can be thought of as,

$$(1 - M^2) \ , \tag{9.60}$$

where M is the local Mach number of the two-phase mixture. For choked conditions, the Mach number is unity and the denominator vanishes. Thus, at the critical condition, the pressure gradient becomes infinite or indeterminate, where in the latter case, the vanishing of the numerator of Eq. (9.59) can be used as a condition to locate the position of the choking plane.

We can solve for the critical mass flux relation by setting the denominator of Eq. (9.59) equal to zero,

$$G_c = \left[-g_c \frac{dp}{d\left(\frac{1}{\langle \rho' \rangle}\right)} \right]^{1/2} . \tag{9.61}$$

It is interesting to note that this is the same result that we would obtain if all terms except the pressure gradient and the spatial acceleration term were neglected in Eq. (9.57).

By assuming that the throat properties are known, Eqs. (5.25), (5.67), and (9.61) can be used to evaluate the critical discharge for any slip model that may be of interest. For instance, $S=1$ and $S=(\rho_f/\rho_g)^{1/3}$ can be used for the equilibrium homogeneous and slip models, respectively.

To illustrate the specific evaluation of Eq. (9.61), we consider the case of the critical flow of a homogeneous, two-phase mixture undergoing an isentropic expansion. First, the derivative in Eq. (9.61) can be expanded as,

$$\frac{d\left(\frac{1}{\langle \rho' \rangle}\right)}{dp} = \left.\frac{\partial\left(\frac{1}{\langle \rho' \rangle}\right)}{\partial p}\right|_{\langle x \rangle} + \left.\frac{\partial\left(\frac{1}{\langle \rho' \rangle}\right)}{\partial \langle x \rangle}\right|_{p} \frac{d\langle x \rangle}{dp} . \tag{9.62}$$

Now, for homogeneous flow in thermodynamic equilibrium,

$$\frac{1}{\langle \rho' \rangle} = v_f + \langle x \rangle v_{fg} .$$

Thus,

$$\left.\frac{\partial\left(\frac{1}{\langle \rho' \rangle}\right)}{\partial p}\right|_{\langle x \rangle} = \frac{dv_f}{dp} + \langle x \rangle \frac{dv_{fg}}{dp} , \tag{9.63}$$

and,

$$\left.\frac{\partial \frac{1}{\langle \rho' \rangle}}{\partial \langle x \rangle}\right|_{p} = v_{fg} . \tag{9.64}$$

For an isentropic process,

$$\langle s \rangle = s_f + \langle x \rangle s_{fg} = \text{constant} .$$

Thus,

$$d\langle s \rangle = 0 = \left.\frac{\partial \langle s \rangle}{\partial p}\right|_{\langle x \rangle} dp + \left.\frac{\partial \langle s \rangle}{\partial \langle x \rangle}\right|_{p} d\langle x \rangle .$$

Hence, for an isentropic process,

$$\frac{d\langle x\rangle}{dp} = -\frac{\left.\frac{\partial\langle s\rangle}{\partial p}\right|_{\langle x\rangle}}{\left.\frac{\partial\langle s\rangle}{\partial\langle x\rangle}\right|_{p}} = -\frac{\left(\frac{ds_f}{dp} + \langle x\rangle\frac{ds_{fg}}{dp}\right)}{s_{fg}} \ . \tag{9.65}$$

By combining Eqs. (9.62) through (9.65),

$$\frac{d\left(\frac{1}{\langle\rho'\rangle}\right)}{dp} = \frac{dv_f}{dp} + \langle x\rangle\frac{dv_{fg}}{dp} - \frac{v_{fg}}{s_{fg}}\left(\frac{ds_f}{dp} + \langle x\rangle\frac{ds_{fg}}{dp}\right) \ . \tag{9.66}$$

Now, by combining Eqs. (9.66) and (9.61), the expression for the critical mass flux is,

$$G_c = \left(\frac{-g_c}{\left\{\frac{dv_f}{dp} - \left(\frac{v_{fg}}{s_{fg}}\right)\frac{ds_f}{dp} + \langle x\rangle\left[\frac{dv_{fg}}{dp} - \left(\frac{v_{fg}}{s_{fg}}\right)\frac{ds_{fg}}{dp}\right]\right\}}\right)^{1/2} \ . \tag{9.67}$$

Thus, the critical mass flux can be evaluated using derivatives obtained from the steam tables. The results for other slip and thermodynamic process assumptions are obtained in a similar manner.

All the models we have discussed so far have been based on the assumption of thermodynamic equilibrium. For some cases of interest, equilibrium conditions are not achieved. Thus, the next section is devoted to a discussion of nonequilibrium effects.

9.2.1.4 *Nonequilibrium Effects*

It is well known that processes involving heat transfer, condensation, evaporation, and bubble expansion involve finite periods of time. It follows that if a small volume of hot water is rapidly decompressed below its saturation pressure, the inherent inertia in the phase change mechanisms causes the fluid state to lag behind equilibrium conditions for a brief period of time. Furthermore, suppose that a hot water particle undergoes decompression during expulsion from a vessel at high pressure through a short flow path to the surroundings. It is possible for the liquid to remain in a nonequilibrium, metastable state and to produce a blowdown rate much higher than that obtained in phase equilibrium flow, even if vessel conditions were saturated. This phenomenon can occur if the time spent in transit through the flow path by the escaping fluid is short compared to the time required for phase change mechanisms to produce an equilibrium state. This situation is shown in Fig. 9-18. Note that for sufficiently high subcoolings,

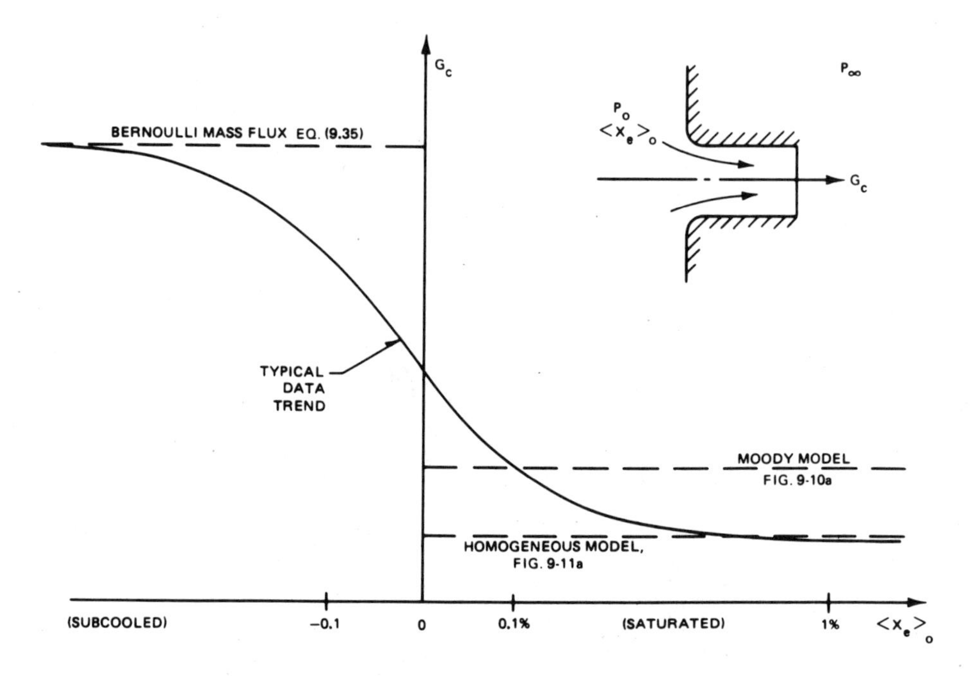

Fig. 9-18 Typical variation of critical mass flux with inlet quality.

single-phase (Bernoulli) discharge is achieved; while for conditions of saturated equilibrium, the observed discharge rate is much lower.

No experimental studies have been conducted that directly measure the time required for a new two-phase equilibrium state to be reached following a change in pressure. However, there are several reported experimental results, which if properly interpreted, provide a means for determining when phasic nonequilibrium is important.

Decompression experiments (Edwards, 1969; Borgartz et al., 1969) were performed with 13-ft-long pipes of 1.0 and 2.88-in. i.d., pressurized initially with 1000 psia, 458°F water (460-psia saturation pressure). Sudden rupture of a glass diaphragm was induced at one end of the pipe, followed by the transit of a decompression wave through the test section. Pressure was measured at approximately 2- to 3-ft intervals along the test section. Figure 9-19 gives the measured pressure trace at 0.25 ft from the discharge end of a 1.0-in.-diam pipe. Note that pressure undershoots the saturation value slightly, but begins to rise again toward the saturation value within 0.5 msec (0.0005 sec). The time duration between the pressure undershoot and lowest pressure point is interpreted as an indication of nucleation time, or time for vapor formation to begin. The nucleation interval clearly is <1.0 msec for hot water.

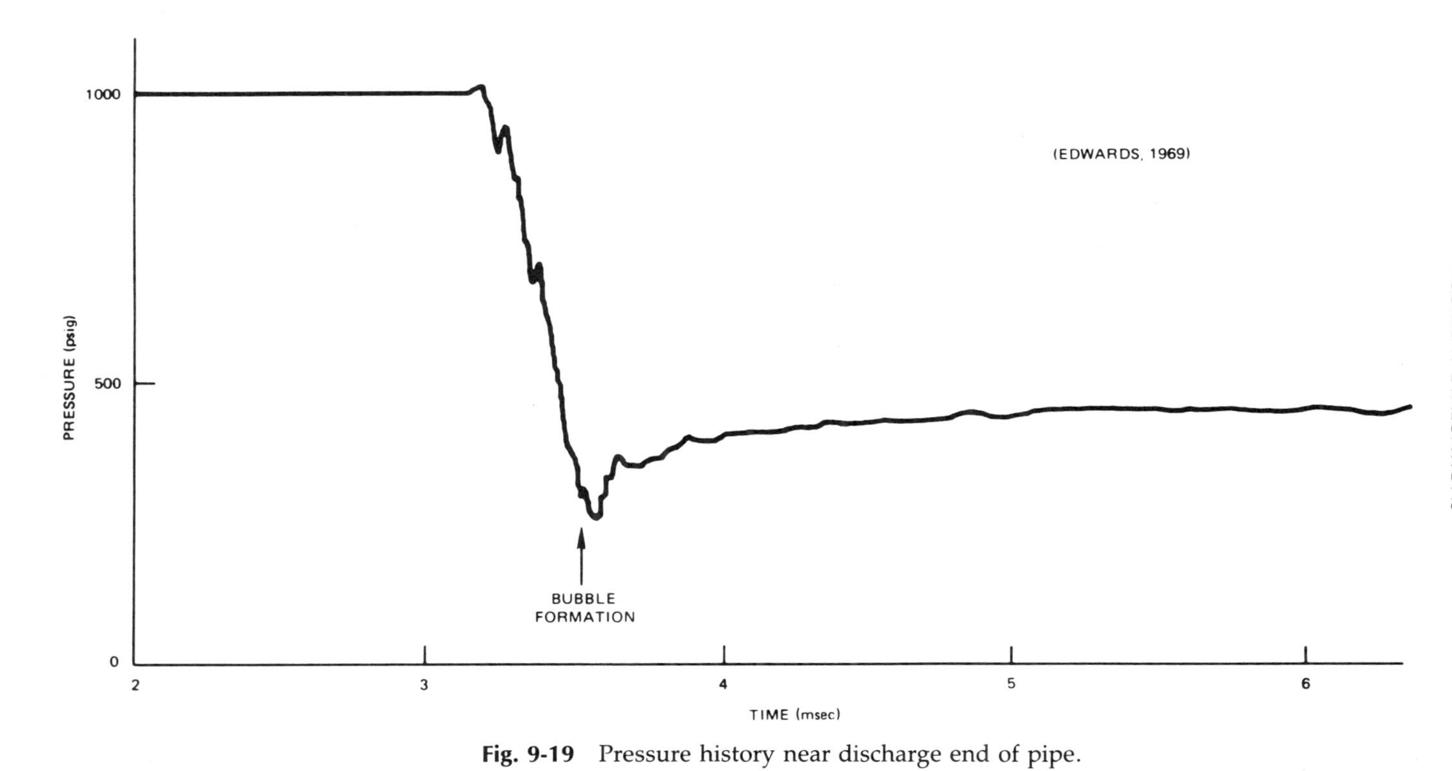

Fig. 9-19 Pressure history near discharge end of pipe.

Other hot water decompression experiments (Zaker and Wiedermann, 1966) similar to that just described have been reported. They found that if nonequilibrium states occurred, they lasted between 0.5 and 1.0 msec.

Similar experiments (Gallagher, 1970) with hot water up to 2000 psia and initial temperature up to saturation have been reported. Gallagher found that if metastable states did occur, they persisted for only ~1.0 msec and were independent of geometry. Therefore, it is concluded that if a fluid particle would be expelled in a time of ~1.0 msec or less, we should expect metastability and must accommodate nonequilibrium phenomenon in the flow rate prediction.

Several investigators have obtained blowdown rates of saturated water from large reservoirs and straight tubes of varying length and diameter. Blowdown rate data (Fauske, 1965; Uchida and Nariai, 1966) in terms of tube length, inside diameter, and reservoir pressure are shown in Fig. 9-20. Note that the blowdown rate decreases rapidly with increasing tube length in the $0 < L < 2.0$-in. interval. Beyond a tube length of 2.0 in., blowdown rate decreases at a much slower rate. If a flow coefficient, $C_f = 0.61$, is applied to the ideal liquid (Bernoulli) flow rate, G_{l_i} of Eq. (9.35), sharp-edged orifice flow is predicted, which corresponds to Fig. 9-20 values at zero length.

The region, $0 < L < 2.0$ in., is interpreted as a region dominated by metastability or nonequilibrium states and two-dimensional flow effects. Once an equilibrium amount of vapor formation is approached, the blowdown stream is expected to fill the flow passage and lend itself to a one-dimensional description.

To include a large-diameter test (Allemann et al., 1970), an initial blowdown rate corresponding to saturated water blowdown from 1600 psia through a 29.0-in-long, 6.8-in.-diam pipe is shown in Fig. 9-20. The mass flux falls below an extension of small-diameter data, suggesting that larger diameters may have the effect of reducing the critical mass flux.

Numbers shown at the extreme right of Fig. 9-20 give ideal critical flow rates of saturated water for the homogeneous equilibrium and equilibrium slip model of Moody. Note that the Moody model overpredicts blowdown rates for $L > 2.0$ in., while the homogeneous model gives fairly good predictions for $L > 10.0$ in. Other independent studies (Simon, 1971; Sozzi and Sutherland, 1975) have produced similar conclusions.

Two-phase blowdown rate models that include the effects of metastability have been proposed (Henry, 1970; Edwards, 1968). These formulations include the additional degree of freedom that local flow quality may lag behind equilibrium quality in a flow passage. Although empirical constants must be evaluated from data in the various flow geometries, these formulations do predict the proper trends in critical flow rate and pressure, particularly in the range where metastability is dominant.

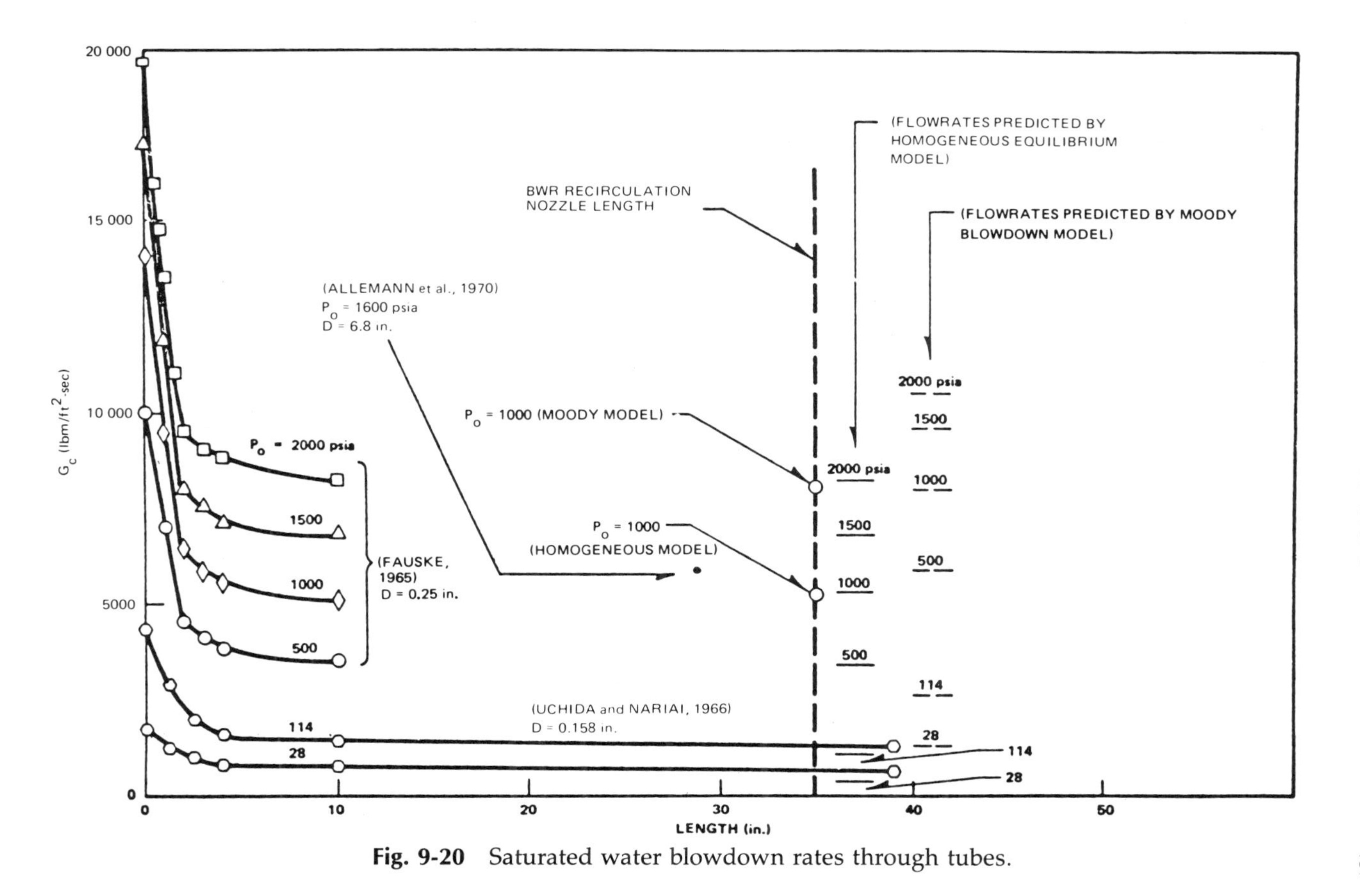

Fig. 9-20 Saturated water blowdown rates through tubes.

9.2.1.5 *The Henry-Fauske Model*

The Henry-Fauske nonequilibrium critical flow model (1970) is often used to predict the critical discharge of subcooled liquids. It was developed from the mixture continuity equation and the mixture momentum equation. In particular, for steady discharges,

Mixture continuity:

$$d(w_l + w_g) = 0 \ , \tag{9.68}$$

where

$$w_l = A_l \langle u_l \rangle_l / v_l \ , \tag{9.69a}$$

$$w_g = A_g \langle u_g \rangle_g / v_g \ . \tag{9.69b}$$

Two-phase momentum conservation (without friction):

$$-g_c A_{x-s} dp = d(w_g u_g + w_l u_l) \tag{9.70}$$

Critical flow is identified as the condition where the mass flux discharged is a maximum with respect to pressure, that is,

$$\frac{dG}{dp} = 0 \ , \tag{9.71}$$

where,

$$G = \frac{w_g + w_l}{A_{x-s}} = \frac{w}{A_{x-s}} \tag{9.72}$$

and,

$$w_g = \langle x \rangle w \ ; \qquad w_l = (1 - \langle x \rangle) w \ . \tag{9.73}$$

If the slip ratio, $s = \langle u_g \rangle_g / \langle u_l \rangle_l$, is introduced, Eqs. (9.68) through (9.73) yield,

$$G_c^2 = -S\left\{ [1 + \langle x \rangle (S-1)] \langle x \rangle \frac{dv_g}{dp} + [v_g\{1 + 2\langle x \rangle (S-1) + Sv_l 2(\langle x \rangle - 1) + S(1 - 2\langle x \rangle)\}] \frac{d\langle x \rangle}{dp} + S[1 + \langle x \rangle (S-2) - \langle x \rangle^2 (S-1)] \frac{dv_l}{dp} + \langle x \rangle (1 - \langle x \rangle)(Sv_l - v_g/S) \frac{dS}{dp} \right\}^{-1} , \tag{9.74}$$

where it has been assumed that v_g, v_l, $\langle x \rangle$, and S can be expressed as functions of pressure only. However, rapid expansion of a mixture through a nozzle is not expected to follow equilibrium states. Therefore, the derivatives are expressed as,

$$\frac{d\phi}{dp} = \left(\frac{d\phi}{dt}\right) \Big/ \left(\frac{dp}{dt}\right) \; ; \qquad \phi = v_g,\ v_l,\ \langle x \rangle,\ S \ . \tag{9.75}$$

The assumptions of negligible slip at critical flow, uniform quality between the stagnation and critical flow states, and isentropic flow yield the following simplifications:

$$\frac{dv_g}{dp} = \frac{v_g}{np} \ , \tag{9.76}$$

where (Tangren, 1949),

$$n = \frac{(1 - \langle x \rangle)(C_l/c_{pg}) + 1}{(1 - \langle x \rangle)(C_l/c_{pg}) + C_v/c_{pg}} \ . \tag{9.77}$$

Furthermore,

$$\frac{dv_l}{dp} \cong 0 \ , \tag{9.78}$$

$$\frac{dS}{dp} \cong 0 \ . \tag{9.79}$$

The equilibrium quality is given by,

$$\langle x_e \rangle = \frac{s_0 - s_{le}}{s_{ge} - s_{le}} \ , \tag{9.80}$$

where subscript e refers to equilibrium conditions at the instantaneous pressure. The actual flow quality, $\langle x \rangle$, is expected to lag behind the equilibrium value so that its rate of change with respect to pressure is given by:

$$\frac{d\langle x \rangle}{dp} = -\left[\frac{(1 - x_0)ds_l/dp + \langle x_0 \rangle ds_g/dp}{s_{g0} - s_{l0}}\right] = N\frac{dx_e}{dp} \ , \tag{9.81}$$

where,

$$N = N(\langle x_e \rangle) = C_l \langle x_e \rangle \ . \tag{9.82}$$

Comparisons with examples of the Marviken time-dependent blowdown data (EPRI, 1982) are shown in Figs. 9-21a and b, with the homogeneous equilibrium model, the Moody model, and the Henry-Fauske model. The blowdowns started with subcooled water in the vessel, which became saturated soon after blowdown began. It is seen that the empirical constant, C_l, can be chosen to match the data closely. Recommended constants are:

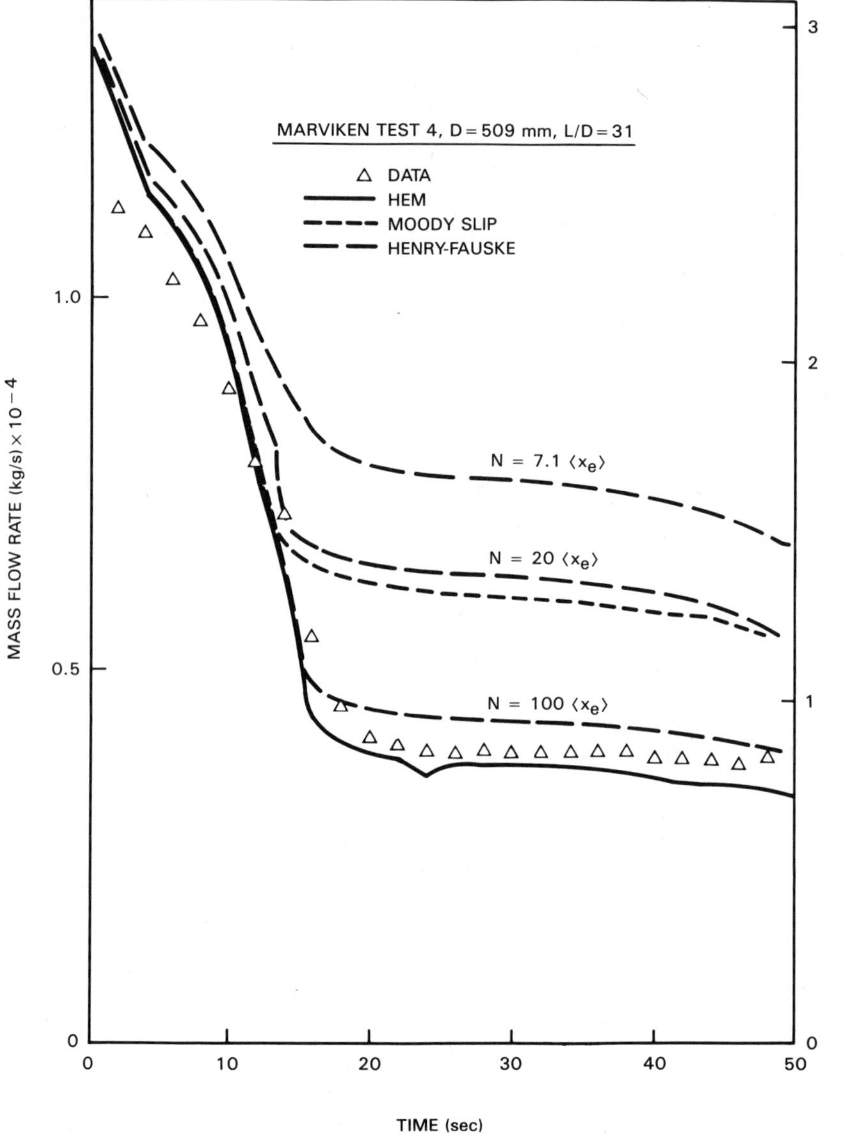

Fig. 9-21a Mass flow history for Marviken test 4.

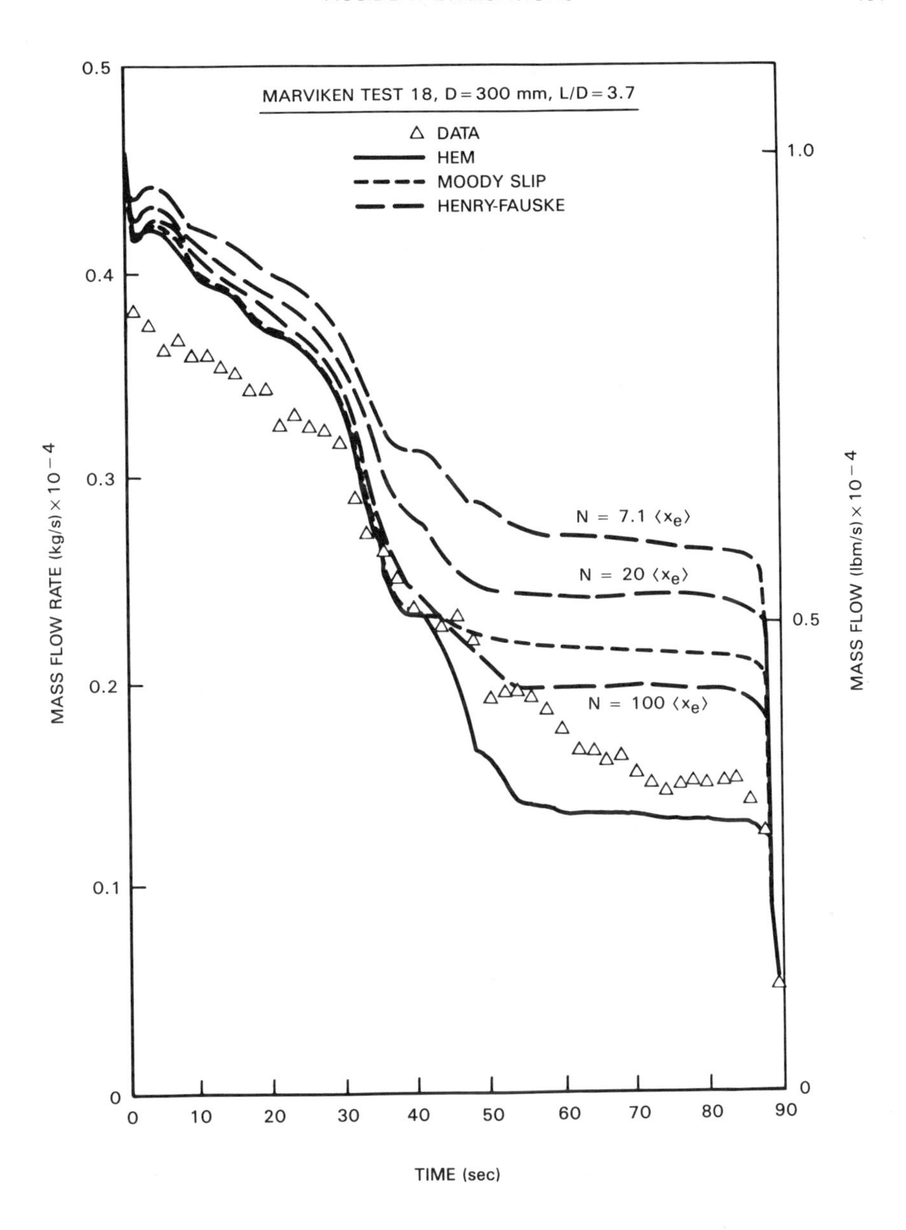

Fig. 9-21b Mass flow history for Marviken test 18.

$C_l = 7.1$ for subcooled blowdown, $L/D \leq 1.5$,

$C_l = 100$ for $L/D \geq 1.5$,

$C_l = 100$ for saturated blowdown .

A more mechanistic approach to nonequilibrium predictions has been suggested (Jones, 1982), whereby three quantities must be adequately specified: the void inception criterion; the interfacial area density; and the interfacial heat transfer rates. Analysis of the nonequilibrium liquid-vapor phase change in decompressive flashing systems has shown that nonequilibrium phase exchange can be approximated by a first-order relaxation process, and is therefore a path-dependent initial value problem, rather than a local phenomenon. Application of this approach is verified by comparison with data (Powell, 1961) in Fig. 9-22.

9.2.1.6 *Effects of Pipe Friction on Two-Phase Maximum Flow Rate*

For blowdown analysis from long pipes, it is necessary to incorporate wall friction effects. The basic model employed is shown in Fig. 9-23. The dotted control volume shown is used for writing the steady conservation equations for an adiabatic pipe in the following form:

Mass:

$$G = G_c = \text{constant} \ , \tag{9.83}$$

Momentum:

$$G^2 \frac{d}{dz}\left(\frac{1}{\langle\rho'\rangle}\right) = -g_c\left(\frac{dp}{dz} + \frac{\tau_w P_f}{A_{x-s}}\right) \ , \tag{9.84}$$

Energy:

$$h_0 = h_f + \langle x\rangle h_{fg} + \frac{G^2}{2g_cJ}\frac{1}{\langle\rho'''\rangle^2} = \text{constant} \ , \tag{9.85}$$

where $\langle\rho'\rangle$ and $\langle\rho'''\rangle$ are, respectively, the so-called momentum and energy-weighted densities, as defined in Eqs. (5.67) and (5.101), respectively.

In this analysis, the slip ratio, $S = (\rho_f/\rho_g)^{1/3}$, is assumed throughout the flow, with corresponding void fraction given by Eq. (5.25). Furthermore, since phase equilibrium is assumed at every location, it follows that for a given G,

$$\frac{1}{\langle\rho'\rangle} = f(p, \langle x\rangle) \ , \tag{9.86}$$

and,

$$h_0 = h_0(p, \langle x\rangle) = \text{constant} \ . \tag{9.87}$$

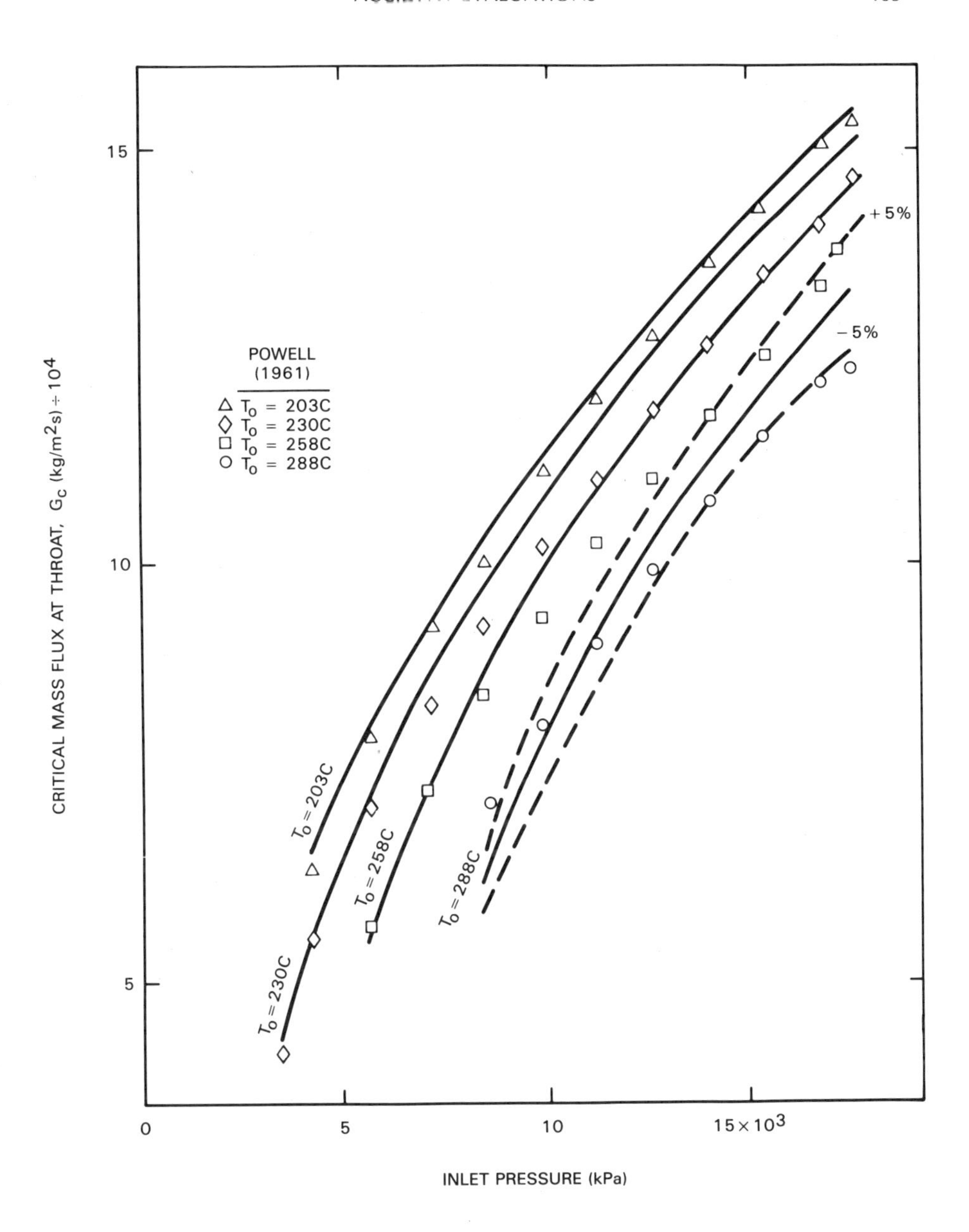

Fig. 9-22 Comparison of critical throat mass fluxes measured by Powell (1961) with predictions for different nozzle inlet pressures and temperatures.

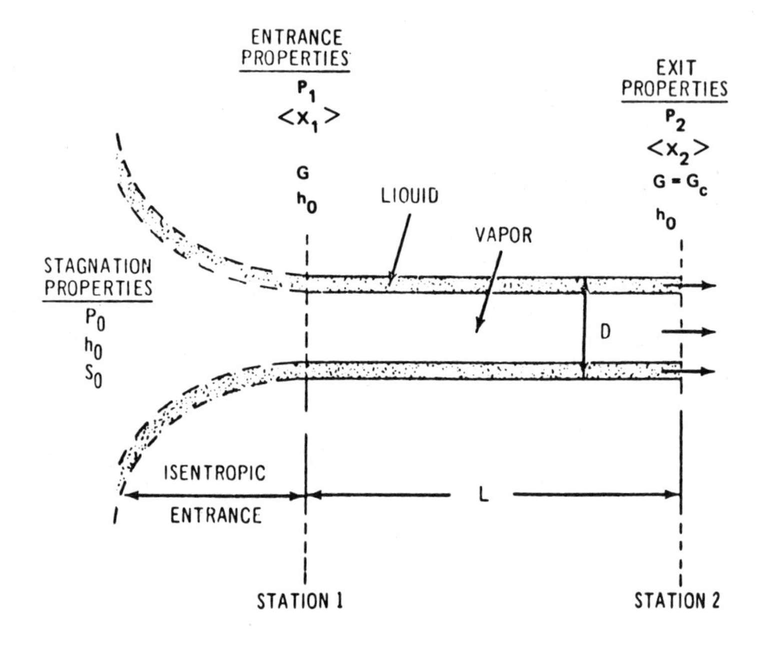

Fig. 9-23 Pipe maximum flow model.

The differential, $d(1/\langle\rho'\rangle)$, in Eq. (9.84) can be expanded to give,

$$G^2\left\{\left[\frac{\partial}{\partial p}\left(\frac{1}{\langle\rho'\rangle}\right)\right]_{\langle x\rangle} dp + \left[\frac{\partial}{\partial\langle x\rangle}\left(\frac{1}{\langle\rho'\rangle}\right)\right]_p d\langle x\rangle\right\} = -g_c\left(dp + \frac{\tau_w P_f}{A_{x-s}}dz\right) . \quad (9.88)$$

Next, for the assumed constant stagnation enthalpy throttling process, the derivative, $d\langle x\rangle/dp$, can be obtained from Eq. (9.87) as,

$$\frac{d\langle x\rangle}{dp} = -\frac{\left(\dfrac{\partial h_0}{\partial p}\right)_{\langle x\rangle}}{\left(\dfrac{\partial h_0}{\partial\langle x\rangle}\right)_p} , \quad (9.89)$$

so that $d\langle x\rangle$ can be eliminated from Eq. (9.88).

There are numerous possible empirical correlations for the wall shear stress in two-phase flow. As discussed in Sec. 5.4, τ_w is normally expressed in terms of wall shear based on an equivalent liquid flow, τ_{wl}, and a two-phase multiplier, ϕ_{l0}^2. Based on the assumption that only liquid contacts the wall (Levy, 1960), ϕ_{l0}^2 can be expressed by Eq. (5.270) with $n=0$. This formulation was chosen (Moody, 1966) because of its simplicity and reasonable accuracy up to 80% quality. Note that for high-quality flows,

$\langle x\rangle \to 1.0$, and a slip ratio of $S=(\rho_f/\rho_g)^{1/3}$, Levy's expression for ϕ_{l0}^2 approaches $(\rho_f/\rho_g)^{4/3}$. Therefore, to apply Levy's formulation at high quality, an adjustment must be made in the friction factor. From Eq. (5.251), with K replaced by fL/D_H, we can write the irreversible pressure loss for gas flow as,

$$\Delta p_g = f_g \frac{L}{D_H}\frac{G^2}{2g_c\rho_g} = f_l \frac{L}{D_H}\frac{G^2}{2g_c\rho_f}\phi_{l0}^2\Big|_{\langle x\rangle\to 1.0} . \tag{9.90}$$

It follows that the friction parameter for Levy's formulation can be expressed in terms of the friction parameter for gas flow as,

$$f_l \frac{L}{D_H} \to f_g \frac{L}{D_H S} ; \quad \langle x\rangle \to 1.0 . \tag{9.91}$$

Thus, from Eqs. (9.88), (9.89), and (5.307),

$$\frac{\rho_f F(p;\, h_0,\, G)dp}{G^2\left[1+\left(\dfrac{1}{S}\dfrac{\rho_f}{\rho_g}-1\right)\langle x\rangle\right]^2} = \frac{f_l}{D_H}dz , \tag{9.92}$$

where f_l is the Darcy friction factor for all liquid flow and,

$$F(p;\, h_0,\, G) \triangleq 2g_c G^2 \left\{ \frac{1}{g_c}\left[\frac{\partial\left(\dfrac{1}{\langle\rho'\rangle}\right)}{\partial\langle x\rangle}\right]_p \frac{\left(\dfrac{\partial\langle h\rangle}{\partial p}\right)_{\langle x\rangle} + \dfrac{G^2}{2g_c J}\left[\dfrac{\partial\left(\dfrac{1}{\langle\rho'''\rangle^2}\right)}{\partial p}\right]_{\langle x\rangle}}{\left(\dfrac{\partial\langle h\rangle}{\partial\langle x\rangle}\right)_p + \dfrac{G^2}{2g_c J}\left[\dfrac{\partial\left(\dfrac{1}{\langle\rho'''\rangle^2}\right)}{\partial\langle x\rangle}\right]_p} - \frac{1}{g_c}\left[\frac{\partial\left(\dfrac{1}{\langle\rho'\rangle}\right)}{\partial p}\right]_{\langle x\rangle} \right\} - 2g_c . \tag{9.93}$$

Properties first must be integrated between Stations 1 and 2 in Fig. 9-23. It is well known that subsonic flow of a compressible fluid approaches critical flow in going through a uniform pipe with friction. Consider a two-phase mass flux, $G<G_c$. In passing through a pipe, friction causes the pressure to decrease. Figure 9-10c shows that for a given h_0, as pressure decreases, the corresponding value of G_c decreases. A condition finally is reached at the pipe exit where $G=G_c$. Therefore, a critical flow state is

specified at Station 2 in Fig. 9-23 and some other state is determined at Station 1, which is distance L upstream from the exit. It follows from Eq. (9.92) that if the pressure at Station 2 is expressed for the critical flow condition as $p_c(h_0, G)$,

$$\int_{p_c(h_0,G)}^{p_1} \frac{\rho_f F(p; h_0, G)dp}{G^2\left[1+\left(\frac{1}{S}\frac{\rho_f}{\rho_g}-1\right)\langle x\rangle\right]^2} = \frac{\bar{f}_l}{D_H}L \; . \tag{9.94}$$

The pressure at Station 1 can be expressed in terms of vessel stagnation properties by idealizing the pipe entrance as an isentropic nozzle such that,

$$s_0(p_0, h_0) = s_f(p_1) + \langle x_1\rangle s_{fg}(p_1) \; . \tag{9.95}$$

Numerical integration involving saturated steam and water properties was carried out to express the maximum two-phase blowdown rate from uniform pipes in either of the forms,

$$G_c = f_1\left(p_0, h_0, \frac{f_l L}{D_H}\right) \tag{9.96}$$

or

$$G_c = f_2\left(p_1, h_0, \frac{f_l L}{D_H}\right) \; . \tag{9.97}$$

Equations (9.96) and (9.97) are plotted in Figs. 9-24a through 9-24e.

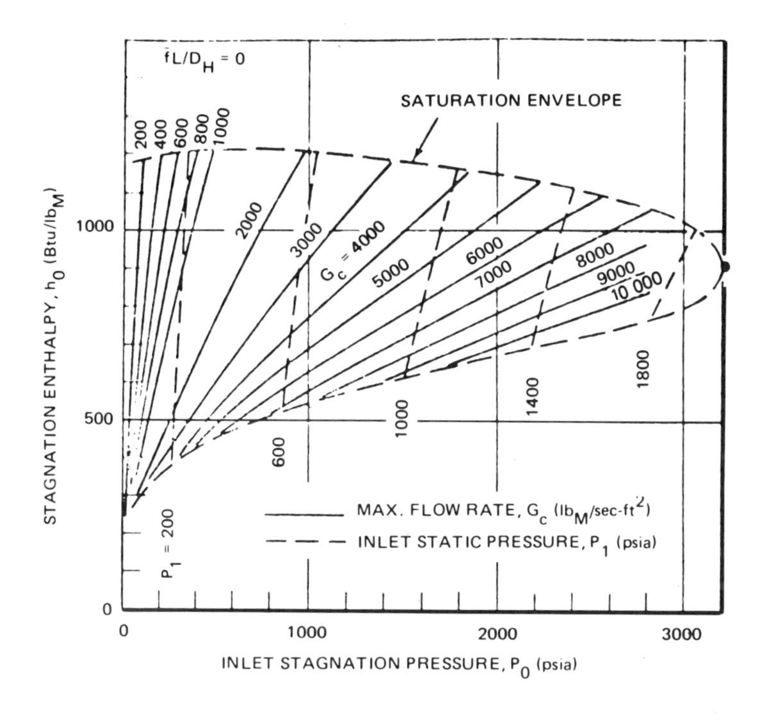

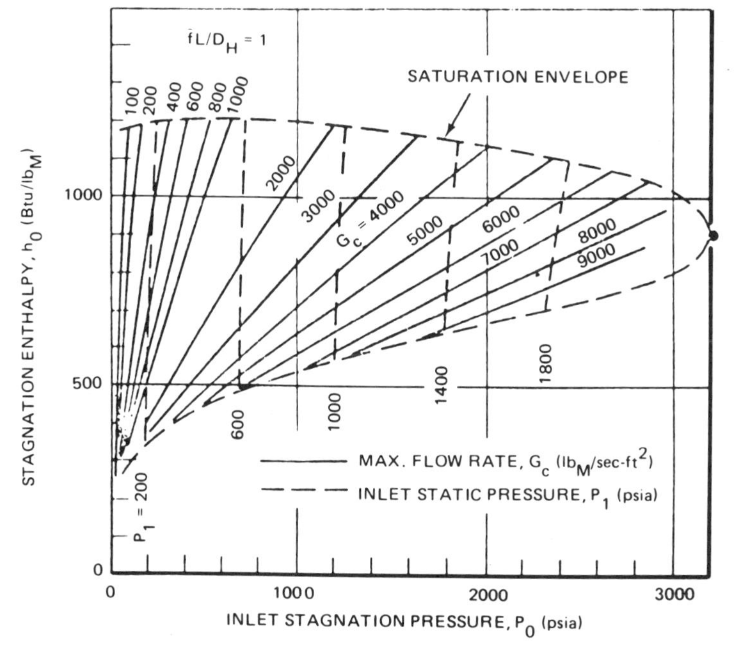

Fig. 9-24a Pipe maximum steam/water discharge rate.

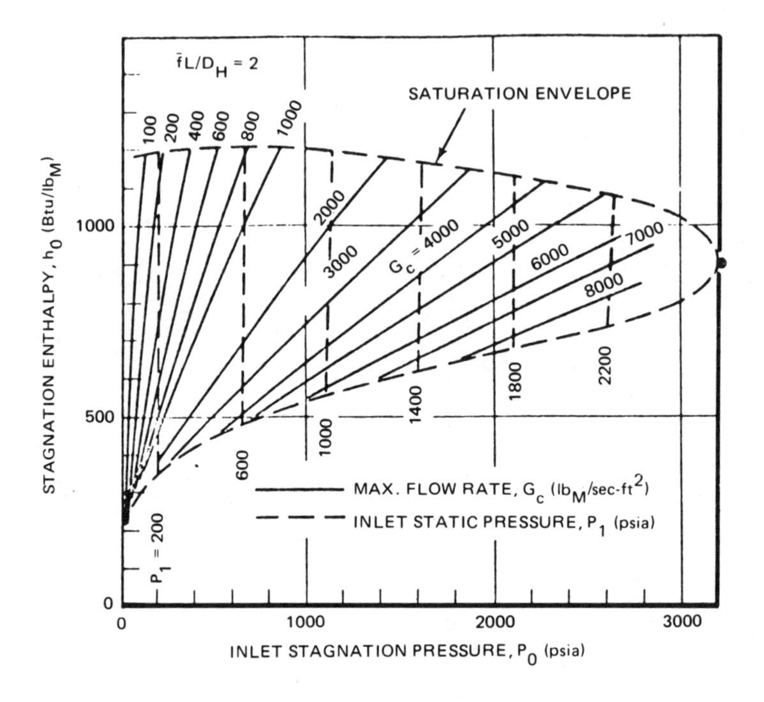

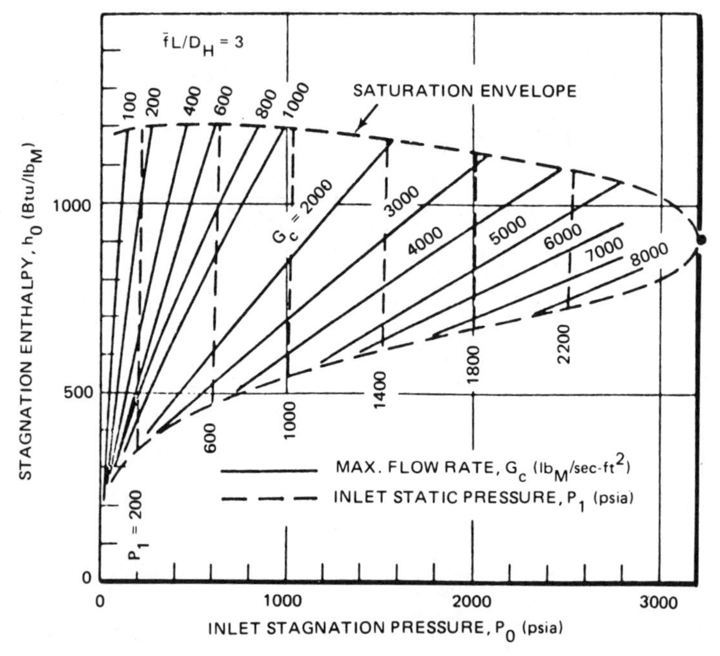

Fig. 9-24b Pipe maximum steam/water discharge rate.

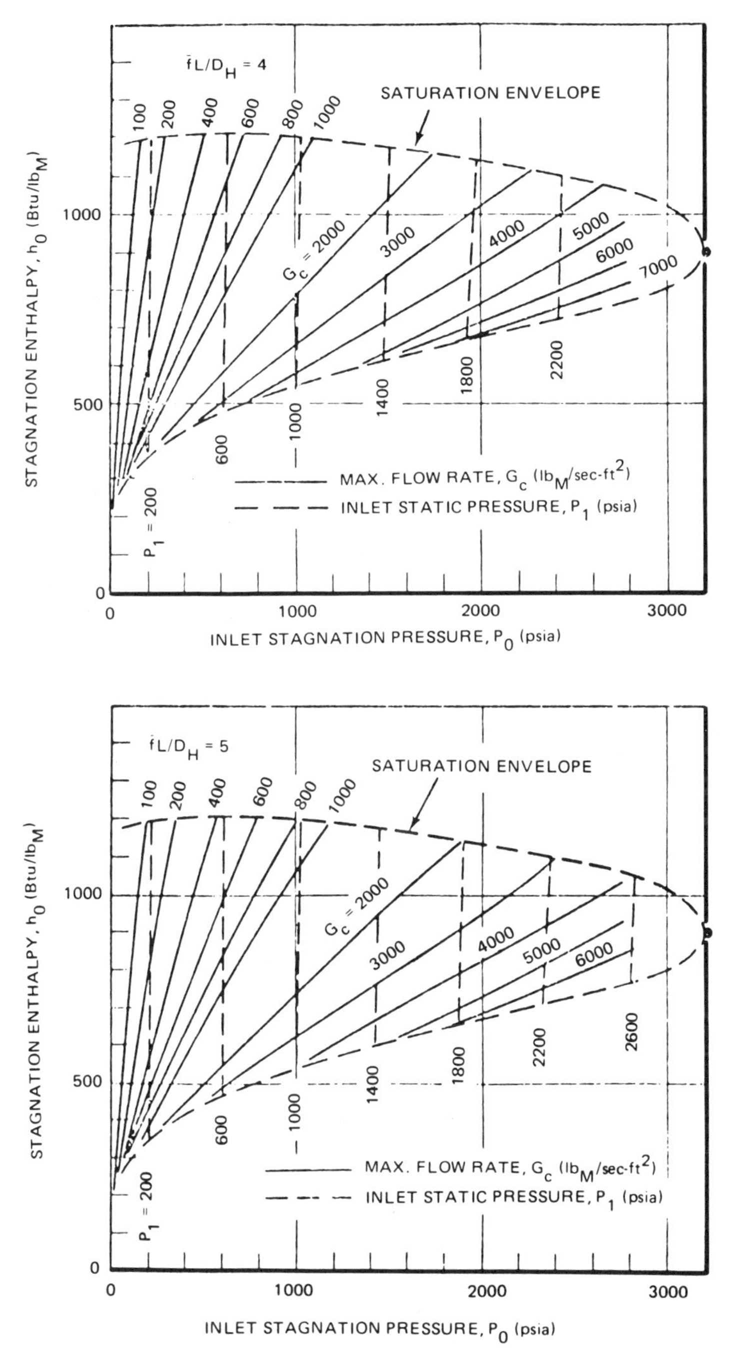

Fig. 9-24c Pipe maximum steam/water discharge rate.

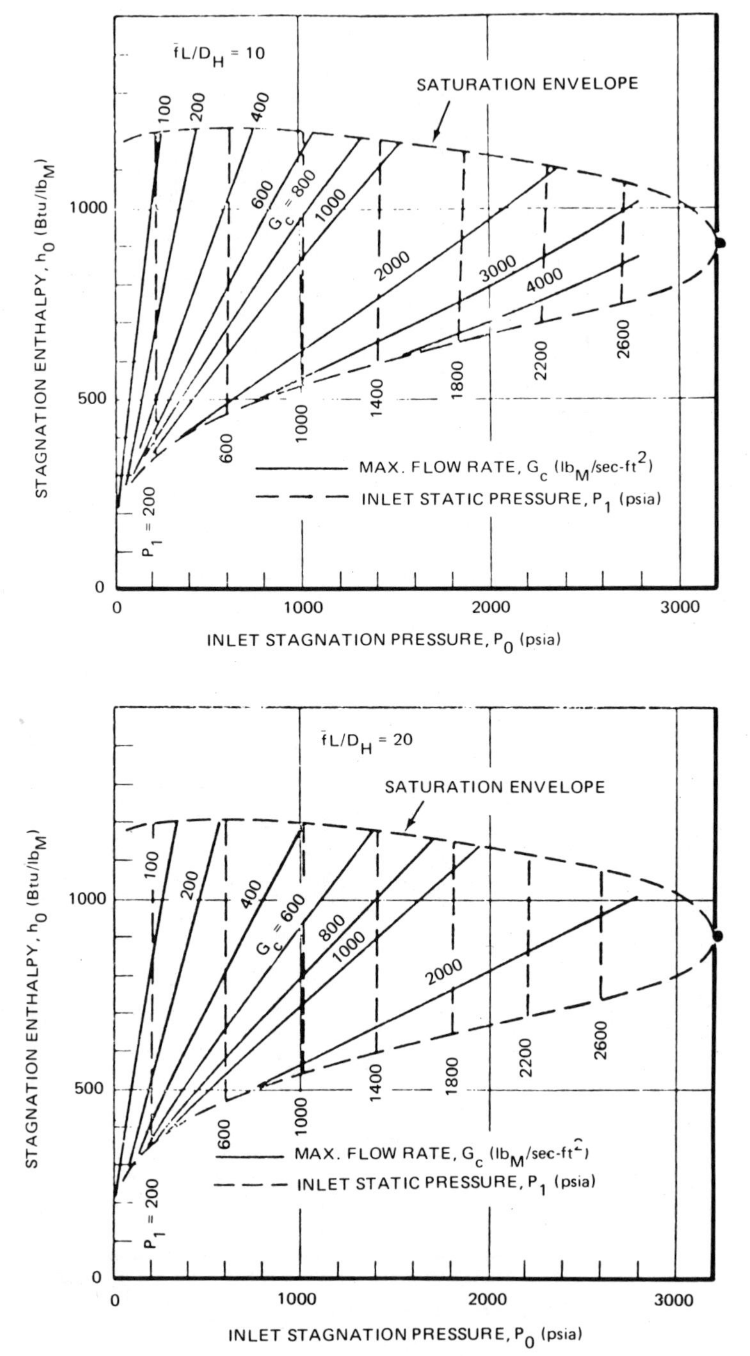

Fig. 9-24d Pipe maximum steam/water discharge rate.

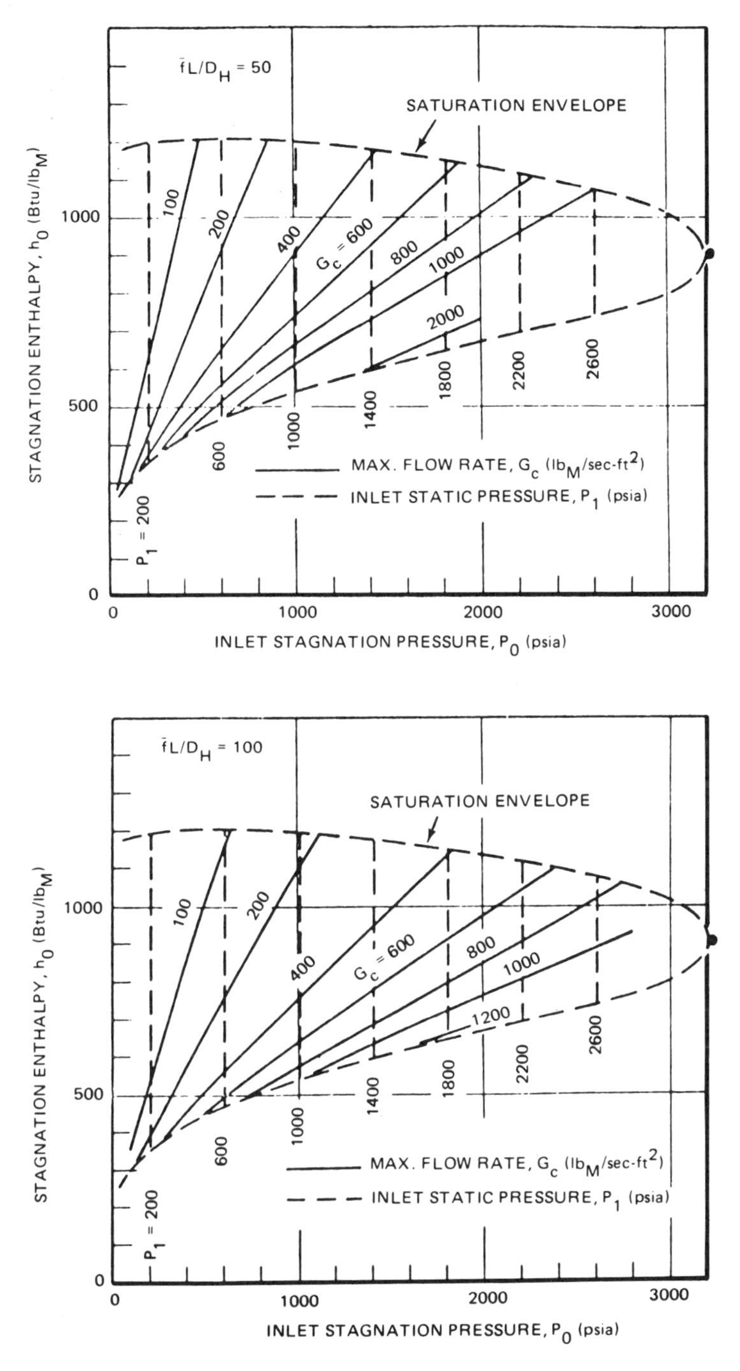

Fig. 9-24e Pipe maximum steam/water discharge rate.

9.2.1.7 *Subcooled Liquid Discharge from Pipes with Friction*

Subcooled water discharge rates from a uniform pipe with friction can be estimated by employing incompressible fluid flow theory from the vessel stagnation state to the location where saturation pressure is reached, and the two-phase flow model of Sec. 9.2.1.6 thereafter to the discharge end of the pipe.

Figure 9-25 shows liquid flow from the vessel to pressure, p_{sat}, at the unknown mass flux for which,

$$p_0 - p_{sat} = \frac{G^2 v_l}{2g_0}\left(1 + \frac{f_1 L_1}{D_H}\right) , \tag{9.98}$$

where f_1 and L_1 are the single-phase liquid friction factor and flow length, respectively. However, neither G nor L_1 is known. The remaining frictional loss coefficient is,

$$\frac{f_2 L_2}{D_H} = \frac{f_2(L - L_1)}{D_H} . \tag{9.99}$$

The stagnation enthalpy, $h_0 \approx h_f(p_{sat})$, and $f_2 L_2 / D_H$ should give a value of G that corresponds to that in Eq. (9.98). The procedure for obtaining consistent values for G, L_1, and L_2 is summarized next.

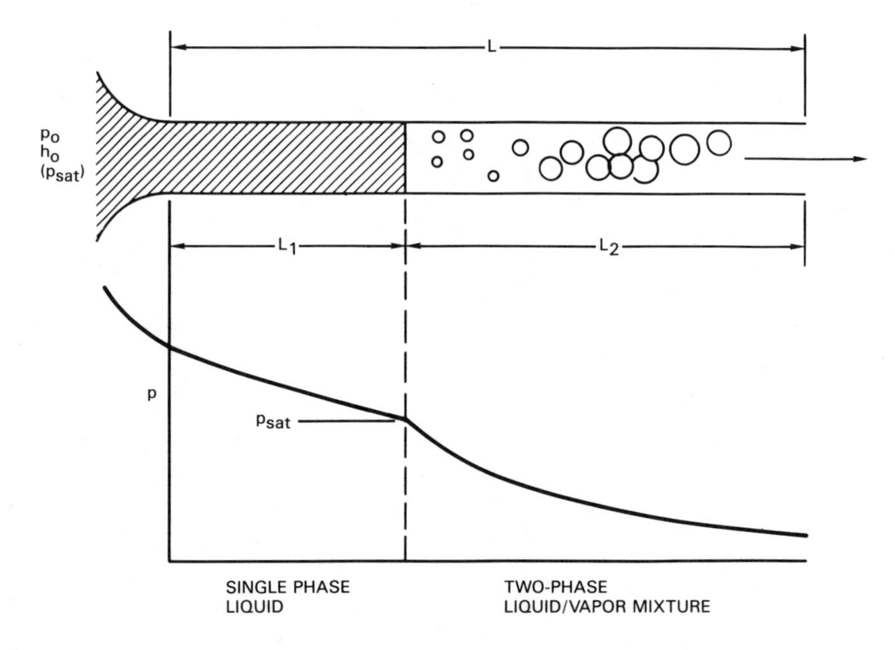

Fig. 9-25 Subcooled liquid blowdown from a pipe with friction.

Calculating Subcooled Blowdown. The following procedure is used for calculating subcooled blowdown from a uniform pipe with friction.

1. Estimate reasonable values for the single and two-phase friction factors f_1 and f_2. List values of p_0, $v_l = v_f(p_0)$, h_0, and p_{sat} from steam tables based on $h_0 = h_f(p_{sat})$, and the total pipe length L.
2. Pick a trial value of the single-phase length, L_1.
3. Calculate the single-phase value of $G_1 = G$ from Eq. (9.98).
4. Calculate the two-phase flow length from $L_2 = L - L_1$, and obtain f_2L_2/D_H.
5. Enter Figs. 9-24 with f_2L_2/D from Step 4 and read the dashed lines of p_{sat} on the saturated water boundary to obtain the two-phase critical mass flux, G_2.
6. Compare G_1 from Step 3 with G_2 from Step 5. A valid solution corresponds to $G_1 = G_2$. If $G_1 > G_2$, the trial value of L_1 was too short, and thus L_2 was too long. Repeat Steps (2) through (6) with a longer L_1. Several iterations may be necessary to obtain a solution for G_c during subcooled discharge.

9.2.2 Vessel Blowdown

The relationships just derived for the critical flow rate of a steam/water mixture can be employed with the conservation of mass and energy for a vessel of fixed volume to determine its time-dependent blowdown properties.

The reactor vessel model is shown schematically in Fig. 9-26. This model can be used to estimate the loss-of-coolant discharge rates of steam/water mass and energy. The model includes both energy and mass inflows and outflows, core heat transfer, heat transfer between the coolant and vessel mechanical components, blowdown discharge, and core sprays. The kinetic and potential energy components of the coolant are neglected since the thermal energy components dominate. A mass inflow rate has a stagnation enthalpy, $h_{0,in}$, determined by its source. A mass outflow rate has the stagnation enthalpy, $h_{0,out}$, determined by vessel fluid properties in the region from which discharge occurs.

Mass and energy conservation equations for Fig. 9-26 are written as:

$$w_{out} - w_{in} + \frac{dM}{dt} = 0 \ , \tag{9.100}$$

and,

$$w_{out}h_{0,out} - w_{in}h_{0,in} + q_{out} - q_{in} + \frac{dU}{dt} = 0 \ . \tag{9.101}$$

The state equations,

$$V = Mv \ , \qquad U = M\mu \ , \qquad \mu = \mu(p,v) \ , \tag{9.102}$$

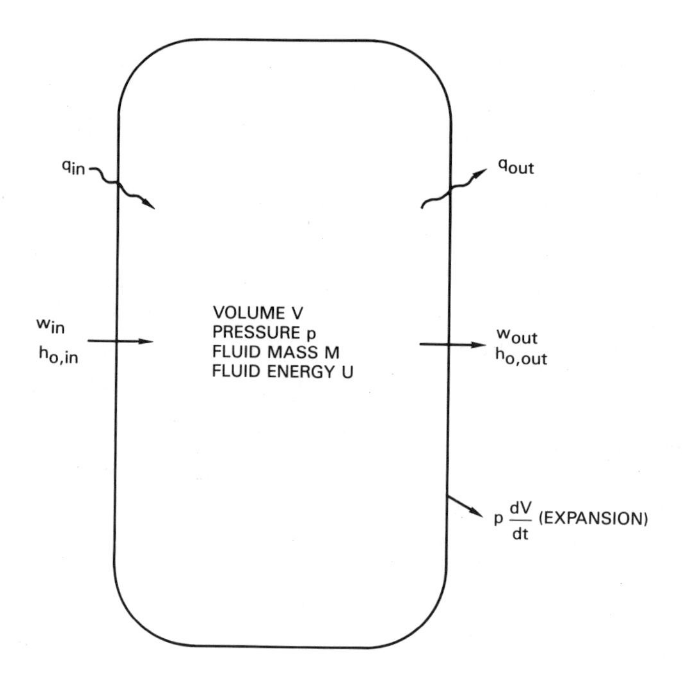

Fig. 9-26 Mass and energy conservation in the fluid region.

can be employed in Eq. (9.101), to give the vessel's rate of pressure change as:

$$\frac{dp}{dt}=\frac{w_{\text{in}}(h_{0,\text{in}}-f_p)-w_{\text{out}}(h_{0,\text{out}}-f_p)+q_{\text{in}}-q_{\text{out}}}{MF(p,V/M)}\ , \tag{9.103}$$

where the functions $f(p)$ and $F(p,V/M)$ are given by:

$$f(p)=e-v\left(\frac{\partial e}{\partial v}\right)\bigg|_p \tag{9.104}$$

and,

$$F(p,V/M)=\left(\frac{\partial e}{\partial p}\right)\bigg|_v \tag{9.105}$$

for any simple compressible substance such as water, steam, or a two-phase liquid/vapor mixture in equilibrium.

In particular, for an incompressible liquid,

$$f(p)=\mu \quad F(p,V/M)=0 \quad \text{(incompressible liquid)}\ , \tag{9.106}$$

while for a perfect gas:

$$f(p)=0 \qquad F(p,V/M)=\frac{v}{(K-1)} \qquad \text{(perfect gas)} \ . \tag{9.107}$$

Finally, an equilibrium two-phase liquid-vapor mixture with a state equation given by,

$$\mu=\mu_f(p)+\frac{\mu_{fg}(p)}{v_{fg}(p)}[v-v_f(p)] \tag{9.108}$$

has the functions,

$$f(p)=\mu(p)-v_f(p)\frac{\mu_{fg}(p)}{v_{fg}(p)} \tag{9.109}$$

and,

$$F(p,V/M)=\mu_f'-\left(v_f\frac{\mu_{fg}}{v_{fg}}\right)'+\frac{V}{M}\left(\frac{\mu_{fg}}{v_{fg}}\right)' , \tag{9.110}$$

where ( )′ denotes $d(\ )/dp$. These functions are plotted in Figs. 9-27 and 9-28.

Heat transfer terms are negligible during the time required for blowdown through a large pipe break. Figure 9-29 shows the calculated coolant mass fraction remaining and pressure-time characteristics for blowdown from an adiabatic vessel initially filled with saturated water at 1000 psia. The blowdown rate was based on the homogeneous equilibrium model (HEM), and core heat transfer was neglected. Vessel blowdown rates do not depend on the drywell pressure p_d as long as the critical discharge pressure, p_c, satisfies the criterion,

$$p_c > p_d \ .$$

The vessel depressurization time, t_{vd}, for either steam or water discharge is proportional to the initial fluid mass, M_i, and inversely proportional to the discharge area, A_b. It can be estimated from the results of Fig. 9-29 as,

$$\frac{A_b t_{vd}}{M_i} \approx 4.0\times10^{-5}\ \text{m}^2\text{/s-kg} \ . \tag{9.111}$$

Initial depressurization rates can be obtained from Eq. (9.103) to show the difference between steam and saturated water discharge. Consider a mass discharge from the vessel at a critical flow rate of,

$$w_{\text{out}}=G_c(p,h_{0,\text{out}})A_b \ . \tag{9.112}$$

The vessel pressure rate is given as,

$$\frac{dp}{dt}=\frac{A_b G_c(p,h_0)}{MF(p,V/M)}[f(p)-h_{0,\text{out}}] \ . \tag{9.113}$$

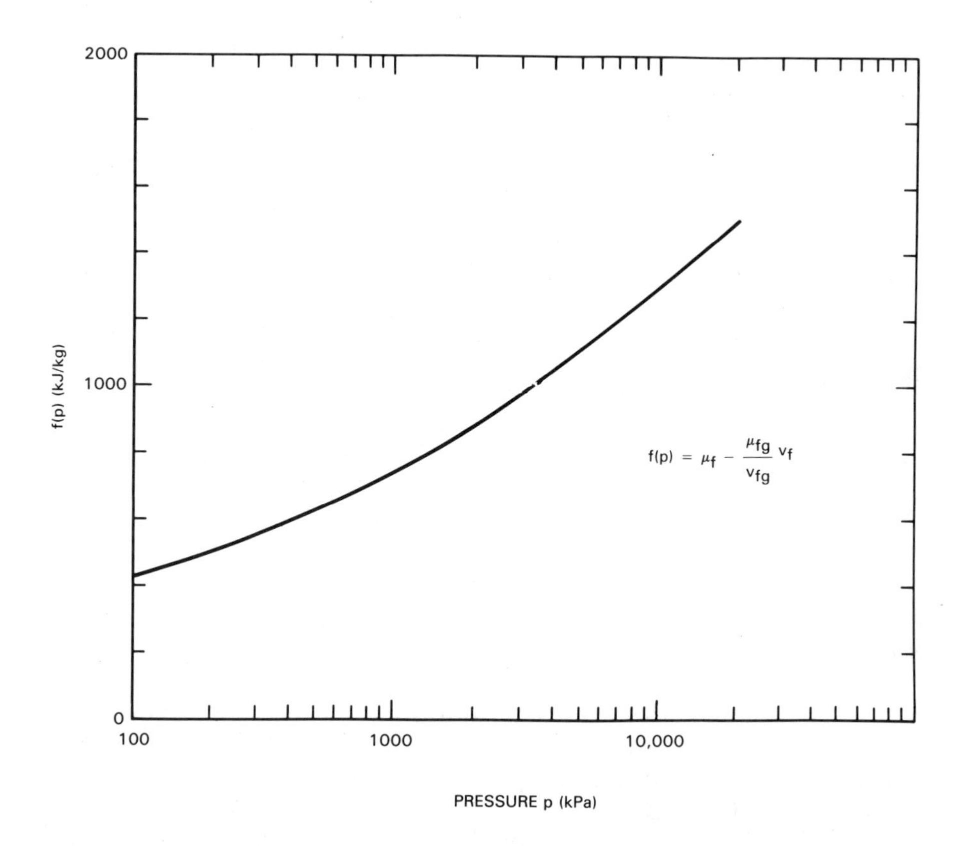

Fig. 9-27 Function $f(p)$, saturated steam/water mixtures.

The critical mass flux for steam and saturated water discharges are G_{cg} and G_{cf} from Sec. 9.2.1.1. The corresponding stagnation enthalpies are $h_{0,g}$ and $h_{0,f}$. The function $f(p)$ and the denominator of Eq. (9.113) depend on the fluid state in the vessel and not the instantaneous blowdown discharge properties. It follows that the ratio of steam-to-water vessel decompression rates is given by:

$$\frac{(dp/dt)_g}{(dp/dt)_f} = \frac{G_c(p,h_{g0})}{G_c(p,h_{f0})} \frac{f(p) - h_{g0}]}{f(p) - h_{f0}]} \; . \tag{9.114}$$

It can be shown that for steam and saturated water blowdowns from 1000 psia, the ratio of decompression rates is

$$\frac{(dp/dt)_g}{(dp/dt)_f} = 7.5 \; .$$

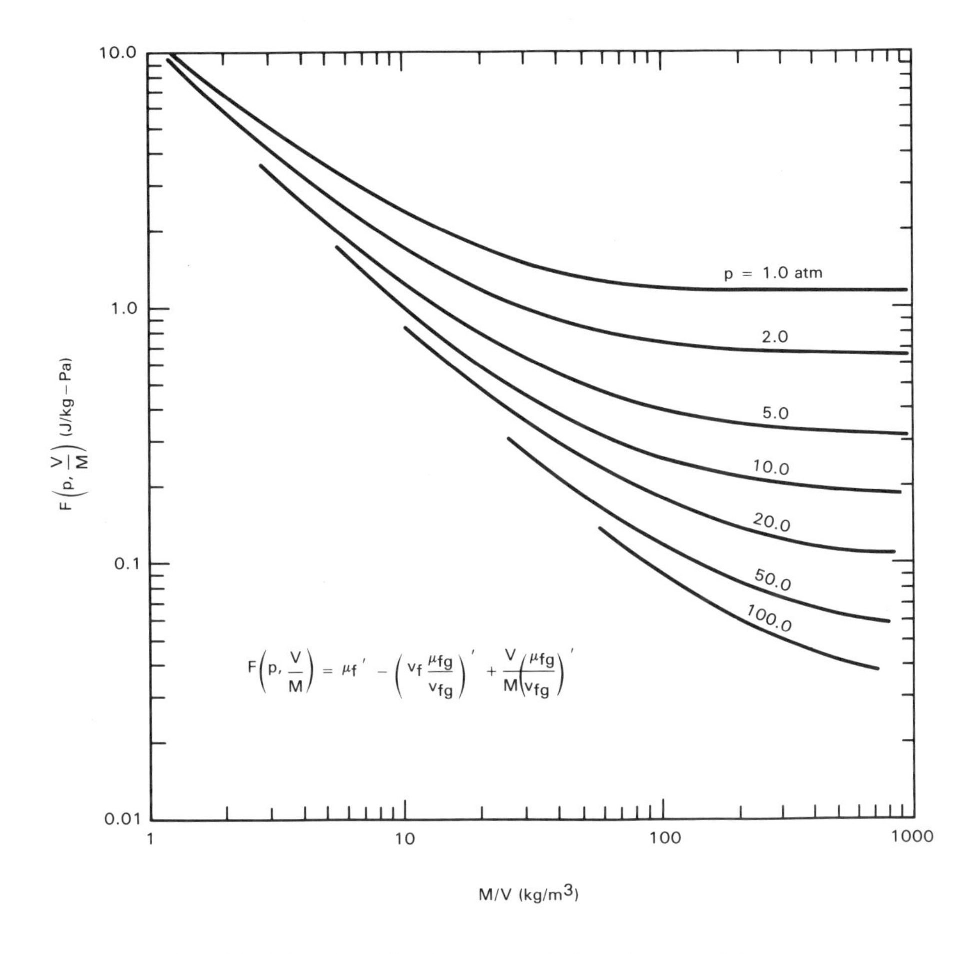

Fig. 9-28 Function $F(p,\rho)$, saturated steam/water mixtures.

That is, the critical discharge of saturated steam gives more than seven times the vessel decompression rate as saturated water from a system at 1000 psia.

Figure 9-30 shows a comparison of typical calculations and data for steam blowdown from a 4.26-m-long, 0.3-m-diam cylindrical pressure vessel through a 0.95-cm orifice. The initial vessel pressure was 68 bars and the starting water level was 3.6 m. The mixture level in the vessel was determined by a computation of vapor formation rate in the liquid during decompression, less that vapor leaving the mixture, as rising bubbles. It can be seen that the agreement between theory and experiment is quite good.

The calculated blowdowns in this section were based on an ideal nozzle of flow area, A_B. For cases where blowdown occurs from a pipe with

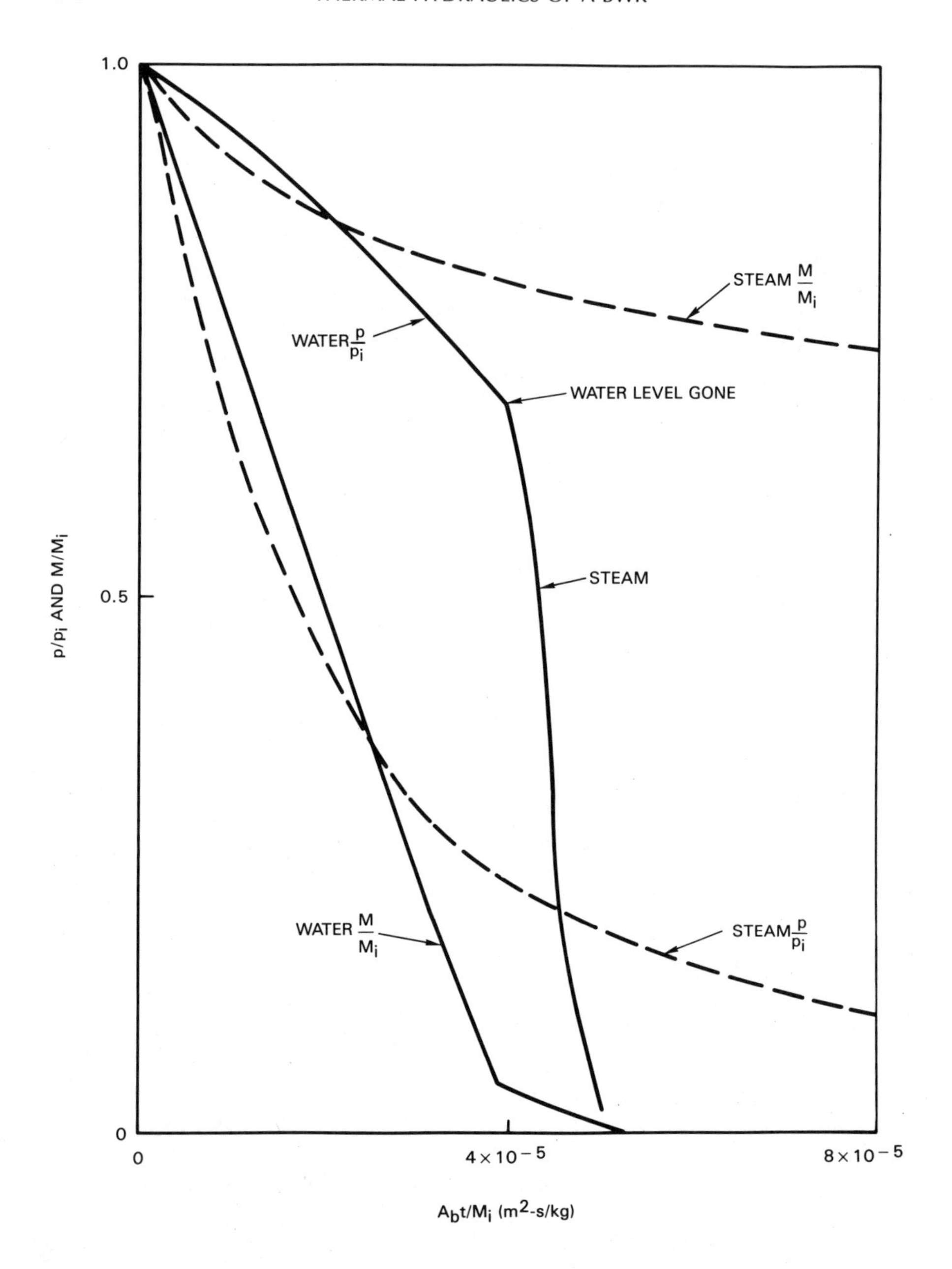

Fig. 9-29 Vessel blowdown calculation, steam/water.

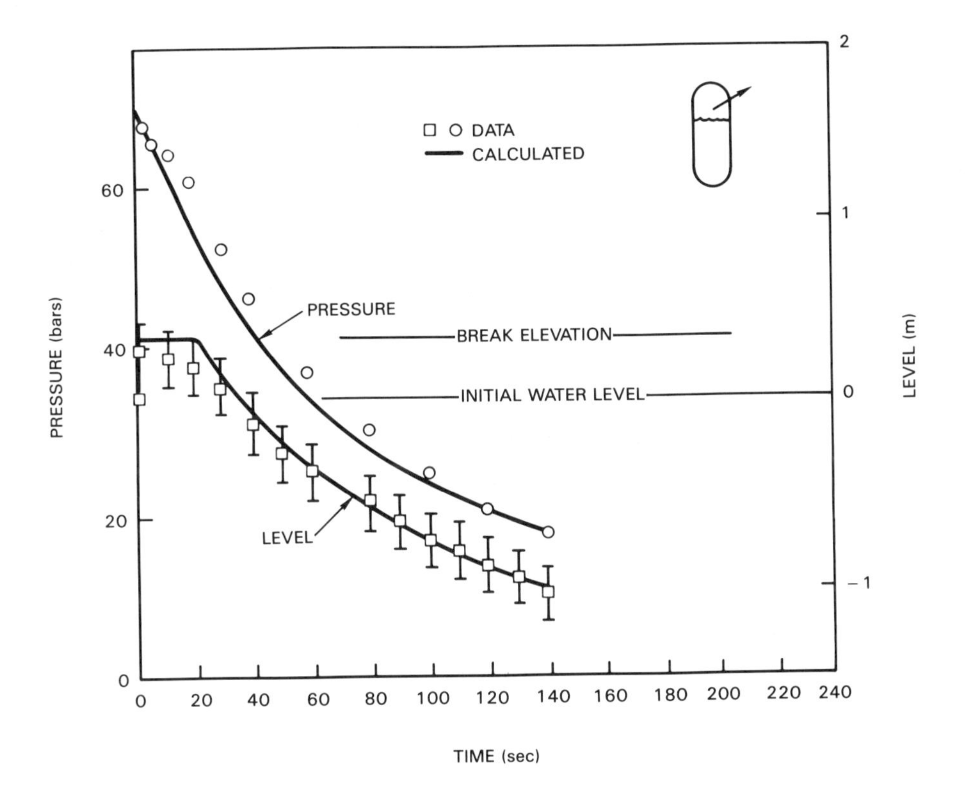

Fig. 9-30 Steam break blowdown test.

appreciable length and associated friction, Figs. 9-11a and 9-24 can be used to estimate an equivalent ideal nozzle area, $A_{B(EQ)}$, which would produce the same blowdown rate as a pipe with appreciable friction and area, A_{pipe}; i.e.,

$$A_{B(EQ)}G_{nozzle}(p_i,\ h_0) = A_{pipe}G\left(p_i,\ h_0;\ \frac{fL}{D_H}\right) \quad . \tag{9.115}$$

To predict the reactor core environment and subsequent thermal behavior, it is necessary to obtain both the pressure and liquid/vapor distribution throughout the vessel. As decompression progresses, vapor formation occurs throughout the bulk of the saturated water. If the vapor formation rate exceeds the rate of vapor separation from the mixture surface, the mixture swells. However, depending on the location of blowdown mass extraction from the vessel, the mixture level also may fall. Prediction of the transient void distribution and level swell during blowdown can be performed readily using the appropriate conservation equations derived in Chapter 5.

References

Allemann et al., "Experimental High Enthalpy Blowdown from a Simple Vessel Through a Bottom Outlet," BNWL-11111, Battelle Northwest Laboratory (1970).

Andersen, J. G. M., Personal Communication, Riso, Denmark (1974).

Andersen, J. G. M., "REMI/HEAT COOL, A Model for Evaluation of Core Heatup and Emergency Core Spray Cooling System Performance for Light-Water-Cooled Nuclear Power Reactors," RISO Report No. 296, Denmark (1973).

Aoki, S., A. Inoue, Y. Kozawa, T. Furubayaki, M. Aritomi, T. Nakajima, and H. Okazaki, "Critical Heat Flux Under Transient Conditions," *Proc. Int. Mtg. Reactor Heat Transfer,* Karlsruhe, Germany (1973).

Baker, L., and L. C. Just, "Studies of Metal-Water Reactors at High Temperature, III. Experimental and Theoretical Studies of the Zircaloy-Water Reaction," ANL-6548, Argonne National Laboratory (1962).

Borgartz, B. O., T. P. O'Brien, N. J. M. Rees, and A. V. Smith, "Experimental Studies of Water Depressurization Through Simple Pipe Systems," CREST Specialist Mtg. Depressurization Effects in Water-Cooled Reactors, Battelle Institute, Frankfurt, Germany, June 10–13, 1969.

Bouré, J., and M. Reocreux, "General Equations of Two-Phase Flows—Applications to Critical Flows and Non-Steady Flows," Fourth All-Union Heat and Mass Transfer Conf., Minsk, U.S.S.R. (1972).

"Critical Flow Data Review and Analysis," EPRI Report NP-2192, Electric Power Research Institute (Jan. 1982).

Dietz, K. A., Ed., "Quarterly Technical Report, Engineering and Test Branch, October 1–December 31, 1967," IDO-17242, Petroleum Company (May 1968).

Edwards, A. R., "Conduction Controlled Flashing of a Fluid and the Prediction of Critical Flow Rates in a One-Dimensional System," AHSB(S)R-1117, Authority Health and Safety Branch, United Kingdom Atomic Energy Authority (1968).

Edwards, A. R., "One-Dimensional Two-Phase Transient Flow Calculations," CREST Specialist Mtg. Depressurization Effects in Water-Cooled Power Reactors, Battelle Institute, Frankfurt, Germany, June 10–13, 1969.

Fabic, S., "Blowdown-2: Westinghouse APD Computer Program for Calculation of Fluid Pressure, Flow, and Density Transients During a Loss-of-Flow Accident," *Trans. Am. Nucl. Soc.,* **12,** 358 (1969).

Faletti, D. W., and R. W. Moulton, "Two-Phase Critical Flow of Steam-Water Mixtures," *AIChE J.,* **9,** 2 (1963).

Fauske, H. K., "Contribution to the Theory of Two-Phase, One-Component Critical Flow," ANL-6633, Argonne National Laboratory (Oct. 1962).

Fauske, H. K., "The Discharge of Saturated Water Through Tubes," *Chem. Eng. Prog. Symp. Ser.,* **61,** 210 (1965).

Gallagher, E. V., "Water Decompression Experiments and Analysis for Blowdown of Nuclear Reactors," IITRI-578-P-21-39 (July 1970).

"General Electric BWR Thermal Analysis Basis (GETAB): Data, Correlation and Design Application," NEDO-10958, General Electric Company (1973).

Gonzalez-Santalo, J. M. and R. T. Lahey, "An Exact Solution for Flow Transients in Two-Phase Systems by the Method of Characteristics," *J. Heat Transfer,* **95** (1973).

Hanson, G. H., "Subcooled-Blowdown Forces on Reactor System Components: Calculational Method and Experimental Confirmation," Report IN-1354, Idaho Nuclear Corporation (June 1970).

Henry, R. E., "Two-Phase Critical Discharge of Initially Saturated or Subcooled Liquid," *Nucl. Sci. Eng.*, **41** (1970).

Jones, O. C., "Towards a Unified Approach for Thermal Nonequilibrium in Gas/Liquid Systems," *Nucl. Eng. Des.*, **69** (1982).

Lahey, R. T., B. S. Shiralkar, J. M. Gonzalez, and L. E. Schnebly, "The Analysis of Transient Critical Heat Flux," GEAP-13249, General Electric Company (1972).

Landau, L. D. and E. M. Lifshitz, *Fluid Mechanics,* Pergamon Press, London (1959).

Levy, S., "Prediction of Two-Phase Critical Flow Rate," *J. Heat Transfer,* **87** (1965).

Levy, S., "Steam Slip-Theoretical Prediction from Momentum Model," *J. Heat Transfer,* **82** (1960).

Moody, F. J., "Maximum Discharge Rate of Liquid Vapor Mixtures from Vessels," *Non-Equilibrium Two-Phase Flows,* ASME Symp. Vol., American Society of Mechanical Engineers (1975).

Moody, F. J., "Maximum Flow Rate of a Single-Component, Two-Phase Mixture," *J. Heat Transfer, Trans. ASME, Ser. C,* **87,** 134 (1965).

Moody, F. J., "Maximum Two-Phase Vessel Blowdown from Pipes," *J. Heat Transfer, Trans. ASME, Ser. C,* **88,** 285 (1966).

Moody, F. J., "Prediction of Blowdown Thrust and Jet Forces," ASME paper no. 69-HT-31, American Society of Mechanical Engineers (1969).

Powell, A. W., "Flow of Subcooled Water Through Nozzles," WAPD-PT-(V)-90, Westinghouse Electric Corporation (Apr. 1961).

Rose, R. P., G. H. Hanson, and G. A. Jayne, "Hydrodynamics Describing Acoustic Phenomena During Reactor Coolant System Blowdown," AEC Research and Development Report TID-4500, U.S. Atomic Energy Commission (July 1967).

Sakurai, A., and M. Shiotsu, "Transient Pool-Boiling Heat Transfer," ASME preprint 74-WA/HT-41, American Society of Mechanical Engineers (1974).

Shapiro, A. H., *The Dynamics and Thermodynamics of Compressible Fluid Flow,* Vols. I and II, The Ronald Press Company, New York (1953).

Shiralkar, B. S., E. E. Polomik, R. T. Lahey, J. M. Gonzalez, D. W. Radcliffe, and L. E. Schnebly, "Transient Critical Heat Flux—Experimental Results," GEAP-13295, General Electric Company (1972).

Siegel, R., and J. R. Howell, *Thermal Radiation Heat Transfer,* McGraw-Hill Book Company, New York (1972).

Simon, U., "Blowdown Flow Rates of Initially Saturated Water," presented at the European Two-Phase Flow Meeting, Risø, Denmark (June, 1971).

Sozzi, G. L., and W. A. Sutherland, "Critical Flow of Saturated and Subcooled Water at High Pressure," NEDO-13418, General Electric Company (1975).

Sparrow, E. M., and R. D. Cess, *Radiation Heat Transfer,* Brooks/Cole Publishing Company, Monterey, California (1967).

Sun, K. H., G. E. Dix, and C. L. Tien, "Effect of Precursory Cooling on Falling-Film Rewetting," ASME paper 74-WA/HT-52, American Society of Mechanical Engineers (1974).

Sun, K. H., J. M. Gonzalez, and C. L. Tien, "Calculations of Combined Radiation and Convection Heat Transfer in Rod Bundles Under Emergency Cooling Conditions," ASME paper 75-HT-64, American Society of Mechanical Engineers (1975).

Tachibana, F., M. Akiyama, and H. Kawamura, "Heat Transfer and Critical Heat Flux in Transient Boiling. I. An Experimental Study in Saturated Pool Boiling," *Nucl. Appl.,* **5,** *3* (1968).

Tangren, R. F., et al., "Compressibility Effects in Two-Phase Flow," *J. Appl. Phys.* **20,** 637 (1949).

Tong, L. S., and J. Weisman, *Thermal Analysis of Pressurized Water Reactors,* American Nuclear Society (1970).

Uchida, H., and H. Nariai, "Discharge of Saturated Water Through Pipes and Orifices," *Proc. 3rd Int. Heat Transfer Conf.,* **5,** 1 (1966).

Wallis, G. B., *One-Dimensional Two-Phase Flow,* McGraw-Hill Book Company, New York (1969).

Wallis, G. B., and J. H. Heasley, "Oscillations in Two-Phase Systems," *J. Heat Transfer* (1961).

Yamanouchi, A., "Effect of Core Spray Cooling in Transient State after Loss-of-Coolant Accident," *Nucl. Appl.,* **5** (1968).

Zaker, T. A., and A. H. Wiedermann, "Water Depressurization Studies," IITRI-578-P-21-26 (June 1966).

Zivi, S. M., "Estimation of Steady-State Void Fraction by Means of the Principle of Minimum Entropy Production," *J. Heat Transfer,* **86** (1964).

CHAPTER TEN

Valve and Piping Transients

10.1 Background

Time- and space-dependent forces are created in piping systems whenever local disturbances occur in the pressure, velocity, or density. For example, valve openings or closures create fluid acceleration and associated pressure changes that propagate through piping geometry, imposing unsteady forces on various pipe segments. Moreover, fluid discharge from a ruptured or otherwise open pipe creates a thrust force, which may change with time as fluid velocity increases. Both the propagating pressure forces and thrust forces can reach magnitudes large enough to require structural design features for the purpose of preventing mechanical overstress or other damage to a piping system or its surroundings.

Increased attention has been focused on transient forces in piping systems during the last several decades, largely motivated by the emphasis on safe design and operation of nuclear power stations (ASCE, 1973, 1979; Haupt and Meyer, 1979). Pipe lengths associated with the nuclear industry often are longer than those in conventional power plants. Consequently, transient loads are exerted for greater time intervals and can reach larger magnitudes on pipe segments in which a disturbance has a longer propagation distance. It follows that the resulting load impulses can create larger pipe motion, leading to higher stresses in the piping system.

Although nuclear containment systems are designed to withstand the consequences of large pipe ruptures and loss-of-coolant accidents, a common design objective has been to prevent pipe ruptures from fluid acceleration loads by specification of appropriate mechanical design. A corollary design objective is to minimize damage even if a pipe failure should occur, by preventing such reactions as pipe whipping, which could cause additional failures.

To provide appropriate mechanical design specifications, it is necessary to predict both steady and transient load magnitudes on the piping system from the fluid response, which could result from various disturbances. Theoretical models and experimental results for pipe reaction forces began to appear in the open literature in the 1960s and 1970s (Fabic, 1967, 1969; Hanson, 1970; Semprucci, 1979; ANSI/ANS, 1980). Solution techniques were introduced for the prediction of unsteady single- and two-phase flow behavior (Haupt, 1979; Wheeler, 1979). Consequently, the state-of-the-art has developed to a sophisticated level at which accurate load descriptions are predictable in terms of valve characteristics or pipe rupture description, piping geometry, and the state properties of the fluid. The predicted loads are subsequently input as forcing functions in dynamic stress analysis programs to determine the pipe system dynamics and to determine design margins for various pipe support and protection features.

This chapter provides a summary of theoretical models and methods employed in the prediction of transient pipe forces. Various studies and experiments are cited that have helped bring the state-of-the-art to its present level of sophistication. Pipe forces resulting from safety/relief valve operation, unsteady gas and liquid-vapor mixture flows, stop valve closure, and pipe rupture pressure transients are discussed.

10.2 Pipe Reaction Forces

If a pipe rupture occurs, thrust forces caused by fluid discharge can cause further damage to piping and other components unless adequate mechanical restraints are employed. Discharging fluid creates a reaction thrust on the ruptured pipe itself, and fluid acceleration inside the pipe generates forces on all segments that are bounded at either end by an elbow or turn. Moreover, the opening or closing of valves also generates fluid acceleration loads on all segments of a connected piping system.

The magnitude, direction, and duration of pipe forces must be known before mechanical restraints can be designed and positioned. Large computer programs (Fabic, 1969; Hanson, 1970; Rose, 1967) can be employed to help predict pipe forces for a postulated fluid transient. However, it is often more efficient and economical to first estimate reaction loads to specify position and capacity of restraints, and then later to check the final design with a detailed analysis.

Figure 10-1 shows a rigid pipe system with N segments of respective lengths L_1, L_2, ..., L_N, attached to a pressure source. Either a rupture of segment N or valve opening or closure is postulated. A disturbance propagates into the fluid contained by the pipe. The disturbance travels through the segments at sonic or shock speed relative to the fluid. As compression

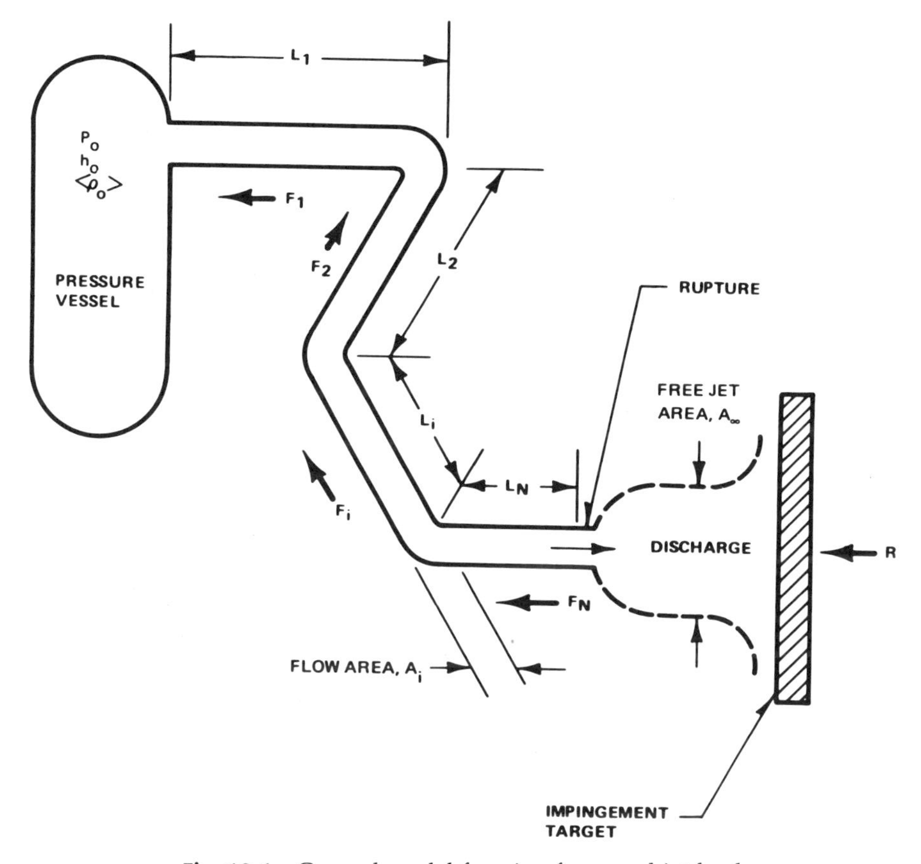

Fig. 10-1 General model for pipe force and jet load.

or decompression waves move through the pipe, the pressure at one end of each segment exceeds the pressure at the other, resulting in net longitudinal forces F_1, F_2, ..., F_N on each segment.

Successive wave transmissions and associated fluid acceleration forces decay as steady discharge is achieved. However, thrust and impingement forces continue to survive after the acceleration forces vanish. Since the discharging fluid pressure inside pipe segment N is generally greater than ambient pressure, p_∞, a free jet of steam or steam/water mixture decompresses and expands to an asymptotic area, A_∞. Impingement forces on structures that overlap the jet depend on the fraction of total forward momentum intercepted.

All reaction forces shown in Fig. 10-1 result from fluid pressure, shear, and momentum. Therefore, these forces are expressed in terms of the fluid properties.

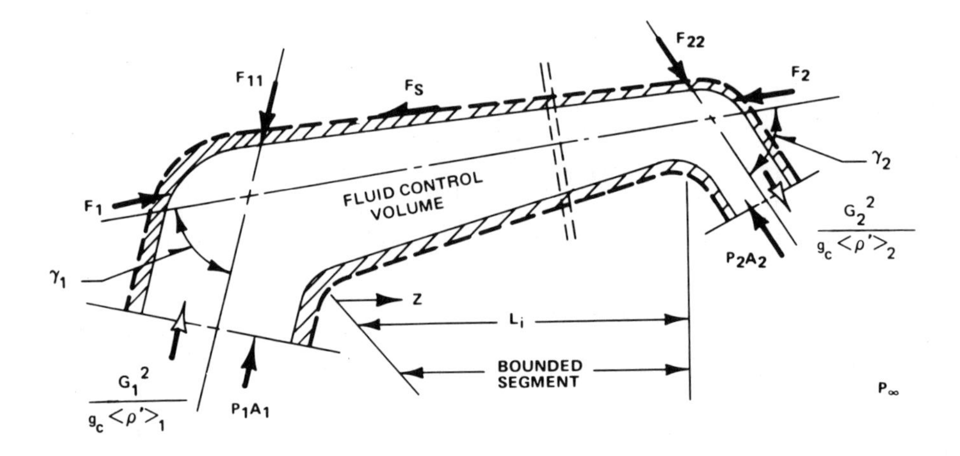

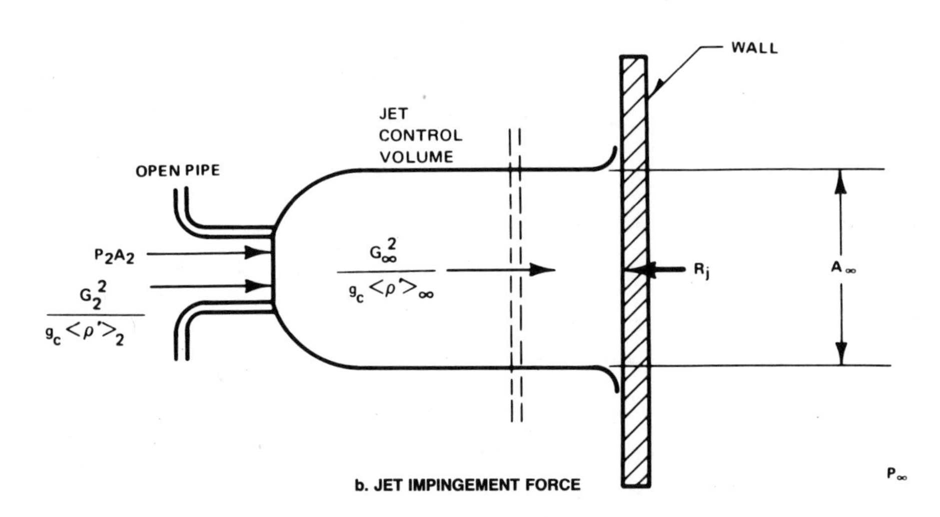

Fig. 10-2 Pipe and jet forces.

Consider the segment of pipe shown between two elbows in Fig. 10-2a. The solid boundary outlines the fluid control volume used for expressing pipe forces in terms of flow properties. It is assumed that the fluid volume contained by the elbows is small compared with segment volumes. The forces, F_{11}, p_1A_1, F_{22}, and p_2A_2 have lines of action parallel to pipe segments joined to each end of the bounded segment. Forces F_1 and F_2 are due to pressure acting over the elbow surface that is formed by projections of the bounded pipe flow area at each end. Ambient pressure, p_∞, acts on the pipe outer surface and cancels everywhere except on the projected areas, A_1 and A_2. The force F_s is caused by wall shear and can include drag caused by orifices or other flow restrictions.

We now consider forces on the bounded segment due to transient wave propagation. Conservation of momentum for a direction parallel to the bounded axis is given by,

$$F_1 - F_s - F_2 = \left[F_{11} - (p_1 - p_\infty) A_1 - \frac{G_1^2 A_1}{g_c \langle \rho' \rangle_1} \right] \cos\gamma_1$$

$$- \left[F_{22} - (p_2 - p_\infty) A_2 - \frac{G_2^2 A_2}{g_c \langle \rho' \rangle_2} \right] \cos\gamma_2 + \frac{1}{g_c} \frac{\partial}{\partial t} \int_0^{L_i} G A dz \ . \tag{10.1}$$

The control volume in Fig. 10-2a can be divided in two by the double dashed line shown normal to the pipe axis. On this line, normal momentum flow rates and pressure forces do not occur. Therefore, momentum conservation normal to the axis can be written for the left and right parts of the control volume as follows,

$$\left[F_{11} - (p_1 - p_\infty) A_1 - \frac{G_1^2 A_1}{g_c \langle \rho' \rangle_1} \right] \sin\gamma_1 = 0 \tag{10.2}$$

$$\left[F_{22} - (p_2 - p_\infty) A_2 - \frac{G_2^2 A_2}{g_c \langle \rho' \rangle_2} \right] \sin\gamma_2 = 0 \ . \tag{10.3}$$

Equations (10.1), (10.2), and (10.3) can be combined to obtain the "wave force," F_{w_i}, on a bounded pipe segment of length, L_i, as,

$$F_{w_i} \stackrel{\Delta}{=} F_1 - F_s - F_2 = \frac{1}{g_c} \frac{\partial}{\partial t} \int_0^{L_i} G A dz \ . \tag{10.4}$$

From Eq. (10.4), note that when steady flow is reached, the wave force vanishes.

Force F_N on the discharging pipe segment can be expressed as a special case of the bounded segment by making $\gamma_2 = 0$. It follows that $F_{22} = 0$ and Eqs. (10.1), (10.2), and (10.3) yield,

$$F_N = F_{W_N} + F_{B_N} \ , \tag{10.5}$$

where,

$$F_{B_N} \stackrel{\Delta}{=} (p_2 - p_\infty) A_2 + \frac{G_2^2 A_2}{g_c \langle \rho' \rangle_2} \ . \tag{10.6}$$

Although the wave force, F_{W_N}, vanishes at steady flow, the "blowdown force," F_{B_N}, survives on the discharge segment.

10.3 Fluid Impingement Forces

Whenever the discharging jet encounters an object in its path, the momentum of some fluid particles is changed and an impingement force is developed. Target shape, projected area, and orientation relative to the jet, as well as jet cross-sectional area and flow properties, make determination of impingement load characteristics difficult. However, the simple model shown in Fig. 10-2b can be used to estimate jet loads on structures encountered in nuclear power systems.

A steady jet discharges from an open pipe with area A_2 and expands to area A_∞ at some distance downstream, where it is assumed to be homogeneous (Moody, 1969). Forward motion of the jet is stopped by the wall shown. The total reaction force, R_j, is obtained from momentum conservation, written for the jet control volume as,

$$R_j = p_2 A_2 + \frac{G_2^2 A_2}{g_c \langle \rho' \rangle_2} + p_\infty (A_\infty - A_2) \ . \tag{10.7}$$

Jet pressure and mass flux are, respectively, p_∞ and $G_2(A_2/A_\infty)$ at the double dashed section drawn normal to the flow direction. It follows that for the part of the control volume to the right, R_j can also be expressed by,

$$R_j = p_\infty A_\infty + \frac{G_\infty^2 A_\infty}{g_c \langle \rho' \rangle_\infty} \ , \tag{10.8}$$

where,

$$G_\infty \triangleq G_2 \left(\frac{A_2}{A_\infty} \right) \ . \tag{10.9}$$

An unbalanced force on the wall is caused by R_j acting on its left side equal and opposite to that shown in Fig. 10-2b, and leftward force, $p_\infty A_\infty$, on its right side. From Eq. (10.7), the net rightward impingement reaction on the wall is, therefore,

$$R \triangleq R_j - p_\infty A_\infty = (p_2 - p_\infty) A_2 + \frac{G_2^2 A_2}{g_c \langle \rho' \rangle_2} \ . \tag{10.10}$$

Comparison of Eqs. (10.6) and (10.10) shows that blowdown thrusts on the discharging pipe segment and impingement reaction of the jet are equal.

Nearby pipes and other targets can intercept only part of the expanded jet area. To estimate the corresponding jet load, an impingement pressure is defined as,

$$p_I \triangleq \frac{R}{A_\infty} \ . \tag{10.11}$$

If Eqs. (10.10) and (10.8) are combined with Eq. (10.9), the expanded jet area can be expressed as,

$$\frac{A_\infty}{A_2} = \frac{G_2^2}{g_c(R/A_2)\langle\rho'\rangle_\infty} \ . \tag{10.12}$$

It follows from Eqs. (10.11) and (10.12) that the impingement pressure is,

$$p_I = \left(\frac{R}{A_2}\right)^2 \frac{g_c\langle\rho'\rangle_\infty}{G_2^2} \ . \tag{10.13}$$

When the discharging jet strikes a mechanical component so that area A_0 intercepts part of the expanded jet, the impingement load, R_0, can be estimated from,

$$R_0 \approx p_I A_0 \ . \tag{10.14}$$

If the target is concave so that jet flow could be turned opposite to its forward direction, R_0 should be doubled. However, pipes and most other possible targets are convex so that jet forward momentum is only decreased rather than stopped. Therefore, Eq. (10.14) should yield higher loads than would actually occur.

The expansion of a fluid jet to asymptotic area A_∞ is an idealization that neglects shear and mixing with the surrounding air at its boundary. These effects tend to increase the expanded area and reduce impingement pressure even more. Saturated water undergoes rapid flashing and expansion in a reduced pressure environment. Furthermore, since shear forces would be absent in the expanded jet, velocities of the liquid and vapor phases would be equal. Therefore, the assumption of a homogeneous expanded jet appears to be reasonable.

The force equations for bounded and open pipe segments and jet impingement loads have been expressed in terms of fluid velocity, pressure, and density. Therefore, it is necessary to obtain time- and space-dependent flow properties from fluid mechanical considerations.

10.4 Unsteady Pipe Flow

Following a postulated pipe rupture, the contained fluid decompresses rapidly. Nonflashing water or steam is treated by methods of single-phase flow. For two-phase blowdown, an appropriate slip assumption must be made. It is well known that two-phase mixtures of liquid and vapor do not always flow in a homogeneous regime. However, the homogeneous flow pattern is probably realistic for a steam-water mixture flow during initial decompression and flashing stages. Later, when steady discharge rates are approached, the homogeneous flow regime may become separated. Effects of separated flow are incorporated in a later consideration of steady thrust forces.

10.4.1 Basic Equations for Pipe Flow

The one-dimensional equations of mass, momentum, and energy conservation that describe homogeneous flow in a rigid flow passage with friction, heat transfer, variable area, and inclination angle, θ, from the horizontal have been given in Eqs. (5.33), (5.38), and (5.45) as,

Mass:

$$\frac{\partial\langle\rho_h\rangle}{\partial t}+\langle j\rangle\frac{\partial\langle\rho_h\rangle}{\partial z}+\langle\rho_h\rangle\frac{\partial\langle j\rangle}{\partial z}+\frac{\langle\rho_h\rangle\langle j\rangle}{A_{x-s}}\frac{\partial A_{x-s}}{\partial z}=0 \tag{10.15}$$

Momentum:

$$\frac{\langle\rho_h\rangle}{g_c}\left(\frac{\partial\langle j\rangle}{\partial t}+\langle j\rangle\frac{\partial\langle j\rangle}{\partial z}\right)=-\frac{\partial p}{\partial z}-\frac{g}{g_c}\langle\rho_h\rangle\sin\theta-\frac{\tau_w P_f}{A_{x-s}}, \tag{10.16}$$

and by neglecting potential energy and internal heat generation contributions,

Energy:

$$\langle\rho_h\rangle\left(\frac{\partial h_0}{\partial t}+\langle j\rangle\frac{\partial h_0}{\partial z}\right)=q''\left(\frac{P_H}{A_{x-s}}\right)+\frac{1}{J}\frac{\partial p}{\partial t}, \tag{10.17}$$

where wall shear stress is given by,

$$\tau_w\triangleq f\frac{\langle j\rangle|\langle j\rangle|}{2g_c}\langle\rho_h\rangle \tag{10.18}$$

and the stagnation enthalpy is defined as,

$$h_0\triangleq\langle h\rangle+\frac{\langle j\rangle^2}{2g_cJ}. \tag{10.19}$$

Simple compressible substances in thermodynamic equilibrium obey the Gibbs equation,

$$Td\langle s\rangle=d\langle h\rangle-\frac{1}{J\langle\rho\rangle}dp, \tag{10.20}$$

where the state equations are of the form,

$$\langle h\rangle=\langle h(p,\langle\rho_h\rangle)\rangle\ ;\qquad p=p(\langle\rho_h\rangle,\langle s\rangle). \tag{10.21}$$

The sonic speed, C, which is used later, is defined by,

$$C^2\triangleq g_c\left(\frac{\partial p}{\partial\langle\rho_h\rangle}\right)\bigg|_{\langle s\rangle}. \tag{10.22}$$

Determination of the flow properties p, $\langle\rho_h\rangle$, and $\langle j\rangle$ in space and time for given initial and boundary conditions permits computation of wave and blowdown forces on the ruptured pipe. However, fluid state equations are required before the flow properties can be determined.

A ruptured piping system in a nuclear reactor initially may contain either subcooled water, saturated water, or steam. Decompression causes saturated water to flash, requiring two-phase state properties. The same is true for subcooled water if the decompression reduces its pressure below saturation. For cases of sufficiently large subcooling, subcooled water properties are used. Saturated steam, if decompressed isentropically, falls into the wet region. However, for the time-dependent part of most practical problems, steam can be treated as an ideal gas.

A general solution for time- and space-dependent values of pressure, velocity, and density from Eqs. (10.15) through (10.22) can be obtained numerically by the method of characteristics (Shapiro, 1953; Anderson, 1982).

10.4.2 The Method of Characteristics

The method of characteristics (MOC) yields three ordinary differential equations for p, $\langle j \rangle$, and $\langle \rho_h \rangle$, which are integrated on two characteristic lines and a particle path line.

Equation (10.19) is first employed to eliminate h_0 from Eq. (10.17). Then the differential dh is obtained from Eq. (10.21) in terms of dp and $\langle d\rho_h \rangle$. Finally, Eqs. (10.20) and (10.22) are used to express the energy equation, Eq. (10.17), in the form:

$$\frac{\partial p}{\partial t}+\langle j\rangle\frac{\partial p}{\partial z}-\frac{C^2}{g_0}\left(\frac{\partial\langle\rho_h\rangle}{\partial t}+\langle j\rangle\frac{\partial\langle\rho_h\rangle}{\partial z}\right)=\frac{f\langle j\rangle^3}{2g_0DT(\partial s/\partial p)_\rho}=F_3(z,t) \qquad (10.23)$$

The dependent variables in Eqs. (10.15), (10.16), and (10.23) are p, $\langle j \rangle$, and $\langle \rho_h \rangle$. The MOC involves a search for paths in the z,t plane on which p, $\langle j \rangle$, and $\langle \rho_h \rangle$ can be integrated as ordinary differential equations.

Let us multiply Eqs. (10.15), (10.16), and (10.23) by unknown constants λ_1, λ_2, and λ_3. The resulting equations can then be added and grouped by their time and space derivatives. The functional forms

$$p=p(z,t) \qquad \langle\rho_h\rangle=\langle\rho_h(z,t)\rangle \qquad \langle j\rangle=\langle j(z,t)\rangle$$

are used to express total derivatives of the dependent variables as:

$$\frac{d(\ )}{dt}=\frac{dz}{dt}\frac{\partial(\ )}{\partial z}+\frac{\partial(\ )}{\partial t}\ , \quad \text{where, } (\ )=p,\ \langle\rho_h\rangle,\ \langle j\rangle\ . \qquad (10.24)$$

The time derivatives, $\partial(\ )/\partial t$, are eliminated next with the last term of Eq. (10.24), and the resulting coefficients of the space derivatives, $\partial(\ )/\partial z$, are set equal to zero, leaving only total derivative terms. This procedure yields:

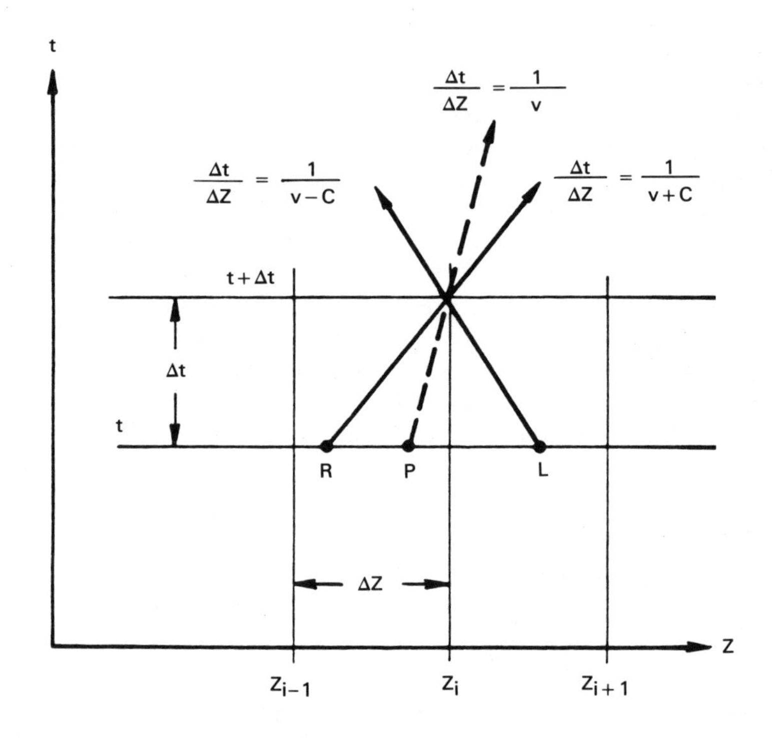

Fig. 10-3 Method of characteristics (MOC) computational mesh.

$$\left.\begin{aligned} dp \pm \frac{\rho C}{g_0} dv &= \left(-\frac{C^2}{g_c} \frac{\rho v}{A} \frac{dA}{dz} \mp \frac{\rho C f v^2}{2 g_0 D} + F \right) dt \ , \quad \text{for } \frac{dt}{dz} = \frac{1}{v \pm C} \\ d\rho - \frac{g_c}{C^2} dp &= -\frac{g_c}{C^2} F dt \qquad\qquad\qquad\qquad , \quad \text{for } \frac{dt}{dz} = \frac{1}{v} \end{aligned} \right\} \qquad (10.25)$$

$$F(z,t) = \left(\frac{f}{D} \frac{v^2 |v|}{2 g_0} + \frac{q'_{in}}{\rho A} \right) \Big/ \left[\left(\frac{\partial h}{\partial p} \right)_\rho - \frac{1}{\rho} \right] \ . \qquad (10.26)$$

The first equation applies on characteristic lines and the second equation applies on a particle path line. Integration can be performed to obtain a solution in either natural or standard Cartesian coordinate systems (Hsiao, 1981). A typical computational mesh is shown in Fig. 10-3 which displays the right- and left-traveling characteristic lines and the fluid path line on which Eq. (10.25) is integrated.

The sound speed, C, determines how fast fluid disturbances propagate through the piping system when shocks are absent.

The sound speed in bubbly, equilibrium steam-water mixtures has the peculiar shape given in Figs. 10-4 and 10-5, which has a minimum value at about 0.5 void fraction for the compressive curve. The decompressive sound speed has no such minimum. The character of the bubbly mixture compressive sound speed has been verified by numerous experimental measurements (Karplus, 1961), such as those shown in Fig. 10-6a for steam-water mixtures. Another experiment (Edwards, 1970) provided a measure of the decompressive sound speed in a 4.0-m-long pipe initially containing subcooled water at 6.8 MPa pressure and 242°C temperature, for which the saturation pressure was 2.8 MPa. A glass diaphragm at one end of the pipe was ruptured, which caused a decompression wave to travel in the subcooled water at 1370 m/sec, suddenly reducing pressure to the saturation value. This high-speed decompression wave was followed by another decompression wave through the saturated water, which decreased the pressure below saturation. Figure 10-6b shows a rapid pressure reduction about 20 msec after the pressure dropped to saturation, signaling arrival of the second decompression at the pressure sensor. This corresponds to a decompression speed of (4 m)/(200 msec) = 20 m/sec. The decompressive sound speed from Fig. 10-5, based on a saturation pressure

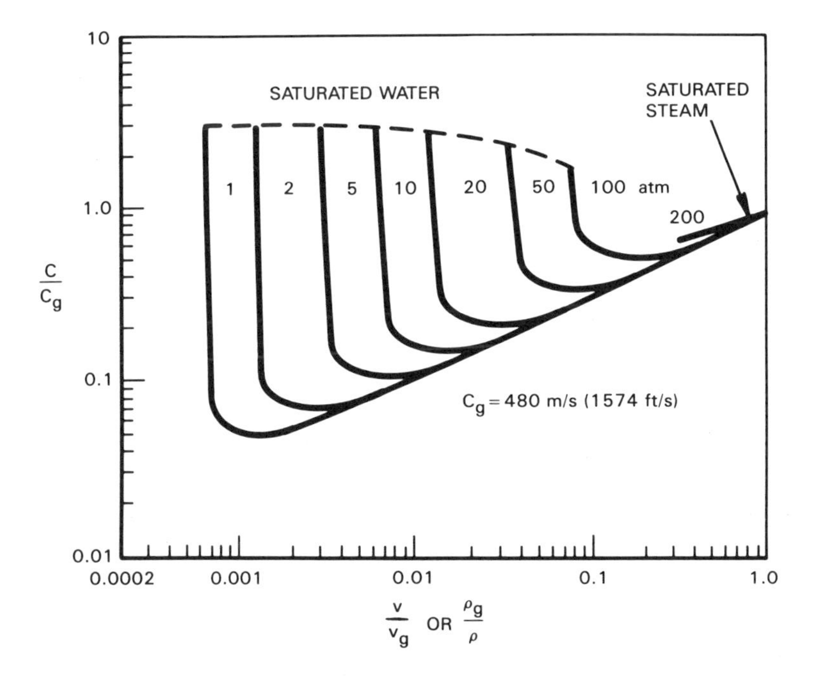

Fig. 10-4 Compressive sound speed, bubbly steam-water mixture.

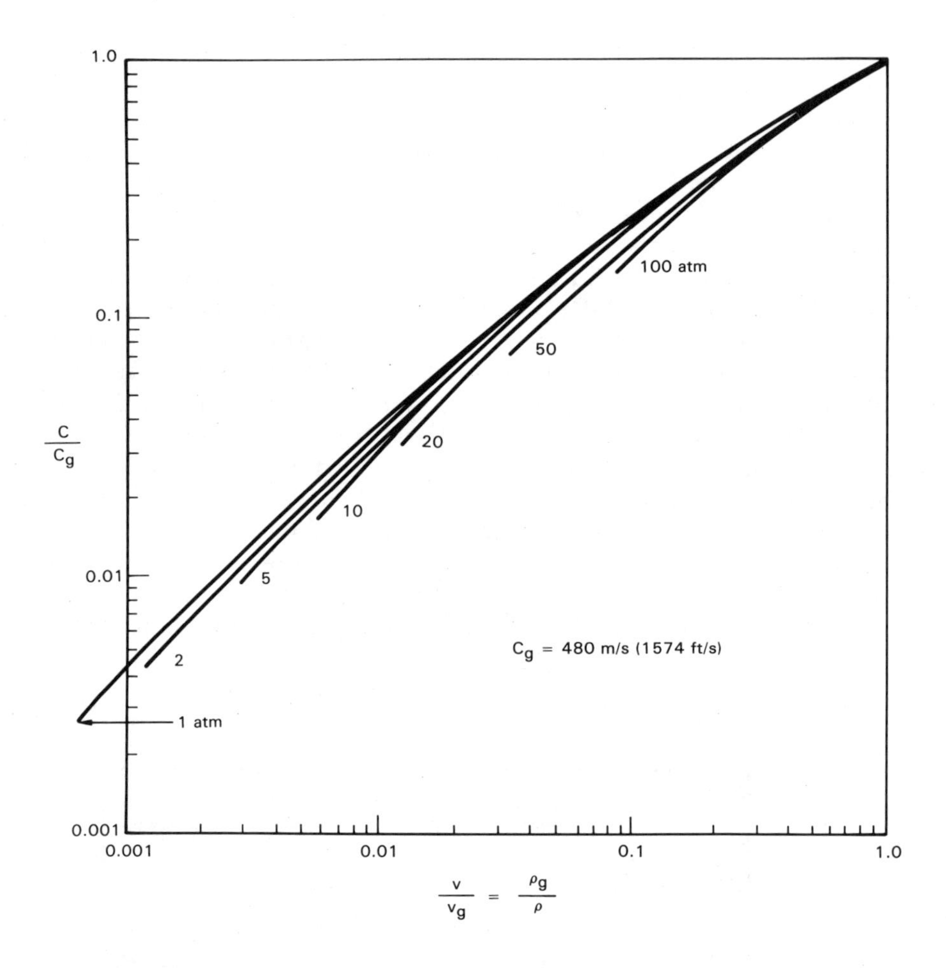

Fig. 10-5 Decompression sound speed, bubbly, saturated equilibrium, steam-water mixture.

of 2.8 MPa (27 atm), 830 kg/m^3 water density, and 13 kg/m^3 steam density, is 19.2 m/sec, which compares favorably with the measured 20 m/sec.

Initial and boundary conditions and fluid state equations are specified to obtain a complete solution for unsteady fluid properties. The boundary conditions may involve sonic or subsonic flow conditions at the discharge end of a ruptured pipe, valve stroke-pressure-flow rate specifications, or relationships for vessel stagnation and pipe entrance flow properties.

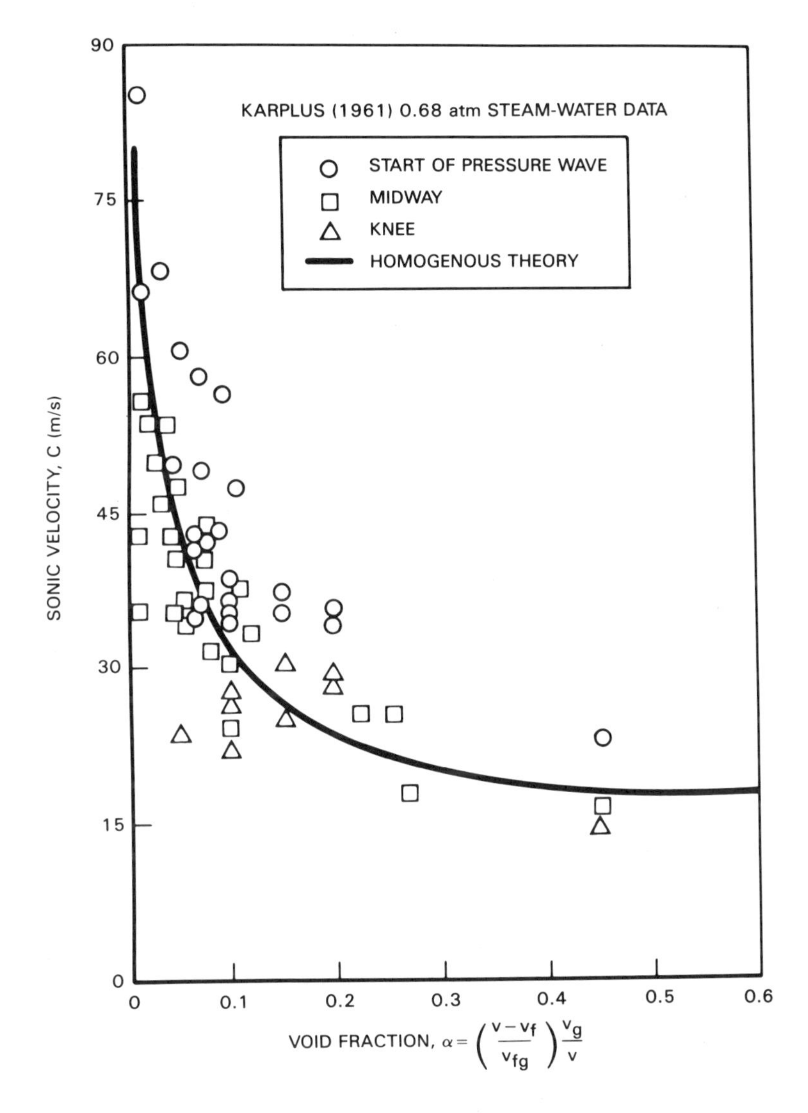

Fig. 10-6a Prediction of sonic velocity data of Karplus (1961).

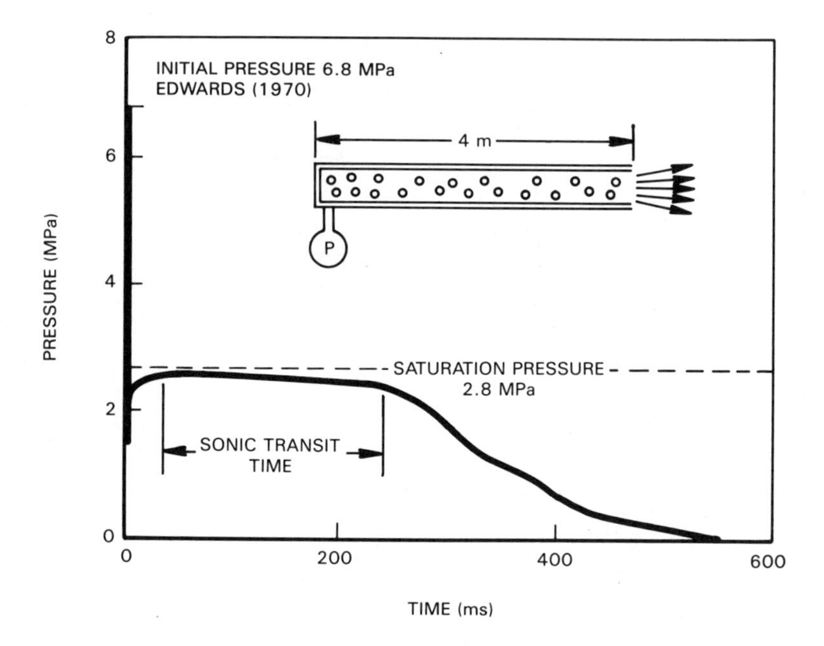

Fig. 10-6b Edwards water decompression experiment (1970).

10.4.3 Isentropic Flows and Simple Waves

When a disturbance propagates through a pipe in one direction without reflections or shock formation, it is called a simple wave. If the flow is also isentropic, it is possible to obtain a closed-form solution for the unsteady flow properties.

For frictionless adiabatic flows that are isentropic, fluid density is only a function of pressure. With the additional simplifications of a constant area flow passage and negligible effects of gravity, Eqs. (10.15) and (10.16) become,

$$\frac{\partial\langle\rho_h\rangle}{\partial t}+\langle j\rangle\frac{\partial\langle\rho_h\rangle}{\partial z}+\langle\rho_h\rangle\frac{\partial\langle j\rangle}{\partial z}=0 \tag{10.27}$$

$$\frac{\langle\rho_h\rangle}{g_c}\left(\frac{\partial\langle j\rangle}{\partial t}+\langle j\rangle\frac{\partial\langle j\rangle}{\partial z}\right)+\frac{\partial p}{\partial z}=0 \ . \tag{10.28}$$

At any location, z, and time, t, there are unique values of p, $\langle\rho_h\rangle$, and $\langle j\rangle$. Since p determines $\langle\rho_h\rangle$ for isentropic flows, $\langle j\rangle$ can be expressed as a function of $\langle\rho_h\rangle$. By employing Eq. (10.22) for the sonic speed, a solution to Eqs. (10.27) and (10.28) can be obtained in the Riemann form (Landau and Lifshitz, 1959),

$$\langle j(z, t)\rangle = \pm g_c \int \frac{dp}{\langle \rho_h(p)\rangle C(p)} + \kappa \tag{10.29}$$

$$z = [\langle j\rangle \pm C]t + f\langle j\rangle \ . \tag{10.30}$$

The $\pm$ in Eq. (10.30) implies right- or left-traveling sonic waves. The arbitrary function, $f\langle j\rangle$, and the constant, κ, are evaluated from appropriate boundary and initial conditions. A further simplification for nonflashing liquid flows, in which $\langle \rho_h\rangle$ and C are constant, permits Eq. (10.29) to be written as,

$$\Delta\langle j\rangle = \pm \frac{g_c}{\rho C} \Delta p \ , \tag{10.31}$$

which can be recognized as the classical "waterhammer" equation, where disturbances are propagated along characteristics in the time-space plane with slope, $dz/dt = \langle j\rangle \pm C$.

As an example of the various boundary and initial conditions required, we assume that the vessel and pipe in Fig. 10-1 are connected by an ideal nozzle through which the flow is isentropic. It follows from Eq. (10.20) that vessel pressure and enthalpy are related to pipe entrance properties by,

$$h_0 = \langle h\rangle + \int_p^{p_0} \frac{dp}{J\langle \rho\rangle} \ , \qquad \text{at } z = 0 \ . \tag{10.32}$$

Further, we assume that an instantaneous circumferential rupture occurs at the other end of the pipe. For a short distance into the pipe, friction effects are negligible. Therefore, Riemann's solution initially applies.

For compressible fluids, sonic discharge flow develops immediately (Shapiro, 1953) providing unique values of discharge pressure, density, and velocity. For the case of nonflashing, incompressible liquid, sonic flow cannot occur and the discharge pressure drops to the ambient value. The exit boundary conditions therefore are expressed as,

$$\langle j(L, t)\rangle = C(L, t);\ p(L, t) > p_\infty \ \text{(compressible fluids)} \tag{10.33}$$

or,

$$p(L, t) = p_\infty \ \text{(incompressible, nonflashing liquid)} \ . \tag{10.34}$$

Normally, it is assumed that the fluid velocity in the pipe before rupture is small compared to discharge flow conditions. Therefore, suitable initial conditions are,

$$p(z, 0) \approx p_0;\ \langle j(z, 0)\rangle \approx 0;\ \langle \rho_h(z, 0)\rangle \approx \langle \rho_h\rangle_0 \ . \tag{10.35}$$

When steady flow conditions are reached following a pipe rupture, wave forces vanish, but the blowdown and jet impingement forces continue to act. The steady-state discharge of steam-water mixtures can be in the form of a homogeneous, separated, or some intermediate flow regime.

As an example solution of fluid properties, we consider a vessel and constant flow area frictionless pipe that contains pressurized ideal steam. Following a postulated rupture, Eqs. (10.15), (10.16), and (10.17) were solved for isentropic vessel attachment and sonic discharge boundary conditions. The resulting solution is shown in Fig. 10-7 as an isometric pressure-space-time surface. Corresponding solution surfaces also exist for velocity and density. Note that immediately following the rupture, discharge pressure drops to ~28% of its initial value and remains constant until a wave is reflected from the vessel; for pipes with friction, the discharge pressure varies during this time interval. When the decompression wave reaches the vessel, pipe entrance pressure begins to drop. Eventually, steady state is reached where the pressure profile becomes flat.

10.5 Evaluation of Reaction and Impingement Forces

Solutions of the space- and time-dependent fluid properties can be used next to obtain wave and blowdown forces from Eqs. (10.4) and (10.6). Typical blowdown and wave forces are shown for ideal steam in Fig. 10-8. The total wave force is obtained from Eq. (10.4), integrated over the entire pipe length, L. From Fig. 10-8a, note that the friction parameter, fL/D, does not affect the initial value of blowdown and wave forces. However, as time progresses, both wave and blowdown forces decrease with increasing friction and the steady blowdown force is strongly dependent on pipe friction. Also note that $(F_B+F_W)/p_0A=1.0$, until the decompression wave reaches the vessel. This feature can be shown to be valid for any fluid in a rigid pipe by considering the rupture as removal of a rightward force, p_0A, from the pipe end. Segmented wave forces in Fig. 10-8b were obtained by integrating Eq. (10.4) from $z=0$ to various fractions of the total pipe length, L. To determine the wave force on a given segment bounded by lengths z_1 and z_2, it is necessary only to subtract values from Fig. 10-8b as follows,

$$\left.\frac{F_W(t)}{p_0A}\right|_{z_1}^{z_2}=\left.\frac{F_W(t)}{p_0A}\right|_{0}^{z_2/L}-\left.\frac{F_W(t)}{p_0A}\right|_{0}^{z_1/L} . \tag{10.36}$$

Presentation of blowdown and segmented wave forces with a wide range of the friction parameter, fL/D, and state properties corresponding to non-flashing water, saturated steam-water mixtures, and steam would be a computational task of staggering magnitude. Fortunately, useful design data can be extracted from initial blowdown and wave forces, steady blowdown and impingement forces, approximate speed of waves in the fluid,

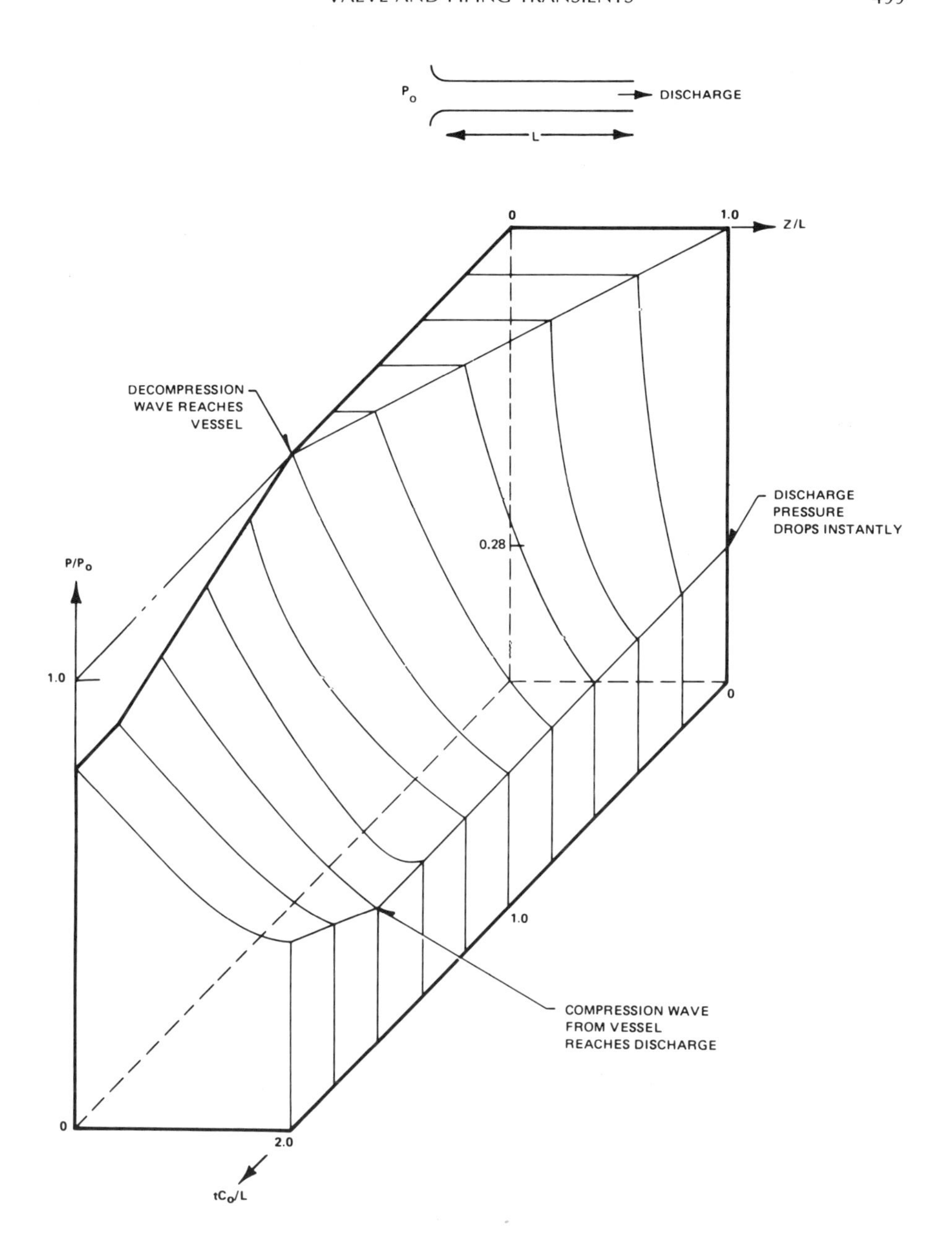

Fig. 10-7 Pressure-space-time surface (steam).

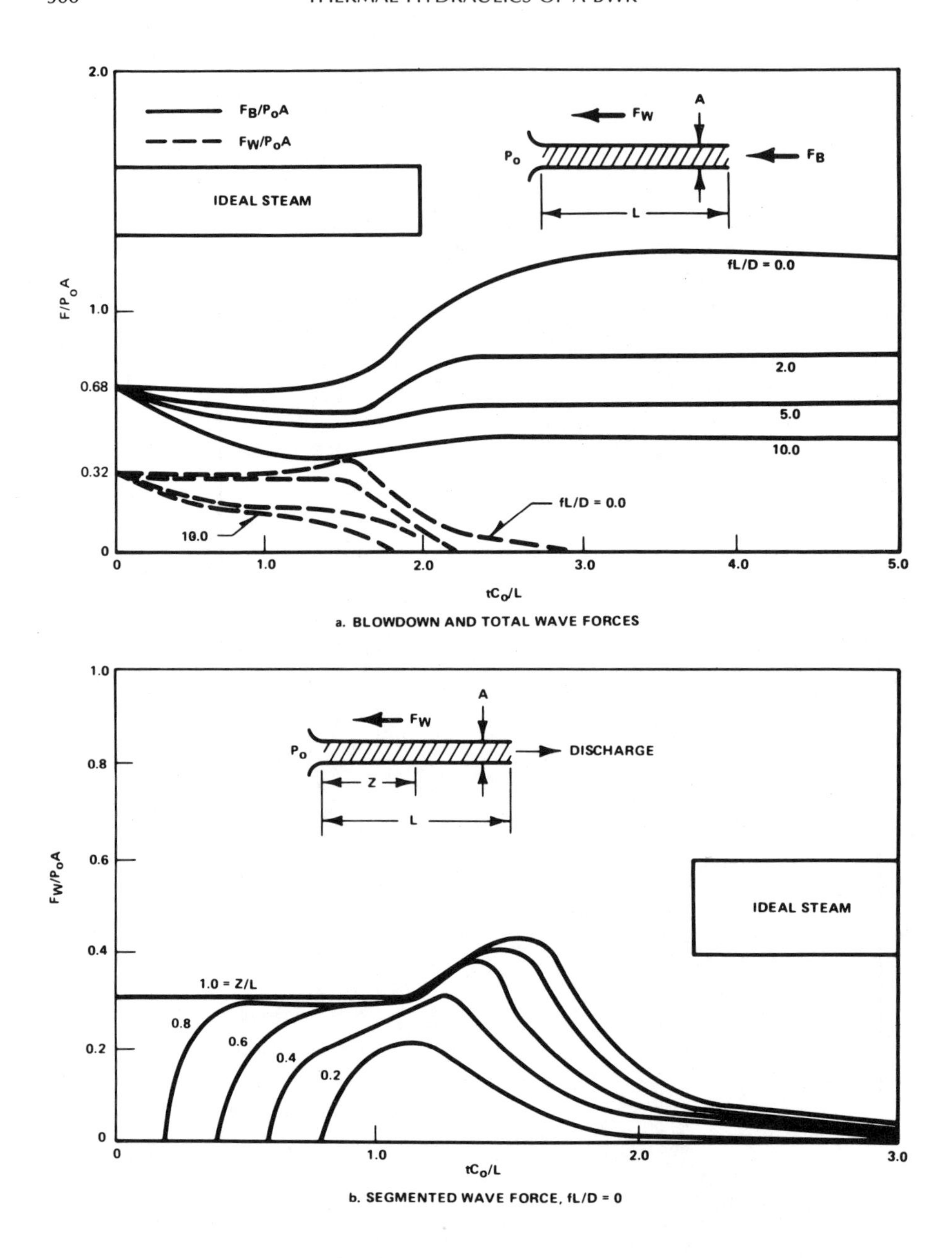

Fig. 10-8 Typical blowdown and wave forces (steam).

and an approximate time required for steady flow to be reached. These design properties are considered next.

By expressing the fluid density and sonic speed in terms of pressure and entropy, and by employing the condition that the discharge velocity is sonic, Eq. (10.29) can be integrated to determine the initial discharge properties. For the case of stagnant ideal steam with $K = 1.3$, initial discharge properties are:

$$\begin{aligned} \frac{p}{p_0} &= \left(\frac{2}{K+1}\right)^{2K/(K-1)} = 0.28 \\ \frac{\langle j \rangle}{C_0} &= \left(\frac{2}{K+1}\right) = 0.86 \qquad \text{(ideal steam)} . \\ \frac{\langle \rho \rangle}{\rho_0} &= \left(\frac{2}{K+1}\right)^{2/(K-1)} = 0.40 \end{aligned} \qquad (10.37)$$

For the case of saturated or subcooled water in the pipe, initial discharge properties are shown in Fig. 10-9. The term C_0 corresponds to the sonic speed in a saturated water mixture at the initial pressure. As an equilibrium-saturated water mixture is decompressed isentropically, its sonic speed increases, explaining why $\langle j \rangle / C_0 > 1.0$ in Fig. 10-9. For nonflashing water, discharge pressure drops to the ambient value and discharge velocity is obtained from Eq. (10.31). At one atmosphere, discharge density is ~ 62.4 lb_m/ft^3, and the initial discharge velocity varies with pressure, p_0, as given in Table 10-1.

Initial discharge properties were used in Eq. (10.6) to obtain the initial blowdown force, which is shown in Fig. 10-10. The solid lines correspond to subcooled water initially at pressure p_0 but whose saturation pressure is p_{sat}. Note that higher subcooling reduces the initial blowdown force, which approaches zero for nonflashing water. In contrast, the initial blowdown force for ideal steam is directly proportional to p_0.

The initial wave force was obtained by subtracting the normalized initial blowdown force from 1.0. These results also are shown in Fig. 10-10.

Note that higher subcooling leads to higher wave forces. It is well known that water can exist in a metastable state before flashing begins, following a decompression. Therefore, following the rupture of a pipe initially containing water, a decompression wave moves at sonic speed in water, roughly 4500 ft/sec. As discussed in Sec. 9.2.1.4 it requires ~ 1.0 msec for flashing to occur. Therefore, a decompression wave could advance into water ~ 4.5 ft before the discharge end flashes, repressurizes, begins sonic outflow, and produces the blowdown and jet forces of Fig. 10-10. It follows that although a brief decompression-recompression wave travels at sonic speed in nonflashing water, a trailing decompression wave advances at the sonic speed corresponding to flashing water. Therefore, a bounded pipe segment

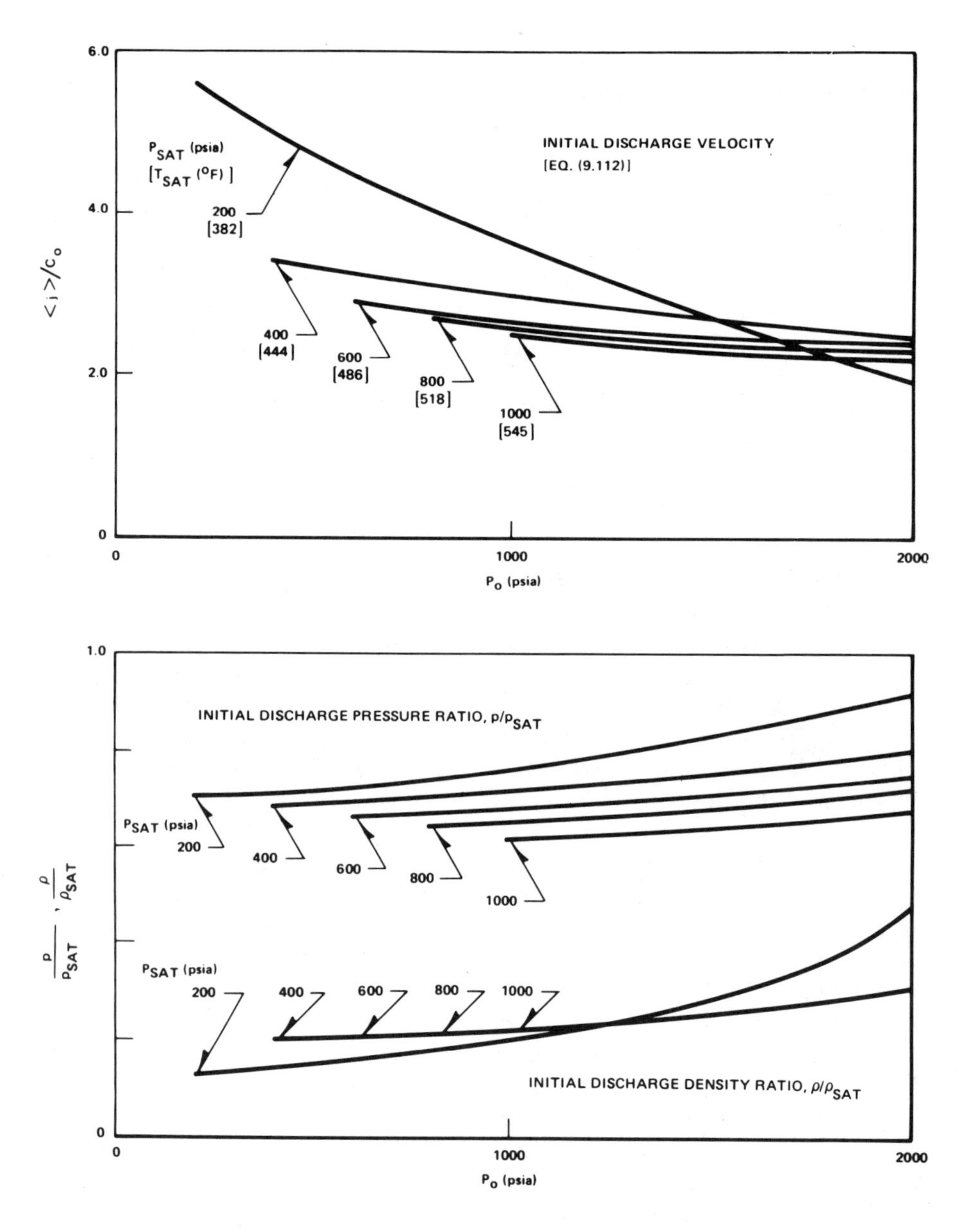

Fig. 10-9 Initial discharge properties, water.

TABLE 10-1
Initial Discharge Velocity, Nonflashing Water

System Pressure, p_0 (psia)	Discharge Velocity, $\langle j \rangle$ (ft/sec)
200	3.7
400	8.4
600	14.3
800	22.0
1000	31.2
1200	42.2
1400	56.6
1600	75.5
1800	96.0
2000	125.0

can be momentarily exposed to vessel pressure, p_0, at one end and a pressure at the other end somewhere between the ambient, p_∞, and saturation pressure, p_{sat}, depending on the degree of metastability. This early wave has a relatively brief period, although its maximum amptitude is between $(p_0 - p_\infty)A$ and $(p_0 - p_{sat})A$. The later decompression wave associated with saturation pressure is of longer duration and normally is the most important from a loading standpoint.

Numerous detailed numerical solutions for time-dependent discharge flow from ruptured pipes lead to the conclusion that steady flow is approached when the initial pipe contents have been expelled. If the average discharge velocity and density are approximated by,

$$\langle j \rangle_{D_{av}} \approx \tfrac{1}{2}(\langle j \rangle_{D,\text{initial}} + \langle j \rangle_{D,\text{final}}) \tag{10.38}$$

$$\langle \rho_h \rangle_{D_{av}} \approx \tfrac{1}{2}(\langle \rho \rangle_{D,\text{initial}} + \langle \rho \rangle_{D,\text{final}}) \ , \tag{10.39}$$

it follows that the time required to discharge the initial contents of a pipe with length, L, and fluid density, ρ_0, is,

$$t_D \approx \frac{\rho_0}{\langle \rho_h \rangle_{D_{av}}} \frac{L}{\langle j \rangle_{D_{av}}} \ . \tag{10.40}$$

For nonflashing liquid flow, $\langle j \rangle_{D_{av}}$ is roughly equal to $\tfrac{1}{2}\langle j \rangle_{D,\text{final}}$, where $\langle j \rangle_{D,\text{final}}$ is obtained from,

$$p_0 - p = \rho_l \frac{\langle j \rangle^2}{2g_0} = \frac{G^2}{2g_0\rho_l} \ . \tag{10.41}$$

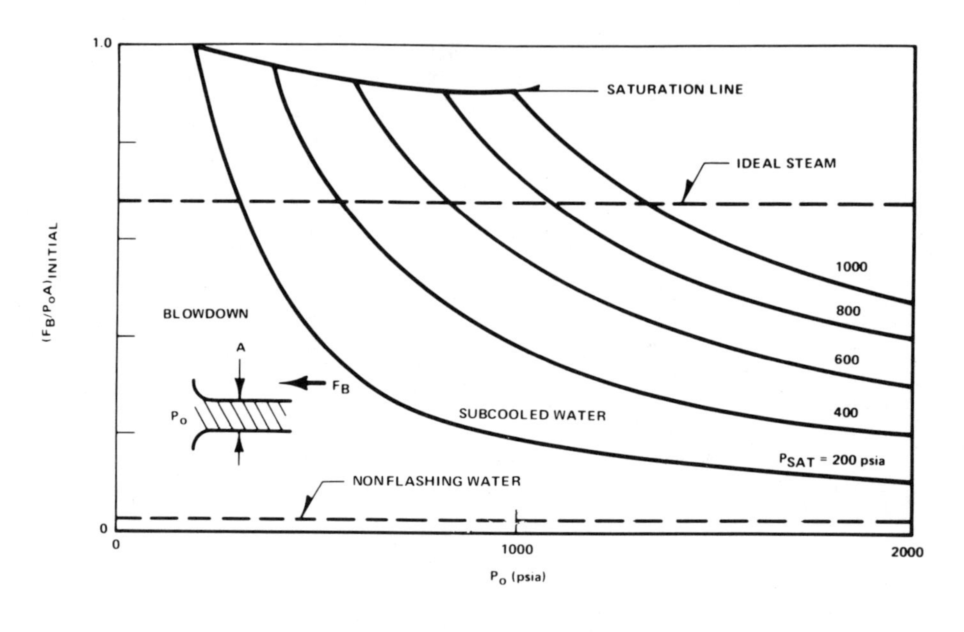

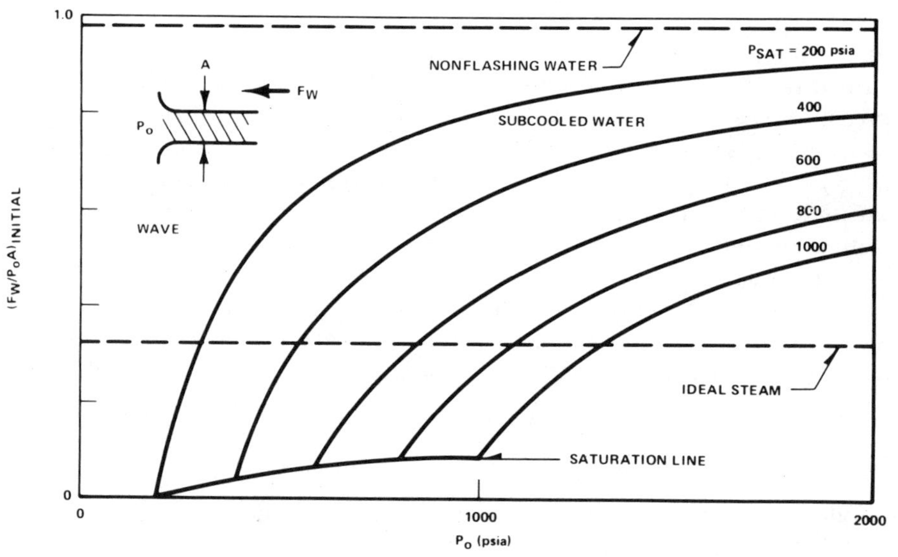

Fig. 10-10 Initial blowdown and wave forces.

The corresponding density ratio, $\rho_0/\langle\rho_h\rangle_{D_{av}}$ is ~ 1.0. Thus, it follows that for nonflashing water,

$$t_D \approx 2L\left[\frac{\rho_0}{2g_c(p_0 - p_\infty)}\right]^{1/2} . \qquad (10.42)$$

For compressible discharge, initial and final density and velocity are roughly the same. Therefore, Eqs. (10.37) or Fig. 10-9 can be employed with Eq. (10.40) in the form,

$$t_D \approx \left(\frac{\rho_0}{\langle\rho_h\rangle_{D_{av}}}\right)\left(\frac{C_0}{\langle j\rangle_{D_{av}}}\right)\left(\frac{L}{C_0}\right) \text{ (compressible fluid) } . \qquad (10.43)$$

Equations (10.42) and (10.43) can be used for estimating the approximate time, t_D, required to reach a steady discharge.

It is readily shown that for separated liquid-vapor flow, the two-phase blowdown and jet impingement forces of Eqs. (10.6) and (10.10) can be written as,

$$F_B = R = (p_2 - p_\infty)A_2 + \frac{G_{max}^2 A_2}{g_c\langle\rho'\rangle_2} , \qquad (10.44)$$

where $\langle\rho'\rangle$ is the momentum density given in Eq. (5.67) as,

$$\langle\rho'\rangle \triangleq \left\{\left[\frac{\langle x\rangle}{\rho_g} + (1-\langle x\rangle)\frac{S}{\rho_f}\right]\left(\langle x\rangle + \frac{1-\langle x\rangle}{S}\right)\right\}^{-1} . \qquad (10.45)$$

Equations (10.44) and (10.45) are used with the formulation for the steam-water critical flow rate expressed functionally by,

$$G_c = G_c\left(p_0, h_0, \frac{fL}{D}\right) . \qquad (10.46)$$

The steady blowdown force, based on both separated and homogeneous flow regimes, is plotted in Fig. 10-11 for frictionless pipes. In the mixture region, vessel pressure has only a slight effect on the nondimensional blowdown force, F_B/p_0A. Outside the saturation boundary, the blowdown force increases more rapidly with increased subcooling, finally reaching the value $F_B = 2.0\ p_0A$, for nonflashing liquid (Moody, 1969). Note from Fig. 10-11 that separated flow gives slightly higher steady blowdown forces than homogeneous flow. Therefore, the separated flow model is conservative for use in subsequent two-phase formulations.

The steady blowdown force for separated flow is graphed in Fig. 10-12 as a function of pipe friction. Also shown is the steady blowdown force for saturated steady and nonflashing water (Moody, 1969). Pressure, p_0, does not strongly affect F_B/p_0A for saturated water in the vessel. The curve for saturated steam was based on 1050-psia vessel pressure, but other values of p_0 would be very close to the dashed line shown.

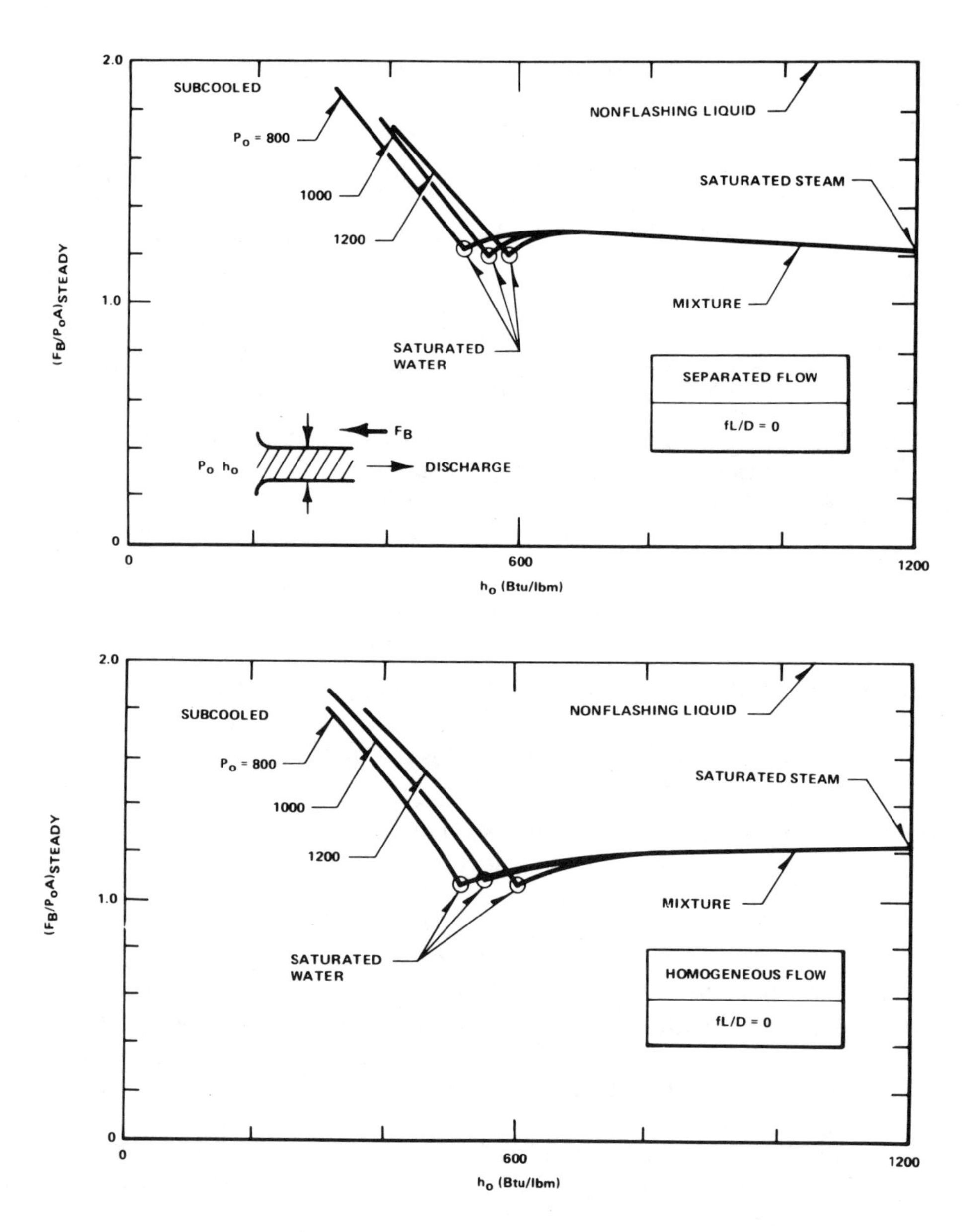

Fig. 10-11 Steady blowdown force.

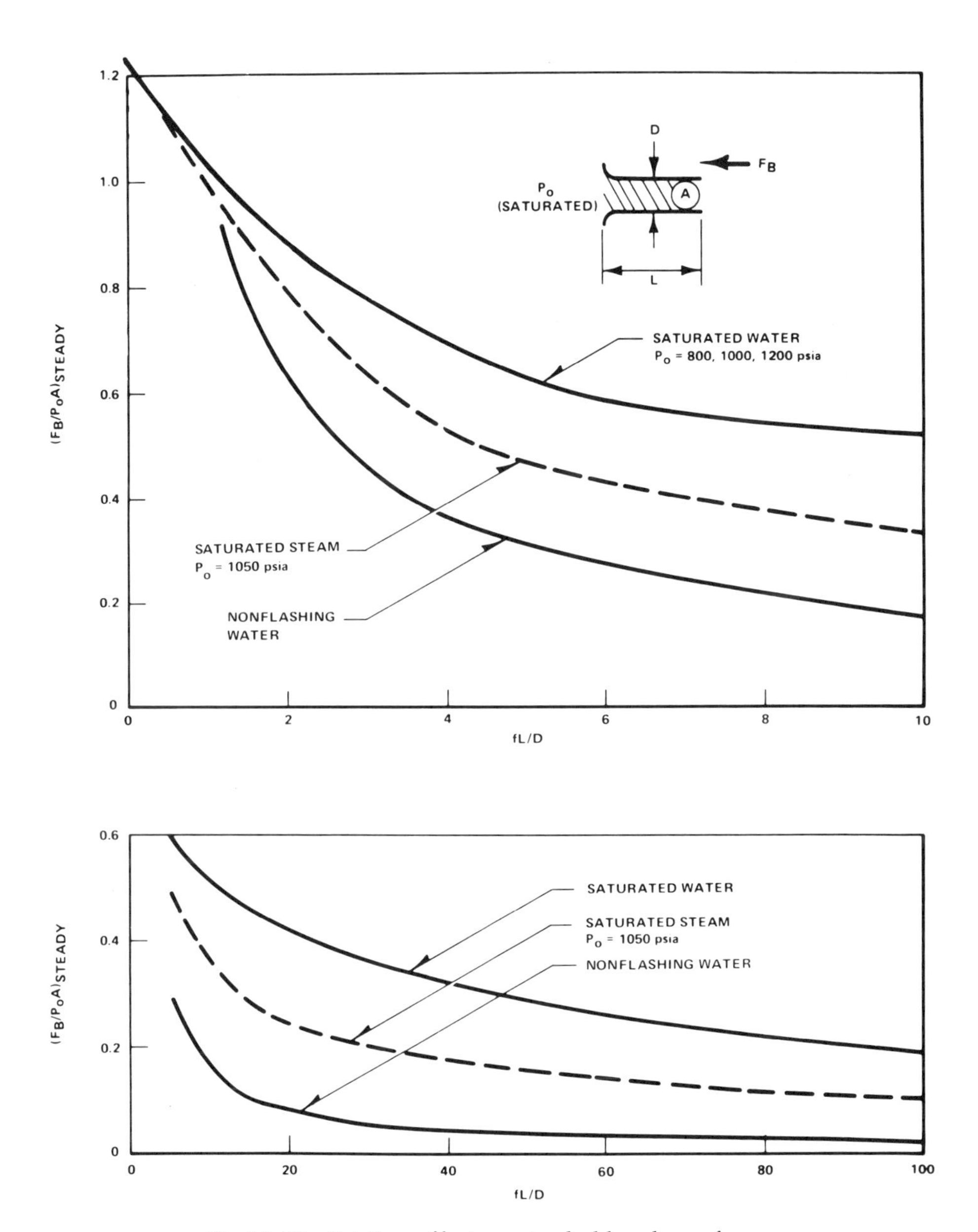

Fig. 10-12 Friction effect on steady blowdown force.

Earlier it was mentioned that the full jet impingement force is equal to the blowdown force. Therefore, the initial blowdown force, $F_{B,\mathrm{initial}}$, of Fig. 10-10 and the steady blowdown force, $F_{b,\mathrm{steady}}$, of Figs. 10-11 and 10-12 can be interpreted, respectively, as the initial and steady full jet impingement forces.

Equations (10.12) and (10.13) were employed with the expanded homogeneous jet density, obtained under the assumption that the stagnation enthalpy equals the static enthalpy of the expanded jet, and the steady blowdown force of Fig. 10-12 was employed to obtain the expanded jet area and impingement pressure. Results are shown in Figs. 10-13 and 10-14.

Previous data (Faletti and Moulton, 1963) for the critical discharge of steam-water mixtures indicate that fully expanded properties are approached closely in less than five discharge pipe diameters of forward travel. Therefore, if a target is more than five pipe diameters away, Figs. 10-13 and 10-14 can be used. For closer targets, the jet boundary should be assumed to expand linearly between the discharge plane and five pipe diameters downstream. Then, if A is the expanding jet area at a target, the normalized jet pressure, p_I/p_0, can be taken as F_B/p_0A.

Sometimes a pipe contains an orifice, valve, or other flow restriction that reduces the steady blowdown force. By equating orifice flow rates for both flashing and nonflashing flows to critical pipe discharge rates, Fig. 10-15 was constructed to show the effect of an upstream flow restriction on the blowdown force. One case is shown for saturated water flow with no flashing in the restriction but subsequent choking in the pipe. Another case is for choked flow in both the restriction and the pipe. A third case is shown for ideal steam flow with choked flow in an upstream Venturi and supersonic discharge flow.

If an upstream restriction exists in a region of nonflashing water, it is likely that choking will not occur in the restriction. However, if vapor begins to form before the restriction, choking probably will occur in both the restriction and pipe. The restriction considered for the upper curves was a sharp-edged orifice with standard pressure loss properties.

To estimate the duration of blowdown and wave forces, it is necessary to know the propagation speed of disturbances in the pipe fluid. Steam and water properties were employed in Eq. (10.22) to determine the propagation (or sonic) speeds shown in Fig. 10-16. Curves for saturated steam are based on the derivative, $(\partial p/\partial \rho_g|_s)$, evaluated on the superheated side of the steam dome. The curve for saturated, flashing water is based on $(\partial p/\partial \langle \rho_h \rangle|_{\langle s \rangle})$, evaluated on the saturated side of the steam dome. For subcooled water, the sonic speed is ~ 4500 ft/sec.

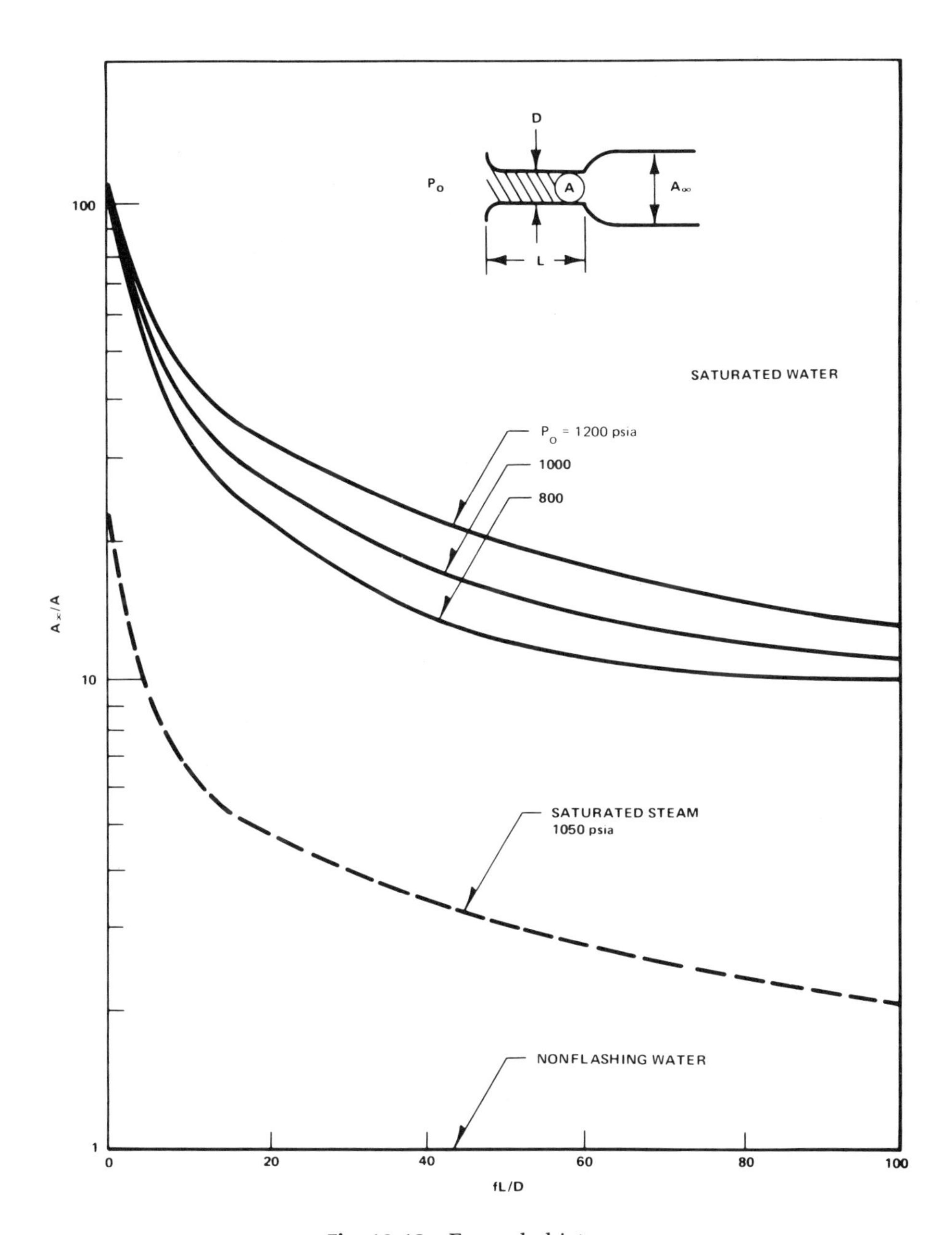

Fig. 10-13 Expanded jet area.

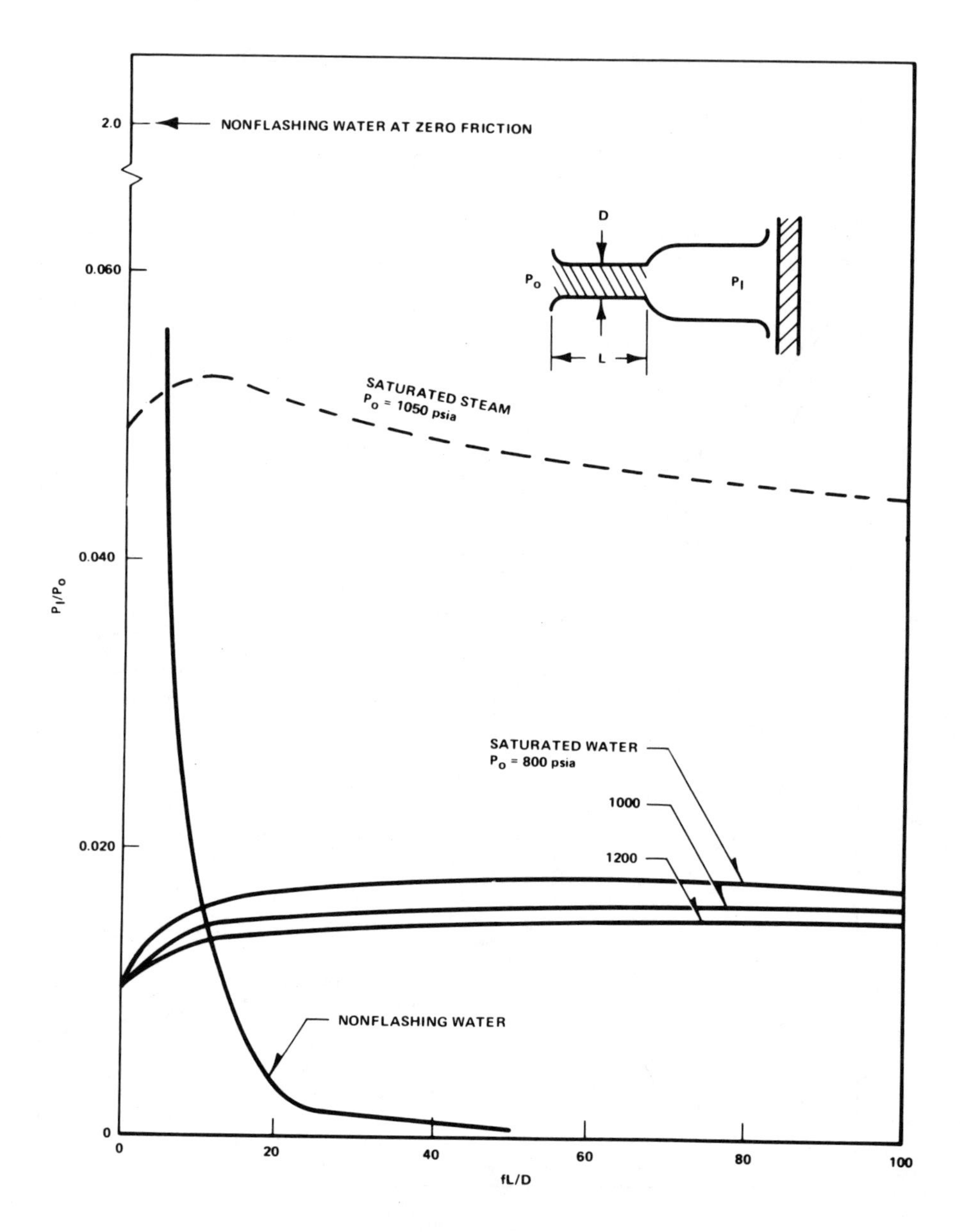

Fig. 10-14 Expanded jet pressure.

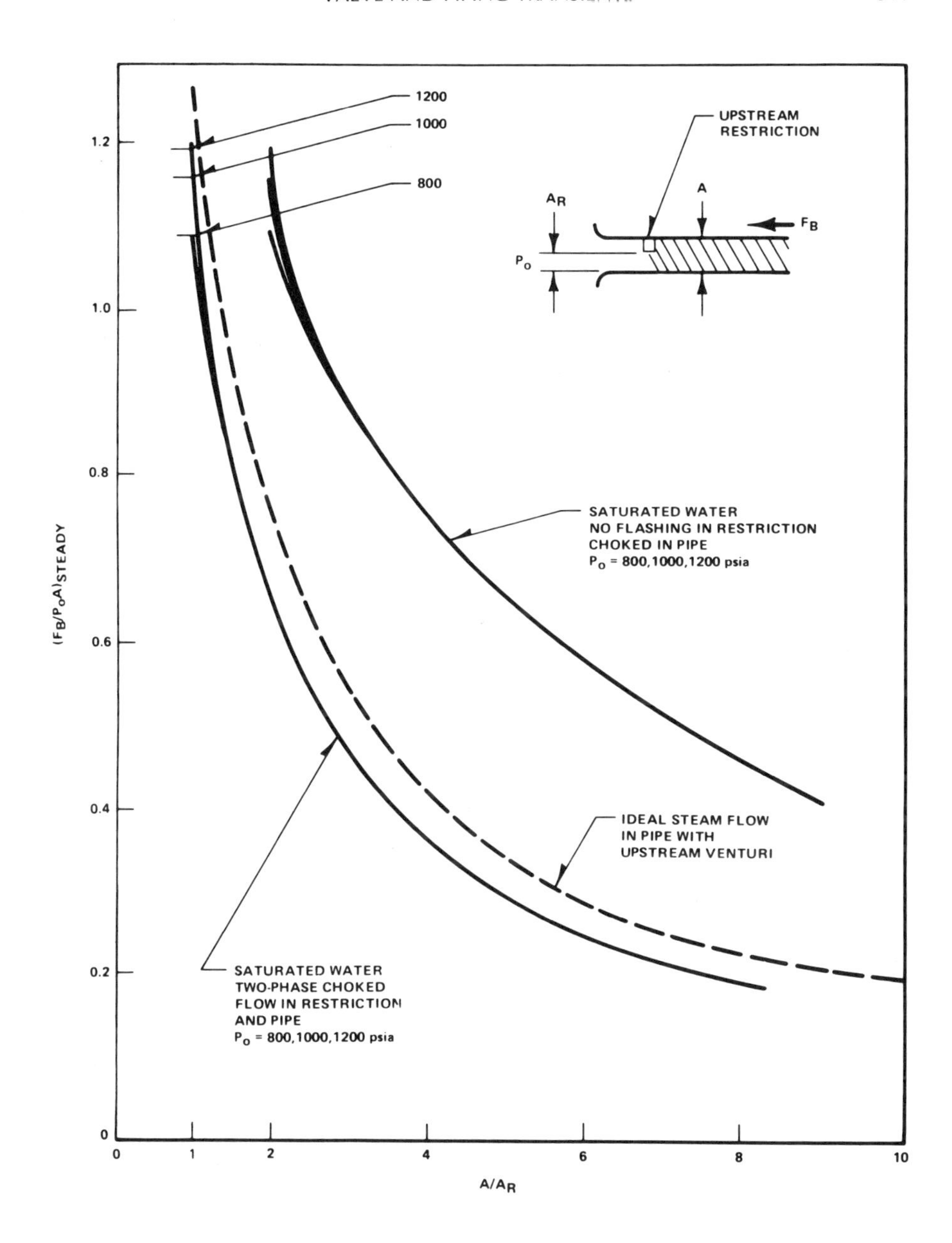

Fig. 10-15 Steady blowdown force with restriction.

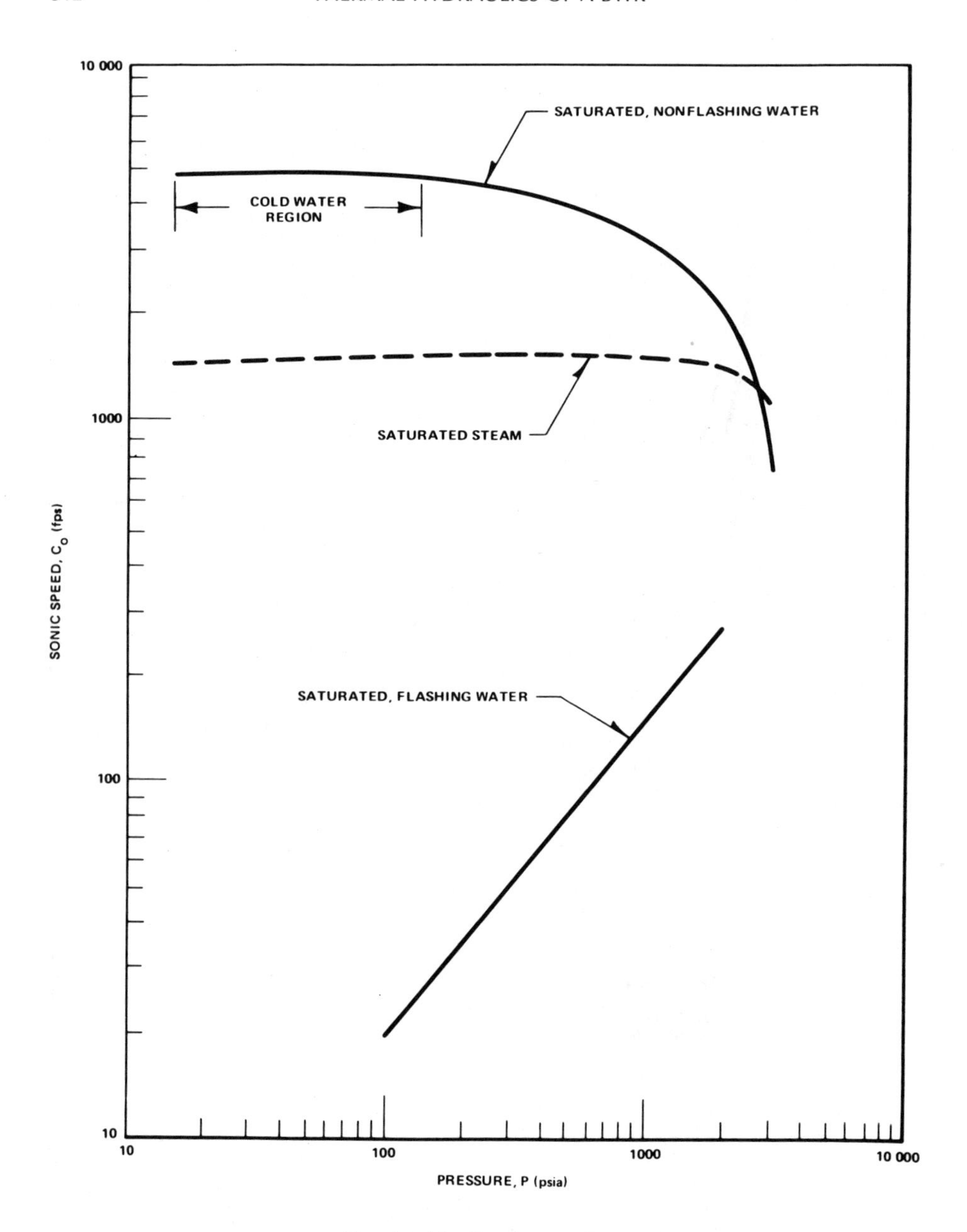

Fig. 10-16 Sonic speed.

10.5.1 Effect of Area Changes

When a decompression or compression wave encounters an abrupt area change like the one shown in Fig. 10-17, flow readjustment causes a change in the wave force. To estimate such an effect, consider an oncoming wave traveling leftward in Sec. I. For the case of nonflashing liquid, further suppose that pressure in the undisturbed fluid is p_{II} and that the oncoming pressure is p_I. Equation (10.31) can be used to show that,

$$\langle j \rangle_I = \frac{g_c}{\rho C}(p_{II} - p_I) \quad . \tag{10.47}$$

On reaching Sec. II, the wave is partly reflected and partly transmitted. For low-velocity liquid, the intermediate pressure, p_i, can be assumed equal in Secs. I and II. Mass conservation in the intermediate region requires,

$$\langle j \rangle_{II_i} A_{II} = \langle j \rangle_{I_i} A_I \quad . \tag{10.48}$$

Equation (10.31) can be employed again to write,

$$\langle j \rangle_{II_i} = \frac{g_c}{\rho C}(p_{II} - p_i) \tag{10.49}$$

$$\langle j \rangle_{I_i} - \langle j \rangle_I = \frac{g_c}{\rho C}(p_i - p_I) \quad . \tag{10.50}$$

The last four equations can be combined to obtain the intermediate properties as,

$$\frac{p_i - p_{II}}{p_I - p_{II}} = 2 \bigg/ \left(\frac{A_{II}}{A_I} + 1\right) \tag{10.51}$$

$$\frac{\langle j \rangle_{II_i}}{\langle j \rangle_I} = 2 \bigg/ \left(\frac{A_{II}}{A_I} + 1\right) \quad . \tag{10.52}$$

$$\frac{\langle j \rangle_{I_i}}{\langle j \rangle_I} = 2 \bigg/ \left(1 + \frac{A_I}{A_{II}}\right) \quad . \tag{10.53}$$

By using the identity, $G = \langle \rho \rangle_l \langle j \rangle$, and Eq. (10.31), Eq. (10.4) can be integrated to show that the total wave force is given by,

$$\left.\begin{aligned} &\frac{F_W}{(p_{II} - p_I)A_I} = 1.0; \text{ before the transition} \\ &\frac{F_W}{(p_{II} - p_I)A_I} = \frac{\left(3 - \dfrac{A_I}{A_{II}}\right)}{\left(1 + \dfrac{A_I}{A_{II}}\right)}; \text{ after the transition} \end{aligned}\right\} \quad . \tag{10.54}$$

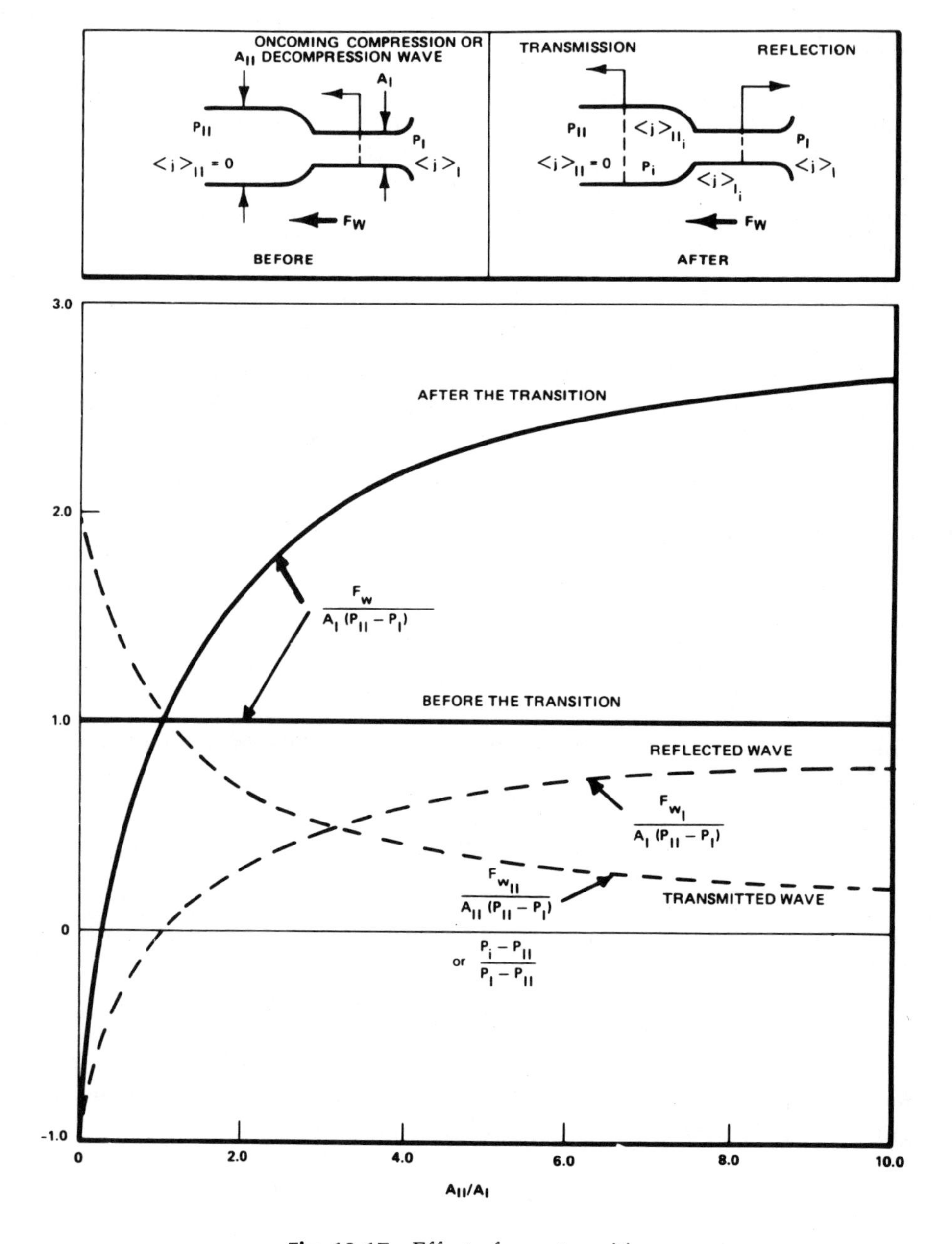

Fig. 10-17 Effect of area transition.

A graph of the wave force before and after transition is shown in Fig. 10-17. Note that for large ratios of A_{II}/A_I, the wave force can triple its magnitude. The total wave force, F_w, is composed of F_{w_I} plus $F_{w_{II}}$, where F_{w_I} is the wave force in Sec. I and $F_{w_{II}}$ in Sec. II. These wave forces are given by,

$$\frac{F_{W_{II}}}{(p_{II}-p_I)A_{II}}=\frac{2}{\left(\dfrac{A_{II}}{A_I}+1\right)} \tag{10.55}$$

$$\frac{F_{W_I}}{(p_{II}-p_I)A_I}=\frac{\left(\dfrac{A_{II}}{A_I}-1\right)}{\left(\dfrac{A_{II}}{A_I}+1\right)}, \tag{10.56}$$

and also are shown in Fig. 10-17.

The graph of Eq. (10.55) is useful in estimating pressure loads imposed on reactor vessel internals following arrival of the decompression wave from a pipe rupture. It should be noted, however, that large decompression waves, which drop pressure below the saturation value, are strongly attenuated in the pipe and decrease about an order of magnitude in intensity when they reach a rigid wall about one pipe diameter into the pressure vessel. Thus, the loads indicated in Fig. 10-17 should be considered as very conservative for estimation of loads on reactor vessel internals.

10.5.2 Comparison with Data

Results from several basic experiments are available to evaluate predictions from the methods just described. Comparison of the steady blowdown force with some Loft data (Dietz, 1968) is shown in Fig. 10-18. Test number 706 did not contain an orifice in the discharge pipe. Only the initial thrust value, based on a 920-psia vessel saturation pressure, was obtained from Fig. 10-10 for a frictionless pipe. Reasonable comparison is shown. Test number 704 contained a 0.027-ft^2 orifice about midway in the 4-ft discharge pipe. The distance from vessel to orifice was sufficient for the saturated water to begin flashing. Therefore, the curve labeled "saturated water, two-phase choked flow in restriction and pipe" in Fig. 10-15 was used to calculate the blowdown force. Good agreement with the test is shown for this comparison also.

The expressions of Eq. (10.54) were employed to predict the wave reaction forces before and after a decompression wave arrived at an abrupt area enlargement. The experimental force measurement (Hanson, 1970) for an area transition from 0.95 to 4.236 $in.^2$ is shown in Fig. 10-19. The pipe system contained cold water initially pressurized to 2175 psia. Discharge

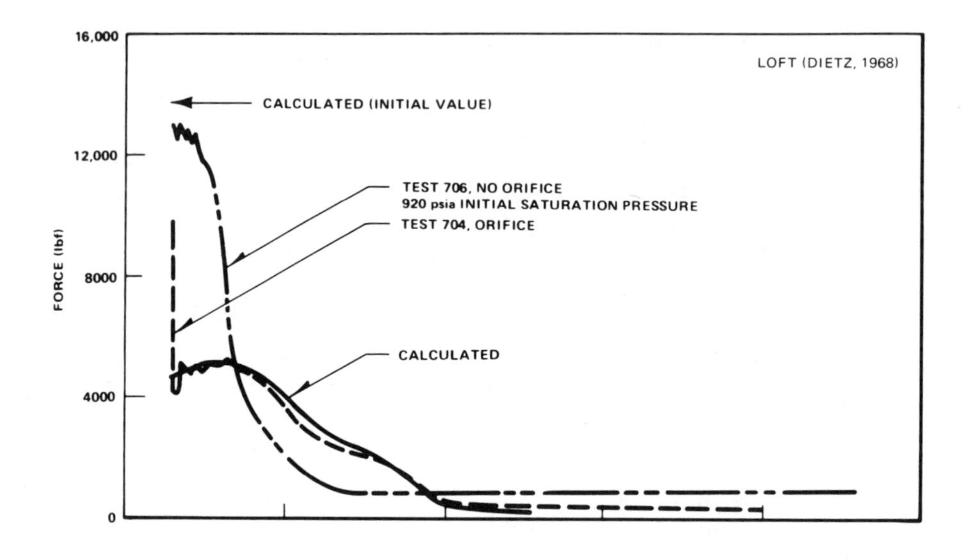

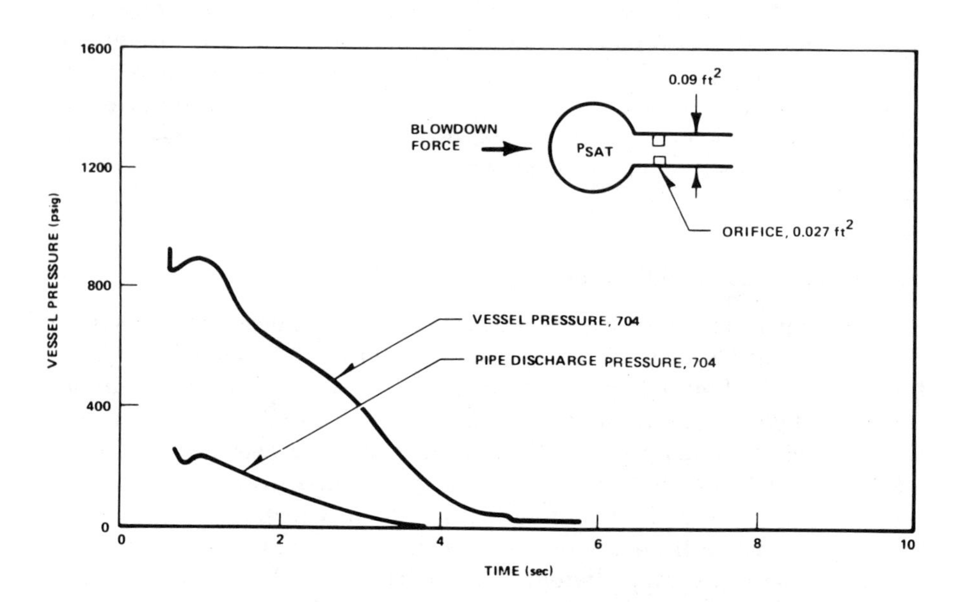

Fig. 10-18 Steady blowdown force comparison.

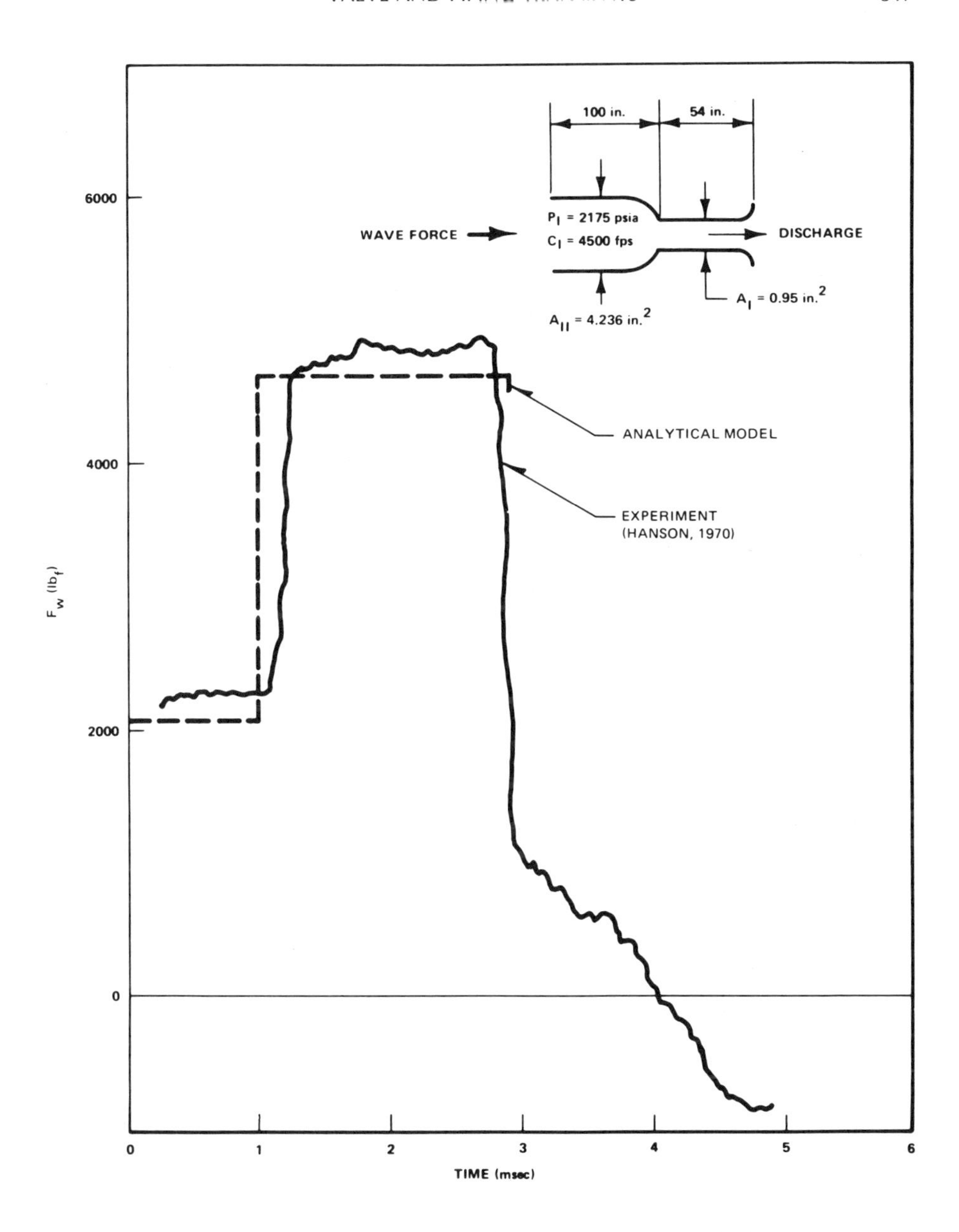

Fig. 10-19 Cold water wave force, comparison with data.

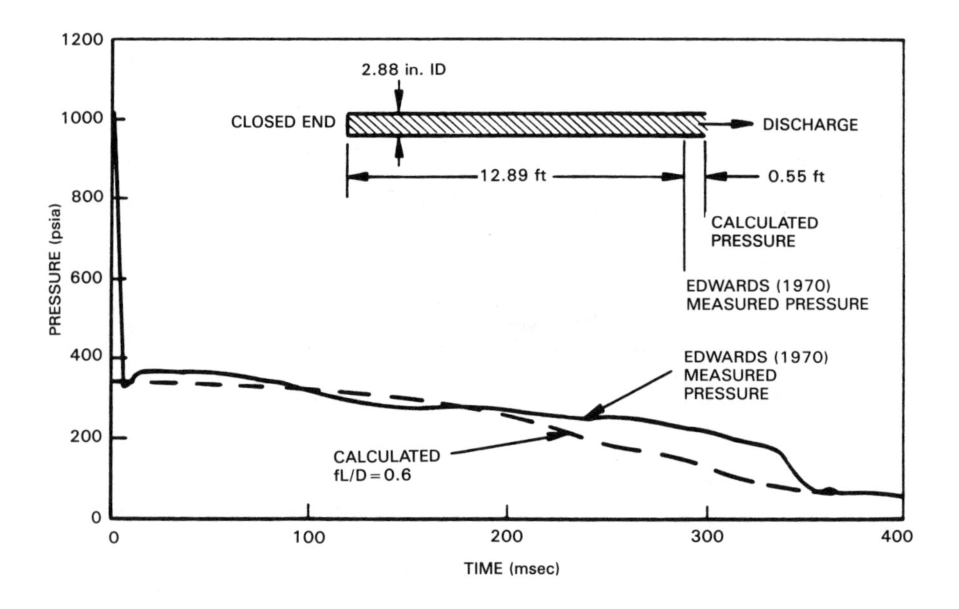

Fig. 10-20 Comparison with Edwards experiment (1970).

was initiated by a rupture disk in the smaller pipe. The analytical model (dashed line) is seen to agree closely with the experimental trace.

The method of characteristics was applied using two-phase homogeneous equilibrium flow theory (Gallagher, 1970) to predict ruptured pipe transients. Figure 10-20 gives a comparison of this theory with the experimental data of Edwards and O'Brien (1970), who ruptured the end of a 13-ft-long, 2.9-in.-diam pipe containing water at 1000 psia with a saturation pressure of 500 psia. It is seen that the predicted and measured pressures are relatively close, which supports the validity of the theory.

Several example problems are presented next to demonstrate the use of the graphs presented so far.

10.5.3 Example Problems

Example Problem #1. The vessel and pipe system shown in Fig. 10-21 contains ideal steam at 1000 psia. The pipe is 160 ft long with segment lengths of 64, 32, and 64 ft, and a flow area of 100 in.^{2} For the postulated rupture, estimate the force-time behavior on the three segments.

The solid curves on Fig. 10-21 were obtained from a detailed analysis. An approximate determination of the forces is obtained by the following procedure.

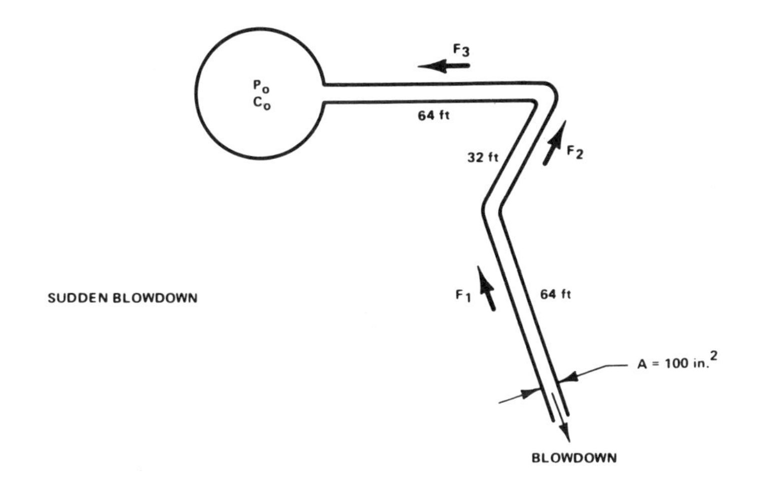

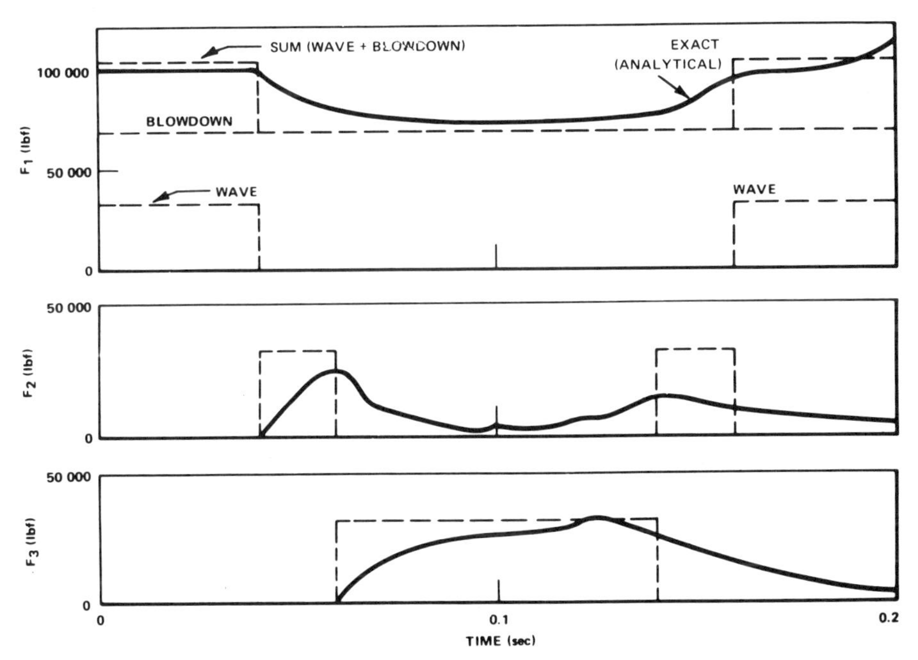

Fig. 10-21 Sample computation.

Use Fig. 10-10 to obtain the initial blowdown and wave forces,

$$F_{B,\text{initial}} = 0.68 p_0 A = (0.68)(1000)(100) = 68\ 000\ \text{lb}_f$$

$$F_{W,\text{initial}} = 0.32 p_0 A = (0.32)(1000)(100) = 32\ 000\ \text{lb}_f\ .$$

The initial blowdown force is constant until a reflected wave returns from the vessel. The wave force is assumed to travel at sonic speed. By using Fig. 10-16 to determine the speed of the sonic wave in saturated steam, $C_0 \approx 1600$ fps.

	Time Interval		Segment Containing Wave
toward vessel	$0 < t < \frac{L_1}{C_0}$	$= \frac{64}{1600} = 0.04$ sec	1
	$0.04 < t < \frac{L_1 + L_2}{C_0}$	$= \frac{96}{1600} = 0.06$ sec	2
	$0.06 < t < \frac{L_1 + L_2 + L_3}{C_0}$	$= \frac{160}{1600} = 0.10$ sec	3
toward rupture	$0.10 < t < \frac{L_1 + L_2 + 2L_3}{C_0}$	$= \frac{224}{1600} = 0.14$ sec	3
	$0.14 < t < \frac{L_1 + 2(L_2 + L_3)}{C_0}$	$= \frac{256}{1600} = 0.16$ sec	2
	$0.16 < t < \frac{2(L_1 + L_2 + L_3)}{C_0}$	$= \frac{320}{1600} = 0.20$ sec	1

The above process is repeated until steady discharge is reached.

To estimate time to reach steady discharge from Eqs. (10.37) and (10.43),

$$\frac{\langle j \rangle}{C_0} = 0.86$$

$$\frac{\langle \rho \rangle}{\rho_0} = 0.40$$

$$t_D \approx \left(\frac{1}{0.40}\right)\left(\frac{1}{0.86}\right)\left(\frac{160}{1600}\right) = 0.29\ \text{sec}\ .$$

The steady blowdown force can be estimated from Fig. 10-11 as,

$$F_{B,\text{steady}} = 1.22 p_0 A$$

$$= (1.22)(1000)(100) = 122\ 000\ \text{lb}_f\ .$$

Example Problem #2. Using the vessel and pipe layout of Example Problem #1, estimate the force-time behavior for a nonflashing cold water system.

Only the basic numbers are obtained here, from which the force-time graphs could be approximated.

The initial blowdown and wave forces can be estimated from Fig. 10-10:

$$F_{B,\text{initial}} \approx 0.0 \text{ lb}_f$$

$$F_{W,\text{initial}} \approx 1.0 p_0 A = (1.0)(1000)(100) = 100\ 000 \text{ lb}_f \ .$$

The wave speed can be estimated from Fig. 10-16,

$$C_0 \approx 4500 \text{ fps} \ .$$

The time to reach steady discharge is obtained from Eq. (10.42) as,

$$t_D \approx 2(160)\left[\frac{(62.4)}{(2)(32.2)(1000)(144)}\right]^{1/2} = 0.75 \text{ sec} \ .$$

The steady blowdown force is obtained from Fig. 10-11 as,

$$F_{B,\text{steady}} = 2.0 p_0 A = (2.0)(1000)(100) = 200\ 000 \text{ lb}_f \ .$$

Example Problem #3. Again, by using the vessel and pipe layout of Example Problem #1, estimate the force-time behavior for a water system whose saturation pressure is 800 psia, but whose initial pressure is 1000 psia. The saturated water enthalpy is 510 Btu/lb_m. Only the basic numbers are obtained for this example.

The initial blowdown and wave forces are estimated from Fig. 10-10,

$$F_{B,\text{initial}} = (0.74) p_0 A = (0.74)(1000)(100) = 74\ 000 \text{ lb}_f$$

$$F_{W,\text{initial}} = (0.26) p_0 A = (0.26)(1000)(100) = 26\ 000 \text{ lb}_f \ .$$

The initial wave speed is estimated from Fig. 10-16 at the saturation pressure. (The large, brief decompression-recompression wave in the liquid is ignored for reasons discussed earlier):

$$C_0 \approx 120 \text{ fps} \ .$$

The time to reach steady discharge is obtained from Eq. (10.43) and Fig. 10-9:

$$\frac{\langle j \rangle}{C_0} = 2.6$$

$$\frac{\langle \rho_h \rangle}{\rho_{sat}} = 0.22$$

$$t_D \approx \left(\frac{1}{0.22}\right)\left(\frac{1}{2.6}\right)\left(\frac{160}{120}\right) = 2.3 \text{ sec} \ .$$

The steady blowdown force is obtained from Fig. 10-11 for separated flow:

$$F_{B,\mathrm{steady}} = 1.32 p_0 A = (1.32)(1000)(100) = 132\ 000\ \mathrm{lb}_f \ .$$

Example Problem #4. Determine the steady blowdown force if a Venturi of 20-in.^{2} area is installed in the 100-in.^{2} pipe of Example Problem #1.

Figure 10-15 is used to obtain the force at a pipe-to-restriction area ratio of 5.0,

$$F_{B,\mathrm{steady}} = (0.34) p_0 A = (0.34)(1000)(100) = 34\ 000\ \mathrm{lb}_f \ .$$

Example Problem #5. A 10-in.^{2} pipe, which contains 1000-psia subcooled water with 800-psia saturation pressure, ruptures. Determine the initial wave force and both transmitted and reflected wave forces at a junction where the 10-in.^{2} pipe joins a 20-in.^{2} pipe. From Fig. 10-9, $p_I = 0.63\ p_{\mathrm{sat}} = 0.63(800) = 504$ psia. Now, use Fig. 10-17 to determine all the forces, noting that $p_{\mathrm{I}} = 504$ psia, $p_{\mathrm{II}} = 1000$ psia, $A_{\mathrm{I}} = 10.0$ in.^{2}, $A_{\mathrm{II}} = 20.0$ in.^{2}, and $A_{\mathrm{II}}/A_{\mathrm{I}} = 2$.

Before reaching the transition,

$$F_W = (1.0)(p_{II} - p_I)A_I$$

$$= (1.0)(496)(10) = 4960\ \mathrm{lb}_f \ .$$

After reaching the transition, the reflected wave force is obtained from,

$$F_{W_I} = (0.3)(496)(10) = 1488\ \mathrm{lb}_f$$

and the transmitted force is,

$$F_{W_{II}} = (0.65)(496)(20) = 6450\ \mathrm{lb}_f \ .$$

Example Problem #6. Consider a break in the 20-in.^{2} pipe of Example Problem #5, and determine wave forces where the section area changes to 10 in.^{2}

Again, use Fig. 10-17 with $p_{\mathrm{I}} = 504$ psia, $p_{\mathrm{II}} = 1000$ psia, $A_{\mathrm{I}} = 20$ in.^{2}, $A_{\mathrm{II}} = 10$ in.^{2}, and $A_{\mathrm{II}}/A_{\mathrm{I}} = 0.5$.

The initial force is,

$$F_W = (1.0)(p_{II} - p_I)A_I$$

$$= (1.0)(496)(20) = 9920\ \mathrm{lb}_f \ .$$

The reflected wave force is given by,

$$F_{W_I} = (-0.35)(496)(20) = -3470\ \mathrm{lb}_f$$

and the transmitted wave force is obtained from,

$$F_{W_{II}} = (1.3)(496)(10) = 6450\ \mathrm{lb}_f \ .$$

In summary, methods for estimating the force-time character of reaction factors caused by the postulated circumferential rupture of a pressurized piping system have been presented in graphic form. The graphs are based on either subcooled water, saturated water, or saturated steam, and can be used to estimate the following:

1. initial blowdown force of the discharging pipe segment
2. transient wave force, which travels back and forth through the pipe fluid until steady blowdown is reached
3. approximate wave speed to determine the transient location of the wave front
4. effect of pipe area change on the wave force
5. approximate time to reach steady-state discharge conditions
6. steady blowdown force, which applies only to the discharging pipe segment and is equal to the full jet impingement force
7. effect of upstream area restriction on the steady blowdown force
8. expanded jet area, to determine which mechanical components overlap the fluid jet and experience impingement forces
9. expanded jet impingement pressure for computing the impingement forces.

The blowdown and wave forces compare favorably with representative experimental data. Therefore, methods presented in this study can be used in the preliminary design and positioning of mechanical restraints to withstand pipe rupture forces resulting from postulated accidents.

10.6 Shocks in Pipes

A shock can form in cases of a rapid valve closing, which decelerates oncoming fluid, or a rapid valve opening, which charges an otherwise stagnant pipe fluid.

If a moving shock forms, it becomes a moving boundary across which fluid properties are assumed to change discontinuously. The MOC can be applied on both sides of a moving shock, but properties across a shock must be obtained from the conservation equations to be developed next (Moody, 1989).

Moving shock relationships can be obtained by analysis of a discontinuity moving rightward at speed S relative to the pipe in Fig. 10-22. Properties in the undisturbed fluid are designated by subscript x. The disturbance has occurred at the left, and shocked properties are designated by subscript y. Mass, momentum, and energy conservation laws across the shock can be written for the dotted CV in which storage terms are negligible. Therefore, we have,

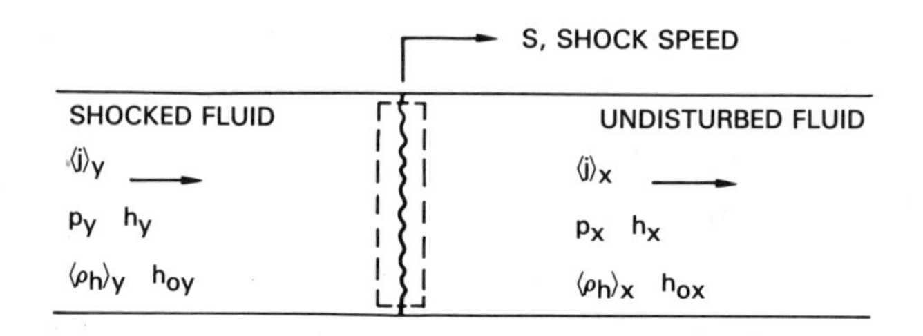

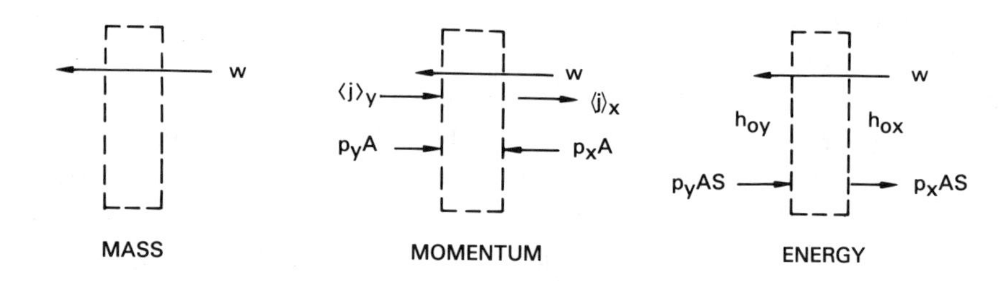

Fig. 10-22 Moving normal shock formulation.

Mass conservation:

$$w = \langle \rho_h \rangle_y A(S - \langle j \rangle_y) = \langle \rho_h \rangle_x A(S - \langle j \rangle_x) \ . \tag{10.57}$$

Momentum conservation:

$$w(\langle j \rangle_y - \langle j \rangle_x) = g_c(p_y - p_x)A \ . \tag{10.58}$$

Energy conservation:

$$w(h_{0y} - h_{0x}) + (p_x - p_y)AS = 0 \ . \tag{10.59}$$

If we substitute $h_0 = h + \langle j \rangle^2/2g_0$, rearrangement yields,

$$\langle \rho_h \rangle_x (S - \langle j \rangle_x) = \langle \rho_h \rangle_y (S - \langle j \rangle_y) \ , \tag{10.60}$$

$$\langle \rho_h \rangle_x (S - \langle j \rangle_x)^2 - \langle \rho_h \rangle_y (S - \langle j \rangle_y)^2 = g_c(p_y - p_x) \ , \tag{10.61}$$

$$h_x - h_y + \frac{1}{2} \frac{p_y - p_x}{\langle \rho_h \rangle_y} \left(\frac{\langle \rho_h \rangle_y}{\langle \rho_h \rangle_x} + 1 \right) = 0 \ . \tag{10.62}$$

Equations (10.60), (10.61), and (10.62) relate properties across a moving normal shock for a simple compressible fluid. State equations for the fluid are required for a complete analysis.

If we employ the perfect gas state properties,

$$h = CpT$$

$$p = \rho RT$$

$$C = \sqrt{K g_0 RT} \ ,$$

then Eqs. (10.60), (10.61), and (10.62) can be written as

$$\frac{\langle \rho_h \rangle_y}{\langle \rho_h \rangle_x} = \left(\frac{k+1}{k-1} \frac{p_y}{p_x} + 1 \right) \left(\frac{k+1}{k-1} + \frac{p_y}{p_x} \right)^{-1} , \qquad (10.63)$$

$$\frac{S - \langle j \rangle_x}{C_x} = \sqrt{\frac{k+1}{2k} \frac{p_y}{p_x} + \frac{k-1}{2k}} , \qquad (10.64)$$

$$\frac{\langle j \rangle_y - \langle j \rangle_x}{C_x} = \left(\frac{p_y}{p_x} - 1 \right) \sqrt{\frac{2}{k(k-1)} \left(\frac{k+1}{k-1} \frac{p_y}{p_x} + 1 \right)^{-1}} . \qquad (10.65)$$

It is difficult to incorporate shocks in a MOC solution that employs standard Cartesian coordinates (Jonsson et al., 1973). Various means of incorporating local dissipation to simulate a moving shock have been proposed, but none have proven to be successful (Von Neumann, 1950). However, it is fortunate that when shocks form, the MOC can be employed to obtain relatively accurate results even for severe pipe flow disturbances without including a moving shock boundary condition. It was found (Wheeler, 1979) that the MOC has an inherent numerical dissipation mechanism when steep pressure profiles form, which closely approximates a shock over three to five spatial mesh points in the numerical computation.

A shock may form immediately and propagate through the piping system in cases of rapid valve opening or closing. When this occurs, a simple solution of the shock equations and prediction of pipe forces is possible.

The discharge mass flow rate w through a valve is usually restricted by the critical flow at the throat such that $w(t) = \langle \rho_h \rangle A \langle j \rangle = G_c A(t)$, where G_c is the critical mass flux, and $A(t)$ is the time-dependent valve throat area. The local critical flow rate for a perfect gas is expressed by:

$$w_{gc} = A \left[k g_c p_0 \langle \rho_h \rangle_0 \left(\frac{2}{k+1} \right)^{(k+1)/(k-1)} \right]^{1/2} , \qquad (10.66)$$

where p_0 and $\langle \rho_h \rangle_0$ are the stagnation pressure and density.

10.7 Safety/Relief Valves

Normally, a safety/relief valve (SRV) would discharge gas or steam. However, there are postulated transients in the power industry that can lead to the discharge of gas-liquid mixtures through a SRV. Figure 9-11a gives the homogeneous equilibrium critical mass flux for saturated steam-water mixtures and subcooled water (Moody, 1975). This model is based on two-phase flow through an isentropic nozzle throat that is 10 cm or longer to permit phase equilibrium.

The homogeneous equilibrium model usually can be employed to approximate steam and high-quality discharge. However, when the water quality drops below about 0.5 steam mass fraction, equilibrium two-phase states are not fully achieved in the relatively short flow length through a valve throat. Other discharge flow models, such as one based on a non-equilibrium formulation (Henry and Fauske, 1971) give better flow rate predictions at low steam mass fractions.

SRV discharge generally occurs into a pipe that contains stagnant air. The entrance boundary condition for SRV discharge makes use of the critical mass flow rate, the quasi-steady, adiabatic energy equation, $h_0 = h + \langle j \rangle^2/2g_0$, written between the reservoir and a pipe section immediately downstream of the valve, and an appropriate state equation to obtain a relationship between entrance values of pressure and velocity. Simultaneous solution with the arriving solution of the characteristic equation yields the fully prescribed entrance condition.

If a pipe is charged at one end and its other end is submerged, the arriving pressure disturbance drives the submerged liquid column out of the pipe. Since the pressure propagation speed in liquid is much faster than it is in gas, all increments of a liquid column tend to respond simultaneously to imposed pressure disturbances. Therefore, the momentum law is written for the entire liquid column, expressing velocity in terms of driving pressure. Simultaneous stepwise solution with the arriving characteristic, or the reflected shock if appropriate, with the condition that equal pressure and velocity exist across the interface, yields the necessary boundary condition. Although a liquid column moves through a distance, the boundary condition normally is applied at one position, provided that its submerged length is much smaller than the total relief pipe length.

SRVs in nuclear plants generally have fast response, going from a closed position to a fully open position in a stroke time of less than 50 msec. Figure 10-23 shows the main disk opening characteristic for a direct acting SRV. The full opening stroke occurs within 24 msec. Most SRVs are close to the pressure vessel. Consequently, the steam flow rate closely follows the valve opening characteristic, as indicated by the mass discharge rate also shown in Fig. 10-23. As high-pressure steam charges from the SRV into the relief line, it creates a compression wave that propagates through an initially stationary gas, which normally is air at 1 atm. The propagating pressure wave changes shape, becoming steeper as it advances through the relief pipe, tending to form a moving shock wave as shown by the pressure profiles in Fig. 10-24. As the developing pressure front passes through various sections of the relief pipe, which are bounded by elbows

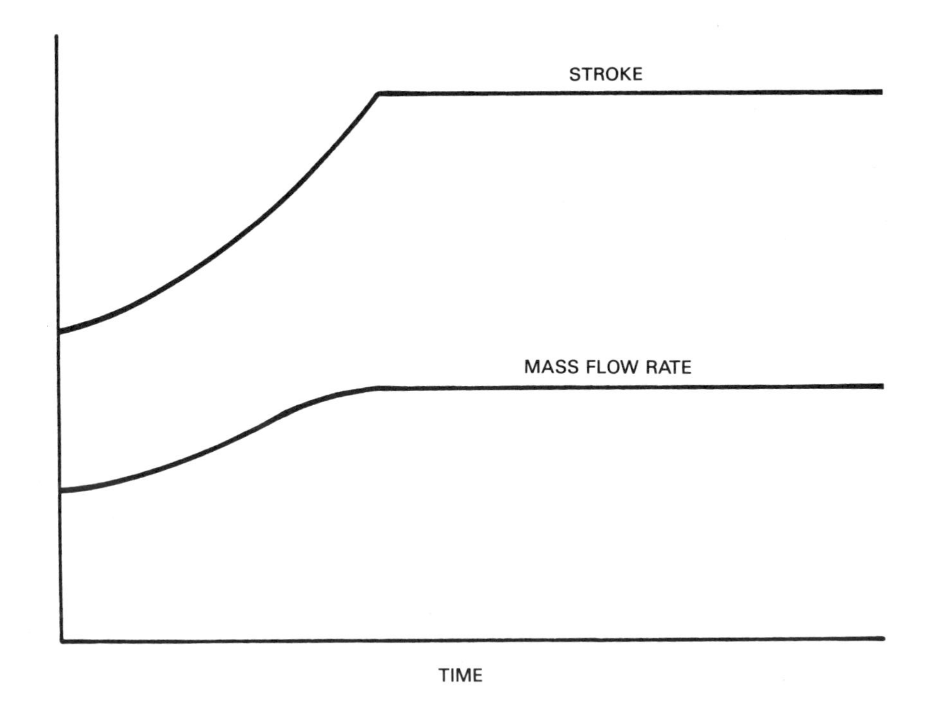

Fig. 10-23 Stroke and mass flow rate, direct acting SRV.

or turns at each end (bounded sections), unbalanced pressures and fluid momentum effects create net longitudinal forces.

The discharging steam pushes the initial gas ahead of it such that a steam-air interface or contact surface follows the propagating pressure front at a slower speed as shown in Fig. 10-24. Although pressure and velocity are continuous across the interface, respective air and steam densities are different. Thus, when the interface enters a bounded pipe section, fluid momentum again creates unbalanced pressures, which result in a net longitudinal pipe force.

If the SRV pipe discharge end is submerged in water, the advancing pressure front eventually arrives at a gas-water interface. The pressure front is reflected at an increased level and simultaneously begins to move the water column, which is accelerated until it moves out of the SRV discharge pipe, causing a decompression wave to propagate backward through the pipe. Ultimately, propagation effects attenuate in the piping system and steady flow is achieved, and only the blowdown thrust force survives on the pipe discharge end. The calculated time-dependent pipe

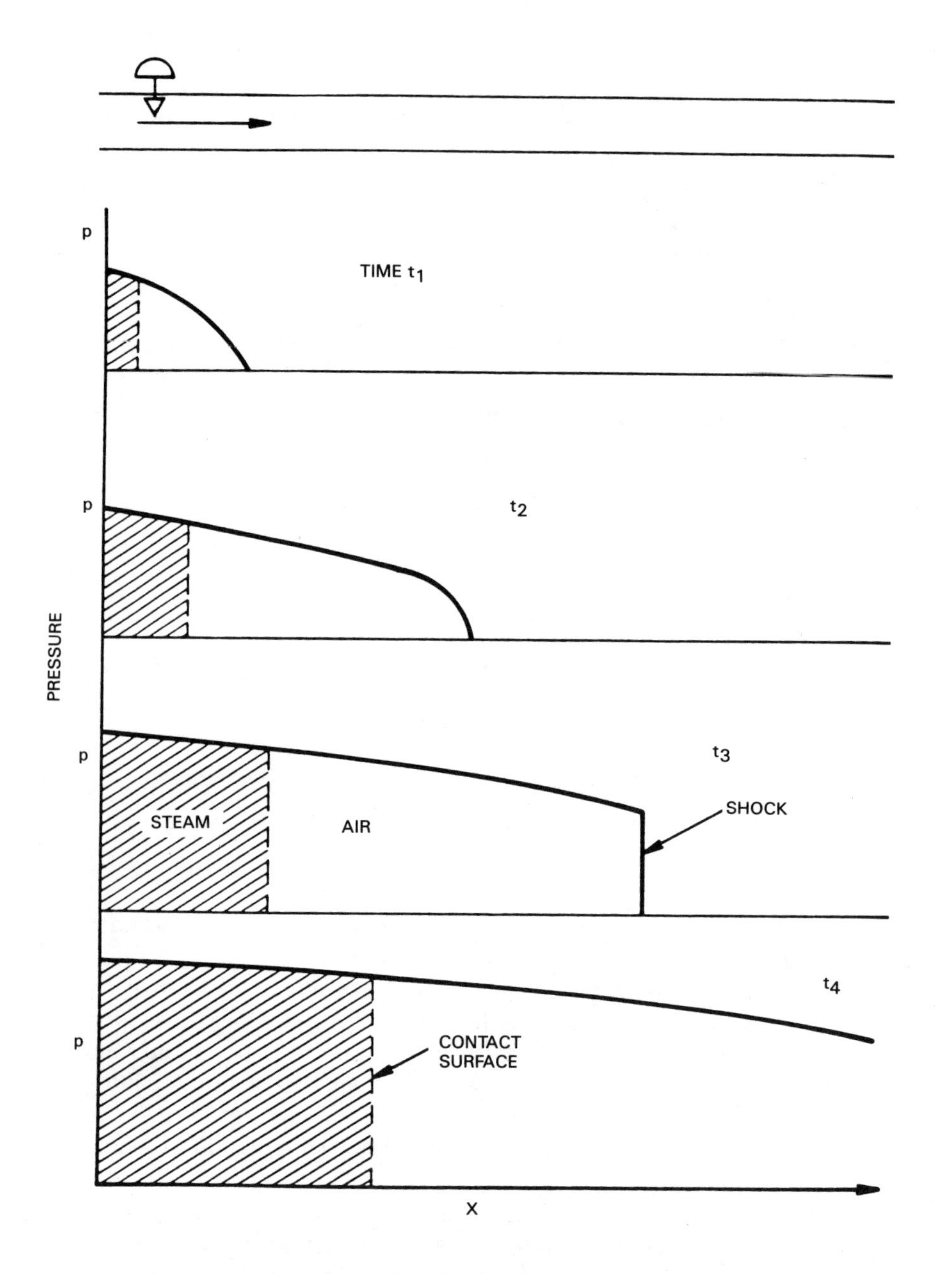

Fig. 10-24 Pressure profile in SRV pipe.

loads on each piping section then are used as input forcing functions for subsequent analysis of the pipe dynamic response.

10.7.1 Safety/Relief Valve Forces, Gas Charging

The following paragraphs present a discussion of the piping forces that are created from SRV gas discharge into a relief line.

Referring to Fig. 10-23, an SRV mass discharge rate usually can be simulated with a ramp-flat approximation according to:

$$w = \begin{cases} w_{\max}\dfrac{t}{\tau} \; ; & t < \tau \\ w_{\max} \; ; & t > \tau \end{cases} ,$$

where $w_{\max}$ is the full valve flow rate and τ is the ramp time. Associated wave forces on various pipe segments correspond to Figs. 10-25a, b, c, and d. Each graph is for a given normalized mass flow rate $w_{\max}/\langle\rho_h\rangle_\infty AC_\infty$ and stagnation properties $(p_0/p_\infty)/\langle\rho_h\rangle_0/\langle\rho_h\rangle_\infty$ for a perfect gas with $k=1.3$ (Moody, 1979). The pipe discharges to ambient air with properties p_∞, C_∞, $\langle\rho_h\rangle_\infty$. The various curves shown on each graph all start at the origin and diverge at

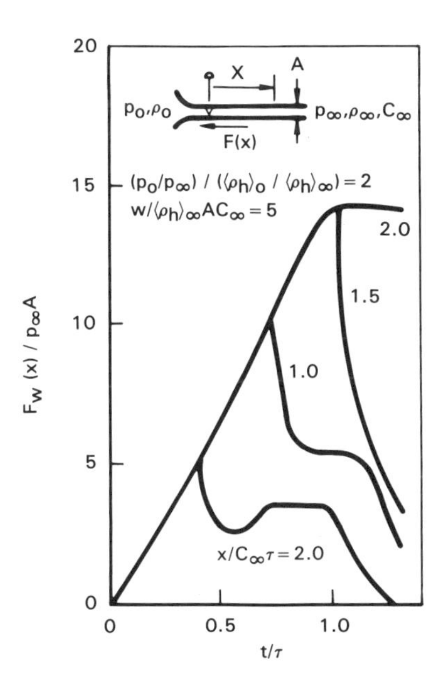

Fig. 10-25a Pipe force with valve flow acceleration.

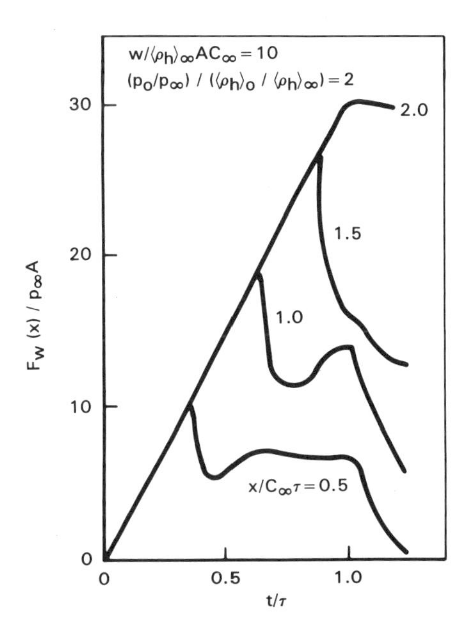

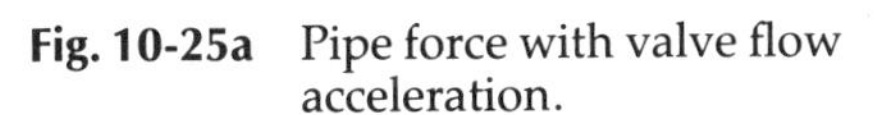

Fig. 10-25b Pipe force with valve flow acceleration.

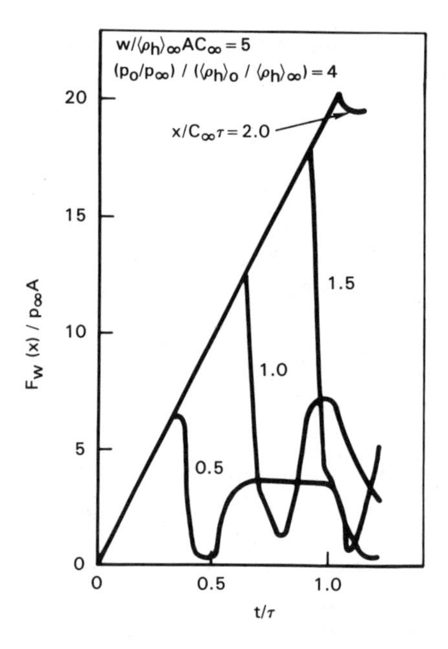

Fig. 10-25c Pipe force with valve flow acceleration.

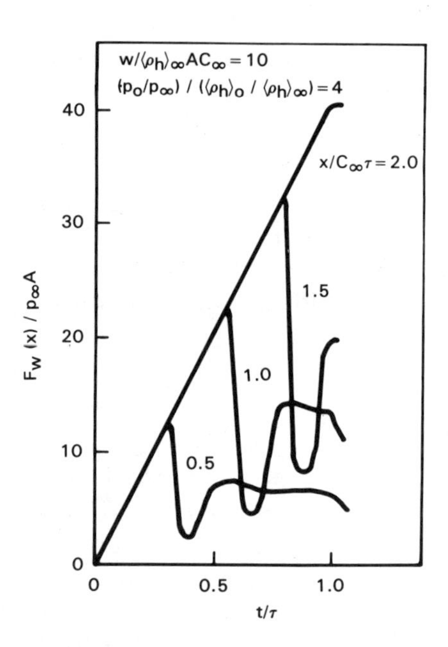

Fig. 10-25d Pipe force with valve flow acceleration.

various times. These individual curves correspond to the total wave force exerted on the relief line between the SRV and a normalized distance $x/C_\infty\tau$. For example, in Fig. 10-25a, the total wave force $F_w/p_\infty A$ between the valve and location $x/C_\infty\tau = 1.0$ rises from 0 to 10, and drops abruptly at a time of $t/\tau = 0.7$ when the pressure front moves past this location. The wave force on a pipe segment between $x/C_\infty\tau = 0.5$ and 1.0 is obtained by the difference of total wave forces corresponding to $x/C_\infty\tau = 1.0$ and 0.5.

10.7.2 Safety/Relief Valves, Instant (Step) Discharge

When an SRV opening time τ and the relief pipe length L are such that $C_\infty\tau/L << 1.0$, pipe forces can be estimated for the idealized case of instant valve opening, which creates a moving normal shock. The approximate location where a shock first forms during a valve opening corresponds to:

$$\frac{x_{\text{shock}}}{L} \approx \frac{C_\infty^2 \tau}{L} \rho(0,t) \bigg/ \left(\frac{w}{A}\right)_\infty = \frac{C_\infty \tau}{L} \frac{\rho(0,t) C_\infty}{(w/A)_\infty} , \tag{10.67}$$

where $\rho(0,t)$ is the estimated density of discharged fluid just downstream of the valve in the relief pipe at initial pipe pressure. For example, the discharge of steam at 6.8 MPa through a 10-msec valve would create a shock at $x_{\mathrm{shock}}/L = 0.2$, which would justify a moving normal shock analysis.

Figures 10-26 and 10-27 give the normalized forces resulting from a shock and steam-air interface moving through a pipe that initially contains gas with properties p_∞, $\langle\rho_h\rangle_\infty$, and C_∞. The pipe suddenly is charged at one end with gas at mass flow rate w coming from a source at pressure p_0 and density $\langle\rho_h\rangle_0$. For example, sudden steam discharge at a mass flow rate of $w = 150$ kg/sec into atmospheric air in a pipe of 0.07 m^2, from a vessel condition of $p_0 = 6.9$ MPa and $\langle\rho_h\rangle_0 = 36$ kg/m^3, creates a shock force of $F_s = 120$ kN and an interface force of $F_i = 2.8$ kN. Usually, the shock force is much greater than the interface force during steam discharge. The opposite may be true if an SRV discharges a two-phase mixture.

The ability to predict unsteady piping loads caused by fluid disturbances is directly related to the ability to predict the time- and space-dependent pressures in a piping system. Figure 10-28 gives the shock pressure for sudden SRV gas discharge. Once the shock pressure is determined for given values of the mass discharge rate and stagnation properties, Fig. 10-29 can be employed to obtain the shock speed and other flow properties. Figure 10-30 shows a comparison of theoretical and measured pressure above the water column near the discharge end of a submerged SRV pipe (McCready et al., 1973). Computations are shown to be in reasonable agreement.

Figure 10-31 shows results from a detailed example calculation of longitudinal forces exerted on various pipe segments during the operation of an SRV. Five straight pipe sections are shown between the SRV and the submerged discharge end. The valve flow rate, w, has been normalized with $\langle\rho_h\rangle_0 C_i A$, where $\langle\rho_h\rangle_0$ is steam density in the pressure vessel, C_i is the sound speed in undisturbed pipe air, and A is the pipe flow area. Time has been normalized with L/C_i, where L is the total pipe length. Resulting forces F_1, F_2, ..., F_n are normalized with $P_i A$, where P_i is initial air pressure in the pipe. Only the ramp opening portion of the SRV is shown, which results in the five time-dependent forces. Force F_1^* in the pipe section connected directly to the SRV begins to rise with the entering pressure front, which pushes against the SRV. When the pressure front moves into the next pipe section, it exerts a force on the elbow that partially counteracts the pressure force on the valve. Simultaneously, F_2^* begins to rise, and then is partially counteracted as the pressure front enters the next pipe section. This process continues until $t^* = 0.5$ when the pressure front enters Sec. 5, reflects on the water interface, and begins acceleration of the submerged water column, at which time F_5^* reverses due to higher pressure on the submerged discharge elbow. When the reflected pressure wave moves backward from the water column into Sec. 4, F_5^* again reverses. The dotted line shows how the reflected compression wave propagates backward through the SRV pipe. At time $t^* = 1.7$, the water plug is fully expelled, pressure at the discharge elbow is reduced, and F_5^* increases.

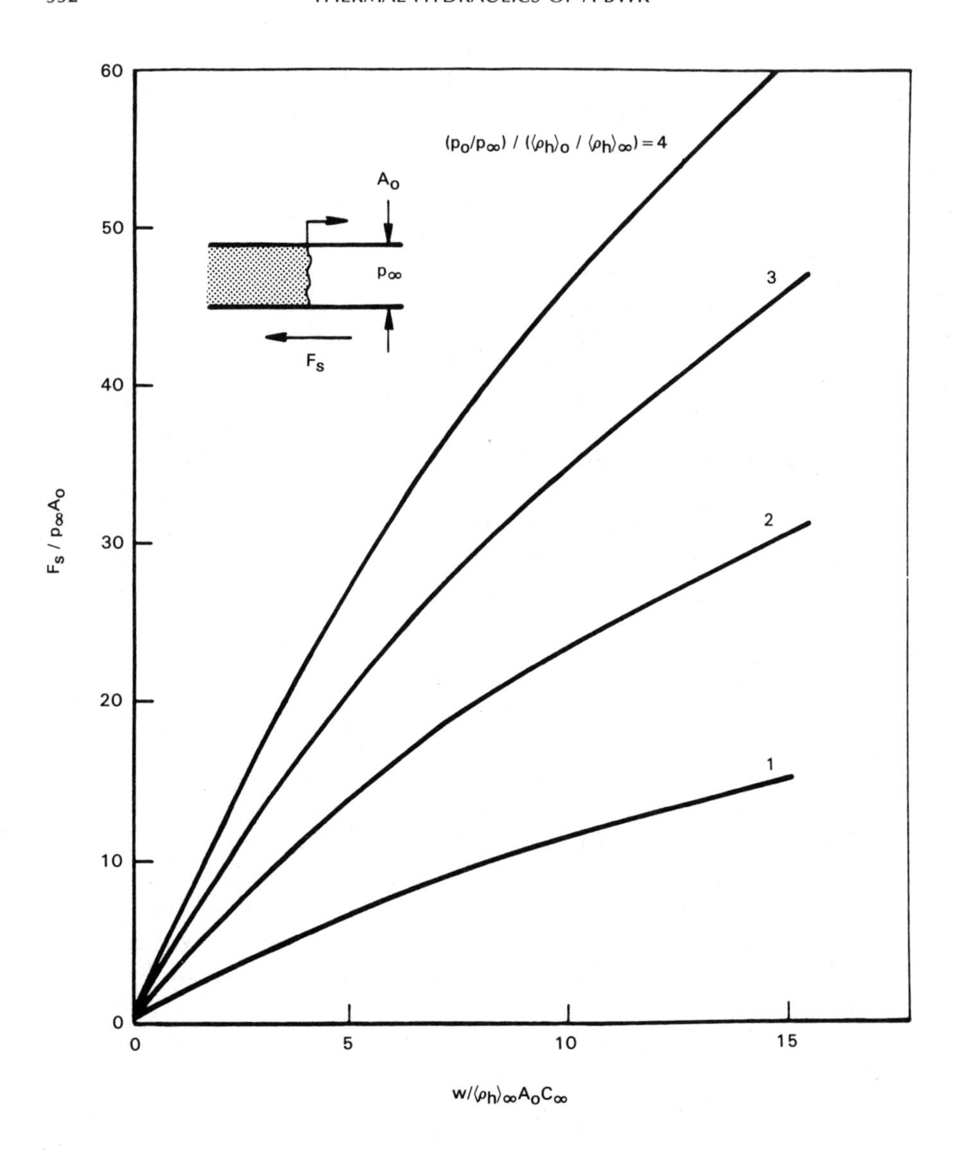

Fig. 10-26 Valve step charging, shock force.

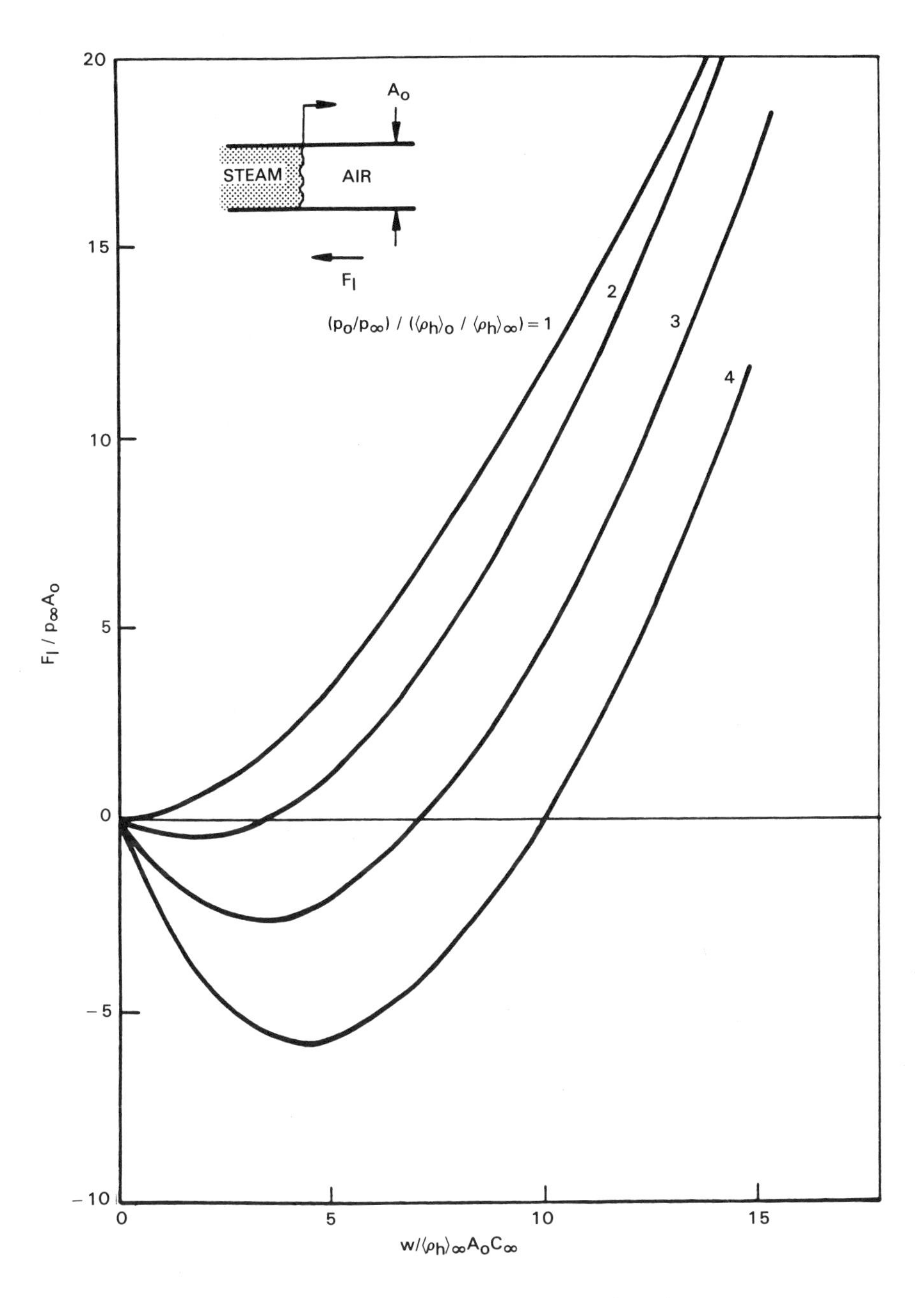

Fig. 10-27 Valve step charging, interface force.

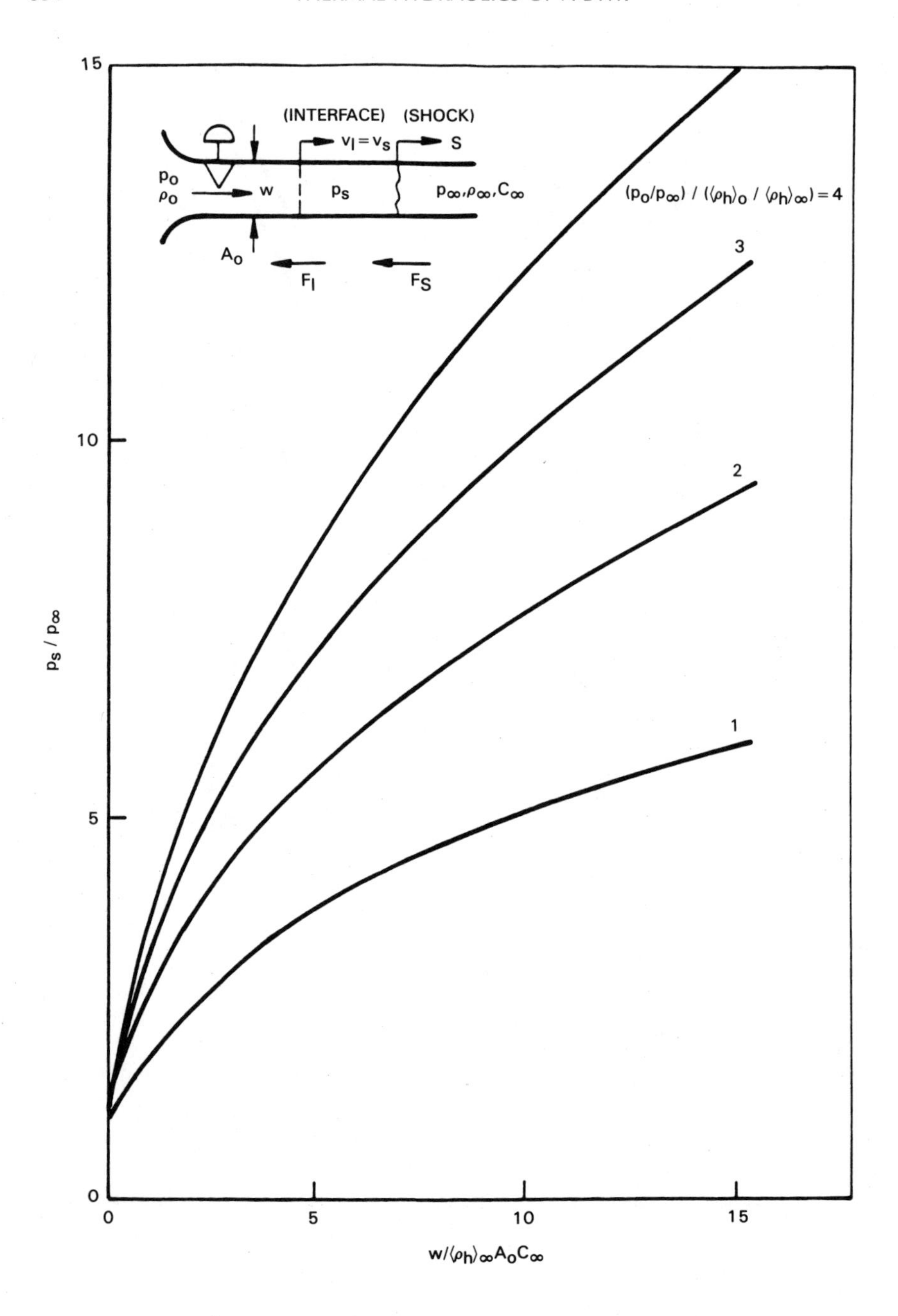

Fig. 10-28 Valve step charging, shock pressure.

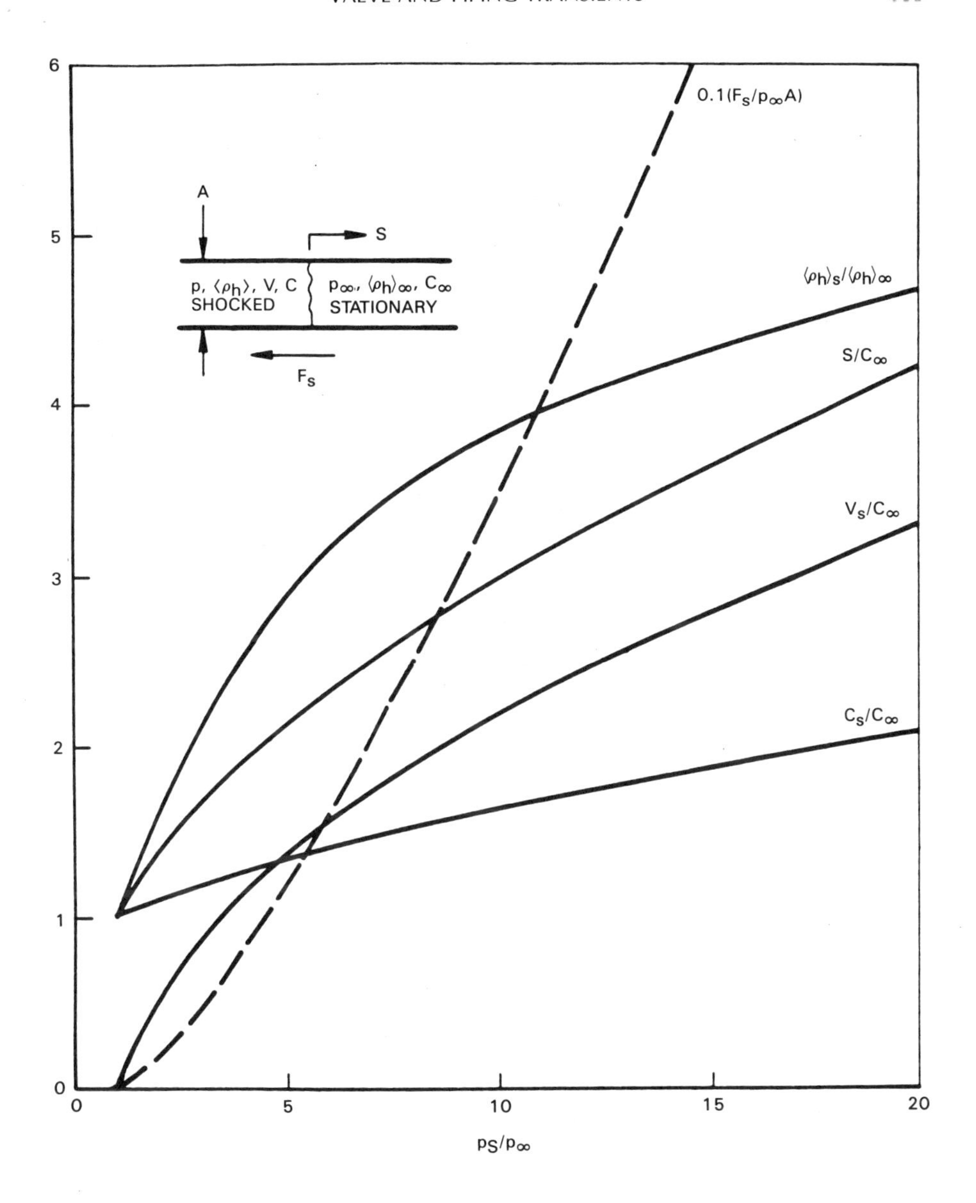

Fig. 10-29 Moving shock properties, $k = 1.4$.

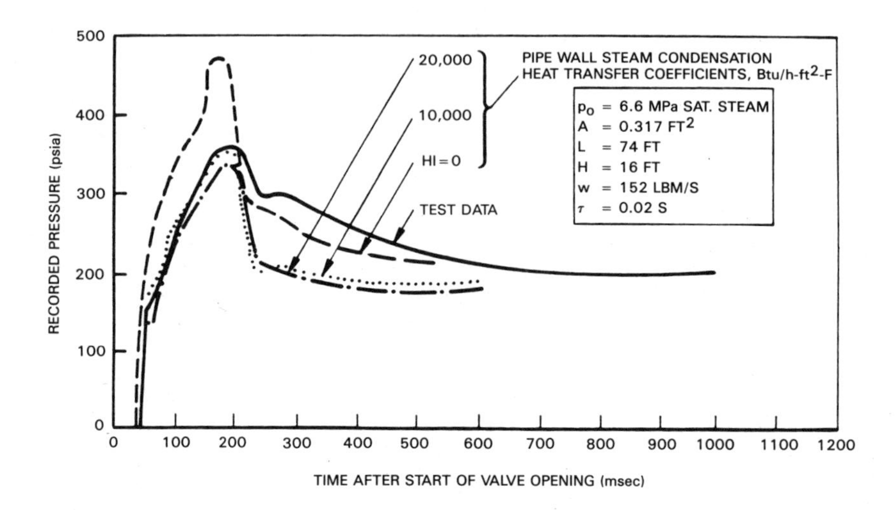

Fig. 10-30 Comparison of model with measured pressure data.

Although subsequent reflections and movement of pressure fronts through the various pipe sections in both directions create increases and decreases in the resulting force components, this discussion should help explain basic pipe loading effects caused by SRV discharge.

10.7.3 Safety/Relief Valve Flooding

Various transient conditions are postulated in the nuclear power industry that involve SRV flooding with steam-water mixtures. The relief pipe forces resulting from two-phase valve discharge must also be accommodated by the design.

Figure 10-32 is based on the perfect gas idealization for air, and gives the calculated shock pressure resulting from sudden steam-water mixture discharge through an SRV for a range of mass flow rate and mixture enthalpy (Moody, 1982). The resulting shock pressure is then employed in Fig. 10-29 to obtain the associated shock force that propagates through the relief pipe. It should be noted that saturated vapor results in the maximum shock force, whereas saturated and subcooled mixture creates smaller shock forces. The main reason for this behavior is that higher enthalpy mixtures undergo greater expansion in the relief line, which results in higher propagating pressure. Figure 10-29 also gives the shock speed and the shocked fluid speed for a range of shock pressures. These speeds are useful in determining both the pipe segment forces, and the time of force application in any segment, based on the shock or interface transit time.

Figure 10-33 gives the calculated mixture-air interface force for two-phase discharge through a SRV (Moody, 1982). It is seen that the interface force

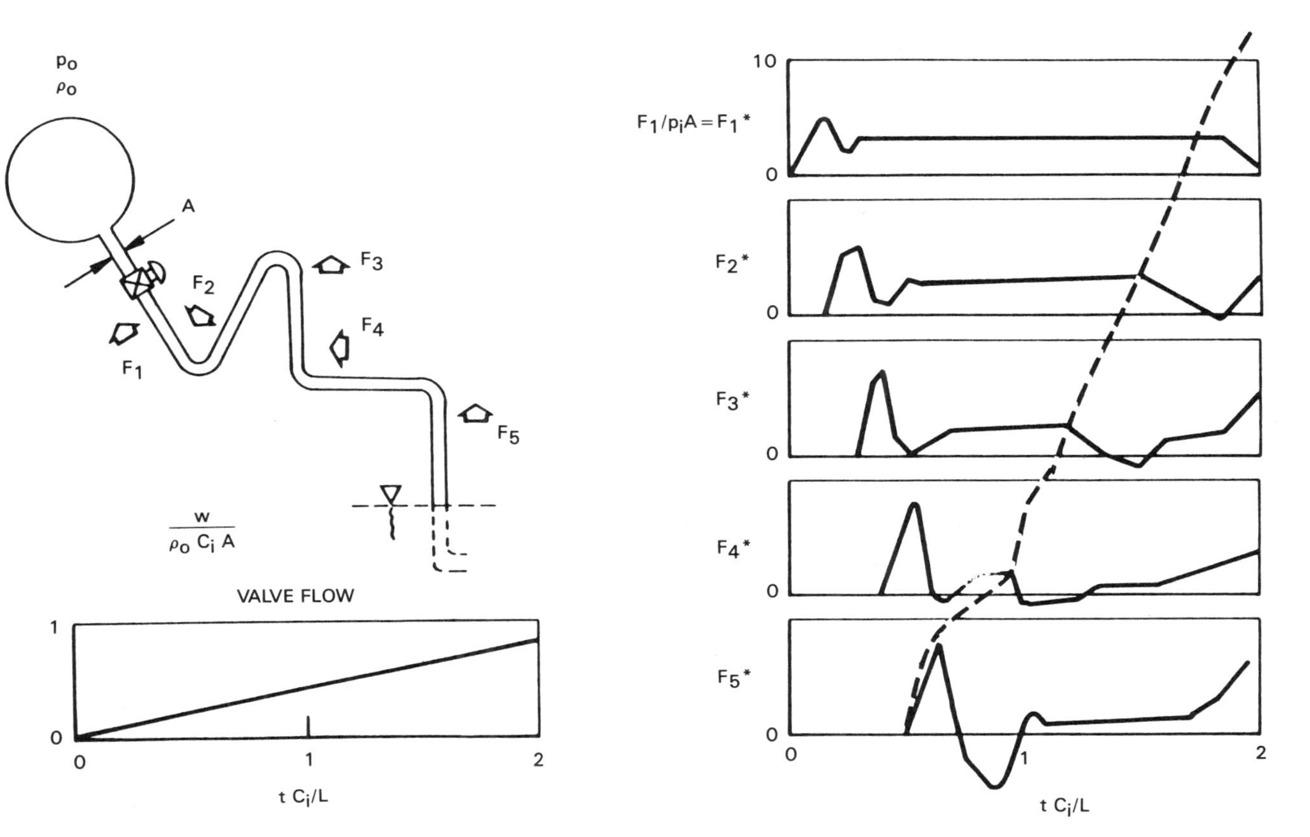

Fig. 10-31 Relief valve opens.

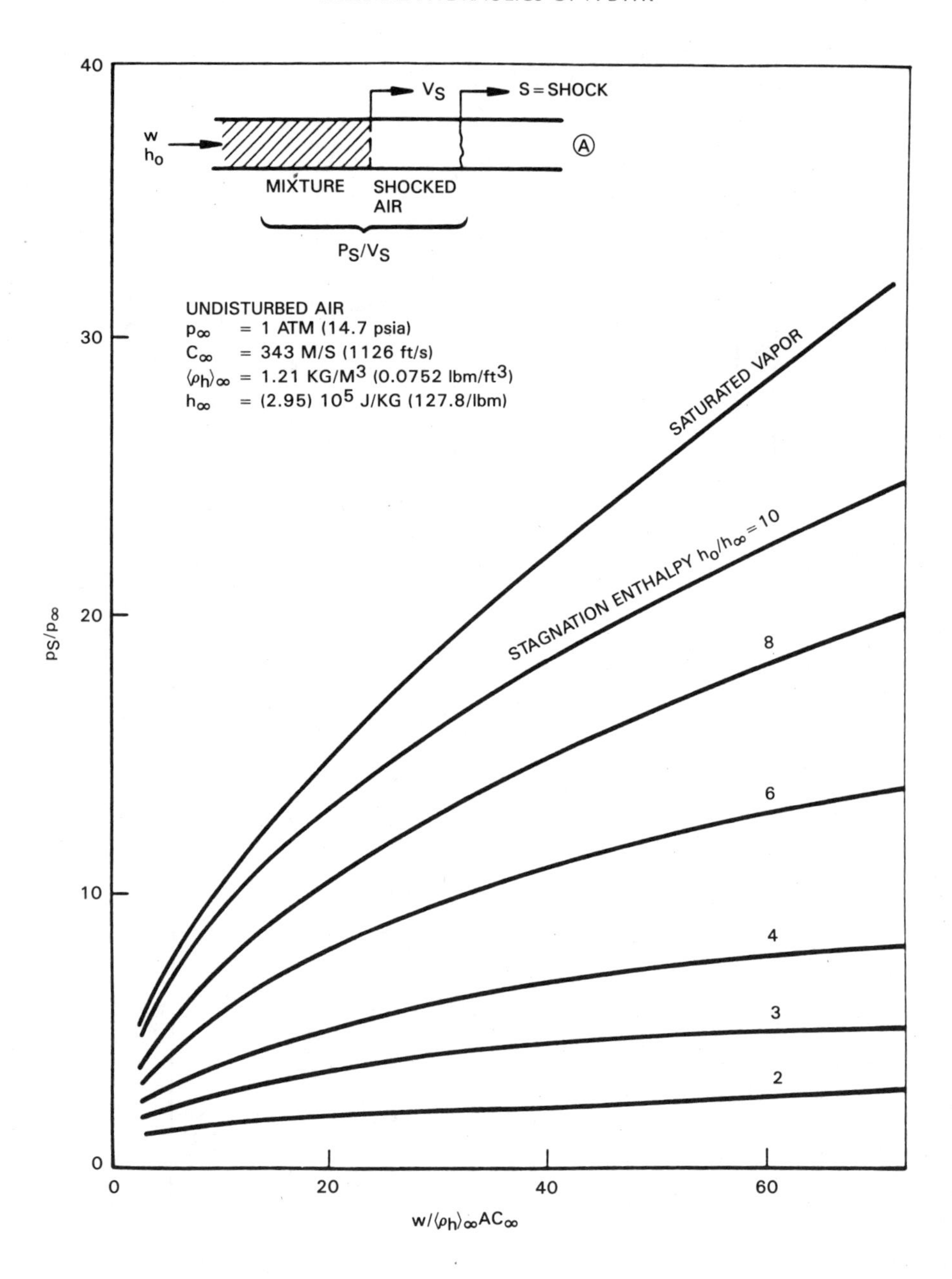

Fig. 10-32 Shock pressure in terms of valve flow rate and fluid enthalpy.

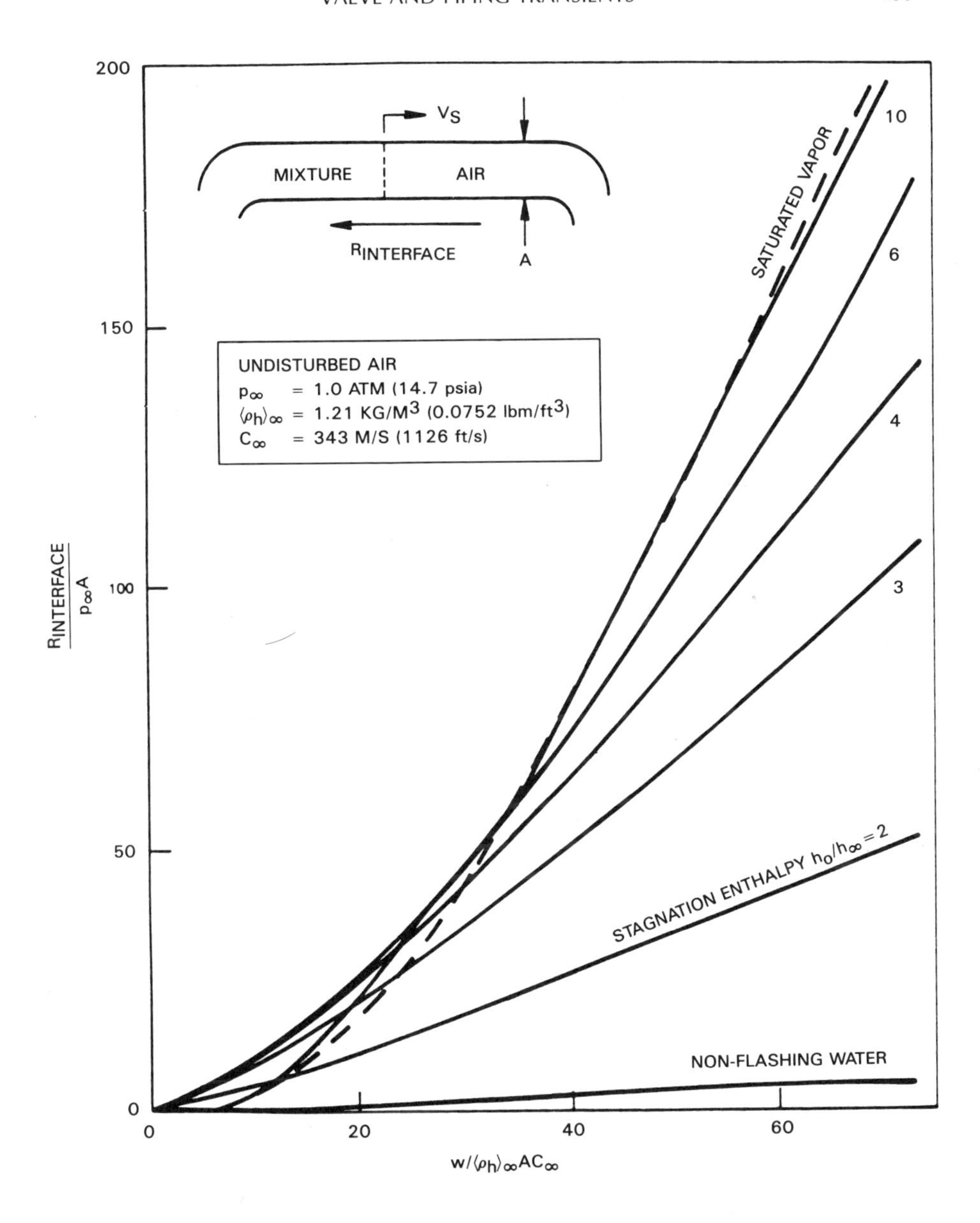

Fig. 10-33 Interface force in terms of valve flow rate and fluid enthalpy.

also increases with mixture enthalpy, and generally is higher than the shock force. For example, 6.8 MPa water, subcooled at 8.8×10^5 J/kg and discharged at a rate of 650 kg/sec into a relief line having 0.03 m^2 flow area would create a shock force of 37 kN and an air-mixture interface force of 220 kN.

Figures 10-29 and 10-33 give conservative force predictions. Pipe friction and other pressure losses in the shocked gas and mixture tend to reduce the shock and interface forces during propagation. This attenuation effect has been formulated in a characteristic solution presented by Hsiao et al. (1981). A typical SRV system is shown in Fig. 10-34. Example reaction force predictions are shown for each straight segment in Figs. 10-35, 10-36, and 10-37 for a 0.025-sec opening SRV discharging: subcooled, 9.9×10^5 J/kg

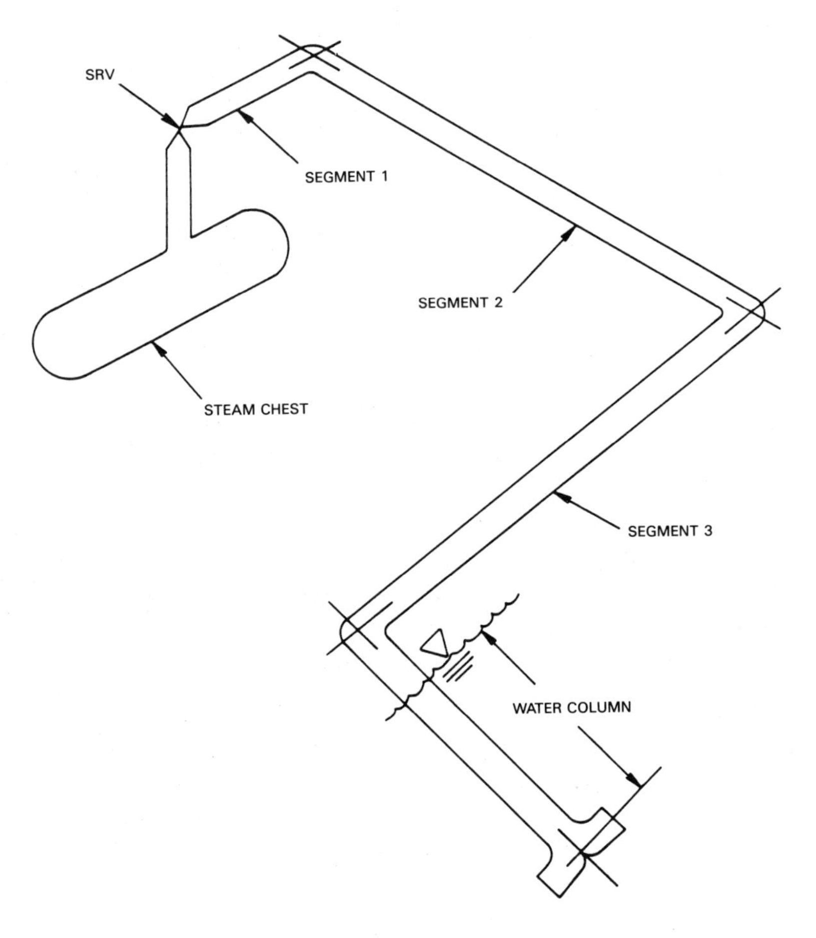

Fig. 10-34 Typical SRV system.

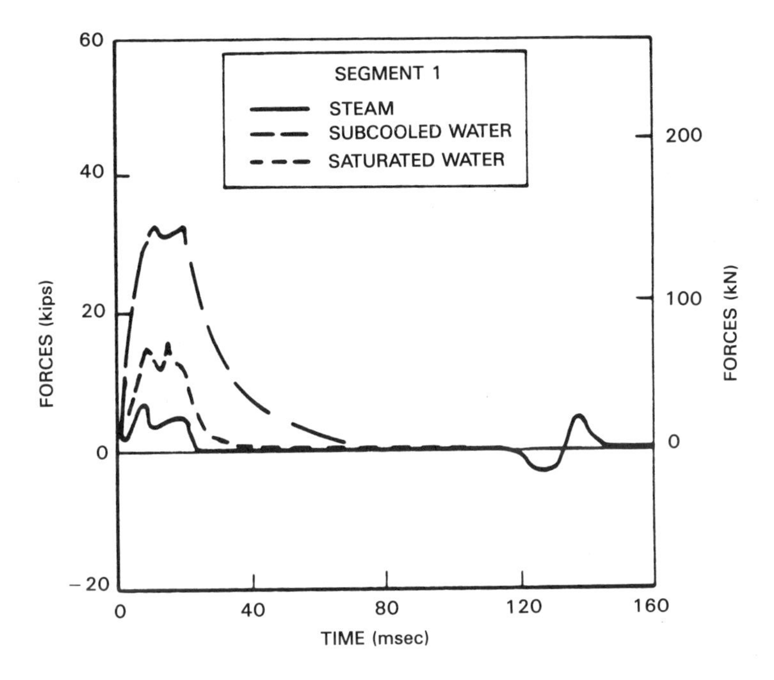

Fig. 10-35 Comparison of forces for segment 1.

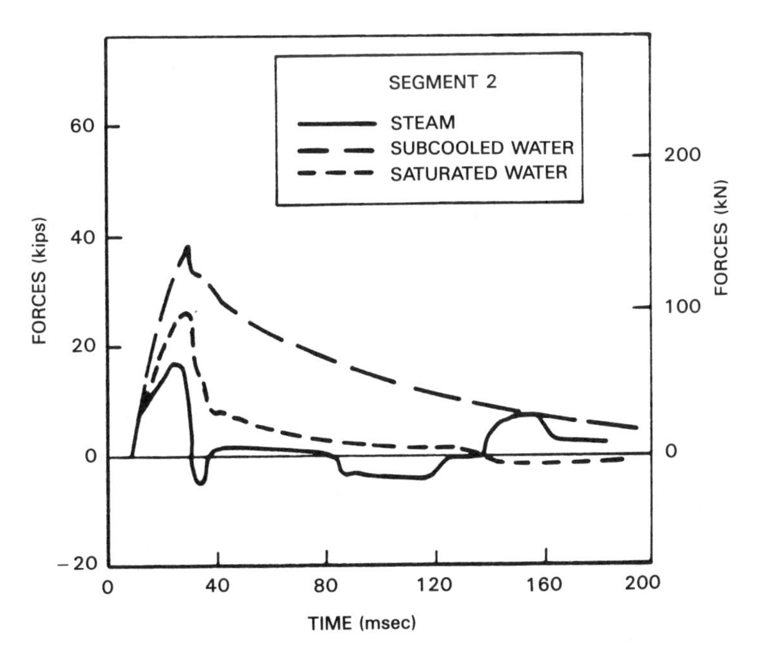

Fig. 10-36 Comparison of forces for segment 2.

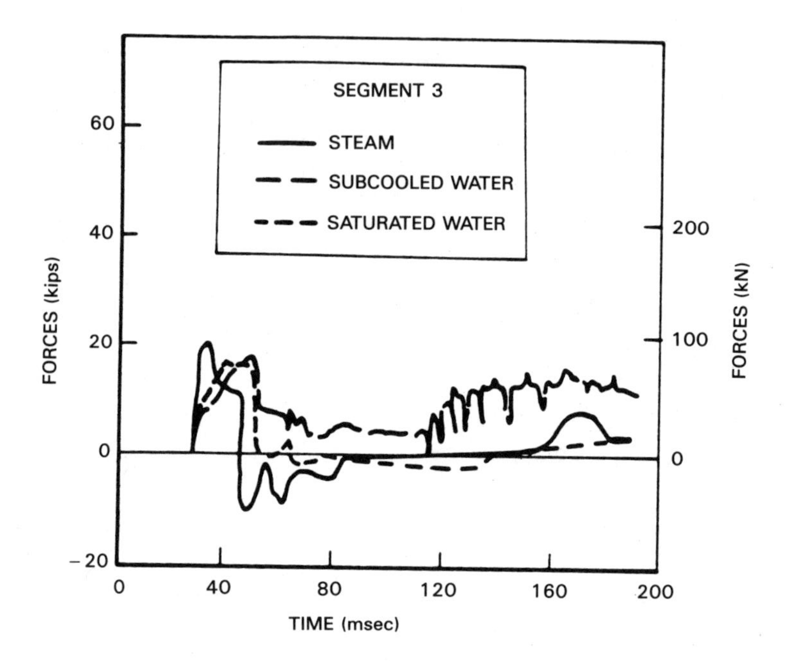

Fig. 10-37 Comparison of forces for segment 3.

water; saturated water; and saturated steam from a vessel at 7.5 MPa into a 0.046-m^2 pipe. Segments 1, 2, and 3 are 3.6, 12, and 12 m long, respectively. The relief line initially is submerged 4.8 m. It is seen that, although subcooled water gives larger forces than saturated water or steam on the first two segments, the resulting forces on segment 3 are much closer together due to higher pressure losses associated with the higher density mixture. In other words, farther away from the SRV flow dissipation reduces pipe forces resulting from saturated and subcooled water discharge to levels of the same magnitude as saturated steam discharge.

The effect of valve opening time on relief pipe force is shown in Fig. 10-38. It is seen that the resulting force is roughly proportional to the reciprocal valve opening time.

10.8 Stop Valve Closure

The pipe loading characteristic from valve closure is different from that associated with valve opening. Figure 10-39 shows the steam velocity reduction associated with a typical turbine stop valve (TSV) closure for which the valve area reduction is approximately linear with time. Steam flow is fully terminated in a time interval τ, which is about 100 msec for some stop

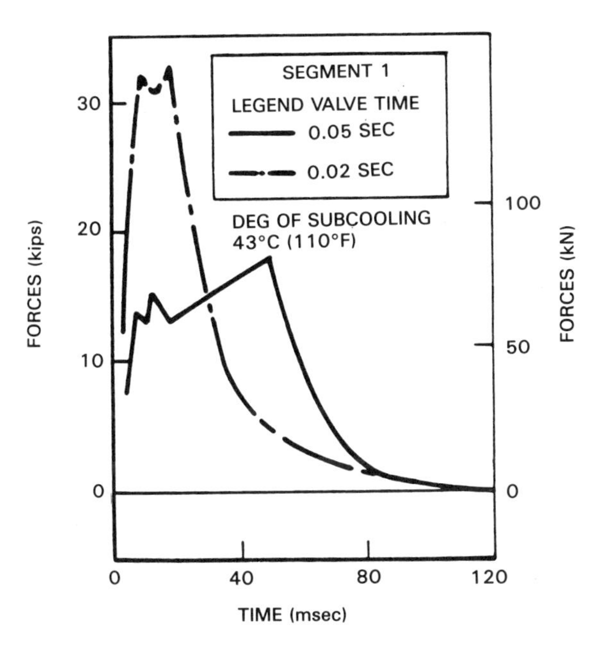

Fig. 10-38 Effect of valve opening time on segment 1 forces.

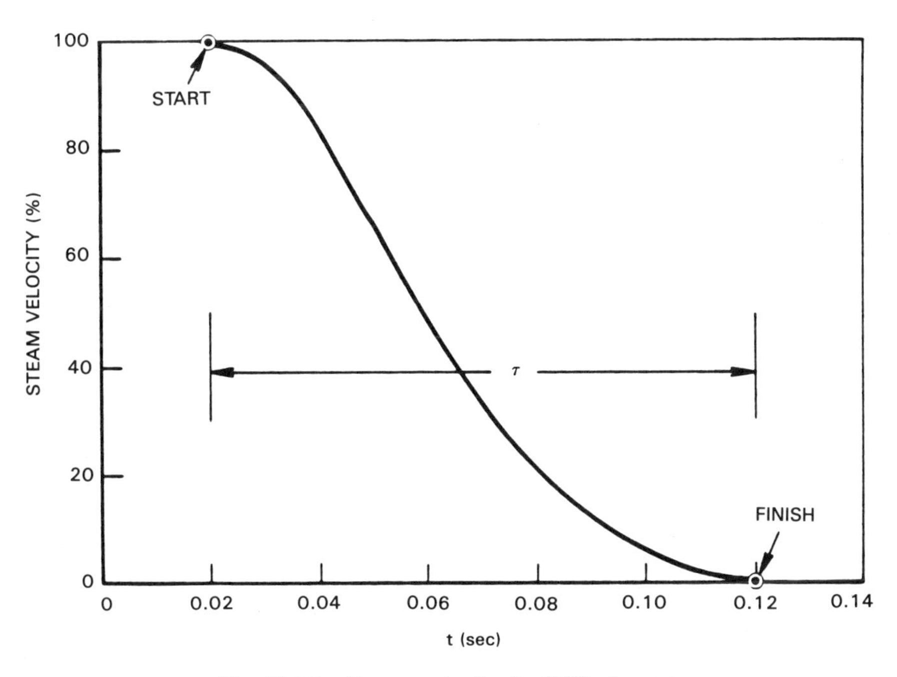

Fig. 10-39 Steam velocity for TSV closure.

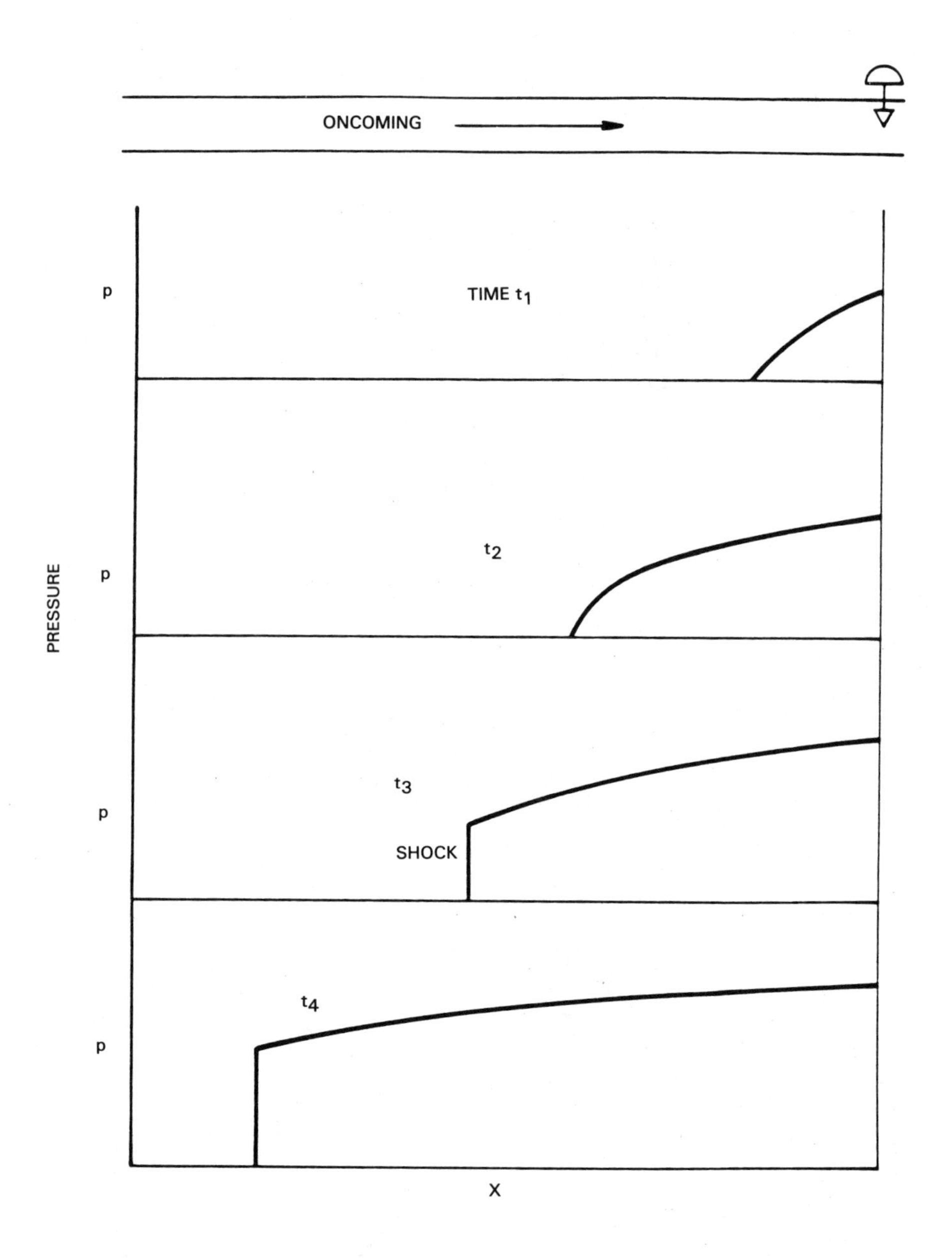

Fig. 10-40 Pressure profile for TSV closure.

valves. As the flow deceleration occurs, an increasing pressure disturbance is propagated into the oncoming steam, tending to steepen while it advances, as shown in Fig. 10-40. Steam pipe lengths usually are long in nuclear plants, making it possible for a shock to form in the pressure front advancing from a TSV closure.

Figure 10-41 shows a sample calculation of pipe section forces resulting from a TSV closure, based on the MOC method already discussed. When valve closure begins, an increased pressure wave moves toward pipe section 1, where it arrives at a normalized time of about 0.8 and exerts force F_1^* toward the valve, as indicated by the negative dip. As the compression wave moves into pipe section 2 at $t^*=1.0$, force F_1^* stops increasing. The same general force profile occurs in sections 2, 3, and 4, acting toward the TSV. The valve is fully closed in a normalized time of 0.8, and the corresponding full pressure achieved at the TSV moves out of section 1 at $t^*=1.8$, causing F_1^* to become zero. The same event occurs in each of the pipe sections in sequence. The compression front arrives at the vessel at $t^*=1.4$, and reflects as a decompression. This causes F_4^* to stop increasing. The decompression travels toward the TSV, moving into section 3 at $t^*=1.6$, causing F_4^* to become zero. The same effect is repeated in sections 3, 2, and 1. There is a delay of about $t^*=1.0$ as the decompression moves from section 1 to the TSV. When the decompression arrives at the TSV, a larger magnitude decompression is reflected. The larger decompression arrives at section 1 at about $t^*=2.8$, and with temporary higher pressure at the elbow of pipe section 1, force F_1^* becomes positive. Reaction forces in the other pipe sections successively rise to positive values. The process continues until propagation dissipates, at which time all net longitudinal forces become zero.

The closure of a valve can be sufficiently fast that a shock forms and propagates through the pipe. The condition for which an idealized instant valve closure analysis is appropriate in a pipe of length L is given by:

$$C_\infty \tau / L << 1.0 \; ,$$

where τ is the closure time. When a saturated bubbly mixture of liquid and vapor flows through a pipe in which a valve closes rapidly, the resulting sudden compression drives the vapor into a superheated state and the liquid into a subcooled state. An approximate fluid state is therefore described by an idealized mixture of incompressible liquid and perfect gas bubbles. Letting $\langle x \rangle$ be the gas mass fraction, Fig. 10-42 gives the pipe force resulting from sudden valve closure on a bubbly mixture flowing at velocity $\langle j \rangle_x$ for several values of the parameter $(1-\langle x \rangle)\rho_g/\langle x \rangle \rho_f$ (Moody, 1973). Also shown are pipe forces for liquid and gas. Shock pressures recorded in a bubbly mixture are compared with predictions in Fig. 10-43. The agreement is reasonable, although predictions of the pressure tend to favor the data upper bound.

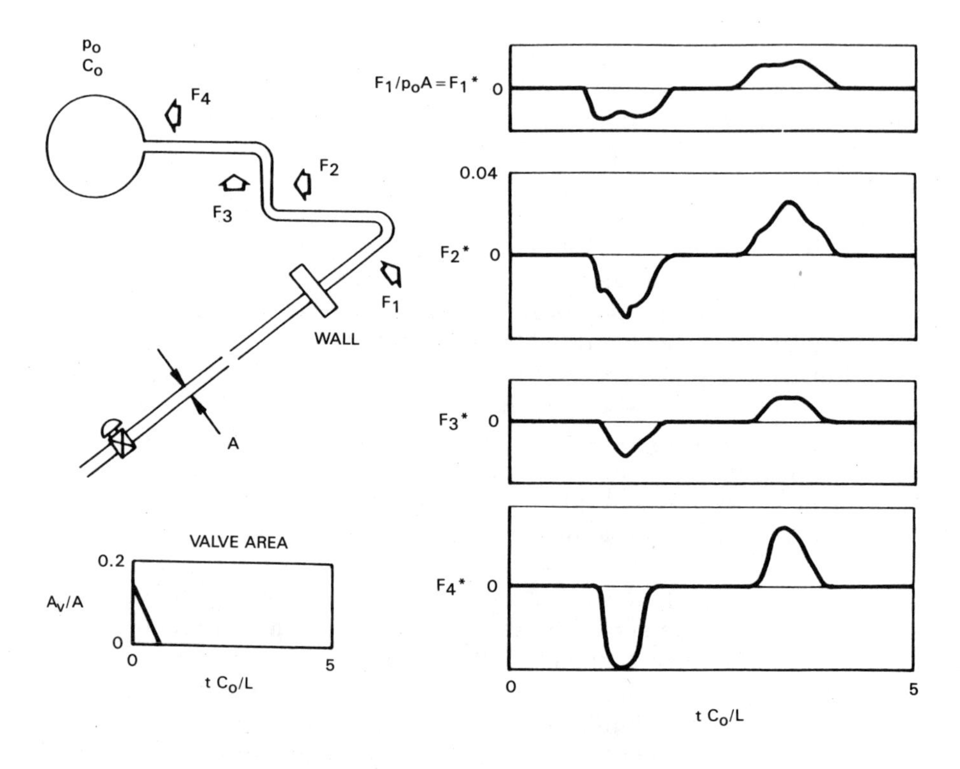

Fig. 10-41 Stop valve closes.

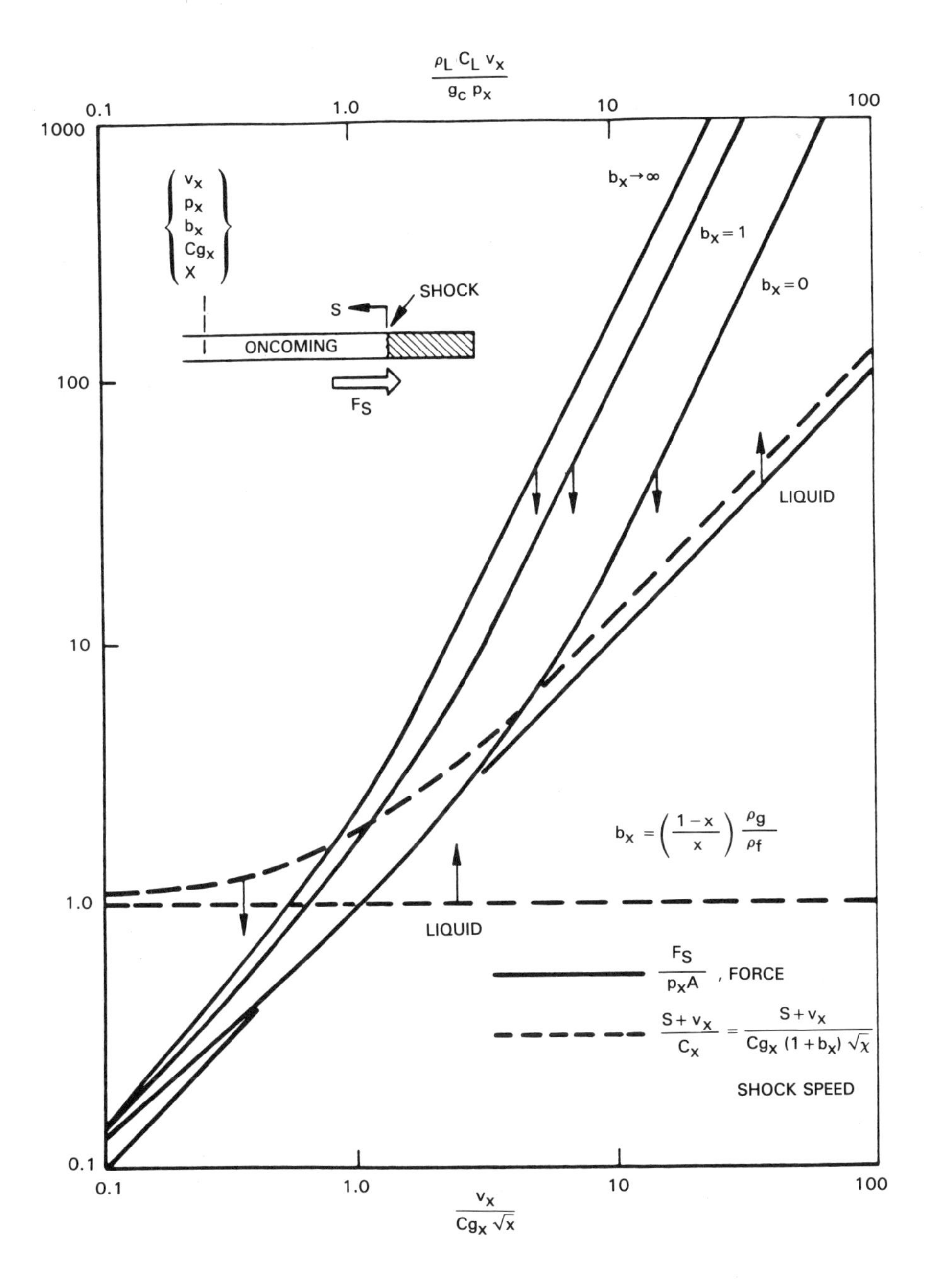

Fig. 10-42 Flow stoppage force and shock speed.

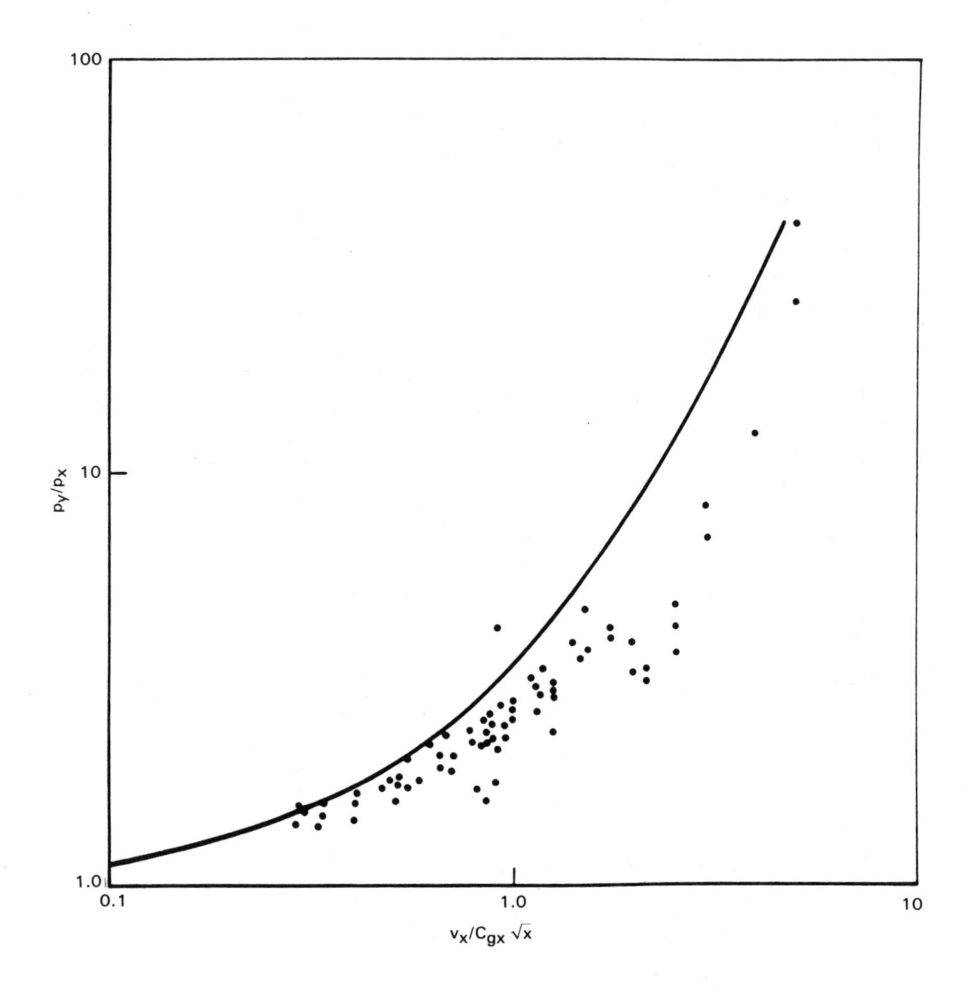

Fig. 10-43 Flow stoppage pressure comparison with data.

References

Anderson, J. D., *Modern Compressible Flow*, McGraw-Hill, New York (1982).

ANSI/ANS-58.2, "Design Basis for Protection of Light Water Nuclear Power Plants against Effects of Postulated Pipe Rupture" (1980).

ASCE, *Structural Design of Nuclear Plant Facilities*, Vols. I, II, III (1973).

ASCE, International Seminar on Probabilistic and Extreme Load Design of Nuclear Plant Facilities (1979).

Dietz, K. A., Ed., "Quarterly Technical Report, Engineering and Test Branch, October 1–December 31, 1967," IDO-17242, Phillips Petroleum Company (1968).

Edwards, A. R., and T. P. Obrien, "Studies of Phenomena Connected with the Depressurization of Water Reactors," *J. Brit. Nucl. Energy Soc.*, **9** (1970).

Fabic, S., "Computer Program WHAM for Calculations of Pressure, Velocity, and Force Transients in Liquid Filled Piping Networks," 67-49-R, Kaiser Engineering (1967).

Fabic, S., "Blodwn-2: Westinghouse APD Computer Program for Calculation of Fluid Pressure, Flow, and Resulting Transients during a Loss of Flow Accident," *Trans. Am. Nucl. Soc.*, **12,** *1,* 358 (1969).

Faletti, D. W., and R. W. Moulton, "Two-Phase Critical Flow of Steam-Water Mixtures," *AIChE J.*, **9,** 2 (1963).

Gallagher, E. V., "Water Decompression Experiments and Analysis for Blowdown of Nuclear Reactor," TID-U500 (1970).

Hanson, G. H., "Subcooled Blowdown Forces on Reactor System Components," IN-1354, Idaho Nuclear (1970).

Haupt, R. W., and R. A. Meyer, Eds., "STEALTH: A LaGrange Explicit Finite Difference Code for Solids, Structural, and Thermo-hydraulic Analysis," Vols. 1–4, EPRI-260, Project 307, August 1976, *Safety Relief Valves,* ASME SP PVP-33 (1979).

Henry, R. E., and H. K. Fauske, "The Two-Phase Critical Flow of One-Component Mixtures in Nozzles, Orifices, and Short Tubes," *Trans. ASME, J. Heat Transfer* (1971).

Hsiao, W. T., P. Valandani, and F. J. Moody, "A Method to Determine Forces Developed during a Time-Dependent Opening of a Relief Valve Discharging a Two-Phase Mixture," ASME Paper No. 81-WA/NE-15 (1981).

Jonsson, V. K., L. Matthews, and D. B. Spalding, "Numerical Solution Procedures for Calculating the Unsteady, One-Dimensional Flow of Compressible Fluid," ASME Paper No. 73-FE-30 (1973).

Karplus, H. B., "Propagation of Pressure Waves in a Mixture of Water and Steam," ARF 4133-12, Armour Research Foundation (1961).

Landau, L. D., and E. M. Lifshitz, *Fluid Mechanics,* Addison-Wesley, Reading, Massachusetts (1959).

McCready, J., et al., "Steam Vent Clearing Phenomena and Structural Response of the BWR Torus (Mark I Containment)," NEDO-10859, General Electric (1973).

Moody, F. J., "Prediction of Blowdown Thrust and Jet Forces," ASME Paper No. 69-HT-31 (1969).

Moody, F. J., "Time-Dependent Pipe Forces Caused by Blowdown and Flow Stoppage," *J. Fluids Eng.* **95,** Ser. 1, No. 3 (1973).

Moody, F. J., "Maximum Discharge Rate of Liquid/Vapor Mixtures from Vessels," in *Non-Equilibrium Two-Phase Flows,* R. T. Lahey and G. B. Wallis, Eds., ASME (1975).

Moody, F. J., "Unsteady Piping Forces Caused by Hot Water Discharge from Suddenly Opened Safety/Relief Valves," *Nucl. Eng. Des.*, **72,** 213–224 (1982).

Moody, F. J., *Introduction to Unsteady Thermofluid Mechanics,* Wiley Interscience, New York (1989).

Moody, F. J., A. J. Wheeler, and M. Ward, "The Role of Various Parameters on Safety Relief Valve Pipe Forces," *Safety Relief Valves,* ASME SP PVP-33 (1979).

Semprucci, L. B., and B. P. Holbrook, "The Application of RELAP 4/REPIPE to Determine Force-Time Histories on Relief Valve Discharge Piping," *Safety Relief Valves,* ASME SP PVP-33 (1979).

Shapiro, A. H., *The Dynamics and Thermodynamics of Compressible Fluid Flow,* The Ronald Press (1953).

Von Neumann, J., and R. D. Richtmyer, "A Method for the Numerical Calculation of Hydrodynamic Shocks," *J. Appl. Phys.,* **21** (1950).

Wheeler, A. J., and F. J. Moody, "A Method for Computing Transient Pressures and Forces in Safety Relief Valve Discharge Lines," *Safety Relief Valves,* ASME SP PVP-33 (1979).

CHAPTER ELEVEN

Pressure Suppression Containment Systems

11.1 General Pressure Suppression System

The major function of a containment system is to protect the environment from an uncontrolled release of radioactive materials in the event that a loss-of-coolant accident (LOCA) should occur. This objective is achieved by designing containment systems to accommodate all combinations of loads generated by the mass and energy releases associated with reactor blowdown from a postulated pipe rupture or safety/relief valve (SRV) discharge. Significant features of pressure suppression containment are passive steam condensation in the pool, lower maximum pressures and rapid pressure reduction, higher capacity for blowdown energy discharge, and insensitivity to the initiation time of heat exchangers for energy removal. The thermal-hydraulic phenomena and associated loads imposed on a pressure suppression containment system are discussed in this chapter.

Major subsystems of the pressure suppression containment concept introduced in Chapter 1 are shown schematically in Fig. 1-7. These include the drywell, vent, water pool, and wetwell air space. Reactor vessel blowdown from a pipe rupture raises the drywell pressure and purges air, or for inerted containments, nitrogen gas, through the vent and pool, where it rises to the wetwell air space in a compressed state. The large volume of steam associated with the reactor pressure vessel blowdown is vented to the pool where it is condensed to a small volume of water. Drywell and wetwell pressures tend to equalize as the vessel blowdown rate diminishes. Emergency core cooling sprays (ECCS) produce additional steam at first, but later can result in drywell pressure reduction when subcooled water flows from the pipe rupture and provides a cool condensing surface for steam. A decreasing drywell pressure causes air backflow through the vacuum breakers and decreases the wetwell pressure.

Pressure suppression systems are designed for a variety of accidents, the most severe being the so-called design-basis accident (DBA), which is a postulated instantaneous circumferential rupture and double-ended blowdown of reactor coolant from that pipe, which causes the highest containment loads. The DBA usually is a main steam line or recirculation line rupture.

The drywell and wetwell volumes, vent flow area, vessel blowdown discharge properties, and vent submergence are important physical parameters, which permit design control of the overall system response. Typical drywell and wetwell pressure-time responses to a DBA are shown in Fig. 11-1. The effect of the pipe break area to vent area ratio on the maximum drywell pressure is given by a correlation of the experimental data shown in Fig. 11-2.

The objective of this chapter is to discuss the state-of-the-art of pressure suppression containment technology. Containment phenomena are summarized in terms of our current understanding, analytical procedures, and simplified analyses, which help to display the qualitative effects of design parameters on containment loads.

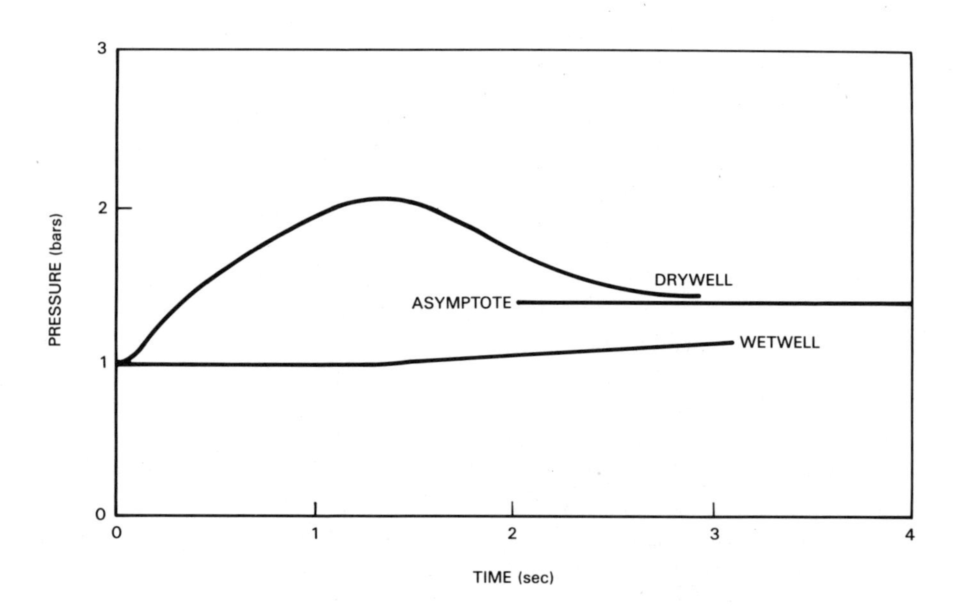

Fig. 11-1 Typical drywell and wetwell pressure response for a DBA.

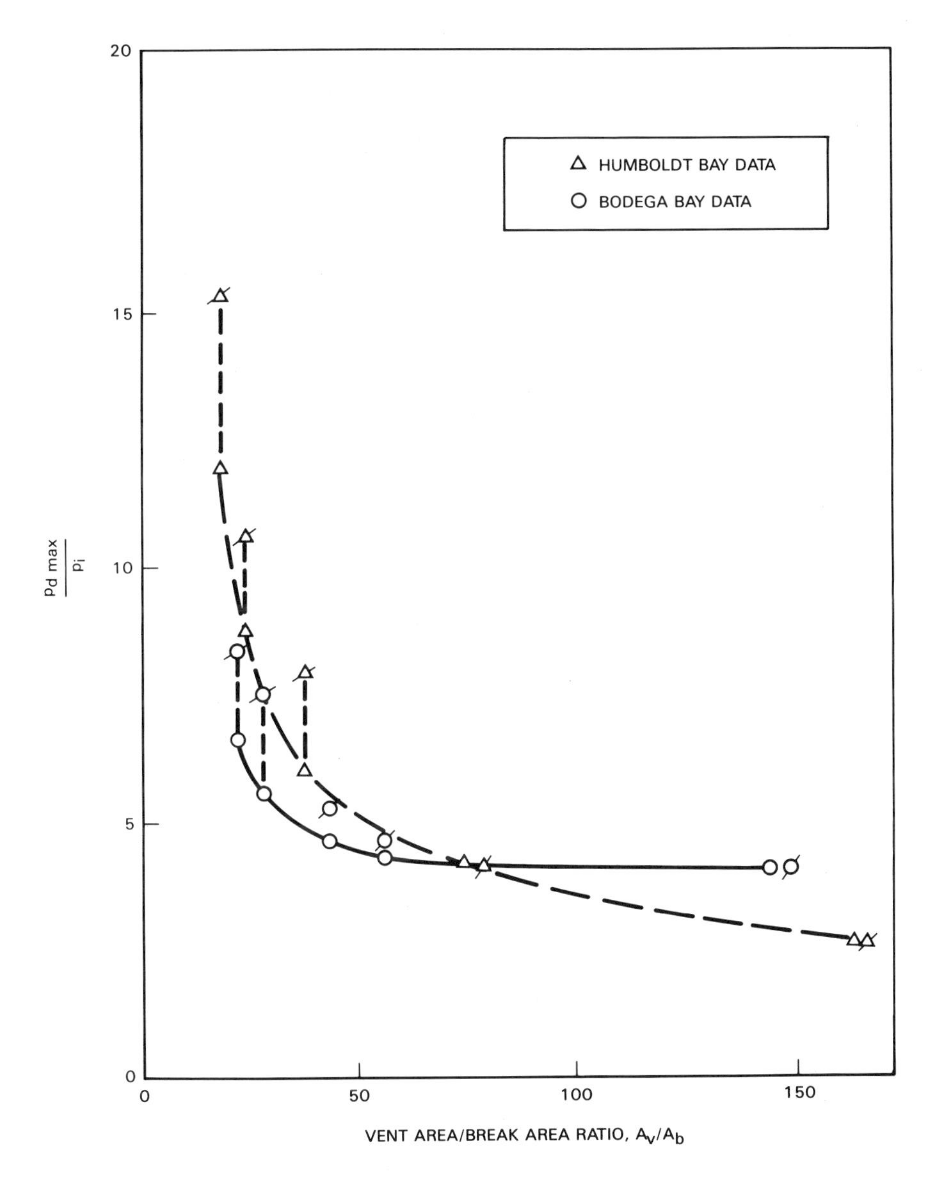

Fig. 11-2 Maximum drywell pressure correlation (PG&E, 1962; Robbins, 1960).

11.1.1 Models of the Individual Subsystems

Basic thermal-hydraulic models of the reactor pressure vessel, drywell, vents, pool, wetwell, and vacuum breakers are discussed in this section. Simplifications also are given to help show the dependence of various containment loads on important parameters.

11.1.1.1 *Reactor Pressure Vessel*

Reactor pressure vessel (RPV) blowdown is discussed in Sec. 9.2.2, where equations and graphs are given for the time-dependent predictions of RPV thermal-hydraulic properties.

11.1.1.2 *Drywell*

The drywell receives coolant discharge from the reactor during a LOCA. Prior to vent clearing, the drywell responds approximately as a closed system, with unrelieved pressure rise. Once the vent water is cleared, drywell venting begins and the pressure rate immediately decreases. The drywell volume, blowdown discharge flow area, and vent flow area determine if the drywell pressure begins to decrease from the maximum nonvented value when venting begins or if it continues to rise at a slower rate to a maximum value.

If the drywell air and reactor steam discharged remain unmixed, Eqs. (9.103), (9.104), and (9.107) yield the initial nonvented drywell pressure rate,

$$\frac{dp}{dt}=\frac{kp_d v_{fg}(p_d)}{h_{fg}(p_d)}[h_0-f(p_d)]\frac{G_c(p,h_0)A_b}{V_d}\ , \tag{11.1}$$

which is inversely proportional to drywell volume, V_d, and proportional to the blowdown mass flow rate.

The initial drywell pressurization rate is important because it plays a major role in the pressure load imposed. Figure 11-3 gives the initial pressurization rate for an initial containment pressure of 101 kPa (1 atm) and a RPV at 1000 psia. It is seen that the pressurization rate has a maximum at some value of blowdown enthalpy in the subcooled region. The pressurization rates are comparable for RPV blowdowns of saturated steam, water, or two-phase mixtures.

When drywell venting begins, the pressurization rate decreases and will tend toward a level where the pressurization rate due to vessel blowdown and the decompression rate due to vent discharge balance to give the quasi-asymptotic condition.

$$\frac{dp}{dt}=0\ . \tag{11.2}$$

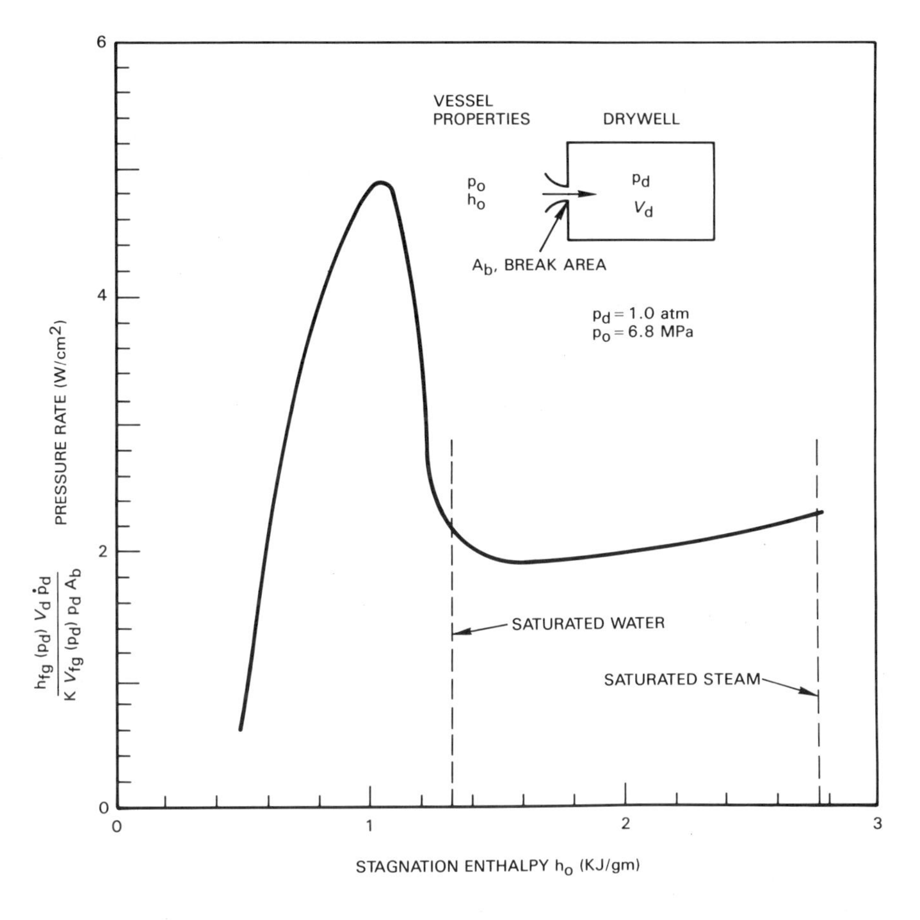

Fig. 11-3 Drywell pressurization rate.

If it is assumed that most of the drywell air has been purged to the wetwell when the condition of Eq. (11.2) is reached, the corresponding drywell pressure (p_d) can be obtained from a solution of Eq. (9.103), written as,

$$\frac{A_v}{A_b} = \frac{G_c(p_{0,\text{in}}, h_{0,\text{in}})}{G_c(p_d, h_{0d})} \frac{[h_{0,\text{in}} - f(p_d)]}{[h_{0d} - f(p_d)]} \tag{11.3}$$

for cases in which the value of p_d is sufficiently large to give critical vent discharge flow. For a vent flow area of A_v and a break area of A_b, Eq. (11.3) was used to obtain Figs. 11-4 and 11-5 for predicting the quasi-asymptotic pressure in a typical drywell after venting begins. Figure 11-4 is based on saturated water blowdown from 1000 psia and either saturated steam or water discharge from the vent. Normally saturated steam will flow into

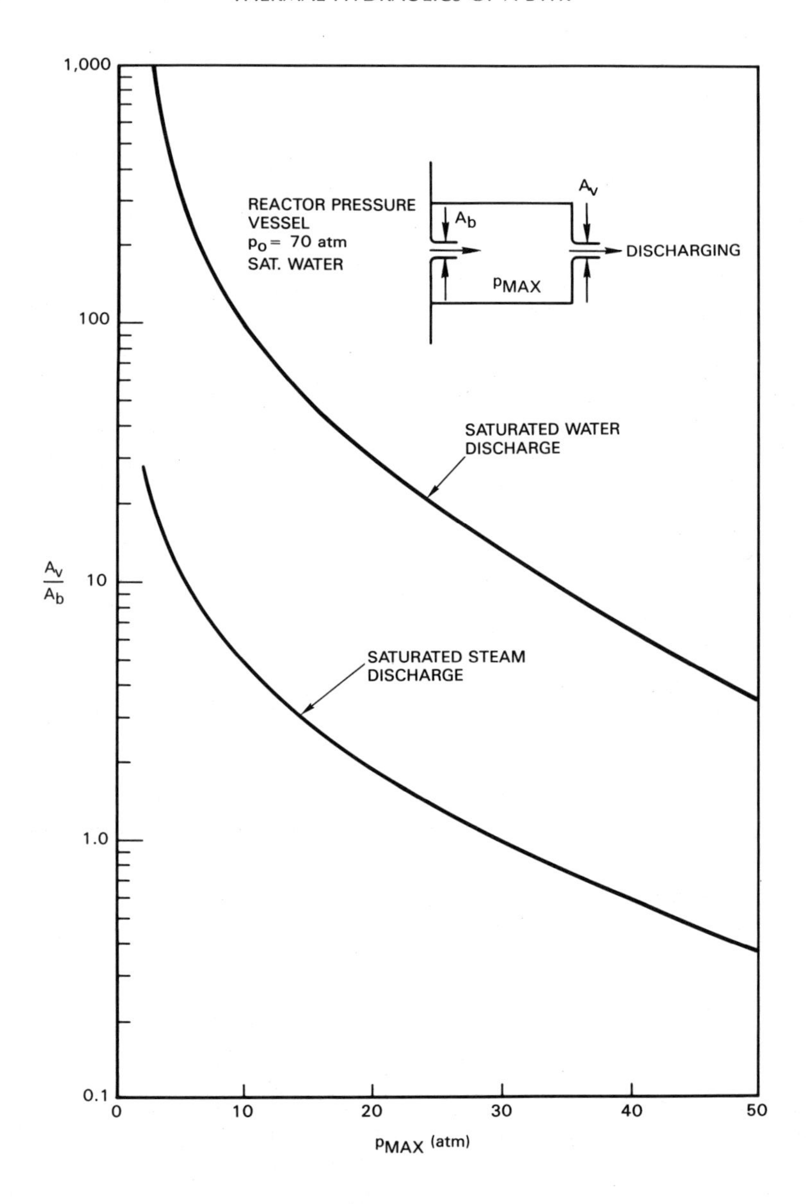

Fig. 11-4 Quasi-asymptotic drywell pressure; saturated water blowdown, steam or water venting.

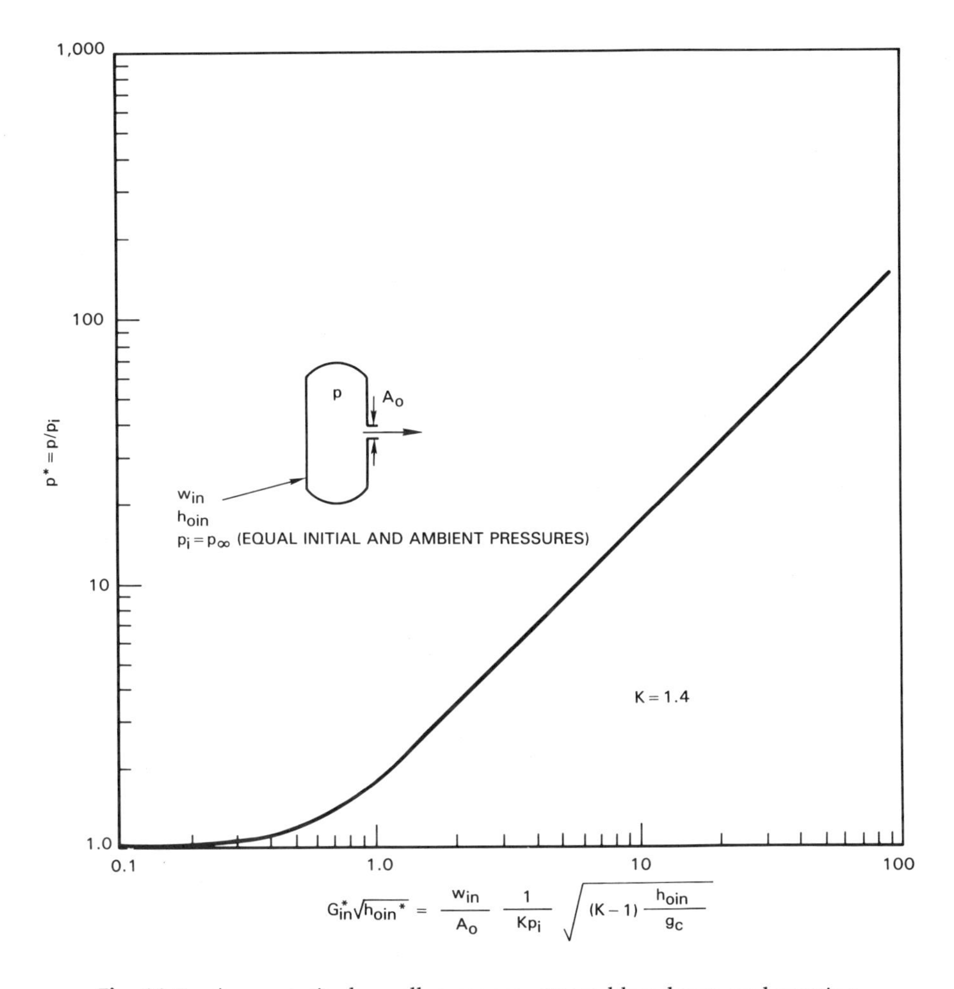

Fig. 11-5 Asymptotic drywell pressure, steam blowdown and venting.

the vents with some carryover of water droplets. Figure 11-5 is based on steam blowdown from the vessel and steam flow in the vents.

11.1.1.3 *Vents*

The vent shown in Fig. 1-8 routes steam and air from the drywell to the pool. Various geometrical vent designs in Figs. 1-9a, b, c, and d include large radial pipes connected to a ring header with downcomers submerged in the pool (MARK-I design); vertical downcomers in the drywell floor, which extend downward into the pool below (MARK-II design); and a weir-vent arrangement with three levels of horizontal vents submerged in the pool (MARK-III design).

Pressure rise in the drywell causes expulsion of the vent's water column, followed by air or nitrogen and steam discharge to the pool. The water expulsion and steam-air discharge are summarized next.

The dynamic equation that governs liquid column expulsion from the submerged vent shown in Fig. 11-6 is given by,

$$(L-y)\frac{d^2y}{dt^2}=\frac{g_c}{\rho_l}[p(t)-p_\infty] \ , \tag{11.4}$$

where friction and gravity are neglected. A more exact treatment of these effects has been given by Khalid and Lahey (1980). Pressure, $p(t)$, is applied to the column surface. A SRV discharge usually forms a shock in the relief pipe, which imposes a step pressure on a submerged water column, whereas a LOCA imposes an almost linearly increasing drywell pressure on a vent water column. Figure 11-6 gives calculated water column transients for step and ramp driving pressures.

A similar water expulsion model for horizontal, parallel vents at different elevations has been formulated and verified experimentally (Bilanin, 1974). Most vent expulsion predictions include a virtual length, equal to about one vent radius, to account for inertia of the surrounding water. Taylor instability of the accelerating vent water surface does not materially affect pressurization loads, and is thus neglected. When the vent water clears, the drywell and vent pipe pressurization rate decreases.

Drywell air, steam-water, or a steam-air mixture begins to discharge from the vents immediately after expulsion of the water column. It is expected that any liquid water will be in the form of small droplets rather than other flow patterns, which require more extensive two-phase flow analysis. Droplet flow is closely approximated by classical methods of compressible flow (Shapiro, 1953), for both the liquid and gas flow rates, w_l and w_g. First, if only gas flow occurs in a vent pipe whose entrance and exit are designated 1 and 2, respectively, the Mach numbers, pressures, and friction parameter are related by:

$$\frac{fl}{D}=\frac{1}{k}\left(\frac{M_2^2-M_1^2}{M_1^2M_2^2}\right)+\left[1+\frac{(k-1)}{2k}\right]\log\left[\frac{\left(1+\frac{k-1}{2}M_2^2\right)M_1^2}{\left(1+\frac{k-1}{2}M_1^2\right)M_2^2}\right] \tag{11.5}$$

and,

$$\frac{p_2}{p_1}=\frac{M_1}{M_2}\sqrt{\frac{1+\frac{k-1}{2}M_1^2}{1+\frac{k-1}{2}M_2^2}} \ . \tag{11.6}$$

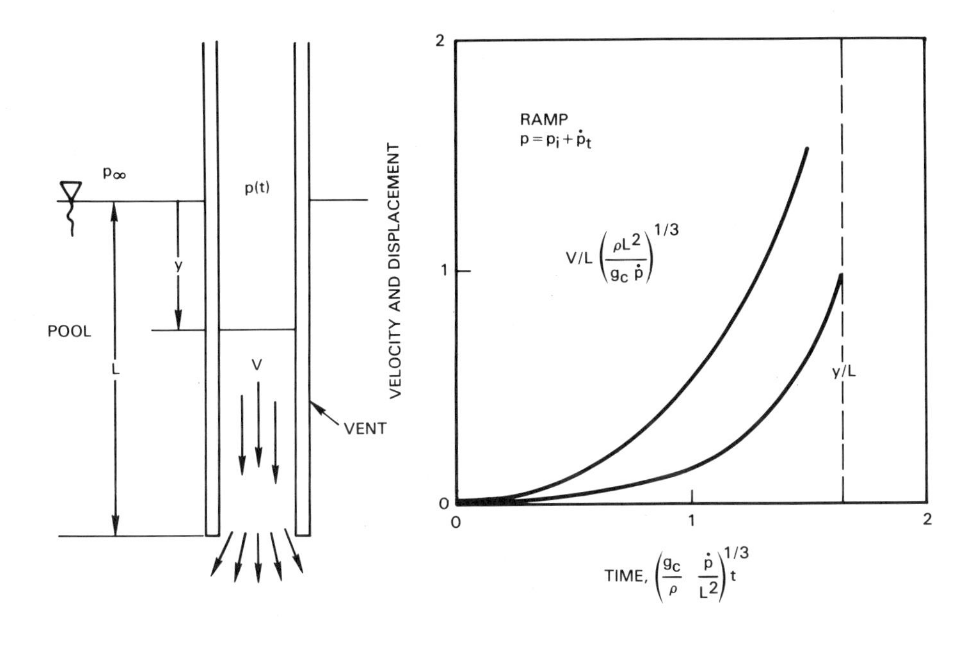

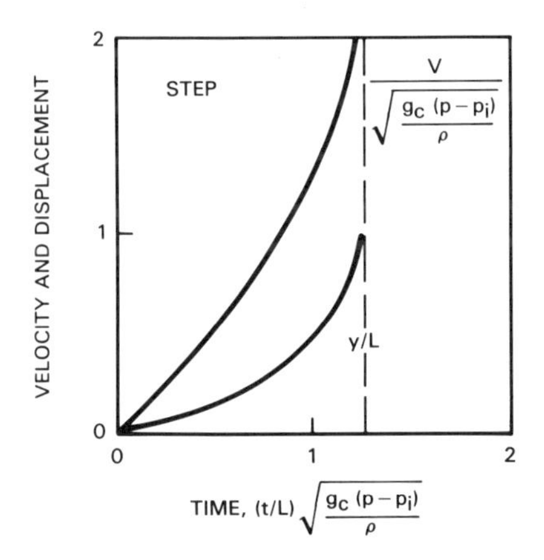

Fig. 11-6 Vent water expulsion, ramp and step pressure.

Also, drywell and vent entrance properties are approximated by an isentropic process,

$$\frac{p_d}{p_1} = \left(1 + \frac{k-1}{2} M_1^2\right)^{k/(k-1)} . \tag{11.7}$$

Equations (11.5), (11.6), and (11.7) were employed to obtain Fig. 11-7, which gives the entrance Mach number in terms of the vent flow resistance and

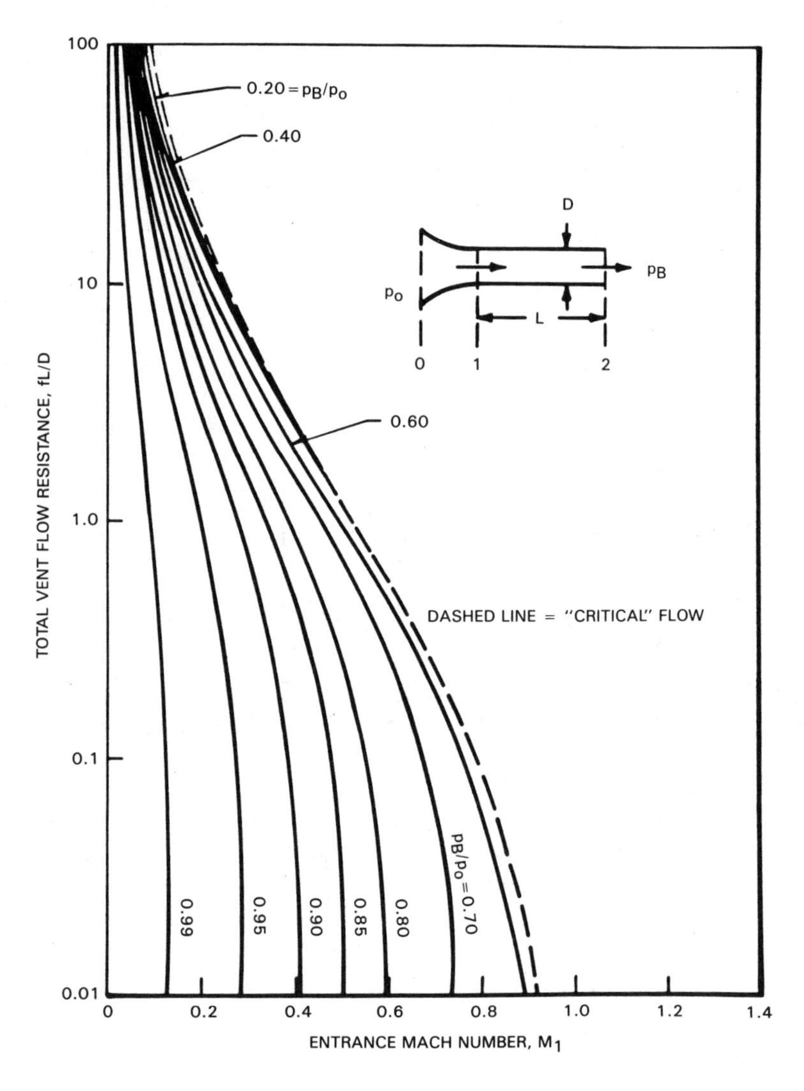

Fig. 11-7 Vent entrance Mach number.

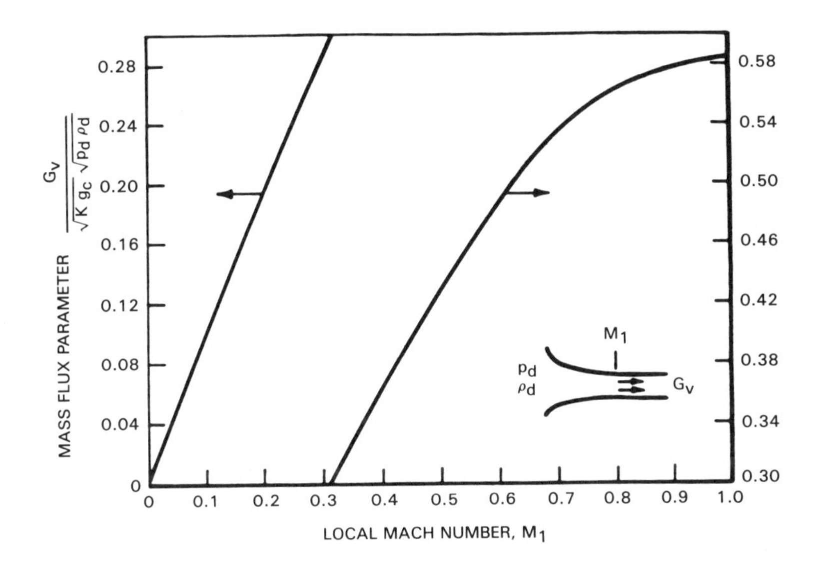

Fig. 11-8 Vent mass flux.

drywell pressure for gas flow. Once M_1 is determined, the vent mass flux G_v for gas flow is obtained from,

$$G_v = \sqrt{k g_c p_d \rho_d}\; M_1 \left(\frac{p_1}{p_d}\right)^{(k+1)/2k} \quad . \tag{11.8}$$

The results of Eq. (11.8) are graphed in Fig. 11-8 for a steam-air mixture.

Equations (11.5), (11.6), and (11.7) can be employed to closely approximate results for a two-phase water droplet-gas mixture of constant quality $\langle x \rangle$, if the quantity $(k-1)$ is replaced by $(k-1)/\langle x \rangle$ whenever it appears. Moreover, when M_1 and p_1 are determined, the mixture mass flux is obtained in terms of G_v in Eq. (11.8) as:

$$G_{\text{mix}} = G_v \left[\langle x \rangle + (1 - \langle x \rangle) \frac{\rho_{gd}}{\rho_l} \left(\frac{p_1}{p_d}\right)^{\frac{1}{k}} \right]^{-1} \quad . \tag{11.9}$$

11.1.1.4 *Pool*

The pool of a pressure suppression containment is designed to provide a large heat sink for absorbing the energy discharged from a reactor vessel during a LOCA. When high-pressure reactor coolant enters the low-pressure drywell, large quantities of saturated steam are generated by flashing, and must be condensed by vent discharge to the pool.

Mass and energy principles for the pool with steam entering at enthalpy, h_g, yield the mass of steam, M_g, which can be condensed in a pool of initial mass, M_{pi}, as:

$$\frac{M_g}{M_{pi}} = \frac{h_f(p_w) - h_{Li}}{h_g(p_d) - h_{Li}} , \tag{11.10}$$

where h_{Li} is the pool's initial enthalpy, $h_f(p_w)$ is saturated water enthalpy at wetwell pressure, and $h_g(p_d)$ is enthalpy of the steam entering from the drywell. For example, a pool at 100°F temperature and between 1 and 2 atm pressure yields,

$$\frac{M_g}{M_{pi}} = 0.1 \ .$$

That is, about 0.1 kg of steam can be condensed in 1.0 kg of water without raising its temperature to the saturated water value. Suppression pools are designed to have ample heat capacity for steam condensation from not only a LOCA, but also continued steam formation from core decay heat.

Reactor heat removal (RHR) heat exchangers in the pool are designed to provide long-term cooling. However, if the RHR systems are unavailable, the pool's heat capacity can absorb LOCA energy or energy release from a stuck-open relief valve (SORV) for about 6 h.

11.1.1.5 *Wetwell*

When air is purged from the drywell during a LOCA, it discharges from the vents, rises through the pool, and enters the wetwell air space. The initial wetwell air mass, M_{awi}, occupies a volume, V_{aw}, at temperature, T_{wi}, and pressure, p_{wi}. If a mass of air, M_{da}, from the drywell enters the wetwell air space at temperature T_∞, a thermodynamic analysis, based on perfect gas equations of state for air, yields the resulting pressure and temperature as,

$$\frac{p_w}{p_{wi}} = 1 + k\frac{T_\infty}{T_i}\frac{M_{da}}{M_{awi}} + \frac{p_g(T_{wa})}{p_{wi}} \tag{11.11}$$

and,

$$\frac{T_{wa}}{T_\infty} = \frac{M_{awi}}{(M_{awi} + M_{da})}\frac{T_{wi}}{T_\infty} + k\frac{M_{da}}{(M_{awi} + M_{da})} , \tag{11.12}$$

where the temperatures are absolute, and the last term of Eq. (11.11) corresponds to the vapor pressure in the wetwell. Consider a wetwell at 1.0 atm and 20°C (293 K). If, for example, the drywell and wetwell air masses were equal, the initial wetwell air and pool temperatures the same, and drywell air entered the wetwell air space at pool temperature so that $T_{wi} = T_\infty$, the pressure and temperature would rise to 2.4 atm and 80°C.

11.1.1.6 *Vacuum Breakers*

Vacuum breakers are employed to permit backflow of water from the wetwell air space to the drywell if the drywell pressure is reduced by steam condensation or other mechanisms. The flow rate through most vacuum breakers can be obtained from incompressible flow theory, based on an orifice-like pressure loss. Thus, if the difference of wetwell and drywell pressures, $p_w - p_d$, is employed as the pressure loss through a vacuum breaker, the flow rate is

$$w_{vb} = \sqrt{(p_w - p_d)\frac{A^2}{K} 2g_c \, \rho_{wa}} \, , \tag{11.13}$$

where ρ_{wa} is the wetwell air-vapor mixture density, and the flow resistance parameter, $A/\sqrt{K}$, is typically about 0.85 m^2.

If the drywell and wetwell pressures are again returned to 1.0 atm, the wetwell temperature considered in the example of Section 11.1.1.5 could, in the limit of an adiabatic wetwell air space, be reduced significantly. However, the pool temperature is elevated during a LOCA, and does not permit an adiabatic limiting temperature reduction during vacuum breaker operation.

11.2 Loss-of-Coolant Scenarios

The containment pressure and thermal loads resulting from a LOCA depend on the discharge rate of reactor coolant. The DBA is a postulated rupture of that largest pipe, which results in the highest containment pressure loads. However, other breaks are postulated that cause containment responses which are different than that of the DBA. Two such other break designations are the intermediate-break accident (IBA) and the small-break accident (SBA).

The containment response to a DBA has been discussed in Sec. 11.1.1, and typical drywell and wetwell pressures are shown in Fig. 11-9.

For discussion of containment response, the IBA is a postulated 0.1-ft^2 (0.0093-m^2) liquid break, which is small enough to prevent rapid depressurization of the reactor pressure vessel, but large enough to prevent the high-pressure coolant injection system from maintaining reactor water level. An SBA is a postulated 0.01-ft^2 (0.00093-m^2) steam break, which is not large enough to depressurize the reactor, and small enough so that the high-pressure coolant injection system is sufficient to maintain reactor water level. Figure 11-9 compares the containment pressurization for an IBA, an SBA, and the DBA postulated for a pressure suppression system with properties of Table 11-1 (Abramson, 1985).

Liquid discharge from an IBA undergoes significant flashing, which fills the drywell with steam. Drywell pressure rises slowly, which depresses the submerged vent water columns and allows air and steam to discharge

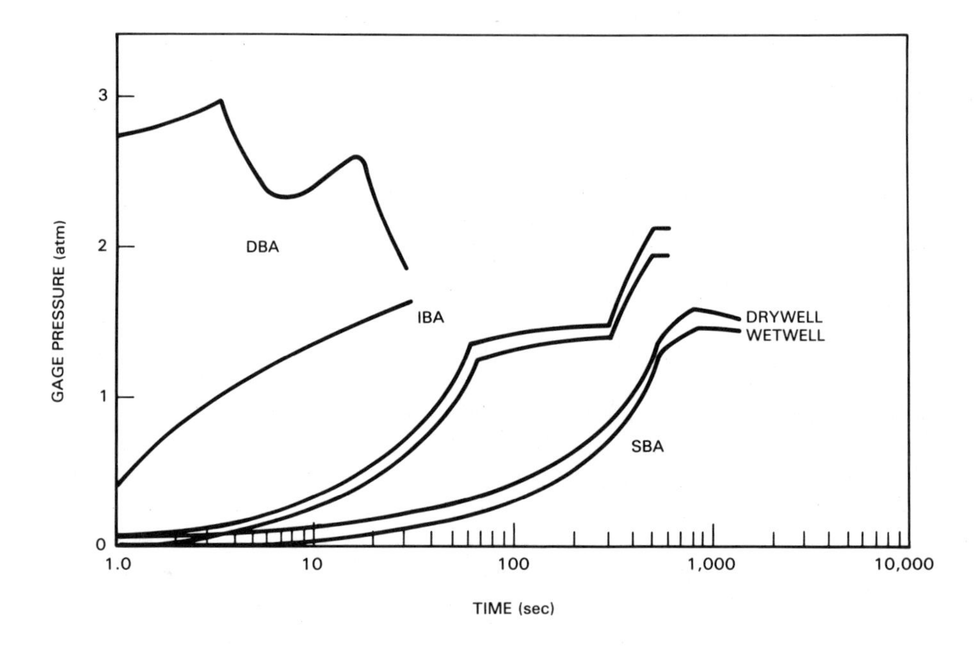

Fig. 11-9 Containment pressure responses.

into the pool. Air transfer from the drywell through the pool increases wetwell air space pressure. The temperature of drywell steam, sensed in the control room, corresponds to the saturated value at drywell pressure. Continued steam discharge to the pool may result in unsteady condensation loads.

If the main steam line isolation valves close due to low vessel water level, SRV operation automatically controls the system pressure. The ADS is actuated on high drywell pressure and low reactor vessel water level resulting from an IBA or SBA and discharges steam directly to the pool

TABLE 11-1
Properties before Pipe Break

Reactor pressure	7.2 MPa (1050 psia)
Drywell air volume	4504 m^3 (159,000 ft^3)
Wetwell air volume	3663 m^3 (129,300 ft^3)
Suppression pool volume	3637 m^3 (128,400 ft^3)
Drywell pressure	110 kPa (16 psia)
Wetwell pressure	101 kPa (14.7 psia)
Pool temperature	27°C (80°F)

through the ADS SRV discharge lines. The core spray systems and LPCI mode of the RHR system will be activated to flood the reactor vessel and cool the core during ADS depressurization. Water cascading into the drywell from the break would cause steam condensation and pressure reduction, resulting in air backflow from the wetwell air space through vacuum breakers, eventually equalizing drywell and wetwell pressures. Since reactor depressurization is less rapid for the IBA than for the DBA, more decay heat is transferred to the suppression pool via the ADS SRV lines and main vents resulting from the break flow. This results in a higher suppression pool temperature at the time of complete reactor depressurization.

An SBA causes drywell pressure to increase at a rate slower than the IBA until the drywell pressure scram set point is reached. Main steam isolation valve (MSIV) closure may occur due to the water level transient in the reactor vessel or low steam line flow. If the main steam lines isolate the reactor system, pressure will increase and intermittent SRV operation will control the system pressure. The steam discharged from the vessel and drywell air is vented to the suppression pool. Steam condensation causes gradual pool heating and continued wetwell pressurization. Unsteady condensation loads may occur during continued steam discharge into the pool.

11.3 Water Expulsion and Air Discharge Loads

The expulsion of vent water and discharge of drywell air may result in loads on submerged wetwell structures. Vent water expulsion in the form of accelerating liquid jets can impose forces on structures in the pool, whereas submerged air discharge can impose pool acceleration loads and pool swell impact forces on structures above the undisturbed water level.

11.3.1 Wetwell Loads at Vent Water Clearing

Approximate vent water expulsion velocity V can be obtained from Fig. 11-6. The expulsion force exerted on surrounding pool water is $\rho A_v V^2$, which reaches a maximum at the instant of water clearing. Subsequent air discharge tends to expand, resembling spherical bubble growth. However, when an approximate hemisphere of air has just emerged whose area is twice that of the vent, the liquid interface velocity reduces to about one-half its expulsion value. The expulsive force, therefore, reaches its maximum just before significant gas bubble growth begins, and decreases thereafter. The time of the maximum vent water clearing loads thus corresponds to that obtained from the clearing times of Fig. 11-6.

11.3.2 Submerged Water Jets

Expulsion of the vent water column creates forces on submerged pool structures in its trajectory. A simplified mass conservation and momentum analysis is formulated for the arbitrary, submerged water jet shown in Fig. 11-10. Since the jet and surrounding liquid have the same density, the buoyancy and weight of the jet cancel, and its motion proceeds as if it were in a zero-gravity environment. Fluid shear forces are neglected, and the surrounding fluid pressure is assumed to be uniform so that no net pressure force acts on the differential control volume shown dotted. The resultant governing equations are written as:

Mass conservation:

$$\frac{\partial w}{\partial x}dx+\frac{\partial M}{\partial t}=0 \tag{11.14}$$

Momentum:

$$\frac{\partial(wv)}{\partial x}dx+\frac{\partial(Mv)}{\partial t}=0 \tag{11.15}$$

Since $M=\rho A dx$ and $w=\rho A v$, Eqs. (11.14) and (11.15) can be written as,

$$\frac{\partial A}{\partial t}+v\frac{\partial A}{\partial x}+A\frac{\partial v}{\partial x}=0 \tag{11.16}$$

and,

$$\frac{\partial v}{\partial t}+v\frac{\partial v}{\partial x}=0\ . \tag{11.17}$$

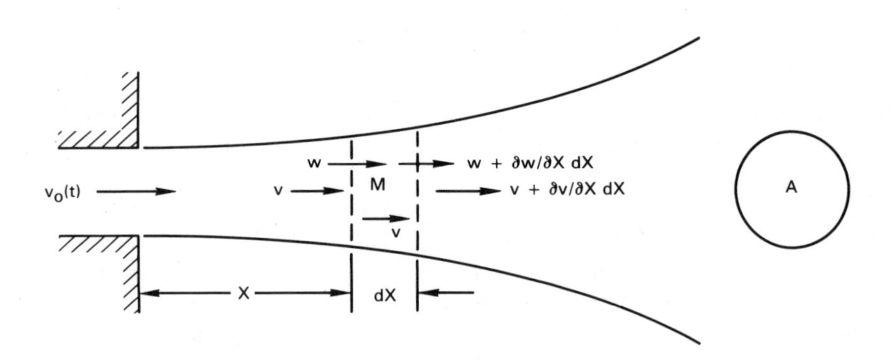

Fig. 11-10 Submerged water jet.

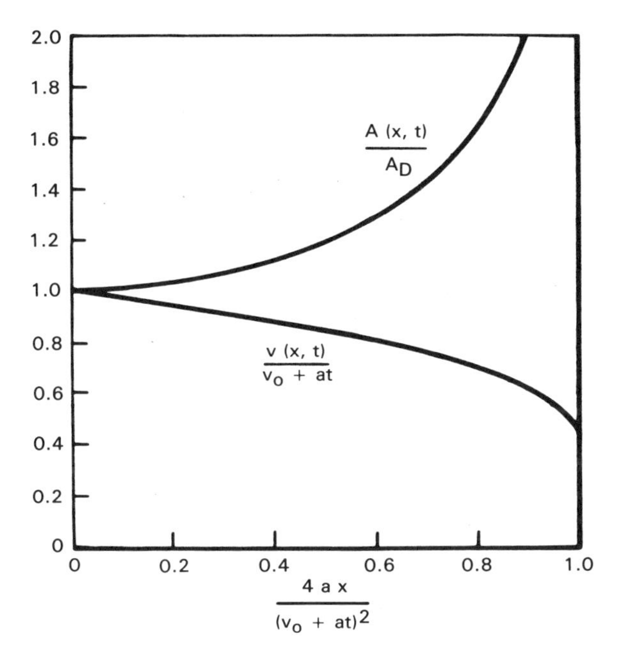

Fig. 11-11 Submerged water jet properties.

The solution for a step-ramp discharge velocity yields local jet velocity and area in the forms,

$$\frac{v(x,t)}{v_D}=\frac{v(x,t)}{(v_0+at)}=\frac{1}{2}\left[1+\sqrt{1-\frac{4ax}{(v_0+at)^2}}\right] \tag{11.18}$$

and,

$$\frac{A(x,t)}{A_D}=\frac{1}{2}\left[\frac{1}{\sqrt{1-\frac{4ax}{(v_0+at)^2}}}+1\right] \tag{11.19}$$

where,

$$v_D=v(0,t)=v_0+at\ . \tag{11.20}$$

These expressions are graphed in Fig. 11-11. Although the velocity magnitude diminishes with distance, it increases with time at a fixed position. Moreover, the jet area increases both with distance and time. The area and velocity formulations are useful in estimating which submerged structures overlap the jet and in predicting the drag forces imposed. Submerged jet discharge properties vary from one design to another, requiring individual predictions for each containment. It can be shown from this simple model that each jet particle proceeds at its expulsion velocity. This permits a jet response time to be defined as the time required for all jet particles to pass a fixed point in space. As a jet particle overtakes the ones expelled ahead of it, the jet front dissipates. Drywell air discharge immediately follows vent water column expulsion.

11.3.3 Submerged Air Discharge

Discharge and expansion of compressed drywell air underneath the pool surface following a LOCA vent water expulsion causes pool swell motion. Also, compressed air discharge from submerged SRV pipes produces bubbles that undergo energetic oscillation. Both cases of air discharge may create dynamic loads on containment structures.

Figure 11-12 shows an experimental bubble expansion and the corresponding pool swell for drywell air discharge from a horizontal (MK-III) vent (McIntyre and Myers, 1974). Calculations based on incompressible, inviscid flow give reasonable predictions of both bubble growth and pool swell (Moody and Reynolds, 1972).

The matter of bubble wall stability during growth or oscillation is largely determined by interface acceleration. Consider a spherical bubble that is being charged with gas at a volume rate, $Q(t)$. The corresponding radius, R, is governed by,

$$Q(t) = \frac{d}{dt}\left(\frac{4}{3}\pi R^3\right) = 4\pi R^2 dr/dt \ . \tag{11.21}$$

If the volumetric flow rate, Q, is constant, the bubble's interfacial acceleration is given by:

$$d^2R/dt^2 = -2(Q/4\pi R^2)^2/R \ , \tag{11.22}$$

which means that as the bubble radius increases, its growth rate decelerates (i.e., it undergoes an acceleration toward the bubble center). Whenever acceleration normal to an interface is from the liquid to the gas, the interface is stable, which is the condition described by Eq. (11.22). It is seen in Fig. 11-12 that the average interface velocity decreases with time, which results in a stable, but increasing, gas volume.

Bubble dynamic behavior implies the dynamic response of the surrounding liquid since the mass of the gas in the bubble is small. Spherical bubble dynamics are governed by the inviscid Rayleigh equation (Lamb, 1945),

$$d^2R/dt^2 + (3/2)(dR/dt)^2 = g_c(p - p_\infty)/\rho_l \ . \tag{11.23}$$

If bubble pressure undergoes adiabatic state changes,

$$pv^k = p_\infty v_\infty^k \quad \text{or} \quad p = p_\infty(R_\infty/R)^{3k} \ , \tag{11.24}$$

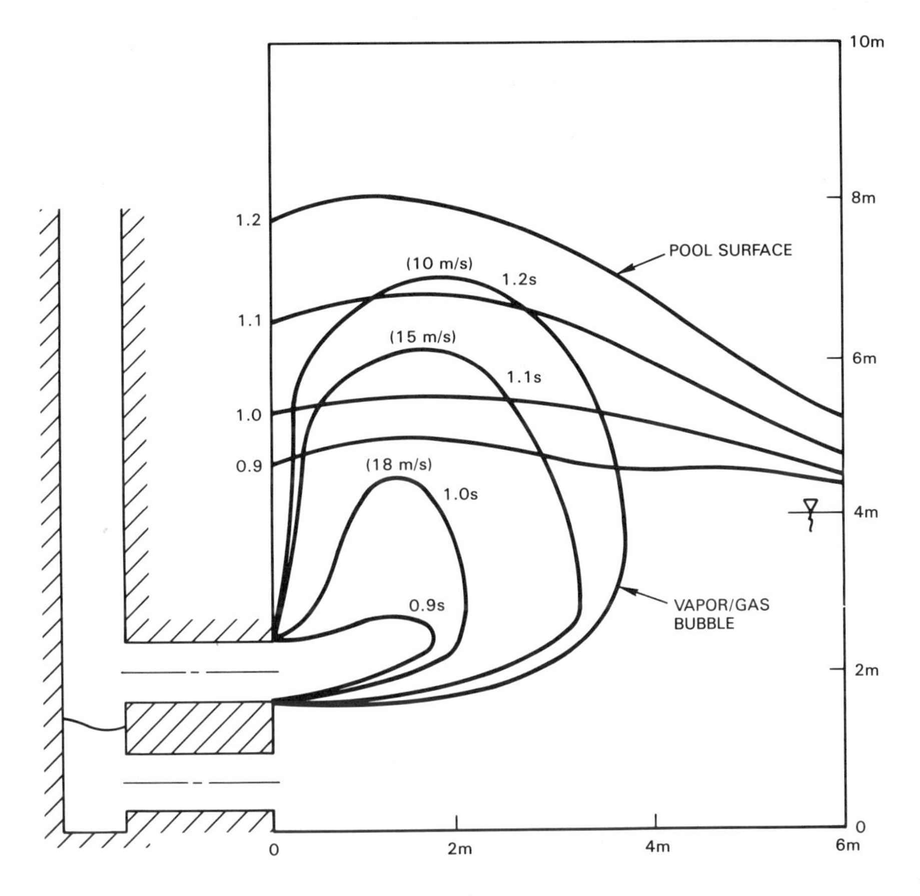

Fig. 11-12 Submerged air discharge, horizontal vents.

where subscript ∞ designates the undisturbed state. Furthermore, linearization of Eq. (11.23) for small-amplitude oscillation about R_∞ yields the equation for a linear oscillator whose frequency is given by,

$$f = \sqrt{3kg_c p_\infty/\rho_l}/2\pi R_\infty \ . \tag{11.25}$$

It is seen that the frequency varies inversely with the radius.

Equation (11.25) gives reasonable estimates of frequency, even for large-amplitude bubble oscillations. The initial pipe air, expelled into the pool after vent clearing from a SRV blow, yields the typical pressure trace shown in Fig. 11-13. The initial frequency is about 8 Hz. The expelled air volume was about 1.0 m^3 at atmospheric conditions, for which Eq. (11.25) gives a frequency estimate of 6.3 Hz, which is in reasonable agreement.

A large-amplitude oscillatory solution for a spherical bubble of maximum pressure $p_{b,\max}$ at minimum radius $R_{\min}$ in liquid surroundings at p_∞ is

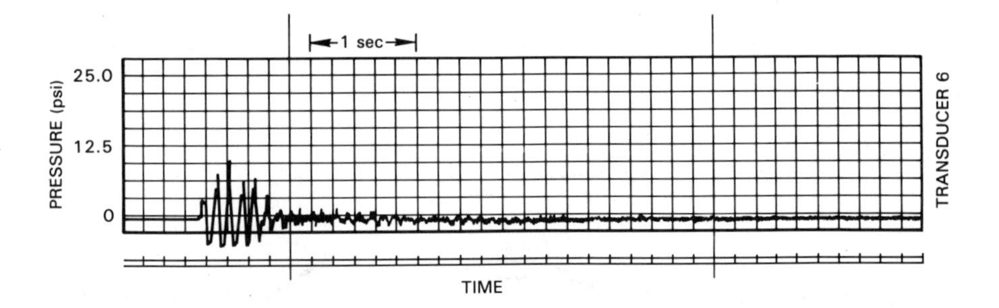

Fig. 11-13 SRV air discharge bubble oscillation pressure.

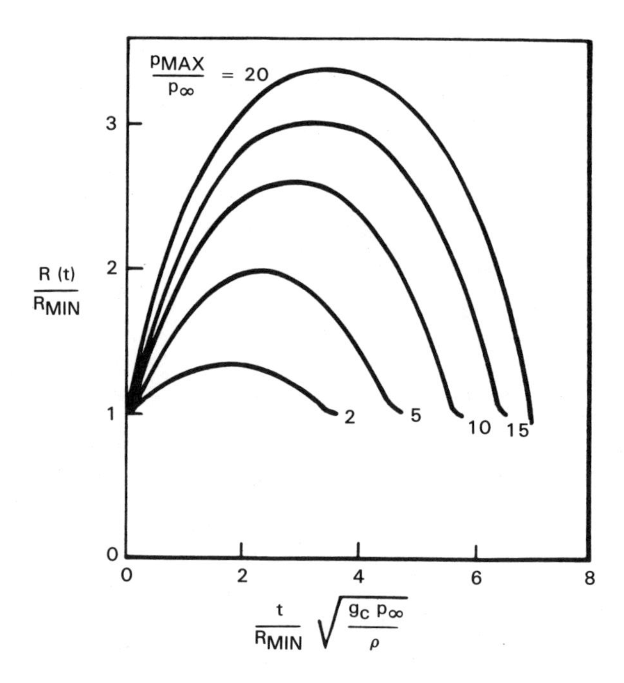

Fig. 11-14 Bubble radius, adiabatic oscillating bubble.

shown for one cycle in Fig. 11-14, which was obtained from a second-order Runge-Kutta solution of Eq. (11.23) for a spherical bubble undergoing adiabatic pressure changes according to Eq. (11.24). This reciprocal dependence of small-amplitude frequency on minimum radius is approximately preserved. Moreover, the maximum radius is roughly proportional to maximum pressure. Submerged oscillating bubbles resulting from safety/relief pipe air discharges are reasonably predicted from the procedures used to obtain Fig. 11-14.

The corresponding bubble interface velocity solutions given in Fig. 11-15 show that the interface is decelerating most of the time during a cycle, which implies a stable interface. The short time intervals during which an unstable condition, $d^2R/dt^2 > 0$, occurs do not permit substantial growth of interface disturbances. Such energetic bubbles usually undergo several oscillations before they reach the pool surface by buoyancy.

Drywell air discharge and bubble growth dynamics can be employed to show approximate trends. If drywell air at an average pressure p_d and density ρ_d discharges through a frictionless nozzle into an expanding bubble at pressure p_b with ambient pressure p_∞, the idealized spherical growth properties correspond to those in Fig. 11-16. The radial growth rate accelerates at first, and then becomes almost linear with time, corresponding to a volume growth rate, dV/dt, proportional to the square of time. This can be estimated from a solution of Eq. (11.23) at constant bubble pressure, which gives $dR/dt = \text{constant}$.

Figure 11-17 shows that bubble pressure decays as soon as radial growth begins. Moreover, surrounding liquid inertia has caused a slight overexpansion of the bubble where $p_b/p_\infty < 1.0$.

The classical analysis of spherical bubble collapse with interior pressure suddenly reduced to zero (Lamb, 1945) is also of interest. Sudden expulsion of steam into cold pool water would cause rapid condensation and almost a step reduction of pressure for an inertia-dominated bubble collapse. The first integral of the Rayleigh equation with $p_b = 0$ is given by

$$\frac{dR}{dt} = \sqrt{\frac{2}{3}\frac{g_c}{\rho_l}p_\infty\left(\frac{R_i^3}{R^3} - 1\right)} \ . \tag{11.26}$$

A second integration yields,

$$\frac{t}{R_i}\sqrt{\frac{g_c p_\infty}{\rho_l}} = 0.915\left[1 - I_{(R/R_i)}\left(\frac{5}{6}, \frac{1}{2}\right)\right] , \tag{11.27}$$

where $I_{(R/R_i)}(5/6, 1/2)$ is an incomplete beta function. Full collapse occurs when

$$\frac{t}{R_i}\sqrt{\frac{g_c p_\infty}{\rho_l}} = 0.915 \ . \tag{11.28}$$

Equations (11.26) and (11.27) are plotted in Fig. 11-18. When noncondensible gas is present or when bubble heat transfer to the liquid is reduced, the bubble's collapse characteristics can be substantially different (Florsheutz, 1965).

An idealized spherical bubble of radius $R(t)$ has the following velocity potential,

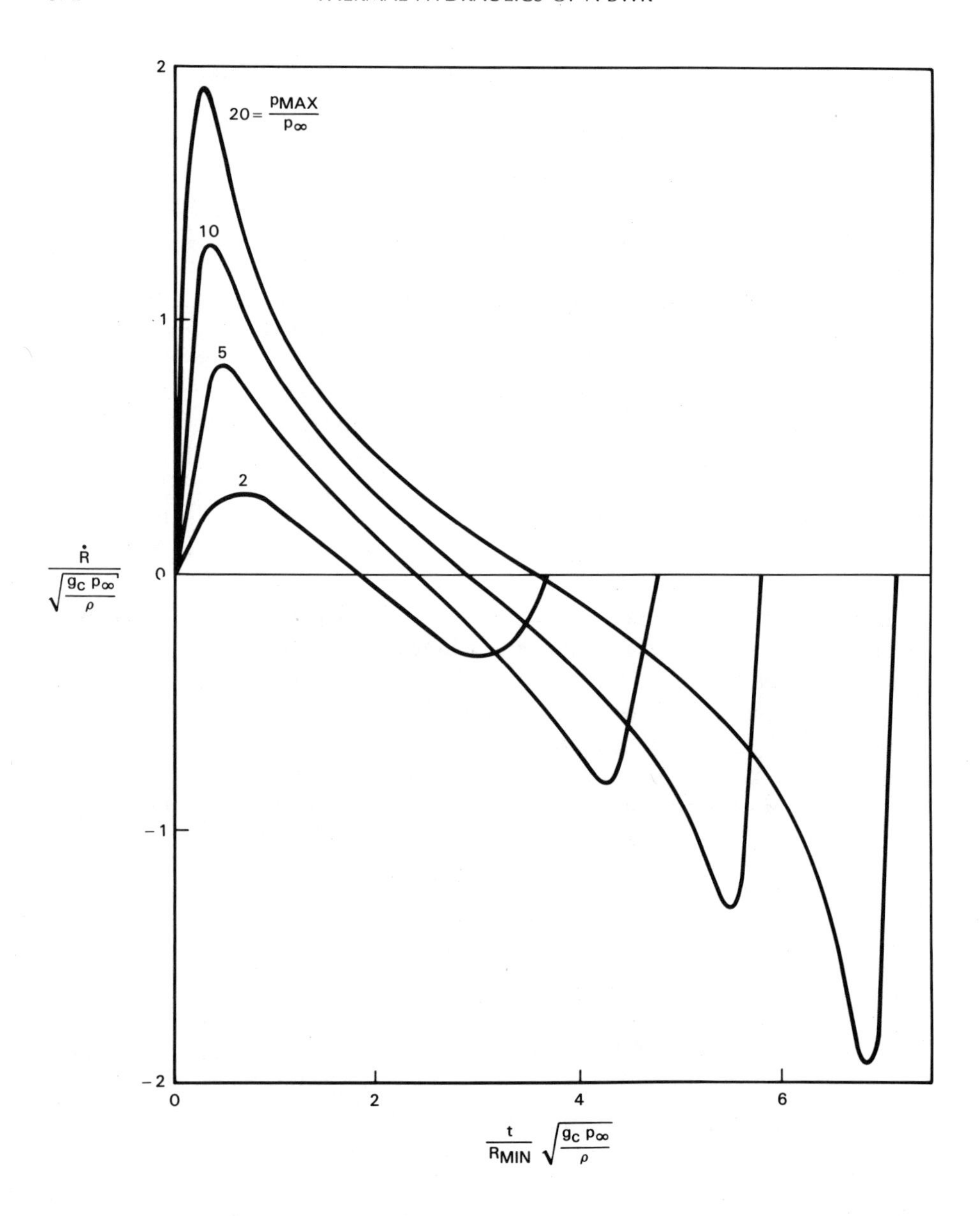

Fig. 11-15 Energetic bubble oscillation velocity.

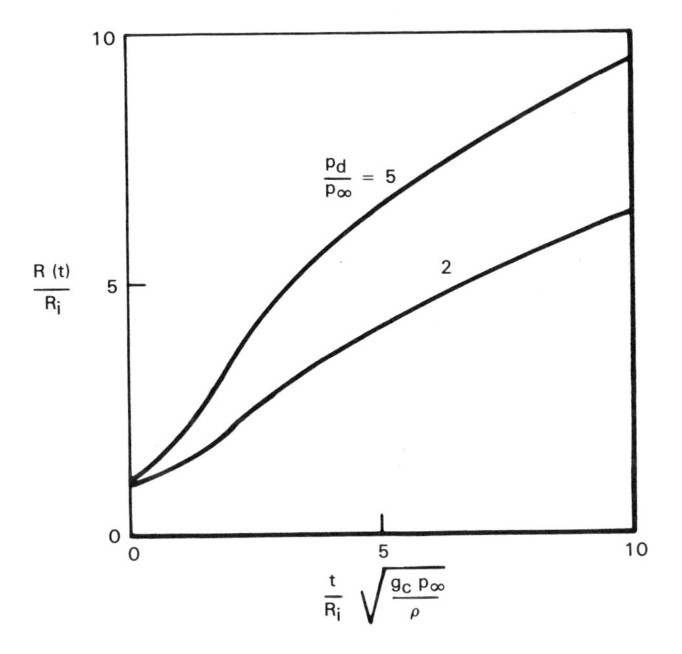

Fig. 11-16 Bubble radius during charging.

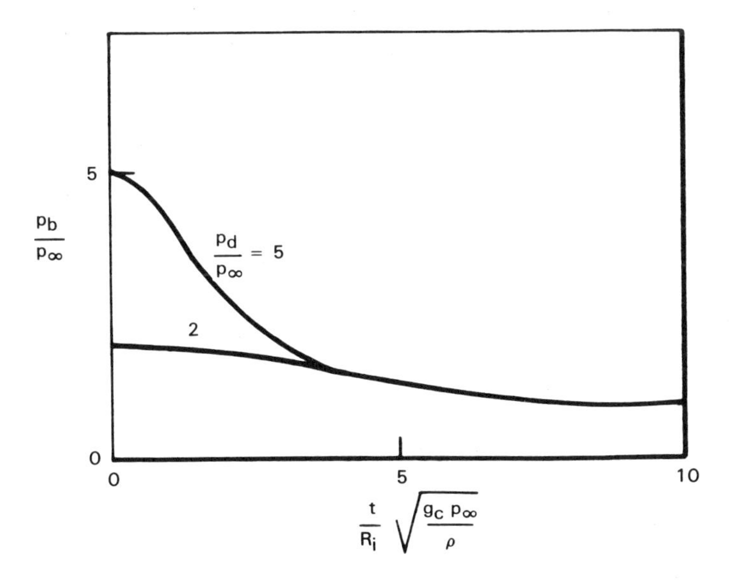

Fig. 11-17 Bubble pressure during charging.

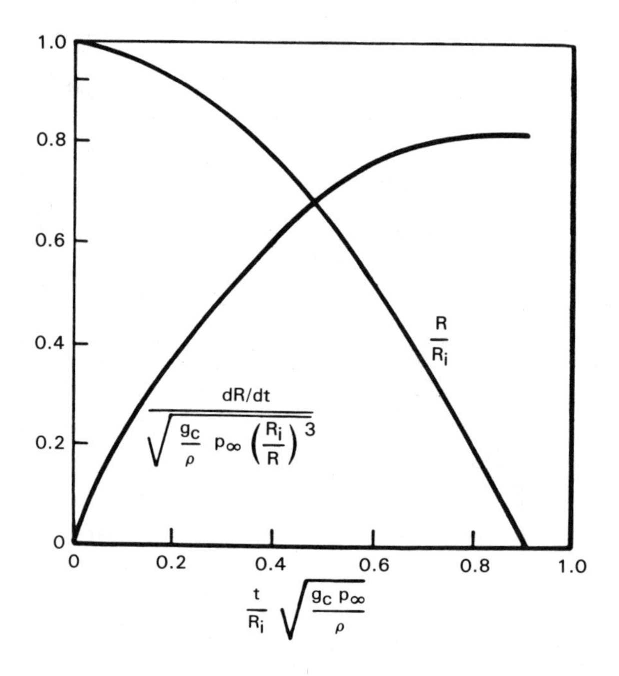

Fig. 11-18 Bubble collapse.

$$\phi = -\frac{R^2\dot{R}}{r} \ . \tag{11.29}$$

The local pressure field in the surrounding pool can be predicted from the unsteady Bernoulli equation,

$$\frac{\partial\phi}{\partial t} + \frac{(\nabla\phi)^2}{2} + \frac{g_c}{\rho_l}p + gy = f(t) \ . \tag{11.30}$$

The bubble's interfacial velocity passes through zero at the maximum and minimum radii during an oscillation. Equation (11.30) shows that at these extremes,

$$\nabla^2 p = 0 \quad \text{whenever } \nabla\phi = V_b = 0 \ . \tag{11.31}$$

That is, a field solution for pressure outside a bubble of radius R at the extremes in its oscillation has the same form as the potential of Eq. (11.29), or

$$p = p_b\frac{R}{r} \quad \text{whenever } \nabla\phi = 0 \ . \tag{11.32}$$

This reciprocal distance pressure dependence is verified from the measured high- and low-pressure fields shown in Fig. 11-19 (McCready et al., 1973). The maximum pressure corresponds to the minimum bubble radius during an oscillation, and the minimum pressure corresponds to its maximum radius.

11.3.4 Pool Wall Pressures

The pressure on a rigid wall during submerged bubble oscillations can be estimated from analyses that employ multiple sources and sinks (Valandani, 1975). A bubble near a stationary wall or corner can be simulated by a point source and images with identical properties placed in a symmetrical pattern. This procedure is based on the property that the velocity potentials, ϕ_n, which are associated with N sources and sinks in an array, can be added, since each satisfies $\nabla^2\phi_n = 0$.

If identical sources of the form given by Eq. (11.29) are placed a distance $2b$ apart, the flow pattern corresponds to a flat wall at the symmetry plane. If the unsteady Bernoulli equation, Eq. (11.30), is employed, the maximum wall pressure on a line joining the two sources, corresponding to a large-amplitude bubble oscillation at the instant when $dR/dt = 0$, is given by:

$$(p_{\text{wall,max}} - p_\infty) = \frac{2R_{\text{min}}}{b}(p_{b,\text{max}} - p_\infty) , \tag{11.33}$$

where p_∞ is the undisturbed pressure. Thus, the maximum wall pressure is proportional to the minimum bubble radius and maximum bubble pressure, and inversely proportional to bubble distance from the wall, b. The factor 2 in Eq. (11.33) is different for bubbles near corners and between parallel walls or below free surfaces. The superposition of point sources has a shortcoming in that calculated wall pressure at instants of zero bubble interface velocity can exceed the bubble pressure. This result is shown to be invalid by setting $V_b = \nabla\phi = 0$ in Eq. (11.30), which occurs at extremes in bubble oscillation. Noting that since $\nabla^2\phi = 0$, the equation $\nabla^2 p = 0$ also must be satisfied when $V_b = 0$, that is, p momentarily satisfies LaPlace's equation, which has the property that no rigid boundary pressure can exceed the highest pressure imposed elsewhere. However, Eq. (11.33) gives infinite pressure if $b \to 0$. The remedy employed is either to restrict p_{wall} to $p_{b,\text{max}}$ or to make use of actual bubble boundary computations without point source simulation. Observations from multiple source analyses are useful in determining how well single-vent steam discharge and collapse experiments represent multivent geometries.

11.3.5 Submerged Structure Loads

Pool motion caused by drywell or SRV air discharge creates forces on submerged structures. These forces are conservatively estimated for rigid structures as a combination of ordinary profile drag,

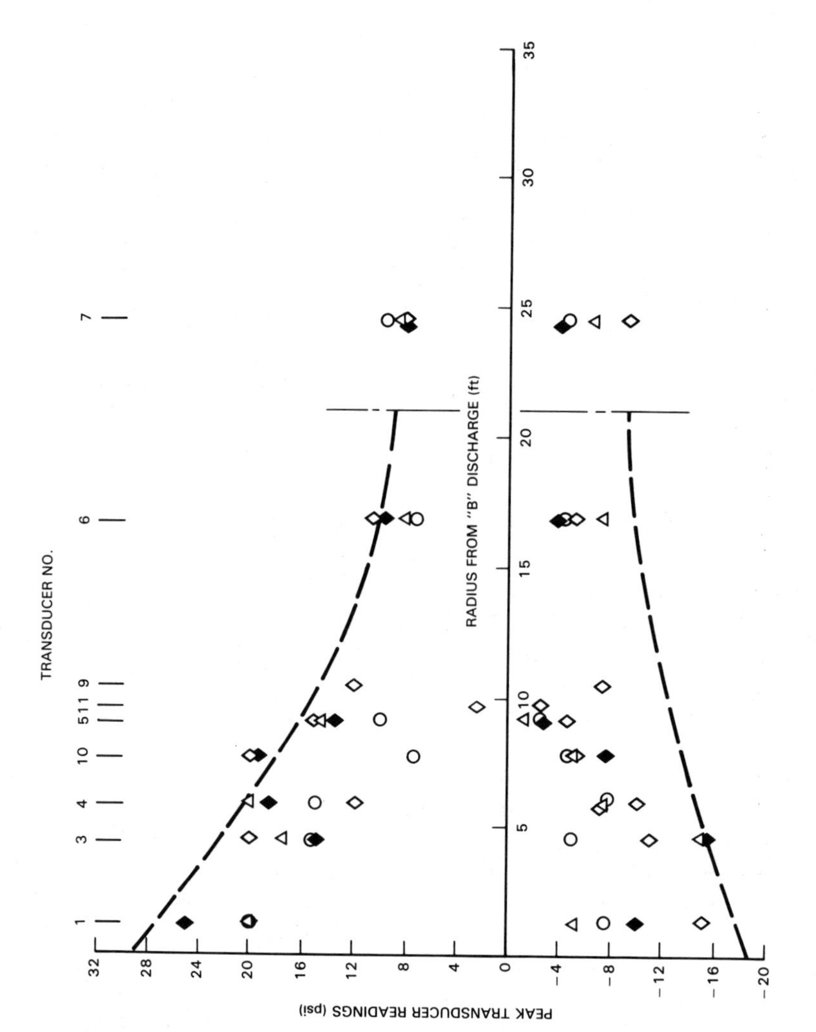

Fig. 11-19 Pressure field from oscillating SRV air bubble discharge (McCready et al., 1973).

$$F_D = C_D A \rho_l \frac{v_l^2}{2g_0} , \tag{11.34}$$

plus a fluid acceleration force (Moody et al., 1977),

$$F_A = \left(V_s + \frac{M_{vm}}{\rho_l} \right) \rho_l \frac{\dot{v}_l}{g_0} , \tag{11.35}$$

where V_s is the structure volume and M_{vm} is the classical hydrodynamic virtual mass.

The force component caused by the structure volume is like a buoyant force in the direction of fluid acceleration. The hydrodynamic virtual mass component is the result of fluid being decelerated and deflected by the structure. Examples of the hydrodynamic virtual mass for various structural geometries found in BWR systems are given in Fig. 11-20 (Patton, 1965).

11.3.6 Pool Swell

The growth rate of a submerged air bubble determines the pool surface velocity at various elevations where structural impact could occur.

Submerged bubble dynamics, at least in the early stages, resembles spherical expansion. In contrast, if bubble growth extends horizontally across the pool, the wetwell geometry resembles a water piston above a flat air cushion. The growth characteristics of spherical and flat air bubbles are different. For example, if p_∞ is the undisturbed liquid pressure and p_b is bubble pressure, the instantaneous radius for a spherical bubble is governed by the Rayleigh equation, Eq. (11.23),where flat bubble growth is governed by,

$$L \frac{d^2y}{dt^2} = \frac{g_0}{\rho_l} (p_b - p_\infty) , \tag{11.36}$$

where L is the depth of the water slug, and y is its upward displacement. The corresponding spherical and flat volume growth rates for constant driving pressure, $\Delta p = p_b - p_\infty$ are given by,

$$\frac{dV_b}{dt} = \begin{cases} 4\pi \left(\dfrac{2}{3} \dfrac{g_0}{\rho_l} \Delta p \right)^{3/2} t^2 \ ; & \text{spherical} \\ A \left(\dfrac{g_0}{\rho_l} \dfrac{\Delta p}{L} \right) t \ ; & \text{flat} \end{cases} , \tag{11.37}$$

where A is the pool surface area. The spherical volumetric growth rate is proportional to t^2, whereas the flat volumetric growth rate is proportional to t. The corresponding pool surface upward acceleration, a_p, is given by,

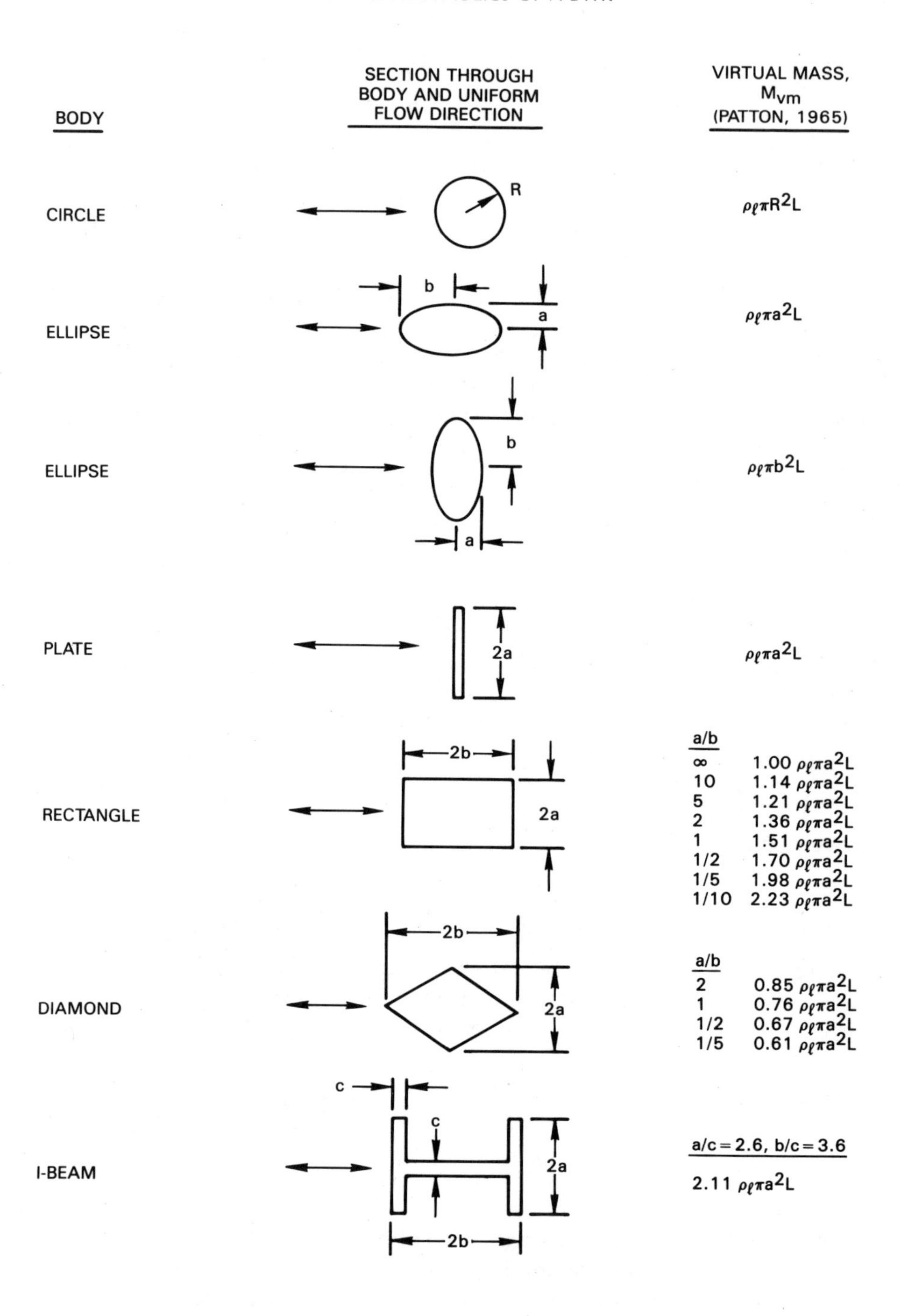

Fig. 11-20a Hydrodynamic virtual mass for two-dimensional structural components (length L for all structures).

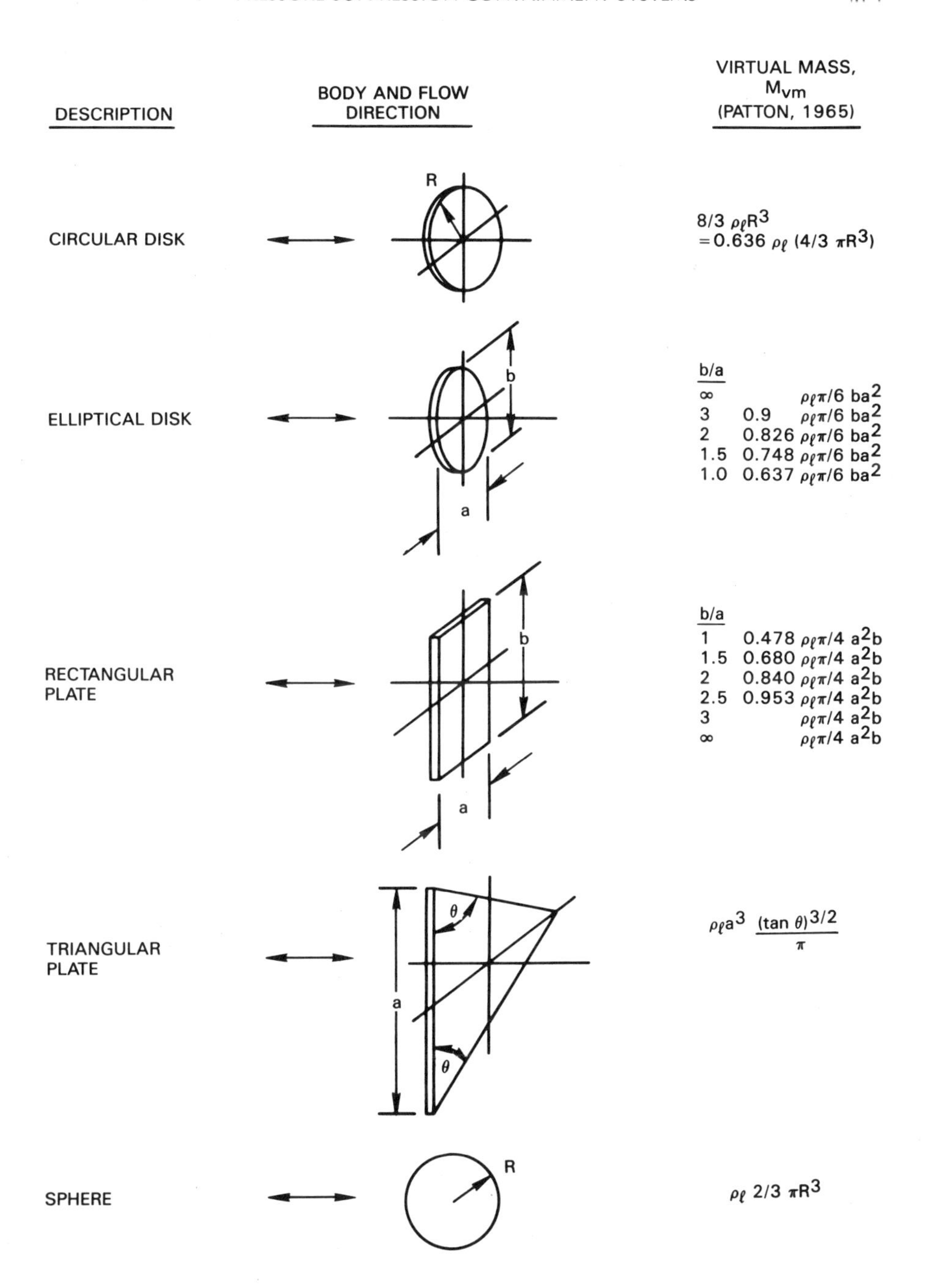

Fig. 11-20b Hydrodynamic mass for three-dimensional structures.

$$a_p = \frac{1}{A}\frac{d^2V_b}{dt^2} = \begin{cases} \dfrac{8\pi}{A}\left(\dfrac{2}{3}\dfrac{g_0}{\rho_l}\Delta p\right)^{3/2} t \; ; & \text{spherical} \\ \dfrac{g_0}{\rho_l}\dfrac{\Delta p}{L} \; ; & \text{flat} \end{cases} . \tag{11.38}$$

It is seen that spherical bubble growth causes a linear change in pool upward acceleration, while flat bubble growth causes constant acceleration for constant p_b. Vertical acceleration largely determines the time required for breakthrough, that is, when the rising bubble overtakes the pool surface.

11.3.7 Wetwell Air Space Compression

Pool swell may be decelerated by compression of the wetwell air prior to breakthrough of the submerged drywell air. The coupling between air bubble charging, pool motion, and wetwell air space compression can be studied by simplified procedures in which a fixed pool mass is accelerated. The charging rate and bubble penetration strongly affect the quantitative results such that simple analyses usually are not accurate. However, these coupled phenomena have been studied by scale models (Torbeck et al., 1976; Anderson et al., 1977; McCauley and Pitts, 1977). The usual procedure has been to deduce appropriate scaling laws, design a small-scale experiment, and scale-up the measurements to predict full-scale behavior. Significant trends have been established from scale model data correlations. For example, the air charging product wh_0 was found to be significant rather than the mass flow rate, w, or stagnation enthalpy, h_0, alone.

Scale model testing has provided insight to other containment phenomena and promises to be useful in further studies. The procedure for scaling thermal-hydraulic phenomena is described in Sec. 11.4.

11.3.8 Pool Swell Structure Impact

A rising pool during submerged air discharge may contact stationary structures in the wetwell above the initial pool surface. When a small structure is fully submerged in the rising pool, the force exerted is characterized by a standard drag load of the type $C_D A \rho_l v_l^2 / 2g_0$, since fluid acceleration is relatively small. Wide structures may overlap a large fraction of the rising pool surface area, imposing a significant pool deceleration, which contributes to the applied force, per Eq. (11.35). Pool surface irregularities and breakthrough strongly affect structural impact loads, and designs typically are based on large-scale experiments (GE, 1980). It has been observed that structural loads applied while the structure is being submerged correspond to the impulsive deceleration of a partial hydrodynamic mass. A study of water impact loads is available that concludes trapped air can play a major role, and that the phenomenon is governed by bulk rather than acoustic effects (Valandani, 1975).

11.3.9 Air Bubble Breakthrough

If drywell air rises as a large bubble with a round top of radius R, it overtakes the pool surface at a relative velocity of (Davies, 1950):

$$v_R \approx 0.6\sqrt{(a+g)R} \ ; \quad a > -g \ . \qquad (11.39)$$

If the rising pool water is decelerating ($a<0$), its top surface develops Taylor instability, during which disturbance wave amplitudes grow downward toward the rising bubble. Small amplitude theory (Taylor, 1950) for waves stabilized only by surface tension σ shows that the early, fastest growing wave length is,

$$\lambda = 2\pi\sqrt{\frac{3g_c\sigma}{\rho_l a}} \ . \qquad (11.40)$$

Pool deceleration is equivalent to downward acceleration of wetwell air toward the pool, and helps explain observed top surface wave growth. A pool swell deceleration of one-tenth gravity corresponds to an estimated surface wave length of 10 cm.

11.3.10 Condensation Loads

Air discharge from the vents during a LOCA is followed by a rapid transition to steam discharge into the pool. The steam discharge rate is high at first, and diminishes over the period of reactor pressure vessel blowdown. Decreasing steam flow passes through several condensation modes, which results in water acceleration and pressure loads in the pool. Figure 11-21 shows a map of condensation modes that have been observed during either a LOCA or SRV discharge. High steam flows result in *steady condensation.* Diminishing steam flows cause condensation to enter the *condensation oscillation* mode, where the steam-water boundary at the vent discharge oscillates somewhat like a flickering candle flame. Further diminished steam flows cause transition to the *chugging* mode during which a column of water rises in the submerged vent, then is blown out by continued steam flow, followed by rapid condensation in the pool, which draws the water column back into the vent and the cycle repeats. Condensation oscillation and chugging phenomena have been identified as the two condensation modes that dominate unsteady containment loads.

11.3.10.1 *Condensation Oscillations*

Figure 11-21 shows that condensation oscillations (CO) can begin when the discharge steam mass flux falls below a pool temperature-dependent value. Observations show that CO can be modeled as an oscillating steam bubble at the exit of the discharge pipe. The bubble may be approximated by part of a sphere, which remains outside the pipe as shown in Fig. 11-22.

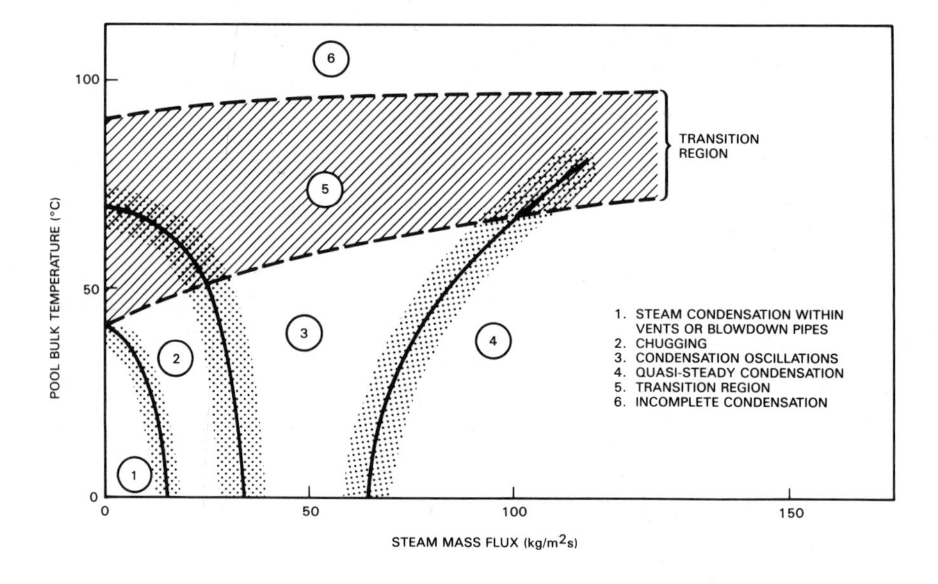

Fig. 11-21 Schematic of typical regions for condensation modes during SRV or LOCA blowdown.

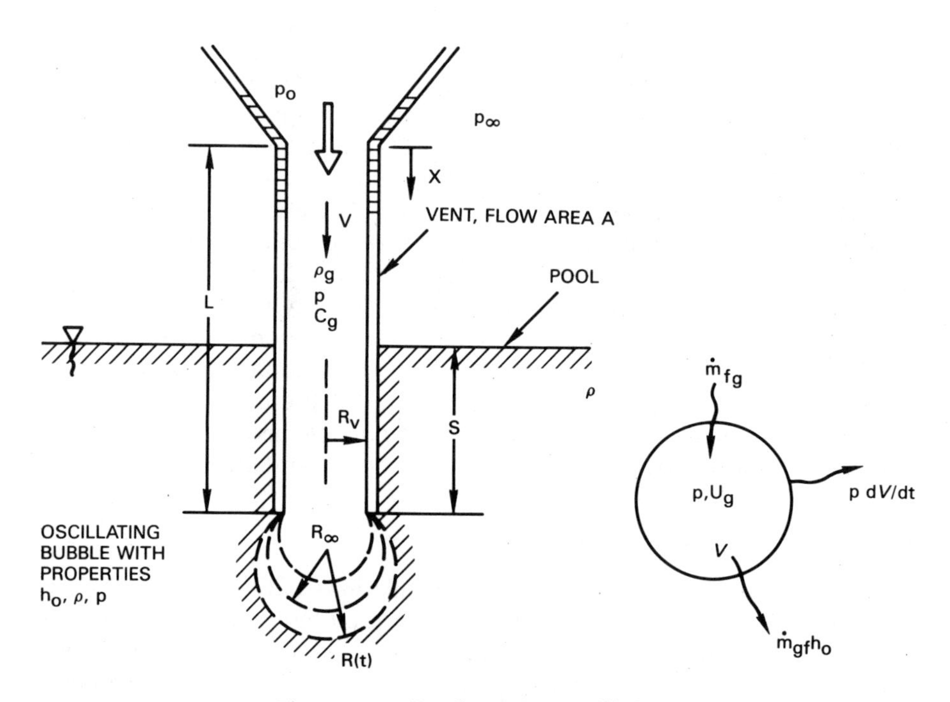

Fig. 11-22 Condensation oscillation.

The bubble is continuously supplied with steam flow, and the oscillation is caused by unsteady condensation on the liquid surface.

Linearized theoretical models, which are based on a heat transfer-controlled condensation rate, yield strongly damped oscillations with reasonable prediction of measured frequencies (Nariai and Aya, 1984). The average bubble size appears to be determined from heat transfer considerations, whereas the oscillation in bubble surface area behaves like a periodic steam sink. When the oscillating area is small, less condensation occurs, bubble pressure increases, and the bubble grows. Overexpansion of the bubble reverses the motion and contraction follows. Acoustic effects in the vent also play a role in some applications (Marks and Andeen, 1979).

The important parameters for CO appear to include the total vent length, L_v; and radius, R_v; the submergence length, L_s; steam flow rate, w; and the steam properties: pressure, p_g; density, ρ_g; stagnation enthalpy, h_{0g}; sonic speed, C_g; and the local steam and water temperatures.

Small-scale experiments were performed to investigate the condensation process and pressure oscillations when steam was discharged into subcooled water (Simpson and Chan, 1982). Pool subcooling exhibited the largest influence on dynamic behavior. The bubble pulsation frequency, f_b, and pressure intensity, $\Delta p/p$, were found to correlate with the Jacob number $(\rho_l c_{pL} \Delta T)/(\rho_g h_{fg})$ and the Reynolds number $(v_g d/\nu_{g0})$ where ρ_l and ρ_g are liquid and steam densities, c_{pL} is the liquid specific heat, ΔT is the pool subcooling, h_{fg} is the vaporization enthalpy, ν_{g0} is the steam jet exit velocity, d is the jet tube diameter, and v_g is the steam kinematic viscosity. Typical correlations are shown in Fig. 11-23 for tube diameters of 0.635, 1.59, and 2.22 cm. The pulsation frequency is seen to increase with both subcooling and discharge velocity. The pressure amplitude increases with discharge velocity, but decreases with subcooling.

It is expected that as subcooling increases and less bubble area is required for condensation, the pulsation frequency would reach a limiting value. Earlier CO tests showed that for higher subcooling, vent acoustics influenced the CO frequency.

A theoretical model (Moody, 1981) was formulated for cases where the vent gas flow is dominated by acoustic effects, for which the governing equations are,

Continuity:

$$\frac{\partial p}{\partial t} + \frac{\rho_g C_g^2}{g_c} \frac{\partial v}{\partial x} = 0 \tag{11.41}$$

Momentum:

$$\frac{\partial v}{\partial t} + \frac{g_c}{\rho_g} \frac{\partial p}{\partial x} = 0 \ . \tag{11.42}$$

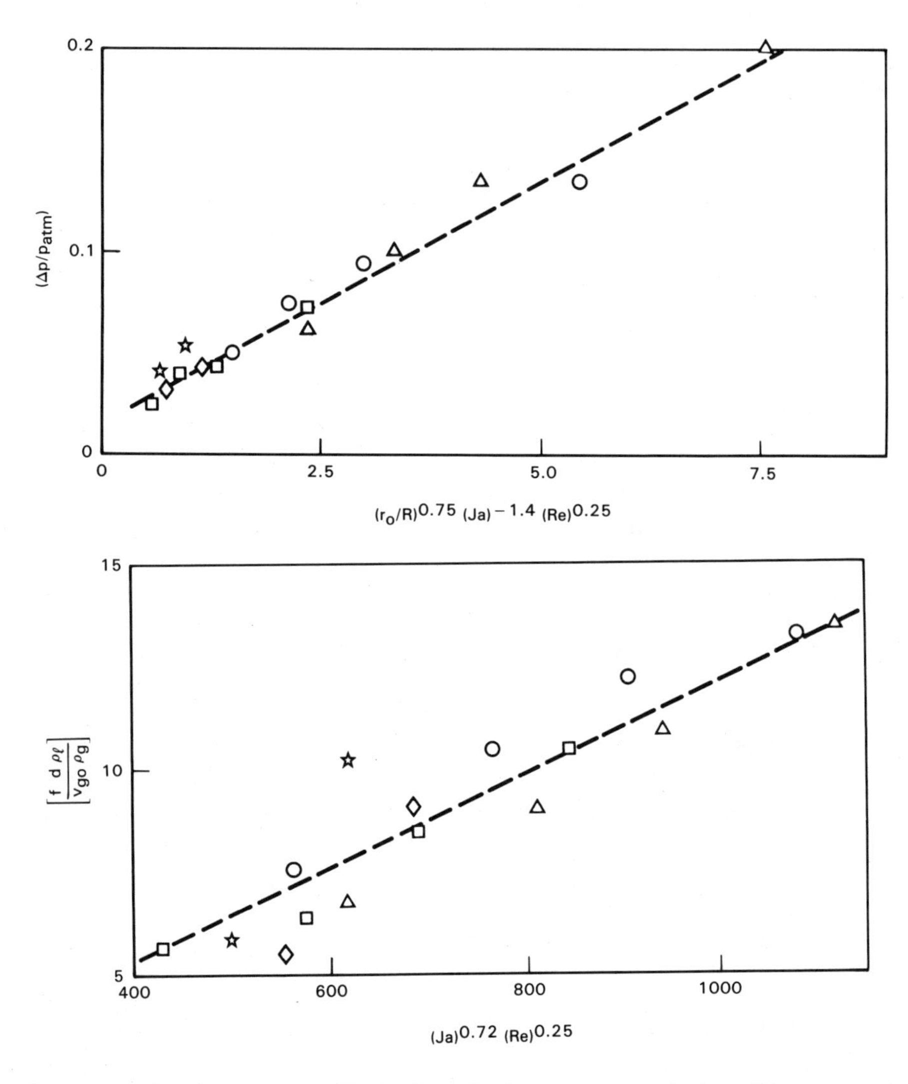

Fig. 11-23 Condensation oscillation impulse frequency correlations (Simpson and Chau, 1982).

The steam was treated as a perfect gas with internal energy $U_g = pV/(k-1)$ and sound speed $C_g = (kg_c p/\rho_g)^{1/2}$. A simplified boundary condition at the submerged end of the vent pipe was considered in which the vapor volume of Fig. 11-22 was assumed to be spherical with radius, $R(t)$. The vent discharge and bubble values of stagnation enthalpy, density, and pressure are assumed to be equal, which yields,

$$h_0(L,t)=\frac{k}{(k-1)}\frac{p(L,t)}{\rho_g(L,t)}+\frac{v(L,t)^2}{2g_0}=h_0 \ , \tag{11.43}$$

$$\rho_g(L,t)=\rho \ , \tag{11.44}$$

$$p(L,t)=p \ . \tag{11.45}$$

The vent discharge flow rate, which charges the bubble at any instant, is expressed by,

$$w=\rho_g(L,t)Av(L,t) \ . \tag{11.46}$$

If vapor condenses on the bubble-liquid interface at a rate $\dot{m}_{gf}$, maintaining the interface liquid at the steam saturation temperature, then interface liquid transports thermal energy by conduction to the surrounding liquid. Condensation is expected to occur primarily where discharged steam strikes the spherical bubble interface directly ahead over a relatively constant area, A, equal to that of the vent. It is assumed that the condensation rate is proportional to the discharge steam density so that the unsteady rate is given by

$$\dot{m}_{gf}=w_\infty\frac{\rho_g(L,t)}{\rho_\infty} \ , \tag{11.47}$$

where w_∞ is the average, or steady steam condensation flow rate at density ρ_∞, namely,

$$w_\infty=\rho_\infty A v_\infty \ . \tag{11.48}$$

The density, $\rho_g(L,t)$, is unsteady because of the surrounding water inertia, which causes the steam bubble and water to respond like a spring-mass oscillator. An energy conservation principle, written for the bubble of Fig. 11-22, yields,

$$p\frac{dV}{dt}+\dot{m}_{gf}h_0-wh_0+\frac{dU_g}{dt}=0 \ . \tag{11.49}$$

If V is replaced by $(4/3)\pi R^3$ and U_g for a perfect gas is employed, Eq. (11.49) becomes

$$p(L,t)R^2\dot{R}+(\dot{m}_{gf}-w)\left(\frac{k-1}{k}\right)\frac{h_0(L,t)}{4\pi}+\frac{R^3}{3k}\dot{p}(L,t)=0 \ . \tag{11.50}$$

Water inertia outside the steam bubble is related to bubble pressure by Eq. (11.23). The entrance boundary is assumed to remain at constant pressure p_0, that is,

$$p(0,t)=p_0 \ . \tag{11.51}$$

If the flow rate or bubble boundary is perturbed from its steady-state values, the vent-bubble-liquid system will respond dynamically in an oscillatory mode. Conditions for steady condensation correspond to the properties p_0, V_∞, ρ_∞, and R_∞. The dynamic behavior is obtained by assuming that some disturbance has been imposed on the system, which causes each property to undergo unsteady response. Thus, each property is written as its steady value plus an unsteady component, that is,

$$\begin{aligned} p &= p_0 + p' \ , \\ v &= v_\infty + v' \ , \\ \rho &= \rho_\infty + \rho' \ , \\ R &= R_\infty + R' \ , \end{aligned} \tag{11.52}$$

where the superscript prime denotes a perturbation. The full perturbed problem formulation is now written from Eqs. (11.41), (11.42), (11.51), (11.50), and (11.23) as

$$\frac{\partial p'}{\partial t} + \frac{\rho_\infty C_\infty^2}{g_0} \frac{\partial v'}{\partial x} = 0 \ , \tag{11.53}$$

$$\frac{\partial v'}{\partial t} + \frac{g_0}{\rho_\infty} \frac{\partial p'}{\partial x} = 0 \ , \tag{11.54}$$

where boundary conditions are given by

$$p'(0,t) = 0 \ , \tag{11.55}$$

$$\dot{R}' + \frac{R_\infty}{3kp_0} p'(L,t) = Bv'(L,t) \ , \tag{11.56}$$

$$R_\infty \ddot{R}' = \frac{g_c}{\rho_l} p'(L,t) \ , \tag{11.57}$$

and B is the constant,

$$B = \frac{A}{4\pi R_\infty^2}\left[1 + \frac{v_\infty^2}{2g_c} \Big/ \left(\frac{k}{k-1}\right)\frac{p_\infty}{\rho_\infty}\right] \ . \tag{11.58}$$

If v' is eliminated from Eqs. (11.53) and (11.54), the wave equation for the disturbed pressure is obtained, namely,

$$\frac{\partial^2 p'}{\partial t^2} - C_\infty^2 \frac{\partial^2 p'}{\partial x^2} = 0 \ . \tag{11.59}$$

An oscillatory solution that satisfies the entrance boundary condition is of the form

$$p'(x,t) = \sin\frac{\omega x}{C_\infty}(c_1 \cos\omega t + c_2 \sin\omega t) \ . \tag{11.60}$$

TABLE 11-2
First Four CO Eigenfrequencies ($f_n L/C_g$)
(high subcooling, vent acoustics dominate)

$R_\infty/L = 0.01$	$R_\infty/L = 0.10$
1.60	1.50
4.71	4.20
7.85	7.10
10.90	10.00

If Eq. (11.56) is differentiated with respect to time, R' is substituted from Eq. (11.57), and $\dot{v}'$ is obtained from Eq. (11.54), we obtain at $x = L$:

$$p' + \left(\frac{\rho_l}{g_0}\frac{R_\infty^2}{3kp_\infty}\right)\ddot{p}' + \frac{\rho_l}{\rho_\infty}B\frac{\partial p'}{\partial x} = 0 \ . \tag{11.61}$$

The bubble and vent radii are assumed to be approximately equal so that $A = \pi R_\infty^2$. Also, $kg_c p_\infty/\rho_\infty$ is replaced with C_∞^2. The solution for p' in Eq. (11.60) is finally substituted into Eq. (11.61) to obtain the following equation for condensation oscillation eigenfrequencies:

$$\tan\left(\frac{\omega L}{C_\infty}\right) = \frac{\frac{1}{4}\left(\frac{R_\infty}{L}\right)\frac{\rho_l}{\rho_\infty}\left(\frac{\omega L}{C_\infty}\right)\left[1 + \frac{k-1}{2}\left(\frac{v_\infty}{C_\infty}\right)^2\right]}{\frac{1}{3}\frac{\rho_l}{\rho_\infty}\left(\frac{R_\infty}{L}\right)^2\left(\frac{\omega L}{C_\infty}\right)^2 - 1} \ . \tag{11.62}$$

The term $(v_\infty/C_\infty)^2(k-1)/2$ is small enough to neglect during condensation oscillation. The first four vent acoustics eigenfrequencies are given in Table 11-2 for water and steam densities ρ_l and ρ_∞ at standard conditions. CO frequencies are inversely proportional to the vent length, and can be controlled by lengthening or shortening the vent.

11.3.10.2 *Vent Chugging*

The phenomenon of vent chugging has been observed during tests when the drywell pressure and vent steam flux are reduced. One form of chugging occurs when a water column rises in a vent, reverses direction because of diminishing condensation, and is expelled into the pool where rapid condensation and steam bubble collapse occur, then repeating the cycle. This process is shown schematically in Fig. 11-24 along with a sample regional map of chugging in vertical vents (Fitzsimmons et al., 1979). A typical trace of a single chug is given in Fig. 11-25 (Marks and Andeen,

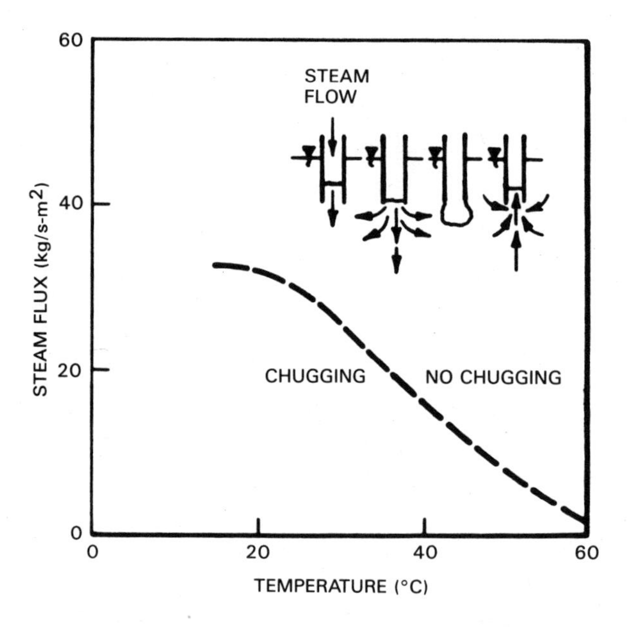

Fig. 11-24 Sample chugging map, vertical vents.

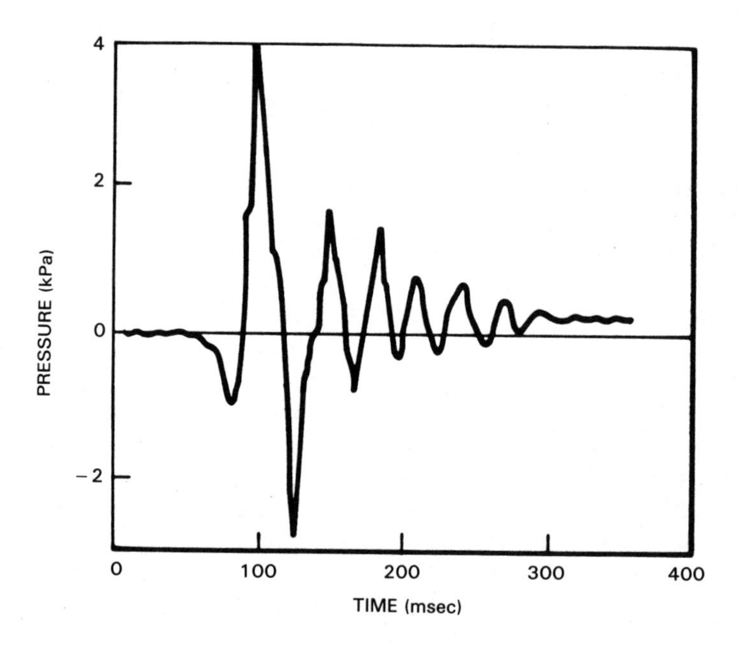

Fig. 11-25 Single chug.

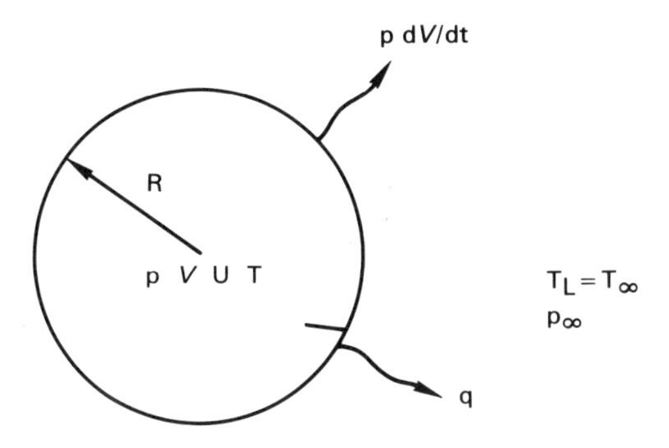

Fig. 11-26 Gas bubble in liquid.

1979), which shows a pressure measurement adjacent to the vent discharge. An initial pressure reduction is seen, followed by an abrupt pressure rise, which is close to the theoretical bubble collapse pressure corresponding to a solution of the Rayleigh bubble equation for a sudden reduction of bubble pressure to zero.

The initial pressure reduction and rapid spike increase in pressure, seen in the chugging trace in Fig. 11-25, are followed by a damped oscillation known as "ringout" (Lahey et al., 1990). It was also noted that the suppression pool water became milky with small bubbles, formed during the initial air discharge into the pool. The presence of these small gas bubbles imposes a strong damping effect on the speed and magnitude of the acoustic pressure waves.

Figure 11-26 shows a gas bubble of radius, $R(t)$, in liquid at temperature, T_∞, and mean pressure, p_∞. The liquid's dynamic response is governed by the Rayleigh equation. The bubble energy balance is given by,

$$pdV/dt + q + dU/dt = 0 \ , \tag{11.63}$$

where the heat transfer, q, is assumed to be dominated by convection according to $q = H_i A(T - T_\infty)$, and the gas internal energy is expressed by, $U = pV/(k-1)$. The perfect gas equation of state, $pV = MR_g T$, and spherical volume, $V = (4/3)\pi R^3$, with surface area, $A = 4\pi R^2$, were employed in Eq. (11.63) to obtain,

$$p\dot{R} + \frac{1}{3k} R\dot{p} + \frac{4\pi H_i(k-1)}{3MR_g k}(pR^3 - p_\infty R_\infty^3) = 0 \ , \tag{11.64}$$

where subscript ∞ designates the equilibrium state. Also, Eq. (11.23) accounts for the water inertia. The equilibrium bubble radius, R_∞, is disturbed to give the initial value,

$$R(0) = R_\infty(1 + \varepsilon) \tag{11.65}$$

at T_∞, where ε is a small number. It follows from Eq. (11.64) that the initial bubble pressure is

$$p(0) = p_\infty \left[\frac{R_\infty}{R(0)} \right]^3 . \tag{11.66}$$

A linearized solution for small amplitude bubble oscillations is obtained by writing $R(t)$ and $p(t)$ as,

$$R(t) = R_\infty + \varepsilon\,\delta R \qquad \text{and} \qquad p(t) = p_\infty + \varepsilon\,\delta p . \tag{11.67}$$

Substitution into Eqs. (11.23) and (11.64), and eliminating δp yields:

$$\frac{d^3\delta R}{dt^3} + a\frac{d^2\delta R}{dt^2} + b\frac{d\delta R}{dt} + c\delta R = 0 , \tag{11.68}$$

with initial conditions given by:

$$t=0, \quad \delta R = 1, \quad \frac{d\delta R}{dt} = 0, \quad \frac{d^2\delta R}{dt^2} = -\frac{3g_c p_\infty}{\rho_l R_\infty^2} , \tag{11.69}$$

and,

$$\begin{aligned} a &= 4\pi(k-1)H_i R_\infty^2 / 3M_g R_g , \\ b &= 3g_c k p_\infty / \rho_l R_\infty^2 \\ c &= 12\pi H_i g_c (k-1) p_\infty / 3M_g R_g \rho_l = a(3p_\infty g_c / \rho_l R_\infty^2) \end{aligned} \tag{11.70}$$

where R_g is the gas constant and M_g is the bubble's mass. An adiabatic bubble ($q=0$) corresponds to $H_i=0$, with an undamped oscillating frequency $(3g_c k p_\infty / R_\infty^2 \rho_l)^{1/2}/2\pi$, which is shown in Fig. 11-27. The isothermal case ($H_i=\infty$) yields another undamped oscillation solution with frequency, $(3g_c p_\infty / R_\infty^2 \rho_l)^{1/2}/2\pi$, which also is shown in Fig. 11-27. However, any other case with a finite value of H_i undergoes a damped oscillation. An example is shown in Fig. 11-27 for

$$H_i = 2\sqrt{g_c p_\infty^3 / \rho_l} / 3(k-1)T_\infty , \tag{11.71}$$

which gives maximum attenuation. It is interesting to note that this damped bubble oscillation closely resembles the damping of a mechanical system by friction. However, there is no frictional dissipation in this formulation. Neither is energy lost from the bubble-liquid system. The damped oscillation for $0 < H_i < \infty$ simply displays an available power loss during a process in which there is entropy production. Thus it appears that the strong attenuation observed in "ringout" involves thermal damping, due to gas bubble heat exchange with the surrounding liquid.

The thermal damping model was extended to a bubbly mixture of noncondensible gas in liquid (Moody, 1983). Equations (5.34) and (5.40) of Chapter 5, which express mass conservation and momentum for one-

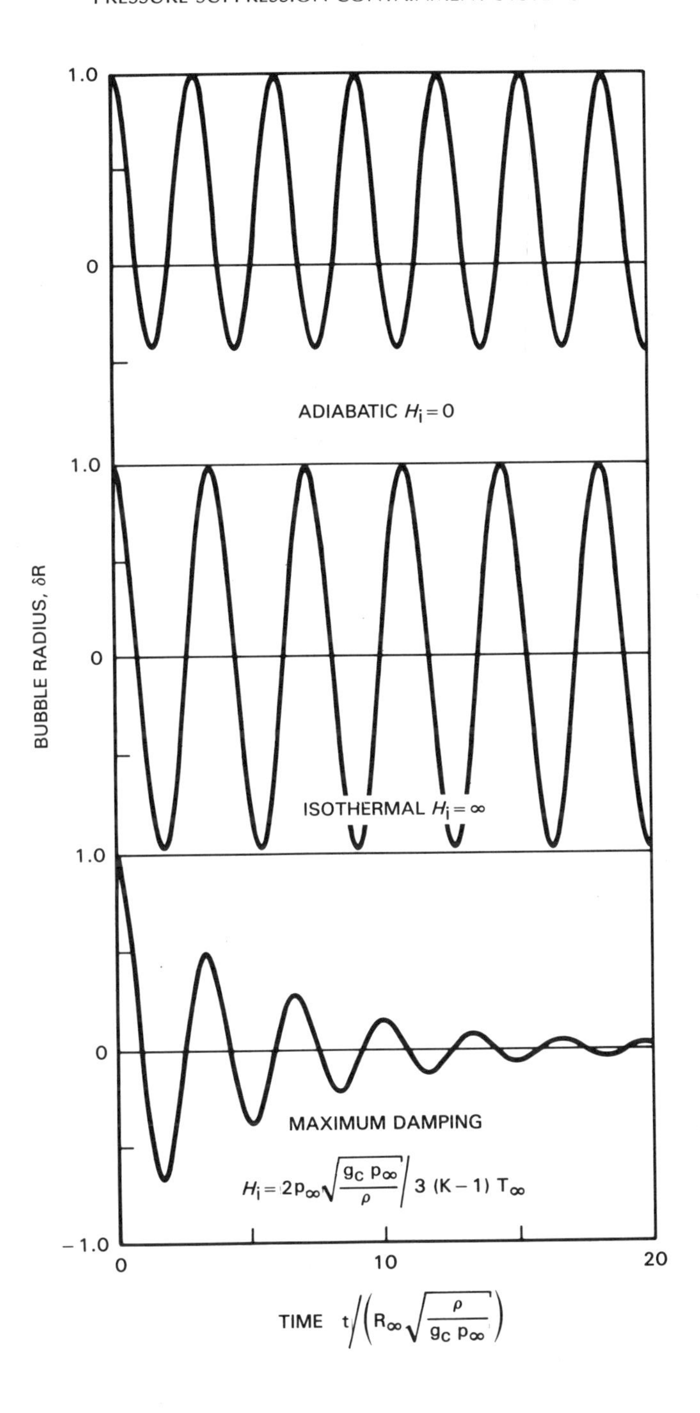

Fig. 11-27 Gas bubble oscillation.

dimensional flow, are written here for horizontal adiabatic flow of a frictionless homogeneous fluid in a uniform passage with $\langle \rho_H \rangle$ and $\langle j \rangle$ replaced by ρ and v:

$$\frac{\partial \rho}{\partial t} + v\frac{\partial \rho}{\partial z} + \rho\frac{\partial v}{\partial z} = 0 \tag{11.72a}$$

$$\frac{\partial v}{\partial t} + v\frac{\partial v}{\partial z} + \frac{g_c}{\rho}\frac{\partial p}{\partial z} = 0 \tag{11.72b}$$

Here, liquid is the continuous phase that contains many small bubbles in an idealized homogeneous pattern with negligible relative motion and interface forces.

Heat transfer between the bubbles and liquid causes nonisentropic state changes. Therefore, mixture density ρ is not a function of pressure only, as in the case of classical acoustics. The mixture density can be written as

$$\rho = [x/\rho_g + (1-x)/\rho_l]^{-1} = \alpha\rho_g + (1-\alpha)\rho_l \ , \tag{11.73a}$$

where quality, x, remains constant in systems without phase change.

Therefore, the density differential is,

$$d\rho = x(\rho/\rho_g)^2 d\rho_g + (1-x)(\rho/\rho_L)^2 d\rho_L \ . \tag{11.73b}$$

The first law for a closed region of fluid is given by

$$q = dU/dt + pdV/dt \ , \tag{11.74}$$

where heat inflow q is considered positive. Writing $V = M/\rho$ and $U = \mu M$, Eq. (11.74) becomes

$$q/M = d\mu/dt - (p/\rho^2)d\rho/dt \ . \tag{11.75}$$

Employing the functional form $\mu(p,\rho)$, its derivative can be written as,

$$d\mu/dt = (g_c c_p/\beta\rho C^2)dp/dt + (p/\rho^2 - c_p/\beta\rho)d\rho/dt \ , \tag{11.76}$$

where,

$$c_p = \left(\frac{\partial h}{\partial T}\right)_p \ ; \quad h = \mu + pv \ ,$$

$$\beta = -\frac{1}{\rho}\left(\frac{\partial \rho}{\partial T}\right)_p \ , \tag{11.77}$$

$$C = g_c\left(\frac{\partial p}{\partial \rho}\right)_s \ .$$

Slight relative motion between the small bubbles and the liquid establishes internal gas circulation so that interphase heat transfer would tend to be limited by convection in the gas. A single bubble at temperature T_g surrounded by liquid at T_l loses thermal energy at a rate

$$q = H_i A(T_g - T_l) \ . \tag{11.78}$$

Since $q > 0$ represents heat inflow to the liquid, $q < 0$ represents heat inflow to the bubble. It follows from Eqs. (11.75) and (11.76) that

$$d\rho_g/dt = (\beta\rho/c_p)_g q/M_g + (g_c/C_g^2)dp/dt \tag{11.79a}$$

for a single bubble, and

$$d\rho_l/dt = -(\beta\rho/c_p)_l q/M_l + (g_c/C_l^2)dp/dt \tag{11.79b}$$

for the liquid in a unit cell of mixture containing the bubble. The total mass $M = M_g + M_l$ is constant in a unit cell, where $M_g = xM$ and $M_l = (1-x)M$. It follows that Eqs. (11.73b), (11.78), (11.79a), (11.79b), and (11.72a) can be written in terms of pressure instead of density derivatives as:

$$p_t + vp_z + (\rho C_b^2/g_c)v_z + C_b^2 F H_i A(T_g - T_l)/g_c M = 0 \ , \tag{11.80}$$

where subscripts t and z designate partial derivatives with respect to time and space and F is the mixture property

$$F \triangleq (\rho/\rho_g)^2(\beta\rho/c_p)_g - (\rho/\rho_l)^2(\beta\rho/c_p)_l \ , \tag{11.81}$$

and the bubbly mixture sound speed is given by,

$$C_b = [x(\rho/\rho_g)^2/C_g^2 + (1-x)(\rho/\rho_l)^2/C_l^2]^{-1/2} \ . \tag{11.82}$$

Gas temperature is expressed from the perfect gas law $T_g = p/R\rho_g$. The liquid temperature is assumed to be at the undisturbed gas temperature such that, $T_l = p_\infty/R\rho_{g\infty}$, which is considered constant, implying that the liquid has much larger thermal capacity than the gas.

Acoustic disturbances result in negligible convective derivatives vp_z and vv_z, and relatively constant derivative coefficients. Therefore, the governing acoustic equations for a bubbly gas-liquid mixture with interphase heat transfer are obtained from Eqs. (11.80) and (11.72b) in the forms,

$$p_t + D(p - p_\infty) + (\rho C_b^2/g_c)v_z = 0 \tag{11.83}$$

and,

$$v_t + (g_c/\rho)p_z = 0 \ , \tag{11.84}$$

where D is the thermal damping coefficient,

$$D = (C_b^2 F H_i A)/(g_c M R \rho_g) \ . \tag{11.85}$$

Either p or V can be eliminated from Eqs. (11.83) and (11.84) to give equations of the form

$$\phi_{tt} + D\phi_t - C_b^2\phi_{zz} = 0 \ ; \quad \phi = p,v \ . \tag{11.86}$$

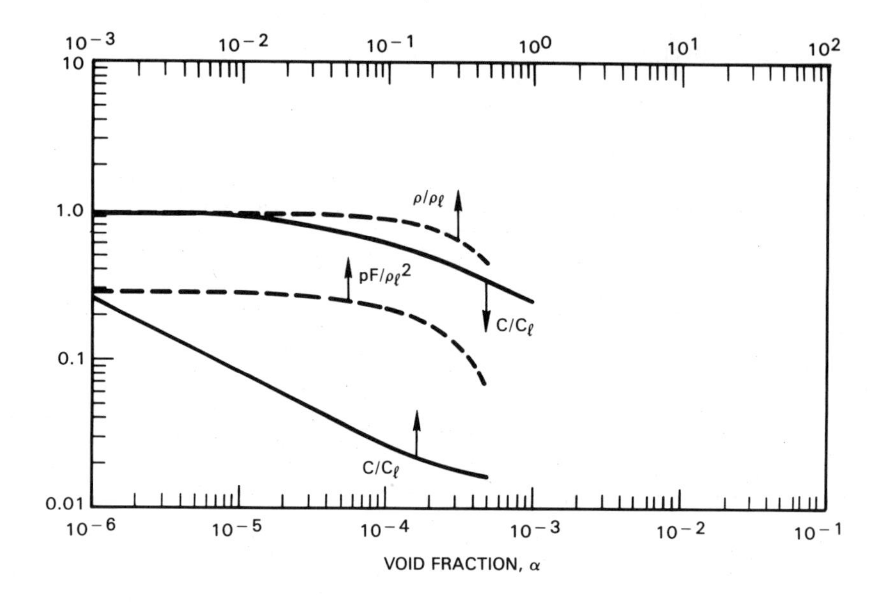

Fig. 11-28 Bubbly mixture properties.

The absence of interphase heat transfer corresponds to $D=0$, for which Eq. (11.86) yields the classical one-dimensional wave equation, $\phi_{tt} - C_b^2\phi_{zz} = 0$. It is interesting to note that Eq. (11.86) also characterizes the behavior of a damped vibrating string.

Figures 11-28 and 11-29 give bubbly mixture properties and the damping coefficient at standard conditions in terms of void fraction. This thermal damping mechanism helps explain the strong damping associated with "ringout."

11.3.10.3 *Fluid-Structure Interaction (FSI)*

Disturbances caused by vent water clearing, air discharge, and unsteady steam condensation create forces that are transmitted through the pool to its boundaries and submerged structures. A simplified procedure for estimating submerged structure response involves a prediction of the pool flow field as if it contained no submerged structures and was bounded by rigid walls and a free surface. Local fluid motion where a structure actually would exist is regarded as a local unsteady uniform flow to be used in a separate analysis, which provides forces acting on the structure as if it were rigid. The force exerted on a rigid structure can be obtained by analysis or experiment. Several considerations are required before the rigid structure force can be employed in a dynamic analysis for the response of the flexible structure.

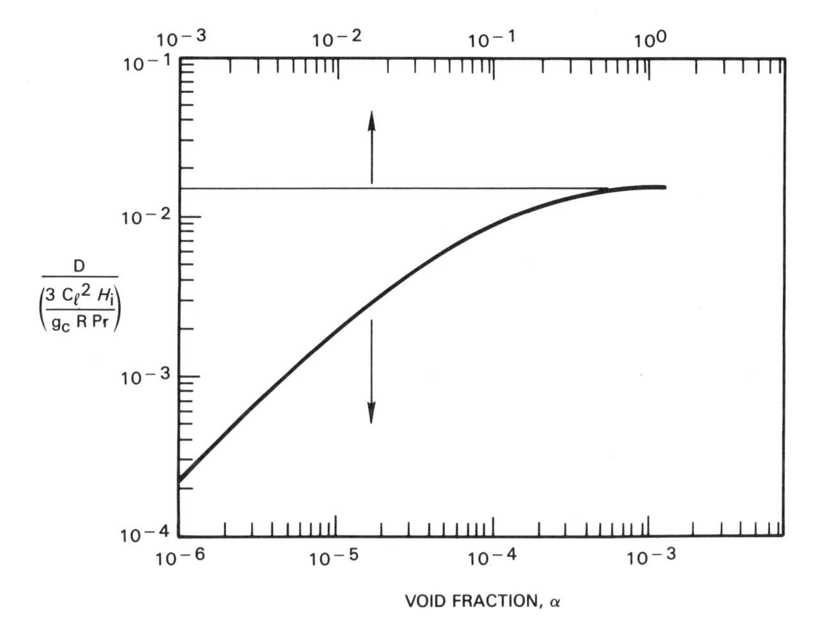

Fig. 11-29 Damping coefficient.

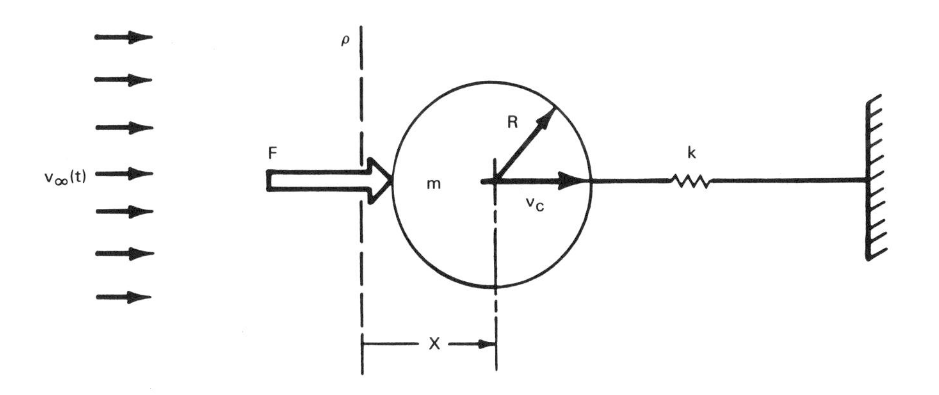

Fig. 11-30 Fluid-structure interaction, bulk flow.

For example, consider a cylinder of length, L, moving rightward at velocity $v_c(t)$ in an unsteady uniform flow of velocity $v_\infty(t)$, as shown in Fig. 11-30. If the disturbance time is long relative to the pool acoustic response time, potential flow methods can be used to predict the resulting fluid force on the cylinder. The potential function for a cylinder moving at velocity, v_c, in the positive x direction through a stationary fluid is,

$$\phi_c = -v_c(t)\frac{R^2}{r}\cos\theta \ , \tag{11.87}$$

where the r,θ coordinate system is fixed at the center of the cylinder. Furthermore, the potential function for uniform flow at unsteady velocity, $v_\infty(t)$, in the positive x direction past a stationary cylinder is

$$\phi_{uf} = v_\infty(t)\left(r + \frac{R^2}{r}\right)\cos\theta \ . \tag{11.88}$$

If gravity effects are neglected, the unsteady Bernoulli equation becomes,

$$\frac{\partial\phi}{\partial t} + \frac{1}{2}\left[\left(\frac{\partial\phi}{\partial r}\right)^2 + \left(\frac{\partial\phi}{r\partial\theta}\right)^2\right] + \frac{g_c}{\rho}p = f(t) \ . \tag{11.89}$$

Equation (11.89) can be used to obtain the pressure distribution on the cylinder. Simultaneous cylinder motion in a uniform unsteady flow is obtained by adding the potential functions of Eqs. (11.87) and (11.88) as

$$\phi = \phi_c + \phi_{uf} \ . \tag{11.90}$$

The differential radial force on a cylinder surface area of $LRd\theta$ is $dF = pLRd\theta$. The x component of this force is $dF_x = \cos\theta dF$, which can be integrated over the cylinder surface to yield,

$$F_x = LR\int_0^{2\pi} pd\theta = \pi R^2 L\frac{\rho}{g_c}(2\dot{v}_\infty - \dot{v}_c) \ . \tag{11.91}$$

If the cylinder were rigid, the applied force would correspond to $V_c = 0$, or,

$$F_R = 2\pi R^2 L\frac{\rho}{g_c}\dot{v}_\infty \ . \tag{11.92}$$

A simple structural dynamic equation for negligible damping is given by,

$$\frac{M}{g_c}\ddot{x} + Kx = F_x \ . \tag{11.93}$$

It follows that the structural dynamics equation can be written as,

$$\frac{(M + \rho\pi R^2 L)}{g_c}\ddot{x} + Kx = 2\pi R^2 L\frac{\rho}{g_c}\dot{v}_\infty = F_R \ . \tag{11.94}$$

The term $\rho\pi R^2 L$ is identified as the virtual mass for a translating cylinder (see Fig. 11-20a).

It follows that for the case of a bulk flow without acoustic effects, the flexible structure analysis can be performed by applying force F_R, obtained for a rigid structure, provided that the structural mass is increased by the virtual mass of the fluid. This result is general and extends to submerged structures of other geometries.

Next, consider a disturbance like a chugging pressure spike whose magnitude varies rapidly over a brief period of time, which is of the same order as the pool acoustic response time (i.e., pool dimension/sound speed). Figure 11-31 shows an acoustic disturbance arriving and reflecting from a flexible structure, which has fluid on one side only. The flexible structure's response for negligible damping can be approximated by,

$$\ddot{x}_s + \frac{Kg_c}{M} x_s = \frac{g_c}{M} p_s A \ . \tag{11.95}$$

The linear wave equations that govern oncoming and reflected pressure and velocity disturbances have the following general solutions for pressure and velocity (Lahey et al., 1990),

$$p(x,t) = F(t + x/C) + G(t - x/C) \ , \tag{11.96}$$

$$v(x,t) = -\frac{g_c}{\rho C}[F(t + x/C) - G(t - x/C)] \ . \tag{11.97}$$

If the initial pressure and velocity are zero, the incident pressure disturbance, $p_0(t)$, corresponds to

$$p_0\left(t + \frac{x}{C}\right) = F_0\left(t + \frac{x}{C}\right) \ , \tag{11.98}$$

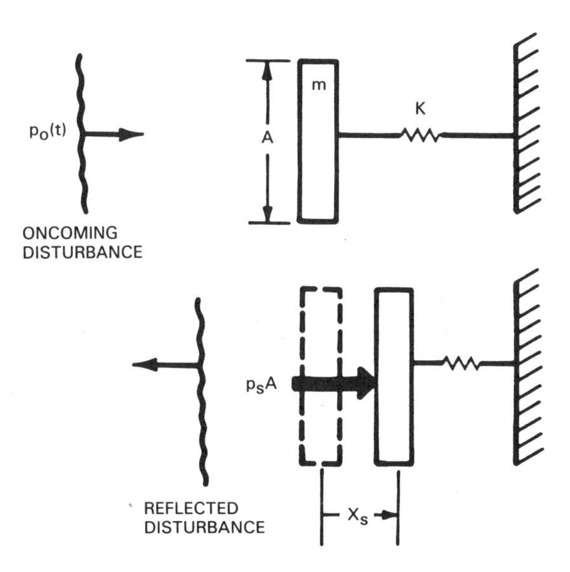

Fig. 11-31 Fluid-structure interaction, acoustic disturbance.

since,

$$G_0\left(t-\frac{x}{C}\right)=0 \ . \tag{11.99}$$

Arrival of the pressure wave at the structure causes it to move with velocity $\dot{x}_s$. The resulting structural motion causes a reflected pressure wave, G_r, which now adds to F_0 so that the pressure and velocity at the structure from Eqs. (11.96) and (11.97) are,

$$p=p_0+G_r \tag{11.100}$$

and,

$$\dot{x}_s=v=-\frac{g_c}{\rho C}(p_0-G_r) \ . \tag{11.101}$$

Substituting into Eq. (11.95) yields,

$$\ddot{x}_s+\left(\frac{\rho AC}{M}\right)\dot{x}_s+\left(\frac{Kg_c}{M}\right)x_s=\frac{2g_c}{M}p_0(t)A \tag{11.102}$$

or, employing Eq. (11.101) in (11.102),

$$\ddot{p}_s+\left(\frac{\rho AC}{M}\right)\dot{p}_s+\left(\frac{Kg_c}{M}\right)p_s=2\left[\ddot{p}_0+\left(\frac{Kg_c}{M}\right)p_0\right] \ . \tag{11.103}$$

The appearance of linear acoustic damping terms, $(\rho AC/M)\dot{x}_s$ and $(\rho AC/M)\dot{p}_s$, accounts for acoustic energy radiation into the fluid. Note that if a structure is rigid (i.e., $K\to\infty$), then Eq. (11.103) yields,

$$p_{s,\text{rigid}}=2p_0 \ . \tag{11.104}$$

This result is consistent with classical waterhammer doubling of the magnitude of a pressure disturbance at a rigid boundary. The forcing function in Eq. (11.102), $2g_cp_0A/M$, can thus be written as,

$$\frac{2g_c}{M}p_0A=\frac{g_c}{M}p_{s,\text{rigid}}A \ . \tag{11.105}$$

It follows that a flexible structure response analysis can be performed without submergence by using the pressure obtained from a submerged rigid structure, $p_{s,\text{rigid}}$, provided that a mechanical damping term is included.

Sophisticated computer programs are available for treating the FSI problem (McCormick, 1979; GE, 1978), which inherently include effects of water inertia and acoustic damping.

11.4 Scale Modeling Procedures

Scale models are useful for predicting certain unsteady thermal-hydraulic behavior in nuclear containments. The design of a scale model requires that all important physical phenomena be preserved. Many researchers, who make extensive use of scale models, have a preferred method of determining the appropriate scaling laws for a given system. One such method is the Buckingham pi theorem (Baker et al., 1973), which is discussed in Sec. 4.3.4. This approach provides model laws, but does not show which effects can be neglected. Another method, fractional analysis (Kline, 1965), which is based on force ratio similitude, does not show how to scale thermodynamic effects. A generalized method of similitude provides model laws from a complete problem formulation and gives the starting point for scale modeling of thermal-hydraulic phenomena.

Two scale-modeling philosophies used extensively in studies of containment phenomena are *segment scaling,* which involves a small section of a full-size system, and *geometric scaling,* which employs a system geometrically similar to full size, but with all dimensions reduced by the same factor. Full-scale thermodynamic properties usually are employed in segment scaling, whereas these properties often must be reduced by a particular fraction in geometric scaling.

A summary of the scale modeling employed in containment technology and important examples of model laws are given in this section.

11.4.1 Segment Scaling

Segment scale modeling was employed in some of the pressure suppression tests mentioned in Chapter 1. A small vessel of saturated steam and water, at prototypical pressure, was discharged through a simulated pipe rupture into a drywell-vent-wetwell system in which the volume ratios were equal to that of full size. Also, the blowdown discharge and vent areas were reduced by the segment volume fraction of full size. Full-size vent flow length, pool depth, and submergence were employed, and the experimental results provided the pressure-time trace expected if a break should occur in the full size system. Figure 11-2 gives a correlation of some of the data from the segment tests conducted for the Humboldt and Bodega containment systems. Although these data were recorded in a small segment test, identical results would be expected for the same volume and area ratios in a full size test.

11.4.2 Geometric Scaling

Geometric scale modeling has also been used successfully in predicting phenomena that include pool sloshing behavior during an earthquake, pool swell during a LOCA, pool circulation and heat up during steam discharge, transient pipe pressures during SRV discharge, and accelerating submerged water jet profiles. It is likely that geometric scaling will also be employed more extensively to study the thermal hydraulics of postulated severe accidents.

A general description of the geometric similarity scale-modeling procedure is given next.

11.4.2.1 *Overall Procedure*

The scale-modeling procedure starts from basic equations of mass, momentum, and energy conservation in a chosen system. Time, space coordinates, and all variable properties are normalized respectively with the response time, characteristic size, and magnitudes of estimated property disturbances so that the resulting nondimensional variables and derivatives are of order 1.0 in magnitude. The relative magnitudes of nondimensional model coefficients appearing in the normalized equations display the physical effects which must be preserved in a scale model. Also, the normalized variables are equal in full and small scales, making it possible to employ data from a small-scale experiment and upscale it to determine full size behavior.

11.4.2.2 *Governing Conservation Equations*

Governing equations of thermal-hydraulic phenomena occurring in severe accidents are summarized below in the integral control volume form (Shapiro, 1953):

Mass conservation:

$$\int_{cs} dw + \frac{d}{dt}\int_{cv} \rho dV = 0 \ . \tag{11.106}$$

Momentum:

$$\int_{cs} \underline{v} dw + \frac{d}{dt}\int_{cv} \rho \underline{v} dV + g_c \int_{cs} p d\underline{A} + g_c \underline{j} \int_{cv} \rho dV - g_c \int_{cs} \underline{\underline{\Gamma}} \cdot d\underline{A} \ . \tag{11.107}$$

Energy conservation (uniform pressure):

$$\int_{cs} (h_0 dw + dq) + \mathcal{P}_{\text{out}} - \mathcal{P}_{\text{in}} - \int_{cv} q''' dV + \frac{d}{dt}\int_{cv} \varepsilon \rho dV = 0 \ , \tag{11.108}$$

where $\mathcal{P}$ designates mechanical power due to expansion, $p(dV/dt)$, or shaft transmission and $\underline{\underline{\Gamma}}$ is the shear stress tensor, defined by,

$$\nabla \cdot \underline{\underline{\Gamma}} = \mu \nabla^2 \underline{V} + \frac{1}{3} \mu \nabla (\nabla \cdot \underline{V}) \ , \tag{11.109}$$

ε is the total stored energy per unit mass, and $d\underline{A}$ is a differential area vector on the surface of the control volume. Integrals identified by *cs* are taken over the control surface, whereas *cv* designates integrals throughout the control volume.

These equations are nondimensionalized with reference variables, which yields parameter groups (model coefficients) to be preserved in a scale model test.

11.4.2.3 *Normalized Variables*

If ϕ is a variable whose initial value is ϕ_i, and its farthest expected departure from ϕ_i is ϕ_r, the normalized form is,

$$\phi^* = \frac{\phi - \phi_i}{\Delta\phi} \; ; \quad \Delta\phi = \phi_r - \phi_i \; , \tag{11.110}$$

and its derivatives are of order 1.0 in magnitude. Space variables x, y, z are normalized to the system space interval L, as $x^* = x/L$, etc. Time is normalized with the response time Δt so that $t^* = t/\Delta t$. Geometric similarity between full-size and small-scale models is usually implied. Table 11-3 gives a list of the normalized variables.

TABLE 11-3
Normalized Variables

$t^* = \dfrac{t}{\Delta t}$	$\varepsilon^* = \dfrac{\varepsilon - \varepsilon i}{\Delta\varepsilon}$
$x^*, y^*, z^* = \dfrac{x, y, z}{L}$	$w^* = \dfrac{w}{w_r}$
$p^* = \dfrac{p - p_i}{\Delta p}$	$h_0^* = \dfrac{h_0}{h_{0r}}$
$\rho^* = \dfrac{\rho - \rho_i}{\Delta\rho}$	$V^* = \dfrac{V - V_i}{\Delta V}$
$\underline{v}^* = \dfrac{\underline{v} - \underline{v}_i}{\Delta v}$	$d\underline{A}^* = \dfrac{d\underline{A}}{A_r}$
$\underline{\underline{\Gamma}} = \dfrac{\underline{\underline{\Gamma}}}{\mu(\Delta v/L)}$	$q^* = \dfrac{q}{q_r}$
$q''^* = \dfrac{q''}{q''_r}$	$q'''^* = \dfrac{q'''}{q'''_r}$
$\mathcal{P}^* = \dfrac{\mathcal{P}}{\mathcal{P}_r}$	

11.4.2.4 *Nondimensional Equations and Model Coefficients*

The nondimensional equations obtained from Eqs. (11.106), (11.107), and (11.108) are

Mass conservation:

$$\int_{cs} \pi_3 dw^* + \frac{d}{dt^*}\int_{cv} (\pi_1 \rho^* + \pi_2) dV^* = 0 \qquad (11.111)$$

Momentum:

$$\int_{cs} \pi_3(\pi_4 \underline{V}^* + \pi_5) d\dot{w}^* + \frac{d}{dt^*}\int_{cv} (\pi_1 \rho^* + \pi_2)(\pi_4 \underline{V}^* + \pi_5) dV^*$$

$$+ \int_{cs} (\pi_6 p^* + \pi_7) d\underline{A}^* + \int_{cv} \pi_8(\pi_1 \rho^* + \pi_2) dV^* - \int \pi_9 \underline{\underline{\Gamma}}^* \cdot d\underline{A}^* = 0 \qquad (11.112)$$

Energy conservation:

$$\int_{cs} \pi_{12} h_0^* dw^* + \int_{cs} \pi_{13} \underline{q}''^* \cdot d\underline{A}^* - \int_{cv} \pi_{14} q'''^* dV^*$$

$$+ \frac{d}{dt^*}\int_{cv} (\pi_1 \rho^* + \pi_2)(\pi_{10} \varepsilon^* + \pi_{11}) dV^* = \pi_{15}(\mathcal{P}_{in}^* - \mathcal{P}_{out}^*) \ , \qquad (11.113)$$

where the resulting nondimensional groups, or model coefficients, are given in Table 11-4. Relative magnitudes of the model coefficients determine which physical effects are negligible and which should be preserved by a scale model.

11.4.2.5 *Reference Parameters for a Postulated Severe Accident*

The nondimensional model coefficients π_j, independent variables t^* and x^*, y^*, z^*, and properties ϕ_k^* provide the information for designing a scale model test and upscaling the results for full size. Although the governing equations are not solved in the modeling procedure, simplified formulations are employed next to estimate the response times, Δt, property disturbances, $\Delta\phi_k$, and reference variables, ϕ_r, to be used in evaluating the π_j.

A postulated severe accident involves the melting of a reactor core from a loss of cooling. Accidents of this type are discussed in Sec. 8.6. As an example of geometric scaling, let us consider a hypothetical severe accident in a BWR having a MARK-I containment. One designs a scale model test that begins with molten core debris draining by gravity from a simulated depressurized reactor vessel break onto a concrete floor (Fig. 11-32). The molten debris spreads and undergoes simultaneous cooling from the top

TABLE 11-4
Model Coefficients

$\pi_1 = \dfrac{\Delta \rho}{\rho_r}$	$\pi_6 = \dfrac{g_c \Delta p \Delta t^2 A_r}{\rho_r V_r L}$
$\pi_2 = \dfrac{\rho_i}{\rho_r}$	$\pi_7 = \dfrac{g_c p_i \Delta t^2 A_r}{\rho_r V_r L}$
$\pi_3 = \dfrac{w_r \Delta t}{\rho_r V_r}$	$\pi_8 = \dfrac{g \Delta t^2}{L}$
$\pi_4 = \dfrac{\Delta V \Delta t}{L}$	$\pi_9 = \dfrac{\nu_r \Delta t^2 \Delta V A_r}{L_2 V_r}$
$\pi_5 = \dfrac{V_i \Delta t}{L}$	$\pi_{10} = \dfrac{\Delta \varepsilon}{\varepsilon_r}$
$\pi_{11} = \dfrac{\varepsilon_i}{\varepsilon_r}$	$\pi_{12} = \dfrac{w_r \Delta t h_{0r}}{\rho_r V_r \varepsilon_r}$
$\pi_{13} = \dfrac{q_r'' \Delta t A_r}{\rho_r V_r \varepsilon_r}$	$\pi_{14} = \dfrac{q_r''' \Delta t v_r}{\rho_r V_r \varepsilon_r}$
$\pi_{15} = \dfrac{\mathcal{P}_r \Delta t}{\rho_r V_r \varepsilon_r}$	

to an overlying water layer, and from the bottom by conduction to the floor, and possibly immobilization by freezing, finally arriving at the dry-well shell. The significant safety question is does the shell melt or not.

Potentially important phenomena that are neglected in this example include nonuniform debris stratification, decay heat from oxidic components, concrete water and gas release, metal-water reaction, containment pressurization by possible hydrogen generation and burning, and boiling of the water layer. Table 11-5 gives example parameters and properties.

11.4.2.6 *Molten Debris Draining*

Reference values associated with gravity draining of molten core debris are:

$$\Delta v = v_r = \sqrt{2gH}\ ; \quad Q = A_b v_r\ ;$$
$$w_r = \rho_d Q\ ; \quad A_r = A_b\ ; \tag{11.114}$$
$$\rho_r = \rho_d\ ; \quad L = H\ ; \quad V_r = V_d$$
$$\Delta p = \rho_d \frac{g}{g_c} H\ .$$

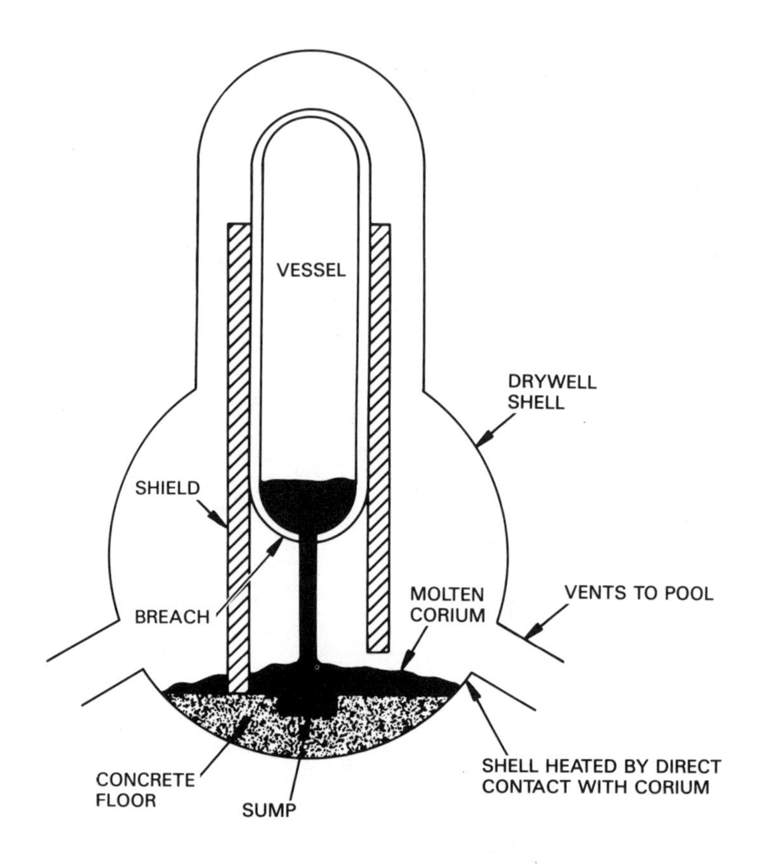

Fig. 11-32 Example system.

TABLE 11-5
Example Parameters and Properties

Vessel branch area A_b	$=16\ \text{cm}^2$	Debris initial temperature T_{di}	$=2600$ K
Debris specific heat c_d	$=480$ J/kg K	Ambient temperature T_∞	$=373$ K
Concrete specific heat c_c	$=835$ J/kg K	Debris solids temperature T_{df}	$=2100$ K
Shield diameter D	$=5$ m	Debris density ρ_d	$=9000\ \text{kg/m}^3$
Heat convection coefficient h	$=400\ \text{W/m}^2$ K	Concrete density ρ_c	$=2300\ \text{kg/m}^3$
Solidification enthalpy h_{LS}	$=250{,}000$ J/kg	Debris thermal diffusivity α_d	$=0.07\ \text{cm}^2/\text{s}$
Debris depth in vessel H	$=2$ m	Concrete thermal diffusivity α_c	$=0.007\ \text{cm}^2/\text{s}$
Debris thermal conductivity k_d	$=30$ W/m K	Steel thermal diffusivity α_s	$=0.14\ \text{cm}^2/\text{s}$
Concrete thermal conductivity k_c	$=1.3$ W/m K	Shell thickness δ	$=3$ cm
Steel thermal conductivity k_s	$=50$ W/m K	Molten debris volume V_d	$=15\ \text{m}^3$

The time response, based on the parameters of Table 11-5, is

$$\Delta t = \frac{V_d}{Q} = 1500 \text{ sec} \ . \tag{11.115}$$

The model design coefficients to be preserved in a scale model, obtained from π_1 through π_9 for negligible heat transfer during core debris discharge show that only geometric similarity should be preserved. The discharge rate and time can be upscaled to predict full-size behavior from

$$w^* = \frac{\rho_d Q}{\rho_d Q_r} = \frac{w}{w_r} = Q^* = \frac{Q}{A_b \sqrt{2gH}} \tag{11.116}$$

$$t^* = \frac{A_b \sqrt{2gH}}{V_d} t \ . \tag{11.117}$$

11.4.2.7 *Molten Debris Spreading on the Floor*

The spreading velocity and depth of molten debris on a horizontal surface are estimated by equating the hydrostatic and velocity heads, which yields $V = (2gY)^{1/2}$. If the volumetric rate of debris pouring onto the surface is Q, and its frontal perimeter is the floor diameter D, then $Q = YD(2gY)^{1/2}$. It follows that reference values for horizontal spreading on the floor are w_r, Q, ρ_r (the same as in Sec. 11.4.2.6):

$$\begin{aligned} Y &= \left(\frac{Q^2}{2gD^2}\right)^{1/3} ; \quad \Delta V = V_r = \left(\frac{2gQ}{D}\right)^{1/3} ; \\ L &= D ; \quad \Delta P = \rho_d \frac{g}{g_c} Y ; \\ A_r &= \frac{\pi D^2}{4} ; \quad V_r = \frac{\pi D^2 Y}{4} \ . \end{aligned} \tag{11.118}$$

The time response for flow a distance D at velocity V_r is, for the parameters of Table 11-5,

$$\Delta t = \frac{D}{V_r} = 15 \text{ sec} \ . \tag{11.119}$$

The model coefficients of Table 11-4 were evaluated, showing that the only nondimensional group to be preserved between a scale-model test and the full-size system is,

$$\pi_6 = \pi_8 = \left(\frac{gD^5}{4Q^2}\right)^{1/3} \ . \tag{11.120}$$

The velocity of spreading and time would be upscaled from

$$\underline{v}^* = \underline{v}\left(\frac{D}{2gQ}\right)^{1/3} \quad ; \quad t^* = t\left(\frac{2gQ}{D^4}\right)^{1/3} . \tag{11.121}$$

11.4.2.8 *Cooling of Molten Debris*

A reference convection heat flux from the top surface is

$$q_r'' = q_h'' = H(T_{di} - T_\infty) . \tag{11.122}$$

If the stored thermal energy is

$$\Delta\varepsilon = \varepsilon_r = c_d(T_{di} - T_\infty) , \tag{11.123}$$

a corium layer of thickness Y, with convection from one surface, will cool according to

$$Y\rho_d c_d \frac{dT}{dt} = q_h'' . \tag{11.124}$$

The solution is exponential with a response time, based on the parameters of Table 11-5,

$$\Delta t_h = \frac{\rho_d c_d Y}{h} = 65 \text{ sec} . \tag{11.125}$$

Conduction from the bottom surface to a concrete floor is obtained from a model of a molten layer of thickness Y and initial temperature T_{di}, suddenly put in contact with a semi-infinite slab at T_∞. The layer and slab temperatures are governed by the layer boundary condition,

$$\rho_d c_d Y \frac{dT_d}{dt} = -k_c \frac{\partial T_c(0,t)}{\partial x} \tag{11.126}$$

and conduction in the slab, governed by,

$$\frac{\partial T_c}{\partial t} = \alpha_c \frac{\partial^2 T_c}{\partial x^2} , \tag{11.127}$$

for which a solution for the debris temperature is,

$$\frac{T_d - T_\infty}{T_{di} - T_\infty} = e^{t/\Delta t} \operatorname{erfc}\left(\frac{t}{\Delta t}\right) , \tag{11.128}$$

with a thermal response time, based on parameters of Table 11-5,

$$\Delta t_c = \frac{Y^2}{\alpha_c}\left(\frac{\rho_d c_d}{\rho_c c_c}\right)^2 = 270 \text{ sec} . \tag{11.129}$$

An approximate heat flux to the concrete is expressed as the quotient of thermal energy $\rho_d c_d(T_{di} - T_\infty)Y$ and the response time, which gives,

$$q_c'' = \frac{\rho_c c_d (T_{di} - T_\infty) Y}{\Delta t_c} \tag{11.130}$$

$$= \frac{k_c}{Y}(T_{di} - T_\infty)\frac{\rho_c c_c}{\rho_d c_d} \ . \tag{11.131}$$

A debris layer of thickness Y, initially at T_{di} with a freezing temperature T_{df}, would have a maximum heat flux

$$q_f'' = \frac{k_d(T_{di} - T_{df})}{Y} \tag{11.132}$$

away from the freezing interface. If the debris is idealized as a pure substance, the heat flux must balance the energy absorption rate of phase change $h_{Ls} v_f \rho_d$ where v_f is the freezing interface velocity. It follows that the freezing response time, based on parameters of Table 11-5, is

$$\Delta t_f = \frac{Y}{v_f} = \frac{Y^2 \rho_d h_{Ls}}{k_d(T_{di} - T_{df})} = 5 \text{ sec} \tag{11.133}$$

after the freezing temperature is reached.

A comparison of thermal response times shows that conduction to the concrete can be neglected from a scale model because it takes an order of magnitude longer than the other heat transfer phenomena. If the debris spreading time is employed as the reference response time, the thermal groups to be preserved in a scale model are:

$$\frac{hD^2}{\rho_d c_d Q} \quad \text{(convection cooling)} \ , \tag{11.134}$$

$$\left(\frac{\rho_c c_c}{\rho_d c_d}\right)^2 \frac{c_d(T_{di} - T_{df})}{h_{Ls}} \quad \text{(freezing)} \ . \tag{11.135}$$

Temperatures would be upscaled by,

$$T^* = \frac{T - T_\infty}{T_{di} - T_\infty} \ . \tag{11.136}$$

11.4.2.9 *Debris Arrival at Drywell Shell*

Sudden arrival of molten debris at the steel shell of a MARK-I containment results in shell heating, which is to be predicted. The separate effect is considered as shell response, and involves only the energy groups of Table 11-4, namely π_{10}, π_{11}, and π_{13}. If the molten debris and shell are treated as two semi-infinite slabs with conduction heat transfer at the contact surface, the contact temperature is (Carslaw and Jaeger, 1947),

$$\frac{T_{con} - T_\infty}{T_{di} - T_\infty} = \left(1 + \frac{k_s}{k_d}\sqrt{\frac{\alpha_d}{\alpha_s}}\right)^{-1} . \tag{11.137}$$

Other reference parameters are

$$\begin{aligned} &A_r = DY \ ; \quad V_r = DY\delta \ ; \quad L = \delta \ ; \\ &\Delta\varepsilon = c_s(T_{con} - T_\infty) \ ; \quad \varepsilon_r = \varepsilon_i = c_s T_\infty \ . \end{aligned} \tag{11.138}$$

The contact heat flux is given by

$$q''_{con} = \frac{k_s(T_{con} - T_\infty)}{\sqrt{\pi\alpha_s t}} . \tag{11.139}$$

The integral of q''_{con} causes shell heatup to T_{con} in a reference time, based on parameters of Table 11-5,

$$\Delta t_s = \frac{\pi\delta^2}{4\alpha_s} = 50 \text{ sec} , \tag{11.140}$$

for which the average shell heat flux is

$$q''_s = \frac{\Delta\varepsilon}{\Delta t_s} = \frac{4k_s}{\pi\delta}(T_{con} - T_\infty) . \tag{11.141}$$

It follows that the nondimensional groups for preserving shell thermal effects in a scale-model test are

$$\begin{aligned} \pi_{10} = \pi_{13} &= \frac{T_{con} - T_\infty}{T_\infty} \\ &= \frac{T_{di} - T_\infty}{T_\infty}\left(1 + \frac{k_s}{k_d}\sqrt{\frac{\alpha_d}{\alpha_s}}\right)^{-1} . \end{aligned} \tag{11.142}$$

The π_{11} was unity, which introduces no model laws. Shell temperature would be upscaled according to Eq. (11.136) and,

$$t^* = \frac{4\alpha_s}{\pi\delta^2} t . \tag{11.143}$$

11.4.2.10 *Multiphenomena Scale Modeling*

The phenomena of molten debris discharge from the vessel, spreading, cooling, and heating of the shell could be studied as separate effects in a scale model. However, it is often desirable to include as many phenomena as possible in an integral scale model. Limitations often are discovered by the inability to satisfy all the model design coefficients that must be preserved. Moreover, phenomena time responses also must be compatible, and this can introduce further limitations.

An integral test should preserve the nondimensional groups of Eqs. (11.120), (11.134), (11.135), and (11.142). The time scale requirements can be obtained by considering two phenomena, j and k. Since nondimensional time t^* is the same in the small-scale model (m) and in full-size (f), for each phenomena,

$$\left(\frac{t_j}{\Delta t_j}\right)_m = \left(\frac{t_j}{\Delta t_j}\right)_f , \tag{11.144}$$

$$\left(\frac{t_k}{\Delta t_k}\right)_m = \left(\frac{t_k}{\Delta t_k}\right)_f . \tag{11.145}$$

To preserve the same relative times for each phenomena in a postulated accident,

$$\left(\frac{t_j}{t_k}\right)_m = \left(\frac{t_j}{t_k}\right)_f , \tag{11.146}$$

which yields,

$$\left(\frac{\Delta t_j}{\Delta t_k}\right)_m = \left(\frac{\Delta t_j}{\Delta t_k}\right)_f . \tag{11.147}$$

If, for example, j and k represent the phenomena of molten debris discharge from the vessel and spreading on the floor, respectively, Eqs. (11.117) and (11.121) yield,

$$\left[\frac{V_d}{DA_b}\left(\frac{A_b}{DH}\right)^{1/3}\right]_m = \left[\frac{V_d}{DA_b}\left(\frac{A_b}{DH}\right)^{1/3}\right]_f , \tag{11.148}$$

which simply implies geometric similarity. Therefore, the time scales for debris discharge and spreading are accommodated by geometric similarity.

Next, consider debris discharge and shell heatup as the j and k phenomena. Eqs. (11.117) and (11.143) yield the requirement,

$$\left(\frac{\delta^2 A_b \sqrt{2gH}}{\alpha_s V_d}\right)_m = \left(\frac{\delta^2 A_b \sqrt{2gH}}{\alpha_s V_d}\right)_f . \tag{11.149}$$

Equation (11.149) shows that in order to include the shell thermal response in an integral scale-model test, the simulated shell thickness should correspond to

$$\delta_m = \delta_f \sqrt{\frac{\alpha_{sm}}{\alpha_{sf}}\left(\frac{D_m}{D_f}\right)^{1/4}} . \tag{11.150}$$

If the shell material is the same in the full size and small scale, a 1/16-scale model of the vessel and floor geometry would require a shell of thickness $\delta_m = 1.5$ cm to preserve the thermal response of a 3-cm-thick full-size shell.

A procedure for obtaining scale-model laws of severe accident phenomena has been described and applied to example events. Nondimensional parameter groups were obtained that should be preserved in both full size and a scale model. Upscaling of data from a scale-model experiment to the full-size system is accomplished with nondimensional variables, which also are preserved. When more than one phenomenon occurs in a system, the relative time-scale ratios in full size and small scale must be equal, which sometimes restricts the range of scale sizes achievable. When restrictions are too cumbersome to manage in a single test, it is desirable to study separate effects.

11.5 Containment Response to Degraded Core Conditions

Let us now consider the response of the containment system to postulated degraded core conditions.

11.5.1 Hydrogen Production During Postulated Core Degradation Accidents

If sufficient loss of core cooling occurs, core overheating and damage would result and some amount of hydrogen production could take place within the RPV. Results from studies on the behavior of hydrogen during hypothetical accidents in light water reactors have been summarized by Sherman et al. (1980). These studies address questions of hydrogen generation, solubility, detection, combustion, and recombination.

Only in highly degraded conditions are large amounts of hydrogen expected to be produced in BWRs. However, if postulated, such accidents lead to core temperatures above 1400°C and steam reaction with Zircaloy cladding, which could produce as much as 2000 kg (about 24,000 m^3) of hydrogen in a BWR. The reaction of steam with vessel stainless steel could also produce hydrogen. Often smaller sources of hydrogen generation include radiolytic decomposition of water, decomposition of paints and galvanized materials, and concrete decomposition by postulated molten core-concrete interaction. If all of the postulated hydrogen burned in the containment without substantial cooling, or if a detonation resulted, it could produce pressures beyond the design values of all existing containments. Thus, hydrogen generation is an important safety issue.

The lower limit of hydrogen flammability in air is about 4% by volume. The possibility of a deflagration (rapid burning), with resulting high pressures, exists if the volumetric concentration of hydrogen is much above about 8% by volume in a dry containment and an ignition source exists. A detonation (burn front moving as a compression shock-wave at supersonic speed into unburned gas) is possible if the hydrogen volume concentration is above 13% by volume. These volume percentages are affected by the presence of steam or other gases and the degree of mixture turbulence. Nevertheless, the amount of hydrogen generated from a severely degraded BWR core could result in a detonatable mixture when released to the containment.

Figure 11-33 gives the calculated and experimental pressure resulting after constant volume, adiabatic combustion of hydrogen and air in terms of the initial hydrogen volumetric concentration. The resulting pressures are much higher for detonations.

Significantly, hydrogen does not separate from air and stratify near the top of the containment once the two are mixed. If hydrogen were introduced near the top or rose to the top as a warm columnar plume without complete mixing during its transit, subsequent mixing with air would be slower, but once mixed, the hydrogen would not stratify, except in the case where it is mixed with a condensible gas such as steam. The introduction of nitrogen, carbon dioxide, or some halogenated hydrocarbons (e.g., Halon 1301) can be employed to make inert an atmosphere containing hydrogen and air. An inert condition implies that a burning front cannot be sustained. Figure 11-34 gives flammability limits of hydrogen in air that is diluted with either carbon dioxide or nitrogen. It can be seen that combustion requires volume concentrations of 60% and 75%, respectively, for carbon dioxide and nitrogen inerting of a hydrogen-air mixture. Moreover,

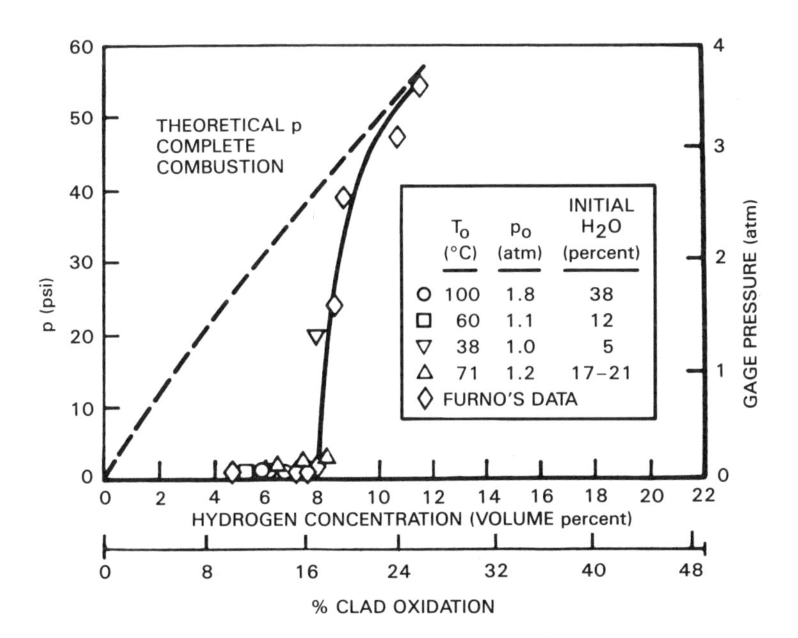

Fig. 11-33 Pressure rise versus hydrogen, spark igniter.

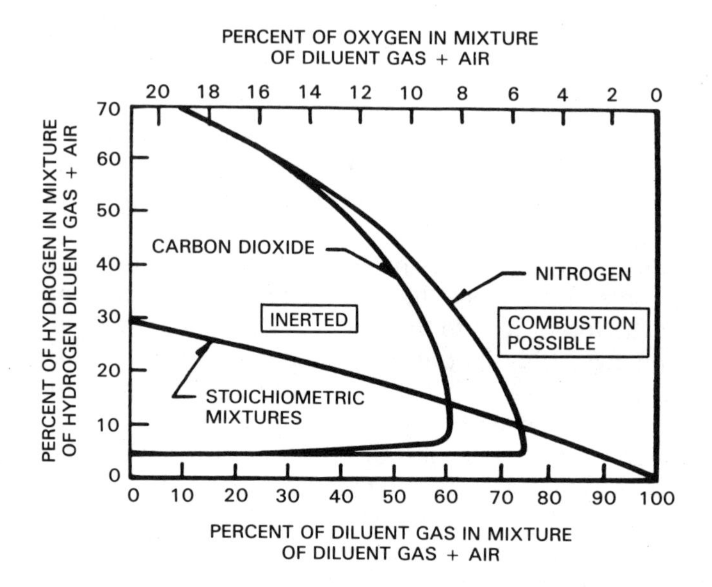

Fig. 11-34 Flammability limits of hydrogen in air diluted with CO_2 and N_2.

it would require a 60% volume percentage of steam for inerting. However, steam condensation in the pool would prevent a steam-inerted condition in the wetwell air space. The inerting property of carbon dioxide, nitrogen, and steam is due to the additional heat capacity, which absorbs sufficient thermal energy to prevent propagation of a flame front. Another possible inerting substance is mono-bromo-tri-fluoro-methane, CF_3Br, of which about 31% volume concentration stops combustion by removal of certain free radicals generated in the burning process. Containment inerting is widely used in MARK-I and MARK-II containments since entry is not required during plant operation.

In contrast, containments that must be entered frequently by workers during power operation (e.g., MARK-III) would normally contain air. Nevertheless, these containments could be rapidly inerted by carbon dioxide or Halon if core degradation were expected following an accident.

An alternative to hydrogen inerting is to burn it with hydrogen igniters as it is generated to prevent concentrations from reaching the detonation limit. Additional cooling of the containment may be required to control the temperature and pressure if this procedure is adopted.

In the unlikely event of hydrogen combustion, it would be most likely to occur outside the drywell above the suppression pool. If a containment overpressurization failure occurred, it would be expected to fail the wetwell air space at the top boundary, rather than the drywell. Consequently, even under extremely degraded conditions, the suppression pool would be retained as a heat sink, a source of water for core cooling, and as a fission product filter.

11.5.2 Fission Product Scrubbing in the Suppression Pool

The pool of a pressure suppression containment is highly effective in absorbing fission products that could be released following a degraded core accident (Rastler, 1981; Moody, 1984). The pool provides a barrier or scrubbing medium for retaining radioiodines and other fission products, except noble gases, which could be generated from a degraded core. Studies have shown that a subcooled pool introduces decontamination factors (i.e., the ratio of incoming to outgoing fission product flows) of at least 100 for elemental iodine and 1000 for cesium iodide and particulates. A saturated pool introduces a decontamination factor of at least 30 for elemental iodine and 100 for cesium iodide and particulates.

If a degraded core condition should occur, discharge of airborne fission products to the pool is accomplished either by SRV discharge or vent discharge. As discussed in detail in Sec. 8.6.8, as bubbles rise through the pool, the fission products are scrubbed by many mechanisms, including sedimentation, diffusiophoresis, and inertial deposition from bubble gas internal circulation. The nature of bubble size, shape, and rise velocity plays a key role in the effectiveness of bubble scrubbing.

11.5.2.1 *Bubble Shattering*

When a relatively constant submerged gas discharge rate occurs into a liquid, single-bubble growth occurs until buoyancy causes it to break loose from the charging source, and another bubble begins to grow in its place. However, when such a bubble is released, it may undergo rapid shattering into a cluster of small bubbles.

It has been observed (Marble, 1983) that large bubble shattering can occur if the initial bubble is charged at nearly the local hydrostatic pressure. Figure 11-35 shows sketches traced from enlarged high-speed movie frames that depict large bubble shattering. Large bubble shattering is relatively complete when the bottom surface of the bubble reaches the top surface. The change in hydrostatic pressure in surrounding water vertically across the bubble is about $\Delta p = \rho_l g D_i / g_c$. Since the bubble gas pressure is essentially uniform, Bernoulli's law requires that the bottom surface overtake the top at an approximate velocity of, $(2g_c \Delta p/\rho_l)^{1/2} = (2gD_i)^{1/2}$. The overtaking time is about $t = D_i/(2gD_i)^{1/2} = (D_i/2g)^{1/2}$. It follows that regardless of the initial bubble size, the early stages of breakup will occur on a normalized time scale of $t(2g/D_i)^{1/2}$. Thus, if D_i is quadrupled, the initial breakup will take twice as long. Since the Taylor rise velocity of a large bubble is about $(gD_i/2)^{1/2}$, the rise distance, Y, for the bottom to overtake the top is $Y = (gD_i/2)^{1/2}(D_i/2g)^{1/2}$.

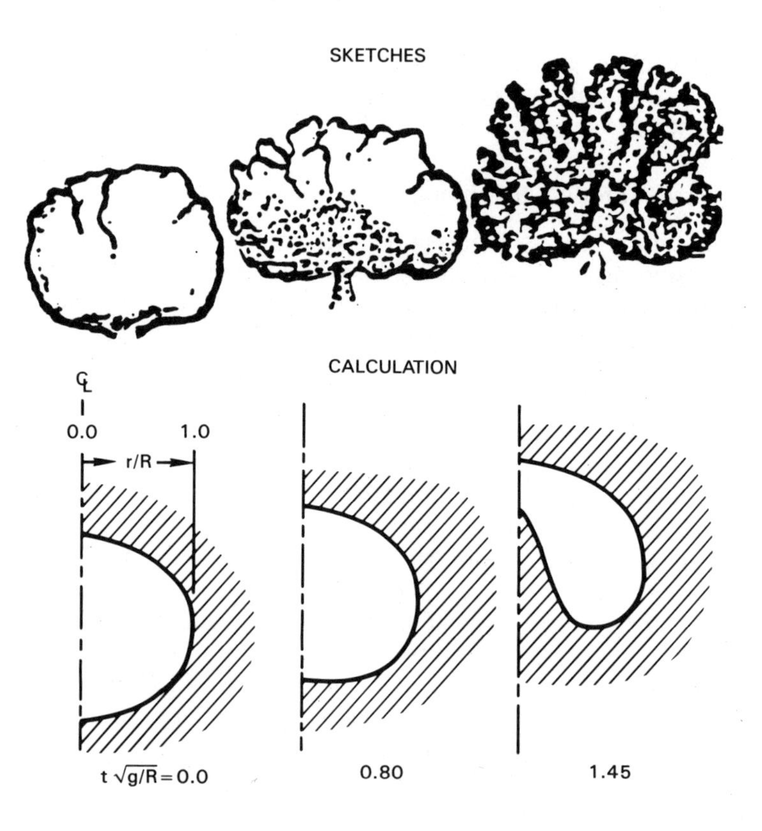

Fig. 11-35 Large bubble shattering.

That is, $Y/D_i = \frac{1}{2}$. Also shown in Fig. 11-35 is a finite difference solution of a gas bubble released in an inviscid liquid. Viscous effects are relatively unimportant during the growth, detachment, and initial bubble breakup process in water. Therefore, the computation is expected to give a reasonable description of how the bubble rises as the bottom surface overtakes the top. It is seen that the bubble shattering distance for all scales is given by $Y/D_i < 0.5$.

11.5.2.2 *Bubble Cluster Rise Velocity*

The effectiveness of fission product scrubbing is directly proportional to transit time of gas through the water pool in a pressure suppression containment system.

Bubbles formed from the shattering of a large bubble rise through the liquid as an approximate spherical cluster of bubbles with internal circulation upward near the vertical axis and downward circulation near the boundary. The observed circulation pattern resembles a Hill's vortex (Panton, 1984). If the mass center of a discharged gas volume V rises a distance Y during the initial breakup process, the energy transferred to the liquid is,

$$E = \rho_l g Y V / g_c \ . \tag{11.151}$$

If the liquid volume entrained in the cluster, V_l, is assumed to be homogeneously mixed with the small bubbles formed, the kinetic energy in the rising Hill's vortex is (Lamb, 1945):

$$E = (10/7)\pi \rho_c v_c^2 R_c^3 / g_c \ , \tag{11.152}$$

where ρ_c is the liquid-bubble mixture density in the cluster, v_c is the cluster rise velocity in stationary liquid, and R_c is the cluster radius. Although fluid inside a bubble cluster circulates, the bubbles are rising by buoyancy relative to the cluster liquid so that the buoyant work on each bubble is continuously dissipated by its drag work. Therefore, the cluster rises with constant kinetic energy and velocity, with a continuous throughflow of liquid. The cluster void fraction is $V_v/(V_v + V_l)$. Since $\rho_v << \rho_l$, the cluster density is, therefore, $\rho_c \cong (1-\alpha)\rho_l$, and the cluster radius is $R_c \cong R_0/\alpha^{1/3}$ where R_0 is the initial bubble radius. If the liquid energy is completely transferred to that in the Hill's vortex, Eqs.(11.151) and (11.152) can be equated, with the discharged gas volume written as $V = (4/3)\pi R_0^3$, to give:

$$v_c = 0.97\sqrt{[\alpha/(1-\alpha)]gY} \ . \tag{11.153}$$

The void fraction of bubble clusters, estimated from slow motion observations (Marble, 1983), appears to be about 0.5, which roughly corresponds to close-packed spheres with $\alpha = 0.52$. This value of void fraction is consistent with idealized shattering by sinusoidal interface wave growth.

Another estimate of the cluster void fraction is obtained by assuming that the initial bubble entrains a volume of liquid equal to $2\pi R_0^3/3$, corresponding to its translational virtual mass, which yields a void fraction of

$$\alpha = \frac{\frac{4}{3}\pi R_0^3}{\left[\frac{4}{3}\pi R_0^3 + \frac{2}{3}\pi R_0^3\right]} = \frac{2}{3} \ . \tag{11.154}$$

This value is more consistent with the actual interface breakup by Taylor instability where round fingers of gas penetrate into the liquid and narrow liquid spikes penetrate into the gas, resulting in less liquid entrainment than an idealized sinusoidal interface.

Measured diameters of the initial bubble released from a 15-cm horizontal vent and the resulting bubble cluster after breakup are given in Fig. 11-36. The dashed correlation line joins the measured initial bubble di-

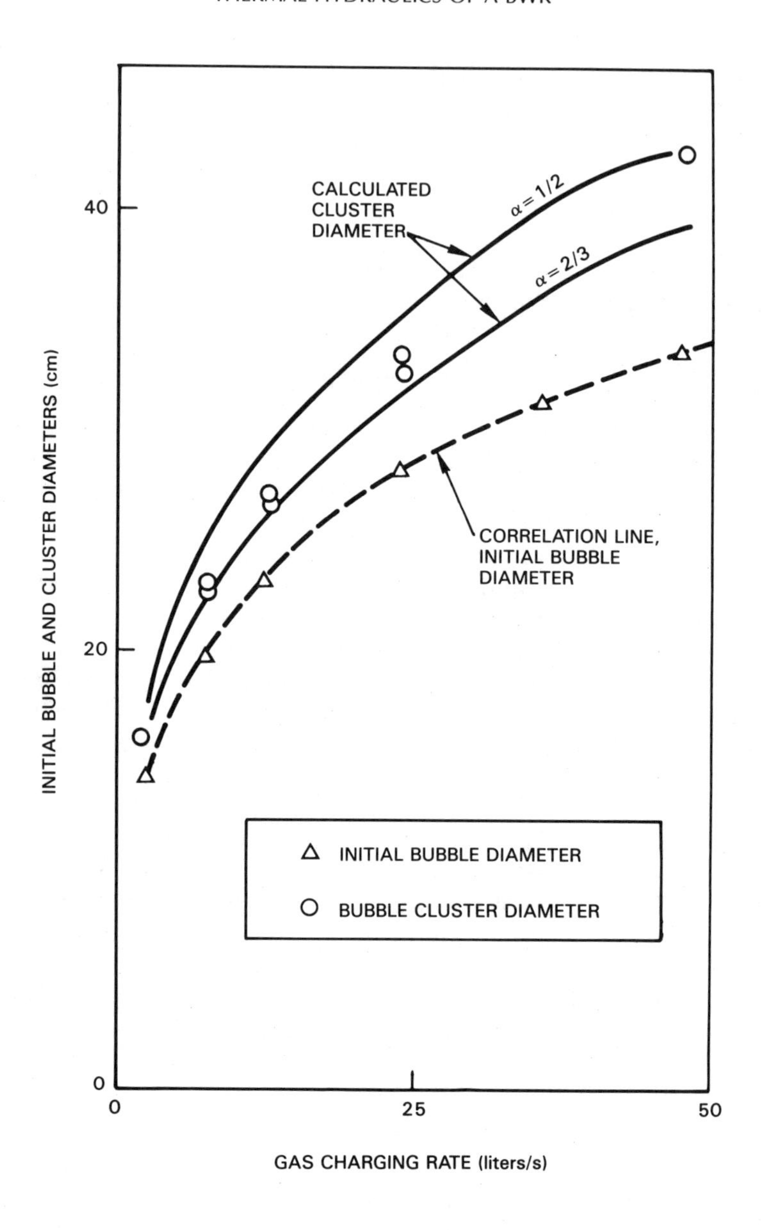

Fig. 11-36 Bubble and resulting cluster diameters.

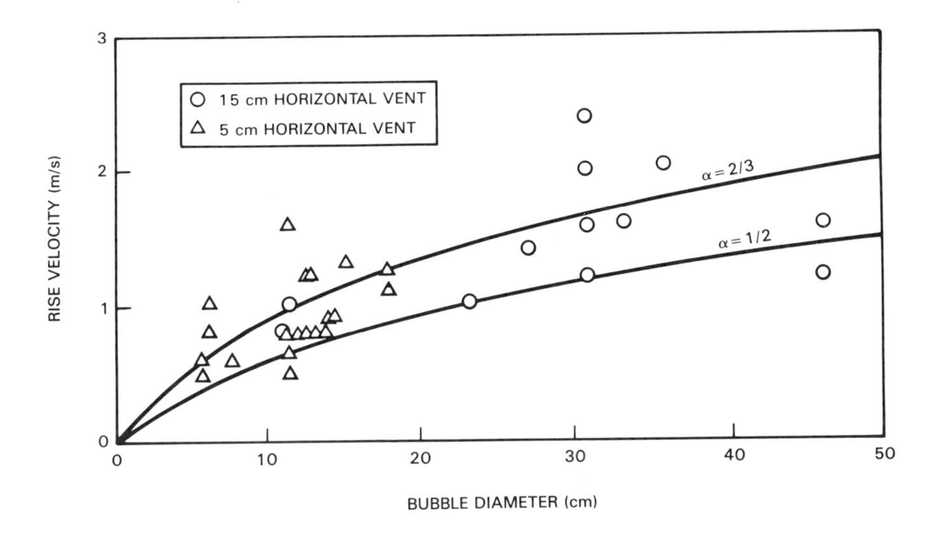

Fig. 11-37 Cluster rise velocity.

ameters. Cluster diameters were calculated for the correlation line, using void fractions of ½ and ⅔, and results shown as solid lines are seen to bound the measured values.

If the estimated breakup distance, $Y = D_i/2$, is employed, it follows from Eq. (11.153) that v_c probably would lie between

$$v_c/\sqrt{gD_i} = \begin{cases} 0.68 & \alpha = 1/2 \\ 0.97 & \alpha = 2/3 \end{cases} . \tag{11.155}$$

The lower limit is a close approximation to the rise velocity of a large spherical gas bubble, $0.707\sqrt{gD_i}$ (Davies and Taylor, 1950). Figure 11-37 shows that a void fraction of ⅔ gives a more uniform prediction of the measured bubble cluster rise velocity, where the close-packed limit of $\alpha = 1/2$ tends to follow the lower boundary of the data.

11.5.2.3 *Bubble Plumes*

Although large bubbles shatter and rise as clusters, a continual discharge of gas from spargers or other distributing devices forms bubble plumes, as shown in Fig. 11-38.

The transit time of bubbles rising in plumes is also important to the fission product scrubbing process. The bubble rise velocity in a plume relative to a laboratory frame of reference can be roughly estimated by considering a column of average area, A, containing bubbles rising through

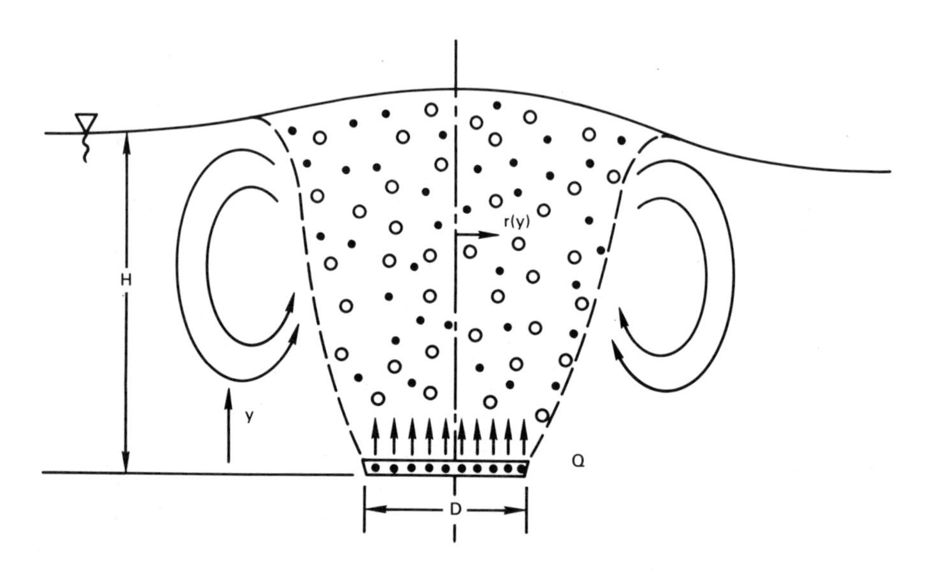

Fig. 11-38 Bubble plume.

a pool of depth, H. The average void fraction of the plume is α. Liquid upward velocity v_l is zero both at the bottom of the plume and again at the pool surface. Thus, the liquid must accelerate to its maximum velocity at approximately $H/2$, and then decelerate in the upper half. The buoyant force of bubbles in the lower half of the plume is given by,

$$F_b = (\rho_l - \rho_v)(g/g_c)A\alpha(H/2) \simeq \rho_l \frac{g}{g_c} A\alpha \frac{H}{2} \ . \tag{11.156}$$

The fluid mass in the lower half of the plume is

$$M_l = \rho_l A(1-\alpha)H/2 \ . \tag{11.157}$$

Thus, Newton's second law yields,

$$a = dv_l/dt = g_c F_b/M_l = g\alpha/(1-\alpha) \ . \tag{11.158}$$

The average liquid velocity, v_l, is obtained by writing $a = v_l dv_l/dy$, and integrating to obtain:

$$v_l = \begin{cases} \sqrt{\dfrac{2g\alpha}{1-\alpha} y} & ; y < \dfrac{H}{2} \\ v_{l,\max} - \sqrt{\dfrac{2g\alpha}{1-\alpha} y} & ; y > \dfrac{H}{2} \end{cases} \tag{11.159}$$

where $v_{l,\max}$ occurs at $y = H/2$. The average liquid velocity, $\bar{v}_l$, is estimated as

$$\bar{v}_l = \sqrt{\alpha/(1-\alpha)}\ \sqrt{gH/2}/2 \ . \tag{11.160}$$

Bubble velocity relative to the laboratory frame is,

$$v_g = v_l + v_{g/l} \ , \tag{11.161}$$

where $v_{g/l}$ is bubble velocity relative to the liquid, which was shown to be nearly constant (Chesters et al., 1980). The total gas volume flow rate is

$$Q = A\alpha v_g \ , \tag{11.162}$$

so that,

$$Q/Av_{g/l} = \frac{(v_l/v_{g/l})^2(1 + v_l/v_{g/l})}{(gH/8v_{g/l}^2) + (v_L/v_{g/l})^2} \ . \tag{11.163}$$

The parameter $Q/Av_{g/l}$ is a Froude number, which governs the coarse structure of a plume.

References

Abramson, P., Ed., *Light Water Reactor Safety Analysis*, Hemisphere, New York (1985).

Anderson, W. G., et al., "Small Scale Modeling of Hydrodynamic Forces in Pressure Suppression Systems: Tests of the Scaling Laws," *Proc. Topl. Mtg. Thermal Reactor Safety*, Sun Valley, ANS CONF 770708, Vol. 3 (1977).

Baker, W. E., Westine, P. S., and Dodge, F. T., *Similarity Methods in Engineering Dynamics*, Hayden Book Company, New Jersey (1973).

Bilanin, W. J., "The General Electric Mark III Pressure Suppression Containment System Analytical Model," NEDO-20533, General Electric (June 1974).

Carslaw, H. S., and J. C. Jaeger, *Conduction of Heat in Solids*, Clarendon Press, Oxford (1947).

Chesters, A. K., et al., "A General Model for Unconfined Bubble Plumes from Extended Sources," *Int. J. Multiphase Flow*, **6**, 499–521 (1980).

Davies, R. M., and G. I. Taylor, "The Mechanics of Large Bubbles Rising through Extended Liquids and through Liquids in Tubes," *Proc. Roy. Soc. London*, **200**, 375 (1950).

Fitzsimmons, G. W., et al., "Mark I Containment Program, Full Scale Test Program Final Report," NEDO-24539, General Electric (Aug. 1979).

Florsheutz, L. W., and B. T. Chao, "On the Mechanisms of Vapor Bubble Collapse," *J. Heat Transfer*, **87**, 2 (May 1965).

Furno, A. L., et al., "Some Observations on Near Limit Flames," *13th Symp. on Combustion*, Pittsburgh, Combustion Inst., pp. 593–599 (1971).

General Electric, "Evaluation of Fluid-Structure Interaction Effects on BWR Mark II Containment Structures," NEDO-21936, General Electric (Aug. 1978).

General Electric, "General Electric Standard Safety Analysis Report (GESSAR II) for BWR-6, Mark III, Appendix 3B," NRC Docket No. STN 50447, General Electric (Mar. 1980).

Khalid, Z. M., and R. T. Lahey, Jr., "The Analysis of Vent Clearing," *Trans. Am. Nucl. Soc.*, **35** (1980).

Kline, S. J., *Similitude and Approximation Theory,* McGraw-Hill, New York (1965).

Lahey, P. M., Lahey, R. T., Jr., and Drew, D. A., "An Analysis of Chugging Loads in Containment Systems," *Trans. Am. Nucl. Soc.*, **62** (1990).

Lamb, Sir Horace, *Hydrodynamics,* Dover, New York (1945).

Marble, W. J., et al., "Preliminary Report on the Fission Product Scrubbing Program," NEDO-30017, General Electric (1983).

Marks, J. S., and G. B. Andeen, "Chugging and Condensation Oscillation Tests," EPRI NP-1167, Electric Power Research Institute (Sep. 1979).

McCauley, E. W., and J. H. Pitts, "Trial Air Test Results for the 1/5 Scale Mark I BWR Pressure Suppression Experiment," UCRL-52371, Lawrence Livermore Laboratory (1977).

McCormick, C. W., Ed., "MSC/NASTRAN Users Manual," MSR-39, MacNeal-Schwendler Corporation (Apr. 1979).

McCready, J. L., et al., "Steam Vent Clearing Phenomena and Structural Response of the BWR Torus (Mk I Containment)," NEDO-10859, General Electric (Apr. 1973).

McIntyre, T. R., and L. L. Myers, "Fifth Quarterly Progress Report: MARK-III Confirmatory Test Program," NEDO-20550, General Electric (July 1974).

Moody, F. J., and W. C. Reynolds, "Liquid Surface Motion Induced by Acceleration and External Pressure," *JBE, Trans. ASME,* **94,** Ser. D, *3* (Sep. 1972).

Moody, F. J., et al., "Analytical Model for Estimating Drag Forces on Rigid Submerged Structures Caused by LOCA and Safety Relief Valve Ramshead Air Discharges," NEDO-21471, General Electric (Sep. 1977).

Moody, F. J., "Unsteady Condensation and Fluid-Structure Frequency Dependence on Parameters of Vapor Quench Systems," ASME SP PVP-46, *Interactive Fluid Structural Dynamics Problems in Power Engineering* (1981).

Moody, F. J., "Interphase Thermal and Mechanical Dissipation of Acoustic Disturbances in Gas-Liquid Mixtures," *Proc. ASME-JSME Thermal Engineering Joint Conference,* Honolulu (1983).

Moody, F. J., "The Importance of Pools, Sprays, and Ice Beds in Fission Product Retention in Containment," *Proc. ANS Topl. Mtg. Fission Product Behavior and Source Term Research* (1984).

Nariai, H., and I. Aya, "Oscillation Frequency of Condensation Oscillation Induced by Steam Condensation into Pool Water," *Basic Aspects of Two-Phase Flow and Heat Transfer*, ASME Publication HTD, Vol. 34 (1984).

Panton, R. L., *Incompressible Flow,* J. Wiley, New York (1984).

Patton, K. T., "Tables of Hydrodynamic Mass Factors for Translational Motion," ASME Paper No. 65-WA/UNT-2 (1965).

PG&E Report, "Preliminary Hazards Summary Report, Bodega Bay Atomic Park, Unit No. 1," Pacific Gas and Electric Company (1962).

Rastler, D. M., "Suppression Pool Scrubbing Factors for Postulated Boiling Water Reactor Accident Conditions," NEDO-24520, General Electric (1981).

Robbins, C. H., "Tests of a Full Scale 1/48 Segment of the Humboldt Bay Pressure Suppression Containment," GEAP-3596, General Electric (1960).

Shapiro, A. H., *The Dynamics and Thermodynamics of Compressible Fluid Flow, Vol. I,* The Ronald Press, New York (1953).

Sherman, M. P., et al., "The Behavior of Hydrogen during Accidents in Light Water Reactors," NUREG/CR-1561, SAND 80-1495 R3, Sandia National Laboratories (Aug. 1980).

Simpson, M. E., and C. K. Chan, "Hydrodynamics of a Subsonic Vapor Jet in Subcooled Liquid," *J. Heat Transfer,* **104,** 271–278 (1982).

Taylor, G. I., "The Instability of Liquid Surfaces when Accelerated in a Direction Perpendicular to their Planes," *Proc. Roy. Soc. A,* **201,** 192–196 (1950).

Torbeck, J. E., et al., "Mark I 1/12 Scale Pressure Suppression Tests," NEDO-13456, General Electric (Sep. 1976).

Valandani, P., "Safety Relief Valve Discharge Analytical Modeling, NEDO-20942, General Electric (May 1975).

INDEX